ICRF Handbook of Genome Analysis

ICRF Handbook of Genome Analysis

THE EDITORS

Nigel K. Spurr
SmithKline Beecham Pharmaceuticals
New Frontiers Science Park
Harlow, Essex

Bryan D. Young
ICRF Department of Molecular Oncology
St Bartholomew's Hospital
London

Stephen P. Bryant
Gemini Research Ltd
Science Park
Cambridge

THE SECTION EDITORS

Stephan Beck
The Sanger Centre
Hinxton, Cambridge

Anna-Maria Frischauf
ICRF
Lincoln's Inn Fields
London

Denise Sheer
Human Cytogenetics Laboratory
ICRF, Lincoln's Inn Fields
London

In two volumes

Volume 2

b

Blackwell
Science

The content of this book expresses the views of the individual authors and not those of the Imperial Cancer Research Fund as a registered UK charity

Editorial Offices:
Osney Mead, Oxford OX2 0EL
25 John Street, London WC1N 2BL
23 Ainslie Place, Edinburgh EH3 6AJ
350 Main Street, Malden
MA 02148 5018, USA
54 University Street, Carlton
Victoria 3053, Australia

Other Editorial Offices:
Blackwell Wissenschafts-Verlag GmbH
Kurfürstendamm 57
10707 Berlin, Germany

Blackwell Science KK
MG Kodenmacho Building
7–10 Kodenmacho Nihombashi
Chuo-ku, Tokyo 104, Japan

First published 1998

A catalogue record for this title
is available from the British Library

ISBN 0-632-03728-8

Library of Congress
Cataloging-in-publication Data

ICRF handbook of genome analysis /
edited by Nigel Spurr.
p. cm.
Includes bibliographical references and index.
ISBN 0-632-03728-8
1. Gene mapping–Laboratory manuals. 2. Nucleotide sequence–Laboratory manuals. I. Spurr, Nigel.
QH445.2. I27 1996
574.87′ 3282–dc20 96–17981
CIP

Set by Setrite Typesetters, Hong Kong
Printed and bound in Great Britain by
MPG Books Ltd, Bodmin, Cornwall

The Blackwell Science logo is a trade mark of Blackwell Science Ltd, registered at the United Kingdom Trade Marks Registry

DISTRIBUTORS

Marston Book Services Ltd
PO Box 269
Abingdon
Oxon OX14 4YN
(Orders: Tel: 01235 465500
Fax: 01235 465555)

USA
Blackwell Science, Inc.
Commerce Place
350 Main Street
Malden, MA 02148 5018
(Orders: Tel: 800 759 6102
781 388 8250
Fax: 781 388 8255)

Canada
Copp Clark Professional
200 Adelaide St West, 3rd Floor
Toronto, Ontario M5H 1W7
(Orders: Tel: 416 597-1616
800 815-9417
Fax: 416 597-1617)

Australia
Blackwell Science Pty Ltd
54 University Street
Carlton, Victoria 3053
(Orders: Tel: 3 9347 0300
Fax: 3 9347 5001)

Contents

List of protocols

Volume 1

List of contributors

G. Argyropoulos *Department of Biochemistry and Molecular Genetics, St Mary's Hospital Medical School, Imperial College of Science, Technology and Medicine, London W2 1PG, UK*

John A.L. Armour *Department of Genetics, School of Medicine, Queen's Medical Centre, Nottingham NG7 2UH, UK*

Mary Berks (deceased) *The Sanger Centre, Wellcome Trust Genome Campus, Hinxton, Cambridge CB10 1SA, UK*

Sandra Birdsall *Molecular Cytogenetics Laboratory, Institute of Cancer Research, 15 Cotswold Road, Sutton, Surrey SM2 5NG, UK*

D. Timothy Bishop *ICRF Genetic Epidemiology Laboratory, St. James's University Hospital, Leeds LS9 7TF, UK*

Martin J. Bishop *HGMP Resource Centre, Hinxton, Cambridge CB10 1SB, UK*

Sharen Bowman *The Sanger Centre, Wellcome Trust Genome Campus, Hinxton, Cambridge CB10 1SA, UK*

S.D.M. Brown *MRC Mouse Genome Centre, Harwell, Oxfordshire OX11 0RD, UK*

Stephen P. Bryant *Gemini Research Ltd, 162 Science Park, Milton Road, Cambridge CB4 4GH, UK*

David Buck *The Sanger Centre, Wellcome Trust Genome Campus, Hinxton, Cambridge CB10 1SA, UK*

Nigel P. Carter *The Sanger Centre, Hinxton Hall, Hinxton, Cambridge CB10 1RQ, UK*

Aravinda Chakravarti *Department of Genetics, Room BRB 721, Case Western Reserve University, 10900 Euclid Avenue, Cleveland, OH 44106, USA*

Carol Churcher *The Sanger Centre, Wellcome Trust Genome Campus, Hinxton, Cambridge CB10 1SA, UK*

Richard Cooke *Laboratoire de Physiologie et Biologie Moléculaire des Plantes, URA 565 du CNRS, Université de Perpignan, 66860 Perpignan, France*

Roger Cox *ICRF, Wellcome Trust Centre for Human Genetics, Nuffield Department of Clinical Medicine, Windmill Road, Headington, Oxford OX3 7BN, UK*

Barbara Czepulkowski *Department of Cytogenetics/ Haematology, Kings College Hospital, Denmark Hill, London SE5 9RS, UK*

Michel Delseny *Laboratoire de Physiologie et Biologie Moléculaire des Plantes, URA 565 du CNRS, Université de Perpignan, 66860 Perpignan, France*

Rosa Maria Diaz *Richard Dimbleby Department of Cancer Research, Rayne Institute, St Thomas' Hospital, Lambeth Palace Road, London SE1 7EH, UK*

Horst Domdey *Laboratorie für Molekulare Biologie, GenZentrum, Ludwig-Maximilian-Universität, Am Klopferspitz, 8033 Martinsried, Germany*

Christine J. Farr *Department of Genetics, University of Cambridge, Downing Street, Cambridge CB2 3EH, UK*

Horst Feldmann *Adolf-Butenandt-Institut für Physiologische Chemie, Physikalische Biochemie und Zellbiologie der Universität, Schillerstrasse 44, D-80336 München, Germany*

Riccardo Fodde *MGC-Department of Human Genetics, Sylvius Laboratory, Faculty of Medicine, University of Leiden, Leiden, The Netherlands*

Fiona Francis *Max-Planck-Institut für Molekulare Genetik, Ihnestrasse 73, D-14195 Berlin, Germany*

Andrei Grigoriev *Max-Planck-Institut für Molekulare Genetik, Ihnestrasse 73, D-14195 Berlin, Germany*

Ilkka Havukkala *Rice Genome Research Program, Institute of Society for Techno-Innovation of Agriculture, Forestry and Fisheries, 446-1, Ippaizuka Kamiyokoba, Tsukuba, Ibaraki 305, Japan*

Peter Heinrich *MediGene GmbH, Lochhamer Strasse 11, D-82152 Martinsreid/München, Germany*

Christoph Heller *Max-Planck-Institut für Molekulare Genetik, Ihnestrasse 73, D-14195 Berlin, Germany*

Prem P. Jauhar *USDA-ARS, Northern Crop Science Laboratory, Fargo, North Dakota 58105-5677, USA*

Monique Jouet *University of Cambridge Clinical School, Addenbrooke's Hospital, Department of Medicine, Hills Road, Cambridge CB2 2QQ, UK*

Lyndal Kearney *MRC Molecular Haematology Department, Institute of Molecular Medicine, John Radcliffe Hospital, Headington, Oxford OX3 9DS, UK*

Bernhard Korn *ICRF, PO Box 123, Lincoln's Inn Fields, London WC2A 3PX, UK*

Nori Kurata *Rice Genome Research Program, National Institute of Agrobiological Resources, 1–2 Kannondai 2-chome, Tsukuba, Ibaraki 305, Japan*

Hans Lehrach *Max-Planck-Institut für Molekulare Genetik, Ihnestrasse 73, D-14195 Berlin, Germany*

Debra M. Lillington *ICRF Department of Medical Oncology, Science Building, St Bartholomew's Hospital Medical College, Charterhouse Square, London EC1M 6BQ, UK*

Yong-Jie Lu *Molecular Cytogenetics Laboratory, Institute of Cancer Research, 15 Cotswold Road, Sutton, Surrey SM2 5NG, UK*

Millicent Masters *Institute of Cell and Molecular Biology, The University of Edinburgh, Darwin Building, King's Buildings, Mayfield Road, Edinburgh EH9 3JR, UK*

Tara Cox Matise *Department of Statistical Genetics, The Rockefeller University, 1230 York Avenue, Box 192, New York, NY 10021, USA*

Sebastian Meier-Ewert *Max-Planck Institut für Molekulare Genetik, Ihnestrasse 73, D-14195 Berlin, Germany*

Aleksandar Milosavljević *CuraGen Corporation, 555 Long Wharf Drive, New Haven, CT 06511, USA*

Yuzo Minobe *Rice Genome Research Program, National Institute of Agrobiological Resources, 1–2 Kannondai 2-chome, Tsukuba, Ibaraki 305, Japan*

Akio Miyao *Rice Genome Research Program, National Institute of Agrobiological Resources, 1–2 Kannondai 2-chome, Tsukuba, Ibaraki 305, Japan*

Simon P. Monard *Aaron Diamond AIDS Research Center, New York NY 10016, USA*

Richard Mott *Bioinformatics Department, SmithKline Beecham Pharmaceuticals, New Frontiers Science Park, Third Avenue, Harlow, Essex CM19 5AW, UK*

Michael A. North *Sequana Therapeutics, 11099 North Torrey Pines Road, Suite 160, La Jolla CA 92037, USA*

Gerald Nyakatura *Institute of Molecular Biotechnology, Department of Genome Analysis, Beutenbergstrasse 11, 07745 Jena, Germany*

Matthias Platzer *Institute of Molecular Biotechnology, Department of Genome Analysis, Beutenbergstrasse 11, 07745 Jena, Germany*

Guy Plunkett III *University of Wisconsin-Madison, Laboratory of Genetics, 445 Henry Mall, Madison WI 53706, USA*

Jaime Prilusky *Bioinformatics Unit, Department of Biological Services, Weizmann Institute of Science 76100, Rehovot, Israel*

Peter Richterich *Genome Therapeutics Corporation, 100 Beaver Street, Waltham MA 02154, USA*

André Rosenthal *Institute of Molecular Biotechnology, Department of Genome Analysis, Beutenbergstrasse 11, 07745 Jena, Germany*

Takuji Sasaki *Rice Genome Research Program, National Institute of Agrobiological Resources, 1–2 Kannondai 2-chome, Tsukuba, Ibaraki 305, Japan*

Robert D.C. Saunders *Department of Anatomy and Physiology, University of Dundee, Dundee DD1 4HN, UK*

Alan J. Schafer *Department of Genetics, University of Cambridge, Downing Street, Cambridge CB2 3EH, UK*

Leonard Schalkwyk *Max-Planck-Institut für Molekulare Genetik, Ihnestrasse 73, D-14195 Berlin, Germany*

Gabriele Senger *Institut für Humangenetik und Anthropologie, Kollegiengasse 10, D-07740 Jena, Germany*

Denise Sheer *Human Cytogenetics Laboratory, ICRF, PO Box 123, Lincoln's Inn Fields, London WC2A 3PX, UK*

Andrew N. Shelling *Department of Obstetrics and Gynaecology, National Women's Hospital, Claude Road, Epson, Auckland, New Zealand*

Janet Shipley *Molecular Cytogenetics Laboratory, Institute of Cancer Research, 15 Cotswold Road, Sutton, Surrey SM2 5NG, UK*

David L. Simmons *Cell Adhesion Laboratory, ICRF, Institute of Molecular Medicine, John Radcliffe Hospital, Headington, Oxford OX3 9DU, UK*

John Sulston *The Sanger Centre, Hinxton Hall, Cambridge CB10 1SA, UK*

Karen Thomas *The Sanger Centre, Wellcome Trust Genome Campus, Hinxton, Cambridge CB10 1SA, UK*

Rob B. van der Luijt *Department of Human Genetics, Faculty of Medicine, Utrecht University, Stratenum, De Uithof, Universiteitweg 100, 3584 CG Utrecht, The Netherlands*

Richard G. Vile *ICRF Laboratory of Cancer Gene Therapy, ICRF Oncology Unit, Hammersmith Hospital, DuCane Road, London W12 ONN, UK*

Sarah P. West *Northern Genetics Service, 19/20 Claremont Place, Newcastle upon Tyne NE2 4AA, UK*

Robert H. Waterston *Department of Genetics and Genome Sequencing Center, Washington University School of Medicine, St Louis, MO 63110, USA*

Kimiko Yamamoto *Rice Genome Research Program, Institute of Society for Techno-Innovation of Agriculture, Forestry and Fisheries, 446-1, Ippaizuka Kamiyokoba, Tsukuba, Ibaraki 305, Japan*

Masahiro Yano *Rice Genome Research Program, National Institute of Agrobiological Resources, 1–2 Kannondai 2-chome, Tsukuba, Ibaraki 305, Japan*

Preface

Genetics as the formal study of inheritance was founded as a field following the rediscovery of Mendel's work at the beginning of this century. This led to the first revolution in our understanding of inheritance, namely of the basic mechanisms of gene transmission, of linkage and of interpretations in terms of the behaviour of chromosomes in meiosis. The second revolution came with the discovery of the Watson–Crick structure of DNA just over 40 years ago, which spelled out the chemical basis for the gene, and then its mode of action. Now, following the development of recombinant DNA technology and many other techniques that enable us to clone and sequence DNA with enormous speed and efficiency, we are entering a third revolutionary phase of genetic analysis as we approach the end of the century. Now is the time when whole genomes are being sequenced and the complete language of organisms is being deciphered.

It was just over 15 years ago that the potential for the complete analysis of the human genome began to be appreciated; it came to be realized that this would provide enormous power for the analysis of all normal biological functions, as well as for the analysis of the basis of essentially all human disease. Thus developed the Human Genome Project, and alongside it many other genome projects.

The rate of advance of the technology and the acquisition of new data could not, I believe, have been predicted even by the wildest speculator. In 1986, I suggested that the project to catalogue and sequence all human genes and place them in their positions along the chromosomes be billed as 'Project 2000'. That prediction we can now see will soon be realized.

Almost daily, new genes are discovered, while many exist and are waiting to be discovered in the databanks of genomic and, especially, partial cDNA sequences. The production and analysis of this extraordinary accumulation of information requires a wide variety of complex techniques; from approaches to the statistical problems of the analysis of complex human pedigrees, to the determination of DNA sequences. This *Handbook* provides an invaluable guide to the wide range of these techniques and is practical and usable. It has required an enormous effort on the part of the authors and, especially, the editors, to put together this most valuable companion and all ought to be congratulated on the achievement.

Only 5 years ago when we were organizing a new form of international Human Gene Mapping Workshop in London, it was hard to convince the pharmaceutical industry that they should be interested. Now, not only is there a huge and burgeoning biotechnology industry, but no major pharmaceutical company can afford any longer not to invest in a major way in genome analysis, and many are accepting that this is where their future lies. The opportunities are enormous but the challenges now are to work with the genes and to understand their functions, and that may take perhaps another century or more to achieve. I am sure that this *Handbook* will make an important contribution towards that end.

Walter Bodmer
ICRF, Laboratory Head

Introduction

The *ICRF Handbook of Genome Analysis* is a combination of protocol manual and informational resource, drawing on the expertise of researchers at ICRF and elsewhere. It describes and evaluates a wide range of techniques pertinent to genome analysis. The first volume comprises a description and evaluation of strategies, techniques and protocols for use in the genetic and physical mapping of the human genome (Chapters 1–19). Genome analysis techniques are also used widely in the study and diagnosis of cancers and other diseases, and some of these applications are also covered. A glossary of abbreviations and acronyms is included at the end of Volume 2.

The second volume includes a comprehensive review section of approaches to DNA sequencing (Chapters 20–25) and reviews of progress in the analysis of the genomes of important model systems (Chapters 26–34). Organisms covered include the mouse, *Drosophila*, *Caenorhabditis elegans*, *Saccharomyces cerevisiae* (the first eukaryote organism to have its genome fully sequenced), *Escherichia coli*, *Arabidopsis thaliana* and rice. The second volume concludes with chapters on information resources and how to access them (Chapters 35–37) and appendices covering materials, preparation of blood samples, suppliers and other useful addresses, extensive tables of mapped human disease genes and mouse knockouts, and tables of chromosomal aberrations associated with cancer. An index to the complete handbook is included at the end of each volume.

One of the main driving forces behind the effort to map and sequence the human genome is the isolation and characterization of human disease genes. The figure on the following page shows the typical stages in such an enterprise and the relevant chapters in the *Handbook* that deal with the techniques involved.

Nigel K. Spurr

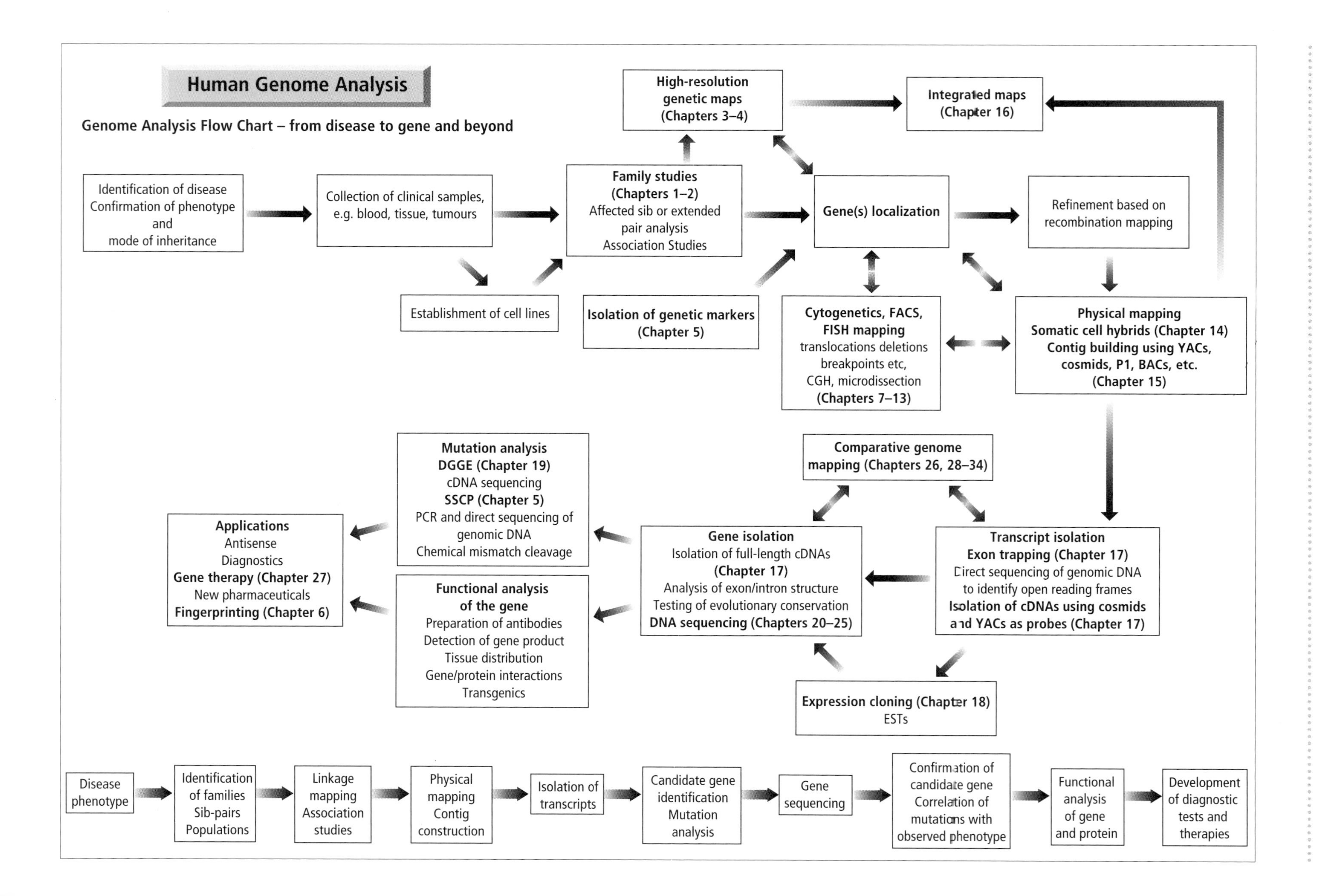
Human Genome Analysis
Genome Analysis Flow Chart – from disease to gene and beyond
Identification of disease
Confirmation of phenotype
and
mode of inheritance
Collection of clinical samples,
e.g. blood, tissue, tumours
Establishment of cell lines
Family studies
(Chapters 1–2)
Affected sib or extended
pair analysis
Association Studies
High-resolution
genetic maps
(Chapters 3–4)
Isolation of genetic markers
(Chapter 5)
Integrated maps
(Chapter 16)
Gene(s) localization
Refinement based on
recombination mapping
Cytogenetics, FACS,
FISH mapping
translocations deletions
breakpoints etc,
CGH, microdissection
(Chapters 7–13)
Physical mapping
Somatic cell hybrids (Chapter 14)
Contig building using YACs,
cosmids, P1, BACs, etc.
(Chapter 15)
Comparative genome
mapping (Chapters 26, 28–34)
Transcript isolation
Exon trapping (Chapter 17)
Direct sequencing of genomic DNA
to identify open reading frames
Isolation of cDNAs using cosmids
and YACs as probes (Chapter 17)
Gene isolation
Isolation of full-length cDNAs
(Chapter 17)
Analysis of exon/intron structure
Testing of evolutionary conservation
DNA sequencing (Chapters 20–25)
Expression cloning (Chapter 18)
ESTs
Mutation analysis
DGGE (Chapter 19)
cDNA sequencing
SSCP (Chapter 5)
PCR and direct sequencing of
genomic DNA
Chemical mismatch cleavage
Functional analysis
of the gene
Preparation of antibodies
Detection of gene product
Tissue distribution
Gene/protein interactions
Transgenics
Applications
Antisense
Diagnostics
Gene therapy (Chapter 27)
New pharmaceuticals
Fingerprinting (Chapter 6)
Disease
phenotype
Identification
of families
Sib-pairs
Populations
Linkage
mapping
Association
studies
Physical
mapping
Contig
construction
Isolation of
transcripts
Candidate gene
identification
Mutation
analysis
Gene
sequencing
Confirmation of
candidate gene
Correlation of
mutations with
observed phenotype
Functional
analysis
of gene
and protein
Development
of diagnostic
tests and
therapies

Section 4 **DNA sequencing**

Section 4 Introduction

Stephan Beck

The Sanger Centre, Hinxton, Cambridge CB10 1SA, UK

The genotype of all living organisms is represented by nucleic acids in the form of either DNA (deoxyribonucleic acid) or RNA (ribonucleic acid). Therefore, it is not surprising that DNA/RNA sequence analysis is so fundamental to genome analysis and the understanding of biological processes in general. The technical breakthrough for DNA sequencing came in 1977 when Maxam and Gilbert described a method for sequencing by base-specific chemical degradation [1] and Sanger and coworkers described a method for enzymatic sequencing using chain-terminating inhibitors [2]. Although the underlying concept of both methods has not changed, many different strategies, modifications and specialized protocols have been developed over the years. These improvements have made large-scale sequencing possible and genome sequencing projects feasible. The largest genome sequenced to date is that of the yeast, *Saccharomyces cerevisiae* (15 Mb) [4], while the longest contiguous sequence is from the nematode *Caenorhabditis elegans* (18 Mb) [5]. In addition, over 8000 of the estimated 100 000 human genes have already been sequenced [5], among them many disease genes.

The main aim of this section of the *Handbook* is to help investigators select the best strategy for a particular project by providing critical reviews as well as protocols of the currently available sequencing methods, and by sharing personal experience and tips. As illustrated in Fig. I.4.1, different codons can code for the same amino acid. By analogy, different methods can be used to sequence a particular DNA fragment. In some cases, the method of choice is just a matter of personal preference, whereas in others, potential problems can be avoided if the right questions are considered early on. Whatever the size and type of the sequencing project, it is well worth spending a little time thinking about the best strategy. A typical sequencing project is subdivided into multiple, individual steps such as cloning, template preparation, labelling, sequencing, electrophoresis, detection, band calling, editing and analysis. The subsequent chapters (Chapters 20–25) discuss various aspects and options for the stages in DNA sequencing, starting with the most basic choices, such as random versus ordered sequencing, *in vivo* versus *in vitro* template amplification, enzymatic versus chemical sequencing, radioactive versus nonradioactive detection.

Factors that may affect the choice of a particular strategy are cost, convenience, accuracy and the complexity of the DNA to be sequenced. Simple clone identification or generation of expressed sequence tags (ESTs), for instance, may not require the same accuracy as the generation of novel sequence (where sequencing of both strands is imperative) and a high GC or repeat content of the target DNA can restrict the choice of applicable strategies. The general choice of sequencing strategies is covered in Chapter 20 (C. Churcher *et al.*), which discusses the advantages and disadvantages in different situations of strategies such as shotgun sequencing, primer-walking, transposon-mediated sequencing and the use of nested deletions. For high-quality and efficient DNA sequencing, all steps of a chosen strategy not only have to work individually but also have to be compatible with each other. A good example in this context is the step of template preparation. It is well known that the ability of DNA templates to sequence well cannot be reliably predicted from agarose gels or optical density measurements, although both methods can be quite helpful in finding out why certain templates do not sequence well. In addition, different strategies have different requirements, and therefore the appropriate template quality is usually determined empirically. A comprehensive selection of 18 different protocols for template amplification and purification is described in Chapter 21 (A. Rosenthal *et al.*).

Chapter 22 (P. Heinrich and H. Domdey) discusses both chemical degradation and dideoxy sequencing chemistries. On the basis of the Science Citation Index (1981–1994), dideoxy DNA sequencing appears to be the most frequently used technique in molecular biology (about 33 500 citations in ref. 2 compared with about 3500 citations for the polymerase chain reaction (PCR) of ref. 7). This enormous success is partly due to the excellent commercial support for DNA sequencing technology and its increasing automation. Chapter 22 concludes with a brief look at some potential future sequencing technologies [8], such as sequencing by hybridization, by mass spectrometry and by high-resolution microscopy. Chapter 23 (C. Heller) discusses the principles and practice of slab gel electrophoresis and provides some basic protocols. In Chapter 24, P. Richterich discusses the advantages and disadvantages of various methods of sequence labelling and detection, including enzyme-linked detection methods. In the final chapter of this section, A. Milosavljevic (Chapter 25) describes the use of the program PYTHIA to detect and characterize repetitive sequences in DNA. Many programs for sequence management and sequence analysis are freely available via WWW, anonymous ftp or e-mail from resource centres such as EMBL, NCBI and the UK-HGMP (see Appendix V and Chapters 35 and 37 for addresses and guidance on access).

References

1. Maxam, A. & Gilbert, W. (1977) A new method for sequencing DNA. *Proc. Natl Acad. Sci. USA* **74**, 560–564.
2. Sanger, F., Nicklen, S. & Coulson, A.R. (1977) DNA sequencing with chain terminating inhibitors. *Proc. Natl. Acad. Sci. USA* **74**, 5463–5467.
3. Bankier, A.T., Beck, S., Bohni, S. *et al.* (1991) DNA sequence of the human cytomegalovirus genome. *DNA Sequence* **2**, 1–12.
4. Wilson, R. *et al.* (1994) 2.2 Mb of contiguous nucleotide sequence from chromosome III of *C. elegans*. *Nature* **368**, 32–38.
5. Schmidtke, J. & Cooper, D.N. (1992) A comprehensive list of cloned human DNA sequences. *Nucleic Acids Res.* **20**, 2181–2198.


```
    *  S  E  L  E  C  T  I  N  G  *  T  H  E  *  R  I  G  H  T
     D  L  N  *  S  V  L  L  M  V  K  P  M  N  S  V  S  D  T  L
      I  *  I  R  V  Y  Y  *  W  L  N  P  *  I  A  Y  R  T  H  *
5'-TGATCTGAATTAGAGTGTACTATTAATGGTTAAACCCATGAATAGCGTATCGGACACACT
            10        20        30        40        50        60
3'-ACTAGACTTAATCTCACATGATAATTACCAATTTGGGTACTTATCGCATAGCCTGTGTGA
    S  R  F  *  L  T  S  N  I  T  L  G  M  F  L  T  D  S  V  S
     I  Q  I  L  T  Y  *  *  H  N  F  G  H  I  A  Y  R  V  C  *
      D  S  N  S  H  V  I  L  P  *  V  W  S  Y  R  I  P  C  V  L

    R  I  T  L  P  S  N  N  Q  R  A  P  R  K  D  I
     G  *  R  F  R  A  I  I  N  A  R  H  G  R  I  L
      D  N  A  S  E  Q  *  S  T  R  A  T  E  G  Y  *
   AGGATAACGCTTCCGAGCAATAATCAACGCGCGCCACGGAAGGATATTGA-3'
            70        80        90       100
   TCCTATTGCGAAGGCTCGTTATTAGTTGCGCGCGGTGCCTTCCTATAACT-5'
    P  Y  R  K  R  A  I  I  L  A  R  W  P  L  I  N
     S  L  A  E  S  C  Y  D  V  R  A  V  S  P  Y  Q
      I  V  S  G  L  L  L  *  R  A  G  R  F  S  I  S
```

Fig. I.4.1 Example of DNA sequence with six-phase amino acid translation in one-letter code. Asterisks (*) indicate stop codons.

6 Robison, K., Gilbert, W. & Church, G.G. (1994) Large scale bacterial gene discovery by similarity search. *Nature Genet.* **7**, 205–214.
7 Saiki, R., Scharf, S., Faloona, F. *et al.* (1985) Enzymatic amplification of β-globin genomic sequences and restriction site analysis for diagnosis of sickle cell anemia. *Science* **230**, 1350–1354.
8 Smith, L. (1993) The future of DNA sequencing. *Science* **262**, 530–531.

Chapter 20 Sequencing strategies

Carol Churcher, Mary Berks, Sharen Bowman, David Buck & Karen Thomas

The Sanger Centre, Wellcome Trust Genome Campus, Hinxton, Cambridge CB10 1SA, UK

This chapter is dedicated to the memory of Dr Mary Berks who died on 12 May 1996.

20.1 Introduction

DNA sequencing has become a widely used tool in many molecular biology laboratories. Its numerous applications range from sequencing tens or hundreds of bases to verify a cloning step, through a few kilobases for many gene-sized projects, to kilobases and megabases for whole-genome projects. The choice of sequencing strategy to be employed can be based to some extent on project size but there are other considerations such as finances, available equipment, and expertise within the laboratory. There are two groups of methods which can be used for the generation of DNA fragments for sequencing: random and ordered. The random methods use restriction enzyme digestion or shearing to produce fragments. The ordered or direct methods commonly used are nested deletions, primer walking, and transposon-facilitated sequencing. Each system is discussed with regard to its applications, advantages and disadvantages. Multiplex sequencing is presented as a method that can utilize DNA fragments generated either randomly or by a direct approach.

20.2 Shotgun sequencing

The shotgun approach (Fig. 20.1) is a random method for DNA sequencing. The DNA is randomly fragmented, each fragment is sequenced, and these 'short reads' are then re-assembled in order, generating the original DNA sequence. With the introduction of more automated methods for sequence data collection, the shotgun approach is gaining popularity and is the method of choice for the majority of large-scale sequencing projects already under way [1–3] (see, for example, Chapter 29).

The stages needed for a shotgun project are described in Sections 20.2.1–20.2.4.

20.2.1 Library preparation

The quality of library production is critical to the success of a shotgun project. It is necessary to

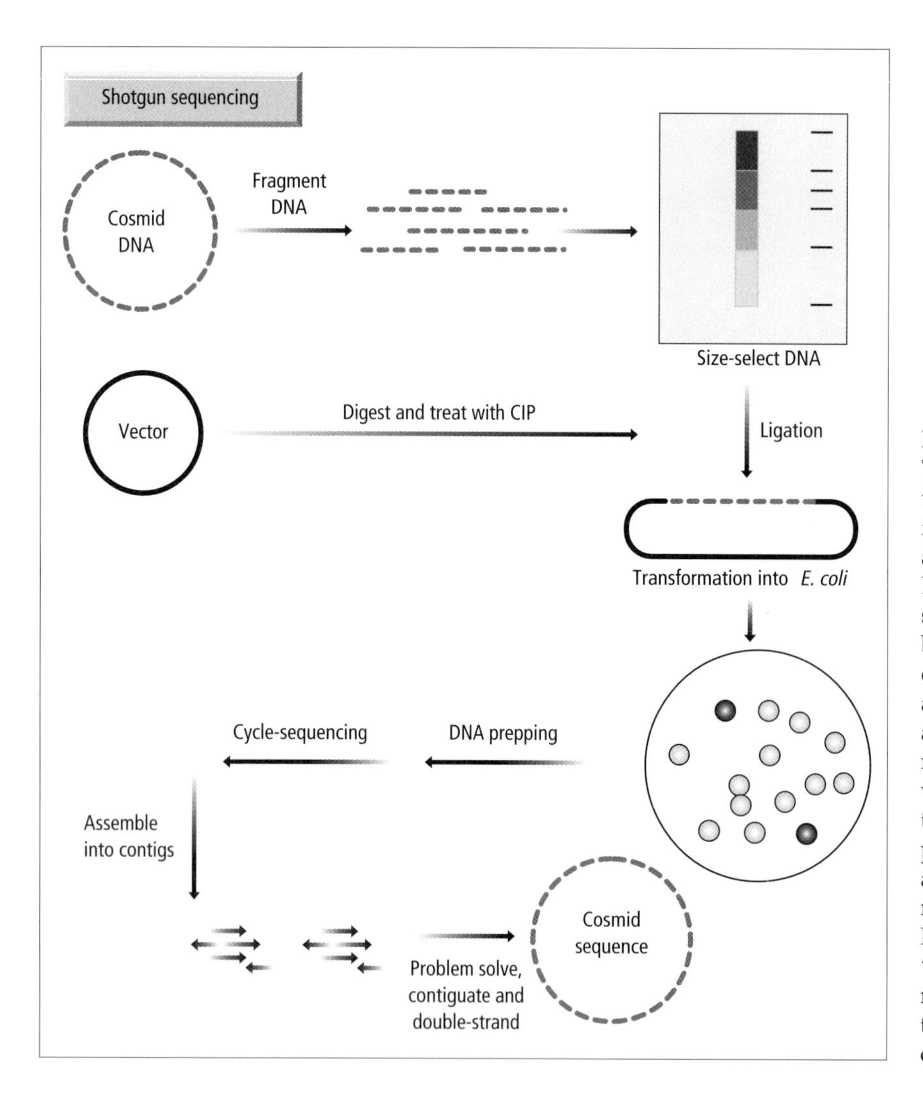

Fig. 20.1 Shotgun sequencing. The target DNA is fragmented using the method of choice, and fragments are size-selected on an agarose gel. The fragmented DNA is then blunt-ended using a suitable enzyme such as mung bean nuclease. Vector DNA is digested with a suitable enzyme and treated with calf intestinal alkaline phosphatase (CIP) to reduce self-ligation. Insert and vector are ligated together and then transformed into *E. coli* and plated out. Individual plaques are prepped, sequenced and the reads assembled to form contigs. Finally, the cosmid is 'contiguated' using directed methods such as primer walking to reconstruct the original cosmid sequence.

fragment the DNA to be sequenced as randomly as possible. If the vector DNA comprises a significant percentage of the total DNA (for example, a lambda-clone) then it is advisable to purify the DNA insert before proceeding any further. For a cosmid project this is not necessary, as the cosmid vector makes up only about 15% of the total DNA, and can provide a convenient internal control to check sequencing accuracy. Several methods can be used to fragment DNA [4–6], but none has been proved to do so in a completely random manner. Sonication is most frequently used, which preferentially breaks DNA in A/T-rich regions or close to the ends of a linear DNA fragment. It is therefore necessary to end-ligate linear DNA before the sonication step to minimize a nonrandom distribution of fragments. To ensure random breakage the DNA solution must be kept as cold as possible (0–4 °C) and sonication restricted to short bursts. Calibration of the sonicator is necessary to achieve the desired size range of fragments. After repairing the damaged ends of the DNA with an enzyme such as mung bean nuclease, the DNA is run on an agarose gel and fragments in the desired size range excised and purified. A fragment range of 1.4–2 kb is optimal: it is greater than the longest achievable read length, is relatively stable in commonly used sequencing vectors, and gives some flexibility for gap closure in the latter stages of a shotgun project. Fragments are then blunt-end ligated into the sequencing vector, usually the single-stranded phage vector M13 and transformed into a suitable *Escherichia coli* host. A test plate of each library should be generated, scored for the ratio of insert containing/no insert clones, and it is advisable, prior to large-scale data production, to sequence some of the clones produced to confirm library quality and randomness.

20.2.2 Sequencing

The difficulties presented in the sequencing phase of a shotgun project are mainly of scale. A large number of sequencing templates of consistently high quality must be prepared. Each template is treated in exactly the same way, so automation of both template preparation and sequencing reactions is possible [7]. A vector-specific universal primer is used for all templates, eliminating the need for costly custom primer synthesis. If 960 sequencing templates are processed for a cosmid project, after sample losses due to reaction failures or presence of cloning (cosmid) or sequencing (M13) vector, an average of 700 useful sequences remain. For a cosmid with an insert size of 35–40 kb, this would give a sequence redundancy of five- to sixfold after assembly. If possible, sequencing reactions should be loaded on a system allowing automatic data collection and computer entry, as it is necessary to process many samples for a single project.

20.2.3 Assembly and editing

Sequences generated by a shotgun project are random, and do not normally exceed 500 bp in length. It is essential to have a sequence assembly software package to process the individual sequences and assemble them into contigs [8]. If repetitive stretches of DNA are present (see Chapter 25), more stringent criteria must be used for data assembly. After a suitable number of reads have been entered, then individual sequences can be compared and edited to remove miscalls and identify sequence-dependent problems.

20.2.4 Directed sequencing

It is rare for the sequence data from the shotgun phase of a project to assemble into one contig of the size expected for the cosmid insert. Usually the data are left in a small number of contigs, and a more directed approach must be applied in order to fill the remaining gaps. If the reason for the remaining gaps was purely due to the random nature of clone distribution intrinsic to this method, then entering more shotgun reads would eventually fill the gaps. However, in practice, gaps seem to occur for other reasons, some due to cloning problems in M13 or regions difficult to sequence. Gaps can be filled by a variety of techniques, including primer walking, or using the polymerase chain reaction (PCR) and reverse primer to sequence the other end of clones lying at the end of contigs, or sequencing a PCR product generated from the original cosmid DNA. Other sequence-dependent problems will also remain, and must be resolved using a directed approach.

The shotgun approach to a sequencing project has several advantages over a directed strategy. Sequence acquisition is rapid: more than 95% of the sequence can be generated rapidly in the initial shotgun phase of the project. A single primer is used, so oligonucleotide synthesis costs are low. The redundancy inherent in this method gives a high level of confidence that the final sequence generated is correct, as each base is sequenced five or six times on average. However, producing redundant data is not an efficient way of generating a sequence—a directed approach would be more efficient in this respect. The library preparation stage of a shotgun project is vital, and can prove difficult. It is necessary to have the ability and equipment to process a large number of samples and automatically enter each sequence into a computer for processing. Problems can arise during data assembly if the DNA sequenced has internal repetitive regions. Even after the shotgun phase, a directed approach must be applied to completely contiguate the project and solve any problems. Nevertheless, the shotgun approach to sequencing has proved to be a dependable method for the rapid, automatable generation of large amounts of sequence.

20.3 Primer walking

In contrast to the random shotgun approach described above, primer walking is a totally directed strategy (Fig. 20.2). It initially involves the use of a custom-made oligonucleotide to prime synthesis from a known site on the DNA template. Subsequent primers are designed according to the sequence that is obtained from successive sequencing reactions until the complete sequence has been elucidated.

In theory, such a totally directed approach to sequencing represents the most efficient strategy, since only a minimal set of sequencing reactions is performed and redundancy is kept relatively low. In practice, however, there are problems. First, the cost of manufacturing even the very small amounts of oligonucleotide required for each sequencing reaction is high. In addition, the method relies on individual primers annealing to unique sites on the DNA template, which presents a problem where DNA is repetitive. There is also a problem in the variation of sequence quality and read length when using custom-made oligonucleotides compared with sequencing methods relying on universal primers which tend to be more consistent.

Primer walking is least suitable for large-scale sequencing projects where many different primers would be required and where problems are most likely to arise from repetitive DNA sequences. It could, however, be the method of choice if the size of cloned DNA to be sequenced is small, especially relative to the vector. In such cases, primers can be designed (or may be commercially available) to known vector sequence flanking the DNA insert, and further primers can be synthesized as described above to complete the sequence of the insert. There is thus no unnecessary sequencing of vector and, with a small insert size of around 2–3 kb, it should be possible to obtain sequence on both strands with as few as 10 different oligonucleotides. Another advantage of primer walking is that there is no problem with assembly (a major consideration in random shotgun approaches) since sequencing always progresses from a known point on the DNA.

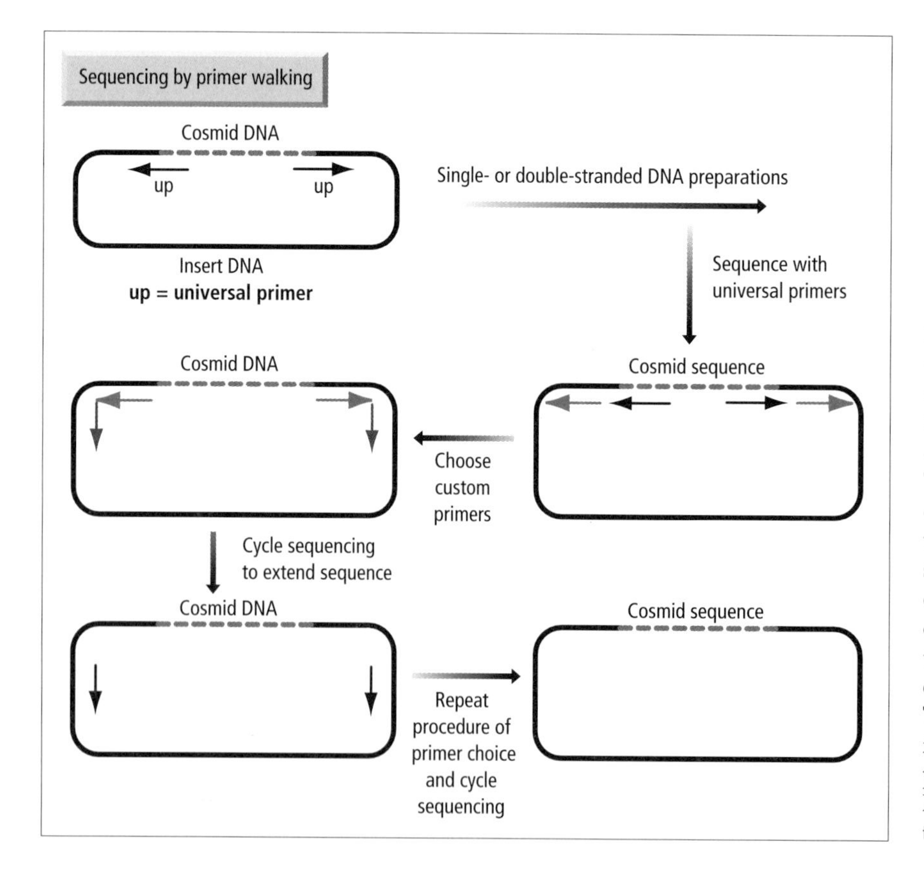

Fig. 20.2 Sequencing by primer walking. Double or single-stranded DNA preparations of the target DNA (in this case a cosmid with insert) are made. The first round of sequencing can be carried out using universal primers to the vector sequence. This yields sequence data for the insert DNA. Custom oligos are then chosen and used to extend the insert sequence data, typically by about 400 bp. This new data is used to choose further custom oligos, and the procedure is repeated until one has 'walked' the entire length of the insert DNA.

Sequencing single- or double-stranded DNA with (unlabelled) custom-made oligonucleotides can be conveniently carried out with fluorescently labelled dideoxy-terminators using either T7 polymerase (Sequenase) or *Taq* polymerase in the sequencing reactions. The labelled products can then be run on ABI 373 A automatic gel readers. One major advantage of these chemistries is that they are efficient at resolving severe compressions and eliminating stops, both of which are common problems in reactions using fluorescently labelled universal primers (as in random shotgun strategies). However, as mentioned above, the quality of the reads may not be sufficiently consistent.

An attempt has been made in recent years to address one of the major drawbacks of a primer walking strategy: the cost of primer synthesis. Two independent groups have reported [9,10] that strings of three adjacent hexamers, and even pentamers, could prime DNA sequencing reactions uniquely without the need for ligation of the adjacent oligonucleotides. In one case, the DNA template was saturated with a bacterial single-stranded binding (SSB) protein which suppresses priming by individual hexamers and most pairs of hexamers but stimulates priming by the 3′ hexamer of most strings of three or more contiguous hexamers. The second case, described by its developers as 'modular primer walking', utilizes pentamers and heptamers without the need for SSB protein. Currently, a pentamer–heptamer–heptamer array has been found to be most successful. In this case, libraries are required of both pentamers and heptamers. Since the two heptamers in the array have two degenerate positions each, the size of the heptamer and pentamer libraries is the same, at 512 sequences each, to cover all possible combinations. In the case of the hexamer library described above, a minimum set of 4096 hexamers is required. Although there has been some success using this strategy for sequencing parts of single-stranded M13 DNA and double-stranded T7 DNA, it remains to be seen whether or not it will be practical in large-scale sequencing projects. The drawbacks of multiple priming to repetitive regions of DNA and possible inconsistent quality of data remain.

Primer walking has been successfully used in conjunction with a primarily random shotgun-based approach in sequencing of *Caenorhabditis elegans* [1,11]. In this large-scale sequencing project, random reads from M13 templates (derived from cosmid DNA) are first carried out to moderate levels of redundancy (fivefold) and this is followed by a directed primer walking strategy to achieve gap closure and to complete double-stranding.

In summary, the main advantages of primer walking are in the sequencing of relatively short stretches of DNA (where the vector is larger than the DNA insert itself) or as a method of completing sequence data following a random shotgun-based strategy in large-scale sequencing projects.

20.4 Transposon-mediated sequencing

Shotgun sequencing has the disadvantage of generating high redundancy and is wasteful of reagents and computer time. In contrast, primer walking is low in redundancy but costly in the requirement for large numbers of oligonucleotides, and is problematic with regard to repetitive sequences. Another approach takes advantage of transposons for directed sequencing of DNA and is both low in redundancy and requires only two universal primers.

Transposons are specialized DNA segments that can move randomly to many sites in a DNA molecule. For use in sequencing, transposons have been engineered to contain the binding sites for the universal sequencing primers from M13. They can be used in two ways, first as simple mobile universal primer-binding sites, and second as a means of generating nested deletions. Both random and ordered transposon-mediated techniques have been developed. The standard mobile site transposon approach is better suited to smaller DNA targets because of the requirement to map the site of insertion, whereas the nested deletion approach is good for large cosmids because of the ease of mapping of the deletion end points.

20.4.1 Transposons as mobile priming sites

The random insertion of a transposon into a target DNA does not disrupt the original linkage among the component parts and can therefore provide access to all positions without recourse to 'shotgun' subcloning or for extensive primer walking [12,13].

The Tn3 family of transposons, which includes Tn3 and γδ, display little sequence specificity for transposition and have been used in plasmid sequencing [12,14,15] as a mobile primer site and for the generation of nested deletions. Their presence can be selected for by a simple bacterial mating. This selection is based on the formation of a cointegrate as the initial product of transposition in the donor cell, and resolution of this cointegrate after it has transferred to the recipient cell (Fig. 20.3). The final product is a simple insertion of the transposon bracketed by a 5-bp direct repeat of target DNA. The *Drosophila* genome sequencing project (see Chapter

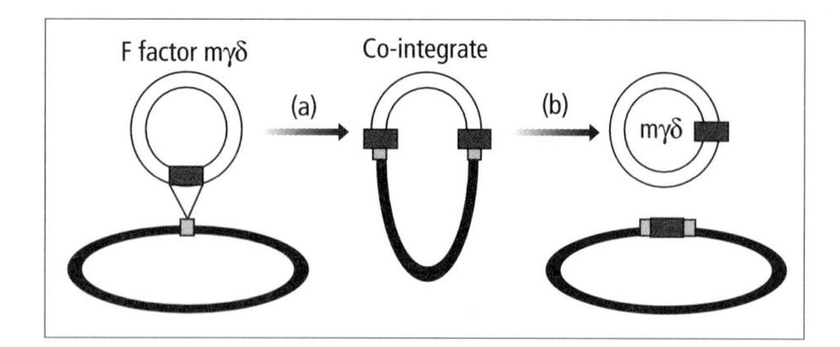

Fig. 20.3 The transposition of γδ and mini-γδ (mγδ). Selection for insertion into a non-conjugative plasmid by mobilization and conjugal transfer. (a) Transposition from donor F factor (for wild-type miniγδ, mγδ) to a non-conjugative plasmid forming an F : plasmid cointegrate. (b) Transfer of cointegrate to a recipient plasmid-free cell via conjugation, followed by resolution to yield a plasmid containing one copy of mγδ, and the donor F factor molecule.

28, Section 28.3.3.4) has developed a transposon-facilitated system [16]. The DNA fragment to be sequenced is subcloned into a minimal plasmid and sites of γδ transpostion mapped by PCR [17] so that a minimal set of sequencing templates can be rapidly obtained. Over 1.25 Mb have been sequenced in a 2-year period using this method.

One of the *E. coli* genome sequencing projects (see Chapter 31) utilizes a complete set of mapped and overlapping λ-phage clones [18]. The sequencing of these clones is being carried out using a strategy employing a Tn5-derived minitransposon developed by Kasai *et al.* [19]. The small minitransposon is necessary for sequencing λ-clones because the size of typical transposons is close to the maximum capacity for the phage head. Tn5 is ideal because all it needs for transposition into λ is a pair of Tn5 inverted repeats (19 bp) and a selectable marker, such as the suppresser tRNA gene *supF* (Tn5*supF* elements are about 300 bp long and contains only *supF*, primer-binding sites and the 19-bp terminal repeats) (Fig. 20.4). Although it inserts less randomly than Tn3 it does not require the formation of a

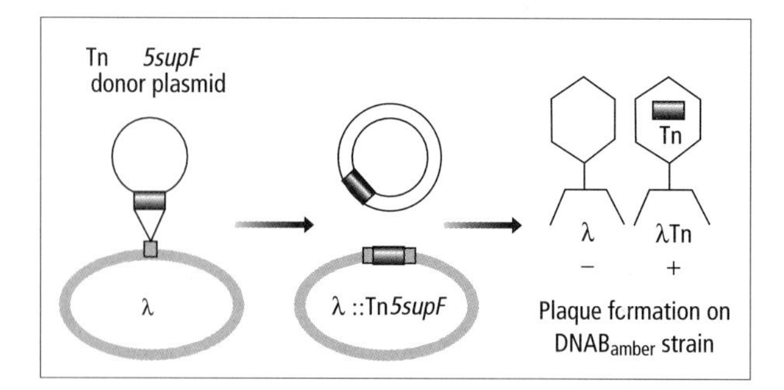

Fig. 20.4 Transposition of Tn5*supF* to phage lambda. Selection for Tn5*supF* insertion in phage lambda (λ) by plaque formation on dnaB$_{amber}$ *E. coli* strain.

cointegrate intermediate (which would necessitate the formation of a λ-lysogen). The transposon Tn5*supF* is delivered in a single cycle of λ-phage infection of donor cells, and transposon containing phage selected by culturing on dnaB$_{amber}$ cells [20]. The transposition product is a simple insertion of Tn5 bracketed by a 9-bp direct repeat of a target DNA.

Although transposons can provide mobile binding sites for sequencing primers, sequence acquisition is random and can be highly repetitive. To reduce redundancy the position of insertion must first be mapped. This can be achieved by restriction digest followed by southern hybridization or by the use of PCR [19,21].

20.4.2 Transposon-generated deletions

The use of transposons to generate deletions in a target DNA was a strategy developed by Ahmed using Tn9 [22]. Deletions are generated by intramolecular transposition to new sites within the same plasmid. This causes a division of the plasmid with only the portion containing the plasmid origin of replication being recoverable. The original Tn9 strategy had the disadvantage of nonrandom insertion and the inability to obtain sequence from both strands. These problems were addressed by Wang *et al.* [23] in the development of a new γδ transposon which transposes more randomly and allows recovery of deletions extending into a cloned fragment in either direction.

The transposition of γδ is replicative, with one entire copy of the transposon ending up in each of the reciprocal deletion derivatives. Both derivatives are made viable by inserting a plasmid origin of replication within the transposon ends. Selection for deletions in either direction is made possible by the incorporation in the transposon vector of cotran-selectable marker genes (*sacB*$^+$, for sucrose sensitivity, and *str*A$^+$ for streptomycin sensitivity) just outside each end of the transposon, and selectable *kan*$^+$ (Kan^r) and *tet*$^+$ (Tet^r) genes between the cloning site and *sacB* and *strA*, respectively. Selection on sucrose tetracycline medium yields deletions extending from one end, while selection on streptomycin kanamycin medium yields deletions in the other direction (Fig. 20.5).

Orientational deletions can be selected, none of which extends beyond the end of the insert DNA. After transposition, one end of the transposon always abuts a deletion end point and can serve as a 'universal' primer-binding site. Deletion end points are mapped by plasmid size, allowing selection of end points in any region.

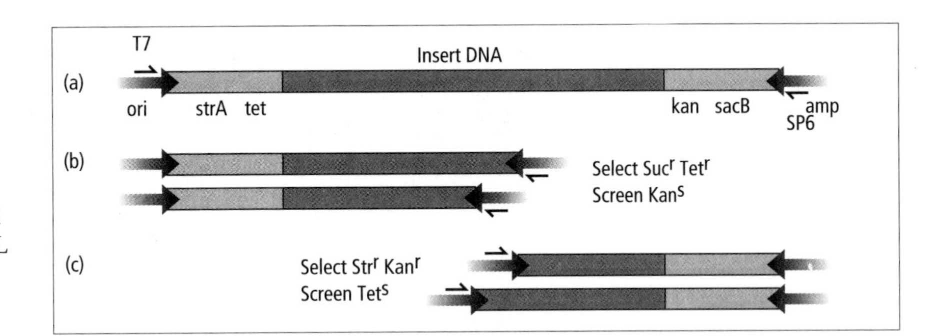

Fig. 20.5 Transposon-generated deletions in pDUAL. (a) pDUAL with cloned fragment of DNA. Selection of clockwise (b) and counterclockwise (c) deletions.

20.5 Nested deletions

Another directed method involves the generation of nested deletions by sequential digestion. The simplest such method begins with the cleavage of double-stranded DNA at a unique site, shortened by enzymatic digestion with *Bal*31 [24] or exonuclease III, followed by either S1 nuclease or exonuclease VII.

*Bal*31 digests double-stranded linear DNA progressively by liberating mononucleotides. The drawback of *Bal*31 is that it will digest both ends of the linearized fragment, necessitating the recloning of the target sequence. Exonuclease III, on the other hand, catalyses the stepwise removal of 5′ mononucleotides from from double-stranded DNA with a protruding 5′ terminus, or blunt end. Ends with protruding 3′ termini are untouched by the enzyme [25]. Judicial choice of restriction enzymes can lead to directed deletions from one end only leaving the vector intact and thus avoiding the necessity for recloning [26,27]. For this reason, exonuclease III is the enzyme of choice for the generation of nested deletions. Figure 20.6 illustrates the procedures involved in the generation of nested deletions using exonuclease III.

An obvious advantage to this approach is the ability to use a universal primer for all the deleted fragments. Thus, sequencing can be carried out using either fluorescent or radioactive sequencing

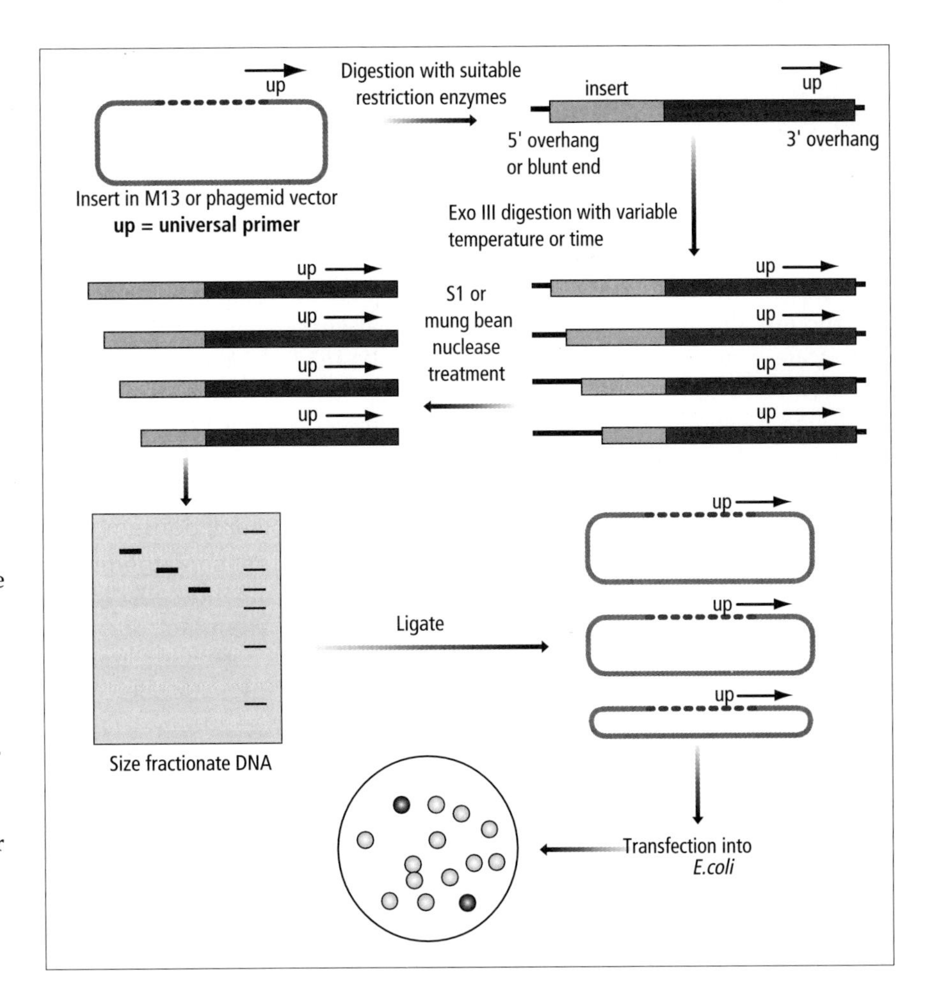

Fig. 20.6 Generation of nested deletions. Double-stranded DNA is digested to leave the insert susceptible to exonuclease III (ExoIII) digestion while the vector remains 'safe'. DNA is treated with ExoIII for variable lengths of time or at different temperatures to yield a range of deleted products. Mung bean or S1 nuclease is used to blunt-end the DNA. Fragments are size-selected and recircularized prior to transfection. The universal primer can then be used to sequence stepwise across the insert.

without the production of costly primers. Thus the method is ideal for small molecular biology labs working on a tight budget.

Exonuclease III is a nonprocessive enzyme with a very stable reaction rate. This allows one to control the rate and extent of deletions by manipulation of reaction temperatures and times (Table 20.1). Thus, progressive deletions of 200–250 bp can be reproducibly achieved.

Because exonuclease III deletions are so controllable, the redundancy can be tailored to meet the demands of the project. In addition, if suitable clones are chosen, both strands can be covered by the use of two sets of deletions.

20.5.1 Applications of nested deletions to sequencing projects

Exonuclease III deletions can be carried out using either M13 or phagemid clones. Thus projects up to about 9 kb are entirely feasible. The limitation of this technique is the distance over which deletions can be obtained. Use of the nested deletion strategy has been reported for a number of cDNA and small-scale genomic sequencing projects, for example the sequencing of the mouse myelin P2 protein gene (4 kb) [28].

Single-stranded DNA can also be used to create exonuclease III deletions, when oligonucleotide hybridization is used to facilitate restriction digest linearization of the target DNA [29,30]. Dale *et al.* [29] reported the generation of deletions for single-stranded M13 clones. They used the method to sequence 2.6 kb of the maize mitochondrial 18 S rDNA and its 5′ flanking region 'in less than a week'.

Although the method would not be ideal for large-scale genome projects, it has obvious advantages when dealing with 'gaps' or repeats. The controlled deletion strategy means that one knows at the assembly stage which copy of the repeat one is dealing with. This 'mapping' data obviously does not exist with a random shotgun approach. Thus, it is a powerful method to be used alongside, or in addition to, a traditional shotgun approach for such projects.

Table 20.1 Exonuclease III deletion rates, based on 20 units per microgram of double-stranded DNA.

Temperature (°C)	**Deletion rate** (bp min^{-1})
37	400
34	375
30	230
23	125

Additionally, nested deletions can been used to facilitate the sequencing of difficult regions, for example very GC-rich sequences [31].

20.6 Multiplex sequencing

Multiplex sequencing is a variant of the shotgun method in which a number of samples are pooled during processing (thereby saving labour) and separated by hybridization detection at the end of the process. Like simple shotgun sequencing it can be used for any size of project, including genomic, YAC, P1, cosmid, plasmid, or cDNA. It can utilize a variety of DNA fragments such as restriction, shotgun, nested deletion, and PCR product (Chapter 21). DNA can be either single or double stranded. Sequencing chemistry (Chapter 22) can be either chemical or dideoxy using a variety of polymerases. Gels can be capillary, electroblotted, or direct transfer electrophoresis (DTE) may be used (see Chapter 23). Sequence detection can be either radioactive or chemiluminescent (Fig. 20.7).

Multiplex sequencing was first described by Church and Kieffer-Higgins [32] as a method of sequencing and electrophoresing many DNA samples in each set of four lanes on a gel and then probing as many times as there are samples in each set. The procedure utilized a set of 20 vectors, each containing two unique oligonucleotide sequences inserted either side of a cloning site. Sonicated genomic DNA fragments were cloned into each of these 20 vectors and transformed. The constructs were then pooled, and DNA preparations carried out by alkaline lysis in groups of 20. Chemical sequencing by the Maxam and Gilbert method [33] was used. After gel electrophoresis the DNA was electro-eluted onto a nylon membrane and cross-linked by UV irradiation. Probing of the sequences was carried out by ^{32}P labelling of oligonucleotides (oligos) complementary to those used in the vector constructs and probing the membranes systematically with each oligo. Bands were visualized by autoradiography using standard X-ray film. This method allowed 40 different sequences to be read from each set of reactions.

Recently, much research has been carried out to find alternative nonradioactive methods of probing membranes. For reasons of safety and storage, chemiluminescent probes are now preferred by many laboratories. Compared with radioactive materials, these nonradioactive labels present no disposal problem, require much shorter exposure times (typically, 10–15 min, whereas radioactive labels require hours or even days) and are easily incorporated into existing protocols without the

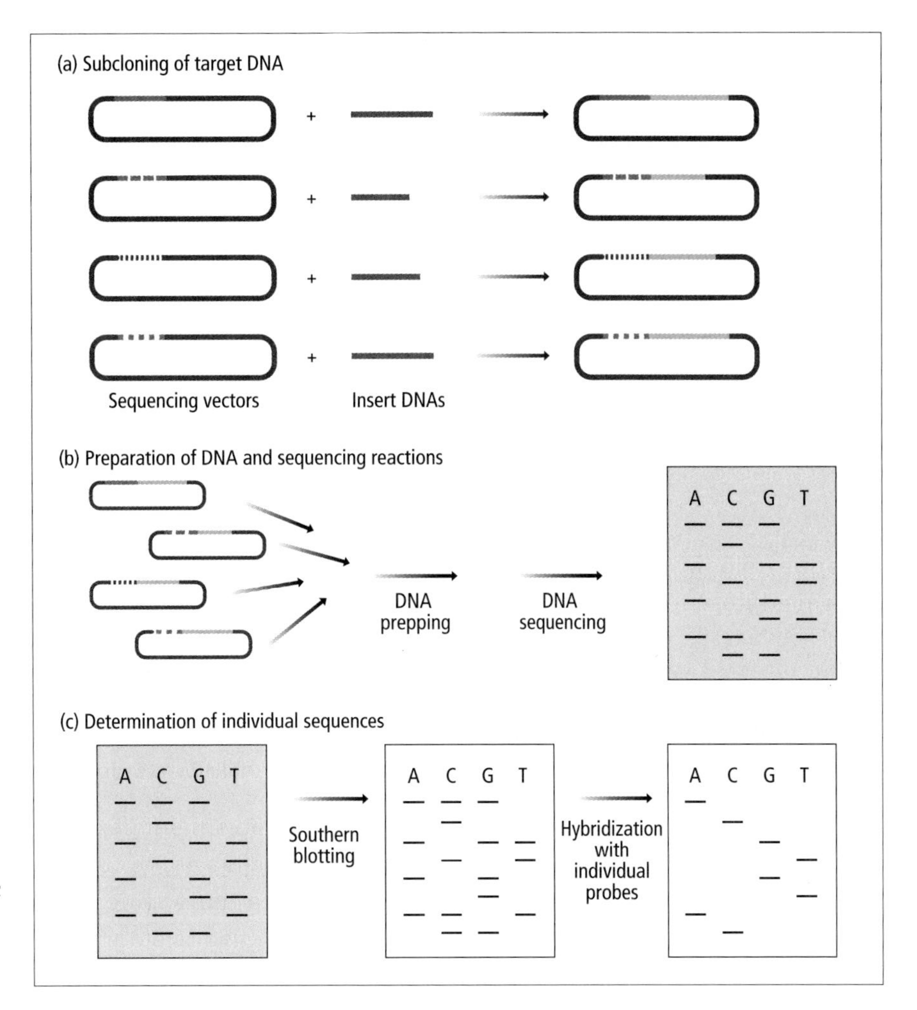

Fig. 20.7 Multiplex DNA sequencing. (a) Subcloning of target DNA. Different DNA fragments are subcloned into a set of different sequencing vectors. (b) Preparation of DNA and sequencing reactions. Templates are pooled, then grown, prepped and sequenced in the same way one would treat individual samples. (c) Determination of individual sequences. Sequencing gels are transferred to membranes, which are hybridized sequentially with probes specific for each different vector, revealing the sequences for the individual clones.

need for expensive equipment. There have been several reports of changes to the original multiplex procedure and protocols are available for the use of:

1 multiplex 'tagged' vectors;
2 single vector, multiplex 'tagged' primers;
3 single vector, multiplex probe labelling.

20.6.1 Multiplex 'tagged' vectors

Multiplex vectors which can be used with the commonly used Sanger dideoxy sequencing technique [34] are now available [35]. Ten-plex vectors based on the original method of incorporating unique oligonucleotide sequences into the vector construct can be used in double-stranded sequencing with forward and reverse primers to produce 20 different sequences to be loaded in each set of four lanes on the gel. Chemiluminescence is used to visualize bands. After the gel has been blotted, biotinylated DNA probes which are complementary to the plex tagging sequences are hybridized.

After hybridization, biotinylated alkaline phosphatase is linked to the bound DNA probe through a streptavidin bridge. The chemiluminescent reaction is performed by adding dioxetane, which produces light at 477 nm and allows the sequence band patterns to be recorded by exposure to X-ray film.

20.6.2 Single vector, multiplex 'tagged' primers

A single vector system using 'tagged' primers has been described [36]. This method uses standard M13/dideoxy sequencing reactions on single-stranded DNA and a set of eight 'tagged' primers. These are 37mers synthesized with the M13 forward primer sequence at the 3′ end and a series of different 20mers at the 5′ end. Sequencing reactions are set up with each of the eight 'tagged' primers then pooled and electrophoresed. The gel is blotted then probed with 20mer oligonucleotide sequences (complementary to each of the 'tagged' primers) which have been labelled with digoxigenin and detected using antidigoxigenin antibody–alkaline phosphatase conjugate and chemiluminescent dioxetane substrate. Sequence is recorded by a 10–15 min exposure to X-ray film.

20.6.3 Single vector, multiplex probe labelling

A method using a single vector and primer with one

Table 20.2 Summary of sequencing strategies.

Method	Advantages	Disadvantages
Random		
Shotgun	Rapid data acquisition Utilizes universal primer Suitable for most sizes of project	High redundancy Rarely completely random
Ordered		
Nested deletion	Low redundancy Useful for repetitive sequence and gaps	Unsuitable for large-scale projects
Primer walking	Low redundancy No subcloning required	Cost of custom primers Unsuitable for large-scale projects
Transposon-mediated	Low redundancy Utilizes universal primer Suitable for any size of project	Large-scale mapping may be required
Random or ordered		
Multiplex	Rapid data acquisition from minimal number of gels Suitable for any size of project	Gel-reading bottleneck

of four different hapten labels attached is described [37]. This method allows the use of familiar vector/primer systems by labelling the primer, this system is comparable with the fluorescent primers used in the ABI373 fluorescent sequencing machines. The hapten labels currently used are biotin, digoxigenin, 2,4-dinitrophenyl and fluorescein. Four separate sequencing reactions are set up with these primers then all T reactions are pooled, all C reactions, etc. Sequence products are simultaneously electrophoresed and blotted by DTE as described by Beck and Pohl [38]. DTE involves the movement of a membrane across the bottom of the gel as it is running, onto which the DNA is transferred as it elutes from the gel. Bands are detected by sequential probing with either hapten-specific alkaline phosphatase or streptavidin–alkaline phosphatase conjugate and a dioxetane solution.

The choice of vector/primer to be used in multiplex sequencing is dependant on the size of the project, available equipment and knowledge of sequencing methodologies. Small projects could quite easily be set up using methods 2 or 3. All protocols can be carried out using commonly available laboratory equipment, although for high throughput automation is desirable, which would increase the cost. The use of commonly available sequencing vectors requires no multiple cloning steps so can be set up in any molecular biology laboratory. Using 'tagged' primers currently allows 8-plexing though one could increase this number by synthesizing more primers. The hapten-labelled primer system allows 4-plexing but work is continuing to find different labels so this number could increase. For large-scale sequencing projects the 'tagged' vector method would seem to be the most efficient and this has been demonstrated in many cases, for example with *Mycoplasma* [39], *Mycobacterium* [40] and *E. coli* [32]. The main disadvantage of these systems would seem to be interpreting and recording the sequence from the autoradiographs. For a small project a sonic digitizer [41] is quite adequate, but for genome-sized projects film reading is a definite bottleneck. An adequate and reliable automated system has yet to be demonstrated although such a system is described by Richterich and Church [42].

A summary of the strategies described in this chapter is given in Table 20.2.

References

1 Wilson, R., Ainscough, R., Anderson, K. *et al.* (1994) 2.2 Mb of contiguous nucleotide sequence from chromosome III of *C. elegans*. *Nature* **368**, 32–38.
2 Johnston, M., Andrews, S., Brinkman, R. *et al.* (1994) Complete nucleotide sequence of *Saccharomyces cerevisiae* chromosome VIII. *Science* **265**, 2077–2082.
3 Bankier, A.T., Beck, S., Bohni, R. *et al.* (1991) The DNA sequence of the human cytomegalovirus genome. *DNA Sequence* **2**, 1–12.
4 Deininger, P.L. (1983) Random subcloning of sonicated DNA: application to shotgun DNA sequence analysis. *Anal. Biochem.* **129**, 216–223.
5 Screifer, L.A., Gebauer, B.K., Qiu, L.Q.Q. *et al.* (1990) Low pressure DNA shearing: a method for random DNA sequence analysis. *Nucleic Acids Res.* **18**, 7455.
6 Bankier, A.T., Weston, K.M. & Barrell, B.G. (1987) Random cloning and sequencing by the M13/Dideoxynucleotide chain termination method. *Meth. Enzymol.* **155**, 51–93.
7 Smith, V., Craxton, M., Bankier, A.T. *et al.* (1993) Preparation and fluorescent sequencing of M13 clones—microtiter methods. *Meth. Enzymol.* **218**, 173–187.

8 Dear, S. & Staden, R. (1991) A sequence assembly and editing program for efficient management of large projects. *Nucleic Acids Res.* **19,** 3907–3911.
9 Kieleczawa, J., Dunn, J.J. & Studier, F.W. (1992) DNA sequencing by primer walking with strings of contiguous hexamers. *Science* **258,** 1787–1791.
10 Kotler, L., Zevin-Sonkin, D., Sabolev, I. *et al.* (1993) DNA sequencing: modular primers assembled from a library of hexamers or pentamers. *Proc. Natl Acad. Sci. USA* **90,** 4241–4245.
11 Sulston, J., Du, Z., Thomas, K. *et al.* (1992) The *C. elegans* genome sequencing project: a beginning. *Nature* **356,** 37–41.
12 Berg, C.M., Wang, G., Strausbaugh, L.D. & Berg, D.E. (1993) Transposon-facilitated sequencing of DNAs cloned in plasmids. *Meth. Enzymol.* **218,** 279–306.
13 Davies, C.J. & Hutchinson, C.A. III (1991) A directed DNA sequencing strategy based upon Tn3 transposon mutagenesis: application to the ADE1 locus on *Saccharomyces cerevisiae* chromosome I. *Nucleic Acids Res.* **19,** 5731–5738.
14 Liu, L., Whalen, W., Das, A. & Berg, C.M. (1987) Rapid sequencing of cloned DNA using a transposon for bidirectional priming sequence of the *Escherichia coli* K-12 autA gene. *Nucleic Acids Res.* **15,** 9461–9469.
15 Berg, C.M., Vartak, N.B., Wang, G. *et al.* (1992) The myd-1 element, a small γδ (Tn1000) derivative useful for plasmid mutagenesis, allele replacement and DNA sequencing. *Gene* **113,** 9–16.
16 Strathmann, M., Hamilton, B.A., Mayeda, C.A. *et al.* (1991) Transposon-facilitated DNA sequencing. *Proc. Natl Acad. Sci. USA* **88,** 1247–1250.
17 Saiki, R., Scharf, S., Faloona, F. *et al.* (1985) Enzymatic amplification of β-globin genomic sequences and restriction site analysis for diagnosis of sickle cell anemia. *Science* **230,** 1350–1354.
18 Kohara, Y., Akiyama, K. & Isono, K. (1987) The physical map of the whole *Escherichia coli* chromosome—application of a new strategy for rapid analysis and sorting of a large genomic library. *Cell* **50,** 495–508.
19 Kasai, H., Isono, S., Kitakawa, M. *et al.* (1992) Efficient large-scale sequencing of the *Escherichia coli* genome: implementation of a transposon and PCR-based strategy for the analysis of ordered λ phage clones. *Nucleic Acids Res.* **20,** 6509–6515.
20 Kurnit, D.M. & Seed, B. (1990) Improved genetic selection for screening bacteriophage libraries by homologous recombination *in vivo*. *Proc. Natl Acad. Sci. USA* **87,** 3166–3169.
21 Strausbaugh, L.D., Bourke, M.T., Sommer, M.T. *et al.* (1990) Probe mapping to facilitate transposon-based DNA sequencing. *Proc. Natl Acad. Sci. USA* **87,** 6213–6217.
22 Ahmed, A. (1987) Use of transposon-promoted deletions in DNA sequence analysis. *Meth. Enzymol.* **155,** 177–204.
23 Wang, G., Blakesley, R.W., Berg, D.E. & Berg, C.M. (1993) pDUAL: a transposon-based cosmid cloning vector for generating nested deletions and DNA sequencing templates *in vivo*. *Proc. Natl Acad. Sci. USA* **90,** 7874–7878.
24 Poncz, M., Solowiejczyk, D., Ballantine, M. *et al.* (1982) Non-random DNA sequence analysis in bacteriophage M13 by the dideoxy chain-termination method. *Proc. Natl Acad. Sci. USA* **79,** 4289–4302.
25 Rogers, S.G. & Weiss, B. (1980) Exonuclease III of *E. coli* K-12 and AP endonuclease. *Meth. Enzymol.* **65,** 201–211.
26 Henikoff, S. (1984) Unidirectional digestion with exonuclease III creates targeted breakpoints for DNA sequencing. *Gene* **28,** 351–359.
27 Henikoff, S. (1987) Unidirectional digestion with exonuclease III in DNA sequence analysis. *Meth. Enzymol.* **155,** 156–165.
28 Narayanan, B., Kaestner, K.H. & Tennekoon, G.I. (1991) Structure of the mouse myelin P2 protein gene. *J. Neurochem.* **57,** 75–80.
29 Dale, R.M.K., McClur, B.A. & Houchins, J.P. (1985) A rapid single-stranded cloning strategy for producing a sequential series of overlapping clones for use in DNA sequencing: application to sequencing the corn mitochondrial 18S rDNA. *Plasmid* **13,** 31–40.
30 Milavetz, B. (1992) Preparation of nested deletions in single-strand DNA using oligonucleotides containing partially random base sequences. *Nucleic Acids Res.* **20,** 3529–3530.
31 Gewain, K.M., Occi, J.L., Foor, F. & MacNeil, D.J. (1992) Vectors for generating nested deletions and facilitating the subcloning of *G* + C-rich DNA between *E. coli* and *Streptomyces* sp. *Gene* **119,** 149–150.
32 Church, G.M. & Kieffer-Higgins, S. (1988) Multiplex DNA sequencing. *Science* **240,** 185–188.
33 Maxam, A.M. & Gilbert, W. (1977) A new method for sequencing DNA. *Proc. Natl Acad. Sci. USA* **74,** 560–564.
34 Sanger, F., Nicklen, S. & Coulson, A.R. (1977) DNA sequencing with chain-terminating inhibitors. *Proc. Natl Acad. Sci. USA* **74,** 5463–5467.
35 Creasey, A., D'Angio, L. Jr, Dunne, T.S. *et al.* (1991) Application of a novel chemiluminescence-based DNA detection method to single-vector and multiplex DNA sequencing. *Biotechniques* **11,** 102–109.
36 Chee, M.S. (1991) Enzymatic multiplex DNA sequencing. *Nucleic Acids Res.* **19,** 3301–3305.
37 Olesen, C.E.M., Martin, C.S. & Bronstein, I. (1993) Chemiluminescent DNA sequencing with multiplex labelling. *Biotechniques* **15,** 480–485.
38 Beck, S. & Pohl, F.M. (1984) DNA sequencing with direct blotting electrophoresis. *EMBO J.* **3,** 2905–2909.
39 Ohara, O., Dorit, R.L. & Gilbert, W. (1989) Direct genomic sequencing of bacterial DNA: the pyruvate kinase I gene of *Escherichia coli*. *Proc. Natl Acad. Sci. USA* **86,** 6883–6887.
40 Smith, D.R., Richterich, P., Rubenfield, M. *et al.* (1993) High throughput multiplex sequencing of mycobacterial genomes. *Genome Sci. Technol.* **1,** 31 (Abstract).
41 Staden, R. (1984) A computer-program to enter DNA gel reading data into a computer. *Nucleic Acids Res.* **12,** 499–503.
42 Richterich, P. & Church, G.M. (1993) DNA sequencing with direct transfer electrophoresis and nonradioactive detection. *Meth. Enzymol.* **218,** 187–222.

Chapter 21 Template amplification and purification

André Rosenthal, Matthias Platzer, Gerald Nyakatura & Monique Jouet*

Institute of Molecular Biotechnology, Department of Genome Analysis, Beutenbergstrasse 11, 07745 Jena, Germany

**University of Cambridge Clinical School, Addenbrooke's Hospital, Department of Medicine, Hills Road, Cambridge CB2 2QQ, UK*

21.1 Introduction

DNA sequencing is a complex process requiring several complicated steps such as DNA cloning, mapping of cloned fragments, subcloning, template generation, sequencing reactions, gel electrophoresis, transferring sequence data into the computer, data analysis and data handling. The availability of high-quality DNA templates in large numbers is important for the success of any DNA sequencing method. It is therefore not surprising that a wide variety of methods and techniques has been developed over the past 15 years to generate templates by bacterial amplification or by the polymerase chain reaction (PCR). Many groups, including those engaged in large-scale sequencing projects, still favour M13, phagemid or plasmid templates which are obtained *in vivo* by bacterial growth. *In vitro* amplification by PCR has significantly changed the way in which DNA can be produced and PCR and DNA sequencing are more and more linked together in modern strategies.

This chapter presents a selection of standard, improved and new protocols for the *in vivo* (Fig. 21.1) and *in vitro* (Fig. 21.2) amplification and purification of templates for DNA sequencing. Most of the protocols presented are used successfully in our laboratories. Table 21.1 gives an overview of the number of templates that can be handled in each protocol.

21.2 In vivo amplification methods

21.2.1 Plasmid templates

21.2.1.1 Standard alkaline lysis miniprep [1–4]
Plasmids are purified from liquid cultures that contain the appropriate antibiotic and have been inoculated with a single bacterial colony picked from an agar plate. Many of the currently used plasmid vectors (e.g. the pUC series) replicate to such high copy numbers that they can be purified in large yield from 2- to 3-ml cultures that have simply been grown to late log phase in standard LB

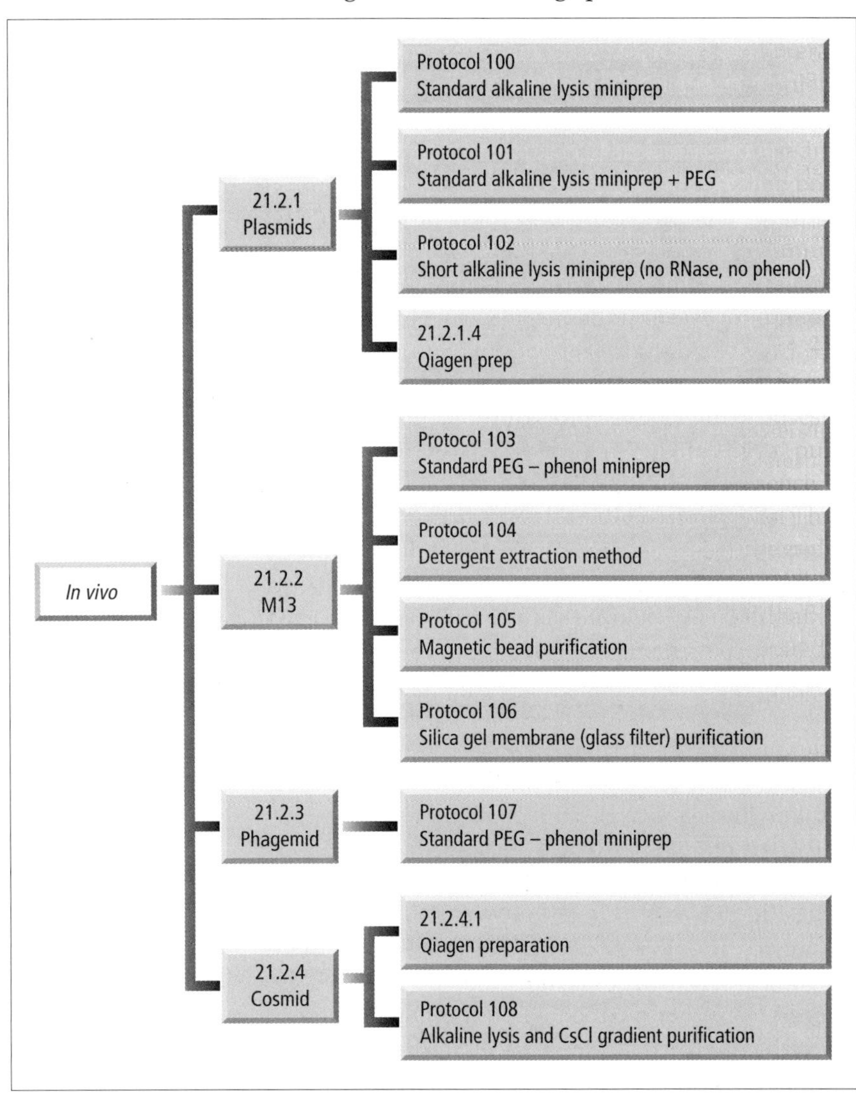

Fig. 21.1 Methods for *in vivo* amplification and purification of templates for DNA sequencing.

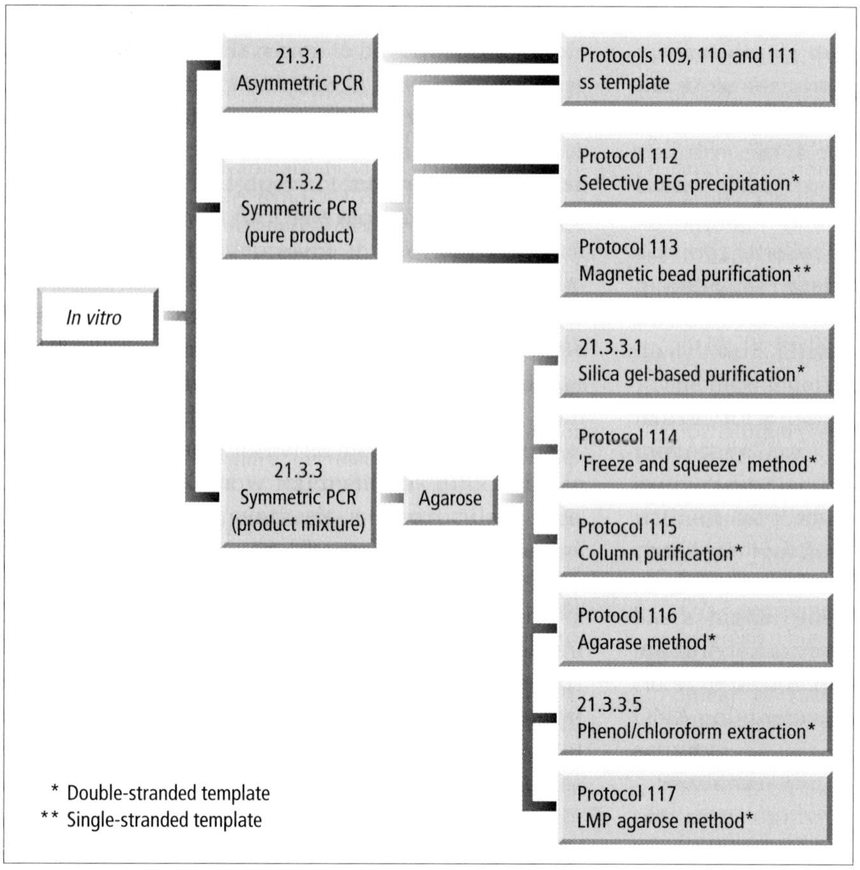

Fig. 21.2 Methods for *in vitro* amplification and purification of templates for DNA sequencing.

medium. The procedure described in Protocol 100 works well for plasmids smaller than 15 kb. Plasmid DNA prepared according to this standard protocol is a good template for radioactive sequencing.

For convenience, 1.5 ml or 2.0 ml snap-cap microfuge tubes are used. Depending on the availability of microcentrifuges the procedure is performed in batches of 12, 18 or 24 samples at a time. The alkaline lysis protocol can also be adapted for the growth and preparation of plasmid DNA from small culture volumes (250–500 μl) utilizing standard 96-well plates. Several hundred plasmids can be prepared simultaneously, yielding sufficient DNA for at least one cycle sequencing reaction.

21.2.1.2 Standard alkaline lysis miniprep followed by polyethylene glycol precipitation [5]

Plasmid DNA prepared according to the standard alkaline lysis protocol still contains residual salt, detergent and some RNase-resistant tRNA species which might interfere with fluorescent sequencing reactions that use dye primer or dye terminator chemistry. In order to obtain high-quality template for fluorescent sequencing, the plasmid DNA should be further purified. We recommend an extra polyethylene glycol (PEG) precipitation (Protocol 101) that effectively removes these impurities.

21.2.1.3 Short alkaline miniprep for plasmid DNA [4,6]

The major difference between this method, which is given in Protocol 102, and the standard method (Protocol 100) is that no RNase digestion and phenol/chloroform steps are used. Alkaline lysis is followed by ethanol precipitation using one Vol. ethanol only. Prior to sequencing, sequencing primer is added and the plasmid DNA is denatured by treatment with NaOH followed by neutralization with HCl. The alkaline treatment also degrades the RNA. The miniprep DNA is an excellent template for radioactive sequencing using α-^{35}S-ATP and Sequenase 2.0. It is not suitable for fluorescent sequencing using dye primer or dye terminator chemistry.

21.2.1.4 QIAGEN preparation

Qiagen is among several suppliers providing fast and convenient DNA purification systems for DNA sequencing. Two basic purification techniques can be used separately (1 and 2) or in combination (3).

(1) *Separation based on anion-exchange chromatography (QIAwell systems)* DNA is a negatively charged biopolymer which can easily bind to a solid support possessing positive charges. After washing, the DNA is removed from the support with simple salt

Table 21.1 Methods for generation of sequencing templates and their scale of application.

Topic	Method	Product	Sample number 1–12	24–48	96
21.2.1	*In vivo*				
21.2.1	**Plasmid**				
21.2.1.1	Standard alkaline lysis miniprep	ds	×	–	–
21.2.1.2	Standard alkaline lysis miniprep + PEG	ds	×	–	–
21.2.1.3	Short alkaline miniprep	ds	×	×	–
21.2.1.4	Qiagen method				
	Anion-exchange based-purification (Qiagen mini/midi/maxi/mega prep QIAwell 8 Kit)	ds	×	×	–
	Silica gel-based purification (QIAprep Spin Kit, QIAprep 8 Kit)	ds	×	×	–
	Anion-exchange + silica gel purification (QIAwell Plus Kit, QIAwell Ultra Kit)	ds	–	×	×
21.2.2	**M13**				
21.2.2.1	Standard PEG-phenol method	ss	×	×	–
21.2.2.2	Detergent extraction method	ss	–	×	×
21.2.2.3	Magnetic bead purification (affinity capture)	ss	–	–	×
21.2.2.4	Silica gel-based membranes (QIAprep)	ss	×	×	–
21.2.3	**Phagemid**				
	Standard PEG-phenol method	ss	×	×	–
21.2.4	**Cosmid**				
21.2.4.1	Anion-exchange-based purification	ds	×	–	–
21.2.4.2	CsCl gradient	ds	×	–	–
21.3	*In vitro*				
21.3.1	**Asymmetric PCR (pure product)**	ss	×	×	×
21.3.2	**Symmetric PCR (pure product)**	ds	×	×	×
21.3.2.1	Selective PEG precipitation	ds	×	×	×
21.3.2.2	Magnetic bead purification by affinity capture	ss	×	×	×
21.3	**Symmetric PCR (product mixture)**				
21.3.3.1	Silica gel-based purification	ds	×	–	–
21.3.3.2	'Freeze and squeeze' method	ds	×	–	–
21.3.3.3	Column purification	ds	×	×	–
21.3.3.4	Agarase method	ds	×	–	–
21.3.3.5	Phenol/chloroform extraction	ds	×	–	–
21.3.3.6	Direct sequencing of DNA in low melting point agarose	ds	×	–	–

buffers followed by ethanol precipitation to remove the salt and to concentrate the DNA. This process is called anion-exchange chromatography and can be used to separate and purify DNA from complex mixtures obtained after alkaline lysis and avoids phenol/chloroform extractions. The effectiveness of the separation process depends largely on the properties of the solid support.

The QIAGEN support is a special silica gel-based resin or membrane which has an optimal particle size of around 100 μm and a special surface coating containing diethyl aminoethyl (DEAE) groups which creates a high surface charge density. It selectively separates DNA from substances such as proteins, carbohydrates and others.

Different kits are available which allow purification of plasmid DNA in microgram to milligram amounts and contain anion-exchange resins in different formats: for example single column (QIAGEN-tip 20–10.000) or strips of eight columns (QIAwell 8 Plasmid Kit).

(2) *Silica gel-based purification (QIAprep systems)* Single- or double-stranded DNA selectively precipitates/adsorbs to silica gel surfaces in the presence of high concentrations of chaotropic salts. Under optimized conditions, carbohydrates, RNA, and proteins do not adsorb and are removed by washing. DNA is then eluted under low salt conditions from the support. QIAprep Spin Plasmid Kits utilize microfuge spin columns and are ideal for 1–12 minipreps. QIAprep 8 Plasmid Kits utilize 8-well strips for higher throughput of up to 48 minipreps.

(3) *Combined anion-exchange chromatography/silica gel-based purification* The quality of the plasmid DNA can be further improved by combining anion-exchange separation with selective binding to silica gel membranes. This avoids the final ethanol precipitation step. Kits based on a combination of these two principles are also available in two different formats: strips of 8 columns (QIAwell 8 Plus Plasmid Kit) and 96-column array (QIAwell 96 Ultra Plasmid Kit). With the help of a vacuum manifold 48–96 samples can be easily processed within 2 h.

The decision about which kit is to be used for a particular project depends on the number and the amount of template that will be needed and on the type of sequencing chemistry. For cycle sequencing with *Taq* polymerase much less plasmid template is needed than for one single sequencing reaction using Sequenase 2.0 or T7 polymerase (e.g. 1 µg of plasmid DNA is needed for cycle sequencing with ABI fluorescent dye terminators, whereas 3–5 µg are needed for the Pharmacia ALF sequencer if dye primer chemistry is used). Also, if a directed sequencing strategy using custom-made primers is adopted, several sequencing reactions with the same template must be performed which results in a need for much more template DNA.

QIAprep plasmid kits are more suitable for small-scale projects based on manual sequencing with radioactive or chemiluminescent labelling. Good results can be obtained with QIAwell plasmid kits and fluorescent T7 DNA polymerase sequencing. For large-scale sequencing projects QIAwell 8 Plus and QIAwell 96 Ultra Kits are favoured because several hundred plasmid templates of high quality can conveniently be prepared in a short period of time. Typical yields are 10–20 µg DNA per sample. The plasmid DNA is suitable for dye primer and dye terminator cycle sequencing using the ABI 373 A system. It can also be used in the Pharmacia ALF system.

For protocols for these methods refer to the instruction manuals supplied with the kits. In order to achieve high yields grow bacteria under optimal conditions. The use of 5 ml overnight cultures in 50-ml Falcon or glass tubes is recommended. Qiagen provides a robotic workstation designed to use the QIAwell 96 Ultra Plasmid kit.

21.2.2 M13 templates

21.2.2.1 Standard PEG–phenol method [7–10]
The growth and purification of M13 DNA from small-volume cultures (2–3 ml) is a straightforward, rapid and easy procedure to perform. M13 phages do not lyse their hosts, but are released from infected cells as the cells continue to grow and divide. Phage particles are readily separated from the bacteria by centrifugation. After precipitation of the phage particles with PEG, the single-stranded DNA is recovered by removal of the phage protein coat with phenol and purified by ethanol precipitation. Small liquid cultures (1.5 ml) can be processed in disposable polypropylene tubes using microcentrifuges and yield pure single-stranded DNA (≈5–10 µg) sufficient for several sequencing experiments. It is possible to grow and purify batches of 12, 18 or 24 minicultures sequentially, so that up to 96 samples can be processed in a day. However, the handling of those numbers is tedious and much time is spent opening and closing tubes and transferring tubes in and out of centrifuges. There are several protocols for M13 minipreps in 96-well microtitre plates. The reader is referred to these procedures if the processing of several hundred templates is being contemplated. Protocol 103 describes a standard PEG–phenol method for recovery of DNA from M13 phage.

21.2.2.2 Detergent extraction method [11,12]
Protocol 104 combines PEG precipitation of M13 phage particles with a nonionic or ionic detergent extraction coupled with heating to denature the M13 protein coat. The whole procedure of growing cultures, PEG precipitation, centrifugation and detergent extraction is performed in special 96-tube boxes allowing the easy handling of hundreds of M13 clones per day. The M13 DNA obtained is an excellent template for fluorescent sequencing using dye primer chemistry.

This detergent extraction method avoids phenol extraction and ethanol precipitation. It is therefore faster and less labour intensive than the standard PEG–phenol method (see Protocol 103).

21.2.2.3 Magnetic bead purification [13–15]
M13 single-stranded DNA can conveniently be purified for large-scale sequencing applications using biotinylated M13-specific oligonucleotides

coupled to streptavidin-coated paramagnetic beads (Protocol 105). First, phage supernatants are lysed. Second, PEG and paramagnetic beads with the oligo probe are added and the oligo is allowed to anneal to the template DNA. Then the beads are separated using a magnet and the supernatant is removed. Finally, the beads are repeatedly washed before the template is released from the beads by heating.

This procedure is carried out in 96-well plates. The yield per sample is ≈0.5 µg DNA which is sufficient for one cycle sequencing reaction. Since centrifugation is not necessary (except for the initial clearing of the bacterial lysates) most of the time is spent waiting for lysis or annealing. Therefore it is possible to process several plates in parallel up to 192 probes in 2 h.

We are currently sequencing cosmids containing genomic DNA by the shotgun approach. We routinely generate 1000–1500 M13 templates per cosmid which then are subjected to cycle sequencing with dye primer chemistry and analysed on 373 A ABI sequencers. The average read length is 400 bases per sample.

The magnetic bead prep procedure can also be adapted to robotic workstations like the Catalyst (Applied Biosystems) and the Biomek 1000 or 2000 (Beckman). However, substantial input is required to alter hardware components and to develop reliable protocols. Also, throughput with these machines is rather limited. New hardware with higher throughput is presently being developed.

Paramagnetic beads with M13-specific oligo probes are commercially available from Promega and Dynal.

21.2.2.4 Silica gel membrane (glass filter) purification [16–18]

M13 phages which previously have been precipitated with acetic acid are applied to silica gel or glass filter membranes. Under these conditions intact phage particles are retained on the membrane. Upon addition of high concentration of chaotropic salt ($NaClO_4$), single-stranded M13 DNA binds (adsorbs) to the membrane, while the phage coat proteins pass through and are efficiently removed. After washing, pure M13 DNA is eluted from the membrane using water or TE.

The method described in Protocol 106 yields a high-quality M13 template DNA which is ideal for any standard manual or automated sequencing application. It is suitable for cycle sequencing using dye primer and dye terminator chemistry and the ABI 373 A as well as for T7 or Sequenase 2.0 extension reactions using dye primer and the Pharmacia ALF system.

21.2.3 Phagemids [19–21]

Several vectors have been developed that combine desirable features of both plasmids and filamentous bacteriophages. These are plasmids containing an origin of replication from a filamentous bacteriophage. They have several attractive features.

1 They provide the same stability and high yields of double-stranded DNA as conventional plasmids.

2 They eliminate the tedious and time-consuming process of subcloning DNA fragments from plasmids to bacteriophage vectors.

3 They are small enough to accommodate segments of foreign DNA up to 10 kb in length that can then be obtained in single-stranded form.

4 Whereas the phagemids can be treated like conventional plasmids, the isolation and purification of phagemid single-stranded DNA, apart from an initial superinfection with helper phage, is as straightforward as for filamentous phages.

Protocol 107 describes the preparation of phagemid DNA.

21.2.4 Cosmids

Cosmids can serve either as a template for sequencing (directed sequencing using custom primers) or as the substrate for generation of shotgun libraries in M13 or plasmids.

21.2.4.1 QIAGEN plasmid kits

A fast and convenient method to generate large amounts (100 µg–10 mg) of cosmid DNA are QIAGEN midi/maxi/mega/giga plasmid kits. Cosmids purified by this procedure are suitable for all kinds of sequencing reactions. But in most cases the read length is considerably shorter than obtained with M13 or plasmid DNA as template.

Cosmid DNA prepared by QIAGEN plasmid kits is often contaminated by considerable amounts of *Escherichia. coli* DNA. Therefore the use of this DNA is not recommended for the generation of subfragment shotgun libraries.

21.2.4.2 CsCl gradient purification [22]

Pure cosmid DNA for library construction can be generated by one round of isopycnic centrifugation over a caesium chloride density gradient (Protocol 108). Using a table top ultracentrifuge (Beckman TL100) this method becomes reasonably fast and easy. Between 5 and 50 µg of supercoiled cosmid DNA can be obtained from a 3-ml CsCl gradient. The *E. coli* contamination of libraries derived from this DNA is less than 5%.

The open circle DNA can also be collected and be

used for direct cosmid sequencing or generation of sequencing templates by PCR.

21.3 In vitro amplification methods

PCR has been used as a means of circumventing bacterial growth in the preparation of DNA templates. The main advantage is that the template preparation becomes a simple biochemical process that can readily be automated if required. There are two major strategies for sequencing of PCR products:

1 direct sequencing of PCR products without cloning;

2 molecular cloning of PCR-amplified material prior to sequence determination.

The direct sequence analysis of PCR products has the advantage that it views an entire mixture of amplified DNA molecules in a single assay and enables rapid and precise determination of sequence identity and variation. For example, direct sequencing of a mixture of two PCR fragments that are heterozygous at one or more base positions will reveal ambiguous signals at these positions. Thus, direct sequencing of PCR fragments is an attractive diagnostic tool for carrier analysis in basic research and routine diagnostics, but can also be used for examining mutations in disease genes. On the other hand, cloning of PCR products allows DNA sequence variants to be separated before sequencing, but a larger number of clones must be analysed to identify polymorphic base positions. Also, errors introduced by the polymerase during PCR amplification are a serious problem if cloning is used prior to sequencing. Thus, direct sequencing of PCR products without an additional cloning step is generally preferable to sequencing cloned material. In addition to the benefit of simplicity, this greatly reduces the potential for errors due to imperfect PCR fidelity, as any random misincorporations in an individual template will not be detectable against the much greater signals of the 'consensus' sequence.

In contrast to plasmid and M13 templates, PCR products are much more difficult to sequence. One problem is that the two strands of a linear double-stranded PCR molecule quickly reanneal, which prevents effective annealing and extension of the sequencing primer. Early protocols suggested the use of organic solvents like DMSO or other detergents to limit template reannealing. Rapid annealing protocols that involved dropping the temperature quickly were also tried to solve this problem.

Several modified PCR protocols were developed to solve this problem in a more general way (Fig. 21.2). In asymmetric PCR (Section 21.3.1) the two primers are added to the reaction mix at different concentrations. Thus, after one primer is exhausted the exponential PCR amplification stops and one of the two strands undergoes further linear amplification. The final product mixture will contain sufficient single-stranded template to be used for sequencing.

In symmetric PCR (Section 21.3.2) the two primers are added to the reaction mix at the same concentration and a double-stranded DNA molecule is formed during amplification. Double-stranded PCR products can easily be sequenced using a linear amplification/sequencing protocol known as cycle sequencing. In this method the sequencing primer is repeatedly annealed to one strand of the PCR molecule and extended by *Taq* polymerase after heat denaturing. Prior to sequencing, excess primers and nucleotides as well as small molecular weight material must be removed from the template. In Section 21.3.2.1 we describe a very effective and easy method for purifying PCR products by selective precipitation with PEG (Protocol 112).

In a different strategy, one of the two strands of a specifically modified PCR product is removed prior to sequence analysis. One method uses the biotin–streptavidin system to affinity purify one of the two PCR strands prior to sequencing (Section 21.3.2.2, Protocol 114). This is achieved by introducing a biotin label into one PCR strand and bind the product to a suitable solid support coated with streptavidin. The free unlabelled strand is then removed using alkaline conditions and the captured strand is sequenced.Another variant is to employ a 5′-phosphorylated primer during PCR to introduce a phosphate group to the 5′ end of one strand. Exonuclease III is then used to 'chew off' this strand leaving the other strand for sequencing.

A major problem with direct sequencing of PCR products, though, is that because of nonoptimal conditions several nonspecific products, in addition to the main PCR product, are often formed during amplification. Also, excess nucleotides and oligonucleotide primers as well as small molecular weight material are still present, leading to a complex mixture. It is important to purify the main PCR product away from nonspecific fragments, nucleotides and primers. The most effective method in our hands is the use of agarose gel electrophoresis for separation of the product mixture followed by isolation of the DNA from agarose. Agarose gel electrophoresis has the advantage of allowing many samples to be handled in parallel. It is cheap and widely distributed. In Sections 21.3.3.1–21.3.3.6 we present several protocols for the isolation of DNA from agarose gels which work well in our laboratories (Protocols 114–117).

Single-stranded PCR templates can be sequenced using conventional primer extension reactions with Sequenase 2.0 or other polymerases. Double-stranded PCR products are effectively sequenced using cycle-sequencing reactions with either labelled primers (radioactive or fluorescence) or with dye terminator chemistry.

21.3.1 Asymmetric PCR (single-stranded template) [23–27]

Asymmetric PCR utilizes an unequal, or asymmetric concentration of the two amplification primers (Protocol 109). During the initial 15–25 cycles double-stranded DNA is generated, but once the limiting primer is exhausted, single-stranded DNA complementary to the limiting primer is accumulated by linear amplification for the next 5–10 cycles. Typical primer ratios for asymmetric PCR are 50:1 or 100:1. The single-stranded template can be sequenced with either the limiting primer or a nested primer.

One problem with asymmetric PCR is that the primer ratio and the thermal cycling conditions must be optimized to ensure reliable generation of single-stranded template suitable for sequencing. This is a tedious process and asymmetric PCR has not proved reliable for routine sequencing. In a slight modification of the original protocol, a regular symmetric PCR with two primers is performed yielding double-stranded DNA. Then, a linear amplification with one of the two original primers is performed using a template containing 1% of the first reaction (Protocol 110).

21.3.2 Symmetric PCR (pure double-stranded template)

During symmetric PCR both primers are added at the same concentration to the reaction mix and a double-stranded PCR product is formed. Total genomic DNA, YACs, cosmids, plasmids or eukaryotic or bacterial cells can be used as a source for template DNA. In order to obtain a pure product in sufficient quantity PCR conditions must be carefully optimized.

Pure double-stranded templates can often be generated from bacterial colonies or cultures (also from phage plaques and stocks) by symmetric PCR using universal primer pairs (e.g. M13–21 forward/M13 reverse, T3/T7, KS/SK) flanking the insertion region. This way plasmid libraries can easily be sequenced in microtitre plates.

The major problem of generating sequencing templates from pure double-stranded PCR products is to remove excess primers and nucleotides as well as larger amounts of truncated amplification products prior to sequencing. In Section 21.3.2.1, we present a method which uses selective PEG precipitation to achieve this (Protocol 112). The pure double-stranded template is then subjected to cycle sequencing.

For diagnostic sequencing (detection of mutations, heterozygous individuals) it is often necessary to use Sequenase 2.0 or T7 polymerase as the sequencing enzyme because they produce a much more even peak distribution than thermostable DNA polymerases like AmpliTaq. In these cases the double-stranded PCR product must be efficiently denatured prior to sequencing. Traditional methods like alkaline treatment or heat denaturing are not efficient because the two PCR strands reanneal very quickly. Several other methods have been published to generate a single-strand template from a double-stranded PCR molecule. One method uses the exonuclease III from λ to chew off one PCR strand which is phosphorylated at its 5′ end. The remaining PCR strand is then sequenced. In Section 21.3.2.2 we present another method of strand separation which is based on biotin capture of one of the PCR strands (Protocol 113).

21.3.2.1 Purification of double-stranded PCR template by selective PEG precipitation [28,29]

Excess PCR primers, nucleotides and truncated PCR products can efficiently be removed by one step precipitation of template with PEG. A special PEG mixture is used to selectively precipitate DNA of more than 150 bp leaving residual primers, nucleotides and small molecular weight PCR products in the supernatant. PEG-precipitated templates can be easily sequenced from both ends using radioactive or fluorescent cycle sequencing methods. Fluorescent dye primer and dye terminator chemistries have been successfully used.

21.3.2.2 Generation of single-stranded template by affinity capture using the biotin–streptavidin system [30–33]

In a very elegant way, biotin is attached to the 5′ end of one strand of the double-stranded PCR product. The biotin label is introduced during PCR by using one biotinylated primer and one unlabelled primer. The biotinylated PCR strand is then captured onto a suitable polymeric support. Paramagnetic beads coated with streptavidin are most suitable for this purpose although streptavidin-coated agarose and plastic surfaces (microtitre wells, pins) have also been used. The captured DNA is then denatured using sodium hydroxide and the second noncaptured strand is removed together with the excess primer and nucleotides by washing. Subsequently, the purified single-stranded template bound to the beads is suitable for sequencing reactions with Sequenase 2.0 or T7 polymerase using radioactive or fluorescent labels. The paramagnetic beads do not interfere with the sequencing reaction.

21.3.3 Symmetrical PCR (product mixture)

Because of nonoptimized amplification conditions and unexplained primer and template variability, the PCR product is often accompanied by one or several minor products. Together with excess primers and nucleotides these have to be removed prior to direct sequencing.

The method of choice to get rid of both is to run the PCR product on an agarose gel, excise the band of interest and recover the DNA. Various procedures have been described for the recovery of DNA from agarose gels. In our hands the following procedures work well.

21.3.3.1 DNA recovery by binding to a silica gel-based matrix

Purification of DNA from agarose slices is based on solubilization of agarose with sodium iodide or perchlorate and selective adsorption of nucleic acids onto silica gel surfaces in the presence of high concentrations of chaotropic salts. Impurities are then washed away and the DNA is eluted with low-salt (e.g. Tris) buffer. Many well-documented, easy-to-use, ready-to-go kits based on this principle are commercially available. For protocols see product guides of the respective companies:

1 GeneClean II Kit (BIO 101 Inc., La Jolla, CA, USA);
2 Qiaex II Kit (Qiagen Inc., Chatsworth, CA, USA);
3 InVisorb Kit (Gesellschaft für Biotechnik mbH, Berlin, Germany);
4 SpinBind Kit (FMC Bioproducts, Rockland, ME, USA);
5 Prep-A-Gene Kit (Bio-Rad Laboratories, Hercules, CA, USA).

Variations of these integrate the more comfortable, albeit usually more expensive, microfuge cartridge systems.

21.3.3.2 DNA recovery by the 'freeze and squeeze' method

In the freeze and squeeze method (Protocol 114), the agarose gel matrix is physically destroyed and the DNA is released together with the water of gelation. As the name implies, the gel slice with the DNA band of interest is frozen and while still frozen squeezed through an appropriate filter (e.g. Millipore's ULTRAFREE-MC 0.45 mm filter unit) by centrifuging at high speed. The filter holds back the dry, powdered agarose. The DNA can then be recovered from the aqueous filtrate by simple ethanol precipitation.

21.3.3.3 DNA recovery by column purification

Purification columns allow the rapid purification of PCR fragments from low melting point agarose gels (Protocol 115). DNA is recovered from the slice of agarose by running the molten gel on a purification column containing a DNA-binding resin (e.g. Magic/Wizard PCR Preps, Promega). After washing the column with isopropanol, DNA is eluted with water. The whole procedure takes about 30 min from start to finish and a clean DNA fragment suitable for radioactive cycle ('fmole', Promega) or fluorescent cycle sequencing with dye terminators (Applied Biosystems) is obtained.

21.3.3.4 Agarase method [34]

DNA can also be recovered from low melting point agarose slices by using an agarose-digesting enzyme (GELase, Cambio Ltd, Cambridge, UK or Agarase, Calbiochem, San Diego, USA) followed by ethanol precipitation of the DNA (Protocol 116).

21.3.3.5 Phenol/chloroform extraction [35]

DNA can easily be recovered from LMP agarose by repeated extraction with phenol and chloroform followed by ethanol precipitation. This traditional method was widely used in the past before DNA binding to a silica gel-based matrix became fashionable. The method is cheap, works very reliably and should also be considered for recovering PCR products from agarose slices for direct sequencing.

21.3.3.6 Direct sequencing of DNA in LMP agarose [36]

Protocols 114–116 and other methods described in Sections 3.3.1–3.3.5 use an additional step to recover the DNA from agarose gels. It has recently been shown that DNA purified in LMP agarose can be directly sequenced in molten agarose (Protocol 117). Therefore, the PCR product can be prepared for sequencing in a single step and sequenced under the same conditions as a normal double-stranded template using techniques already available in most laboratories. DNA purified according to Protocol 117 can be sequenced using Sequenase 2.0 and AmpliTaq. Single primer extension reactions and cycle sequencing protocols can be applied.

Protocol 100 Standard alkaline lysis miniprep of plasmid DNA

For details of solutions, media and materials, see Appendix I. For suppliers and contact addresses see Appendix III.

Materials

- GET buffer: 50 mM glucose, 25 mM Tris-HCl (pH 8.0), 10 mM EDTA
- lysis buffer: 0.2 M NaOH, 1% SDS
- RNase A (200 μg ml^{-1})
- potassium acetate (KOAc) (3 M)
- sodium acetate (NaOAc) (3 M, pH 5.2)
- phenol
- chloroform

Method

1 Spin for 30 s to pellet the cells from 1.5 ml overnight culture. Resuspend cells in 100 μl GET buffer.

2 Add 200 μl freshly prepared lysis buffer, mix gently by inverting a few times (do not vortex), leave on ice for 5 min.

3 Add 150 μl 3 M KOAc solution, mix gently by inverting, leave on ice for 5 min. Spin for 5 min to pellet the chromosomal DNA and cell debris. Transfer 400 μl supernatant to a clean tube.

4 Add 1 ml absolute ethanol and mix by vortexing. Leave at room temperature for 5 min to precipitate the DNA.

5 Pellet the DNA by centrifugation for 5 min at room temperature. Decant supernatant and wash pellet once with 1 ml 70% ethanol. Dry briefly under vacuum.

6 Resuspend the DNA pellet in 100 μl RNase A (200 μg ml^{-1}) and incubate at 37 °C for 1 h.

7 Extract once with 50 μl phenol followed by 50 μl chloroform. Transfer aqueous phase to a new tube and precipitate DNA by addition of $\frac{1}{10}$ vol. 3 M NaOAc (pH 5.2) and 2 vol. ethanol. Incubate for 5 min at room temperature.

8 Pellet the plasmid DNA by centrifugation for 15 min at room temperature. Wash once with 1 ml 70% ethanol, and briefly dry the DNA under vacuum. Dissolve pellet in 100 μl water.

Protocol 101 PEG precipitation of plasmid DNA

For details of solutions, media and materials, see Appendix I. For suppliers and contact addresses see Appendix III.

Materials

- PEG solution: PEG 8000 (26.2%), $MgCL_2$ (6.6 mM), NaOAc (0.6 M, pH 5.2)

Method

1 Precipitate DNA obtained by Protocol 100 (step 8) with 100 μl PEG solution. Mix thoroughly and incubate at room temperature for 5 min.

2 Pellet the plasmid DNA by centrifugation for 15 min at room temperature. Wash once with 1 ml 70% ethanol, and briefly dry the DNA under vacuum. Dissolve pellet in 100 μl water.

Protocol 102 Short alkaline miniprep for DNA

For details of solutions, media and materials, see Appendix I. For suppliers and contact addresses see Appendix III.

Materials

- GET buffer (see Protocol 100)
- lysis buffer (see Protocol 100)
- KOAc (3 M)

Materials

1 Decant 2 ml overnight culture into a 2-ml snap-cap tube. Spin for 30 s to pellet the cells and resuspend pellet in 200 μl GET buffer.

2 Add 400 μl freshly prepared lysis buffer, mix gently by inverting a few times (do not vortex), leave on ice for 5 min.

3 Add 300 μl KOAc solution, mix gently by inverting, leave on ice for 5 min. Spin for 5 min to pellet chromosomal DNA and cell debris. Transfer 800 μl supernatant to a clean 2-ml snap-cap tube.

4 Add 1 vol. (900 μl) absolute ethanol and mix by vortexing and immediately pellet the DNA by centrifugation for 5 min at room temperature. Decant supernatant and wash pellet once with 2 ml 70% ethanol. Dry briefly under vacuum.

5 Resuspend the DNA pellet in 40 µl sterile water. Two microlitres of the DNA are then used for sequencing. Add sequencing primer, 3 µl 0.4 M NaOH and incubate at 65 °C for 5 min. Neutralize with 3 µl 0.4 M HCl.

Protocol 103 Standard PEG–phenol method for recovery of DNA from M13 phage

For details of solutions, media and materials, see Appendix I. For suppliers and contact addresses see Appendix III.

Materials

- TY medium
- PEG/NaCl solution (20%/2.5 M)
- TE buffer (10 mM)
- NaOAc (3 M, pH 5.2)
- phenol
- ethanol

Method

1 Toothpick a white plaque into 1.5 ml of a 1 : 100 dilution of an overnight culture of TG1 cells in TY medium. Shake at 300 r.p.m. for 5–6 h at 37 °C.

2 Transfer the culture to a 1.5-ml microfuge tube and spin for 5 min at top speed. Transfer supernatant into a fresh 1.5-ml microfuge tube containing 100 µl 20% PEG/2.5 M NaCl solution without disturbing the cell pellet. Vortex well and incubate at room temperature for 10 min.

3 Spin for 10 min to pellet the phage particle. Remove carefully all of the supernatant by aspiration. Spin again for 1 min and aspirate off supernatant to remove traces of PEG.

4 Resuspend phage pellet in 100 µl 10 mM TE buffer. Add phenol, vortex thoroughly, and incubate at room temperature for 5 min. Vortex again and spin for 2 min to separate the phases. Transfer 60 µl of the aqueous phase, avoiding the interphase, into fresh 1.5-ml microfuge tube containing 6 µl 3 M NaOAc (pH 5.2) and 150 µl ethanol. Vortex thoroughly and precipitate DNA at –20 °C.

5 Pellet the M13 DNA by centrifugation for 10 min at room temperature. Decant supernatant and wash pellet once with 1.0 ml 70% ethanol. Dry briefly under vacuum.

6 Resuspend DNA in 10–50 µl sterile water to a final concentration of 100 µg ml^{-1}.

Protocol 104 Detergent extraction method for M13 DNA (for 96 samples)

For details of solutions, media and materials, see Appendix I. For suppliers and contact addresses see Appendix III.

Materials

- TY medium
- PEG/NaCl solution (20%/2.5 M)
- Triton-TE extraction buffer: 0.5% Triton X-100, 10 mM Tris-HCl (pH 8.0), 1 mM EDTA (pH 8.0), or
- ionic extraction buffer:, e.g. Tris-HCl (pH 8.0), 1 mM EDTA, 125 mM KI (potassium iodide), 0.16 mM KDS (potassium lauryl sulphate)
- 96-tube boxes (Beckman)
- 96-tube cap (Beckman)
- 1.2-ml centrifuge tubes (Beckman) or
- 96 deep well microtitre plate (Beckman)
- rotor #362349 (Beckman)
- 3M silver tape (R.S. Hughes Company)
- 96-well microtitre plate (Corning)

Method

1 Use a 96-tube box that holds strips of 12 1.2-ml tubes. A 96 deep-well microtitre plate can also be used. Add 800 µl of a 1 : 100 dilution of an overnight culture of JM101 or TG1 in TY medium.

2 Clear plaques are picked using toothpicks which are dropped into single tubes of the 96-tube box. When each tube contains a toothpick, all are removed and discarded. The box is incubated with lid taped on securely for 12–16 h at 37 °C and 300 r.p.m.

3 Centrifuge the 96-tube box at 3500 r.p.m. for 15 min using a special rotor (# 362349, Beckman).

4 Prepare a second 96-tube box by placing 120 µl 20% PEG/2.5 M NaCl solution into each tube using a 12-channel pipetter. Transfer 600 µl of phage supernatant from first 96-tube box to the PEG-containing tube box. A 96-tube cap is placed over all tubes and the box is inverted several times to mix. Incubate for 15 min at room temperature.

5 Centrifuge the 96-tube box at 3500 r.p.m. for 15 min, and decant the supernatant by inverting the box over a sink. Leave the box in an inverted position on a paper towel for 2 min.

6 To completely remove the PEG from the tube walls, a paper towel is placed underneath the lid of the box and the entire box is centrifuged in an inverted position at 200–250 r.p.m. for 2 min.

7 Phage pellets are resuspended by adding 20 µl nonionic Triton-TE extraction buffer to each tube. (In a modified protocol, an ionic

extraction buffer, e.g. Tris-HCl (pH 8.0), 1 mM EDTA, 125 mM KI, 0.16 mM KDS is used.) All tubes are covered with a piece of 3M silver tape and vortexed on a floor model for 30–45 s with pulsing between vortex speed levels 2 and 6. Centrifuge briefly to collect phage solution.

8 Remove bottom of the 96-tube box to check all tubes for complete resuspension of the phage pellets. Place box in water bath at 80 °C (for non-ionic extraction buffer) or at 90 °C (ionic extraction buffer) and incubate for 10 min to achieve phage lysis. Place 96-tube box into a 4 °C ice slurry for 5 min. Centrifuge briefly to collect condensate.

9 Remove silver tape and transfer resulting solution to 96-well microtitre plate. Add 20–40 µl water to each well to dilute DNA. Store at –20° C.

Protocol 105 Magnetic bead purification of M13 DNA (for 96 samples) and bead re-use

For details of solutions, media and materials, see Appendix I. For suppliers and contact addresses see Appendix III.

Materials

- TY medium
- PEG/NaCl (20%/2.5 M)
- 15% SDS
- 0.1 × SSC
- streptavidin-coated paramagnetic beads with linked biotinylated M13-specific oligos (DYNAL)
- 2-ml microfuge tubes (Eppendorf)
- 96-well microtitre plate (Falcon)
- 96-well magnet
- multipipette with 0.5-ml adaptor (Eppendorf)
- 12-channel pipette

Method

1 Pick 96 white plaques into 2-ml sterile microfuge tubes without lids containing 1 ml of a 1 : 100 dilution of an overnight culture of TG1 cells in TY medium. Place tubes in plastic racks (Eppendorf; each rack holds 10 tubes) and shake for 5–6 h at 37 °C.

2 Prior to sample preparation wash beads as follows: take 4 ml of paramagnetic beads with M13-specific oligo (10 mg ml^{-1}), wash beads three times in 4 ml water and resuspend in 1 ml sterile water.

Add 10 µl washed beads to each well of the microtitre plate using a multipipette with a 0.5-ml adaptor.

3 Spin racks with tubes from step 1 at top speed for 5 min (Eppendorf centrifuge 5416 or 5413).

4 Transfer 190 µl supernatant to each well of a microtitre plate.

5 Add 10 µl 15% SDS to each sample using a multipipette with a 0.5-ml adaptor and put plate on heat block at 80 °C for 5 min.

6 Add 50 µl PEG/NaCl solution per well and put microtitre plate on heat block at 45 °C for 20 min.

7 Collect beads by placing microtitre plate on a 96 well magnet. Aspirate off supernatants from all wells.

8 Wash beads three times with 100 µl 0.1 × SSC and mix by pipetting. Collect beads with magnet. Aspirate off supernatants from all wells.

9 Add 20 µl water to each well, place microtitre plate on heat block at 80 °C for 3 min to release M13 DNA from beads.

10 Collect beads by placing microtitre plate on a 96-well magnet. Transfer DNA to new microtitre plate using 12-channel pipette.

Note: Paramagnetic beads can be re-used. After washing, old beads are combined with new beads.

BEADS WORK-UP PROTOCOL

Additional materials

- re-use solution: 0.15 M NaOH, 0.001% Tween 20
- 50-ml Falcon tubes
- 10 × PBS, 0.01% BSA

Method

1 Add 50 µl re-use solution to each well of the microtitre plate.

2 Pool beads in 50-ml Falcon tube. Wash beads twice with an equal volume of re-use solution. Collect beads with magnet.

3 Wash beads once with an equal volume of 10 × PBS, 0.01% BSA, collect beads with magnet and resuspend beads in half the original volume of 10 × PBS, 0.01% BSA.

4 Mix used beads with new beads in a 1 : 1 ratio.

Protocol 106 Silica gel membrane (or glass filter) purification of M13 DNA (24 samples)

For details of solutions, media and materials, see Appendix I. For suppliers and contact addresses see Appendix III.

Materials

- TY medium
- chaotropic salt ($NaClO_4$)
- TE buffer (0.1 M)
- acetic acid
- 70% ethanol
- 1.5-ml microfuge tubes
- Whatman GF/C filter
- filtration unit

Method

1 Toothpick a white plaque into 1.5 ml of a 1 : 100 dilution of an overnight culture of TG1 cells in TY medium. Shake for 5–6 h at 37 °C.

2 Transfer the culture to a 1.5-ml microfuge tube and spin for 5 min at top speed. Transfer supernatant into a fresh 1.5-ml microfuge tube containing 15 µl acetic acid.

3 Put a Whatman GF/C filter into a spin-X filter carrier and place the spin-X filter carriers into a filtration unit. Add the supernatant from step 2 onto the Whatman GF/C filter using gentle suction. The precipitated phages will stick onto the glass filter.

4 Add 1 ml 4 M $NaClO_4$ in TE to the filter.

5 Wash the filter with 1 ml 70% ethanol and dry the filter for 5 min.

6 Place filter onto a 1.5 ml-microfuge tube. Add 20 µl 0.1 M TE to the filter and spin tube for 30 s in a centrifuge.

Following this protocol 24 samples can be processed within 30 min. Alternatively, Qiagen offers commercial kits for M13 DNA preparation based on the same principle in two formats: microspin columns (QIAprep Spin M13 Kit), and 8-well strips (QIAprep 8 M13 Kit). Up to 48 samples can be processed in parallel in less than 30 min using a vacuum manifold and a multichannel pipette. A QIAprep 96 M13 Kit is also available.

Protocol 107 Preparation of phagemid DNA

For details of solutions, media and materials, see Appendix I. For suppliers and contact addresses see Appendix III.

Materials

- pBluescript (Stratagene)
- phage M13KO7 (or VCSM13, Stratagene)
- TY medium
- kanamycin
- PEG/NaCl solution (20 %/2.5 M)
- TE buffer (10 mM)
- NaOAc (3 M, pH 5.2)
- phenol
- ethanol

Method

1 Suspend a fresh bacterial colony containing a phagemid (e.g. pBluescript) in a sterile 15-ml culture tube with 2–3 ml 2 × TY medium containing the appropriate antibiotic. Add M13KO7 (or its commercial derivative VCSM13 from Stratagene) to a final concentration of 2×10^7 PFU ml^{-1}. Incubate for 1–1.5 h at 37 °C with strong agitation.

2 Add kanamycin to a final concentration of 70 μg ml^{-1}. Continue incubation for a further 4–5 h at 37 °C.

3 Prepare single-stranded DNA as described in Protocol 103, step 2.

Protocol 108 CsCl gradient purification of cosmid DNA

For details of solutions, media and materials, see Appendix I. For suppliers and contact addresses see Appendix III.

Materials

- TY medium
- GET buffer (see Protocol 100)
- TE buffer (10 mM)
- 10 mM Tris-HCl (pH 8.0), 0.1 mM EDTA
- 0.2 M NaOH/1% SDS
- 3 M NaOAc, (pH 5.2)
- 96% ethanol, 70% ethanol
- CsCl
- ethidium bromide (10 mg ml^{-1})
- isobutanol
- polyallomer Quick Seal tubes (Beckman)

- 500-ml centrifuge bottles (Sorvall)
- 30-ml centrifuge bottles
- 50-ml centrifuge tubes

Method

1 Streak cosmid directly from a –70 °C stock to obtain single colonies.

2 Inoculate 4 × 200 ml TY containing the appropriate antibiotic with a single colony each (avoid very tiny or large colonies). Shake at 37 °C for 18–24 h.

3 Spin at 10 000 *g* for 5 min in 500-ml Sorvall centrifuge bottles. Decant the medium and place the bottles on ice for 2 min. Aspirate off any remaining medium and add 6 ml GET buffer.

4 Resuspend pellet and transfer with a 10-ml pipette to a 30-ml centrifuge bottle.

5 Add 8 ml 0.2 M NaOH/1% SDS, mix by inverting and incubate on ice for 15 min.

6 Add 6 ml 3 M NaOAc (pH 5.2), mix by inverting and incubate on ice for 30 min.

7 Spin at 17 000 *g* for 10 min and transfer the supernatant (15–18 ml) to a 50-ml tube.

8 Add 2 vols 96% ethanol, mix and spin at 900 *g* for 5 min. Decant the supernatant and drain.

9 Wash the pellet with 20 ml 70% ethanol, spin at 6000 *g* for 5 min, decant the supernatant, drain for 5 min and vacuum dry for 2 h.

10 Add 2.5 ml 10 mM TE buffer and allow DNA to dissolve overnight.

11 Add 2.9 g CsCl and 0.25 ml 10 mg ml^{-1} ethidium bromide, mix and spin at 6000 *g* for 10 min. Transfer the supernatant to polyallomer Quick Seal tubes. Ensure the tubes are full, then seal and place in a Beckman TL 100.3 rotor with spacers.

12 Spin at 20 °C at 70 000 r.p.m. for 17 h or at 83 000 r.p.m. for 6 h.

13 Collect the cosmid DNA with a 20-G needle on a 1-ml syringe by piercing the wall of the tube about 2 mm below the lower supercoiled cosmid band. Collecting 200 μl yields about 90% of the cosmid. Do not try to collect more.

14 Add 300 μl water and extract with isobutanol until the organic layer is colourless.

15 Add water to a final volume of 400 μl and precipitate DNA with 800 μl 96% ethanol.

16 Dissolve cosmid DNA in 20 μl 10 mM Tris-HCl (pH 8.0), 0.1 mM EDTA.

Cosmid clones tend to be unstable, therefore it is recommended to pick several colonies for independent large-scale growth. The different preparations should be checked for rearrangements by restriction enzyme digestion or fingerprinting.

Protocol 109 Generation of single-stranded DNA sequencing template by asymmetric PCR (one-step)

For details of solutions, media and materials, see Appendix I. For suppliers and contact addresses see Appendix III.

Materials

- 10 mM Tris-HCl (pH 8.0), 0.1 mM EDTA
- 1×PCR buffer: 10 mM Tris-HCl (pH 8.3), 50 mM KCl, 1.5 mM $MgCl_2$, dNTPs (250 μM each)
- universal forward primer 1 (0.5 μM)
- universal reverse primer 2 (0.01 μM)
- Taq polymerase (AmpliTaq, 1.0 units, Perkin Elmer)
- 3 M NaOAc, pH 5.2
- isopropanol
- 70% ethanol
- 0.5-ml microfuge tubes
- PCR tubes
- light mineral oil
- materials and equipment for agarose gel electrophoresis

Method

1 Pick a fresh phage plaque with the tip of a pasteur pipette into 100 μl of 10 mM Tris-HCl (pH 8.0), 0.1 mM EDTA.

2 Prepare a mix for 24, 48 or 96 asymmetric PCR reactions (or multiples) containing 1×PCR buffer, universal forward primer 1 (0.5 μM), universal reverse primer 2 (0.01 μM), and Taq polymerase (AmpliTaq, 1.0 units). Dispense 95 μl of the asymmetric PCR mix into 0.5-ml microfuge tubes using a multichannel pipette. Transfer 5 μl culture (phage stock) into PCR tubes. All reactions are then overlaid with 100 μl light mineral oil and asymmetric PCR is performed for 35 cycles (a typical cycle is 30 s at 95 °C, 30 s at 50–55 °C and 1–2 min at 72 °C).

3 A 5-μl aliquot of the PCR product is examined by agarose gel electrophoresis. Single-stranded DNA runs slower than double-stranded DNA and can be visualized by staining with ethidium bromide, although the fluorescence is much reduced relative to an equivalent amount of double-stranded DNA.

4 The reaction mixture is carefully removed from under the mineral oil and transferred to a clean 1.5-ml microfuge tube. 10 μl NaOAc and 100 μl isopropanol are added, the mixture is vortexed and the DNA precipitated by incubation at room temperature for 10 min. DNA is pelleted by centrifugation at 13 000 *g* for 10 min, washed once with 400 µl 70% ethanol and briefly dried. The DNA is dissolved in 25 µl water.

Protocol 110 Generation of single-stranded DNA sequencing template by asymmetric PCR (two-step)

For details of solutions, media and materials, see Appendix I. For suppliers and contact addresses see Appendix III.

Materials

- as Protocol 109

Method

1 Perform symmetric PCR as described in steps 1–2 of Protocol 111.

2 A 5-µl sample of the reaction mixture is removed from under the mineral oil and transferred to a new tube (or microtitre plate) containing 45 µl water.

3 A linear amplification mix for 24, 48 or 96 PCR reactions is prepared containing 1 × PCR buffer, one universal primer (0.5 µM) and Taq polymerase (AmpliTaq, 0.5 units). Dispense 50–100 µl of the PCR mix into 0.5-ml microfuge tubes using a multichannel pipette. Transfer 1–2 µl diluted double-stranded PCR product from step 2 into new tube. All reactions are then overlaid with 50–100 µl light mineral oil and linear amplification is performed for 35 cycles (a typical cycle is 30 s at 95 °C, 30 s at 50–55 °C and 1–2 min at 72 °C).

4 A 5-µl aliquot of the PCR product is examined by agarose electrophoresis.

5 The reaction mixture is carefully removed from under the mineral oil and transferred to a clean 1.5-ml microfuge tube. 10 µl NaOAc and 100 µl isopropanol are added, the mixture is vortexed, and the DNA precipitated by incubation at room temperature for 10 min. DNA is pelleted by centrifugation at 13 000 *g* for 10 min, washed once with 400 µl 70% ethanol and briefly dried. The DNA is dissolved in 25 µl water.

Protocol 111 Generation of double-stranded DNA sequencing template by symmetric PCR (for 24, 48 or 96 templates)

For details of solutions, media and materials, see Appendix I. For suppliers and contact addresses see Appendix III.

Materials

- TB broth or 2×TY broth
- PCR buffer (see Protocol 109)
- dNTPs (250 µl each)
- universal primers (0.3 µl each)
- Taq polymerase (AmpliTaq, Perkin Elmer)
- microtitre plate (Corning)
- 0.25-ml microfuge tubes or heat-stable microtitre plate (Techne)
- 96-pin hedgehog device

Method

1 Recombinant colonies are toothpicked into separate wells of a microtitre plate containing 100 µl of TB or 2×TY broth with the appropriate antibiotic. The plates are incubated with lids on at 37 °C for 12–24 h without shaking. Culture microtitre plates are stored at 4 °C for several weeks until sequencing is finished. Replica plates containing glycerol are stored at –70 °C.

2 Prepare a mix for 24, 48 or 96 PCR reactions (or multiples of this) containing 1×PCR buffer, universal primers (0.3 µM each) and Taq polymerase (0.5 units per well). Dispense 20–30 µl of the PCR mix into 0.25-ml microfuge tubes or into wells of a heat-stable microtitre plate using a multichannel pipette. Transfer a small amount of culture (0.5 µl) into PCR tubes. A 96-pin hedgehog device is used for simultaneous transfer of many samples from the culture plate to the PCR plate. All reactions are then overlaid with 20 µl of light mineral oil and PCR is performed in a thermal cycler (e.g. MW-1, PHC-3, Techne). After an initial denaturation period at 95 °C for 150 s, in order to free some template DNA, 35 cycles are carried out including denaturing at 95 °C for 30 s, annealing at 50–55 °C for 30 s and extension at 72 °C for 1–2 min (depending on insert size).

3 A 5-µl aliquot of the PCR product is examined by agarose gel electrophoresis to estimate insert size and check purity of PCR product.

Protocol 112 Purification of PCR-generated double-stranded DNA sequencing template by selective polyethylene glycol (PEG) precipitation

For details of solutions, media and materials, see Appendix I. For suppliers and contact addresses see Appendix III.

Materials

- PEG mix: 26.2% PEG 8000, 6.6 mM $MgCl_2$, 0.6 M NaOAc (pH 5.2)
- ethanol

Method

1 A large portion of the aqueous phase of the PCR product from step 2 of Protocol 111 is transferred to 0.5-ml microfuge tubes containing an equal volume of PEG mix. It is mixed thoroughly and then incubated at room temperature for 5 min. The PEG mix is dispensed in advance using a multiple pipetter, e.g. Eppendorf 4780.

2 Spin the samples at 13 000 *g* for 5 min and remove supernatant carefully with a yellow tip, avoiding the usually invisible DNA pellet.

3 Wash pellets once with ethanol, dry tubes and redissolve DNA in water.

Protocol 113 Generation of single-stranded DNA sequencing template by PCR followed by affinity capture using the biotin–streptavidin system

For details of solutions, media and materials, see Appendix I. For suppliers and contact addresses see Appendix III.

Materials

- streptavidin-coated magnetic beads (Dynabeads M-280, Dynal)
- materials for PCR (see Protocol 111)
- biotinylated universal primer 1
- unbiotinylated universal primer 2
- 0.15 M NaOH

Method

1 PCR amplification is performed under the usual conditions (see Protocol 111) but using 0.3 μM biotinylated universal primer 1 and 0.3 μM unbiotinylated universal primer 2.

2 For each amplified template, 30 μl of washed streptavidin-coated magnetic beads are added directly to the reaction tube or microtitre well including the oil overlay. The biotinylated PCR product is allowed to bind to the beads by incubating at room temperature for 10 min.

3 Beads are pelleted towards the side of the tube or well using a suitable magnet and the supernatant is removed.

4 Denature captured DNA by adding 20 µl 0.15 M NaOH and incubate the beads for 5 min at room temperature. Beads are pelleted again and the supernatant is removed.

5 Wash the beads once with 20 µl 0.15 M NaOH, followed by three washes with 40 µl water. During each wash pellet the beads with the magnet and remove supernatant.

6 Beads are resuspended in water and used for sequencing reaction.

Protocol 114 Recovery of PCR product by the 'freeze and squeeze' method

For details of solutions, media and materials, see Appendix I. For suppliers and contact addresses see Appendix III.

Materials

- Millipore ULTRAFREE-MC 0.45 mm microfuge cartridge
- 3 M NaOAc, pH 5.2
- absolute ethanol, 70% ethanol

Method

1 Excise DNA band of interest with a clean razor from agarose gel.

2 Stuff the slice into the filter unit of Millipore ULTRAFREE-MC microfuge cartridge.

3 To freeze the agarose slice, put the assembled cartridge (filter unit with slice in the provided microfuge tube) at –70 °C for 15 min or at –20 °C for 30 min.

4 Immediately (while the agarose is still frozen) spin at 13 000 *g* for 5 min.

5 Discard the filter unit with the dry powdered agarose. To the aqueous filtrate add 0.1 vol. NaOAc and 3 vols cold absolute ethanol to precipitate the DNA. Incubate at room temperature for 5–10 min, spin at 13 000 *g* for 5 min, discard supernatant, wash once with 70% ethanol, air dry and dissolve DNA in an appropriate volume of water.

A cheaper way of doing this is to punch a small hole at the bottom of a 0.5-ml tube and to stuff the bottom of the tube with siliconized glass wool to hold back the agarose. This can then be inserted into a 1.5-ml tube and be used instead of the filter unit.

The quality of DNA obtained by the freeze and squeeze method can be improved by subjecting the filtrate to a phenol/chloroform extraction prior to ethanol precipitation or by doing a PEG precipitation (see Protocol 101) instead of the ethanol precipitation.

Protocol 115 Recovery of PCR product by column purification

For details of solutions, media and materials, see Appendix I. For suppliers and contact addresses see Appendix III.

Materials

- low-melting-point (LMP) agarose gel (SeaPlaque or NuSieve GTG agarose, FCM Bioproducts)
- ethidium bromide
- Magic PCR Preps resin (Promega)
- 80% isopropanol
- equipment for electrophoresis
- 1.5-ml microfuge tubes
- Magic minicolumn (Promega)

Method

1. Separate the PCR reaction products by electrophoresis in an LMP agarose gel containing ethidium bromide using standard procedures.
2. Visualize band under long wavelength UV light and excise the desired DNA band using a clean razor blade.
3. Transfer agarose slice to a 1.5-ml microfuge tube and incubate sample at 70 °C until agarose is completely molten.
4. Add 1 ml Magic PCR Preps resin to the molten agarose slice and vortex for 20 s, then leave on bench for 5 min for DNA to bind to the resin.
5. For each PCR product, prepare one Magic Minicolumn. Remove and set aside the plunger from a 3-ml disposable syringe. Attach the syringe barrel to the luer-lock extension of each minicolumn.
6. Pipette the resin/DNA mix from step 4 into the syringe barrel. Insert the syringe plunger slowly, and gently push the slurry into the minicolumn with the plunger.
7. Detach syringe from minicolumn, and remove the plunger from the syringe. Re-attach the syringe barrel to the minicolumn. Pipette 2 ml

80% isopropanol into the syringe to wash the column. Insert the plunger into the syringe and gently push the isopropanol through the column.

8 Remove the syringe and transfer minicolumn to a 1.5-ml microcentrifuge tube. Centrifuge the minicolumn for 20 s at 13 000 *g* to dry the resin.

9 Transfer minicolumn into new microfuge tube. Apply 50 μl water to the column and wait 3–5 min. Centrifuge column for 20 s at 13 000 *g* to elute the bound DNA.

10 Repeat step 9 with another 50 μl water.

11 The amount recovered is estimated by running $\frac{1}{10}$ of each fragment on an agarose gel. Approximately 30 ng double-stranded DNA of about 250–400 bp is used per 'fmole' cycle sequencing reaction and 500 ng for cycle sequencing with dye terminators.

The use of a vacuum manifold allows for the processing of 20 fragments at a time and the whole procedure takes less than 1 h.

Protocol 116 Recovery of PCR product by the agarose method

For details of solutions, media and materials, see Appendix I. For suppliers and contact addresses see Appendix III.

Materials

- 10 mM Tris-HCl (pH 7.6), 5 mM EDTA (pH 8), 0.1 M NaCl
- GELase (Cambio Ltd)
- ethanol
- 3 M NaOAc

Method

1 Incubate a gel segment of low-melting-point (LMP) agarose containing the DNA of interest at room temperature for 30 min in 20 vols 10 mM Tris-HCl (pH 7.6), 5 mM EDTA (pH 8.0) and 0.1 M NaCl.

2 Remove excess buffer carefully, transfer the gel segment to a clean tube and incubate at 70 °C until gel is completely molten.

3 Equilibrate the molten gel carefully to 45 °C. Centrifuge tube briefly to collect all the material in the bottom of the tube.

4 Add 1 unit GELase per 600 mg of 1% agarose gel and incubate at 45 °C for 1 h. During this time the agarose is digested to oligosaccharides.

5 DNA is further purified by ethanol precipitation in the presence of 3 M NaOAc, washed once with 70% ethanol, dried and resuspended in a suitable volume of water for sequencing.

The LMP agarose gel must be completely molten for the agarose bonds to be accessible to GELase.

GELase is rapidly inactivated at temperatures above 45 °C, while LMP agarose begins to resolidify below 45 °C.

Protocol 117 Direct sequencing of DNA in low-melting-point agarose

For details of solutions, media and materials, see Appendix I. For suppliers and contact addresses see Appendix III.

Materials

- low-melting-point (LMP) agarose (SeaPlaque or NuSieve GTG agarose, FCM Bioproducts)
- 1 × TBE/0.5 mg ml^{-1} EDTA

Method

1 Prepare LMP agarose gels in 1 × TBE containing 0.5 mg ml^{-1} EDTA.

2 Mix 20 µl aliquots of the PCR reaction mixture with 4 µl of loading dye. Load samples in individual lanes of the gel. Run gels long enough to yield well-resolved bands. Excise bands using a razor under long-wavelength UV light. It is important to trim away all excess agarose from the product band.

3 Melt the DNA-containing gel slice by heating for 5 min at 68 °C. Use 10 µl DNA for sequencing.

One-half to 2.0% agarose gels can be used and are compatible with the method. SeaPlaque agarose should be used for separation of fragments smaller than 1 kb and NuSieve for bands greater than 1 kb.

References

1 Birnboim, H.C. & Doly, J. (1979) A rapid alkaline extraction procedure for screening recombinant plasmid DNA. *Nucleic Acids Res.* **7**, 1513–1523.

2 Sambrook, J., Fritsch, E.F. & Maniatis, T. (1989) Small-scale preparations of plasmid DNA. In *Molecular Cloning: A Laboratory Manual*, 1.25–1.39 (2nd edn, Cold Spring Harbor Laboratory Press, Cold Spring Harbor, NY).

3 Gibson, T. & Sulston, J.E. (1987) Preparation of large numbers of plasmid DNA samples in microtitre plates by the alkaline lysis method. *Gene Anal. Technol.* **4**, 41–44.

4 Rosenthal, A. & Stephen Charnock-Jones, D. (1993) Linear amplification sequencing with dye terminators. In *Methods in Molecular Biology*. Vol. 23: *DNA Sequencing Protocols* (Griffin, H. & Griffin A., eds), 281–296 (Humana Press Inc., Totowa, NJ).

5 Perkin, E. *Taq Dye Deoxy Terminator Cycle Sequencing Protocol* (Applied Biosystems GmbH Weiterstadt, Germany).

6 Jones, D.S.C. & Schoefield, J.P. (1990) A rapid method for isolating high quality plasmid DNA suitable for DNA sequencing. *Nucleic Acids Res.* **18**, 7463–7464.

7 Bankier, A.T., Weston, K.M. & Barrell, B.G. (1987) Random cloning and sequencing by the M13/dideoxynucleotide chain termination method. *Meth. Enzymol.* **155**, 51–93.

8 Sambrook, J., Fritsch, E.F. & Maniatis, T. (1989) Small-scale preparation of single-stranded bacteriophage M13 DNA. In *Molecular Cloning: A Laboratory Manual* (2nd edn, Cold Spring Harbor Laboratory Press, Cold Spring Harbor, NY), 4.29–4.30.

9 Eperon, I.C. (1986) Rapid preparation of bacteriophage DNA for sequence analysis in sets of 96 clones, using filtration. *Anal. Biochem.* **156**, 406–412.

10 Smith, V., Craxton, M., Bankier, A.T. *et al.* (1993) Microtitre methods for the preparation and fluorescent sequencing of M13 clones. *Meth. Enzymol.* **218**, 173– 187.

11 Mardis, E.R. (1994) High-throughput detergent extraction of M13 subclones for fluorescent DNA sequencing. *Nucleic Acids Res.* **22**, 2173–2175.

12 Beck, S. & Alderton, R.P. (1993) A strategy for the amplification, purification and selection of M13 templates for large scale DNA sequencing. *Anal. Biochem.* **212**, 498–505.

13 Hawkins, T. (1992) M13 single-strand purification using a biotinylated probe and streptavidin coated magnetic beads. *DNA Sequence* **3**, 65–69.

14 Fry, G., Lachenmeier, E., Mayrand, E. *et al.* (1992) A new approach to template purification for sequencing applications using paramagnetic particles. *BioTechniques* **13**, 124–131.

15 *Promega Technical Bulletin* **183**.

16 Kristensen, T., Voss, H. & Ansorge, W. (1987) A simple and rapid preparation of M13 sequencing templates for manual and automated dideoxy sequencing. *Nucleic Acids Res.* **15**, 5507–5516.

17 Zimmermann, J., Voss, H., Kristensen, T. *et al.* (1989) Automated preparation and purification of M13 templates for DNA sequencing. *Meth. Mol. Cell. Biol.* **1**, 29–34.

18 Ansorge, W., Zimmermann, J., Erfle, H. *et al.* (1993) Sequencing reactions for ALF (EMBL) automated DNA sequencer. In *Methods in Molecular Biology*. Vol. 23: *DNA Sequencing Protocols* (Griffin, H. & Griffin, A., eds), 317–356 (Humana Press Inc., Totowa, NJ).

19 Sambrook, J., Fritsch, E.F. & Maniatis, T. (1989) Production of single-stranded phagemid DNA. In *Molecular Cloning: A Laboratory Manual* (2nd edn, Cold Spring Harbor Laboratory Press, Cold Spring Harbor, NY), 4.44–4.50.

20 Vieira, J. & Messing, J. (1987) Production of single-stranded plasmid DNA. *Meth. Enzymol.* **153**, 3–11.

21 Alting-Mees, M.A., Sorge, J.A. & Short, J.M. (1992) pBluescript II: multifunctional cloning and mapping vectors. *Meth. Enzymol.* **216**, 483–495.

22 Craxton, M. (1993) Cosmid sequencing. In *Methods in Molecular Biology*. Vol. 23: *DNA Sequencing Protocols* (Griffin, H. & Griffin, A., eds), 149–167 (Humana Press Inc., Totowa, NJ).

23 Gyllenstein, U.B. & Erlich, H.A. (1988) Generation of single-stranded DNA by the polymerase chain reaction and its application to direct sequencing of the *HLA-DQA* locus. *Proc. Natl Acad. Sci. USA* **85**, 7652–7656.

24 McCabe, P.C. (1990) Production of single-stranded DNA by asymmetric PCR. In *PCR Protocols: A Guide to Methods and Applications* (Innis, M.A., ed.), 76–83 (Academic Press, New York).

25 Wilson, R.K., Chen, C. & Hood, L. (1990) Optimization of asymmetric polymerase chain reaction for rapid fluorescent sequencing. *BioTechniques* **8**, 184–189.

26 Gibbs, R.A., Nguyen, P.N., McBride, L.J. *et al.* (1989) Identification of mutations leading to Lesch–Nyhan syndrome by automated direct DNA sequencing of in-vitro amplified cDNA. *Proc. Natl Acad. Sci. USA* **86**, 1919–1923.

27 Allard, M.W., Ellsworth, D.L. & Honeycutt, R.L. (1991) The production of single-stranded DNA suitable for sequencing using the polymerase chain reaction. *BioTechniques* **10**, 24–26.

28 Rosenthal, A., Coutelle, O. & Craxton, M. (1993) Large-scale production of DNA sequencing templates by microtitre format PCR. *Nucleic Acids Res.* **21**, 173–174.

29 Du, Z., Hood, L. & Wilson, R.K. (1993) Automated fluorescent DNA sequencing of polymerase chain reaction products. *Meth. Enzymol.* **218**, 104–121.

30 Hultman, T., Stahl, S., Hornes, E. & Uhlén, M. (1989) Direct solid phase sequencing of genomic and plasmid DNA using magnetic beads as solid support. *Nucleic Acids Res.* **17**, 4937–4946.

31 Jones, D.S.C., Schofield, J.P. & Vaudin, M. (1991) Fluorescent and radioactive solid phase dideoxy sequencing of PCR products in microtitre plates. *DNA Seq.* **1**, 279–283.

32 Hultman, T., Bergh, S., Moks, T. & Uhlén, M. (1991) Bidirectional solid-phase sequencing of *in vitro*-amplified plasmid DNA. *BioTechniqes* **10**, 84–93.

33 Schofield, J.P., Jones, D.S.C. & Vaudin, M. (1993) Fluorescent and radioactive solid-phase dideoxy sequencing of polymerase chain reaction products in microtitre plates *Methods Enzymol.* **218**, 93–103.

34 Burmeister, M. & Lehrach, H. (1989) Isolation of large DNA fragments from agarose gels using agarase. *Trends Genet.* **5**, 41–48.

35 Sambrook, J., Fritsch, E.F. & Maniatis, T. (1989) Small-scale preparations of plasmid DNA. In *Molecular Cloning: A Laboratory Manual* (2nd edn, Cold Spring Harbor Laboratory Press, Cold Spring Harbor, NY), 6.30–6.31.

36 Kretz, K.A. & O'Brien, J.S. (1993) Direct sequencing of polymerase chain reaction products from low melting temperature agarose. *Meth. Enzymol.* **218**, 72–79.

Chapter 22 Sequencing chemistries: chemical vs. enzymatic DNA sequencing and conventional and novel technologies

Peter Heinrich & Horst Domdey*

MediGene GmbH, Lochhamer Strasse 11, D-82152 Martinsreid/München, Germany

**Laboratorie für Molekulare Biologie, GenZentrum, Ludwig-Maximilian-Universität, Am Klopferspitz, 8033 Martinsried, Germany*

22.1 Introduction

Determination of the sequence of DNA is one of the most important aspects of modern molecular biology. Since the development in the late 1970s of the currently used DNA sequencing techniques, dideoxy chain termination [1] and chemical degradation [2], newer and faster methods have been devised and developed. Now, over 150 million bases are available, stored in international data banks such as EMBL, GenBank, and the DDBJ, making them fully accessible to the scientific community. Many viral genomes have been completely sequenced, including the genome of cytomegalovirus, which consists of 200 000 bp [3]. The complete sequences of the bacterium *Escherichia coli* (see Chapter 31), and the yeast *Saccharomyces cerevisiae* (see refs 3–5 and Chapter 30) are now available. Projects are also under way to determine the DNA sequence of the fruitfly (*Drosophila melanogaster*) (Chapter 28), the mouse (*Mus musculus*) (Chapter 26), the nematode worm (*Caenorhabditis elegans*) (Chapter 29), a plant (*Arabidopsis thaliana*) (Chapter 33), and the human (*Homo sapiens*) genome.

The two original methods of DNA sequencing that were described in 1977 differ considerably in principle. The enzymatic (or dideoxy chain termination) method of Sanger *et al.* [1] involves the synthesis of a DNA strand from single-stranded template by a DNA polymerase. The Maxam and Gilbert (or chemical degradation) method [2] involves chemical degradation of the original DNA. Both methods produce populations of radioactively labelled polynucleotides that begin from a fixed point and terminate at points dependent on the location of a particular base in the original strand. The fragment sets of the four reactions are loaded on adjacent lanes on a polyacrylamide slab gel and resolved by polyacrylamide gel electrophoresis. Autoradiographic imaging of the pattern of the labelled DNA bands in the gel reveals the relative sizes, corresponding to band mobilities, of the fragments in each line, and the DNA sequence is deduced from this pattern [6]. Using the high-resolution denaturing polyacrylamide electrophoresis procedures, one can resolve single-stranded oligodeoxynucleotides of up to 700 bases long.

Although both these techniques are still employed today, there have been many modifications and improvements to the original protocols. The chemical degradation method is still in use, but the enzymatic chain termination method is by far the most established and widely used technique for sequence determination. Fundamental to dideoxy sequencing are three major tools:

- DNA polymerases;
- deoxynucleotide analogues;
- polyacrylamide gel electrophoresis (PAGE) [7].

The most common enzymes currently in use are the Klenow fragment of DNA polymerase I [1], modified T7 DNA polymerase [8], and the thermophilic DNA polymerase (*Taq* polymerase) of *Thermophilus aquaticus* [9]. The usually used Klenow enzyme has been increasingly replaced by the T7 and *Taq* DNA polymerases. The basic chemistry of deoxynucleotides has remained unchanged, but new analogues, including deoxyinosine (dITP) [10] and the 7-deaza analogues, c7dATP and c7dGTP [11], have been synthesized for special purposes in sequencing. There have been fewer changes in PAGE technique compared to the DNA polymerases; however, the introduction of thin gels [12], wedge-shaped gels [13], and buffer gradient gels [14] has significantly improved the data obtainable from PAGE both in quality and quantity.

The largest number of bases resolved from a gel run of a single sequencing reaction is around 500–750. For longer stretches to be analysed, additional strategies have to be pursued. For example, 'walking' primers are used for sequence lengths between 500 and 2000 bp. Between 2000 and 500bp, an ordered deletion method like the *Bal*31 exonuclease approach [15] or the unidirectional exonuclease III method [16] is employed. Both methods involve a controlled progressive degradation of parts of the DNA insert from a fixed point. For DNA fragments longer than 5000 bp, the random shotgun approach is used [17–19]. Using this approach, the DNA insert is enzymatically or sonically cut at non-specific points into smaller fragments and subcloned into a plasmid vector suitable for performing sequencing reactions. The data obtained from the individual sequencing reactions are compiled with the aid of a computer, resulting in the final DNA sequence.

Another point to be adressed in sequencing is the type of DNA being used as template (see Chapter 21). In general, there are different sources of DNA in use. Both the single-stranded M13 phage [20] and phagemids, which are plasmids containing the intergenic region of M13 which can be packed into single-stranded form using helper phage [21], are widely used. Double-stranded DNA from circular plasmids, cosmids, or linear phage lambda DNA can also be used [13]. Finally, the products of polymerase chain reaction (PCR) are routinely used as DNA templates for sequencing [22]

To automate the process, the use of hazardous and expensive radioisotopes has been eliminated and replaced by fluorescence methods; a chemistry has been devised for attaching DNA molecules to fluorophores, which can be detected by laser technology.

Additional recent innovations include the use of PCR technology to enable the sequencing reaction to be used repetitively to generate a sequencing ladder [23,24]. A recently developed detection method that is comparable in sensitivity to traditional radiolabelling is sequencing by chemiluminescence, in which detection of the sequencing products occurs by a chemiluminescent reaction that can be monitored by autoradiography [25,26].

Another innovative approach is multiplex sequencing, which uses hybridization to a specific probe to detect an individual sequencing ladder in a mixture of ladders (see Chapter 20). In this method, DNA samples are mixed and sequenced using either the enzymatic or the chemical method, and the products are fractionated on a sequencing gel which is transferred to a membrane, and hybridized with a probe specific for one template, a process which can be repeated up to 40 times employing different probes. Thus, the amount of sequence information available from one gel can be multiplied by the number of times the membrane can be rehybridized [27]. Multiplex sequencing originally used radioactive probes and chemical sequencing technology [27], but can be extended beyond the Church approach to include chemiluminescent label instead of a radioisotope [25] and to use Sanger dideoxy chemistry instead of Maxam–Gilbert [28].

Another recent innovation that is applicable to both manual and automated DNA sequencing is solid-phase DNA sequencing [29,30]. In this procedure one strand of a double-stranded DNA molecule is biotinylated. The hemibiotinylated DNA molecule is then bound to streptavidin–ferromagnetic beads. Sequencing reactions can be performed using the denatured biotinylated strand preparation as the template.

Commercial efforts are being made to automate parts of the sequencing process, from sample preparation through DNA analysis, by using robotic work stations [31,32]. All commercially available automated sequencers are designed for enzymatic sequencing reactions with manual gel preparation and sample loading, but with automatically controlled electrophoresis and data analysis.

In future, employing mass spectrometry or tunnelling electron microscopy for DNA sequencing might dramatically increase the rate at which DNA can be sequenced.

22.2 Chemical DNA sequencing (Maxam–Gilbert method)

22.2.1 Overview

In the chemical method of DNA sequencing developed by Maxam and Gilbert [2], the target DNA is radioactively labelled at one end (3′- or 5′-end). This label is the reference point for determining the positions of the nitrogenous bases. In four separate reactions, the labelled DNA is cut with a base-specific chemical reagent under limiting conditions and the reaction products are separated on a sequencing gel. Because only end-labelled fragments are observed following autoradiography of the sequencing gel, the DNA sequence can be read from the four DNA ladders. The sequencing reaction consists of two stages:

1 the chemical modification step, which is carried out in such a way that only one base of one type, such as guanine, is modified once in every 500–1000 bases;

2 chain cleavage at the modification sites, which is taken to completion.

The modification reactions use very toxic and fairly unstable reagents (dimethyl sulphate, hydrazine, potassium permanganate, etc.), which is one of the reasons for the rapid decrease in the popularity of this method. Another reason is the development of simple and improved methods for enzymatic DNA sequencing based on the Sanger method [1].

Although the chemical procedure for DNA sequencing is not as widely used as the Sanger method [1], it has some advantages and can be very useful in certain situations. Sequencing can be performed from any point in the clone where a suitable restriction site occurs, obviating the need for further subcloning. The sequence thus obtained can be used to design oligonucleotide primers, and further sequence can then be obtained by Sanger sequencing. The Maxam–Gilbert method is also very useful for resolving regions of DNA that yield poor results in Sanger sequencing owing to secondary structures in the DNA [32].

22.2.2 Chemical sequencing in practice

Chemical sequencing requires a DNA fragment that is labelled at only one end. It is a disadvantage, but a prerequisite for chemical DNA sequencing to have a detailed restriction map for the fragment to be sequenced. This knowledge permits the generation of a series of subfragments that can be enzymatically labelled at both ends. These labelled fragments have to be cut asymmetrically to produce two fragments labelled at each end, which can be purified by gel electrophoresis. But, with the aid of specially constructed vectors [33,34] the requirement for purifying and labelling individual restriction fragments can be circumvented, a fact which greatly facilitates chemical sequencing projects. These vectors enable DNA that is labelled at only one end to be sequenced. This strategy is analogous to the universal priming site in enzymatic dideoxy sequencing. These vectors allow the cloning of a set of nested deletions, generated, for example, by the unidirectional *Bal*31 deletion strategy. This nested deletion strategy is to be recommended for large chemical sequencing projects [33].

22.2.3 Vectors for chemical sequencing

Recently, the construction of specialized vectors for rapid and simultaneous sequence analysis of a large number of samples by chemical sequencing has been described [33,34]. These vectors, pSP64CS and pSP65CS, are high-copy-number plasmids with a synthetic polylinker containing two *Tth*111I sites flanking a *Sma*I site. The two *Tth*111I sites cleaves at the sequence GACNNNGTC leaving a single protruding 5′ base. The way the two *Tth*111I sites are devised makes it possible to selectively label one end of a *Tth*111I fragment using the Klenow fragment of the DNA polymerase I in an end-labelling reaction. The labelled fragment can be sequenced directly, without prior gel purification.

The use of these vectors has several advantages. First, they allow for rapid and simultaneous sequencing of a large number of samples by the chemical cleaving method. Second, the enzymatic end-labelling step is simply and easily to perform, Third, owing to the high copy number of these plasmids, a small plasmid preparation yields enough DNA to repeat the sequencing reactions several times, once the DNA is radiolabelled. Fourth, the vectors are versatile, and allow for the sequencing of a large number of fragments bordered by any of the unique restriction sites in the polylinker. Fifth, each plasmid contains the SP6 promoter element which can be utilized to synthesize RNA complementary to the cloned DNA. This feature can be useful in exon/intron mapping or other types of nuclease protection/mapping experiments, which can be carried out simultaneously with sequencing using the same plasmid [33,34].

22.2.4 Direct DNA sequencing of PCR products

PCR amplification has enormously simplified and accelerated the analysis of DNA that is only available in small amounts. For example, mutations in genes of patients suffering from genetic diseases can be diagnosed routinely by PCR. Various methods of detecting mutations in PCR-amplified products have been described, including allele-specific oligonucleotide hybridization [36], restriction enzyme digestion, sequencing of subcloned PCR products, transcript sequencing of amplified DNA [37], and direct DNA sequencing using single-stranded or double-stranded PCR products employing the chain termination method of Sanger *et al.* [1]. But direct sequencing of single or double-stranded DNA by the enzymatic approach gives often poor resolution, especially if shorter DNA templates are used.

During their study of mutations in patients with propionic acidemia, Kraus and Tahara [35] developed a protocol that allows the direct sequencing of PCR amplified genomic DNA by the Maxam–Gilbert method. This procedure yields clearly readable DNA sequences, 100–400 bp in length, derived from human genomic DNA, in 4 days. Essentially, the procedure can be divided into four steps:

1 the radiolabelling of primers with [γ-^{32}P]ATP prior to amplification;

2 PCR, which generates radiolabelled amplified DNA;

3 removal of excess primers by spin dialysis;

4 chemical cleavage of the PCR product with some modifications.

22.2.5 Outlook

The original method of chemical DNA sequencing [2] has been modified and improved over the years [38]. Additional chemical cleavage reactions have been devised [39], new end-labelling techniques developed [34], and shorter, simplified protocols have been described [40,41]. The main advantage of chemical degradation sequencing is that the sequence is obtained from the original DNA molecule and not from an enzymatic copy, in which wrong bases could have been incorporated. Therefore, with this method it is possible to analyse DNA modifications such as methylation, and to study protein–DNA interactions. Being confronted with strong secondary structures, which cannot be resolved by enzymatic sequencing, the chemical approach will be the method of choice. There are several advantages of the Maxam–Gilbert sequencing method for analysis of PCR products [38]. The first is its consistency, yielding clear DNA sequences of fragments 100–400 bp long, whereas either single or double-stranded sequencing by the chain termination methods often yields unreadable sequences. Second, it allows examination of all sequences present in the PCR-amplified products. Finally, because the sequence starts with the first (5′) nucleotide of the primer, there is no loss of sequence information immediately adjacent to the primer and no modifications to the sequencing protocol are required to obtain these data.

22.3 Enzymatic DNA sequencing (Sanger method)

22.3.1 Overview

In the conventional dideoxy sequencing reaction, an oligonucleotide primer is annealed to a single-stranded DNA template and extended by *Escherichia coli* DNA polymerase I to synthesize a complementary copy of a single-stranded DNA in the presence of four deoxyribonucleoside triphosphates (dNTPs), one of which is ^{35}S-labelled. DNA polymerases are not able to initiate DNA chains. Therefore, chain elongation occurs at the 3′-end of a short complementary primer which is annealed adjacent to the DNA segment to be sequenced. Chain growth involves the formation of a phosphodiester bridge between the 3′-hydroxyl group at the growing end of the primer and the 5′-phosphate group of the incorporated deoxynucleotide. Thus, overall chain growth is in the 5′→3′ direction. The reaction mixture contains one of four dideoxyribonucleoside triphosphates (ddNTPs) that terminate elongation when incorporated into the growing DNA chain. The enzymatic sequencing method is based on the ability of DNA polymerases to use both 2′-deoxynucleotides and 2′,3′-dideoxynucleotides as substrates. When a dideoxynucleotide is incorporated at the 3′-end of the growing primer chain, the elongation is terminated selectively at A, C, G or T owing to the missing 3′-hydroxyl group of the primer chain. After completion of the sequencing reactions, the products are subjected to electrophoresis on a high-resolution denaturing polyacrylamide gel and then autoradiographed to visualize the DNA sequence.

Since intact *E. coli* DNA polymerase I also has 5′→3′ exonuclease activity, the large fragment (Klenow fragment) of *E. coli* DNA polymerase I, which can still carry out the elongation reaction, has historically been used. Alternatively, reverse transcriptase (either from Moloney murine leukaemia virus or from avian myeloblastosis virus) can also be employed. In addition, the use of T7 bacteriophage DNA polymerase (Sequenase) [8] has improved and simplified DNA sequence analysis both in respect to quality of resolution and quantity of data obtainable. The use of *Taq* DNA polymerase from the thermophilic bacterium *T. aquaticus*, which has a temperature optimum for polymerization of 75–80 °C, is particularly advantageous for sequencing DNA templates that exhibit strong secondary structures at lower temperatures [42], since a higher reaction temperature will disrupt DNA secondary structures and inhibit reannealing of denatured, double-stranded templates.

Another variation to conventional DNA sequencing involves substitution of dITP or 7-deaza-dGTP for dGTP in the nucleotide mixes to destabilize secondary structures that can otherwise form in the sequencing products during electrophoresis to cause gel 'compressions'. In addition, manganese can be substituted for magnesium in the labelling/termination reaction. Manganese increases the band uniformity exhibited by Sequenase and can increase the intensity of the sequencing ladder near the primer.

In practice, enzymatic DNA sequencing involves the following steps.

1 The DNA to be sequenced is prepared as single-stranded molecules.

2 A short, chemically synthesized oligonucleotide primer is annealed to the 3′-end of the region to be sequenced. The annealed oligonucleotide serves as primer for DNA polymerase.

3 The hybrid molecules are divided into four aliquots. Each contains all four dNTPs, one of which is ^{35}S-labelled, and also contains one of the four 2′,3′-ddNTPs. The dNTP/ddNTP concentration is adjusted such that termination of the elongation primer occurs at each base in the template resulting in a population of radiolabelled extended primer chains, which have a fixed 5′-end determined by the annealed primer and a variable 3′-end terminating at a specific base.

4 The radiolabelled reaction products are denatured by heating and then separated on a sequencing gel in adjacent lanes. The DNA sequence can be read directly from the autoradiograph of the gel.

Most protocols for enzymatic DNA sequencing utilize [α-^{35}S]dATP to label the nascent chain. Alternative protocols include the use of [α-^{32}P]dATP or [α-^{32}P]dCTP or [α-^{33}P]dATP and the use of primer radiolabelled at the 5′-end for the sequencing reaction.

Most enzymatically based sequencing studies have used bacteriophage M13 DNA as cloning vector; this replicates as a double-stranded DNA molecule, but is packaged as single-stranded DNA in the virus [43]. This method, however, usually requires the subcloning of DNA fragments from plasmids into the double-stranded M13 replicating form, in which the stability of some DNA inserts has, at times, been problematic [7]. To eliminate subcloning steps, Wallace *et al.* [44] introduced a procedure using linearized plasmid DNA as template for sequencing. In this method, sequencing primers were synthesized to align chain elongation next to the vector cloning site and were hybridized to heat-denatured DNA. Chen and Seeburg [7] demonstrated that alkaline denaturation of supercoiled plasmid DNA is more efficient than heat denaturation.

DNA sequencing of alkali-denatured supercoiled plasmid DNA can produce sequencing gels of the same quality as those obtained with M13 vectors. The past few years have witnessed several efforts to make the direct sequencing of plasmid DNA feasible by the chain termination method. One significant advantage of plasmid sequencing over the M13 system is that production of unidirectional overlapping deletions and the DNA sequencing can be performed on a single vector, avoiding subdividing steps into M13. A recent vector development, combining the characteristics both of the M13 phage and the plasmids, concerns the so-called phagemids [21]. This vector system is widely used for cloning, mutagenesis and DNA sequencing.

More recent innovations in sequencing chemistries include the development of nonradioactive detection methods and the automation of some of the sequencing stages [45]. The use of methods that replace radiolabelled DNA in sequencing reactions with fluorescently labelled DNA is growing rapidly. The use of fluorescent, rather than radioactive, material has the advantages of greater safety, less expensive waste disposal, generation of machine-readable data, and greater reagent stability. Detection of fluorescently labelled material has the disadvantage of being rather less sensitive than detection of radiolabelled materials and requires expensive equipment.

22.3.2 Vectors for dideoxy sequencing

Dideoxy sequencing requires a single-stranded template to which the primer can anneal. Single-stranded templates can be easily generated using specialized vectors derived from M13 [43]. Dideoxy sequencing can also be readily carried out using double-stranded DNA, which has to be denatured by heat [44] or alkali [7] prior to the sequencing reaction. Dideoxy sequencing of a double-stranded template is the only rapid method available for verifying a particular plasmid construction, but can also be used for large-scale sequencing projects [33]. The products of PCR can also be sequenced by the dideoxy method.

22.3.2.1 Filamentous phages

The dideoxy sequencing method has been greatly facilitated by the development of the filamentous *E. coli* phage M13 as a cloning vector [43]. Analysis of the life cycle of phages M13, f1, and fd revealed that they have a biological system which enables them to separate the strands of a DNA molecule [46]. Only one particular strand of the double-stranded replicative DNA is packaged into the viral capsid and secreted into the culture fluid from the infected cells. Phage-infected cells grow more slowly than uninfected ones, but do not lyse. Thus, cells infected with these phages can be grown as normal colonies and as plaques, which are defined as regions of slowed growth on a continuous lawn of uninfected bacteria.

The M13 vectors most widely used for dideoxy sequencing are a series called M13mp constructed by J. Messing *et al.* [43,47–49]. The M13mp series contains the *lacZ* promoter and a partial *lacZ* gene, encoding the α-fragment of β-galactosidase. After infection of an *E. coli* F'host containing another partial *lacZ* gene encoding the ω-fragment of β-galactosidase and induction by isopropyl-β-D-thiogalactoside (IPTG), M13mp phage produce blue plaques on Xgal-containing agar plates. In addition, each vector in the M13mp series contains a synthetic polylinker inserted into the fifth codon of *lacZ* without changing the *lacZ* reading frame, enabling them to produce blue plaques on Xgal agar. Insertion of a DNA fragment into one of the unique polylinker cloning sites has a high probability of disrupting the lacZ reading frame, thus generating a recombinant phage producing colourless plaques. Because the polylinker is inserted into the same site in *lacZ* in all M13mp derivatives, a synthetic oligonucleotide primer, which is complementary to a region of *lacZ* adjacent to the 3′ side of the polylinker, is used as a 'universal' primer for all sequencing reactions.

Vectors based on filamentous phages have a major advantage, because they produce a high yield of recombinant single-stranded DNA from a 1 ml volume of bacterial culture. Isolation of hundreds of DNA preparations from M13 clones can be done by hand or using automated equipment. However, the size of DNA fragments to be inserted into the vector should be limited to about 2000 bp, because phages harbouring large foreign DNA fragments grow poorly and have a tendency to undergo deletions [13]. Routinely, one can sequence 350–500 nucleotides in M13 vectors in a single set of reactions.

22.3.2.2 Phagemids

Plasmids containing the origin of replication from filamentous phages can be packaged into the phage capsid when the bacterial culture is superinfected with a phage. A mixture of single-stranded helper phage DNA and single-stranded plasmid DNA can thus be isolated from the culture [50]. The principle of the phagemid system is that it combines the properties of plasmid and M13 vectors in a single vector. However, when the helper phage and the plasmid replicate simultaneously, the latter appears to the phage as a defective interfering particle [50], which reduces the DNA yields of both the helper phage and the single-stranded form of the plasmid in stationary cultures. The unpredictability of the ratio in yield between recombinant phagemid DNA and the helper phage DNA in the DNA prepared from phage particles is a problem with using recombinant phagemids. But in recent innovations, new defective helper phages M13K07 and M13R408 have been developed [51,52], which provide almost no competition with empty vectors, since more than 95% of the yield is single-stranded vector DNA. This ratio can drop when recombinant DNA is packaged.

22.3.2.3 Plasmid sequencing vectors

Numerous plasmid vectors, most available commercially, can be used for double-stranded dideoxy sequencing [33]. Most of these plasmids contain an origin of replication derived from plasmids pMB1 or ColE1 [53,54], modified to weaken copy number control. These plasmids allow a blue-vs.-white screen for inserts, and for most of them, primers are commercially available for sequencing both strands of insert DNA. The design of the polylinker is the most important feature of the commercially available vectors. The polylinker of pUC18/19 [49] is found in many of these vectors. Several other vectors, including Bluescript, have polylinkers containing several useful sites not present in the pUC18/19 polylinker, as well as a different configuration of these sites. To determine which plasmid vector is best suited for a particular sequencing project, one should check the catalogues of the many commercial suppliers (see Appendix III).

One of the disadvantages of plasmid sequencing is that template renaturation during the polymerase reaction occurs relatively rapidly. But this drawback can be greatly compensated for by using fast-acting polymerases such as the phage T7 polymerase (Sequenase) or the appropriate DNA polymerases derived from thermophilic organisms, such as *Taq* polymerase. Using these enzymes much better results can be obtained in plasmid sequencing than when using the Klenow fragment or reverse transcriptase.

22.3.3 Basic techniques of enzymatic DNA sequencing

22.3.3.1 Difference between labelling/termination and the Sanger procedure

Labelling/termination sequencing reactions are carried out in two steps. In the first step, the primer is extended in the presence of low concentrations of dNTPs, including [α-^{35}S]dATP, until one or more of the dNTP pools is depleted (labelling step). The limiting levels of dNTPs and a low reaction temperature reduce the processivity of Sequenase, increasing the number of chains that are extended and labelled. At the end of the labelling reaction, the uniformly labelled fragments range in size from a few to several hundred nucleotides. In the second step, synthesis resumes in the presence of additional dNTPs and ddNTPs (termination step). In the termination reaction, the high dNTP concentration and an increased reaction temperature render Sequenase processive, ensuring that the polymerase extends each chain without dissociation until the incorporation of a dideoxynucleotide [33]. The labelling/termination procedure is mostly used for Sequenase, but can also be applied to other polymerases. However, because each polymerase has different buffer and Mg^{2+} concentration optimum, and each discriminates to a different extent against ddNTPs, the concentration of these components must be modified in each case.

Two steps are also employed in the Sanger dideoxy sequencing reaction, but the purposes of the steps are different. In the first step, a pulse reaction extends the primer in the presence of [α-^{35}S]dATP, unlabelled dNTPs, and ddNTPs. Thus, both labelling and termination occurs in the pulse. The second step, a chase, employs a high concentration of all four dNTPs and ensures that all extended primers that have not incorporated a ddNMP are extended past the region to be sequenced.

22.3.3.2 Sequencing with Klenow fragment

The traditional Sanger method is based on two steps. First, on the primed DNA synthesis, which occurs in the presence of a mixture of [α-^{35}S]dATP, unlabelled dNTPs and ddNTPs and in which termination takes place when a dideoxynucleotide is incorporated into the growing chain. Thus, both labelling and termination occur in the pulse. Second, a chase with high concentrations of dNTPs makes sure that oligonucleotides that have not specifically terminated by incorporation of a dideoxynucleotide are elongated past the region to be sequenced [33].

The Klenow fragment has a 3′→5′ exonuclease activity, which shows a significantly lower activity than that described for the native T7 DNA polymerase [55] and does not interfere with the use of Klenow fragment for DNA sequencing. The 5′→3′ exonuclease activity present in the native *E. coli* DNA polymerase I is absent from Klenow fragment. The Klenow enzyme is relatively non-processive and has an intermediate elongation rate. The low processivity contributes to a somewhat higher background and lower signal-to-noise ratio than that found in reactions performed using Sequenase. In some cases, Klenow fragment has difficulties in reinitiating DNA synthesis at particular sites, resulting in 'ghost' or 'shadow' bands. Klenow fragment discriminates strongly against ddNTPs, having a several thousandfold preference for a dNTP over the corresponding ddNTP. In addition, it exhibits a sequence dependent variability in discrimination. Although the discrimination against ddNTPs is reduced over 100-fold by inclusion of Mn^{2+} into the reaction mix [56], a significant degree of sequence-dependent discrimination persists, producing bands that are still more variable in intensity than those resulting from reactions with Sequenase. Nucleotide analogues such as 7-deaza-dGTP and dITP can be used in sequencing reactions with Klenow fragment [33].

In the labelling/termination protocol, extension lengths can be modulated by altering the dNTP concentrations in the first step. Thus, this method can produce longer products, on average, than the Sanger protocol. This is an advantage when trying to maximize the amount of sequence information obtained from each template. It can be a disadvantage, however, when only the first few nucleotides of sequence information after the primer are desired.

For most sequencing projects, where maximizing the amount of sequence information obtained per template is desired, the labelling/termination approach is recommended. For situations where limited amounts of sequence information are required, the Sanger protocol is appropriate.

22.3.3.3 Sequencing with Sequenase

The labelling/termination sequencing protocol involves two steps [55,57]. In the labelling step, primed DNA synthesis is initiated in the presence of limiting concentrations of all four dNTPs, including [α -^{35}S]dATP, and continues until one of the dNTP pools is depleted. At this point, the uniformly labelled DNA chains have a random length distribution ranging from a few nucleotides to hundreds of nucleotides. In the second step, synthesis resumes in the presence of additional dNTPs and one ddNTP. Elongation of the DNA chains in this step is rapid and processive until termination occurs at specific bases after incorporation of the corresponding dideoxynucleotide. The average length of the radioactively labelled oligonucleotide products can be modified by altering the concentration of dNTPs in the first step. It can also be regulated by altering the dNTP/ddNTP ratio in the termination reaction.

The high level of 3′→5′ exonuclease activity in the native form of T7 DNA polymerase makes it ineffective for DNA sequencing. However, the exonuclease activity can be selectively removed, without affecting the polymerase activity, by a chemical reaction [55], or by genetically based modification [56]. The genetically modified enzyme has a higher specific activity than the chemically modified enzyme and the resulting dideoxynucleotide-terminated fragments are more stable. T7 DNA polymerase does not have a 5′→3′ exonuclease activity.

Sequenase synthesizes DNA at a rapid elongation rate by a highly processive mechanism. Under the termination reaction conditions, it discriminates against ddNTPs by about fourfold compared to dNTPs. Thus, although Sequenase produces bands of higher uniformity than those produced by Klenow fragment or *Taq* DNA polymerase, there is about a 10-fold variation in the intensity of adjacent bands. When Mn^{2+} is included in the reaction mixture, Sequenase incorporates dideoxynucleotides at the same rate as deoxynucleotides and the bands are almost completely uniform [57].

Sequenase utilizes dITP and 7-deaza-dGTP efficiently. It has a tendency to stall in sequencing reactions using dITP, resulting in sequencing ladders with bands in all four lanes. These terminated products are described to be removable by employing terminal deoxynucleotidyl transferase [58]. Labelling reactions using modified T7 DNA polymerase should be kept below 25 °C, both to reduce the processivity of the enzyme and to maintain its activity in the termination reaction. However, termination reactions can be performed from 37 °C to 55 °C.

22.3.3.4 Sequencing with Taq DNA polymerase

Taq DNA polymerase can be used in DNA sequencing employing both the termination/labelling procedure and the Sanger protocol [33]. The use of *Taq* DNA polymerase is indicated for templates exhibiting secondary structures that may inhibit elongation of polymerase. Using *Taq* polymerase, the reactions can be performed at temperatures high enough to destabilize many secondary structures.

Taq DNA polymerase lacks a 3′→5′ exonuclease activity. Native *Taq* DNA polymerase has 5′→3′ exonuclease activity; however, a number of genetically engineered or post-translationally modified versions of *Taq* DNA polymerase have recently been made commercially available in which the 5′→3′ exonuclease activity has been removed. *Taq* DNA polymerase synthesizes DNA at an intermediate elongation rate with a moderate degree of processivity. It discriminates strongly against ddNTPs and requires a ddNTP/dNTP ratio comparable to Klenow fragment. It exhibits better band uniformity than does Klenow fragment, but not as good as that exhibited by Sequenase. Sequencing reactions carried out with *Taq* DNA polymerase give a very clean background on the sequencing gel. For total breakage of potential secondary structure in the template to be sequenced, 7-deaza-dGTP or 7-deaza-dATP can each be used in sequencing reactions with *Taq* DNA polymerase. However, dITP is usually not recommended for use with *Taq* DNA polymerase because of an unacceptable high frequency of inappropriately terminated chains. The sequencing reaction using *Taq* DNA polymerase can be performed at 55–70 °C in low salt buffer, conditions which destabilize many template secondary structures [9,33,59].

A variety of other thermophilic DNA polymerases are being developed for use in sequencing reactions, including *Bst* DNA polymerase [60], *Tth* DNA polymerase [61], and Vent DNA polymerase [62].

22.3.3.5 Sequencing with reverse transcriptase

Avian myeloblastosis virus (AMV) reverse transcriptase is a RNA-dependent DNA polymerase that uses single-stranded RNA or DNA as a template to synthesize the complementary strand. This polymerase can be used for synthesizing long cDNA molecules and for generating high-quality chain termination sequence data. AMV reverse transcriptase provides excellent sequencing band resolution, but it is not widely used for DNA sequencing.

This enzyme lacks both 3′→5′ and 5′→3′ exonuclease activities. Compared to the other polymerases used for sequencing, it has an intermediate level of processivity and a low rate of elongation, which is disadvantageous for DNA sequencing since there is not as much incorporation of radioactivity into the fragments. In addition, background bands due to sites where the polymerase has paused are more frequent. AMV reverse transcriptase uses ddNTPs efficiently and requires ddNTP/dNTP ratios comparable to Sequenase. It exhibits slightly better band uniformity than does Klenow fragment. 7-deaza-dGTP can be used in sequencing reactions with AMV reverse transcriptase. Sequencing reactions can be performed at 37–42 °C [33,63,64].

22.3.3.6 Sequencing with 5′-end-labelled primers

Primers labelled on the 5′-end are used primarily for sequencing very long double-stranded DNA fragments or for templates that have given ambiguous results with nascent chain labelling. Nicks in DNA templates can act as priming sites in the labelling reaction, resulting in a high background on the sequencing gel. This can be eliminated when a 5′-end-labelled primer is used; only clear products from elongation of the primer will be detected by autoradiography.

Because the primer is labelled prior to the sequencing reaction, only one step is required for extension of the primer in the presence of ddNTPs. Nucleotide mixes vary depending on the DNA polymerase used in the reaction. Sequenase, Klenow fragment and thermophilic DNA polymerases can be used for the one-step protocol as long as the optimum reaction conditions for the individual enzyme are chosen. Sequencing ladders derived from 5′-end-labelled primers generally have a clean background and band uniformity which is limited by variations caused by polymerase-specific artefacts. Primers may be 5′-end-labelled either with [γ-^{35}P]ATP or [γ-^{35}S]ATP using T4 polynucleotide kinase [33].

22.3.3.7 Choosing between chemical and enzymatic DNA sequencing

There is no doubt that dideoxy DNA sequencing is simpler and quicker than chemical sequencing. A large number of single- or double-stranded samples can be prepared for sequencing simultaneously. The method also offers excellent band resolution if ^{35}S-labelled nucleoside triphosphates are used in labelling the DNA. The primer-annealing and sequencing reactions can be completed within an hour. Therefore, the enzymatic approach has become used more frequently for DNA sequencing. But it also has disadvantages, which relate to the property of the various DNA polymerases sometimes to terminate chain elongation prematurely owing to secondary structures in the templates to be sequenced. Although the use of various DNA polymerases with their different properties, and the application of various nucleotide analogues may help in overcoming some of these problems, there often remain DNA regions which are poorly resolved by the chain termination method.

Problems associated with polymerase-mediated synthesis of chain elongation may be eliminated by utilizing the chemical approach. Employing chemical sequencing, premature termination due to DNA sequence or structure does not occur, permitting sequencing of DNA stretches which cannot be sequenced enzymatically.

To obtain the sequence of short stretches of DNA using the chemical method, there is no need to subclone into an appropriate sequencing vector. In addition, sequencing of small oligonucleotides is only possible by chemical degradation.

22.3.3.8 Choosing a DNA polymerase for DNA sequencing

The decision about which DNA polymerase to use for a sequencing project should be based on individual enzyme characteristics. In general, more than one polymerase can give reliable sequence information when used in a good protocol with clean DNA templates. Since any of the DNA polymerases can produce artefactual sequence data under certain conditions, an effective strategy will be to choose one enzyme to generate the bulk of the sequence data and then switch to another enzyme and/or protocol to resolve remaining ambiguities.

For large sequencing projects, Sequenase with the labelling/termination protocol is the method of choice because of its high degree of band uniformity, low background, and efficient use of ddNTPs and other nucleotide analogues. Because of Sequenase's lack of thermostability, one should switch to *Taq* DNA polymerase in situations where regions with secondary structure have to be resolved.

Klenow fragment can be used both in the labelling/termination and the Sanger protocol. It is the most widely used DNA polymerase in the Sanger protocol and has a long track record of use in DNA sequencing, but its popularity is dropping significantly. *Taq* DNA polymerase is the alternative of choice when template secondary structure causes premature termination of Klenow fragment. Template secondary structure most commonly occurs in stretches of DNA rich in G + C or A + T or which are extensively palindromic. Reverse transcriptase has been shown to be effective at sequencing through G + C-rich regions while Klenow fragment is effective for A + T-rich regions [33].

Thermostable polymerases are required for thermal cycle sequencing protocols and they are useful for sequencing templates which are generated by PCR. In this case, the high temperature of the sequencing reaction not only destabilizes template secondary structure but also provides increased priming specifity.

22.3.4 Novel DNA sequencing techniques

22.3.4.1 Cycle sequencing

For primer extension by polymerase to occur in the course of a chain termination sequencing protocol, DNA within the region to be sequenced must be in a single-stranded form. With double-stranded templates, this is generally accomplished using either alkaline [13] or heat [44] denaturation. During polymerization, each molecule of template is copied once as the complementary primer-extended strand. Since a finite number of template molecules are present in a 'one-cycle' sequencing reaction, the purity and concentration of the input template DNA are critical to the generation of satisfactory sequence data. Under these conditions, the use of highly purified templates is necessary to prevent false priming. The amount of template added to the reaction dictates signal strength of the data generated. For very large templates, such as λ or cosmid DNA, it is difficult to introduce sufficient template to achieve a good signal. Templates which exhibit secondary structure or which undergo rapid reannealing also present special sequencing challenges.

The introduction of thermostable DNA polymerases has made it feasible to repeatedly cycle a sequencing reaction through alternating periods of heat denaturation, primer annealing, extension and dideoxy termination. This cycling process effectively amplifies small amounts of input double-stranded DNA template to generate sufficient template for sequencing. By annealing and extending the primer at elevated temperatures, problems inherent to traditional sequencing protocols are eliminated. Double-stranded templates remain in a denatured state for longer periods, thus increasing the amount of template available for primer annealing.

Thermal cycle sequencing is a relatively simple process whose success depends on the use of a thermostable DNA polymerase that is functional at temperatures which will denature the template. A thermostable polymerase, such as *Taq* DNA polymerase [9], *Bst* DNA polymerase [60], *Tth* DNA polymerase [61] or Vent DNA polymerase [62], need only be added once to the reaction, thus making programmed cycling possible. Primer is annealed and then extended by the polymerase at temperatures sufficiently high to minimize the secondary structure of the template and strand reannealing. The newly synthesized strand is then dissociated from the template by heating. When the reaction is cooled, more primer anneals and a second round of synthesis occurs. Each cycle increases the amount of product available for sequencing, with the theoretical yield roughly equal to the number of cycles performed.

Since so little material is required, template purity is less critical. Even slightly impure templates can be diluted such that impurities which may be present are at non-interfering levels. Because the template is amplified, it is even possible to sequence DNA extracted from a single colony or plaque [24], if efficient lysis conditions are used. The linear amplification achieved by cycling compensates for reduced template amount and can generate detectable sequence where standard sequencing methods fail.

With cycle sequencing, even traditionally 'difficult' templates such as λ or cosmid DNA, PCR fragments, GC-rich templates or single-stranded templates with difficult stretches can yield satisfactory sequence data. The high annealing and polymerization temperatures that are used are advantageous in several respects. First, since rapid strand reannealing is inhibited, the template remains denatured longer and template utilization is thus more efficient. Second, template secondary structure is minimized so that the polymerase is less likely to dissociate from the template. Even highly structured regions can be satisfactorily sequenced using thermal cycling conditions. And, third, since hybridizations are more stringent at higher temperatures, nonspecific primer annealing is reduced, resulting in less background.

Cycle sequencing can be performed with either end-labelled primers or with primers that incorporate label through extension reactions. When false priming is a problem, the use of end-labelled primers offers several advantages. Since only the specific sequencing primer is labelled prior to annealing, chains extended from primers binding nonspecifically to other sites do not contribute to background on the sequencing gel. Only those sequences derived from end-labelled primer will be detected. False signals sometimes seen when sequencing impure DNA with the label-incorporating protocol will thus be avoided. Moreover, the use of end-labelled primers is generally less expensive, since labelled primer can be prepared in bulk for use with many templates over a period of several days. End-labelled primers are especially useful for generating sequence data very close to the primer, and for sequencing large double-stranded DNA templates. Since each molecule contains only a single radiolabel, this gives an essentially uniform band intensity throughout the sequencing ladder. Furthermore, degradation of the sequencing products by radiolysis simply results in unlabelled fragments which are not detected on the autoradiogram.

With its inherent advantages, cycle sequencing is quickly becoming a standard protocol [24,65–71]. Unlike traditional sequencing methods, it is less affected by the nature and concentration of the template. In addition, it eliminates the need for tedious denaturing protocols and produces data which generally exhibit less background, few if any strong stops and readable sequence up to 500 bases from the primer.

22.3.4.2 Primer-directed DNA sequencing

The partial DNA sequences of the multiple DNAs cloned in the same vector can be obtained using a 'universal' primer. The sequence of this primer is selected to be complementary to a known region of the cloning vector near the multiple cloning site. This primer is therefore universal in the sense that it can be used to obtain sequence information from any unknown insert that has been cloned into the particular vector from which the primer was derived.

Specific primers can be used to fill gaps between contiguous stretches of extended sequence (contigs) obtained using random sequencing methods. In general, random sequencing will provide about 90% of the complete sequence of a large clone before the effort needed to further extend the contigs reaches an unacceptably high level. The remaining 10% usually consists of several small gaps between the contigs. These gap sequences may be obtained by choosing a specific primer from the end of a known contig sequence and using it to obtain sequence across the immediately adjacent gap region. By sequentially applying this strategy a limited number of times, the entire gap sequence can be obtained. Closure is realized when this process produces the end sequence from the adjacent contig.

Extended sequence of a particular cloned DNA can be obtained using sequence information generated from a universal primer to select a new insert-specific primer for a further round of sequence analysis. Successive cycles of sequencing and new primer generation yield the complete sequence of interest. The advantage of the approach lies in the large amount of sequence information that may be obtained from a single clone. However, the speed with which the entire insert sequence is acquired is a function of several factors: the rate at which new specific primers can be selected, synthesized and purified; the rate of new sequence acquisition for each primer; the amount of sequence information sufficiently accurate to pick a succeeding primer obtained per run; and the percentage of selected primers giving rise to interpretable sequence. For sizable clones, a large number of primers are required and the gene-walking process can be quite slow and expensive, the cost of oligonucleotide synthesis may even be prohibitive [72].

22.3.4.3 DNA sequencing by primer walking with short oligomers

In most strategies of DNA sequencing, such as shotgun or nested deletion subcloning (see Chapter 20), the major bottleneck in efficiency is not the stage of sequencing, but rather, the subcloning and template preparation ('front end') and/or the integration of sequences from individual shotgun runs ('back end'). In contrast, sequencing by primer walking minimizes the front and back end problems, using the same template many times in a processive manner, with a new primer for each run. But it presents its own bottleneck in terms of time and cost of the primer synthesis step. This step produces a huge excess of the synthesized primer (0.2–1.0 μmol) over the amount needed for a typical sequencing reaction (0.5 pmol) [73].

A proposed solution for the inconvenience and expense of having to synthesize a primer for each sequencing reaction is to use primers that are short enough that a manageable library of primers would allow any DNA molecule to be sequenced entirely with primers selected from the library. So, Studier [74] has suggested eliminating the need for synthesis of walking primers by building a library of presynthesized short oligonucleotides (8-mers and/or 9-mers, the shortest primers expected to be unique for plasmid-size template), but the size of such a potential library proved to be problematic.

Szybalsky [75] proposed an improvement by performing a template-directed ligation, a technique which is compatible with current sequencing procedures. Efficient ligation requires that oligonucleotides pair at adjacent sites in the template DNA. He proposed that two hexamers be ligated on the template into a unique 12-mer primer, which means a 64-fold reduction in the library size (6-mers vs. 9-mers); but to make the procedure suitable for routine use, the complete ligation of hexamers on the template has to be shown.

Kieleczawa *et al.* [76] discovered that saturating the template DNA with single-stranded DNA binding protein (SSB) stimulates strings of three or more unligated hexamers to prime specifically at the position of the string and at the same time suppressed priming by individual hexamers or by many pairs of contiguous hexamers. When template DNA is saturated with SSB, strings of three or four contiguous hexanucleotides can cooperate through base–stacking interactions to prime DNA synthesis specifically from the 3′-end of the string. Under the same conditions, priming by individual hexamers is suppressed. Strings of three or four hexamers representing more than 200 of the 4096 possible hexamers, primed easily readable sequence ladders at more than 75 different sites in single-stranded or denatured double-stranded templates. A synthesis of 1 μmol of hexamer supplies enough material for thousands of primings, so multiple libraries of all 4096 hexamers could be distributed at a reasonable cost, allowing rapid and economical DNA sequencing.

Kotler *et al.* [73] published a protocol in which they describe a remarkable manyfold increase in the sequence specifity of priming by short oligonucleotides, such as hexamers or pentamers, when tandemly annealed to the DNA template, as compared when each is annealed separately. In DNA sequencing reactions this phenomenon results in unique priming by what they term a 'modular primer', a tandem string of two or three short oligonucleotides, with no ligation required. In contrast, the same pentamers or hexamers show nonunique, multiple priming when used 'one at a time' without adjacent partners. This effect is interpreted as resulting in part from the increase in the affinity of the oligonucleotides for the template caused by their base-stacking, as they anneal to the template next to them, in comparison with their annealing alone, with no neighbours. Modular primers showed a 91% success rate in sequencing reactions, which is comparable to the performance of conventional 17-mer primers. A complete oligonucleotide library of all possible pentamer or hexamer sequences comprises only 1024 or 4096 samples, respectively, and would remove the need for synthesis of new primers for each walking step. Not only time but also cost per walking step is thus reduced, since the scale of oligonucleotide synthesis is sufficient to produce thousands of libraries for users.

22.3.4.4 Sequencing by hybridization

The success of the human genome project will depend on whether DNA sequencing approaches can greatly increase throughput and decrease cost. Strategies that may help to accomplish this task include a greatly improved method of sequencing based on the conventional automated fluorescent DNA sequencers and the advent of new sequencing technologies [77].

Any linear sequence is an assembly of overlapping, shorter sequences. Sequencing by hybridization (SBH) [78,79] is based on the use of oligonucleotide hybridization to determine the set of constituent subsequences (such as 8-mers) present in a DNA fragment. Unknown DNA samples can be attached to a support and sequentially hybridized with labelled oligonucleotides (format 1); alternatively, the DNA can be labelled and sequentially hybridized to an array of support-bound oligonucleotides (format 2). Highly discriminative hybridization is required to distinguish between perfect DNA fragment and oligonucleotide complementarity and all hybridizations exhibiting one or more nucleotide mismatches [78]. Reliable conditions for this discrimination have recently been determined for format 1 [78].

The sets of n-mer oligomers used as hybridization probes can vary in number from several hundred to all possible combinations (65 536 for octamers), depending on the type of sequence information required. The completeness of the probe set and its design, which can vary according to such parameters as length of probe and internal or flanking positioning of unspecified bases, determines the kind of sequence information that can be extracted from individual DNA fragments or libraries of fragments. Mapping information that determines clone overlap can be obtained with 100–200 probes. The positioning and identification of genome structural elements (partial sequencing) requires 500–3000 probes [78], and complete sequencing requires data from 3000 or more septamer probes on three to five related genomes.

By combining SBH and traditional gel sequencing methods, overall throughput and accuracy can be improved by at least an order of magnitude. SBH can be used to fill in gaps and check errors. It is possible that SBH could itself become a primary method of genomic sequencing except for segments consisting of multiple tandem repeats. In addition, because of the combinatorial and parallel features of SBH, several similar genomes could be readily sequenced simultaneously [78]. Furthermore, SBH is an ideal method for identification of genes, repeats, and motifs in chromosomes and cDNAs as well as for studying the differences between related populations and species.

It is expected that SBH will contribute to the development of rapid and inexpensive techniques for mapping clones, facilitate DNA sequence analysis, and extend the applicability of DNA fingerprinting for diagnostic use [78].

22.4 Direct sequencing of PCR products

22.4.1 Overview

PCR has gained widespread application, allowing sequences of interest to be amplified from a complex sample of nucleic acid [9,42,80]. PCR is based on the use of two oligonucleotides to prime DNA-polymerase-catalysed synthesis from opposite strands across a region flanked by the priming site of the two oligonucleotides. By repeated cycles of DNA denaturation, annealing of oligonucleotide primers, and primer extension, an exponential increase in copy number of a discrete DNA fragment can be achieved. Although the cloning of amplified DNA is relatively straightforward, direct sequencing of PCR products facilitates and speeds the acquisition of sequence information. As long as the PCR reaction produces a discrete amplified product, it will be amenable to direct sequencing.

In contrast to methods where the PCR product is cloned and a single clone sequenced, the approach in which the sequence of PCR products is analysed directly is generally unaffected by the comparatively high error rate of *Taq* DNA polymerase. Errors are likely to be stochastically distributed throughout the molecule, thus, the overwhelming majority of the amplified product will consist of the correct sequence. The only exceptions to this rule may be those cases where only a very small number of template molecules are present at the outset of the reaction. In such cases, an error occurring in the first cycle of amplification may be exponentially amplified [81].

Templates for sequencing can advantageously be produced by PCR from any DNA-containing source. It is even useful for amplifying sequence from already cloned material, since it avoids having to grow clones and isolate DNA. PCR products can be sequenced by either the dideoxy (Sanger) [1] or the chemical (Maxam–Gilbert) [2] approach.

22.4.2 Generation of sequencing template

By using PCR, templates for sequencing can be generated more efficiently than with cell-dependent methods either from genomic targets or from DNA inserts cloned into vectors. Amplification of cloned inserts can be achieved using oligonucleotides that are priming inside, or close to, the polylinker of the cloning vector [82].

Sequencing the PCR products directly has two advantages over sequencing cloned PCR products. First, it is readily standardized because it is a simple enzymatic process that does not depend on the use of living cells. Second, only a single sequence needs to be determined for each sample.

22.4.2.1 Generation of single-stranded DNA templates

Most of the difficulties arising in sequencing double-stranded DNA are derived from strand reassociation during the sequencing reaction. These problems can be avoided by preparing single-stranded DNA templates by any of the following number of methods.

Asymmetric PCR In this approach, an excess of one amplified strand (relative to its complement) is generated by the addition of one primer in vast excess over the other. The resulting excess of single-stranded product is then used as a template for the production of the dideoxy-terminated chains from which the sequence is derived.

During the first 20–25 cycles, double-stranded DNA is generated, but when the limiting primer is exhausted, single-stranded DNA is produced for the next 5–10 cycles by primer extension. Using an initial ratio of 50 pmol of one primer to 0.5 pmol of the other primer, the amount of double-stranded DNA accumulates exponentially to the point at which the primer is almost exhausted, and thereafter stops. The single-stranded DNA generation starts at about cycle 25, the point at which the limiting primer is almost depleted. In general, a ratio of 50 : 1–0.5 pmol for a 100-μl PCR reaction will result in about 1–3 pmol single-stranded DNA after 30 cycles of PCR [82].

Low yields of single-stranded DNA using asymmetric PCR may reflect either too little of the limiting primer, preventing the accumulation of enough double-stranded DNA as a template for the primer-extension reaction, or it reflects too much of the limiting primer, saturating the reaction with double-stranded DNA before any single-stranded DNA is produced [82].

The single-stranded DNA generated can then be sequenced using either the PCR primer that is limiting, or any complementary sequence internal to the 3′-end of the single-stranded template.

The use of asymmetric primer ratios does not always result in reproducible high yields of single-stranded product. An alternate protocol entails isolation of double-stranded PCR products followed by reamplification, which is performed in the presence of a single primer. The asymmetric PCR has the advantage that, because the limiting primer is exhausted, there is no need to remove excess primers prior to initiating the sequencing reaction. Protocols for generation of templates by asymmetric PCR are described in refs 81–84 (see also Chapter 21, Protocols 109 and 110).

Lambda exonuclease-generated single-stranded DNA An alternative approach for generating single-stranded products, which does not require the use of unequal primer concentrations is described by Higuchi and Ochmann [85]. In this procedure, one of the oligonucleotide primers is treated with polynucleotide kinase to introduce a 5′-phosphate prior to the PCR. After a symmetric PCR, the products are exposed to the double-strand-specific 5′→3′ exonuclease, which is only active if a phosphate is present at the 5′-position. Only the strand flanked by the phosphorylated primer will be degraded. The single-stranded DNA is then purified from the reaction mix and used for sequencing.

Solid-phase DNA sequencing Solid-phase methods of producing sequencing templates from PCR products for conventional, including fluorescent, methods of nucleic acid sequencing are gaining widespread use (see Chapter 21). They produce quality templates at a high rate, particularly when automated, and avoid centrifugation, phenol extraction, ethanol precipitation, or column chromatography [29,30,86,87].

Strongly binding a single-strand from a PCR reaction to a solid phase allows the remainder of the reaction components to be removed by thorough washing. Streptavidin-coated magnetic beads have been successfully used as the solid phase to pioneer the approach [30] (see Chapter 21, Protocol 113). Streptavidin has an extremely high affinity ($K_D = 10^{-15}$ mol^{-1}) and specifity for biotin [88]. The strand to be immobilized on streptavidin-coated magnetic beads must therefore contain biotin, which is conventionally achieved by biotinylating the specific primer.

The beads on which a strand has been immobilized can be washed repeatedly. They are collected by a magnet, other reaction components aspirated, and then the beads resuspended in fresh liquid, following removal of the magnet.

Each strand of a PCR can be prepared separately for sequencing because in non-denaturing conditions the non-biotinylated strand remains hydrogen bonded to the immobilized strand but can be removed by strong alkaline denaturing conditions without affecting the biotin–streptavidin binding.

Protocols for generation of single-stranded templates by solid-phase technique are described in refs 82, 83 and 89.

22.4.2.2 Generation of double-stranded DNA templates
Although double-stranded, closed circular DNA templates can be alkaline-denatured and sequenced using dideoxy chain termination protocols, this method of template preparation gives poor results with linear PCR products. Many problems associated with direct sequencing of PCR products result from the ability of the two strands of the linear

amplified product to reassociate rapidly after denaturation. This leads either to blocking the extension of the primer–template complex or to preventing the sequencing oligonucleotide from annealing efficiently [82,83]. This problem is more severe for longer PCR products. To circumvent the strand reassociation of double-stranded DNA, a number of alternative methods have been developed.

Precipitation of denatured DNA The template DNA is denatured in 0.2 M NaOH, neutralized by adding 0.4 vol. of 5 M ammonium acetate (pH 7.5), and DNA is immediately precipitated with 4 vols ethanol. The DNA is resuspended in sequencing buffer and primer at the desired annealing temperature [82].

Snap-cooling of template DNA The template DNA is denatured by heating (95 °C) for 5 min and quickly freezing the tube by putting it in a dry-ice–ethanol bath to slow down the reassociation of strands. The sequencing primer is added either prior to or after denaturation and brought to the appropriate annealing temperature [82,90].

Cycle sequencing of PCR Another choice for generating enough sequencing template is to cycle the sequencing reaction, using appropriate thermophilic DNA polymerases, such as *Taq* DNA polymerase I, as enzymes for both amplification and sequencing. Even though only a small fraction of the templates will be utilized in each round of extension and termination, the amount of specific terminations will accumulate with the number of cycles [65,66,82,90]. A more detailed description of cycle sequencing is given in Section 22.3.4.1.

22.4.3 Sequence analysis of PCR products in practice

Both the dideoxy and the chemical sequencing approaches can be used to analyse PCR products directly. The general differences, advantages and shortcomings of the two sequencing approaches are discussed in Sections 22.3.3.7 and 22.3.3.8. In general, the dideoxy method involves fewer steps than chemical sequencing. Where thermophilic DNA polymerases are used for sequencing reactions, the same buffer will serve for the amplification and sequencing steps, and products do not have to be cleaned and isolated repeatedly.

22.4.3.1 Enzymatic approach

The technology and procedures used in dideoxy sequencing are by now well standardized; both T7 DNA polymerase and Klenow fragment as well as *Taq* DNA polymerase or other appropriate thermophilic DNA polymerases can be employed. A number of commercially available dideoxy 'kits' yield good results. Provided that single-stranded templates of good quality are used, the dideoxy methods permit rapid sequencing of amplified products [81].

22.4.3.2 Chemical approach (genomic sequencing)

The genomic sequencing approach of sequencing PCR products combines the PCR amplification method [9,36] with the genomic sequencing technique [2,91]. Following PCR, the amplified DNA is chemically sequenced, transferred by electroblotting, and covalently bound by UV crosslinking onto a nylon filter that can be repeatedly probed with short, sequence-specific oligonucleotides. This method has proved particularly useful for sequencing large regions. In such cases, the sequences of interest can be amplified simultaneously in a single PCR reaction, or separately as a set of discrete adjacent or overlapping fragments. Several amplified fragments can be mixed, simultaneously sequenced, run on gels, and the sequence of the different fragments successively visualized by the use of appropriate end-labelled probes. In addition, sequences from both strands can be derived from a single filter, with a consequent increase in sequence accuracy. The direct chemical sequencing of a labelled strand is a fast alternative if only a single product and a single strand is being sequenced. The utility of this direct approach is thus limited to those cases when the PCR yields a single DNA species or when the product of interest can be readily purified [82].

Detailed protocols for direct sequencing of PCR products based on the enzymatically mediated chain termination method are described in several publications [59,81,83,84]. Chemical sequencing of PCR products is described in refs 81, 92 and 93.

22.5 Automation in DNA sequencing

One of the major advances in sequencing technology in recent years is the development of automated DNA sequencers, which automate the gel electrophoresis step, detection of DNA band pattern, and analysis of bands. These machines are based on the chain termination method and utilize fluorescent rather than radioactive labels. The fluorescent dyes can be attached to the sequencing primer, to the dNTP, or the ddNTPs, and are incorporated into the DNA chain during the strand synthesis reaction mediated by a DNA polymerase, such as Sequenase, Klenow fragment or *Taq* DNA polymerase. The four sets of oligonucleotides generated by the sequencing reactions are loaded onto the gel manually and electrophoresis is then controlled automatically.

Detection occurs at a point near the bottom edge of the gel in one of two ways. In one method, applicable for either fluorescently or radioactively labelled DNA, the bands of DNA moving sequentially past a detector are recorded. In the second method, the banding pattern of fluorescently labelled DNA is detected using an imaging camera. All automated sequencers include data-collection capabilities and include either further analysis programs or allow the data to be taken to external data-analysis software programs. A number of such sequencing machines are now commercially available and are becoming increasingly popular [77,94–96].

Automation by robotics of template preparation and purification and of the sequencing reaction is under development [33,98–103].

Research is under way to develop the technology of mass spectrometry for DNA sequencing [104], a technique that could replace the gel electrophoresis step in DNA sequencing. Alternatively, resonance ionization spectroscopy combined with mass spectrometry could enable much faster analysis of isotopically labelled DNA bands [32,104].

22.6 Future techniques in DNA sequencing

The demand for improved DNA sequencing methodologies posed by the Human Genome Project has spurred the development of both conventional and unconventional approaches [79,105].

22.6.1 Gel electrophoresis

The slow step in the operation of automated DNA sequencers is electrophoresis. Typically, the separation requires at least 12–14 h, during which the expensive instrument is fully occupied. If the speed of the separation could be substantially increased, the throughput of the instrument could be increased correspondingly, decreasing the cost of the sequencing process.

Two recent developments have the potential to significantly increase the throughput of electrophoresis-based sequencing instruments: ultrathin gel electrophoresis in ultrathin (50 μm) gels increases heat transfer efficiency, which permits higher electric fields to be applied without deleterious thermal effects resulting in correspondingly more rapid separations. Employing this technique, separation of fragments up to 600 bases in length can be achieved in 2 h. This separation rate is a factor of four to five greater than can be achieved in conventional slab-gel electrophoresis. Work is in progress on such systems, both with arrays of capillaries [106] and ultrathin slabs [107].

Several efforts have been undertaken to make capillary electrophoresis usable for DNA sequencing [108]. Currently, the principal advantage of this method is its speed, which can be up to 25-fold faster than conventional analysis [108]. To date, most capillary electrophoresis systems have used a single capillary. To compete with slab-gel electrophoresis, many capillaries must be used simultaneously. The presently available systems allow separation of fragments up to 320 bases in length. Sequence length must be increased to be useful in large-scale sequencing efforts.

22.6.2 Scanning microscopy

In this approach, atomic force or scanning tunnelling microscopy (STM) would be used to generate a high-resolution image of individual DNA molecules. The goal is to identify individual bases in single-stranded DNA; the sequence is determined by imaging along the length of the strand. Early work with atomic-scale imaging led to high hopes for this sequencing method. The best achievable resolution of DNA fragments is between 2 and 5 nm. Although it is possible to observe hints of helical structure in double-stranded DNA with the current generation of microscopes, more than one order of magnitude improvement in resolution will be required to produce images with sufficient resolution to determine DNA sequence accurately [109, 110].

22.6.3 Mass spectrometry

Mass spectrometric approaches range from simple replacement of the fluorescence detector in gel electrophoresis with a mass spectrometric detector to ambitious approaches for sequence determination on a single large DNA molecule in an ion trap. However, before mass spectrometry can be used for sequencing, three issues need to be addressed. First, the sequencing sample must be introduced to the gas phase without damage. Second, the sequencing fragments must be separated on the basis of mass. Last, the massive DNA fragments must be detected efficiently.

An intermediate approach being pursued in several laboratories is the replacement of the gel electrophoresis separation of Sanger fragment sets with mass spectrometric separation and detection.

This possibility arises because of the relatively new technique of matrix-assisted laser desorption (MALDI), which permits singly charged ions from proteins as large as 300 000 Da to be produced and mass analysed [105]. With the best matrix for mixed sequence oligomers found to date, 3-hydroxypicolinic acid, the largest oligomer determined was 67 bases, with about 10 pmol of each component analysed [105].

A factor of 10 in mass range and more than a factor of 10 in resolution will be required to achieve useful sequencing accuracy. If the progress in this field over the last three to four years is continued in the future, separations of Sanger mixtures may be performed in seconds, which compares well with the hour or so required even in the ultrafast gel electrophoresis system.

22.6.4 Sequencing by hybridization

This up-and-coming technology is described in more detail in Section 22.3.4.4. Sequence information is obtained by hybridization of small probes to a target to be sequenced. Two formats have been proposed for the method. In one, many targets are arranged on a support and hybridized successively with every possible oligonucleotide probe. In the case of 8-mer probes, this could require as many as 65 000 successive hybridizations, a somewhat intimidating prospect. In the other format, this scenario is inverted by arraying the 65 000 probe oligomers on a support and hybridizing the target sequence to the array; the hybridization pattern then determines the sequence. Several technical issues arise in practice: A particularly thorny one is the effect of repetitive DNA upon sequence reconstruction. Because of this, the method is unlikely to serve as a primary sequencing tool for complex genomes. But these arrays could be very powerful for the sequence analysis of short non-repetitive DNA fragments and used in conjunction with primary sequence data derived by other methods to provide a rapid means of confirming and correcting sequence data [77–79,105].

References

1 Sanger, F., Nicklen, S. & Coulson, A.R. (1977) DNA sequencing with chain-terminating inhibitors. *Proc. Natl Acad. Sci. USA* **74**, 5463–5467.
2 Maxam, A.M. & Gilbert, W. (1977) A new method for sequencing DNA. *Proc. Natl Acad. Sci. USA* **74**, 560–564.
3 Blattner, F.R., Plunkett, G., Bloch, C.A. *et al.* (1997) The complete genome sequence of *Escherichia coli* K-12. *Science* (in press).
4 Oliver, S.G. *et al.* (1992) The complete DNA sequence of yeast chromosome III. *Nature* **357**, 38–46.
5 Dujon, B. *et al.* (1994) Complete DNA sequence of yeast chromosome XI. *Nature* **369**, 371–378.
6 Sambrook, J., Fritsch, E.F. & Maniatis, T. (1989) *Molecular Cloning: A Laboratory Manual* (2nd edn, Cold Spring Harbor Laboratory Press, Cold Spring Harbour, NY).
7 Chen, E.Y., Kuang, W.-J. & Lee, A.L. (1991) In *Methods: A Companion to Methods in Enzymology* Vol. 3, 3–19.
8 Tabor, S. & Richardson, C.C. (1989) Effect of manganese ions on the incorporation of dideoxynucleotides by bacteriophage T7 DNA polymerase and Escherichia coli DNA polymerase I. *Proc. Natl Acad. Sci. USA* **86**, 4076–4080.
9 Innis, M.A., Myambo, K.B., Gelfand, D.H. & Brow, M.A.D. (1988)DNA sequencing with Termus aquaticus DNA polymerase and direct sequencing of polymerase chain reaction-amplified DNA. *Proc. Natl Acad. Sci. USA* **85**, 9436–9440.
10 Mills, D.R. & Kramer, F.R. (1979) Structure-independent nucleotide sequence analysis. *Proc. Natl Acad. Sci. USA* **76**, 2232– 2235.
11 Mizusawa, S., Nishimura, S. & Seela, F. (1989) Improvement of the dideoxy chain termination method of DNA seqencing by use of deoxy-7-deazaguanosine triphophate in place of dGTP. *Nucleic Acids Res.* **14**, 1319–1324.
12 Sanger, F. & Coulson, A.R. (1978) The use of thin acrylamide gels for DNA sequencing. *FEBS Lett.* **87**, 107–110.
13 Chen, E.Y. & Seeburg, P.H. (1985) Supercoil sequencing: a fast and simple method for sequencing plasmid DNA. *DNA* **4**, 165–170.
14 Biggin, M.D., Gibson, T.J. & Hong, G.F. (1983) Buffer gradient gels and 35S labels as an aid to rapid DNA sequence determination. *Proc. Natl Acad. Sci. USA* **80**, 3963–3965.
15 Misra, T.K. (1987) DNA sequencing: a new strategy to create ordered deletions, modified M13 vector, and improved reaction conditions for sequencing by dideoxy chain termination method. *Methods Enzymol.* **155**, 119–139.
16 Henikoff, S. (1984) Unidirectional digestion with exonnuclease III creates targeted breakpoints for DNA sequencing. *Gene* **28**, 351–359.
17 Sanger, F., Coulson, A.R., Hong, G.F. *et al.* (1982) Nucleotide sequence of bacteriophage lambda DNA. *J. Mol. Biol.* **162**, 729–773.
18 Bankier, A.T., Weston, K.M. & Barrell, B.G. (1987) Random cloning and sequencing by the M13/dideoxynucleotide chain termination method. *Methods Enzymol.* **155**, 51–93.
19 Anderson, S. (1981) Shotgun DNA sequencing using cloned DNase-I-geneerated fragmants. *Nucleic Acids Res.* **9**, 3015–3027.
20 Yannish-Perron, C., Vieira, J. & Messing, J. (1985) Improved M13 phage cloning vectors and host strains: nucleotide sequences of the M13mp18 and pUC19 vectors. *Gene* **33**, 103–119.
21 Vieira, J. & Messing, J. (1987) Production of single-standed plasmid DNA. *Methods Enzymol.* **153**, 3–11.
22 Olsen, D.B., Wunderlich, G., Uy, A. & Eckstein, F.

(1993) Direct sequencing of polymerase chain reaction products. *Methods Enzymol.* **218**, 79–92.
23 Murray, V. (1991) Improved double-stranded Dna sequencing using the linear polymerase chain reaction. *Nucleic Acids Res.* **17**, 8889.
24 Krishnan, B.R., Blakesly, R.W. & Berg, D.E. (1991) Linear amplification DNA sequencing directly from single phage plaques and bacterial colonies. *Nucleic Acids Res.* **19**, 1153.
25 Beck, S., O'Keefe, T.O., Coull, J.M. & Koster, H. (1989) Chemiluminescent detection of DNA: application for DNA sequencing and hybridization. *Nucleic Acids Res.* **17**, 5115–5123.
26 Martin, C., Bresnick, L., Juo, R.-R., Voyta, J.C. & Bronstzin, I. (1991) Improved chemiluminescent DNA sequencing. *BioTechniques* **11**, 110–113.
27 Church, G.M. & Kieffer-Higgins, S. (1988) Multiplex DNA sequencing. *Science* **24**, 185–188.
28 Chee, M. (1991) Enzymatic multiplex DNA sequencing. *Nucleic Acids Res.* **19**, 3301–3305.
29 Hultman, T., Bergh, S., Moks, T. & Uhlen, M. (1991) Bidirectional solid-phase sequencing of in vitro-amplified plasmid DNA. *BioTechniques* **10**, 84–93.
30 Kaneoka, H., Lee, D.R., Hsu, K.-C., Sharp, G.C. & Hoffman, R. (1991) Solid-phase direct DNA sequencing of allele-specific polymerase chain reaction-amplified HLA-DR genes. *BioTechniques* **10**, 30–40.
31 Smith, V., Brown, C.M., Bankier, A.T. & Barrell, B.G. (1990) Semiautomated preparation of DNA templates for large-scale sequencing projects. *J. DNA Seq. Map.* **1**, 73–78.
32 Griffin, H.G. & Griffin, A.M. (1993) DNA sequencing. Recent innovations and future trends. *Appl. Biochem. Biotechnol.* **38**, 147–159.
33 Ausubel, F.M., Brent, R., Kingston, R.E. *et al.* (eds) (1987) *Current Protocols in Molecular Biology*, Chapter 7 (John Wiley, New York).
34 Eckert, R. (1987) New vectors for rapid sequencing of DNA fragments by chemical degradation. *Gene* **51**, 245–252.
35 Kraus, J.P. & Tahara, X. (1993) Direct DNA sequencing of polymerase chain reaction-amplified genomic DNA by Maxam-Gilbert method. *Methods Enzymol.* **218**, 227–233.
36 Saiki, R.K., Bugawan, T.L., Horn, G.T. *et al.* (1986) Analysis of enzymatically amplified beta-globin and HLA-DQ alpha-DNA with allele-specific oligonucleotide probes. *Nature* **324**, 163–166.
37 Stoflet, E.S., Koeberl, D.D., Sarkar, G. & Sommer, S.S. (1988) Genomic amplification with transcript sequencing. *Science* **239**, 491–494.
38 Maxam, A.M. & Gilbert, W. (1980) *Methods Enzymol.* **65**, 499–560.
39 Ambrose, B.J.B. & Pless, R.C. (1987) DNA sequencing: chemical methods. *Methods Enzymol.* **152**, 522–538.
40 Pichersky, E. (1993) Chapter 32. In *Methods in Molecular Biology. DNA Sequencing: Laboratory Protocols* (Griffin, H.G. & Griffin, A.M., eds) (Humana Press, Totowa, NJ).
41 Rosenthal, A. (1993) Chapter 33 In *Methods in Molecular Biology. DNA Sequencing: Laboratory Protocols* (Griffin, H.G. & Griffin, A.M., eds), (Humana Press, Totowa, NJ).
42 Erlich, H.A., Gelfand, D.H. & Saiki, R.S. (1988) *Nature* **331**, 461–462.
43 Messing, J. (1983) New M13 vectors for cloning. *Methods Enzymol.* **101**, 20–78.
44 Wallace, R.B., Johnson, M.J., Suggs, S.V. *et al.* (1981) A set of synthetic oligodeoxyribonucleotide primers for DNA sequencing in the plasmid vector pBR322. *Gene* **16**, 21–26.
45 Edwards, A., Voss, H., Rice, P. *et al.* (1990) Automated DNA sequencing of the human HPRT locus. *Genomics* **6**, 593–608.
46 Rache, I. & Oberer, E. (1986) Ff coliphages: structural and functional relationships. *Microbiol. Rev.* **50**, 401–427.
47 Messing, J. & Vieira, J. (1982) A new pair of M13 vectors for selecting either DNA strand of double-digest restriction fragments. *Gene* **19**, 269–276.
48 Messing, J. (1988) *Focus (BRL)* **10**, 21–26.
49 Yanisch-Perron, C., Vieira, J. & Messing, J. (1985) Improved M13 phage cloning vectors and host strains: nucleotide sequences of the M13mp18 and pUC19 vectors. *Gene* **33**, 103–109.
50 Dente, L., Cesareni, G. & Cortese, R. (1983) pEMBL: a new family of single-stranded plasmids. *Nucleic Acids Res.* **11**, 1645–1655.
51 Vieira, J. & Messing. (1987) Production of single-stranded plasmid DNA. *Methods Enzymol.* **153**, 3–11.
52 Russell, M., Kidd, S. & Kelley, M.R. (1986) An improved filamnetous helper phage for generating single-stranded plasmid DNA. *Gene* **45**, 333–338.
53 Bolivar., F., Rodriguez, L., Greene, P.J. *et al.* (1977) Construction and characterization of new cloning vehicles. II. A multipurpose cloning system. *Gene* **2**, 95–113.
54 Kahn, M., Kolter, R., Thomas, C. *et al.* (1979) Plasmid cloning vehicles derived from plasmids ColE1, F, R6K and RK2. *Methods Enzymol.* **68**, 268–280.
55 Tabor, S. & Richardson, C.C. (1987) DNA sequencing analysis with a modified bacteriophage T7 DNA polymerase. *Proc. Natl Acad. Sci. USA* **84**, 4767–4771.
56 Tabor, S. & Richardson, C.C. (1989) Selective inactivation of the exonuclease activity of bacteriophage T7 DNA polymerase by in vitor mutagenesis. *J. Biol. Chem.* **264**, 6447–6458.
57 Tabor, S. & Richardson, C.C. (1990) DNA sequence analysis with a modified bacteriophage T7 DNA polymerase. Effect of pyrophosphorolysis and meta ions. *J. Biol. Chem.* **265**, 8322–8328.
58 Fawcett, T.W. & Bartlett, S.G. (1990) An effective method for eliminating 'artifact banding' when sequencing double-stranded DNA templates. *BioTechniques* **9**, 46–48.
59 Gyllenstein, U.B. (1989) *BioTechniques* **9**, 46–48.
60 McClary, J., Ye, S., Hong, G.F. & Whitney, F. (1991) Sequencing with the large fragment of DNA polymerase I from *Bacillus stearothermophilus*. *J. DNA Seq. Map.* **1**, 173–176.
61 Myers, T.W. & Gelfand, D.H. (1991) Reverse transcription and DNA amplification by a *Thermus thermophilus* DNA polymerase. *Biochemistry* **30**, 7661–7666.
62 Mattila, P., Korpela, J., Tenkänen, T. & Pitkänen, K. (1991) Fidelity of DNA synthesis by the *Thermococcus litoralis* DNA polymerase—an extremely heat stable

enzyme with proofreading activity. *Nucleic Acids Res.* **19**, 4967–4973.
63 Stoflet, E., Koeber, L., Sarkar, G. & Sommer, S. (1988) Genomic amplification with transcript sequencing. *Science* **239**, 491–494.
64 Krawetz, S. (1987) *BioTechniques* **5**, 620–627.
65 Craxton, M. (1993) In *Methods in Molecular Biology. DNA Sequencing: Laboratory Protocols* (Griffin, H.G. & Griffin, A.M., eds) (Humana Press, Totowa, NJ).
66 Craxton, M. (1991) Chapter title. In *Methods: A Companion to Methods in Enzymology* Vol. **3**, 20–26.
67 Adams, S.M. & Blakesley, R. (1991) *Focus* (BRL) **13**, 56–58.
68 Manfioletti, G. & Schneider, V. (1988) A new and fast method for preparing high quality lambda DNA suitable for sequencing. *Nucleic Acids Res.* **16**, 2873–2884.
69 Kim, B.S. & Jue, C. (1990) Direct sequencing of lambda-gt11 recombinant clones. *BioTechniques* **8**, 156–159.
70 Gal, S. & Hohn, B. (1990) Direct sequencing of double-stranded DNA PCR productes via removing the complementary strand with single-stranded DNA of an M13 clone. *Nucleic Acids Res.* **18**, 1076.
71 Green, A., Roopra, A. & Vaudin, M. (1990) Direct double-stranded sequencing from agarose of polymerase chain reaction products. *Nucleic Acids Res.* **18**, 6163.
72 Kaiser, R., Hunkapiller, T., Heiner, C. & Hood, L. (1993) Specific primer-directed DNA sequence analysis using automated fluorescence detection and labeled primers. *Methods Enzymol.* **218**, 122–153.
73 Kotler, L.E., Zevin-Sonkin, D., Sobolev, I.A., Beskin, A.D. & Ulanovsky, L.E. (1993) *Proc. Natl Acad. Sci. USA* **90**, 4241–4245.
74 Studier, F.W. (1989) a strategy for high-volume sequencing of cosmid Dnas: random and directed priming with a library of oligonucleotides. *Proc. Natl Acad. Sci. USA* **86**, 6917–6921.
75 Szybalski, W. (1990) Proposal for sequencing DNA using ligation of hexamers to generate sequential elongation primers. *Gene* **90**, 177–178.
76 Kieleczawa, J., Dunn, J.J. & Studier, W. (1992) DNA sequencing by primer walking with strings of contiguous hexamers. *Science* **258**, 1787–1791.
77 Hunkapiller, T., Kaiser, R.J., Koop, B.F. & Hood, L. (1991) Large-scale and automated DNA sequence determination. *Science* **254**, 59–67.
78 Drmanac, R., Drmanac, Z., Strezoska, Z. *et al.* (1993) DNA sequence determination by hybridization: a strategy for efficient large-scale sequencing. *Science* **260**, 1649–1652.
79 Strezoska, Z., Paunesku, T., Radasavljevic, D. *et al.* (1991) DNA sequencing by hybridization: 100 bases read by a non-gel-based method. *Proc. Natl Acad. Sci USA.* **88**, 10089–10093.
80 Innis, M.A., Gelfand, D.H., Sninsky, J.J. & White, T.J. (eds) (1990) *PCR Protocols* (Academic Press, New York).
81 Ausubel, F.M., Brent, R. & Kingston, R.E. (eds) (1987) Chapter 15. In *Current Protocols in Molecular Biology* (John Wiley, New York).
82 Gyllenstein, U.B. & Allen, M. (1993) Sequencing of in vitro amplified DNA. *Methods Enzymol.* **218**, 3–15.
83 Gyllenstein, U.B. (1989) In *PCR Technology* (Erlich, H.A., ed.), 45–60 (Stockton Press, New York).
84 Kreitman, M. & Landweber, L. (1989) A strategy for producing single-stranded DNA in the polymerase chain reaction. A direct method for genomic sequencing. *Gene Anal. Technol.* **6**, 84–88.
85 Higuchi, R.G. & Ochman, H. (1989) Production of single-stranded DNA templates by exonuclease digestion following the polymerase chain reaction. *Nucleic Acids Res.* **17**, 5865.
86 Greene, A., Roopra, A. & Vaudin, M. (1990) Direct single stranded sequencing from agarose of polymerase chain reaction products. *Nucleic Acids Res.* **18**, 6163–6164.
87 Rosenthal, A. & Jones, D.S.C. (1990) Genomic walking and sequencing by oligo-cassette mediated polymerase chain reaction. *Nucleic Acids Res.* **18**, 3095–3096.
88 Pahler, A., Hendrickson, W.A., Gawinowicz-Kolks, M.A. *et al.* (1987) Characterization and crystallization of core streptavidin. *J. Biol. Chem.* **263**, 13933–13937.
89 Schofield, J.P., Jones, D.S.C. & Vaudin. (1993) Fluorescent and radioactive solid-phase dideoxy sequencing of polymerase chain reaction products in microtiter plates. *Methods Enzymol.* **218**, 93–103.
90 Kusukawa, N., Uemori, T., Asada, K. & Kato, I. (1990) Rapid and reliable protocol for direct sequencing of material amplified by the polymerase chain reaction. *BioTechniques* **9**, 66–72.
91 Church, G.M. & Gilbert, W. (1984) Genomic sequencing. *Proc. Natl Acad. Sci. USA* **81**, 1991–1995.
92 DiMarzo, R., Dowling, C.E., Wong, C. *et al.* (1988) The spectrum of beta-thalassaemia mutations in Sicily. *Br. J. Haematol.* **69**, 393–397.
93 Ohara, O., Dorit, R.L. & Gilbert, W. (1989) One-sided polymerase chain reaction: the amplification of cDNA. *Proc. Natl Acad. Sci. USA* **86**, 5673–5677.
94 Smith, L.M., Sanders, J.Z., Kaiser, R.J. *et al.* (1986) Fluorescence detection in automated DNA sequence analysis. *Nature* **321**, 674–679.
95 Ansorge, W., Sproat, B., Stegemann, J. & Schwager, C. (1986) A non-radioactive automated method for DNA sequence determination. *J. Biochem. Biophys. Meth.* **31**, 315–323.
96 Prober, J.M., Trainor, G.L., Dam, R.J. *et al.* (1987) A system for rapid DNA sequencing with fluorescent chain-terminating dideoxynucleotides. *Science* **238**, 336–341.
97 Bankier, A. (1993) Chapter 38. In *Methods in Molecular Biology. DNA Sequencing: Laboratory Protocols* (Griffin, H.G. & Griffin, A.M., eds) (Humana Press, Totowa, NJ).
98 Frank, R., Bosserhoff, A., Boulin, C. *et al.* (1988) *Biotechnology* **6**, 1211–1213.
99 Mardis, E.R. & Roe, B.A. (1989) Automated methods for single-stranded DNA isolation and dideoxynucleotide DNA sequencing reactions on a robotic workstation. *BioTechniques* **7**, 840–850.
100 Wilson, R., Yuen, A., Clark, S. *et al.* (1988) Automation of dideoxynucleotide DNA sequencing reactions using a robotic workstation. *BioTechniques* **6**, 776–787.

101 Fujita, M., Usui, S., Kiyama, M. *et al.* (1990) Chemical robot for enzymatic reactions and extraction processes of DNA in DNA sequence analysis. *BioTechniques* **9**, 584–591.
102 Smith, V., Brown, C.M., Banker, A.T. & Barrell, B.G. (1990) Semiautomated preparation of DNA templates for large-scale sequencing projects. *J. DNA Seq. Map.* **1**, 73–78.
103 D.'Cunha, X., Berson, B.J., Brumly, R.L. *et al.* (1990) An automated instrument for the performance of enzymatic DNA sequencing reactions. *BioTechniques* **9**, 80–90.
104 Jacobson, K.B., Arlinghaus, H.F., Buchanan, M.V. *et al.* (1991) Applications of mass spectrometry to DNA sequencing. *Genet. Analysis: Techniques & Applic.* **8**, 223–229.
105 Smith, L.M. (1993) *Science* **262**, 530–531.
106 Mathies, R.A. & Huang, X.C. (1992) *Nature* **359**, 167–169.
107 Kostichka, A.J., Marchbanks, M.L., Brumley, R.L. Jr *et al.* (1992) High speed automated DNA sequencing in ultrathin slab gels. *Bio/Technology* **10**, 78–81.
108 Luckey, J.A., Drossman, H., Kostichka, T. & Smith, L.M. (1993) *Methods Enzymol.* **218**, 154–172.
109 Beebe, T.P., Wilson, T.E., Ogletree, O.F. *et al.* (1989) Direct observation of native DNA structures with the scanning tunneling microscope. *Science* **243**, 370–374.
110 Weaver, J.M.R., Walpita, L.M. & Wickramasinghe, K. (1989) *Nature* **342**, 738.

Chapter 23 Electrophoresis

Christoph Heller

Max-Planck-Institut für Molekulare Genetik, Ihnestrasse 73, D-14195 Berlin, Germany

23.1 Introduction

Before the products of sequencing reactions (see Chapter 22) can be analysed, they have to be separated. For this purpose, slab gel electrophoresis is generally used, although in recent years, capillary electrophoresis has been developed as a possible alternative.

For large sequencing projects, this separation step is currently the bottle-neck in the whole sequencing procedure. In order to read a sequence, single base resolution is needed. The sequencing reactions produce DNA fragments of up to a thousand bases or more, but it is not yet possible to separate molecules of that size at high resolution. Some effort has been made to improve the readability of the gels, but the average reading in real sequencing projects is still only a few hundred bases [1–3].

It has to be pointed out that the limits of electrophoretic separation are intrinsic to this method. The resolution (R) that can be achieved depends on different factors: generally, it is defined as the band separation ΔX (distance between the centres of two bands) divided by the average of the two band widths W_1 and W_2 (Fig. 23.1):

$$R = \frac{\Delta X}{\frac{1}{2}(W_1 + W_2)} \quad (23.1)$$

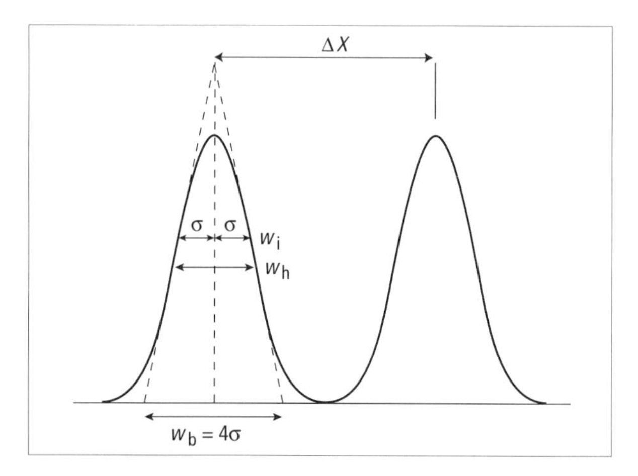

Fig. 23.1 Schematic concentration profile of two bands of gaussian shape, showing the parameters influencing the resolution. ΔX, interband distance; W_i, band width at the inflection point ($W_i = 2\sigma$); W_h, band width at half-height ($W_h = 2\sigma$ (2ln 2)); W_b, band width at the baseline, estimated by drawing the tangents to the peak at the inflection point and extrapolating to the baseline ($W_b = 4\sigma$). The resolution between the two bands is defined as $R = 2\Delta X/(W_{b1} + W_{b2}) = (2\ln 2\Delta X/(W_{h1} + W_{h2})$, using the band width at the baseline or at half-height, respectively.

Therefore, we can distinguish between two main factors, that is, 'band-spacing effects' and 'band-broadening effects', which influence the resolution.

The interband spacing is given by the length of the migration path and the velocity difference between the two species, which in turn is determined by the migration mechanism. In sequencing gels, different separation mechanisms can be identified [4,5]. Small DNA fragments are separated by a sieving mechanism [6,7] and their mobility is proportional to exp(–molecular weight). With increasing molecular weight, they become larger than the pore size and they begin to 'reptate' [8–10]. As a consequence, their mobility becomes inversely proportional to their size. Both mechanisms give a good band separation. With a further increase in molecular weight, however, the mobility reaches a plateau and separation fails. From theoretical considerations [8–10] and experimental studies of double-stranded DNA in agarose [11,12], this effect is known to be due to orientation of the molecules. This upper separation limit decreases with increasing electric field, which means that higher electric fields are counterproductive. However, recent data indicate that at very high fields a different migration process might occur, which seems to preserve good band separation up to a thousand bases [13].

Experimental studies [4,14] and numerical estimates [15] show that in sequencing gels under the usual conditions (40–50 V cm^{-1}, 6% polyacrylamide), molecular orientation only takes place at a fragment size of about 1.5–2 kb, which is well above the current practical limit of gel readings. Therefore, molecular orientation cannot be the main reason for the limited readability.

The second limiting factor is the band width, that is, the resolution limit is reached when bands become so broad that they overlap. The band width can be influenced by the migration mechanism (e.g. ref. 16), but is mainly due to dispersion effects independent of the migration mechanism. Assuming that the bands have a gaussian shape, their width is determined by the variance of the concentration distribution. The band width can be taken as the width at half peak height (W_h) or as the width at the base, which is four times the square root of the variance ($W_b = 4\sqrt{\sigma^2}$; see Fig. 23.1). The total variance is the sum of the individual contributions of different (presumably independent) factors, for example diffusion (dif), Joule heating and temperature profile (ΔT), adsorption (ads), initial band width (ibw) and other possible sources [17]:

$$\sigma^2_{tot} = \sigma^2_{dif} + \sigma^2_{\Delta T} + \sigma^2_{ads} + \sigma^2_{ibw} + \sigma^2_{oth} \quad (23.2)$$

Among these, probably the most important contributors are diffusion and Joule heating. Band broadening due to diffusion increases as the square root of time (e.g. ref. 18), which means that reducing the run time is advantageous. Assuming a parabolic temperature profile, Joule heating causes the band width to increase with the square of the gel thickness and the square of the electric field [18,19] (other authors estimate a third-power dependence [5,20]), which is a strong motivation to use thin gels. Therefore, regarding the electric field, a compromise has to be found between the reduction of run time (i.e. to reduce the diffusion) on one hand and the increase in band width due to thermal effects on the other. In fact, there is an optimal field strength, where the band spreading is minimized [5,18,19].

In practice, the band width is also strongly dependent on the 'quality' of the gel and the loading. In order to achieve optimum results, any factor that could increase the band width has to be minimized or avoided. This includes for example impurities or air bubbles in the gel, badly formed wells, urea in the wells, etc. Depending on the gel concentration, the mobility of the sample is slowed down when entering the gel. Therefore it becomes concentrated and the initial band width is reduced, but irregularities on the surface of the gel are 'imprinted' into the band shape.

In classical slab gel electrophoresis, it is a band-spacing effect that limits the readability. As explained above, the mobility of the (non-orientated) DNA molecules is inversely proportional to their length. As the run time is the same for all fragments, the relationship between distance travelled and fragment size is a hyperbolic function, which means that there are widely spaced bands at the bottom and crowded bands at the top of the gel. In other words, the large fragments do not travel far enough to be well separated. It is this reduction in interband spacing that limits the readability of classical sequencing gels [15]. Possible solutions to this problem are the use of very long gels or the use of gradient gels (see below).

However, there exists another solution to this problem. Instead of recording the band pattern in the whole gel at a fixed time, the passing bands can be detected at a fixed position in the gel over a long time period. As all fragments have to travel the same distance, the relation between retention time and fragment size becomes a linear function, and in consequence, constant interband or peak spacing is obtained in the nonorientated regime (e.g. ref. 21). This principle has been realized in direct blotting electrophoresis as well as in electrophoresis with on-line detection (automated sequencers, capillary electrophoresis), and in this case it is the band broadening that has been identified as the limiting factor [15].

If the band width or the reduction in interband spacing at the top of the gel could be minimized, the separation would eventually be limited by a zero velocity difference due to molecular orientation [22]. Recently, the use of high gel concentrations and high electric fields for the optimization of sequencing electrophoresis within that limit has been proposed [23].

Pulsed electric fields, which can successfully cancel the molecular orientation of large double-stranded DNA molecules in agarose gels and therefore expand the range of separation, have been shown to give only slight improvements in polyacrylamide gels at usual field strengths [4,24–27]; however, they might be more effective at higher field gradients [15]. Trapping electrophoresis [14] has been proposed, but band broadening seems to be a major problem.

So far, the idea of sequencing thousands of bases from a single sample on a polyacrylamide gel remains an illusion. The most promising way to enhance the amount of information obtained from a single gel seems to be the multiplex sequencing method [28] (see Chapter 20). However, enhancing the readability of a gel is still important, as it would reduce the number of clones needed in shotgun sequencing and would also reduce the amount of work needed if ordered strategies are employed.

The factors limiting sequencing gel electrophoresis are listed in Table 23.1 along with their remedies.

23.2 Slab gel electrophoresis

Slab gel electrophoresis is the 'classical' method for DNA sequencing. This process can be automated, and there are a few commercially available instruments on the market, all of them based on fluorescence labelling and detection. However, as these instruments are still rather expensive, the manual method is still in use in many laboratories (see Protocol 118).

23.2.1 Manual sequencing

23.2.1.1 Gel matrix

In free solution, nucleic acids of different sizes have the same electrophoretic mobility and cannot be separated in an electric field. A matrix is therefore needed, which on one hand serves as an anticonvective medium and on the other has to have good 'sieving' properties. At present, polyacrylamide gels are the only matrices which have a high enough

Table 23.1 Limiting factors in sequencing gel electrophoresis.

	Technique			
Limiting factor	**Classical slab gel**	**DTE/Gel readers**	**CE**	**Remedy**
Loss of separation due to molecular orientation	*	**	***	Reduce electric field Use pulsed fields Use trapping electrophoresis
Bandwidth, due to:	**	***	–	
Initial band width				Load small sample volume Wash wells
Diffusion				Increase electric field Use longer gels Reduce gel thickness
Joule heat				Reduce electric field Control gel temperature
Loss of separation due to reduced interband spacing at the top	***	–	–	Use gradient gels Use long gels

CE, capillary electrophoresis.
***Main factor.
*,**Secondary factors.

resolving power for DNA sequencing. The gel is prepared by polymerizing acrylamide monomers in the presence of a crosslinker, which forms covalent bridges between the polyacrylamide chains, resulting in a three-dimensional network (Fig. 23.2). There are a number of different crosslinkers (see refs 29 and 30 for reviews), but *N,N′*-methylene-*bis*-acrylamide ('*bis*-acrylamide') is by far the most popular.

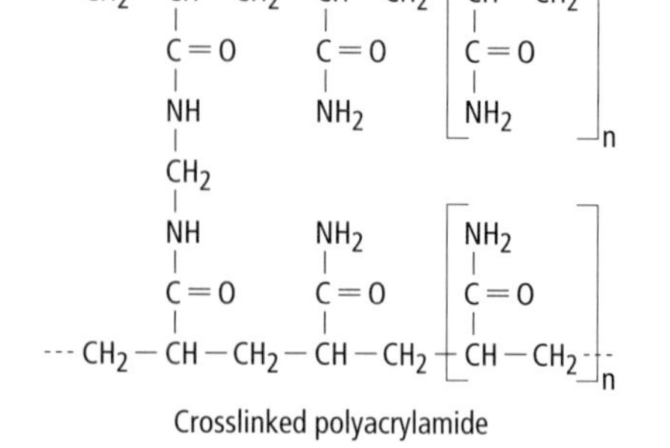

$CH_2{=}CH{-}C({=}O){-}NH_2$

Acrylamide

$CH_2{=}CH{-}C({=}O){-}NH_2{-}CH_2{-}NH_2{-}C({=}O){-}CH{=}CH_2$

N,N′-methylene *bis*-acrylamide

Crosslinked polyacrylamide

Fig. 23.2 Structure of polyacrylamide.

The following terminology for describing gel composition is very useful and has been widely adopted: the gel concentration is given as the total monomer concentration (acrylamide plus crosslinker), for example '6%T', whereas the crosslinker concentration is given in percentage of the total concentration ('%C').

The pore size of the gel is dependent on both the crosslinker and the total concentration, but from electron micrographs [31] and experimental studies [32,33] it is known that an acrylamide : *bis*-acrylamide ratio of 19:1 (i.e. 5%C) gives the smallest pore size and therefore the highest resolution for any given total concentration, %T [34].

The choice of the gel concentration is determined by the competition between the maximum interband spacing that can be obtained and the reduction of the spacing that occurs in classical slab gels, as described above. Concentrated gels (e.g. 8 or 10%T) have a high resolving power, but this is drastically reduced with increasing molecular weight (more than about 200 bases). Diluted gels (e.g. 4%) can separate much larger molecules, but at lower resolution [34]. Therefore, for standard sequencing gels, a concentration of 6%T is the best compromise.

To initiate the polymerization reaction, a catalyst that produces free radicals is needed. As for the crosslinker, there are a number of catalyst systems for polyacrylamide (see ref. 35 for review), but for DNA sequencing gels only one catalyst-redox system is generally in use: ammonium persulphate (APS), which acts as radical donor, in combination with *N,N,N′,N′*-tetramethyl-1,2-diaminoethane (TEMED) as the catalyst.

After addition of the initiator, gelation should take place within 20–30 min; the time can be tested using an aliquot of the gel solution in a closed vessel (i.e. reaction tube). However, as the polymerization reaction continues, it is better to prepare the gel well before the start of the electrophoresis (for the dynamics of the polymerization of acrylamide, see, e.g. ref. 36). If a sequencing run is to be done within one day, we recommend preparing the gel first thing in the morning. It will then be ready for use when the sequencing reactions are completed (around 2–3 h in the case of enzymatic sequencing starting from a purified template).

As oxygen is a 'trap' for free radicals, it inhibits the polymerization reaction. Therefore, in order to achieve higher reproducibility, we recommend that the gel solution be degassed prior to pouring. Care has to be taken that the gel edges are not exposed to the air, that is proper sealing of the glass plates and a fitting well-former (comb) are essential.

All substances mentioned above have to be handled with care, as acrylamide and *bis*-acrylamide are toxic and are absorbed through the skin. TEMED is corrosive and APS is a strong oxidizing agent. Both catalysts are hygroscopic and have only a limited shelf life.

Reagents should be of the highest quality and purity. Poor quality acrylamide and *bis*-acrylamide can contain the following:

1 Acrylic acid, the hydrolysis product of acrylamide, will polymerize with acrylamide or *bis*-acrylamide, and therefore change the properties of the gel. The degradation reaction of acrylamide and *bis*-acrylamide is catalysed by light; storage in the dark is therefore recommended.

2 Linear polyacrylamide, which is caused by catalytic contaminations in the dry monomer, will affect the polymerization of the gel and the effective acrylamide concentration, leading to loss of reproducibility.

3 Metal ions as contaminants can inhibit or accelerate the polymerization or affect the mobility of the DNA.

APS is very hygroscopic and decomposes almost immediately when dissolved in water. The result is loss of activity. As this compound affects the rate of polymerization, it is important to prepare it fresh daily in order to achieve reproducible results. Alternatively, aliquots of a stock solution can be frozen and discarded after use.

TEMED is very reactive and subject to oxidation. The oxidized form is yellow and less reactive. As TEMED is also hygroscopic, it will accumulate water, which again accelerates the oxidative decomposition. Therefore, only water-free TEMED, greater than 99% pure, should be used.

Recently, new monomers, such as *N,N*-dimethylacrylamide (DMA) and similar alkyl-substituted acrylamides have been introduced (see refs 37 and 38 for reviews). A commercial product containing these formulations (HydroLink™) has been shown to be useful for DNA sequencing [39]. The authors claim an increase in readability.

Another novel monomer, *N*-acryloylamino-ethoxyethanol (AAEE) has been synthesized, which is much more hydrophilic than DMA and therefore better suited for serving as an electrophoretic matrix [38]. Both poly(AAEE) and poly(DMA) have much better resistance to hydrolysis under both acidic and alkaline conditions than polyacrylamide [37,38], which would allow the use of buffers with higher pH, which would in turn help keep the DNA fully denatured during electrophoresis.

23.2.1.2 Buffer

For DNA sequencing gels, 1×TBE buffer (89 mM Tris, 89 mM boric acid, 2.5 mM EDTA, pH 8.3) is normally used. However, a precipitate can form during prolonged storage of concentrated stock solutions. The remedy is to use a modified TBE (133 mM Tris, 44 mM boric acid, 2.5 mM EDTA, pH 8.8) which does not precipitate in concentrated form. Such a buffer with a higher pH has been reported to give a better resolution [34].

Again, only reagents of high quality should be used with a low amount of metal or non-buffer ion contaminants.

23.2.1.3 Denaturants

Intramolecular base pairing can occur within a DNA strand, resulting in the formation of small loop structures within the molecule. As such conformations can alter the mobility of DNA fragments, a molecule with intramolecular base pairing can travel at the same speed as a shorter one without a loop. On sequencing gels, this effect manifests as a 'compression', where the band separation is drastically reduced because of the mobility change.

In order to avoid such intramolecular base pairing, the gel has to be run under denaturing conditions. Therefore, a denaturing agent, such as urea or formamide has to be added during polymerization. Alkali cannot be used as it deaminates acrylamide and methylmercuric hydroxide inhibits polymerization. Most researchers use urea (at a concentration of 7–8 M) as it does not have to be deionized as do most batches of formamide. However, the addition of formamide (up to 40%) increases the denaturing capacity of the gel.

The denaturing power of these agents alone is not sufficient and the gels have to be run at an elevated temperature. Generally, a temperature of 50–60 °C is high enough to keep the DNA fragments denatured.

23.2.1.4 Gel dimensions and apparatus

The gel is poured between two glass plates, one of them being notched to ensure contact with the buffer. Its thickness is determined by the thin plastic strips that are used as spacers between the glass plates. Standard sequencing gels are 0.3–0.4 mm thick, but the use of thin (0.1–0.2 mm) and ultrathin (0.05 mm) gels has been reported [34,40]. Pouring of these gels requires special methods, such as the sliding technique [41], or the clapping technique [42,43]. Thinner gels generate less Joule heat and can therefore be run at higher voltages, resulting in lower run times and less diffusion. The temperature gradient across the gel is smaller, also causing less band spreading. However, thin gels are very fragile, and sample loading is difficult, whereas thicker gels accept larger sample volumes, but take longer to fix and dry.

As the distance between bands increases linearly during the run and band broadening only increases with the square root of time, the number of resolved bases should increase with longer runs. To prevent the short fragments running out of the gel, long gels are required and indeed such gels (100–120 cm) have been reported to give a resolution up to 600 bases [34]. However, large gels are difficult to handle and the gain of information often does not justify the effort. The 'standard' gel length of 40–50 cm is much more convenient and matches the common size of gel dryers, X-ray films and cassettes.

The gel width is very much a matter of personal preference, but should also be chosen according to the film size (e.g. 20 or 40 cm). As the very edges of a gel should not be used because of the temperature gradient, the actual loading width of a gel 20 cm wide is about 15–16 cm, which is enough to load 40 samples (10 clones).

There are a number of devices for manual sequencing with different features. We obtained very good results with a simple two-tank design as described in ref. 44.

23.2.1.5 Sample wells and loading

The slots that accommodate the sample are formed by introducing a comb at the top of the gel immediately after pouring. The size and number of the wells is again a matter of personal preference. A comb with teeth 2 mm wide and 3 mm deep and a space of 1 mm between them, for example, allows at least 10 clones to be loaded on a gel 20 cm wide.

Another type of slot former is the 'sharkstooth' comb. The gel is polymerized with a precomb (a rectangular piece of spacer material) in place, to ensure a flat surface. After polymerization, the precomb is replaced by the sharkstooth comb so that its teeth, which serve as barriers between the samples, are just penetrating the gel. The comb rests in place during electrophoresis.

Sharkstooth combs have the advantage that the sample tracks are immediately adjacent, making gel reading easier. On the other hand, a perfectly flat surface is essential, and introducing the comb needs some experience in order to avoid leaks or deformations of the surface.

In both cases, a perfect polymerization is important. Therefore, exposure to air should be avoided, and the combs should be absolutely proper. A 'trick' to ensure proper polymerization at the wells consists of applying tiny amounts of the APS solution onto the comb with a paper tissue, just before inserting into the gel (R. Reinhard, personal communication 1995).

Before loading, the wells must be flushed to remove unpolymerized acrylamide and urea which diffuses out of the gel. The (denatured) samples can be loaded with the help of a glass capillary or special thin pipette tips. One has to keep in mind that the final width of the bands in the gel is also dependent on the initial band width and that the concentration effect taking place at the gel surface might not be strong enough to compensate for dilute samples. Therefore, only small volumes should be loaded and as quickly as possible to minimize diffusion (and renaturation). Properly washed wells with an even surface are essential. Loading the sample directly onto the surface is better than letting it trickle down the well, which causes dilution with the buffer.

23.2.1.6 Field conditions

As pointed out above, the influence of the electric field on the resolution is severalfold. The optimum field strength that gives the minimum band width depends on different factors, but increases with decreasing gel thickness, which is why in thin gels higher electric fields can be used. The empirically determined optimum field strength of 30–50 V cm^{-1} [34] coincides well with the estimated one [19].

If the gel is not actively thermostatted, the electric power also has the task of heating the gel. An electric field should be chosen which generates enough power to bring the gel temperature to 50–60 °C (measured on the outside). Too strong an electric field, however, will result in overheating, which in turn enhances the conductivity of the gel. Current and power will increase, leading to even more heating and so forth ('thermal runaway'). Therefore, to ensure a constant heat input, the gel should be run at constant power rather than at constant voltage. For routine sequencing, an active temperature control is not necessary, but we recommend the use of a metal plate in contact with one of the glass plates, to achieve a more even heat distribution.

For a 20 × 48 cm gel 0.4 mm thick, a power of 40 W is sufficient to generate the required temperature. This results in a current of about 20–25 mA and an electric field of about 1.6–2 kV corresponding to 33–42 V cm^{-1}. Under these conditions, molecular orientation does not occur below fragment lengths of 1 kb, which is far above the actual reading limit.

23.2.1.7 Gradient gels

As explained above, for resolving large fragments, long gels have to be used in order not to loose the small molecules. An alternative method is to slow down the fragments in the lower part of the gel, thus preventing them from being electrophoresed out of the gel. This can be achieved with a nonuniform electric field across the gel, that is the field in the lower part has to be weaker than in the upper part.

One way of creating such a gradient, is by preparing a wedge-shaped gel, which is thicker at the bottom (0.6–0.75 mm) than at the top (0.25 mm) [45–47]. In the thicker part, the gel has a lower resistance, leading to a lower voltage drop across that part of the gel.

However, there is another way to produce a voltage gradient. Increasing the ionic strength by increasing the buffer concentration in the lower part of the gel also results in a lower resistance and therefore a lower voltage drop [48]. Such a buffer concentration gradient can be made with two gel solutions containing different TBE concentrations (e.g. 0.5 and 5 × TBE, see Fig. 23.3a).

A similar effect can be achieved by adding sodium acetate to the lower chamber buffer. The salt diffuses into the lower part of the gel, thus increasing the ionic strength [49].

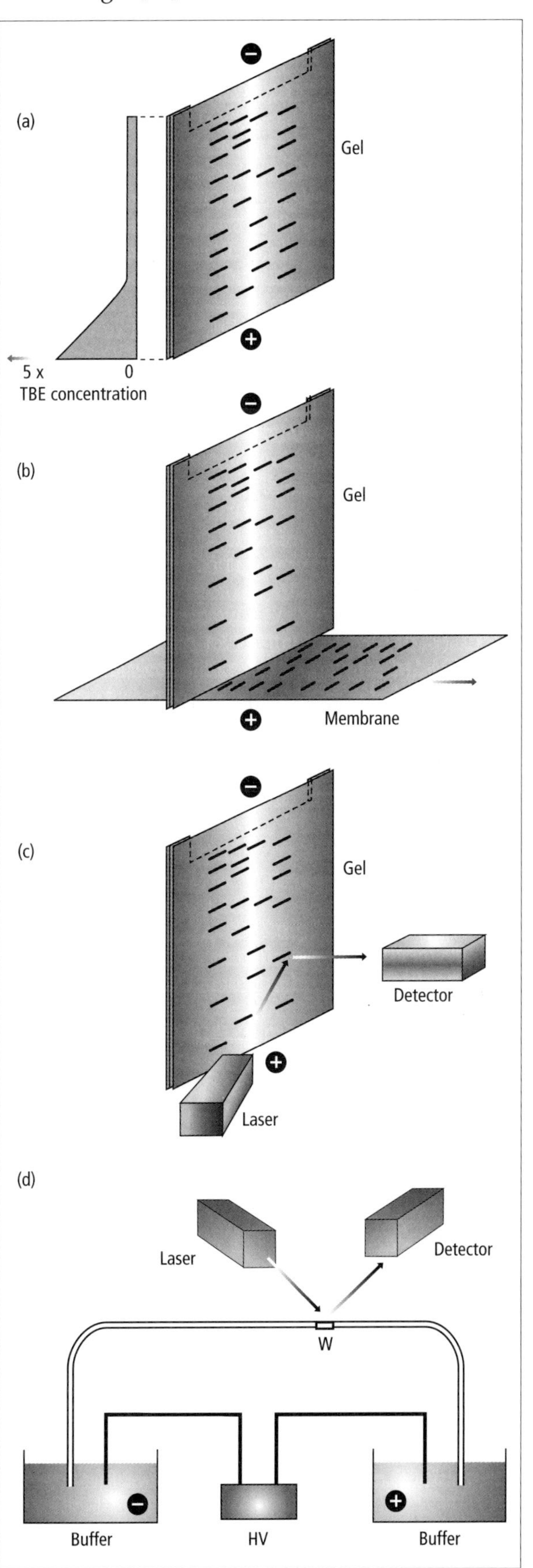

Fig. 23.3 Schematic illustration showing the principles of the different electrophoretic techniques described in the text. (a) Slab gel with gradient profile; (b) direct transfer electrophoresis; (c) automated gel reader; (d) capillary electrophoresis. W, detection window; HV, high voltage power supply.

Due to the field gradient, the DNA molecules are gradually slowed down, i.e. the bands are compressed and band width and interband spacing are reduced. If the gradient is carefully chosen, these gels give a nearly even band spacing, thus making gel reading easier. Because of these advantages, the use of a gradient gel is strongly recommended. Which kind of gradient is used remains a matter of personal choice. Wedge-shaped gels can be prepared with a single gel solution, but they take much longer to dry. Without distorting the glass plates, the magnitude and flexibility of the gradient remains limited. In buffer gradients, on the other hand, uniformity across the gel can be difficult to obtain and distortion of the lanes can occur. See also Chapter 19, Protocol 99 for the preparation of a denaturing gradient gel.

23.2.1.8 End of the run and autoradiography

In a 6% gel, the bromophenol blue marker approximately runs at a rate equivalent to that of a DNA fragment 25 nucleotides long and the xylene cyanol marker runs equivalent to a fragment of about 115 nucleotides. Depending on the vector and the primer used, the start of the unknown sequence is about 50 nucleotides from the 3′-end of the primer. Under the conditions described above (20×48× 0.04 cm buffer gradient gel, 40 W power input), molecules of this length will reach the end of the gel after about 4 h.

In classical sequencing gels, radioactive labels are used. To detect the bands, an X-ray film is placed in close contact to the gel (exposure) and processed afterwards. In order to obtain sharp bands, a low-energy radioactive source, such as ^{35}S is preferred. (^{32}P generally gives more diffuse bands). To prevent quenching of the weak signal, urea has to be washed out and the gel has to be dried.

23.2.1.9 Blotting

There are some applications where it is necessary to transfer the DNA fragments from the gel onto a membrane (blotting). These include the nonradioactive detection of the sequencing pattern with colorimetric [42,50] or chemiluminescent methods [51,52] as well as multiplex sequencing [28] (see Chapter 20). In the case of sequencing gels, there are two main methods in use: electroblotting and capillary blotting are 'off line' methods, which means that after electrophoresis, the gel has to be dismantled and the fragments transferred to a membrane with the help of capillary forces or a transverse electric field. Direct blotting, or direct transfer electrophoresis (DTE) is an 'on line' method: a membrane is moved across the bottom of the sequencing gel, and the DNA molecules are immobilized onto that matrix as they are eluted (Fig. 23.3b) (see Protocol 119).

After the transfer, the membranes can be used for hybridization (multiplex sequencing) or can be developed to visualize the DNA fragments. If blotting is used for routine nonradioactive sequencing, it might be a good idea to try to automate the developing step (see ref. 42, for example) as this can become very time consuming.

Electroblotting and capillary blotting Sequencing gels can be blotted in the same way as agarose gels by using capillary forces [53]. Electroblotting is much faster, but unfortunately, most commercially available blotting devices have been built for blotting standard agarose or protein gels and are too small to accommodate a large sequencing gel. Therefore, those interested in this technique might be obliged to construct an apparatus on their own. (The construction and use of such a device is described in ref. 54.)

The principle is essentially the same as for blotting agarose or SDS–polyacrylamide gels; however, some experience and extra care is needed in order to avoid air bubbles being trapped between the gel and the membrane.

We have successfully used a home-made 'wet blotting' device. With 0.25×TBE transfer buffer and an electric field of about 50 V cm^{-1}, transfer is complete within 15–20 min.

Direct transfer electrophoresis More than 10 years ago, a new technique was introduced that combines the separation and the blotting into a single step. This direct blotting electrophoresis or direct transfer electrophoresis (DTE) has proved useful for DNA sequencing [21,42,55] and the device is also commercially available. DTE allows nonradioactive sequencing at relatively low cost, but some experience is needed in order to achieve good results. In DTE, the DNA fragments travel the same distance but over different time spans (in contrast to normal gels); thus they are evenly spaced on the membrane over a wide size range [21]. As explained above, in this case band spreading is the limiting factor, therefore the following strategy should be adopted: thin gels (0.1 mm) should be used, as they minimize the band width. The loading capacity is reduced, however, which can be alleviated by the use of 'inverse wedge' gels [42]. An air-bubble-free, homogeneous gel with even edges in the wells, as well as a clean and even lower edge is also very important. The gel is bound to one glass plate with 3-methacryloxy-propyltrimethoxysilane [41] to avoid slipping. Longer gels increase the resolution, but lead to long run times. As the gels are thinner, stronger electric fields can be used to compensate, but this might eventually lead to a domination of the molecular orientation.

Good results have been obtained with 30-cm long gels and an electric field of 1800 V (60 V cm^{-1}) [42,55]. The optimum speed of the membrane is then 8–20 cm h^{-1}. Further improvements can be made by obeying the theoretical considerations outlined above: by using longer gels (60 cm) and higher electric fields (70 V cm^{-1}), but lower gel concentrations (3.5%), sequence readings up to 800 bases have been achieved [43]. When even higher electric fields (100 V cm^{-1}) are applied, the distances between the bands of the long fragments are reduced (Fig. 8 of ref. 43), which shows the onset of molecular orientation and indicates that the improvements have pushed the method close to its inherent limits.

23.2.2 Automated sequencing gel electrophoresis

In the past few years automated DNA sequencers have been developed [56–61]. This name is somewhat misleading, as in fact these devices are 'on-line' gel readers. The principle of slab gel electrophoresis remains the same and the gels still have to be poured and loaded manually. Therefore, what has been said above remains valid, however, more care has to be taken when preparing the gel and the gel solutions, as dirt, dust and fluorescent contaminants in the gel or on the glass plates can disturb the detection (see Protocol 120).

The existing devices are based on fluorescence detection. They require reaction products with a fluorescent group attached either to the primer, to the dideoxy nucleotide analogues or to the deoxynucleotides. The products are detected directly within the gel, by excitation of the fluorescent labels with a laser and detection of the signals emitted at the respective wavelengths. The laser beam can either be directed onto the gel surface (e.g. ref. 56) or by 'side excitation' [59,62]. During electrophoresis, the bands are passing the laser and detector, which have a fixed vertical position, leading to 'on-line' and 'real-time' detection (see Fig. 23.3c). The bands that have passed the detector are no longer needed as they have already been processed, and are electrophoresed off the gel.

For labelling and detection, different strategies can be adopted: if different fluorescent tags are used for each of the four terminator reactions, all four reaction sets can be electrophoresed in the same lane. This avoids problems due to possible mobility variations in different lanes. However, the different fluorophores change the mobility of the DNA fragments and the different shifts have to be corrected by a computer program.

Alternatively, a 'one-dye/four-lane' approach can be used, which avoids the need of four different labels, but problems of misalignment between lanes can occur and computational algorithms are needed to compensate [63]. Of course, fewer samples can be loaded on a gel of the same size. The advantages and disadvantages of the different strategies are discussed in detail in ref. 64.

As in DTE, all DNA bands travel the same distance but over increasing time spans in proportion to their size and pass the detector at regular time intervals. Again, the band width is the limiting factor and all means to reduce band broadening will enhance the performance of sequencers with on-line detection.

However, the first generation of commercially available sequencing automates did not seem to obey these rules and unsurprisingly, their performance was no better than manual sequencing (e.g. ref. 3). This is mainly due to technical limitations: to obtain a stronger signal, relatively thick gels (0.3–0.5 mm) are used. Detection systems that scan the gel have a long data acquisition time. In order to be identified, the DNA fragments have to travel at low speed, which means long run times and enhanced band width by diffusion.

Recently, theoretical considerations concerning resolution have been confirmed by empirical studies (e.g. ref. 19) and instrumental improvements have been made. These include the use of thin (0.1–0.25 mm) or long (50–90 cm) gels, stronger electric fields (up to 80 V cm^{-1}), reduced laser beam diameter, faster detection systems (e.g. without filter wheels) or simultaneous detection of all lanes [5,65–67]. These measures improve resolution and/or speed and therefore the throughput of obtainable sequence data.

Automated sequencers are not suitable for multiplexing, unless a sophisticated, multispectral labelling technique is developed.

23.3 Capillary electrophoresis

During the past 10 years, capillary electrophoresis has been developed into a powerful analytical method. Separation takes place in a thin fused silica capillary, coated with polyimide on the outside. A small window in the coating allows detection by absorption or fluorescence (Fig. 23.3d).

Capillary electrophoresis can also be used to separate oligonucleotides and DNA sequencing reaction products [68–72]. For labelling and detection, the same sequencing chemistries can be used as for the automated slab gels (ref. 73; see ref. 74 for review). However, despite a number of publications concerning this technique it has, to our knowledge, not yet been used for a larger sequencing project. One obvious reason for the low acceptance of this method for DNA sequencing was the lack of a suitable commercially available apparatus. However, in the meantime, several prototypes have been described [75–77], and by the time of writing, one device, specialized for DNA sequencing and analysis (based on a single capillary system) is on the market. There have also been some technical problems like the filling of the capillary and the stability of the gel [78]. However, the field is developing fast, and capillary electrophoresis might soon become a real alternative to the conventional method (see Protocol 121).

In principle, capillary electrophoresis has several advantages over slab gel electrophoresis. Because of the small diameter of the capillaries, heat dissipation is very effective and band broadening due to Joule heating is minimized. Strong electric fields (up to 400 V cm^{-1}) can be used, therefore reducing run time and diffusion. Capillaries are available in a variety of diameters (about 10–300 μm) and their length can be chosen in a wide range. The sensitivity is very high and minute amounts of sample can be analysed, which probably could make the amplification of the template (*in vivo* or *in vitro*) unnecessary and offer the opportunity of sequencing DNA directly isolated from plaques or colonies.

The biggest advantage, however, is the potential for full automation of the separation process for DNA sequencing. The samples are injected by pressure or electrokinetic injection. After each run, the separation matrix in the capillary can be replaced by rinsing with pressure. Both processes, injection and rinsing, can be done automatically, therefore avoiding the time-consuming tasks of gel pouring and gel loading. The commercially available instrument, for instance, has a capacity of 48 samples, which can be analysed in a fully automated way.

However, in contrast to slab gels, only one sample can be loaded at a time, which means a total analysis time of 125 h (140 min separation time, 15 min for refilling and prerun each) for the 48 samples. This is still much longer than the ≈10 h (2 h for gel preparation and loading, 8 h for prerun and run), needed for the separation of 36 samples in an automated slab gel sequencer.

Therefore it will only be possible to exploit the full potential of capillary electrophoresis for DNA sequencing if many capillaries in parallel (capillary arrays) are used. Several prototypes have been described [79–83], but a 'ready to use' system is not yet on the market. These prototypes mainly differ in the way the DNA is detected (sheath flow cuvette, confocal system or direct observation), as this is not a simple task: the detection must be sensitive and fast at the same time.

Because of the reduced band width, capillary electrophoresis has potentially a very high separation efficiency (several million theoretical plates per meter) and we should expect a better readability compared to the automated slab gel systems. So far, this has not been achieved. This could be owing to the molecular orientation, which becomes a limiting factor for separation at high field strengths [5], or to enhanced diffusion, which has been predicted to occur above a certain molecular size [16].

23.3.1 Capillaries

In capillary electrophoresis (CE), columns made of fused silica are used. The surface of untreated fused silica is comprised of silanol groups, which are negatively charged at any pH above 2. These fixed charges are balanced by positive ions in the bulk solution, which form a thin sheet of charged fluid close to the capillary wall. When an electric field is applied, these positively charged ions will move towards the cathode, dragging with them the bulk solution. This flow is called electro-osmotic flow (EOF) and in many applications is used as a 'pump' for the separation process. In CE separation of DNA, the capillary is filled with a polymeric matrix, which suppresses this EOF to a certain extent. Therefore a coating of the inner capillary surface (to suppress the EOF) is not necessarily needed, but it has been found that in many cases such a coating is advantageous for reproducibility and quality of the separation. The coatings often consist of polymers which are absorbed or covalently bonded (see ref. 84 for review). A number of coatings (e.g. methylsilicone, phenyl, polyethylene glycol, trifluorpropyl, polyacrylamide or polyvinylalcohol) are commercially available and have been successfully used in the separation of DNA.

23.3.2 Gel matrix

For separating oligonucleotides and sequencing reaction products, the same matrix as in slab gels (i.e. crosslinked polyacrylamide) is used. The gels are prepared in the same manner by adding the catalysts to the monomer solution which is then pumped into the capillary, where the polymerization takes place. The capillary is treated before with

3-methacryloxypropyltrimethoxysilane, which fixes the gel to the inner wall and prevents it from being extruded by electro-osmotic forces. Gel concentrations vary from 3 to 6%T and 3–5%C, and the same buffer and denaturants as in slab gels are used. Prefilled capillaries have also become commercially available.

However, gel-filled capillaries have several disadvantages. First, the capillary has to be filled extremely carefully in order to avoid introducing air bubbles. Shrinkage of the gel during polymerization can also be a source of bubbles [85]. It has been observed that during repeated use, bubbles can form at the sample-injection end of the capillary [78,86]. Drying out of the gel at the ends can also be a problem [78]. However, when carefully handled, gel-filled capillaries can be re-used several times, but about 50 separations seem to be the maximum.

A solution to these problems is to replace the rigid nonflowable gel with a flowable network, which in return requires noncovalent crosslinks. One type of such interactions are polymer entaglements: if a solution is sufficiently concentrated, the polymer chains become entangled, forming a transient network that has good 'sieving' properties. The first separations of short DNA molecules were done in polymer solutions such as linear (uncrosslinked) polyacrylamide. For separating oligonucleotides and sequencing reaction products, solutions of about 8–10% polyacrylamide are needed [75,87,88]. However, such solutions are highly viscous and the capillaries cannot be filled manually or with the existing capillary electrophoresis devices [89]. Therefore, as in the case of crosslinked gels, polymerization has to take place within the capillary. Again, as the capillaries cannot be rinsed or refilled, they have a rather limited life and only electrokinetic injection of the sample can be used. Recently, the use of polyacrylamide with low molecular weight has been proposed to circumvent the problem [76,90,91]. These formulations contain polymer chains that are small enough to give a low viscosity, but long enough to be still entangled. Beside polyacrylamide, other polymers such as polyethyleneoxide (PEO) have been successfully used for DNA sequencing [92].

Another type of noncovalent crosslinks are hydrophobic associations. These can be obtained if hydrophobic end groups are attached to hydrophilic polymer backbones. When such copolymers are dissolved in water, the hydrophobic ends will associate into micelles. Above a certain concentration, these micelles will form continuous superstructures ('self-assembling gels'). The mesh size and viscosity of such networks can be influenced by choosing appropriate hydrophilic backbones and different hydrophobic end groups. Such a flowable gel is successfully used as a matrix for separating DNA sequencing fragments in capillary electrophoresis [93].

23.3.3 Field conditions

For DNA sequencing, electric fields between 100 and 465 V cm^{-1} have been used. Apparently, the readability does not change very much with the electric field, but seems to get worse above 400 V cm^{-1}. These voltages were probably used to obtain a high speed rather than long readings.

Clearly, at such high electric fields, the loss of separation due to molecular orientation will occur very early, but the exact limit is not yet known. Again, there is a trade-off between the reduction of diffusional band broadening and the reduction of the thermal gradient. The optimal electric field strength is also dependent on the fragment size, but a value of about 150–250 V cm^{-1} has been found to be a good compromise [94,95]. Under these conditions, up to 300–350 bases can be read in run times of only 30–60 min (e.g. refs 72,76) and up to 450 bases in 140 min [95]. Recently, separation of sequencing fragments over 550 bases in length in about 2 h have been reported [96].

23.4 Final notes

The theoretical considerations described in this chapter concerning readability are only valid for an 'ideal' sample. It has to be pointed out that the quality of the sample, which is dependent on different factors like the DNA preparation and the sequencing chemistry used, plays an important role. In capillary electrophoresis with electrokinetic injection for example, the sample must be desalted. The sequence of the sample itself (GC content, self-complementary sequences, stretches of the same nucleotide) influences the readability. The existing automated sequencers cannot cope with large differences in band intensities and are therefore more sensitive to the type of chemistry used [97]. (However, this problem has been partially circumvented recently, with the introduction of new polymerases with a more uniform incorporation of nucleotides [98]). Finally, the software used for detection and 'band calling' in film readers and automated sequencers might produce errors.

Scientific and commercial publications often claim readings of several hundred bases, mainly obtained with M13 DNA, but the average reading length in real sequencing projects can be much lower [1–3,99].

Protocol 118 How to set up and run a standard sequencing gel

For details of solutions, media and materials, see Appendix I. For suppliers and contact addresses see Appendix III.

This protocol describes the preparation and running of a 'standard' sequencing gel, using a buffer gradient. It is, of course, a compromise between the length of the reading which can be achieved, the 'robustness' of the method and the ease of handling. Good and detailed descriptions of how to prepare and run a sequencing gel are also given in refs 44 and 100.

For the other techniques described in this chapter, see the instructions of the suppliers or the publications given in the references.

Materials

- 10×TBE buffer (pH 8.8): 162 g Tris, 27.5 g boric acid, 9.2 g Na_2EDTA; make up to 1 litre with H_2O
- 40% acrylamide solution: 380 g acrylamide, 20 g *bis*-acrylamide; make up to 1 litre with H_2O and dissolve. Add 20 g mixed bed resin (e.g. Amberlite MB1 or equivalent). Stir carefully for 20 min (this step removes metal ions and acrylic acid). Store in the dark at 4 °C. *Always wear gloves and work in the hood*
- 0.5×TBE 6% gel solution: 460 g urea, 150 ml 40% acrylamide solution, 50 ml 10×TBE buffer; make up to 1 litre with H_2O and dissolve. Filter through sintered glass funnel and store in the dark at 4 °C up to several weeks
- 5×TBE 6% gel solution: 115 g urea, 37.5 ml 40% acrylamide solution, 125 ml 10×TBE buffer, 10 mg Bromophenol blue (optional); make up to 250 ml with H_2O and dissolve. Filter through sintered glass funnel and store in the dark at 4 °C up to several weeks
- 25% ammonium persulphate: make up 2.5 g ammonium persulphate to 10 ml with H_2O and store for up to several weeks at 4 °C
- TEMED (*N,N,N′,N′*-tetramethyl-1,2-diaminoethane)
- repel-silane (dimethylchlorosilane solution)

Method

1. Thoroughly wash a set of glass plates with detergent and warm water. Rinse with deionized water and let them air-dry.

2. Treat one glass plate (e.g. the notched one) with repel-silane by spreading 1 ml of solution with a paper tissue all over the inner surface (work in the fume hood).

3. Wipe the inner surface of both plates with a few millilitres of ethanol using a paper tissue.

4. Put the spacers (0.5–1 cm wide) onto the edges of one plate and assemble both plates. Form a liquid- and air-tight seal on both sides

and at the bottom with polyester tape. Take particular care at the bottom corners of the plates.

5 For a 20 × 50 × 0.04 cm gel use 45 ml 0.5 × TBE gel solution and 7 ml 5 × TBE gel solution. Add to both solutions 2 µl ammonium persulphate and 2 µl TEMED for each ml of gel solution. Mix the solutions by swirling.

6 Immediately, take up about 39 ml of the 0.5 × TBE mix into a 50-ml syringe and put it aside. Take up the rest into a 25-ml glass pipette fitted with a pipette controller.

7 Draw the 5 × TBE solution into the same pipette and allow a few air bubbles to pass upwards through the gel solutions in order to establish a rough gradient.

8 Slowly pour the solution into the mould (held at an angle), either down one side (easier to perform) or down the centre of the mould (gives a more even gradient).

9 Lower the mould, pick up the syringe containing the 0.5 × TBE gel mix and continue pouring. Control the flow rate by altering the angle at which the mould is held.

10 Examine the gel for air bubbles. Often, air bubbles can be driven out of the gel by lifting the plates and slightly knocking against the mould. Alternatively, a thin spacer can be introduced to push a bubble aside.

11 Lower the mould to a nearly horizontal position and insert the comb. Clamp the plates together over the side spacers and leave it to polymerize. If some solution remains in the syringe, pour it into a reaction tube and close it. This way the polymerization can be monitored. The gelification should take place after about 15–20 min, but polymerization continues for a much longer time.

12 When polymerization is completed, wash away any dried gel from the outside of the plates and carefully remove the slot former (works best under an overlay of water).

13 Remove the tape from the bottom and attach the gel to the electrophoresis apparatus. Fill the buffer chambers with 1 × TBE.

14 Denature the samples for 20 min at 80 °C. Thoroughly flush the wells and immediately load 1–2 µl of the sequencing reactions (see Section 23.2.1.5).

15 When the gel is loaded, close the apparatus and connect it to the power supply (see Section 23.2.1.6 for field conditions).

16 At the end of the run (see Section 23.2.1.8), disconnect the power supply. Discard the buffer (radioactive!) and remove the gel from the apparatus. Peel off the tape and separate the glass plates, using a spatula. The gel should stick to the nonsilanized plate.

17 Transfer the gel (on the plate) into a 10% acetic acid solution and leave it for 15 min.

18 Carefully remove the glass plate with the gel from the acetic acid and let the liquid drain off. Let it dry in a nearly vertical position for 15 min.

19 Lay the glass plate and the gel in a horizontal position, cut the gel to size with a 'pizza cutter' and place a piece of Whatman 3MM paper on top of it. Peel off the paper with the gel stuck to it.

20 Put a sheet of Saran plastic wrap on top of the gel, trim the whole and dry the gel on a gel dryer.

21 After drying, peel away the Saran wrap, place the gel into a film cassette and position a sheet of X-ray film in direct contact with it. Expose overnight and develop the film.

Protocol 119 How to set up and run a direct blotting gel

For details of solutions, media and materials, see Appendix I. For suppliers and contact addresses see Appendix III.

Running a direct blotting gel requires some extra care and experience. One point that needs special attention is the lower surface of the gel. The gel must be uniformly polymerized, the glass plates must have a flat and smooth surface over the whole width of the gel, and they must be carefully aligned. Air bubbles between the gel and the membrane must be absolutely avoided. We therefore refer the reader to detailed protocols [42,43] and the instructions of the supplier [101].

Method

1 Thoroughly wash a set of glass plates with detergent and warm water. Rinse in deionized water and let them air-dry.

2 Treat both glass plates with bind-silance by spreading 1 ml of solution with a paper tissue all over the inner surfaces. Let stand for 10–15 min.

3 Wipe the inner surfaces of both plates with a few millilitres of ethanol using a paper tissue.

4 Put the spacers onto the edges of one plate but do not assemble both plates.

5 Prepare 4% gel solution by adding 43.3 ml urea diluent (84 g urea, 18.7 g 10×TBE, 73.4 g water) to 6.7 ml 30% acrylamide stock solution (for other concentrations, adjust volumes accordingly).

6 For a 32-cm long gel, take 17 ml degassed gel solution and add 75 µl 10% APS and 17 µl TEMED. Mix the solutions by swirling.

7 Immediately, take up into a 20-ml syringe and pour the gel by either the sliding technique [101] or the clapping technique [43].

8 Examine the gel for air bubbles and ensure parallel alignment of the lower glass plate edges.

9 Insert the comb. Clamp the plates together over the side spacers and leave it to polymerize. If some solution remains in the syringe, pour it into a reaction tube and close. This way the polymerization can be monitored. Gelification should take place after about 15–20 min, but polymerization continues for a much longer time.

10 Fill the lower buffer chamber with 1 × TBE and attach the membrane to the conveyor belt.

11 When polymerization is completed, wash away any dried gel from the outside of the plates and carefully remove dried gel from the lower edges without touching the gel itself. Remove the clamps and carefully put the plates into the apparatus. Attach an aluminium plate to the front glass plate with two clamps.

12 Fill 1 × TBE into the upper buffer chamber and remove the comb (or precomb, respectively).

13 Perform pre-electrophoresis for 30 min. For exact conditions see refs 42 and 101.

14 Switch off and disconnect the electrodes. Flush the wells and load the samples. (If a sharkstooth comb is used, flush the gel pocket, carefully introduce the comb and load the samples.)

15 When the gel is loaded, close the apparatus and reconnect it to the power supply (see Section 23.2.1.9 for field conditions).

16 When the bromophenol blue front is close to the bottom edge, switch off the power supply and move the membrane under the gel. Set the speed of the conveyor belt to 8–20 cm h^{-1}. Switch on the power again and continue the run.

17 At the end of the run, disconnect the power supply and remove the gel from the apparatus. Carefully hold the membrane with tweezers and detach from the conveyor belt.

18 Dry the membrane, crosslink with UV light and expose to film (when radioactive label was used) or 'develop' with the non-radioactive detection system.

Protocol 120 How to set up and run a sequencing gel for automated sequencing

For details of solutions, media and materials, see Appendix I. For suppliers and contact addresses see Appendix III.

Preparing a gel for automated sequencing requires more care than for standard gels, to ensure absence of fluorescence contaminants and to exploit fully the theoretically possible longer read lengths. For more details see the instructions of the suppliers (e.g. refs 102 and 103).

Method

1. Thoroughly wash a set of glass plates with detergent (nonfluorescent!) and warm water. Rinse with deionized water and let them air-dry.
2. Wipe the inner surface of both plates with a few millimetres of ethanol or methanol using a paper tissue.
3. Assemble the glass plates according to the instructions of the supplier. (Alternatively, use a 'clapping' technique.)
4. To prepare a 25×48×0.02 cm gel: to 28.8 g urea add 8 ml 40% acrylamide solution, 35 ml distilled water and 1 g mixed-bed ion exchange resin. (This will result in a 4% gel. If a gel of different concentration, e.g. 5% or 6%, is needed, adjust the volumes accordingly.)
5. Stir until the urea is dissolved and filter through a 0.22 μm pore size filter.
6. Transfer into a cylinder, add 8 ml 10×TBE and pure water up to 80 ml.
7. Degas for a few minutes.
8. Add 400 μl of a freshly made 10% APS solution and 55 μl TEMED. Swirl gently and take up into a syringe.
9. Introduce the gel solution into the assembled glass plates (or use the clapping technique).
10. Examine the gel for air bubbles. Often, air bubbles can be driven out of the gel by lifting the plates and slightly knocking against the mould. Alternatively, a thin spacer can be introduced to push a bubble aside.
11. Lower the mould to a nearly horizontal position and insert the comb or precomb. Clamp the plates together over the side spacers and leave it to polymerize. If some solution remains in the syringe, pour a small quantity into a reaction tube and close it. This way the

polymerization can be monitored. Gelification should take place after 15–20 min, but polymerization continues for a much longer time.

12 When polymerization is completed (1–2 h), wash away any dried gel from the outside of the plates and carefully remove the slot former. (When using a sharkstooth comb, remove the precomb and replace with the sharkstooth comb.)

13 Clean the plates and mount the gel into the electrophoresis apparatus according to the instructions of the supplier. Fill the buffer chambers with 1 × TBE.

14 Attach the heat transfer plate (if available) and connect the electrode cables.

15 Perform a prerun for about 20 min.

16 Pause the prerun, load the samples, and start the run.

Protocol 121 How to set up and run a capillary electrophoresis gel for sequencing

For details of solutions, media and materials, see Appendix I. For suppliers and contact addresses see Appendix III.

Sequencing with capillary electrophoresis requires only a few preparation steps as the whole procedure is fully automated. For more details see the instructions of the supplier [95].

Method

1 Equilibrate the plastic syringe containing the separation matrix to room temperature and install into the pump block.

2 Install the sequencing capillary according to the instructions of the supplier.

3 Install the glass syringe (serving as a reservoir for the gel).

4 Push the gel out of the plastic syringe into the glass syringe, either manually or with the help of the machine. Be careful not to introduce air bubbles.

5 Fill the vials with buffer and water according to the instructions.

6 Prepare the samples, cap the sample vials with septa and load the autosampler.

7 For running parameters (injection time and voltage, run time, voltage and temperature) use programmed values or adjust according to your needs.

8 Start the run.

The samples are then automatically injected by electrokinetic injection and separated in the gel-filled capillary. After each run, a small amount of gel is automatically pushed out of the glass syringe into the capillary, replacing the used gel.

References

1 Smith, V., Brown, C.M., Bankier, A.T. & Barrell, B.G. (1990) Semiautomated preparation of DNA templates for large-scale sequencing projects. *DNA Sequence* **1**, 73–78.

2 Davison, A.J. (1991) Experience in shotgun sequencing a 134 kilobase pair DNA molecule. *DNA Sequence* **1**, 389–394.

3 Khurshid, F. & Beck, S. (1993) Error analysis in manual and automated DNA sequencing. *Anal. Biochem.* **208**, 138–143.

4 Heller, C. & Beck, S. (1992) Field inversion gel electrophoresis in denaturing polyacrylamide gels. *Nucleic Acids Res.* **20**, 2447–2452.

5 Grossman, P.D., Menchen, S. & Hershey, D. (1992) Quantitative analysis of DNA-sequencing electrophoresis. *Genet. Anal. Tech. Appl.* **9**, 9–16.

6 Ogston, A.G. (1958) The spaces in a uniform random suspension of fibers. *Trans. Faraday. Soc.* **54**, 1754–1757.

7 Rodbard, D. & Chrambach, A. (1970) Unified theory of gel electrophoresis and filtration. *Proc. Natl Acad. Sci. USA* **4**, 970–977.

8 Lerman, L.S. & Frisch, H.L. (1982) Why does the electrophoretic mobility of DNA in gels vary with the length of the molecule? *Biopolymers* **21**, 995–997.

9 Lumpkin, O.J. & Zimm, B.H. (1982) Mobility of DNA in gel electrophoresis. *Biopolymers* **21**, 2315–2316.

10 Slater, G.W. & Noolandi, J. (1985) New biased-reptation model for charged polymers. *Phys. Rev. Lett.* **55**, 1579–1582.

11 Hurley, I. (1986) DNA orientation during gel electrophoresis and its relation to electrophoretic mobility. *Biopolymers* **25**, 539–554.

12 Jonsson, M., Akerman, B. & Norden, B. (1988) Orientation of DNA during gel electrophoresis studied with linear dichroism spectroscopy. *Biopolymers* **27**, 381–414.

13 Mayer, P., Slater, G.W. & Drouin, G. (1993) Exact behaviour of single stranded DNA electrophoretic mobilities in polyacrylamide gels. *Appl. Theor. Electrophoresis* **3**, 147–155.

14 Ulanovsky, L., Drouin, G. & Gilbert, W. (1990) DNA trapping electrophoresis. *Nature* **343**, 190–192.

15 Slater, G.W. & Drouin, G. (1992) Why can we not sequence thousands of DNA bases on a polyacrylamide gel? *Electrophoresis* **13**, 574–582.

16 Slater, G.W. (1993) Theory of band broadening for DNA gel electrophoresis and sequencing. *Electrophoresis* **14**, 1–7.

17 Giddings, J.C. (1965) *Dynamics of Chromatography* (Marcel Dekker, New York).

18 Hjertén, S. (1990) Zone broadening in electrophoresis with special reference to high performance electrophoresis in capillaries. *Electrophoresis* **11**, 665–690.

19 Nishikawa, T. & Kambara, H. (1991) Analysis of limiting factors of DNA band separation by a DNA sequencer using fluorescence detection. *Electrophoresis* **12**, 623–631.

20 Reijenga, J.C. & Kenndler, E. (1994) Computational simulation of migration and dispersion in free capillary zone electrophoresis. *J. Chromatogr. A* **659**, 403–415.

21 Beck, S. & Pohl, F.M. (1984) DNA sequencing with direct blotting electrophoresis. *EMBO J.* **3**, 2905–2909.

22 Issaq, H.J., Atamna, I.Z., Muschik, G.M. & Janini, G.M. (1991) The effect of electric field strength, buffer type and concentration on separation parameters in capillary zone electrophoresis. *Chromatographia* **32**, 155–161.

23 Slater, G.W., Mayer, P. & Drouin, G. (1993) On the limits of near-equilibrium DNA gel electrophoretic sequencing. *Electrophoresis* **14**, 961–966.

24 Lai, E., Davi, N.A. & Hood, L. (1989) Effect of electric field switching on the electrophoretic mobility of single-stranded DNA molecules in polyacrylamide gels. *Electrophoresis* **10**, 65–67.

25 Birren, B.W., Simon, M.I. & Lai, E. (1990) The basis of high resolution separation of small DNA by asymmetric-voltage field inversion electrophoresis and its application to DNA sequencing gels. *Nucleic Acids Res.* **18**, 1481–1488.

26 Daniels, D.L., Marr, L., Brumley, R.L. & Blattner, F. (1990) Field inversion gel electrophoresis as applied to DNA sequencing. In *Structure and Methods* (Sarma, R.H. & Sarma, M.H., eds), Vol. 1, 29–35 (Adenine Press, New York).

27 Brassard, E., Turmel, C. & Noolandi, J. (1992) Pulsed

field sequencing gel electrophoresis. *Electrophoresis* **13**, 529–535.
28 Church, G.M. & Kieffer-Higgins, S. (1988) Multiplex DNA sequencing. *Science* **240**, 185–188.
29 Hochstrasser, D.F., Patchornik, A. & Merril, C.R. (1988) Development of polyacrylamide gels that improve the separation of proteins and their detection by silver staining. *Anal. Biochem.* **173**, 412–423.
30 Gelfi, C. & Righetti, P.G. (1981) Polymerization kinetics of polyacrylamide gels. I. Effect of different cross-linkers. *Electrophoresis* **2**, 213–219.
31 Rüchel, R., Steere, R.L. & Erbe, E.F. (1978) Transmission-electron microscopic observations of freeze-etched polyacrylamide gels. *J. Chromatogr.* **166**, 563–575.
32 Morris, C.J.O.R. & Morris, P. (1971) Molecular-sieve chromatography and electrophoresis in polyacrylamide gels. *Biochem. J.* **124**, 517–528.
33 Chrambach, A. & Rodbard, D. (1971) Polyacrylamide gel electrophoresis. *Science* **172**, 440–451.
34 Ansorge, W. & Barker, R. (1984) System for DNA sequencing with resolution of up to 600 base pairs. *J. Biochem. Biophys. Meth.* **9**, 33–47.
35 Hochstrasser, D.F. & Merril, C.R. (1988) 'Catalysts' for polyacrylamide gel polymerization and detection of proteins by silver staining. *Appl. Theor. Electrophoresis* **1**, 35–40.
36 Gelfi, C. & Righetti, P.G. (1981) Polymerization kinetics of polyacrylamide gels. II. Effect of temperature. *Electrophoresis* **2**, 220–228.
37 Righetti, P.G., Chiari, M., Nesi, M. & Caglio, S. (1993) Towards new formulations for polyacrylamide matrices, as investigated by capillary zone electrophoresis. *J. Chromatogr.* **638**, 165–178.
38 Righetti, P.G. (1995) Macroporous gels: facts and misfacts. *J. Chromatogr. A* **698**, 3–17.
39 Gelfi, C., Canali, A., Righetti, P.G. *et al.* (1990) DNA sequencing in HydroLink matrices. *Electrophoresis* **11**, 595–600.
40 Brumley, R.L. & Smith, L.M. (1991) Rapid DNA sequencing by horizontal ultrathin gel electrophoresis. *Nucleic Acids Res.* **19**, 4121–4126.
41 Garoff, H. & Ansorge, W. (1981) Improvements of DNA sequencing gels. *Anal. Biochem.* **115**, 450–457.
42 Richterich, P., Heller, C., Wurst, H. & Pohl, F.M. (1989) DNA sequencing with direct blotting electrophoresis and colorimetric detection. *Biotechniques* **7**, 52–59.
43 Richterich, P. & Church, G.M. (1993) DNA Sequencing with direct transfer electrophoresis and non-radioactive detection. *Meth. Enzymol.* **218**, 187–222.
44 Bankier, A.T. & Barrell, B.G. (1989) Sequencing single-stranded DNA using the chain-termination method. In *Nucleic Acids Sequencing: A Practical Approach* (Howe, C.J. & Ward, E.S., eds), 117–135 (IRL Press, Oxford).
45 Ansorge, W. & Labeit, S. (1984) Field gradients improve resolution on DNA sequencing gels. *J. Biochem. Biophys. Meth.* **10**, 237–243.
46 Olsson, A., Moks, T., Uhlen, M. & Gaal, A.B. (1984) Uniformly spaced banding pattern in DNA sequencing gels by use of field-strength gradient. *J. Biochem. Biophys. Meth.* **10**, 83–90.
47 States, J.C., Patel, L.R. & Li, Q. (1991) A gel electrophoresis system for resolving over 500 nucleotides with a single sample loading. *Biotechniques* **11**, 46–48.
48 Biggin, M.D., Gibson, T.J. & Hong, G.F. (1983) Buffer gradient gels and ^{35}S label as an aid to rapid DNA sequence determination. *Proc. Natl Acad. Sci. USA* **80**, 3963–3965.
49 Sheen, J.-Y. & Seed, B. (1988) Electrolyte gradient gels for DNA sequencing. *Biotechniques* **6**, 942–944.
50 Beck, S. (1987) Colorimetric-detected DNA sequencing. *Anal. Biochem.* **164**, 514–520.
51 Beck, S., O'Keefe, T., Coull, J.M. & Köster, H. (1989) Chemiluminescent detection of DNA: Application for DNA sequencing and hybridization. *Nucleic Acids Res.* **17**, 5115–5123.
52 Karger, A.E., Weiss, R. & Gesteland, R.F. (1992) Digital chemiluminescence imaging of DNA sequencing blots using a charge-coupled device camera. *Nucleic Acids Res.* **20**, 6657–6665.
53 Chee, M. (1991) Enzymatic multiplex sequencing. *Nucleic Acids Res.* **19**, 3301–3305.
54 Saluz, H.P. & Jost, J.P. (1987) *A Laboratory Guide to Genomic Sequencing* (Birkhäuser, Basel).
55 Pohl, T.M. & Maier, E. (1995) Sequencing 500 kb of yeast DNA using a GATC 1500 direct blotting electrophoresis system. *Biotechniques* **19**, 482–486.
56 Smith, L.M., Sanders, J.Z., Kaiser, R.J. *et al.* (1986) Fluorescence detection in automated DNA sequence analysis. *Nature* **321**, 674–678.
57 Ansorge, W., Sproat, B.S., Stegemann, J. & Schwager, C. (1986) A non-radioactive automated method for DNA sequence determination. *J. Biochem. Biophys. Meth.* **13**, 315–323.
58 Prober, J.M., Trainor, G.L., Dam, R.J. *et al.* (1987) A system for rapid DNA sequencing with fluorescent chain-terminating dideoxynucleotides. *Science* **238**, 336–341.
59 Kambara, H., Nishikawa, T., Katayama, Y. & Yamaguchi, T. (1988) Optimization of parameters in a DNA sequenator using fluorescence detection. *Bio/Technology* **6**, 816–821.
60 Brumbaugh, J.A., Middendorf, L.R., Grone, D.L. & Ruth, J.L. (1988) Continuous on-line DNA sequencing using oligodeoxynucleotide primers with multiple fluorophores. *Proc. Natl Acad. Sci. USA* **85**, 5610–5614.
61 Middendorf, L.R., Bruce, J.C., Bruce, R.C. *et al.* (1992) Continuous, on-line DNA sequencing using a versatile infrared laser scanner/electrophoresis apparatus. *Electrophoresis* **13**, 487–494.
62 Chen, D., Peterson, M.D., Brumley, R.L. *et al.* (1995) Side excitation of fluorescence in ultrathin slab gel electrophoresis. *Anal. Chem.* **67**, 4305–3411.
63 Fujii, H. & Kashiwagi, K. (1992) Compensation for mobility inequalities between lanes from band signals in on-line fluorescence DNA sequencing. *Electrophoresis* **13**, 500–505.
64 Hawkins, T.L., Du, Z., Halloran, N.D. & Wilson, R.K. (1992) Fluorescence chemistries for automated primer directed DNA sequencing. *Electrophoresis* **13**, 552–559.
65 Stegemann, J., Schwager, C., Erfle, H. *et al.* (1991) High

speed on-line DNA sequencing on ultrathin slab gels. *Nucleic Acids Res.* **19**, 675–676.
66 Kostichka, A.J., Marchbanks, M.L., Brumley, R.L. *et al.* (1992) High speed automated DNA sequencing in ultrathin slab gels. *Bio/Technology* **10**, 78–81.
67 Nishikawa, T. & Kambara, H. (1992) High resolution-separation of DNA bands by electrophoresis with a long gel in a fluorescence-detection DNA sequencer. *Electrophoresis* **13**, 495–499.
68 Cohen, A.S., Najarian, D.R., Paulus, A. *et al.* (1988) Rapid separation and purification of oligonucleotides by high-performance capillary gel electrophoresis. *Proc. Natl Acad. Sci. USA* **85**, 9660–9663.
69 Cohen, A.S., Najarian, D.R. & Karger, B.L. (1990) Separation and analysis of DNA sequence reaction products by capillary gel electrophoresis. *J. Chromatogr.* **516**, 49–60.
70 Swerdlow, H. & Gesteland, R. (1990) Capillary gel electrophoresis for rapid, high resolution DNA sequencing. *Nucleic Acids Res.* **18**, 1415–1420.
71 Luckey, J.A., Drossman, H., Kostichka, A.J. *et al.* (1990) High speed DNA sequencing by capillary electrophoresis. *Nucleic Acids Res.* **18**, 4417–4422.
72 Karger, A.E., Harris, J.M. & Gesteland, R.F. (1991) Multiwavelength fluorescence detection for DNA sequencing using capillary electrophoresis. *Nucleic Acids Res.* **19**, 4955–4962.
73 Williams, D.C. & Soper, S.A. (1995) Ultrasensitive near-IR fluorescence detection for capillary gel electrophoresis and DNA sequencing applications. *Anal. Chem.* **66**, 1021–1026.
74 Swerdlow, H., Zhang, J.Z., Chen, D.Y. *et al.* (1991) Three DNA sequencing methods using capillary gel electrophoresis and laser-induced fluorescence. *Anal. Chem.* **63**, 2835–2841.
75 Pentoney, S.L.J., Konrad, K.D. & Kaye, W. (1992) A single-fluor approach to DNA sequence determination using high performance capillary electrophoresis. *Electrophoresis* **13**, 467–474.
76 Ruiz-Martinez, M.C., Berka, J., Belenkii, A. *et al.* (1993) DNA sequencing by capillary electrophoresis with replaceable linear polyacrylamide and laser-induced fluorescence detection. *Anal. Chem.* **65**, 2851–2858.
77 Luckey, J.A., Drossman, H., Kostichka, T. & Smith, L.M. (1993) High-speed DNA sequencing by capillary gel electrophoresis. *Meth. Enzymol.* **218**, 154–172.
78 Swerdlow, H., Dew, J.K.E., Brady, K. *et al.* (1992) Stability of capillary gels for automated sequencing of DNA. *Electrophoresis* **13**, 475-483.
79 Zagursky, R.J. & McCormick, R.M. (1990) DNA sequencing separations in capillary gels on a modified commercial DNA sequencing instrument. *Biotechniques* **9**, 74–79.
80 Huang, X.C., Quesada, M.A. & Mathies, R.A. (1992) Capillary array electrophoresis using laser-excited confocal fluorescence detection. *Anal. Chem.* **64**, 967–972.
81 Takahashi, S., Murakami, K., Anazawa, T. & Kambara, H. (1994) Multiple sheath-flow gel capillary array electrophoresis for multicolour fluorescent DNA detection. *Anal. Chem.* **66**, 1021–1026.
82 Ueno, K. & Yeung, E.S. (1994) Simultaneous monitoring of DNA fragments separated by electrophoresis in a multiplexed array of 100 capillaries. *Anal. Chem.* **66**, 1424–1431.
83 Lu, X. & Yeung, E.S. (1995) Optimization of excitation and detection geometry for multiplexed capillary array electrophoresis of DNA fragments. *Appl. Spectrosc.* **49**, 605–609.
84 Chiari, M., Nesi, M. & Righetti, P.G. (1996) Surface modification of silica walls. In *Capillary Electrophoresis in Analytical Biochemistry* (Righetti, P.R, ed.), 1–36 (CRC Press, Boca Raton, FL).
85 Dolnik, V., Cobb, K.A. & Novotny, M. (1991) Preparation of polyacrylamide gel-filled capillaries for capillary electrophoresis. *J. Microcol. Sep.* **3**, 155–159.
86 Dubrow, R.S. (1992) Capillary gel electrophoresis. In *Capillary Electrophoresis* (Grossman, P.D. & Colburn, J.C., eds), 133–156 (Academic Press, San Diego).
87 Huang, X.C., Quesada, M.A. & Mathies, R.A. (1992) DNA sequencing using capillary array electrophoresis. *Anal. Chem.* **64**, 2148–2154.
88 Manabe, T., Chen, N., Terabe, S. *et al.* (1994) Effects of linear polyacrylamide concentrations and applied voltages on the separation of oligonucleotides and DNA sequencing fragments by capillary electrophoresis. *Anal. Chem.* **66**, 4243–4252.
89 Chiari, M., Nesi, M., Fazio, M. & Righetti, P.G. (1992) Capillary electrophoresis of macromolecules in 'syrupy' solutions. *Electrophoresis* **13**, 690–697.
90 Heller, C. & Viovy, J.-L. (1994) Electrophoretic separation of oligonucleotides in replenishable polyacrylamide-filled capillaries. *Appl. Theor. Electrophoresis* **4**, 39–41.
91 Grossman, P.D. (1994) Electrophoretic separation of DNA sequencing extension products using low-viscosity entangled polymer networks. *J. Chromatogr. A* **663**, 219–227.
92 Fung, E.N. & Yeung, E.S. (1995) High-speed DNA sequencing by using mixed poly (ethylene oxide) solutions in uncoated capillaries. *Anal. Chem.* **67**, 1913–1919.
93 Menchen, S.M., Winnik, M.A. & Johnson, B.F. (1995) Viscous electrophoresis polymer medium and method. United States Patents No. **5**, 468, 365 and **5**, 290, 418.
94 Luckey, A.J. & Smith, L.M. (1993) Optimization of electric field strength for DNA sequencing in capillary gel electrophoresis. *Anal. Chem.* **65**, 2841–2850.
95 ABI (1995) *Prism, 310 Genetic Analyzer User's Manual.*
96 Best, N., Arriga, E., Chen, D.Y. & Dovichi, N.J. (1994) Separation of fragments up to 570 bases in length by use of 6%T non-crosslinked polyacrylamide for DNA sequencing in capillary electrophoresis. *Anal. Chem.* **66**, 4063–4067.
97 Sanders, J.Z., MacKellar, S.L., Otto, B.J. *et al.* (1990) Peak height variability and accuracy in automated DNA sequencing. In *Structure and Methods* (Sarma, R.H. & Sarma, M.H., eds), Vol. 1, 89–102 (Adenine Press, New York).
98 Tabor, S. & Richardson, C.C. (1995) A single residue in

DNA polymerases of the *E. coli* DNA polymerase I family is critical for distinguishing between deoxy- and dideoxyribonucleotides. *Proc. Natl Acad. Sci. USA* **92**, 6339–6343.

99 Naeve, C.W., Buck, G.A., Niece, R.L. *et al.* (1995) Accuracy of automated DNA sequencing: a multi-laboratory comparison of sequencing results. *Biotechniques* **19**, 448–453.

100 Sambrook, J., Fritsch, E.F. & Maniatis, T. (1989) *Molecular Cloning* (Cold Spring Harbor Laboratory Press, Cold Spring Harbor, NY).

101 GATC (1993) *1500 Direct Blotting Electrophoresis System: Operating Instructions.*

102 ABI (1995) *Prism, 377 DNA Sequencer User Manual.*

103 ALFDNA (1994) *Sequencer User Manual.*

Chapter 24 Sequence labelling and detection

Peter Richterich

Genome Therapeutics Corporation, 100 Beaver Street, Waltham MA 02154, USA

24.1 Introduction

The goal of this chapter is to give a newcomer to the field of DNA sequencing an overview of labelling and detection methods used. In addition, guidelines are provided to help in the choice of a method for a sequencing project, along with example protocols.

For at least 10 years after the original description of DNA sequencing by the dideoxy sequencing method [1] and the chemical sequencing method [2] (see Chapter 22), virtually all sequencing was done using radioactive labelling and autoradiographic detection. Today, however, a scientist has the choice between a number of different labelling-detection methods for DNA sequencing. The most commonly used methods can be grouped into three categories:

1 radioactive isotope methods;
2 fluorescence-based machines [3–6];
3 enzyme-linked methods using colorimetric, chemiluminescent and fluorigenic substrates.

Two other methods to detect DNA sequence patterns are less commonly used, hybridization [7,8] and silver staining [9,10].

A typical band in a DNA sequence pattern consists of 0.01–0.5 femtomoles (fmol) of DNA; for comparison, a typical band in an agarose gel contains 10–200 fmol DNA. To detect these minute amounts of DNA, some kind of signal amplification mechanism is used by all detection methods. With radioactive labelling and detection, the amplification takes place when a few decay events lead to a chemical chain reaction in the film which produces a visible silver grain. With fluorescence detection, one takes advantage of the fact that each fluorescent dye molecule can be excited and then emit light many thousands of times.

Finally, enzyme-linked methods utilize the high substrate turnover of the enzymes used to obtain a signal amplification of several orders of magnitude, thereby creating a signal which can be seen by the naked eye. Enzyme-linked methods typically employ labelling molecules such as biotin [11] or digoxigenin [12]. These are detected through bridge molecules such as streptavidin or antibodies that are labelled with enzymes such as alkaline phosphatase or horseradish peroxidase. Upon incubation with colorimetric, chemiluminescent, or fluorigenic substrates, the product molecules of the enzymatic reaction give rise to a coloured precipitate, emit light, or become fluorescent.

A variety of different combinations of labels, enzymes and substrates have been successfully used for DNA sequencing with enzyme-linked detection; similarly, a variety of different isotopes and fluorescent sequencers are available for radioactive and fluorescent sequencing. This chapter gives an overview of the most commonly used enzyme-linked and radioactive methods. Fluorescent methods, on the other hand, will be treated only briefly, mainly because the initial cost of fluorescence-based sequencers will rule out their use for many small laboratories.

In addition, hybridization-based detection methods will be discussed. The concept was originally developed for methylation studies [7] and later extended to multiplex sequencing [8]. Radioactive hybridization probes [8] as well as enzyme-linked [13–15] and fluorescent [16] detection schemes have been used for multiplex sequencing (see Chapter 20). This section may be of interest even if the reader is not considering multiplex sequencing as an option, since the use of enzyme-labelled hybridization probes can be a convenient alternative to labels like biotin or digoxigenin.

24.2 Choosing a detection method

The choice of the labelling and detection method depends on a number of factors. These include:

- local experience and equipment;
- personal preferences;
- health concerns;
- time constraints;
- project size;
- regulatory restrictions;
- reagent and material costs;
- sequencing strategy.

The relative importance of these factors will vary from case to case, and individual factors will often make the choice obvious. To give an example, results might be needed very quickly in a laboratory where only experience with radioactive sequencing exists, making radioactive labelling the method of choice; or, as is the case for the author's laboratory, personal preference and experience can tilt the decision the other way, towards enzyme-linked methods.

In less obvious cases, one can give each factor a weight and all of the possible methods a score. By summing over all of the scores multiplied by the weights, a rational choice can be made for the method with the higher weighted score. Table 24.1 compares the major advantages and disadvantages for radioactive and enzyme-linked detection and may be helpful for this task.

Unless regulatory or monetary restriction dictate the choice of a detection system, personal preferences will often play a major role. For best results with enzyme-linked detection, good protocols and the help from experienced users can be of critical importance. The single most determining factor of

Table 24.1 Comparison of radioactive and enzyme-linked detection methods for DNA sequencing.

Radioactive	Enzyme-linked
Advantages	*Advantages*
Established in most laboratories	Stable reagents and reactions (can be stored for months)
Reagent costs can be lower than for enzyme-linked detection	No training for use of radioisotopes required
No transfer to membranes required	No special licences or work areas for radioactivity needed
Most straightforward method for 'primer-walking' strategies	Multiple exposures can be obtained within 1–3 h
	Results can be obtained within 2 h of electrophoresis
	Easy switch to label multiplexing for higher efficiency
	Very high spatial resolution [29] and multicolour detection [64] with colorimetric substrates
	Well suited for users of direct transfer electrophoresis [19,20,29]
Disadvantages	*Disadvantages*
Potential radiation hazards	Transfer from gels onto membranes required
Training in handling of radioisotopes required	Reagent costs can be higher, especially with chemiluminescent detection
Unstable reagents and reactions	Background problems can result from bacterial contamination of buffers and handling errors
Long exposure times, especially with ^{33}P and ^{35}S	Additional 'hands on' time (0.5–2 h) required for detection procedures
Usage may be restricted by local and state regulations	

success, however, is often how careful and meticulous the person performing the experiments is.

24.2.1 Cost considerations

In many cases, the higher cost of nonradioactive detection kits is viewed as a strong argument for radioactive detection. For typical chemiluminescent detection kits, costs for membranes, buffers, antibody–enzyme complexes, substrates, and films come to approximately $40–$50 per membrane (15×40 cm). With 10 clones and an average read length of 200–250 bases, this amounts to 2 cents per raw base. These costs can easily be reduced by using lower concentrations of chemiluminescent substrates, or by using colorimetric or fluorigenic substrates.

Cost for radioisotopes, on the other hand, can be as high or higher. The isotope ^{33}P, which is often used because of its lower radiation hazards and easier handling [17], for example, has a current list price of $7 per reaction. For a gel with 10 clones, the resulting isotope costs of $70 would be higher than for enzyme-linked detection with chemiluminescent substrates. For occasional users of radioactivity who do not use an entire vial before decay, costs can be even higher.

However, the cost contribution of labelling and detection reagents to the overall costs in DNA sequencing is generally very small; this is generally true for both enzyme-linked and radioactive detection. For most sequencing projects, the total cost is typically between $1 and $3 per finished base pair, or between 10 and 50 cents per 'raw' base. Thus, costs of labelling and detection typically contribute less than 5–10% of the overall costs in DNA sequencing projects.

24.2.2 Project size

Another misconception is that enzyme-linked detection methods may be appropriate for small projects, but not for medium-sized or large projects. Quite the opposite is true: within the last three years, at least three projects have used different enzyme-linked approaches to generate more than 600 000 bases of raw sequence data each. In one case, more than 4 million bases of raw sequence for 500 kb finished sequence [18] have been generated. Two of these projects were run under severe cost constraints, illustrating the fact that enzyme-linked detection can be cost-efficient.

24.2.3 Handling and time considerations

The fact that DNA needs to be transferred onto

nylon membranes for enzyme-linked detection may keep some from switching to nonradioactive systems. However, two different transfer methods can make this task very straightforward: direct transfer electrophoresis (DTE) [19,20] and 'contact' capillary transfer (see Section 24.4.9).

In enzyme-linked detection, the transfer step as well as incubations and washes during development can make the overall procedure more time-intensive than radioactive detection. However, this can be offset by convenience gains. Reactions can be done on any bench without special safety precautions, no radioisotopes need to be ordered and disposed, and reagents and reactions can be stored without any impact on the quality of results, thus allowing for more flexible work schedules.

24.3 Radioactive labelling

24.3.1 Isotope choices

Classically, radioactive DNA sequencing has been done with ^{32}P-labelled deoxynucleotides. ^{32}P has a half-life of 14 days; since it is a strong β-emitter (maximum at 1.71 MeV [21]), it necessitates the use of plastic and/or lead shields to minimize health hazards. In addition, the high-energy radiation tends to increase the width of bands on autoradiographs.

^{35}S-labelled nucleotides [22] have several advantages over ^{32}P. The longer half-life (87 days) and weaker emission [21] (maximum: 0.16 MeV) give longer shelf lives, allow for storage of sequencing reactions for at least one week, and eliminate the need for shields. Furthermore, the weaker emissions lead to sharper bands and longer sequence reads. On the other hand, exposure times are several times longer than with ^{32}P, and gels have to be dried before film exposure. However, fixation in acetic acid/ methanol mixtures and direct contact of gel and film, as originally described [22], is not essential [17,23].

Often, the isotope of choice [17] in radioactive sequencing is ^{33}P rather than ^{35}S or ^{32}P. The emission energy of ^{33}P is about sixfold lower than of ^{32}P (maximum: 0.248 MeV vs. 1.7 MeV), and the half-life is about twice as long [21] (25 days vs. 14 days). It therefore offers the same advantages as ^{35}S: no shielding is required, band patterns are sharp, reactions can be stored for a week, and the shelf life is longer. However, unlike ^{35}S, it does not require drying of gels before exposure, and exposure times are only about 1.5–3 times as long as with ^{32}P. Furthermore, contamination is easier to detect with ^{33}P than with ^{35}S.

The one major drawback of ^{33}P is the higher cost: current prices are three- to sixfold higher than for ^{32}P, and it can exceed the cost of enzyme-linked labelling and detection. Furthermore, new licences for handling this isotope may be required, and on-site storage times before radioactive waste can be disposed of as nonradioactive waste are twice as long as for ^{32}P.

24.3.2 Incorporation vs. end-labelled primers

Conventional radioactive labelling procedures are based on the incorporation of labelled nucleotides into the synthesized DNA strand. Alternatively, the primer oligonucleotides can be 5′-labelled with γ-labelled dNTPs; this is most often done when cycle sequencing protocols are used. End-labelled primers can often lead to cleaner sequences, for example when RNA contaminations lead to false priming which results in increased background.

24.4 Enzyme-linked detection

24.4.1 Choosing a label

In DNA sequencing with enzyme-linked detection, biotin and digoxigenin are currently the most commonly used labelling molecules. Typically, oligonucleotide primers are chemically labelled at the 5′-end during or after synthesis. However, the use of hapten-labelled ribo- and deoxynucleotide in enzymatic labelling reactions has also been described [24–27]. Enzymatic labelling protocols are especially attractive when the primer is used only in one or a few sequencing reactions, for example with primer walking strategies. For universal primers, on the other hand, the extra cost and/or time to end-label oligonucleotides chemically tends to be negligible since primers are used many times. For biotin as well as digoxigenin, reagents for chemical labelling (*N*-hydroxysuccinimidyl-compounds and phosphoramidites) and enzymatic labelling (ribo- and deoxynucleotide) are readily available from several commercial sources. In addition, many companies that specialize in custom oligonucleotide synthesis offer biotin- and digoxigenin-modified oligonucleotides.

In addition to biotin and digoxigenin, fluorescein and 2,4-dinitrophenyl (DNP) have also been used for enzyme-linked detection of DNA sequence patterns [28]. These labels are attractive for label multiplexing strategies (see Section 24.4.5 below), but their general use may be limited by the availability of highly active antibody–enzyme conjugates.

In our experience, biotin-based detection systems

tend to give shorter exposure times than digoxigenin-based detection systems. However, cost and convenience considerations may be more important. The lowest costs per development can be obtained by using biotinylated primers, streptavidin, biotinylated alkaline phosphatase, and colorimetric detection as described [15,29,30]. For maximum convenience, however, the use of streptavidin–enzyme or antibody–enzyme complexes can lead to protocols with fewer incubation and washing steps.

24.4.2 Biotin: one- and two-component systems

The first publications on enzyme-linked detection of DNA sequences used a two-component system: membranes were first incubated with streptavidin, followed by incubation with biotin-labelled alkaline phosphatase. After several washes, the sequence patterns were detected by incubation with the colorimetric substrate/enhancer combination BCIP/NBT [30]. Background problems with the original procedure were reduced by using high concentrations of SDS to block nonspecific binding of streptavidin to nylon membranes [29]. For fast results, the total time for incubation and wash steps can be reduced to less than 30 min [31].

Alternatively, one-component systems, that is preformed complexes of streptavidin and alkaline phosphatase, can be used. This reduces the number of incubation and wash steps and therefore the total hands on time. The elapsed time, however, tends to be longer due to longer incubation times. Furthermore, 100-fold differences in signal intensity as well as limited shelf lifes have been observed with different streptavidin–phosphatase complexes (P.R., unpublished; C.M. Martin, unpublished). Therefore, it is advisable to use streptavidin–phosphatase complexes specifically tested for detection of DNA sequence patterns. Even for highly active streptavidin–phosphatase conjugates, sequences within 10–30 bases of the primer tend to be weaker and may not be readable.

Finally, biotin is sufficiently stable to be used as a label in chemical sequencing [32]; this allows the use of enzyme-linked detection in chemical sequencing as well as in footprinting studies of DNA–protein interactions.

24.4.3 Digoxigenin and other haptens

Digoxigenin has also been used successfully for enzyme-linked DNA sequencing [16,18,28]. Development protocols and results are similar to one-component biotin detection systems; in fact, the same protocol can be used for digoxigenin detection with antidigoxigenin alkaline phosphatase-labelled antibodies and for biotin detection with streptavidin–phosphatase complexes [18] (see Section 24.4.4). Results are similar, although exposure times with digoxigenin tend to be longer.

In addition to alkaline phosphatase-labelled antibodies, peroxidase-labelled antibodies to digoxigenin in conjunction with enhanced chemiluminescence [33,34] have been used for DNA sequencing, allowing for simple protocols for 'label duplexing' [20].

For other labels like fluorescein and DNP, detection systems are currently not as widely available as systems for biotin and digoxigenin. Therefore, the use of these haptens will typically be restricted to label multiplexing procedures [20,28] (see Section 24.4.5). The main advantage of fluorescein and other fluorescent haptens is that purification of labelled primers by gel electrophoresis is simplified, since the labelled primer can easily be seen by eye during electrophoresis. However, precautions to minimize exposure to light and photo bleaching have to be taken when fluorescent haptens are used.

24.4.4 Enzymes and substrates

24.4.4.1 Alkaline phosphatase vs. horseradish peroxidase

The most commonly used enzyme by far for nonradioactive detection of DNA sequencing patterns is alkaline phosphatase, typically from calf intestine (CIP). The high specific activity and the large number of available substrates make CIP the enzyme of choice. In addition, the high stability of CIP enables prolonged signal developments of up to several days.

Peroxidase from horseradish, another enzyme which has been used for DNA sequencing, is less stable in the presence of substrates. Signal intensities with peroxidase-based chemiluminescent detection decrease rapidly after 1 h [34]. Therefore, this system is less well suited when signal intensities are very low, or when multiple exposures are desired. Furthermore, film exposure have to be taken as soon as possible after addition of peroxidase substrates, thus making peroxidase-based protocols less convenient and flexible than alkaline phosphatase-based protocols.

24.4.4.2 Colorimetric substrates

Use of colorimetric substrates for DNA sequencing with end-labelled primers has only been reported for alkaline phosphatase [30], not for peroxidase. The commonly used substrate is BCIP (5-bromo-4-chloro-3-indolyl phosphate; also called X-Phos) with the enhancer NBT (nitro blue tetrazolium). One advantage of colorimetric detection is the low substrate cost (about $1 per membrane, compared to $10–20 for chemiluminescent substrates). Furthermore, colorimetric detection gives very high resolution of sequencing band patterns [29]. Colorimetric detection is well suited for manual sequence reading with a digitizer tablet, or for scanning with flat-bed scanners which are less costly than film scanners ($800–$1500 as opposed to $10 000–$30 000).

For long-term storage of colorimetric sequence patterns, efficient washing of membranes after the colour development is essential. Washing with nonionic detergent solutions for 30 min or longer, followed by rinsing with water, removes excess substrate and enhancer and allows storage for several years with minimal degradation of the pattern quality [16].

Compared with chemiluminescent substrates, colorimetric detection has three limitations.

1 The detection step takes longer, from 1 h to overnight.

2 Multiple exposures cannot be obtained.

3 Removal of precipitated product for reprobing in multiplex experiments requires washes with hot dimethylformamide, *which is highly toxic* [35].

For typical in-house sequencing projects, however, limitations 2 and 3 will not matter, and the longer signal development times will be tolerable under most circumstances.

24.4.4.3 Chemiluminescent substrates

Chemiluminescent enzyme substrates emit light after enzymatic modification (therefore, the more correct term would be 'chemiluminogenic substrates'). The use of chemiluminescent enzyme substrates avoids the three limitations for colorimetric substrates mentioned above: exposure times are fast, typically from several minutes to 1 h; multiple exposures can be obtained within one or a few hours; and the substrate as well as the product of the enzymatic reaction can easily be removed from the membrane to allow successive developments. Furthermore, researchers used to radioactive sequencing may be more comfortable with obtaining results in a familiar form, on X-ray film.

On the down side, costs for chemiluminescent substrates tend to be significantly higher than for colorimetric substrates. However, these costs can easily be reduced, at least for alkaline phosphatase-based detections, by using lower substrate concentrations. Suggested concentrations in most kits are close to the K_m of the enzyme, and they can typically be lowered by a factor of 2–10. Exposure times will be longer by a similar factor, but this may often be acceptable.

One aspect of chemiluminescent detection which can be of interest for high-throughput sequencing projects is that the sequence patterns can be detected directly by light-sensitive cameras [36]. However, exposure times range from several to 30 min, necessitating the use of rather expensive equipment for direct detection.

Chemiluminescent substrates for alkaline phosphatase

1,2-dioxetane-based substrates for alkaline phosphatase [37,38] are currently available from many distributors under names such as AMPPD, CPD, and CPD-Star. The substrates are stable in alkaline solution, but become unstable upon dephosphorylation, and eventually break apart, emitting light with an emission maximum at 460 nm. In addition, the dephosphorylation leads to binding of the product molecule to nylon membranes. This binding to nylon membranes increases the product half-life to several hours, compared to minutes in aqueous solution [14,39].

The long half-life of the product molecule leads to accumulation of product and to an increase in signal strength over time. As a result, exposures taken after several hours' preincubation are much darker than exposures taken shortly after substrate addition. Product accumulation can also lead to band broadening, most noticeably when exposures are taken the next day.

Two improved dioxetane substrates for alkaline phosphatase, named CSPD [40] and CPD-Star, have been developed. The dephosphorylated product of CSPD shows a reduced half-life of 40 min on nylon membranes, leading to faster exposures and reduced band broadening effects [40]. An example of a sequencing pattern detected with CSPD is shown in Fig. 24.1.

The newest substrate, CPD-Star, offers signal intensities that are ≈10-fold higher than the intensities from CSPD, making it the ideal choice for direct detection with light-sensistive cameras. For film-based detection, exposure times of less than 1 min can easily be achieved with CPD-Star; alternatively, much lower substrate concentrations can be used for exposure times comparable to those typical for CSPD or AMPPD.

Enhanced chemiluminescence for horseradish peroxidase

Chemiluminescent substrates for peroxidase show quite different characteristics from phosphatase substrates. Light emission is maximal shortly after addition of the substrate and decays rapidly [34]; after an overnight incubation, hardly any signal

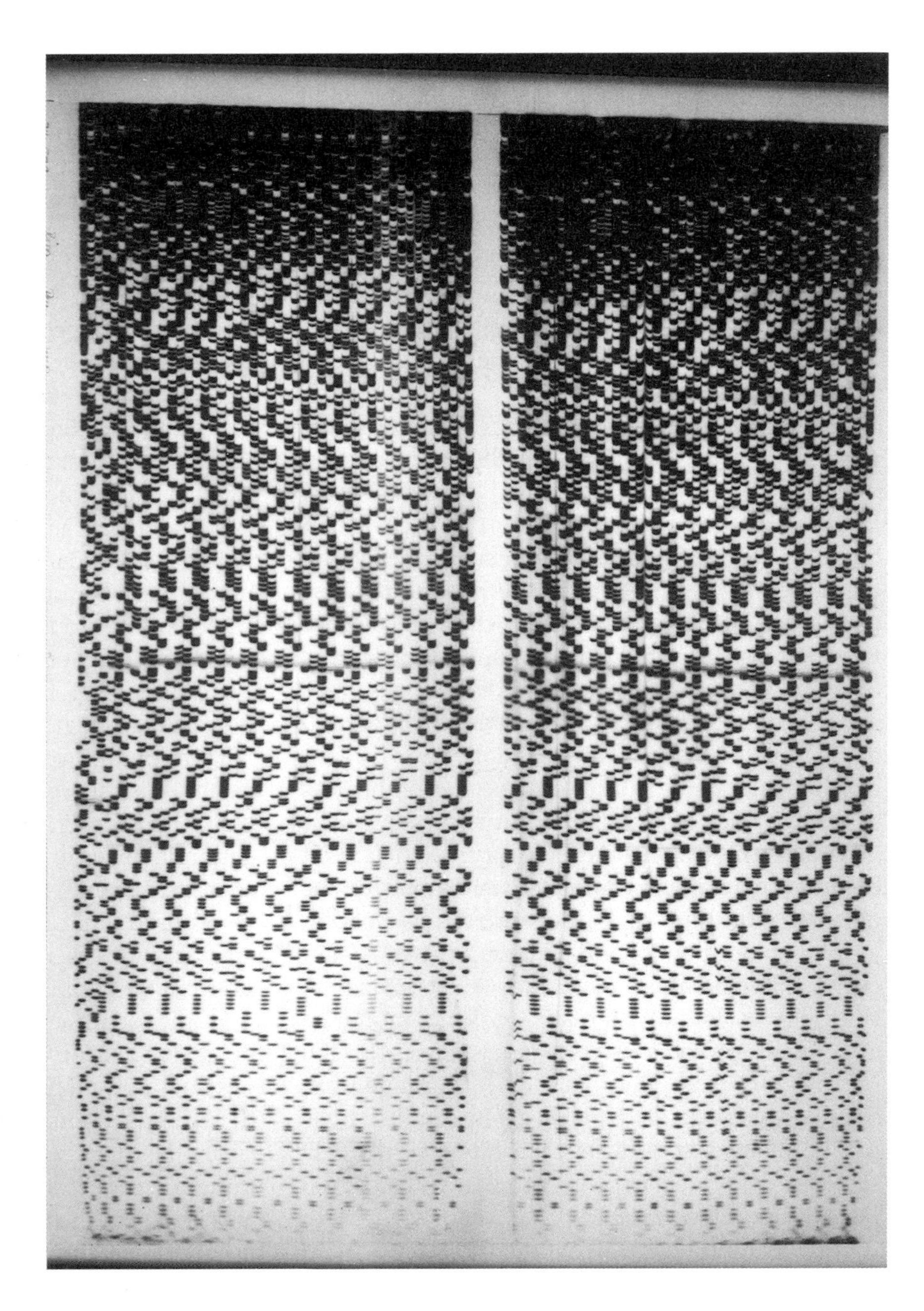

Fig. 24.1 Enzyme-linked detection of DNA sequencing patterns. Sequencing was done from recombinant M13-clones with 5′-biotin labelled primer, modified T7 DNA polymerase and manganese buffers. Reactions were separated on 6% acrylamide wedge gels and transferred onto Biodyne A membranes by contact transfer as described in the text. Detection was done with streptavidin, biotinylated alkaline phosphatase, and CSPD as described [20]. Exposure time, 1 h.

intensity is left. This poses some restrictions on the timing and number of exposures; more important, it limits sensitivity. While the sensitivity of phosphatase-based detections is typically limited by nonspecific binding of enzymes and other components to the nylon membrane, peroxidase-based detection is often limited by the total signal intensity. However, the signal levels are high enough for typical sequencing experiments.

At least theoretically, peroxidase-based detection can be less contamination-sensitive than phosphatase-based detection. Alkaline phosphatase is ubiquitous, and buffer contamination by bacteria, for example, can lead to very high background levels when phosphatase-based detection is used. To avoid such problems, many protocols suggest the use of bacteriostatic stock solutions and/or preparation of buffers shortly before use.

Exposure considerations Exposure to X-ray film can be done directly in hybridization bags, which may have been used for the preceding incubations and washes, as suggested by several distributors of detection kits. However, most hybridization bags are thicker than Saran wrap (25–50 μm vs. 11 μm), and band sharpness suffers somewhat. This is especially noticeable in regions where bands are closer together, towards the top of the gel. When maximum read lengths are desired, it is preferable to wrap membranes in Saran wrap or very thin Mylar sheets (optionally on top of a thicker plastic backing). In addition, weights on top of exposure holders should be used to insure good contact between membrane and film.

24.4.4.4 Fluorigenic substrates

Besides colorimetric enzyme substrates, fluorigenic substrates have been used for a long time in enzyme-linked detection methods [41]. Ideal fluorigenic substrates show no (or minimal) fluorescence; upon enzymatic modification, for example dephosphorylation by alkaline phosphatase, the product molecules show strong fluorescence.

A new fluorigenic substrate for alkaline phosphatase, called Atto-Phos, has been introduced recently [42] and shows this desirable characteristic. Furthermore, the product of the enzymatic reaction binds strongly to nylon membranes, thus enabling the detection of sequencing patterns [15]. Because of the enzymatic amplification, the fluorescence intensity is so high that patterns can be seen simply by illumination with hand-held UV lamps. Very strong sequence patterns can even be seen in normal daylight without special illumination.

While fluorigenic substrates are considerably less expensive than chemiluminescent substrates, their general use for DNA sequencing is currently limited because permanent records cannot easily be obtained. Results can be documented by photography under illumination with long-wave UV, but only small sections of membranes can be imaged at sufficient resolution. Alternatively, detection can be done with cooled charge-coupled device (CCD) cameras; exposure times for Atto-Phos range from one to several seconds. However, only small sections can be imaged at sufficient resolution with CCD cameras, unless very expensive cameras (2000×2000 pixels or more) are used (see Chapter 13).

We have observed two other problems with fluorigenic substrates. First, the dephosphorylated product diffuses somewhat at higher signal intensities. This makes bands appear more blurry, and also limits the dynamic range. Second, the product is difficult to remove completely for multiple reprobings, requiring either very long washes or alkaline stripping buffers. Alkaline stripping conditions, however, lead to DNA loss and thus limit the number of successful reprobings. Therefore, chemiluminescent substrates will be a better choice than fluorigenic substrates in most instances.

24.4.5 Label multiplexing

To increase overall efficiency of enzyme-linked protocols, combinations of several haptens can be used instead of single haptens. To give a simple example, two sets of sequencing reactions, one with biotin-labelled primer and the other with digoxigenin-labelled primer, can be done individually and then combined before electrophoresis. After transfer to membranes, detection of the biotin- and digoxigenin-labelled sequences can be done successively [20]. Compared to using just one label, only half as many gels have to be run and transferred, giving significant savings in time. Furthermore, costs for membranes are halved.

The efficiency gains can be even higher when two differently labelled primers are used in the same sequencing reaction. This reduces the number of sequencing reactions by a factor of two, simultaneously reducing the costs for sequencing reagents. This approach is well suited to cycle sequencing protocols. With double-stranded templates, sequence from both strands can be obtained in the same reaction. However, the distance between the primers (the insert size) may need to be larger than the desired read lengths to obtain optimal results.

The concept of 'label multiplexing' has been extended to the use of four labels per membrane [28]. This further increases efficiency and reduces costs. However, exposure times with chemiluminescent substrates will increase with the number of labels used, unless the total volume of the sequencing reactions is reduced, for example by ethanol precipitation.

If the same enzyme is used in subsequent detections of different labels, the enzyme from previous detections must be removed or irreversibly denatured. With alkaline phosphatase, this can easily be done either by incubation in low-pH buffers at room temperature [28], or by inactivation with hot buffers containing SDS-EDTA [20]. For horseradish peroxidase-based detection, incubation in peroxide-containing substrate solutions may be sufficient between subsequent detections, since horseradish peroxidase is inactivated by peroxide [34]. An even simpler approach is the use of different enzymes. Then, incubation with different antibody–enzyme or streptavidin–enzyme conjugates can be done simultaneously, and only the substrate solutions need to be changed between film exposures [20].

24.4.6 Detection with oligonucleotide–enzyme conjugates

The idea of using hybridization with oligonucleotide–enzyme conjugates to detect sequence patterns [14] might seem odd. However, the convenience and efficiency of such an approach is at least comparable to conventional hapten-based approaches owing to several factors. First, stringency requirements are low and probes are short, so that hybridization and posthybridization washes can be done at room temperature. Second, hybridization times can be very short, from 10 to 30 min with probe concentrations from 0.5–2 nM. And third, a large number of protocols for the efficient preparation of oligonucleotide–enzyme conjugates has been described [43–46]. Directions for the preparation and use of such conjugates are given in Protocol 123.

While hybridization-based detection can be used for universal primer-based sequencing strategies, it can also be adapted for 'primer walking' strategies. In this case, walking primers with a 5′-tag sequence are used, and oligonucleotides complementary to the tag sequence are used for detection. The tag sequences lead to some additional costs per primer; however, these costs are typically lower than costs for custom labelling with biotin or digoxigenin, and are likely to go down further in the future.

24.4.7 General procedures for enzyme-linked detection

24.4.7.1 Labelling of oligonucleotides

To introduce labels like biotin or fluorescein into oligonucleotide primers for DNA sequencing, chemical as well as enzymatic methods can be used. Chemical labelling is typically used for universal primers which are be used many times, while enzymatic labelling is convenient if walking primers or existing, unmodified primers are to be used.

Long spacer arms between the hapten and the DNA improve signal intensities [47,48]. Longer spacers maximize the accessibility of the hapten to streptavidin antibodies. Therefore, compounds with the longest possible spacer arm should be chosen when different choices are available.

24.4.7.2 Chemical end-labelling

The most convenient labelling method is the use of hapten-labelled phosphoramidites during chemical synthesis of oligonucleotides. Phosphoramidites labelled with biotin [49,50], digoxigenin, and fluorescein have been described.

Labelling during synthesis, however, can be expensive. Occasionally, we have also observed poor efficiencies, most likely due to prolonged storage and use of modified phosphoramidites. Therefore, postsynthesis labelling of amino-modified oligonucleotides by reaction with amino-reactive haptens is often preferred. Most commonly, oligonucleotides are synthesized with a 5′-amino group on a C6-spacer and reacted with a 10–50-fold molar excess of *N*-hydroxysuccinimidyl (NHS)- or isothiocyanate (ITC)-modified haptens. Modification efficiencies are typically between 40 and 90%. Protocol 122 describes labelling with NHS-LC-biotin. However, it can be used without changes for the labelling with NHS-digoxigenin, fluorescein-ITC, ITC-infrared dyes, and other NHS- or ITC-modified haptens.

To label unmodified oligonucleotides with biotin or other haptens, enzymatic phosphorylation followed by reaction with diaminohexane can be used [51]. However, enzymatic labelling with terminal transferase (TdT) or polymerases, as described in Section 24.4.7.3, is simpler and preferable.

24.4.7.3 Enzymatic labelling

Introduction of haptens by enzymatic reactions before or during sequencing reactions provides a convenient alternative to chemical labelling, especially for primer-walking strategies. A number of different approaches have been described:

1 labelling by TdT and ribonucleotides [25];
2 the use of labelled dideoxynucleotides in Sanger sequencing protocols [4];
3 end-filling reactions for chemical sequencing [32];
4 the incorporation of labelled nucleotides into the growing DNA chain [24,26,27].

Methods 1 and 4 are most generally useful. With method 1, a sufficient amount of primer for several hundred sequencing reactions can be generated by 3′-labelling with terminal transferase. However, the primer has to be chosen so that the first nucleotide to be incorporated into the growing strand is the labelled nucleotide.

Method 4, labelling by incorporation of one hapten-modified deoxynucleotide during synthesis, has been used successfully for fluorescein [24,52], biotin [27], and infrared fluorescence [26]. To obtain consistent results, some protocols use a two-step reaction. In the labelling step, the absence of at least one deoxynucleotide limits the elongation, leading to the incorporation of a single labelled nucleotide. In the following elongation reaction, an excess of all four unmodified dNTPs is used. This approach can be used with modified T7 DNA polymerase [27] as well as with cycle sequencing protocols [26].

24.4.8 Purification of labelled oligonucleotides

Purification of oligonucleotides labelled with biotin, digoxigenin or fluorescent dyes can easily be done by polyacrylamide gel electrophoresis (PAGE) or by reverse phase HPLC (RP-HPLC). In PAGE purification, the hapten leads to an apparent size increase corresponding to 1–2 nucleotides over the unlabelled oligonucleotide. In RP-HPLC, the hydrophobic nature of the haptens leads to longer retention times. An example of an RP-HPLC purification is shown in Fig. 24.2; protocols for PAGE purification have been described [20,51].

While purification often-used primers may be advisable, it is not necessary in every case. The main negative effect of incomplete labelling is a proportionate reduction in signal intensity; with typical labelling efficiencies of 50% and higher, this can often be tolerated.

24.4.9 Transfer to membranes

Before enzyme-linked detection can be done, DNA sequence patterns need to be transferred from gels onto nylon membranes. This transfer can be done by capillary blotting, electrophoretic transfer, or direct transfer electrophoresis (DTE).

The two most convenient methods are contact capillary blotting and DTE. In DTE, a membrane is moved along the lower edge of the gel during electrophoresis; DNA is immobilized on the membrane as it leaves the gel. Transfer by DTE is virtually complete, and DTE gives longer read lengths than conventional gels [19,20,29]. However, special sequencing machines are required for DTE.

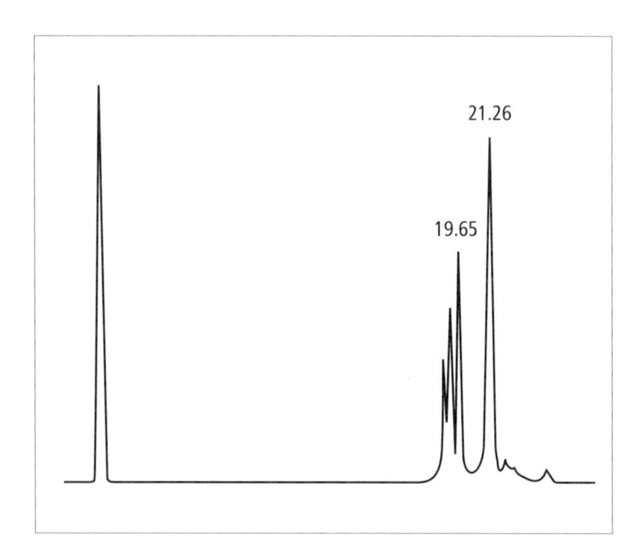

Fig. 24.2 Purification of biotin-labelled oligonucleotides by reverse phase HPLC. A 5′-amino-labelled oligonucleotide was chemically labelled with NHS-LC-biotin, and excess biotin was removed by gel filtration as described in the text. A quarter of the sample (0.25 ml) of the biotinylated oligo was loaded onto a RP-HPLC column equilibrated with buffer A (5% acetonitrile, 95% 100 mM triethyl ammonium acetate). After 5 min, a gradient of 0–40% buffer B (65% acetonitrile, 35% TEA) over 20 min was started. The biotinylated oligonucleotide eluted in the last major peak at 21.26 min.

Contact capillary transfer, on the other hand, can be done without any special equipment and from all typically used gel formats. Before electrophoresis, glass plates need to be treated so that they can be separated easily after electrophoresis, with the gel sticking to only one glass plate. This can be achieved by using very clean glass plates, treating both glass plates with a hydrophobic solution like Sigmacote or Rainex, or by treating one glass plate with Sigmacote and the other glass plate with bind silane [9].

After electrophoresis, glass plates are pried apart, leaving the gel on the lower glass plate. A nylon membrane, cut to size and prewetted in electrophoresis buffer (TBE), is placed onto the gel, and air bubbles are squeezed out (always wear gloves when handling membranes!). Two sheets of dry Whatman 3MM paper are placed on top of the membrane and pressed on to eliminate air bubbles. The other glass plate is put on top of the Whatman paper, and a weight (2–4 kg) is placed on top of this glass plate. Transfer times are typically one hour, but shorter times and overnight transfer has also been used successfully. Before development, membranes are crosslinked by UV irradiation at 150 mJ cm^{-2}.

24.4.10 Development procedures

Many protocols for enzyme-linked detection of sequencing patterns have been described in detail in the literature. Therefore, we will just point out a few general considerations for enzyme-linked detection. An example protocol which has successfully been used for biotin- and digoxigenin-based detection in large scale applications [18] is given in Section 24.7.

Development of sequencing membranes can be done in hybridization bags [30], large trays, or in large cylinders [29,53]. Accordingly, all the equipment needed for enzyme-linked detection is either a shaker or an instrument to rotate large drums. Details of the different procedure have been discussed [20]. Each of the three approaches offers some advantages, but we have used them interchangeably, typically dependent on what kind of equipment was available.

Overlaps between different membranes or different parts of one membrane during development are avoided in all of the above approaches, whenever possible. While rolling up membranes tightly works well for radioactive hybridizations, it tends to give low signal intensities and background problems with enzyme-linked detection protocols. This can be explained by several factors, such as limited diffusion on enzymes through membrane pores and higher concentrations of detection system components when compared to radioactive probes.

One major advantage of large, rotating drums for nonradioactive detection is that the wash and incubation steps can be easily automated [32]. At least one machine for automated nonradioactive detection is currently commercially available. With chemiluminescent detection, membranes need to be taken out of the drums for exposures; with colorimetric detection, however, all steps can be done in the drums.

Recently, a variation of the drum approach has been described [54]. Membranes are attached to the outside of large drums which rotation in slightly larger half-cylinders. This method allows for direct detection of the sequence patterns by cameras, making this scheme very attractive for multiplex and other high-throughput projects.

24.4.11 Adapting sequencing protocols for enzyme-linked detection

Sequencing protocols for radioactive sequencing will typically work with minimal or no changes for enzyme-linked detection. Sequencing kit protocols for end-labelled primers can generally be used directly in enzyme-linked detection. Alternatively, two-step protocols with incorporation of radioactive dNTPs in the first step can be also followed if the radioactive nucleotide is replaced by its nonradioactive equivalent [30].

Slight protocol modifications, however, can sometimes be useful to simplify procedures or to optimize results. For example, one-step reactions instead of two-step reactions can be used [29]. If nucleotide mixes are changed, higher dNTP to ddNTP ratios are often used to even out band intensities for long reads, especially if direct transfer electrophoresis is used to generate the sequence patterns [20].

24.5 Fluorescent detection

24.5.1 Systems for on-line detection during electrophoresis

A number of machines for fluorescent DNA sequencing, based on the detection of fluorescently labelled DNA molecules during electrophoresis, are currently available. These machines eliminate manual steps for the visualization of sequence patterns completely, and they offer soft-ware for the automatic reading of sequence patterns, thereby potentially increasing the efficiency of DNA sequencing. Furthermore, they simplify the scaling up of sequencing projects and are often the method of choice for large-scale projects.

However, the costs of automated DNA sequencers as well as operating costs may be prohibitive for occasional sequencing needs. This may change in the future when the use of automated DNA sequencers in central facilities, similar to the use of oligonucleotide synthesizers and protein sequencers, is likely to become more common. We will therefore discuss some aspects of automated DNA sequencers briefly.

Automated sequencers can be divided into single-dye [6,55,56] (marketed by Pharmacia, Li-Cor, Millipore and others) and four-dye systems [3,4] (marketed by Applied Biosystems). Single-dye systems use one fluorescent label and four lanes per sequencing reaction, similar to radioactive sequencing. Four-dye systems use different fluorochromes for each of the four nucleotides, so four colours instead of four lanes are used per sequencing reaction. As a result, more sequencing reactions can be loaded per gel, currently 36 compared with 10–12 on single-dye systems. Therefore, four-dye systems are typically used when high throughput is important, for example in large-scale sequencing projects [57].

A relatively recent development in fluorescent sequencers is the availability of machines which can accommodate gels up to 60 cm long. Previously, gels around 30 cm long were the only option. Longer gels lead to significant increases in electrophoretic resolution and read length; typically, more than 700 bases can be resolved to single base resolution on 60-cm gels, compared to 350–400 bases on 30-cm gels [20,52]. In addition, the better electrophoretic resolution of longer gels can also lead to improved base-calling accuracy, in particular for the first 400 bases.

Longer read lengths can reduce the number of sequencing reactions, templates and walking primers needed to sequence a given stretch of DNA, thus increasing efficiency and reducing costs. However, electrophoresis run times tend to be longer, and throughput numbers (in bases per hour) tend to be lower for longer gels.

Labelling procedures for single-dye automated DNA sequencers are very similar or identical to procedures for biotin and other haptens as described above. Chemical as well as enzymatic labelling can be used, and primer walking strategies can be pursued by incorporation of fluorescent deoxynucleotides into the growing chain (see Section 24.4.7). Most of the commercially available instruments use visible fluorescence; only one instrument uses fluorescence in the near-infrared region [56]. The main advantages of using the near-infrared region are significantly reduced background fluorescence and the availability of cost-efficient, highly reliable lasers and detectors.

Multiple dye sequencing machines, on the other hand, are often preferred when maximum throughput is important. However, labelling strategies are more complicated, since four different dyes are used, and mobility differences between the dyes can have a negative impact on results if new primers are chosen. These problems can be circumvented by the use of dye-labelled dideoxynucleotides, but reagent costs can be higher than for incorporation strategies with a single label.

24.5.2 Detection after electrophoresis: fluorescence scanning

While fluorescence detection during electrophoresis is convenient, the throughput is limited by the speed of electrophoresis. Typically, just one or two runs per day can be obtained per machine. Scanning after electrophoresis, on the other hand, can be significantly faster. Most film scanners, for example, scan 35×43 cm X-ray films in less than 10 min. Therefore, fluorescence scanners could be an attractive alternative, especially in situations where equipment resources are shared between different groups, and when equipment funds are limited.

The use of a fluorescence scanner for DNA sequencing has been described [58]. A 532-nm laser was used for excitation, and glass plates for sequencing gels had to be nonfluorescent. For detection from membranes, a detection limit of 10 fmol was reported. It has been observed that high intrinsic fluorescence of membranes can limit detection sensitivity [59].

As mentioned above, background fluorescence is much lower in the near-infrared region of the spectrum. Detection limits of 60 attomoles have been obtained with an infrared scanner prototype [16]. This sensitivity is two orders of magnitude better than in the visible region, and is sufficient to allow detection of DNA sequence patterns from nylon membranes. This, in turn, allows for the use of DTE, enabling longer reads than with conventional gels. An example is shown in Fig. 24.3. The sensitivity is also sufficient to allow the detection of multiplex sequence patterns with infrared-labelled hybridization probes [16]. However, general use of these methods is limited at the time of this writing since infrared scanners are not yet commercially available. We also encountered problems with the reproducibility of infrared fluorescence detection from membranes, which were tentatively attributed to fluorescence quenching by minute amounts of contaminants.

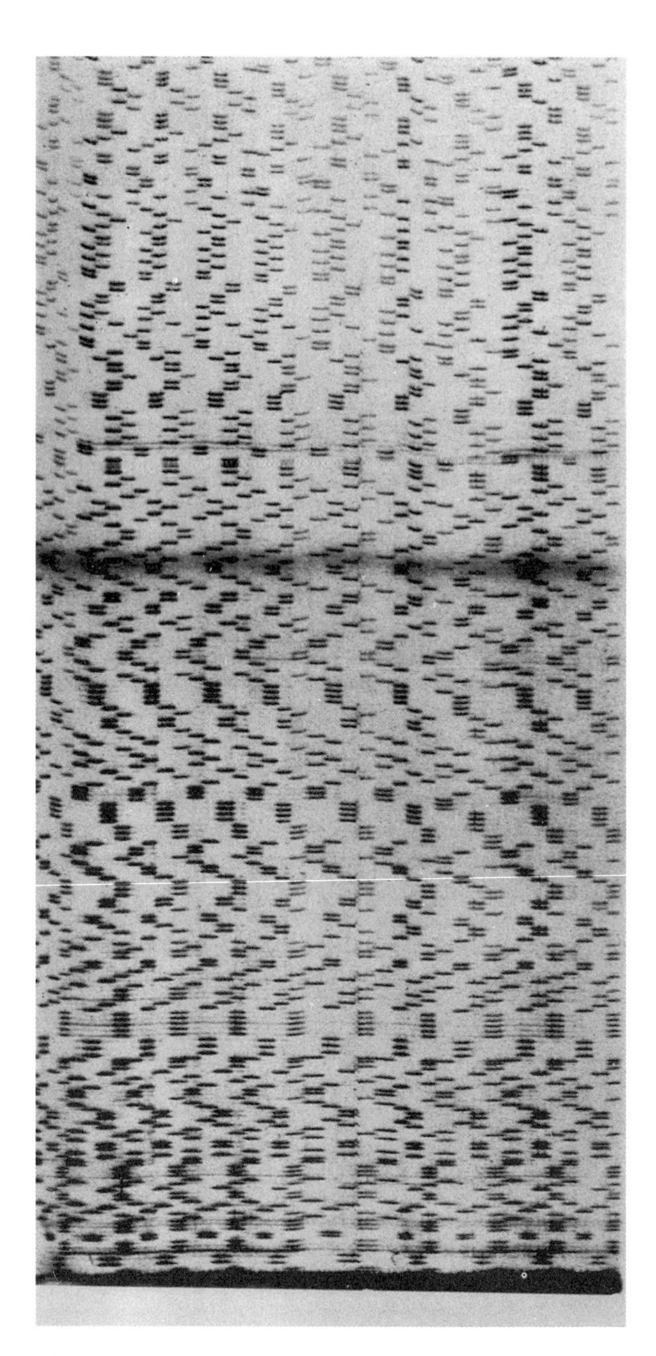

Fig. 24.3 Detection of DNA sequence patterns by fluorescence end-labelling and scanning. Sequencing was done from recombinant M13-clones with modified T7 DNA polymerase and manganese buffers, as described [20]. The primer was 5′-labelled with IRD40 (Li-Cor, Lincoln, Nebraska). Reactions were separated by direct transfer electrophoresis onto Biodyne A nylon membranes. After drying, the membrane was sandwiched between two glass plates and scanned with an prototype scanner [16] based on the detection optics of a Li-Cor DNA sequencer [56].

24.6 Silver staining

Silver staining of DNA in polyacrylamide gels can be sufficiently sensitive to visualize DNA sequence patterns [9,10]. No special labelling of primers is required, and costs for chemicals are lower than for many enzyme-linked procedures. Covalent fixation of the gel to one glass plate is required, and the overall procedure takes about the same time as enzyme-linked developments, thus being faster than radioactive detection. Multiple exposures can be obtained by a procedure similar to the taking of contact prints from black-and-white negatives.

However, the amount of DNA per sequence band is very close to the detection limit of silver staining procedures. The supplier suggests cycle sequencing protocols which tend to give intense sequence patterns, and results may not be satisfactory with weak sequences, especially close to the primer. For comparison, the detection sensitivity for chemiluminescent detection is much higher than required for most sequence patterns, and patterns with less than 2% of the typical intensity can be visualized [20].

24.7 Hybridization-based detection and multiplex sequencing

Conceptually, multiplex sequencing [8] is very similar to DNA sequencing with enzyme-linked detection. In enzyme-linked detection, the sequence is labelled by a hapten such as digoxigenin, and detected through incubation with a reagent that specifically recognizes the hapten, for example an antibody–enzyme conjugates; in multiplex sequencing, a short 'sequence tag' is used instead of a hapten, and specific recognition is achieved by hybridization with a cDNA instead of the antibody–enzyme conjugate. In the simplest case, the sequence tag can be a universal primer, and the probe the reverse complement of the primer, conjugated to an enzyme such as alkaline phosphatase. Then, almost identical protocols can be used to visualize the sequence pattern, as Table 24.2 shows.

All incubation and wash steps are done at room temperature, and identical buffers are used for steps 3–9 (see Table 24.2). For the hybridization buffer, a high salt concentration is used for maximum speed. Alternatively, buffers containing 5–7% SDS [7,20] and hybridization times of 30–60 min can be used if background minimization is more important than short hybridization times. Time for steps 1 and 2 may be increased to 45 min for digoxigenin, and reduced to 15 min for oligo-enzyme conjugates. Additional wash steps may be added if the background is high. Buffers used are:

- steps 1 and 2, digoxigenin: 1.5% casein-based blocking reagent (Boehringer Mannheim) in maleate buffer (1.16% w/v maleic acid, 0.876% w/v NaCl, pH adjusted to 7.5 with NaOH);
- steps 1 and 2, oligo-enzyme conjugate: 2% casein-based blocking reagent (Boehringer Mannheim) in 750 mM NaCl, 50 mM Tris-HCl (pH 8);
- steps 3–5: maleate buffer, pH 7.5 (see step 1);
- steps 6–8: 0.1 M diethanolamine-HCl, pH 9.5, 1 mM $MgCl_2$.

While hybridization-based detection can be efficiently used for the detection of single, non-multiplexed sequencing patterns, higher efficiency can be gained by combining sequencing reactions with different tags before electrophoresis, and reading out the individual patterns through successive hybridizations. Many different approaches can be used to introduce the multiplex sequence tags: reactions done with different primers can be combined, primers which differ only in a 5′-tag sequence can be combined [60], or the DNA of interest can be subcloned so that the insert is flanked by different tag sequences, either by using dedicated multiplex vectors [8], or by using tagged linkers for subcloning [61].

The more clones are pooled together, or — in other words — the higher the multiplex factor is, the more efficiency is gained by multiplexing. Similarly, maximum efficiency gains can be realized when pooling is done early in the protocol. To give an

Table 24.2 Comparison of protocols for hapten-based and hybridization-based detection.

Step number	Digoxigenin	Oligo-enzyme conjugate
1	Block 30 min	Prehybridize 30 min
2	Incubate with 30 ml antibody–alkaline phosphatase complex (1 : 5000), 30 min	Hybridize with 30 ml oligo-enzyme (1 nM) for 30 min at room temperature
3–5	Wash 3 × 10 min, 250 ml	Wash 3 × 10 min, 250 ml
6–7	Wash 2 × 5 min, 250 ml	Wash 2 × 5 min, 250 ml
8	Incubate with 30 ml substrate (CSPD 0.05 mM) for 5 min	Incubate with 30 ml substrate (CSPD 0.05 mM) for 5 min
9	Expose to X-ray film for 10–60 min	Expose to X-ray film for 10–60 min

example, pooling clones from 20 different libraries before the DNA preparation [8] is much more efficient than pooling four sequencing reactions just before electrophoresis [60].

Multiplex sequencing can be done in conjunction with all commonly used labelling and detection methods: radioactive [8], enzyme-linked [13,15,31, 60], and—at least in principle—fluorescent detection [16]. Multiplex sequencing has been used in various large-scale sequencing projects [62], for example to obtain more than 760 000 bases of sequence from the genomes of *Mycobacterium leprae* and *M. tuberculosis* [63].

While multiplex sequencing is extremely well suited for large-scale sequencing projects, the hybridization-based detection can also be the method of choice for smaller projects, since protocols for hybridization with oligonucleotide–enzyme conjugates are as convenient as other enzyme-linked detection protocols. A number of protocols to conjugate enzymes to oligonucleotides have been described [43–46], and numerous refinements of the original protocols have been developed. Protocol 123 describes a time-efficient conjugation and purification method. The protocol can be done simultaneously for 2–4 conjugates in less than 2 h, excluding purification. Yields are typically between 10 and 20%, enough for 10–20 full-size sequencing membranes. Conjugates are stable for at least 6 months when stored at 4 °C.

To summarize, developments in the past 10 years have given scientists the opportunity to choose between a variety of different labelling/detection methods for DNA sequencing. These include different radioactive isotopes, automated fluorescent DNA sequencing machines, and a variety of enzyme-linked protocols with colorimetric, chemiluminescent, and fluorigenic substrates. Each of these methods has been applied to large-scale as well as small-scale sequencing projects. Kits and machines are commercially available from a number of different sources, giving the potential user the flexibility to adapt to local restrictions and personal preferences.

A number of other factors which may influence the decision for or against a given system have been outlined in this chapter, and a few example protocols have been given. Admittedly, the views presented might be biased by the author's long positive experience with enzyme-linked detection and multiplex sequencing. If this is so, it may serve to counteract a very common reason for choosing radioactive labelling—the 'we have always done it this way' syndrome.

Protocol 122 Labelling oligonucleotides with NHS- or ITC-haptens

For details of solutions, media and materials, see Appendix I. For suppliers and contact addresses see Appendix III.

Materials

- NHS-LC-biotin (40 mM) (Pierce)
- dimethylformamide
- sodium carbonate (1 M, pH 9.0)
- Tris-HCl (0.1 mM, pH 8.0)
- Sepharose G25 column
- buffer: 100 mM triethylammonium acetate, pH 7 (Applied Biosystems), or
- TE (10 mM Tris-HCl, pH 8.0, 0.1 mM EDTA)

Method

1 Resuspend deprotected and lyophilized, amino-modified oligonucleotide in water to a nominal concentration of 2 mM (100 μl for a 0.2 μmol synthesis; do *not* resuspend in Tris- or other amine-buffers!).

2 Prepare a fresh 40 mM solution of NHS-LC-biotin in dimethylformamide (DMF; dimethyl sulphoxide can also be used; use high-quality, water-free reagents).

3 In an Eppendorf tube, mix: 45 μl oligonucleotide, 5 μl sodium carbonate, and 50 μl DMF.

4 Add 100 μl NHS-LC-biotin solution. A precipitate may form, but can be ignored. Incubate at room temperature for 2–24 h (shorter incubation times can be used, but labelling efficiencies may be reduced). For fluorescent haptens, incubate in a dark place and minimize exposure to light in all steps.

5 Add 300 μl 0.1 M Tris-HCl (pH 8.0), and incubate for 10–60 min.

6 Purify on a Sephadex G25 column (NAP5 column, Pharmacia). If purification by reverse phase-HPLC or polyacrylamide electrophoresis is intended, use 100 mM triethylammonium acetate (pH 7.0) as buffer and lyophilize eluate. For direct use without further purification, use TE as buffer.

Protocol 123 Preparation and purification of oligonucleotide–alkaline phosphatase conjugates

For details of solutions, media and materials, see Appendix I. For suppliers and contact addresses see Appendix III.

Materials

- sodium carbonate (1 M, pH 9.0)
- DMSO
- disuccinimidyl suberate (DSS), 25 mg ml^{-1} in DMSO (Pierce)
- calf intestinal alkaline phosphatase (CIP) 10 mg ml^{-1} in 3 M NaCl, 1 mM $MgCl_2$, 0.1 mM $ZnCl_2$, 30 mM triethanolamine (pH 7.6) (Boehringer Mannheim)
- Sephadex spin columns (Centri-Sep, Princeton Separations)
- ProteinPAK 300 SW column (Millipore/Waters)

Method

1 Dissolve deprotected, lyophilized and amino-modified oligonucleotide in water to a concentration of 2 mM.

2 To 8 μl oligo (16 nmol), add 2 μl sodium carbonate (1 M, pH 9.0) and 10 μl DMSO.

3 Add 10 µl of a fresh solution of DSS to each oligo mix, and incubate for 5 min.

4 Add 50 µl water, mix by vortexing. This precipitates most of the excess DSS. Spin at maximum speed in an Eppendorf centrifuge for 2 min.

5 Carefully take off 60 µl and load onto Sephadex spin columns pre-equilibrated in sodium carbonate according to the manufacturer's instructions. Spin for 3 min at 4000 r.p.m. Recovery of 20-mers is about 40%.

6 Add 25 µl CIP to each oligo, mix by pipetting; incubate at room temperature for 2–16 h, then store at 4 °C until purification.

7 Purify conjugate by gel filtration on a ProteinPAK 300 SW column using 20 mM Tris-HCl, 100 mM NaCl as running buffer. Characterize fractions by $OD_{260/280}$ ratios. Conjugate peaks may overlap with free enzyme for oligonucleotides shorter than 20 bases, but good separation to free oligonucleotides is generally achieved. Alternatively, gel filtration on Biogel P60, ion-exchange chromatography, or other methods can be used [45,46].

References

1 Sanger, F., Nicklen, S. & Coulson, A.R. (1977) DNA sequencing with chain-terminating inhibitors. *Proc. Natl Acad. Sci. USA* **74**, 5463–5467.
2 Maxam, A.M. & Gilbert, W. (1977) A new method for sequencing DNA. *Proc. Natl Acad. Sci. USA* **74**, 560–564.
3 Smith, L.M., Sandes, J.Z., Kaiser, R.J.P. *et al.* (1986) Fluorescence detection in automated DNA sequence analysis. *Nature* **321**, 674–679.
4 Prober, J.M., Trainor, G.L., Dam, R.J. *et al.* (1987) A system for rapid DNA sequencing with fluorescent chain-terminating dideoxynucleotides. *Science* **238**, 336–341.
5 Ansorge, W., Sproat, B., Stegemann, J. *et al.* (1987) Automated DNA sequencing, ultrasensitive detection of fluorescent bands during electrophoresis. *Nucleic Acids Res.* **15**, 4593–4602.
6 Brumbaugh, J.A., Middendorf, L.R., Grone, D.L. & Ruth, J.L. (1988) Continuous, on-line DNA sequencing using oligodeoxynucleotide primers with multiple fluorophores. *Proc. Natl Acad. Sci. USA* **85**, 5610–5614.
7 Church, G.M. & Gilbert, W. (1984) Genomic sequencing. *Proc. Natl Acad. Sci. USA* **81**, 1991–1995.
8 Church, G.M. & Kieffer-Higgins, S. (1988) Multiplex DNA sequencing. *Science* **240**, 185–188.
9 Promega (1993) Silver Sequence Kit, 135.
10 Alain, S., Mazeron, M.C., Pepin, J.M. *et al.* (1995) Value of a new rapid non-radioactive sequencing method for analysis of the cytomegalovirus UL97 gene in ganciclovir-resistant strains. *J. Virol. Methods* **51**, 241–251.
11 Langer, P.R., Waldrop, A.A. & Ward, D.C. (1981) Enzymatic synthesis of biotin-labelled polynucleotides: novel nucleic acid affinity probes, *Proc. Natl Acad. Sci. USA* **78**, 6633–6637.
12 Martin, R., Hoover, C., Grimme, S. *et al.* (1990) A highly sensitive, nonradioactive DNA labelling and detection system. *BioTechniques* **9**, 762–768.
13 Beck, S., O'Keeffe, T., Coull, J.M. & Kπster, H. (1989) Chemiluminescent detection of DNA, application for DNA sequencing and hybridization. *Nucleic Acids Res.* **17**, 5115–5123.
14 Tizard, R., Cate, R.L., Ramachandran, K.L. *et al.* (1990) Imaging of DNA sequences with chemiluminescence. *Proc. Natl Acad. Sci. USA* **87**, 4514–4518.
15 Cherry, J.L., Young, H., DiSera, L.J. *et al.* (1993) Enzyme-linked fluorescent detection for automated multiplex DNA sequencing. *Genome Sequencing and Analysis Conference V*, Hilton Head, SC (Abstract), 65.
16 Richterich, P., Imrich, J., Mao, J. & Middendorf, L. (1993) Use of infrared fluorescence for membrane-based procedures (Abstract). *Genome Sequencing and Analysis Conference V*, Hilton Head, SC, 55.
17 Zagursky, R.J., Conway, P.S. & Kashdan, M.A. (1991) Use of ^{33}P for Sanger DNA sequencing. *BioTechniques* **11**, 36–38.
18 Pohl, T.M. & Maier, E. (1995) Sequencing 500 kb of yeast DNA using a GATC 1500 Direct Blotting Electrophoresis System. *BioTechniques* **19**, 482–486.
19 Beck, S. & Pohl, F.M. (1984) DNA sequencing with direct blotting electrophoresis, *EMBO J.* **3**, 2905–2909.

20 Richterich, P. & Church, G.M. (1993) DNA sequencing with direct transfer electrophoresis and nonradioactive detection. *Meth. Enzymol.* **218**, 187–222.
21 Kocher, D.C. (1981) *Radioactive Decay Data Tables* (National Technical Information Service, DOE/TIC-11026, Springfield).
22 Biggin, M.D., Gibson, T.J. & Hong, G.F. (1983) Buffer gradient gels and ^{35}S label as an aid to rapid DNA sequence determination. *Proc. Natl Acad. Sci. USA* **80**, 3963–3965.
23 Yang, D. & Waldman, A.S. (1994) Production of autoradiographs from unfixed ^{35}S-labelled DNA sequencing gels. *BioTechniques* **16**, 224–226.
24 Voss, H., Schwager, C., Wirkner, U. *et al.* (1991) A new procedure for automated DNA sequencing with internal labelling by fluorescent dUTP. *Meth. Mol. Cell. Biol.* **3**, 153–155.
25 Igloi, G.E. & Schiefermayr, E. (1993) Enzymatic addition of fluorescein- or biotin-riboUTP to oligonucleotides results in primers suitable for DNA sequencing and PCR. *BioTechniques* **15**, 486–497.
26 Steffens, D., Sutter, S., Jang, G. & Mühlegger, K. (1993) An infrared labelled dATP for labeling DNA sequencing fragments (Abstract). *Genome Sequencing and Analysis Conference V*, Hilton Head, SC, 53.
27 US Biochemical (1994) *Sequenase images non-isotopic DNA sequencing system* (US Biochemical/Amersham Life Sciences), 11 (Amersham Life Science Inc, Cleveland, OH).
28 Olesen, C.E.M., Martin, C.S. & Bronstein, I. (1993) Chemiluminescent DNA sequencing with multiplex labelling. *BioTechniques* **15**, 480–485.
29 Richterich, P., Heller, C., Wurst, H. & Pohl, F.M. (1989) DNA sequencing with direct blotting electrophoresis and colorimetric detection. *BioTechniques* **7**, 52–59.
30 Beck, S. (1987) Direct transfer electrophoresis used for DNA sequencing. *Anal. Biochem.* **164**, 514–520.
31 Creasy, A., D'Angio, L., Dunne, T.S. *et al.* (1991) Application of a novel chemiluminescence-based DNA detection method to single-vector and multiplex DNA sequencing. *BioTechniques* **11**, 102–109.
32 Richterich, P. (1989) Non-radioactive chemical sequencing of biotin labelled DNA. *Nucleic Acids Res.* **17**, 2181–2186.
33 Thorpe, G.H.G., Kricka, L.J., Gillespie, E. *et al.* (1985) Enhancement of the horseradish peroxidase-catalyzed chemiluminescent oxidation of cyclic diacyl hydrazides by 6-hydroxybenzothiazoles. *Anal. Biochem.* **145**, 96–100.
34 Durrant, I., Benge, L.C.A., Sturrock, C. *et al.* (1990) The application of enhanced chemiluminescence to membrane-based nucleic acid detection. *BioTechniques* **8**, 564–570.
35 Gebeyehu, G., Rao, P.Y., SooChan, P. *et al.* (1987) Novel biotinylated nucleotide-analogs for labelling and colorimetric detection of DNA. *Nucleic Acids Res.* **15**, 4513–4534.
36 Karger, A.E., Weiss, R. & Gesteland, R.F. (1992) Digital chemiluminescence imaging of DNA sequencing blots using a charge-coupled device camera. *Nucleic Acids Res.* **20**, 6657–6665.
37 Schaap, A.P., Sandison, M.D. & Handley, R.S. (1987) Chemical and enzymatic triggering of **1**, 2-dioxetanes. 3 alkaline phosphatase-catalyzed chemiluminescence from an aryl phosphate-substituted dioxetane. *Tetrahedron Lett.* **28**, 1159–1162.
38 Bronstein, I. & McGrath, P. (1989) Chemiluminescence lights up. *Nature* **338**, 559–560.
39 Bronstein, I., Juo, R.R., Voyta, J.C. & Edwards, B. (1991) Novel chemiluminescent adamantyl 1, 2-dioxetane enzyme substrates. In *Bioluminescence and Chemiluminescence, Current Status* (Stanley, P.E. & Kricka, L.J., eds), 73 (Wiley, Chichester).
40 Martin, C., Bresnick, L., Juo, R.-R. *et al.* (1991) Improved chemiluminescent DNA sequencing. *BioTechniques* **11**, 110–113.
41 Burstone, M.S. (1960) Postcoupling, noncoupling, and fluorescence techniques for the demonstration of alkaline phosphatase. *J. Natl Cancer Inst.* **24**, 1199–1207.
42 Cano, R.J., Torres, M.J., Klem, R.E. & Palomares, J.C. (1992) DNA hybridization assay using ATTOPHOS, a fluorescent substrate for alkaline phosphatase. *BioTechniques* **12**, 264–268.
43 Renz, M. & Kurz, C. (1984) A colorimetric method for DNA hybridization. *Nucleic Acids Res.* **12**, 3435–3444.
44 Jablonski, E., Moomaw, E.W., Tullis, R.H. & Ruth, J.L. (1986) Preparation of oligodeoxynucleotide-alkaline phosphatase conjugates and their use as hybridization probes. *Nucleic Acids Res.* **14**, 6115–6128.
45 Murakami, A., Tada, J., Yamagata, K. & Takano, J. (1989) Highly sensitive detection of DNA using enzyme-linked DNA-probe. 1. Colorimetric and fluorometric detection. *Nucleic Acids Res.* **17**, 5587–5595.
46 Farmar, J.G. & Castaneda, M. (1991) An improved preparation and purification of oligonucleotide-alkaline phosphatase conjugates. *BioTechniques* **11**, 588–589.
47 Leary, J.J., Brigati, D.J. & Ward, D.C. (1983) Rapid and sensitive colorimetric method for visualizing biotin-labeled DNA probes hybridized to DNA or RNA immobilized on nitrocellulose: bio-blots. *Proc. Natl Acad. Sci. USA* **80**, 4045–4049.
48 Forster, A.C., McInnes, J.L., Skingle, D.C. & Symons, R.H. (1985) Non-radioactive hybridization probes prepared by the chemical labelling of DNA and RNA with a novel reagent, photobiotin. *Nucleic Acids Res.* **13**, 745–761.
49 Misiura, K., Durrant, I., Evans, M.R. & Gait, M.J. (1990) Biotinyl and phosphotyrosinyl phosphoramidite derivatives useful in the incorporation of multiple reporter groups on synthetic oligonucleotides. *Nucleic Acids Res.* **18**, 4345–4354.
50 Pieles, U., Sproat, B.S. & Lamm, G.M. (1990) A protected biotin containing deoxycytidine building block for solid phase synthesis of biotinylated oligonucleotides. *Nucleic Acids Res.* **18**, 4355–4360.
51 Chollet, A. & Kawashima, E.H. (1985) Biotin-labeled synthetic oligodeoxyribonucleotides: chemical synthesis and uses as hybridization probes. *Nucleic Acids Res.* **13**, 1529–1541.
52 Voss, H., Wiesmann, S., Wirkner, U. *et al.* (1992) Automated DNA sequencing system resolving 1000 bases with fluorescein-15-dATP as internal label. *Meth.*

Mol. Cell. Biol. **3**, 30–34.

53 Thomas, N., Jones, C.N. & Thomas, P.L. (1988) Low volume processing of protein blots in rolling drums. *Anal. Biochem.* **170**, 393–396.

54 Ferguson, M., Kimball, A., DiSera, L.J. *et al.* (1993) An automated hybridization/imaging device for fluorescent multiplex DNA sequencing (Abstract). *Genome Sequencing and Analysis Conference V*, Hilton Head, SC, 65.

55 Ansorge, W., Sproat, B., Stegemann, J. & Schwager, C. (1986) A non-radioactive automated method for DNA sequence determination. *J. Biochem. Biophys. Meth.* **13**, 315–323.

56 Middendorf, L.R., Bruce, J.C., Bruce, R.C. *et al.* (1992) Continuous, on-line DNA sequencing using a versatile infrared laser scanner/electrophoresis apparatus. *Electrophoresis* **13**, 487–494. .

57 Wilson, R., Ainscough, R., Anderson, K. *et al.* (1994) 2.2 Mb of contiguous nucleotide sequence from chromosome III of *C. elegans*. *Nature* **368**, 32–38.

58 Ishino, Y., Mineno, J., Inoue, T. *et al.* (1992) Practical applications in molecular biology of sensitive fluorescence detection by a laser-excited fluorescence image analyser. *BioTechniques* **13**, 936–943.

59 Chu, T.J., Caldwell, K.D., Weiss, R.B. *et al.* (1992) Low fluorescence background electroblotting membrane for DNA sequencing. *Electrophoresis* **13**, 105–114.

60 Chee, M. (1991) Enzymatic multiplex sequencing. *Nucleic Acids Res.* **19**, 3301–3305.

61 Blum, H., Mallok, S., Goesswein, C. & Domdey, H. (1993) New perspectives in DNA-multiplex-sequencing (Abstract). *Genome Sequencing and Analysis Conference V*, Hilton Head, SC, 63.

62 Church, G.M., Gryan, G., Lakey, N. *et al.* (1994) Automated multiplex sequencing. *Automated DNA sequencing and analysis techniques*. (Craig Venter, L., ed.), 11–16 (Academic Press, San Diego).

63 Smith, D., Richterich, P., Lundstrom, R. *et al.* (1993) High throughput multiplex sequencing of mycobacterial genomes (Abstract). *Genome Sequencing and Analysis Conference V*, Hilton Head, SC, 55.

64 Höltke, H.J., Ettl, I., Finken, M. *et al.* (1992) Multiple nucleic acid labelling and rainbow detection. *Anal. Biochem.* **207**, 24–31.

Chapter 25 Repeat analysis

Aleksandar Milosavljević

CuraGen Corporation, 555 Long Wharf Drive, New Haven, CT 06511, USA

25.1 Introduction

Repetitive elements are DNA fragments that occur repeatedly within a genome. This broadest definition includes gene families such as rRNA and tRNA. A more limited definition [1] includes tandem arrays of repeats (satellites, microsatellites, and minisatellites), telomeric and subtelomeric repeats, retroposons (short interspersed repetitive elements such as L1), medium and low reiteration frequency repeats, endogenous retroviruses, and viral retrotransposons. As much as 60% of the human genome may consist of repetitive elements, even if the less inclusive definition is adopted [1].

The REPBASE database [2] contains prototypical interspersed repetitive elements from primates, rodents, mammals, vertebrates, invertebrates, and plants, as well as a collection of prototypical simple DNA sequences of Alu, L1, MIR and THE repetitive elements. REPBASE is available in directory 'repository/repbase' via ftp from ncbi.nlm.nih.gov.

With the advent of large-scale DNA sequencing, the analysis of the newly sequenced DNA for the presence of repetitive elements is becoming a frequent practice. Basic repeat analysis consists of the following three steps (we here assume human DNA is being analysed).

1 *Recognition of known repeats.* Occurrences of known interspersed repetitive elements are discovered. The DNA sequence is separated into repeat and nonrepeat regions.

2 *Repeat subfamily identification.* Alu sequences, and perhaps other repeats, discovered in the first step are assigned to subfamilies.

3 *Recognition of internal repeats.* Nonrepeat regions that are produced in the first step are analysed for local or global repetitions that are either direct or inverted.

The non-repeat regions that remain after the third step are used for comparisons against DNA or protein data bases in search of biologically interesting similarities.

In the following section we follow the three-step protocol using PYTHIA programs and REPBASE data base. We then review other methods and programs for individual steps of the protocol. In the appendix to this chapter, we discuss in more detail SMPL, a core program of PYTHIA. While we focus on the analysis of human genomic sequences, the analysis of sequences of other organisms may be performed similarly.

25.2 Repeat analysis via PYTHIA

PYTHIA is currently available on an electronic mail server at the address pythia@anl.gov. To get an update on the current status of the server, send to it the word 'help' in the subject line.

In the following we describe in detail the three-step protocol for repeat analysis using the human tissue plasminogen activator gene as an example.

The body of an e-mail message containing a request to PYTHIA consists of sequences in Intelligenetics format. An example of input is in Fig. 25.1. Although in our example we assume that a single locus is analysed, it should be noted that PYTHIA accepts multiple loci as well. The subject line contains one of the keywords: 'RPTS' (recognition of known repeats), 'ALU' (Alu subfamily identification), or 'SMPL' (recognition of internal repeats).

25.2.1 RPTS: recognition of known repeats

HUMREP is a file within REPBASE that contains prototypical human interspersed repeats. A version of HUMREP that is augmented by repeats in oppo-

```
;   HUMTPA Length: 36594 June 4, 1994 19:29 Type: N Check: 8313 ..
HUMTPA
TTCACACAACTGGTGCTGTTACCACCATGGGCGTCTAGTCTGGATCAGTGGTCCTCAGTCTTTTTTGCAC
CAGGGACCAGTTTTGTAAAGATAGCTTTTCCACGGACAGAGGGAGGGGAGATAGTTTCGGGATGATTCAA
...
...

AAAGGGCAGACTTGTAGTAGAATTCAGTTGCAAGAGGGATTGGGGAATCTTAAGGAAAAAATAGAATCTT
AAGGAAAAAATAACTGGGTGAGACGTGGACTGTGGACAGGCGCGGAAAAGGCAC1
```

Fig. 25.1 Human tissue plasminogen activator sequence in Intelligenetics format. Only the first two and the last two lines of sequence are shown here. Intelligenetics format requires that every sequence be preceded with at least one line starting with a semicolon (the rest of the letters in these lines do not matter) and by a locus name consisting of a contiguous sequence of letters at the beginning of a separate line. The sequence itself consists of uppercase letters A, G, C, and T and ends with a 1. Blank lines are not allowed, but empty spaces within the sequence itself are tolerated. The maximum length of a line accepted by PYTHIA programs is 99 letters. Every request consists of one or more sequences.

site orientation can be obtained by sending the word 'repbase' in the subject line. The names of repeats in opposite orientation are derived by appending 'c' at the end of the original name (e.g. an L1 sequence in opposite orientation is denoted L1c).

In order to search sequences against HUMREP, send them to pythia@anl.gov with term 'RPTS' in the subject line. As an example, in the following we assume that we have sent as an input the human tissue plasminogen activator (HUMTPA) gene sequence [3], GenBank [4] accession number K03021, containing 36 594 bases, as illustrated in Fig. 25.1. PYTHIA responds with a message consisting of four parts, as described below.

1 *Occurrences of repeats.* A partial listing for our example is in Fig. 25.2.

2 *Listing of Alu regions.* A few Alu sequences from our example are shown in Fig. 25.3. These Alu regions can be excised using a test editor and then sent directly back to pythia@anl.gov with 'ALU' in the subject line for Alu subfamily identification, an analysis option described in Section 25.2.2.

3 *Local alignment of repeats.* A local alignment of a repeat fragment from our example is shown in Fig. 25.4. There is no rigorous significance theory of alignment scores yet; a threshold for significant alignment scores is set based on empirical testing. An additional test homology, suggested by J. Jurka [5], is to count the ratio of transitions vs. total point mutations; a significant ratio of about 1:3, the expected value for random sequences, indicates true homology.

4 *Listing of nonrepeat fragments.* Some of the nonrepeat fragments from our example are in Fig. 25.5. The sequence returned by PYTHIA is formatted so that it can be immediately mailed to pythia@anl.gov with 'SMPL' in the subject line for the purpose of discovering internal repetitive patterns, or it can be directly used for database searches. Searching for internal repetitive patterns is recommended because it may further reduce uninteresting similarities (e.g. see Fig. 25.6) produced by the standard similarity search algorithms.

```
ALU :
HUMTPA 739 1022
HUMTPA 8862 9165

...

HUMTPA 32921 33210
HUMTPA 34234 34503

ALUc :
HUMTPA 5671 5960
HUMTPA 6319 6463
HUMTPA 6483 6746
HUMTPA 7224 7512
HUMTPA 10513 10938
HUMTPA 11728 11869
HUMTPA 12700 13143
HUMTPA 18640 18796
HUMTPA 21651 21940

...

MER1 :
HUMTPA 40 578
HUMTPA 26355 26380

MER12c :
HUMTPA 17555 17769

MER1c :
HUMTPA 301 408

...
```

Fig. 25.2 Some occurrences of repeats in the human tissue plasminogen activator sequence that are identified by PYTHIA.

```
;
HUMTPA[739->1022](0,0)
GGCTGGGCGCGGTGGCTCACACCTATAATCCCAGCACTTTGGGAGGCTGAGGCAGGTGGATCACGAGGTC
GGGGGTTTGAGACCAGCCTGACCAACATGGTGAAACCCCGTCTCTACTAAAATACAAAAAATTAGCTGGG
CGTGGTGGCGGGCACCTGTAATCTCAGCTACTCAGGAGGCTGAGGCAGGAGAATTGCTTGAACCTGGTGG
AGGTTGCAGTGAGCCGAGATCACACCACTGCACTCTAGCCTGGGCGACAGAGCAAGACTCTGTCTCAAAA
AAAA1
;
HUMTPA[8862->9165](0,0)
GGCCGGGCACACAGCTCCTGCCTGTAATCCCAGCACTTTGGGAGCCCGAGGTGGGCGGGTTGCTTGAGCC
AAGGAGTTTGAAACCAGCCCGGGTCTTGAACATAGCGAAGACTCTGTCTCTACAAAAAAATGAAAAAAAA
AAAAAAATTAGCCAGACATGGTGGCACGCACCTGTAGTCCCAGCTACTTGAGAAGCTGAGGTGAGAGGAT
CACTTGAGCCAGGGAGGTTGAAACTGCAGTGAGCTGTGATCACGCCACTGCACTCCAGTCTGGGTGACTG
GGCGAGACCCTGTCTCAAAACAAA1

...
```

Fig. 25.3 Partial listing of Alu regions.

```
score: -44


top: locus: ALUc beginning: 150 end: 290 length: 141
bottom: locus: HUMTPA beginning: 11728 end: 11869 length: 142
local_indels: 1 mismatches: 39 transitions: 25

@150      @160       @170       @180       @190       @200       @210
CCGGCTAATTTTTGTATT-TTTAGTAGAGACGGGGTTTCACCATGTTGGCCAGGCTGGTCTCGAACTCCT
**:*..**.*.**.**** ***.*.***.*:*****:*::*::******.*:**********:********
CCAGGAAAATATTCTATTCTTTTGAAGACATGGGGTCTTGCTATGTTGCCTAGGCTGGTCTTGAACTCCT
  @11730    @11740    @11750    @11760    @11770    @11780    @11790

 @220      @230       @240       @250       @260       @270       @280
GACCTCAGGTGATCCGCCCGCCTCGGCCTCCCAAAGTGCTGGGATTACAGGCGTGAGCCACCGCGCCCGG
.:*****.*****.*.**:*****:******::*:..***:***:******:*::*:*****:*:***:*
CGCCTCAAGTGATGCTCCTGCCTCAGCCTCCTGAGTAGCTAGGACTACAGGTGCAAACCACCACACCCAG
  @11800    @11810    @11820    @11830    @11840    @ 11850    @11860

@290
CC
*  :
CT
```

Fig. 25.4 A local alignment of a repeat region. Note that the ratio of transitions to total point mutations significantly exceeds the expected value of 1 : 3, thus providing additional evidence for true homology.

```
;
HUMTPA[1->738](0,0)
TTCACACAACTGGTGCTGTTACCACCATGGGCGTCTAGTCTGGATCAGTGGTCCTCAGTCTTTTTTGCAC
CAGGGACCAGTTTTGTAAAGATAGCTTTTCCACGGACAGAGGGAGGGGAGATAGTTTCGGGATGATTCAA

...

CGGGGATTCCCAGTCTAGATGGAGACCTAGACAAGGCGTGCGACAATAACACCGATTTTAGATCCATCAT
GACATTTACCCCATCCCCTGCAAAGCCAGATGGCTACCAAAATTAAATCTTAGTTTAGACACAGAATGTC
CGTCTTCTGGTCCAAAACATCCTTGTCATAAGTTTCTT1
;
HMTPA[1023->8861](0,0)
AAAAAAGAAAATAAAAAAAAAAAAAACAAGTTTCTTGCCCACTCTTCCTTTCTCTGAGTTTCCAGAGACAT
CACATCATTTCTTACCCAGCTGAGCAGAGTCCCAGCATGGCTCTCGTTCGAATACCCATCCTGCCACCTG

...
```

Fig. 25.5 A sample of nonrepeat fragments. Loci names are augmented by fragment location: `HUMTPA [1023 => 8861]` (0, 0) denotes subfragment 1023–8861 within locus HUMTPA in direct orientation; the first 0 within the parentheses means that sequence is not reversed (1 means reversed) while the second 0 means that sequence is not complemented (1 means complemented). The same fragment in opposite orientation would be denoted HUMTPA `[1023 => 8861] (1, 1)`.

25.2.2 ALU: subfamily identification of Alu sequences

In order to identify subfamily membership of Alu sequences, send them with the word 'ALU' in the subject line. In the following, we assume that the Alu sequences obtained in the previous section are sent as the input. PYTHIA responds with a message consisting of two parts, as described below.

1 *Alignment of Alu sequences against Alu consensus.* An example of an alignment is shown in Fig. 25.7. The alignment enables exact localization of diagnostic positions within the Alu sequence. The diagnostic positions are important for subfamily identification, which is described next.

2 *Alu subfamily identification.* The bases in diagnostic positions determine Alu subfamily membership. The output of PYTHIA describes the identification procedure; an example of the output is in Fig. 25.8. Alu subfamilies and diagnostic positions are described in more detail in ref. 6. The method that was used for discovering Alu subfamilies is described in ref. 7.

```
ATTAACAGTAACTGCTTCATAGATAGA-AGATAGATAGATTAGATAGATAGATAG
**.:**:***:.**..*.*:******** **** *****:*  *********** *
ATAGACGGTAGATGGATGACAGATAGACAGAT-GATAGGT--GATAGATAGAT-G
```

Fig. 25.6 Sequence similarity due to shared repetitive structure.

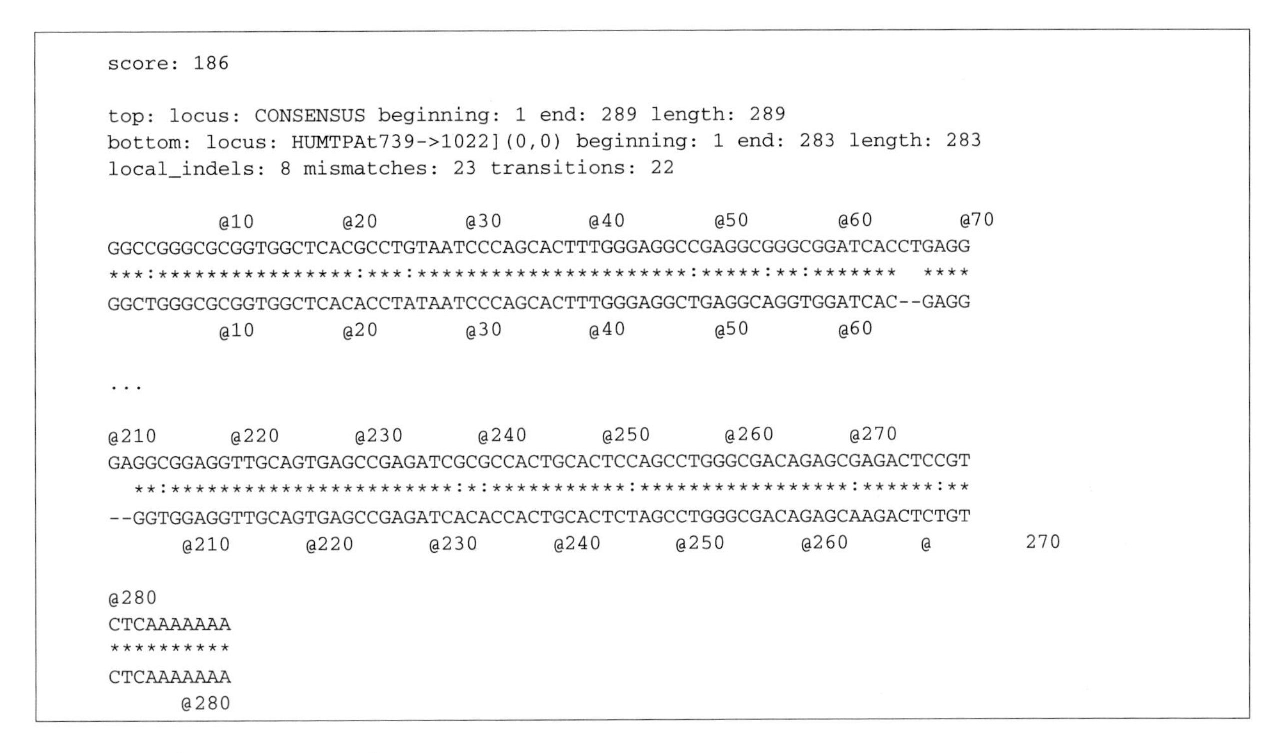

```
score: 186

top: locus: CONSENSUS beginning: 1 end: 289 length: 289
bottom: locus: HUMTPAt739->1022](0,0) beginning: 1 end: 283 length: 283
local_indels: 8 mismatches: 23 transitions: 22

         @10       @20       @30       @40       @50       @60       @70
GGCCGGGCGCGGTGGCTCACGCCTGTAATCCCAGCACTTTGGGAGGCCGAGGCGGGCGGATCACCTGAGG
***:***************:***:************************:*****:**:*******  ****
GGCTGGGCGCGGTGGCTCACACCTATAATCCCAGCACTTTGGGAGGCTGAGGCAGGTGGATCAC--GAGG
         @10       @20       @30       @40       @50       @60

...

@210      @220      @230      @240      @250      @260      @270
GAGGCGGAGGTTGCAGTGAGCCGAGATCGCGCCACTGCACTCCAGCCTGGGCGACAGAGCGAGACTCCGT
  **:***********************:*:***********:*****************:******:**
--GGTGGAGGTTGCAGTGAGCCGAGATCACACCACTGCACTCTAGCCTGGGCGACAGAGCAAGACTCTGT
      @210      @220      @230      @240      @250      @260      @        270

@280
CTCAAAAAAA
**********
CTCAAAAAAA
      @280
```

Fig. 25.7 An alignment against Alu consensus.

25.2.3 SMPL: recognition of internal repeats

Even after small occurrences of known repetitive elements are eliminated from a genomic sequence, the sequence may still contain internal repetitions of different kinds: satellites, microsatellites, and minisatellites consisting of tandem arrangements of short oligomers (see Chapter 5); self-complementary sequences; pseudogenes that arose by gene duplication; or perhaps two or more occurrences of as yet undiscovered repetitive elements in the same or in the opposite orientation. Such structures can be recognized by sending the sequences with the word 'SMPL' in the subject line.

The SMPL program, the format of its output, and the method for establishing significance are discussed in more detail in the appendix to this chapter. The output of SMPL consists of five parts, each corresponding to a combination of the following two criteria: that repetitive structures can be local (within a window of 128 bases) or global (perhaps even occurring across different loci), and that repetitions may occur in the same or in opposite orientation.

1 *Simple regions.* Regions consisting of repetitions of words are identified and parsed, as shown in Fig. 25.9.

2 *Listing of complex regions.* The sequence that remains after the regions identified in step 1 are 'censored' may now be used for standard database search.

3 *Local reverse complementarity.* Fragments that contain a significant number of long words in both orientations are detected. If a long word is identical to its own reverse complement, it may be detected. This kind of symmetry may be a simple consequence of tandem repetitiveness (e.g. $(AT) * N$ sequence), or it may indicate the presence of secondary structure in the transcribed RNA or in DNA. Regions detected in our example are listed in Fig. 25.10.

4 *Global repeats.* Repeats can be due to duplications of large genomic segments or due to the presence of repetitive elements. Figure 25.11 contains the output for our example. Note that SMPL can analyse multiple loci at once, thus serving as a method for fast identification of repetitive structure in data bases of DNA sequence.

5 *Global inverted repeats.* The list of global inverted repeats in our example is empty, but if we did not

```
An Alu sequence is identified by performing                   Alu
a series of decisions as illustrated                         / #\
on the right. Each decision leads to                      J       S
the placement of the sequence in a more                          /# \
specific subfamily. As an example, the decisions              Sbc     Spqx
leading to the identification of an Sbl sequence             /# \     / \
are marked by '#'. (SbO denotes the members                 Sb   Sc  Sx  Spq
of Sb that are neither Sbl nor Sb2.)                       /# \          / \
                                                      Sb0Sb1  Sb2      Sp   Sq
                                                        /# \
                                                      Sb0   Sb1

     pos:  57  65  70  71  94 101 120 163 194 204 214 215 220 233 275 277
       J:                                           *                   *
       S:           *   *   *   *   *   *   *   *       *   *   *   *
  weight:   5  18   5   9  20   2   5  12          22  20   7   1   7   6   6   8

     pos:  78  88  95 153 163 197 219
     Sbc:               *
    Spqx:   *   *   *       *       *
  Weight:  36  28   5   3  13   2   1

     pos: 155 244 262 272
     Spq:
      Sx:   *   *   *   *
  Weight:   2   2   4  38

 locus contains an Alu-Sx
```

Fig. 25.8 Alu subfamily identification.

```
;
; 41 HUMTPA 7103 7230
HUMTPA[7103=>7230](0,0){41}1self
T-C-T-T-C-C-G-A-T-A-G-T-G-G-C-T-C-A-G-T-T-T-C-T-A-C-T-T-A-C-A-T-A-A-
A-A-A-G-A-C-A-G-C-A-C-A-T-T-C-T-C-T-T-A-G-C-A-A-T-A-T-G-T-G-T-T-T-G-T-
A-T-G-TGTGTGTGTGTGTGTGTGTGTGTGTGTGT-A-TATATATATATATATATATATA-A-T-T-T-A1

...

; 8 HUMTPA 21251 21278
HUMTPA[21251=>21278](0,0){8}1self
A-A-G-A-A-A-A-A-G-AAAAGAAAAGAAAA-A-A-T-T-A1
;
; 42 HUMTPA 23879 24006
HUMTPA[23879=>24006](0,0){42}1self
A-A-T-A-C-A-G-G-A-T-G-G-A-T-A-GATGGATAGATG-A-T-A-G-A-C-A-G-A-T-A-ATAGA
TGATAG-G-T-GATAGATGATAGA-T-TGATAGATGATAGAT-GATAGGTGATAGAT-T-A-G-A-T-A-
AATAGATGATA-C-A-T-A-C-ATGATAGAT-A-G-A1
;

...
```

Fig. 25.9 Simple regions. Dashes are inserted to indicate parsing, as described in the appendix to this chapter. A locus name is augmented by the redundancy information (the method for computing encoding lengths and redundancy is explained in the appendix): `HUMTPA [7103 = > 7230] (0, 0) {41} | self` means that the fragment can be encoded in 41 bits less than required by the strainghtforward encoding (2 bits per letter), indicating repetitions of words within the fragment at the significance level 2^{-41}.

eliminate two occurrences of Alu sequences in opposite orientation, they would have been detected at this point. While the occurrences of a repeated region in identical orientation may be explained by gene duplications, occurrence in opposite orientation is more likely to indicate the presence of a

```
;
; 21 HUMTPA 7167 7230 HUMTPA 7167 7230
HUMTPA[7167=>7230](0,0){21}1HUMTPA[7167=>7230](1,1)
G-T-T-T-G-T-A-T-G-T-G-T-G-T-G-T-G-T-G-T-G-T-G-T-G-T-G-T-G-T-G-T-G-T-G-
TATATATATATATATATATATATA-A-T-T-T-Al
;
; 21 HUMTPA 7199 7262 HUMTPA 7199 7262
HUMTPA[7199=>7262](0,0){21}1HUMTPA[7199=>7262](1,1)
G-T-G-TATATATATATATATATATATATA-A-T-T-T-A-G-A-G-A-C-A-A-G-G-T-C-T-G-A-C
-T-C-C-A-T-C-A-C-C-C-A-G-G-C-T-G-Gl
;
; O HUMTPA 7935 7998 HUMTPA 7935 7998
HUMTPA[7935=>7998](0,0){0}1HUMTPA[7935=>7998](1,1)
A-G-G-A-T-T-G-A-T-C-A-G-A-A-G-A-T-C-T-G-A-T-T-C-C-ACCTGGA-G-C-C-T-C-T-
GAAGTGATCACTTC-C-A-G-G-T-T-A-G-G-C-T-Gl
;
; 2 HUMTPA 27229 27292 HUMTPA 27229 27292
HUMTPA[27229=>27292](0,0){2}1HUMTPA[27229=>27292](1,1)
T-A-C-A-T-A-A-ATATGTGT-G-T-G-G-G-T-G-T-G-T-G-TATATATATA-T-G-T-A-A-T-AC
ACATAT-A-T-TAAATTTA-T-A-T-Al
;
; O HUMTPA 32792 32855 HUMTPA 32792 32855
HUMTPA[32792=>32855](0,0){0}1HUMTPA[32792=>32855](1,1)
T-A-A-A-C-T-G-C-A-GGAAATTTCC-C-C-A-G-G-A-T-C-T-G-C-A-C-A-G-G-C-A-A-C-T
-C-C-A-CCTGTACAGG-ATTTCCT-T-T-C-Al
```

Fig. 25.10 Local reverse complementarity. Locus name is augmented by the encoding information: `HUMTPA[7167=>7230](0,0){21}|HUMTPA[7167=>7230](1,1)` means that 21 bits can be saved by encoding the fragment 7167–7230 relative to its own reverse complement, indicating reverse complementarity at the significance level 2^{-15}. The words that are not interrupted by dashes occur in opposite orientation within the same region, as determined by the parsing procedure that is described in the appendix to this chapter.

```
;
; 36 HUMTPA 25479 25618 HUMTPA 3071 3326
HUMTPA[25479=>25618](0,0){36}1HUMTPA[3071=>3326](0,0)
ATTCCAG-T-CCACAC-CTTGTCA-A-T-T-T-GGCACC-C-A-TGTGCATC-TCCTT-AAACC-ATCCT
T-CACCTCC-A-A-G-TAAACAC-A-G-G-A-ACAAA-A-T-C-A-T-A-C-TCCTGCCT-A-A-C-A-T
-G-A-TAGAA-CTACC-AGTGT-A-CAACC-A-A-A-A-A-C-G-CACTCCCl
;
; 84 HUMTPA 24583 24838 HUMTPA 5375 5630
HUMTPA[24583=>24838](0,0){84}1HUMTPA[5375=>5630](0,0)
CAGGA-G-G-TCCTGAGGACAT-G-T-G-C-CCAAGGT-TGTCAG-G-G-C-A-C-A-G-C-T-T-GCCT
TT-A-G-A-C-G-T-T-T-T-AGGGAG-T-CATGAGACAT-C-AATCAA-CATGTG-T-G-AGATGT-A-
C-A-TCGGT-TTGGT-C-G-GGAAAG-T-T-G-G-G-A-T-AACTCGAAG-C-A-A-GGGCTTCCAGG-C
-CATAGGTAGATAAGAGA-C-A-AAAGGC-T-G-T-A-TTCTGAGTC-C-T-T-G-A-TCAGC-T-TTTC
ACTGAA-C-ACACAATT-GAGTCT-G-G-C-T-C-A-G-TTCAT-C-T-G-C-A-T-T-TTTACATA-A-
A-A-Al

...
```

Fig. 25.11 Global repeats. Locus name indicates regions compared and gives relative encoding length: `HUMTPA[25479 => 25618] (0, 0) {36} | HUMTPA[3071 => 3326] (0, 0)` means that 36 bits can be saved by encoding the fragment 25479–25618 relative to fragment 3071–3326 in direct orientation, indicating similarity at the significance level 2^{-36}. The words that are not interrupted by dashes occur in both fragments and are determined by the parsing procedure that is described in the appendix to this chapter.

repetitive element, which typically integrates in either orientation. This is how many repeats from REPBASE have been found.

25.3 Repeat analysis by methods other than PYTHIA

Many methods for repeat analysis have been proposed and implemented. In the following, we discuss a representative but not comprehensive sample of such programs, grouping them according to the three steps of our repeat analysis protocol.

25.3.1 Recognition of known repeats

The CENSOR program [8] and GenQuest server [9] also provide searches against REPBASE. The XBLAST program [10] provides a search against a data base of Alu sequences.

For the purposes of comparison, the same sequence that was sent to PYTHIA with 'RPTS' in the subject line (as described in Section 25.2) was sent to the GenQuest server [9]; the response is in Fig. 25.12. While all the Alu occurrences in direct orientation were correctly identified by both programs, only PYTHIA identified all the occurrences of Alu in reverse orientation. GenQuest also missed a number of occurrences of other repetitive elements that are not listed in Fig. 25.2 due to lack of space. In some other cases, GenQuest may also detect elements not recognized by PYTHIA. Rather than indicating superiority of one program over the other, this single example only illustrates the fact that best analysis may be performed by combining the independent results of a number of programs.

One should mention that occurrences of Alu elements can be censored prior to sequence similarity searches using XBLAST [10]. Unlike PYTHIA and GenQuest, which both employ alignment algorithms that tolerate insertions and deletions, XBLAST is essentially a parser of BLAST output and inherits its insensitivity to these two kinds of mutations. Since repetitive elements are subject to insertions and deletions, the methodological disadvantage of the latter method is clear, especially in the case of older and more decayed elements.

The sensitivity of CENSOR and RPTS are comparable. However, CENSOR is more than a hundred times faster than the current inefficient implementation of RPTS. Preparations are in progress to either improve RPTS or replace it with CENSOR.

In addition to the methods discussed so far, a number of other methods and programs for analysis and visualization of repeats and for large-scale sequence comparisons have been proposed [11–14].

25.3.2 Repeat subfamily identification

The only programs other than PYTHIA for identifying Alu and other repeat subfamilies are available from Jerzy Jurka (jurka@gnomic.stanford.edu).

```
>ALU
 len = 290
forward strand
hit 1:        239      739-   1015
hit 2:        203     8862-   9165

...

hit 16:        235  32922- 33189
hit 17:        186  34234- 34464
reverse strand
hit  1:       234   21661- 21941
hit  2:       153   12863- 13144
hit  3:       167   10680- 10939
hit  4:       164    7259-  7514
hit  5:       193    6506-  6747
hit  6:       222    5695-  5961
>MER1
 len = 539
forward strand
hit 1:        539      40-    578

...

>MER12
 len = 240
reverse strand
hit 1:       127   17556- 17786
```

Fig. 25.12 Some occurrences of repeats in the human tissue plasminogen activator sequence that are identified by GenQuest.

25.3.3 Recognition of internal repeats

Simple sequences that have biased sequence composition can also be detected by the SEG program [15,16], which is based on the concept of compositional complexity [17,18], and is currently also available individually or as '- filter seg' option in BLAST. Program XNU [10,19] finds short tandem repeats by performing self-alignment. XNU is also available as XBLAST postfilter of the BLASTN program.

SEG [15,16] works well in cases with repetitive pattern bias in the base frequencies (as with poly(A) tails), but it fails to recognize longer tandem repeats of balanced base composition. XNU and XBLAST [10,19] partly rectifies this problem by performing self-alignments, a computationally expensive operation that may discover repeats of longer periodicity. The SEG and XNU programs have been judged complementary, each being sensitive to a particular kind of repetitiveness [20]. As described in the appendix, SMPL requires only a linear time computation, in contrast to the quadratic pattern, irrespective of its periodicity. SMPL exactly measures redundancy in terms of the number of bits, d, which directly implies significance of 2^{-d}.

In addition to the methods discussed so far, a number of other methods and programs for analysis and visualization of internal repetitive structures have been proposed [21–24].

Finally, we should mention Sequence Landscapes [25], a pioneering program for analysis of internal repetitive structures, still rarely superseded in its generality and clarity. The directed acyclic word graph data structure [26] that underlies Sequence Landscapes has also been employed in the PYTHIA SMPL program.

Acknowledgements

I thank David Nadziejka for editorial assistance. This work was supported in part by the US Department of Energy, Office of Health and Environmental Research, under Contract W-31-1009-ENG-38.

References

1 Jurka, J. (1995) Human repetitive elements. In *Molecular Biology and Biotechnology: A Comprehensive Desk Reference* (Meyers, R.A., ed.), 438–441 (VCH Publishers, New York).
2 Jurka, J., Walichiewicz, J. & Milosavljević, A. (1992) Prototypic sequences for human repetitive DNA. *J. Mol. Evol.* **35**, 286–291.
3 Friezner-Degen, S.J., Rajput, B. & Reich, E. (1986) The human tissue plasminogen activator gene. *J. Biol. Chem.* **261**, 6972–6985.
4 Benson, D., Lipman, D.J. & Ostell, J. (1993) GenBank. *Nucleic Acids Res.* **21**, 2963–2965.
5 Jurka, J. (1994) Approaches to identification and analysis of interspersed repetitive DNA sequences. In *Automated DNA Sequencing and Analysis Techniques* (Venter, J.C., ed.), 294–298 (Harcourt Brace Jovanovich, New York).
6 Jurka, J. & Milosavljević, A. (1991) Reconstruction and analysis of human Alu genes. *J. Mol. Evol.* **32**, 105–121.
7 Milosavljević, A. & Jurka, J. (1993) Discovery by minimal length encoding: a case study in molecular evolution. *Machine Learning J.* **12** (Spec. Issue, *Machine Discovery*), 69–87.
8 Jurka, J., Klonowski, P., Dagman, V. & Pelton, P. (1996) Censor—a program for identification and elimination of repetitive elements from DNA sequences. *Comp. Chem.* **20**, 119–121.
9 Uberbacher, E., Xu, Y. & Mural, R.J. (1996) Discovering and understanding genes in human DNA sequence using GRAIL. *Meth. Enzymol.* **266**, 259–281.
10 Claverie, J.-M. (1993) Effective large-scale sequence similarity searches. *Meth. Enzymol.* **266**, 212–227.
11 Agarwal, P. & States, D. (1994) The repeat pattern toolkit (rpt): analyzing the structure and evolution of the *C. elegans* genome. In *Proceedings of the 2nd International Conference on Intelligent Systems for Molecular Biology* (Altman, R., Brutlag, D., Karp, P. *et al.*, eds) (AAAI Press, Menlo Park, USA).
12 Madden, T.L., Tatusov, R.L. & Zhang, J. (1996) Applications of network BLAST server. *Meth. Enzymol.* **266**, 131–141.
13 Parsons, J.D. (1995) Micropeats: graphical DNA sequence comparisons. *Comp. Appl. Biosci.* **11**, 615–619.
14 Quentin, Y. & Fichant, G.A. (1994) Fast identification of repetitive elements in biological sequences. *J. Theor. Biol.* **166**, 51–61.
15 Wootton, J.C. & Federhen, S. (1996) Analysis of compositionally biased regions in sequence databases. *Meth. Enzymol.* **266**, 554–571.
16 Wootton, J.C. & Federhen, S. (1993) Statistics of local complexity in amino acid sequences and sequence databases. *Comp. Chem.* **17**, 149–163.
17 Salamon, P. & Konopka, A.K. (1992) A maximum entropy principle for distribution of local complexity in naturally occurring nucleotide sequences. *Comp. Chem.* **16**, 117–124.
18 Salamon, P., Wootton, J.C., Konopka, A. & Hansen, L. (1993) On the robustness of maximum entropy relationships for complexity distributions of nucleotide sequences. *Comp. Chem.* **17**, 135–148.
19 Claverie, J.-M. & States, D.J. (1993) Information enhancement methods for large scale sequence analysis. *Comp. Chem.* **17**, 191–201.
20 Altschul, S.F., Boguski, M.S., Gish, W. & Wootton, J.C. (1994) Issues in searching molecular sequence databases. *Nature Genet.* **6**, 119–129.
21 Benson, G. & Waterman, M.S. (1994) A method for fast database search for all k-nucleotide repeats. *Nucleic Acids Res.* **22**, 4828–4836.
22 Hancock, J.M. & Armstrong, J.S. (1994) SIMPLE34: an improved and enhanced implementation for VAX and Sun computers of the SIMPLE algorithm for analysis of clustered repetitive motifs in nucleotide sequences. *Comp. Appl. Biosci.* **10**, 67–70.
23 Sonhammer, E.L.L. & Durbin, R. (1994) A workbench for large-scale sequence homology analysis. *Comp. Appl. Biosci.* **10**, 301–307.
24 Mrázek. & Kypr, J. (1993) UNIREP: A microcomputer program to find unique and repetitive nucleotide sequences in genomes. *Comp. Appl. Biosci.* **9**, 355–360.
25 Clift, B., Haussler, D., McConnell, R. *et al.* (1986) Sequence landscapes. *Nucleic Acids Res.* **14**, 141–158.
26 Blumer, A., Blumer, J., Haussler, D. *et al.* (1985) The smallest automaton recognizing the subwords of a text. *Theor. Comp. Sci.* **40**, 31–55.
27 Milosavljević, A. (1995) The discovery process as a search for concise encoding of observed data. *Foundations Sci.* **1**, 212–218.
28 Batzer, M.A., Deininger, P.L., Hellman-Blumberg, U. *et al.* (1996) Standardized nomenclature for alu repeats. *J. Mol. Evol.* **42**, 3–6.
29 Milosavljević, A. & Jurka, J. (1993) Discovering simple DNA sequences by the algorithmic significance method. *Comp. Appl. Biosci.* **9** (4), 407–411.
30 Milosavljević, A. (1995) Discovering dependencies via algorithmic mutual information: a case study in DNA sequence comparisons. *Machine Learning J.* **21**, 35–50.
31 Chen, E.Y., Liao, Y.C., Smith, D.H. *et al.* (1989) The human growth hormone locus: nucleotide sequence,

biology and evolution. *Genomics* **4**, 479–497.

32 Storer, J.A. (1988) *Data Compression: Methods and Theory* (Computer Science Press, Rockville, USA).

33 Milosavljević, A. (1993) Discovering sequence similarity by the algorithmic significance method. In *Proceedings of the 1st International Conference on Intelligent Systems for Molecular Biology* (Hunter, L., Searls, D. & Shavlik, eds), 284–291 (AAAI Press, Rockville, USA).

34 Milosavljević, A. (1994) Sequence comparisons via algorithmic mutual information. In *Proceedings of the 2nd International Conference on Intelligent Systems for Molecular Biology* (Altman, R., Brutlag, D., Karp, P. *et al.*, eds), 303–309 (AAAI Press, Rockville, USA).

Appendix: encoding and parsing

In the following we discuss in more detail the method employed by SMPL. We also present the method for determining significance of patterns discovered by SMPL.

The SMPL program is based on the general premise that the process of inference and pattern recognition can be viewed as a search for concise encoding of data; for a general argument in support of this premise, see ref. 27. Every aspect of the analysis of repetitive DNA elements perfectly fits this premise. In fact, some of the currently accepted Alu subfamilies [28] were discovered by computing concise encodings [6, 7].

Repetitive patterns can best be defined via encoding length: a completely random DNA sequence that does not resemble any other known sequence requires 2 bits per letter; however, if a sequence contains long repeated words, then the repeated occurrences can be replaced by short pointers to earlier occurrences within the same sequence, thus reducing the total number of bits. The newly developed algorithmic significance method [29] states that d bits of information can be saved with probability 2^{-d}. In other words, a random sequence is unlikely to be compressed by chance.

The problem of determining significance of similarities between objects is best captured using the concept of algorithmic mutual information [30], which is defined as the difference between the sum of individual encoding lengths of objects and their joint encoding length. This general formulation enables search for sequences of low complexity that exhibit enough similarity to prove their homology. This kind of analysis may be applied to track mutations that abound in sequences of low complexity. Figure 25.13 contains one example obtained by the analysis of the 66 495 bp human genomic locus that contains growth hormone and chorionic somatotropin (HUMGHCSA) genes [31], GenBank Accession No. J03071. The approaches that rely on 'censoring' [8, 19] of simple DNA sequence would clearly miss these homologies.

We now turn to the encoding algorithms employed by the SMPL program. SMPL and the algorithmic significance method are described in detail in two earlier papers [29,30]; for the sake of completeness, some of the material is reviewed here.

The number of bits needed to encode sequence t by itself is denoted by $IA_0(t)$. The encoding length of sequence t relative to sequence s is denoted $I_A(t \mid s)$. An encoding of a sequence can in either case be represented by a parsing, which we describe below.

When encoding a sequence by itself, a repeated occurrence of a word is replaced by a pointer to its previous occurrence within the same sequence. We assume that a pointer consists of two positive integers: the first integer indicates the beginning position of a previous occurrence of the word while the second integer indicates the length of the word. For example, sequence

`AGTCAGTTTT`

may be encoded as

`AGTC(1,3)(7,3),`

where `(1,3)` points to the occurrence of `AGT` from position 1 to position 3, and `(7,3)` points to the occurrence of `TTT` from position 7 to position 9 in the original sequence.

The decoding algorithm A_0 consists of the following two steps:

1 replace each pointer by a sequence of pointers to individual letters;

2 replace the new pointers by their targets in the left-to-right order.

Continuing our example, the first step would yield

`AGTC(1,1)(2,1)(3,1)(7,1)(8,1)(9,1),`

and the second step would yield the original sequence. From this decoding algorithm it should be obvious that the original sequence can be obtained despite overlaps of pointers and their targets, as is the case with the pointer `(7,3)` in our example.

When encoding a target sequence relative to a source sequence, the pointers point to the occurrences of the same words in the source.

Consider an example where the target sequence is

`GATTACCGATGAGCTAAT`

and the source sequence is

`ATTACATGAGCATAAT`

The occurrences of some words in the target may be replaced by pointers indicating the beginning and the length of the occurrences of the same words in the source, as follows:

`G(1,4)CCG(6,6)(13,4)`

The decoding algorithm A is very simple: it only needs to replace pointers by words.

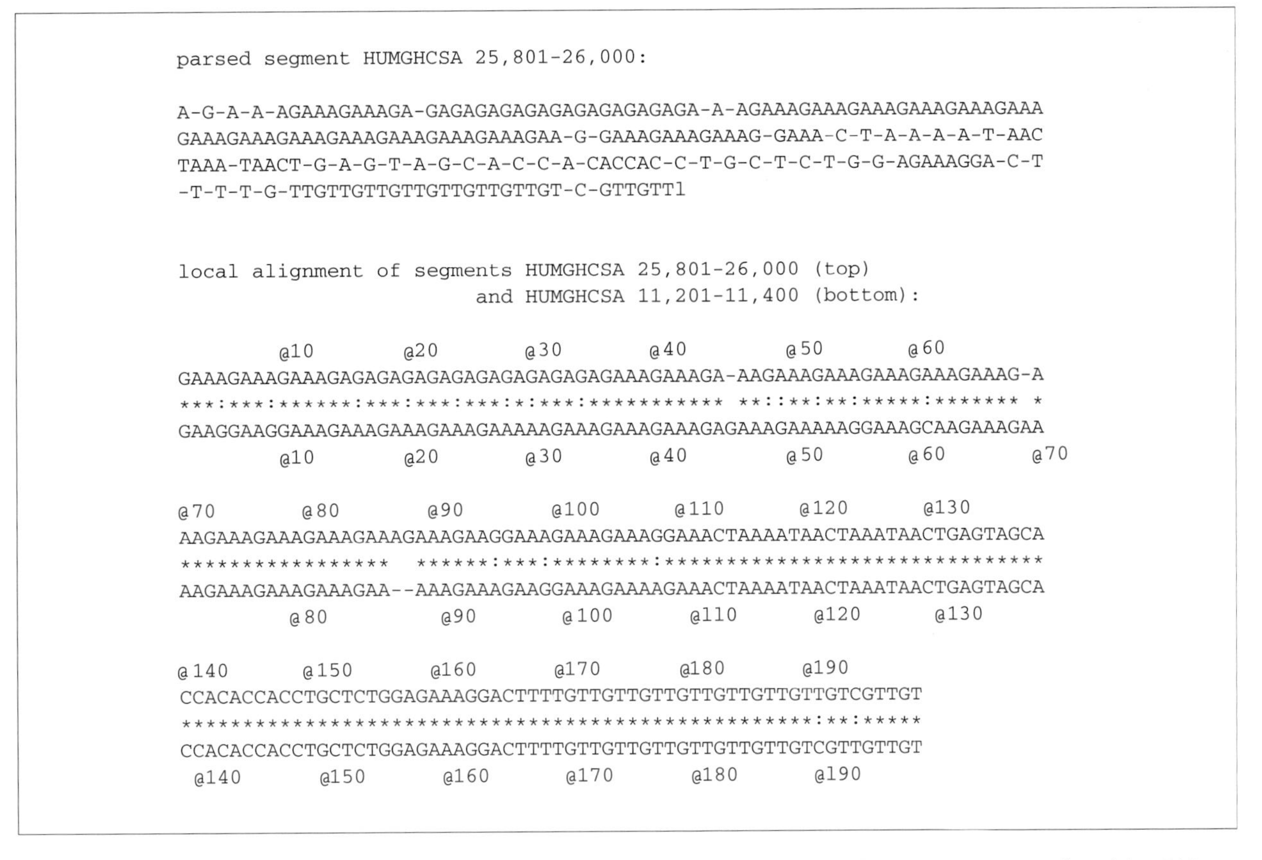

```
parsed segment HUMGHCSA 25,801-26,000:

A-G-A-A-AGAAAGAAAGA-GAGAGAGAGAGAGAGAGAGAGA-A-AGAAAGAAAGAAAGAAAGAAAGAAA
GAAAGAAAGAAAGAAAGAAAGAAAGAAAGAA-G-GAAAGAAAGAAAG-GAAA-C-T-A-A-A-A-T-AAC
TAAA-TAACT-G-A-G-T-A-G-C-A-C-C-A-CACCAC-C-T-G-C-T-C-T-G-G-AGAAAGGA-C-T
-T-T-T-G-TTGTTGTTGTTGTTGTTGT-C-GTTGTT1


local alignment of segments HUMGHCSA 25,801-26,000 (top)
                        and HUMGHCSA 11,201-11,400 (bottom):

         @10       @20       @30       @40        @50       @60
GAAAGAAAGAAAGAGAGAGAGAGAGAGAGAGAGAGAAAGAAAGA-AAGAAAGAAAGAAAGAAAGAAAG-A
***:***:******:***:***:***:*:***:*********** **::**:**:*****:******* *
GAAGGAAGGAAAGAAAGAAAGAAAGAAAAAGAAAGAAAGAAAGAGAAAGAAAAAGGAAAGCAAGAAAGAA
         @10       @20       @30       @40        @50       @60       @70

@70       @80       @90        @100      @110      @120      @130
AAGAAAGAAAGAAAGAAAGAAAGAAGGAAAGAAAGAAAGGAAACTAAAATAACTAAATAACTGAGTAGCA
****************  *****:***:********:*********************************
AAGAAAGAAAGAAAGAA--AAAGAAAGAAGGAAAGAAAAGAAACTAAAATAACTAAATAACTGAGTAGCA
          @80        @90       @100       @110      @120      @130

@140      @150      @160      @170       @180      @190
CCACACCACCTGCTCTGGAGAAAGGACTTTTGTTGTTGTTGTTGTTGTTGTCGTTGT
********************************************:**:*****
CCACACCACCTGCTCTGGAGAAAGGACTTTTGTTGTTGTTGTTGTTGTCGTTGTTGT
 @140      @150      @160      @170       @180      @190
```

Fig. 25.13 HUMGHCSA genomic region: parsing of segment 25,801–26 000 and its local alignment with segment 11,201–11 400. This homology between simple regions was discovered by the SMPL program. Note the (*GA*) * *N* pattern that is present around position 20 in the top segment but does not occur in the bottom segment. This pattern may have occurred by duplication of the *GA* dimer, a frequently occurring mutation.

In either kind of encoding, one can think of the encoded sequence as being parsed into words that are replaced by pointers and into the letters that do not belong to such words. One may then represent the encoding of a sequence by inserting dashes to indicate the parsing. In the self-encoding example, the parsing is

A-G-T-C-AGT-TTT

while in the relative-encoding example the parsing is

G-ATTA-C-C-G-ATGAGC-TAAT

Note that there are many possible encodings. We will be particularly interested in the shortest ones. With every saved bit we improve significance twofold: the algorithmic significance method states that d bits can be saved by chance with probability at most 2^{-d} (refs 29 and 30). Concise self-encoding indicates that sequence is not random while concise relative encoding indicates homology.

To apply the algorithmic significance method, we need to count the number of bits that are needed for a particular encoding. We may assume that the encoding of a sequence consists of units, each of which corresponds either to a letter or to a pointer. Every unit contains a (log 5)-bit field that either indicates a letter or announces a pointer. A unit representing a pointer contains two additional fields with positive integers indicating the position and length of a word. These two integers do not exceed n, the length of the source sequence. Thus, a unit can be encoded in log 5 bits in case of a letter or in $\log 5 + 2 \log n$ bits in case of a pointer.

If it takes more bits to encode a pointer than to encode the word letter by letter, then it does not pay off to use the pointer. Thus, the encoding length of a pointer determines the minimum length of common words that are replaced by pointers in an encoding of minimal length.

Note that we do not need to actually construct encodings—it suffices to estimate the encoding lengths. Thus, we may assume that we have even more powerful decoding algorithms that would enable smaller pointer sizes. For further details on pointer sizes, see ref. 29.

The encodings of minimal length can be computed efficiently by a classical data compression algorithm [32]. We here focus on the algorithm for encoding one sequence relative to the other. The case when a sequence is self-encoded requires only a slight modification.

The minimal length encoding algorithm takes as an input a target sequence t and the encoding length p of a pointer and computes a minimal length encoding of t for a given source s. Since it is only the ratio between the pointer length and the encoding length of a letter that matters, we assume, without loss of generality, that the encoding length of a letter is 1.

Let n be the length of sequence t and let tk denote the $(n-k+1)$-letter suffix of t that starts in the kth position. Using a suffix notation, we can write t_1 instead of t. By $I_A(t_k \mid s)$ we denote the minimal encoding length of the suffix t_k. Finally, let $l(i)$, where $l \leq i \leq n$, denote the length of the longest word that starts at the ith position in target t and that also occurs in the source s. If the letter at position i does not occur in the source, then $l(i) = 0$. Using this notation, we may now state the main recurrence:

$$I_A(t_i \mid s) = min(1 + I_A(t_{i+1} \mid s), p + I_A(t_{i+l(i)} \mid s)$$

Proof of this recurrence can be found in ref. 32.

Based on this recurrence, the minimal encoding length can now be computed in linear time by the following two-step algorithm. In the first step, the values $l(i)$, $l \leq i \leq n$ are computed in linear time by using a directed acyclic word graph data structure that contains the source s [34]. In the second step, the minimal encoding length $I_A(t \mid s) = I_A(t_1 \mid s)$ is computed in linear time in a right-to-left pass using the recurrence above.

Section 5 **Other model systems**

Section 5

Introduction

Nigel K. Spurr

SmithKline Beecham Pharmaceuticals, New Frontiers Science Park, Harlow, Essex CM19 5AW, UK

Why study the genomes of other organisms? This question is often asked by clinicians and other medical researchers, and carries the implication that humans are the only species on which the limited funding available for medical research should be spent. However, as the recent completion of the DNA sequence of the genome of the yeast *Saccharomyces cerevisiae* shows, much information can be gained from the study of other organisms that will have direct applications to the study of human diseases.

Apart from the fact that some model organisms are of commercial importance in their own right, the major reasons for analysing the genomes of other organisms can be summarized as follows.

- Gene function can usually be more easily determined in a simpler model organism, especially where transgenic techniques and controlled breeding can be implemented, and this may throw light on the function of the gene in humans.
- The regulation of gene expression can also be more easily studied through transgenic techniques and induced mutations.
- Comparison of the genomes of different organisms will throw light on the evolution and conservation of gene function.
- Some organisms can provide models for human disease.

One of the key reasons for studying simpler model organisms is to determine a basic set of eukaryotic functional genes. These can then act as a reference set of genes against which those of other plants and animals and of humans can be compared. Such comparative evolutionary studies will also reveal whether gene organization is conserved between particular species and, in addition, the role and position of introns and other chromosome elements can be determined and compared. This may give a greater insight into the role of such noncoding sequences in the control of transcription and gene expression. Finally, the pattern and timing of gene expression during development can be more easily studied in simpler organisms.

Chapters 26–34, which make up this section of the *Handbook*, give an insight into the progress of genome projects in a wide range of species that are either important as laboratory models or of commercial importance. The model organisms include the mouse (Chapter 26, G. Argyropoulos & S.D.M. Brown), the fruit fly (*Drosophila melanogaster*) (Chapter 28, R.D.C. Saunders), the nematode worm (*Caenorhabditis elegans*) (Chapter 29, J. Sulston, R. Waterston and members of the *C. elegans* Genome Consortium), and the bacterium *Escherichia coli* (Chapter 31, M. Masters & G. Plunkett). Model plants are represented by *Arabidopsis thaliana*

(Chapter 33, M. Delseny & R. Moore). Organisms of commercial importance include the brewing and baking yeast (*Saccharomyces cerevisiae*) (Chapter 30, H. Feldmann) and rice (*Oryza sativa*) (Chapter 34, T. Sasaki *et al.*). In addition, Chapter 27 (R. Vile) reviews progress in somatic gene therapy in humans.

There are various reasons for the concentration of the genome communities on these organisms. The mouse is a mammal like ourselves, with many genes homologous to those of humans, and the study of its genome has closely paralleled the Human Genome Project. There are numerous models of human inherited diseases in the mouse, and the ability to construct transgenic mice with precisely targeted gene knockouts now makes it even more useful.

The yeast *S. cerevisiae* is of commercial importance and has also been one of the most favoured model organisms for studying the regulation and control of the cell cycle in eukaryotes. The recently completed genome DNA sequence is the first complete genome sequence of a eukaryote. The genetics and development of the nematode worm *C. elegans* have also been intensively analysed and this simple multicellular organism will be the first multicellular eukaryote to have its genome sequence completed (by 1997). The fruit fly *Drosophila* has been used as a genetic model for more than 100 years, and there is a great wealth of information available on its genes and mutations. These have provided important insights into the genetic basis of development of multicellular organisms in general. Mutant and transgenic flies can readily be generated and techniques are available for the rapid identification of the genes affected.

The bacterium *E. coli* was for many years the principal experimental organism for most molecular biologists. In recent years its dominance in the field has lessened as it has become possible to study and manipulate eukaryotic genes at the molecular level, and interest has shifted from prokaryotes to eukaryotes. However, its genome sequence is still of great interest because of the wealth of knowledge on gene function and expression in *E. coli*, and for comparison with other prokaryote genomes now being sequenced. *E. coli* is also still of interest as a pathogenic bacterium, as witnessed in Japan in 1996 with the outbreak of food poisoning contracted from strain 0157 and affecting over 6000 schoolchildren.

Chapter 32 (P. Jauhar) outlines the particular problems associated with studying plant genomes—for example, the hybrid origin and polyploidy characteristic of many crop species such as wheat. This is a valuable insight into an area unfamiliar to many genome scientists. Of the two plant species discussed in detail in this section, *A. thaliana* (Chapter 33) is a small insignificant weed, but unlike many plants it has a small compact genome and has been chosen by the plant community as an ideal organism for developmental genetics, gene mapping and genome sequencing. It is simple and quick to grow and many mutants are available for study. Rice, the other plant species represented here (Chapter 34), is an important staple crop, and has been chosen by Japan and other countries as a key organism for study due to its importance as a food and as a model for the family of grassses as a whole, to which many staple crops belong.

Chapter 26 Mouse genome mapping

G. Argyropoulos & S.D.M. Brown*

Department of Biochemistry and Molecular Genetics, St Mary's Hospital Medical School, Imperial College of Science, Technology and Medicine, London W2 1PG

**MRC Mouse Genome Centre, Harwell, Oxfordshire OX11 0RD, UK*

26.1 Introduction

As the Human Genome Project progresses towards the determination of the complete DNA sequence of the human genome, a parallel project, with similar aims, is under way for the mouse genome [1]. The role of the mouse as a model organism for studying genetic disorders in humans is long established, and the creation of detailed genetic and physical maps of its chromosomes will facilitate its future use.

A number of so-called model organisms have played an important role in the Genome Project. The construction of complete genetic and physical maps of model organisms presents an unparalleled opportunity for comparisons of related genomes that will contribute significantly to our knowledge of:

1 the function of genes and genomes and;

2 the mechanics of evolutionary changes of genomes.

The mouse, with its short breeding cycle of around 8 weeks (gestation time, 21 days; length of time to sexual maturity, 4–6 weeks), represents an ideal organism for the study of mammalian genetics and genome mapping.

This chapter introduces the mouse as a laboratory organism, its capacity as a tool for the study of complex mammalian genetic systems, and the role it is playing in the dissection of genome organization. In Section 26.2 we present a short background to the laboratory mouse and its potential as an experimental model in mammalian genetics. In Sections 26.3 and 26.4, we examine the strategies used to map the mouse genome, providing examples of specific genome mapping approaches.

26.2 The laboratory mouse

26.2.1 Mutations in the laboratory mouse

A wide range of mutations have been identified in the laboratory mouse including gene defects affecting skin texture, coat colour, tail shape and length, the skeleton, the eye, the inner ear, neurology and neuromusculature as well as genes affecting behaviour and reproduction (Table 26.1). Over 100 mutations have been identified as having some neurological or neuromuscular effect and another 100 mutations have been shown to affect the skeleton. In addition, a large number of mutations have been identified affecting the function of a variety of proteins including enzymes and cell-surface antigens. Many mutations that have been characterized appear to be monogenic. On the other hand, some strains of mice appear to be carrying defects that are polygenically determined. Such polygenic mice include strains of mice predisposed towards diabetes and obesity. The vast array of mouse mutations represents a powerful collection of animal models for the study of human disease. As well as spontaneous and induced mutations, many additional mutations have been produced by gene knock-out (see Appendix VIII).

26.3 Strategies for mapping the mouse genome

For most mutations the underlying gene carrying the altered DNA sequence is not known. The Mouse Genome Project aims to provide the resources needed to map and identify the mutated gene. The

Table 26.1 Mouse mutants and loci grouped according to a variety of phenotypic classes.

Phenotype	Number of mutant loci
Colour and white spotting	81
Skin and hair texture	94
Skeleton	134
Tail and other appendages	94
Eye	87
Inner ear and circling	45
Neurological and neuromuscular	126
Other behavioural	19
Haematological	56
Endocrinological, hormonal, growth and obesity	98
Reproductive organs, sterility	45
Defects of viscera	48
Immune defects	21
Homozygous lethality or sublethality	115

The number of mutants known in each class is given. Adapted from ref. 14.

overall strategy for the provision of complete genetic and physical maps of the mouse genome involves a number of elements including:

1 isolation and characterization of cloned DNA fragments—DNA markers—from the mouse genome;

2 establishing suitable genetic crosses with high resolution that allow us to complete high-density genetic maps of DNA fragments across the entire mouse genome;

3 linking our mapped DNA fragments into a complete physical map of the mouse genome that gives us access to all of its incumbent DNA sequences.

26.3.1 Genetic mapping of the mouse genome: methodologies

26.3.1.1 DNA markers

The pivotal feature of DNA markers used in mouse genetic mapping is their ability to detect some kind of sequence variation between parental strains that allows us to follow the segregation of the polymorphic locus in the appropriate genetic cross (see below). The methods used for detecting sequence variation vary according to the type of marker employed. The main approaches used are as follows.

1 *Analysis of the segregation of a restriction fragment length variant (RFLV) detected between the parental strains used in the cross (see Fig. 26.1b).* This analysis is applicable to both coding sequences as well as random, nongenic cloned fragments, though in the former case it may be more difficult to detect an RFLV in a conserved, coding region of the genome.

2 *Analysis of segregation of simple sequence length polymorphisms.* Mammalian genomes contain frequent short tandem repeat sequences called *microsatellites* (refs 2–4, see also Chapter 5). These microsatellites are often composed of a short array of dinucleotide repeats. Dinucleotide repeats have been shown to vary in length even between closely related inbred laboratory strains of mice. Length variation in a microsatellite is known as a simple sequence length polymorphism (SSLP). Such SSLPs can be used as markers to detect the segregation pattern of a locus of interest (see Fig. 26.1c). Moreover, microsatellite sequences are present at frequent intervals in the mouse genome with one microsatellite every 20 kb or so, representing a total of 150 000 microsatellites in the entire mouse genome [3].

3 *Analysis of the segregation of variants of interspersed repeat sequence polymerase chain reaction (IRS-PCR) products.* PCR of genomic DNA using primers to interspersed repeat sequences in mammalian genomes allows the recovery of PCR products to the sequences between closely spaced repeat sequences (see Chapters 10 and 14). In the mouse, two major short interspersed repeat sequence families have been used for the generation of IRS-PCR products—the B1 and B2 repeat families, both present in around 50 000–100 000 copies per genome [5]. In humans, the short interspersed repeat sequence Alu (equivalent to B1 in the mouse) has been used to generate IRS-PCR markers in a similar fashion (see Chapters 9 and 10 for protocols). IRS-PCR product length can vary between different strains or species of mice, allowing us to score their segregation in genetic crosses based upon a length polymorphism. Alternatively, depending upon variation in repeat location between strains or species of mice, presence/absence polymorphisms may be observed that may be amenable to segregation analysis depending upon the nature of the genetic cross employed (see ref. 6 and below). Presence/absence polymorphisms are the most common variant observed between species.

26.3.1.2 Interspecific and intersubspecific genetic crosses

The interspecific back-cross [7] has become one of the most important genetic tools for the construction of genetic maps spanning the mouse genome (Fig. 26.1a). A laboratory strain of mouse is crossed to a wild species of mouse, *Mus spretus*. Fertile female progeny can be produced which are then back-crossed to either the parental laboratory strain or the *M. spretus* strain. Thus, back-cross progeny are produced that segregate DNA sequence variation derived from the laboratory strain or *M. spretus*.

Sometimes a subspecies (*M. castaneus*) related to the laboratory mouse is used, in which case the cross is known as intersubspecific [3]. The laboratory strain used in the back-cross may be carrying an interesting mutation and the back-cross progeny produced will also be segregating the mutation of interest. Thus, interspecific back-crosses not only produce the resources for mapping of DNA markers but also may provide the necessary resources for mapping and localizing mutations (see below).

26.3.1.3 Advantages of an interspecific back-cross

Laboratory strains of mouse and the wild species *M. spretus* separated some 2–3 million years ago. *M. castaneus* and laboratory strains diverged somewhat later. However, the DNA sequence of the two parental strains in both interspecific and intersubspecific crosses is highly diverged, making it relatively easy to identify sequence variation between the parental genomes for any DNA marker.

Over 90% of microsatellites show SSLPs between laboratory mice and *M. spretus* and this percentage is not much lower when *M. castaneus* and other laboratory strains are compared [3]. In addition, there is a remarkably high rate of variation between standard laboratory inbred strains of mice. Around 50% of microsatellites demonstrate scorable SSLPs between inbred strains of mice [2,3]. Thus, it is possible to carry out considerable genetic analysis of DNA markers even in genetic crosses employing standard laboratory strains. When mapping mouse mutations, it is important to be aware that the penetrance or expressivity of the mutation (see Chapter 1) may vary according to the genetic background of the mouse strains employed in the cross. For some mutations, it may be prudent to employ a back-cross strategy that does not involve a wild mouse species.

26.3.1.4 Analysing a mouse back-cross

DNA from the back-cross progeny is analysed for a number of types of sequence variation including RFLVs or SSLPs, as described previously. For RFLVs and SSLPs, each DNA marker can be defined as a sequence-tagged site (STS) since at least a portion of its sequence is usually known. The sequence data are readily transferred to another laboratory and the DNA marker can be reproduced by manufacturing primers for each STS and the application of PCR. The high DNA sequence variation that characterizes the parental strains of interspecific and intersubspecific crosses means that it is feasible to analyse interspecific back-cross progeny for all available DNA markers. The ability to analyse many DNA markers in one back-cross in a multipoint fashion allows us to order our markers by minimizing the number of observed recombination events [8]. This

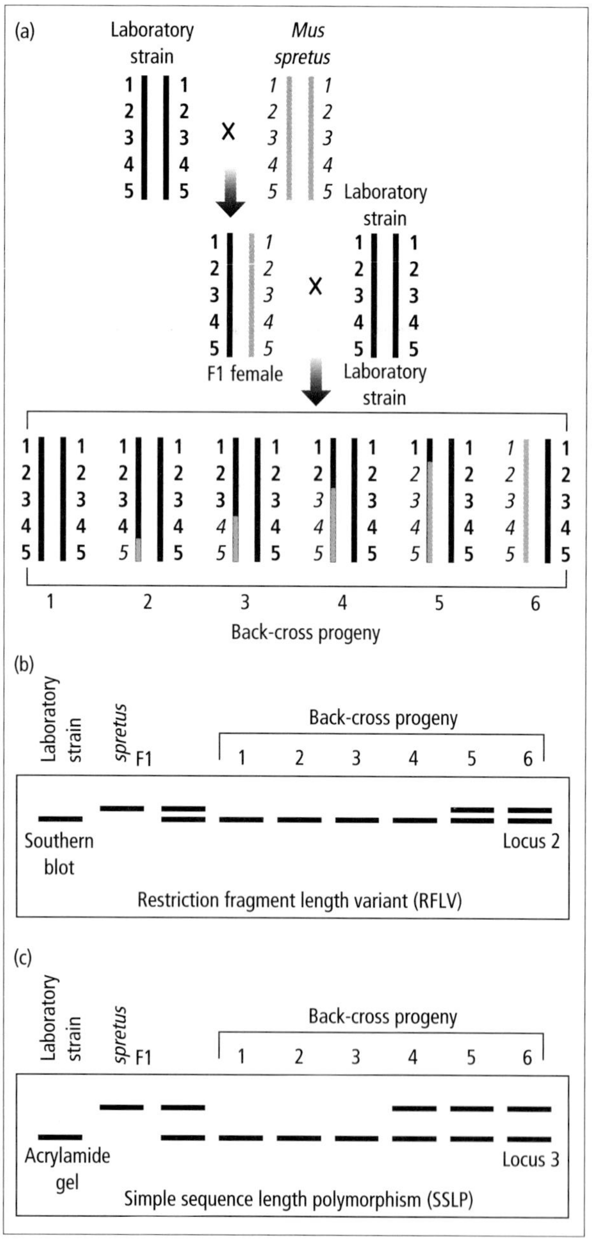

Fig. 26.1 Genetic mapping using a mouse interspecific back-cross. (a) Construction of a genetic map of five markers by multipoint analysis on one mouse chromosome. Laboratory strain sequence variants are shown in bold type and *Mus spretus* strain sequence variants are shown in italics. A number of back-cross progeny are recovered derived from recombination between the laboratory strain chromosome and the *M. spretus* chromosome in the F1 female. The genetic order of markers on the chromosomes is determined by a haplotype analysis that minimizes the number of observed recombination events across the back-cross progeny set. Changing the order of markers would necessitate an overall increase in recombination events observed and, in addition, the appearance of triple recombinants, which are usually very rare. (b, c) Diagrammatic representation of the analysis of two markers from our interspecific back-cross (loci 2 and 3, see part a). Locus 2 (part b) has been analysed using a RFLV. DNA from each back-cross progeny is digested with the appropriate restriction enzyme, fragments separated on an agarose gel and transferred to a nylon membrane. The membrane is then hybridized with radiolabelled marker 2 and following exposure of the membrane to autoradiographic film, the segregation of the RFLV can be read. Locus 3 (part c) is analysed by following the segregation of an SSLP. The SSLP is visualized by amplification of the variant microsatellite using radioactively labelled primers and PCR followed by running the products on an acrylamide gel. The gel is exposed to autoradiographic film allowing the segregation of the SSLP can be read. For locus 2, back-cross progeny 4 has not inherited a *spretus* variant from the F1 female but has inherited a *spretus* variant for locus 3. Thus, back-cross progeny 4 is recombinant between locus 2 and locus 3 (see part a).

is sometimes known as haplotype or pedigree analysis (see Chapter 1).

The methodology used in the interspecific back-cross means that mouse genetic mapping differs qualitatively from that usually appropriate to the human genome and human genetic mapping (see Chapters 1–3). The emphasis is on multipoint mapping within a single pedigree, which allows the mouse geneticist to construct high-integrity, high-resolution and ordered genetic maps of DNA markers across most of the mouse genome.

26.3.1.5 Alternatives to a back-cross analysis

One alternative mapping resource in the mouse that has been extensively used in the past, but less so recently, are the recombinant inbred (RI) strains [9]. RI strains are recovered by mating two inbred laboratory progenitor strains to obtain first an F1, and then an F2 generation. This is followed by rounds of brother–sister matings (Fig. 26.2). A number of inbred strains are established that carry one or more recombination events between the parental mouse chromosomes. Analysis of parental sequence variation for DNA markers from each chromosome in a number of RI strains establishes a strain distribution pattern (SDP). Analysis of new DNA markers through a variety of RI strains, comparison of the SDPs and computation of linkage to already mapped markers enables the mouse geneticist to arrive at an accurate chromosomal assignment (Fig. 26.2). The analysis is formally similar to the haplotype or pedigree analysis carried out for interspecific and intersubspecific back-crosses. However, there are two limitations. First, RI strains carry a limited number of recombinants on each chromosome and therefore limited resolution. Second, their derivation from closely related laboratory inbred strains hinders the discovery of sequence variation for some DNA markers (see above).

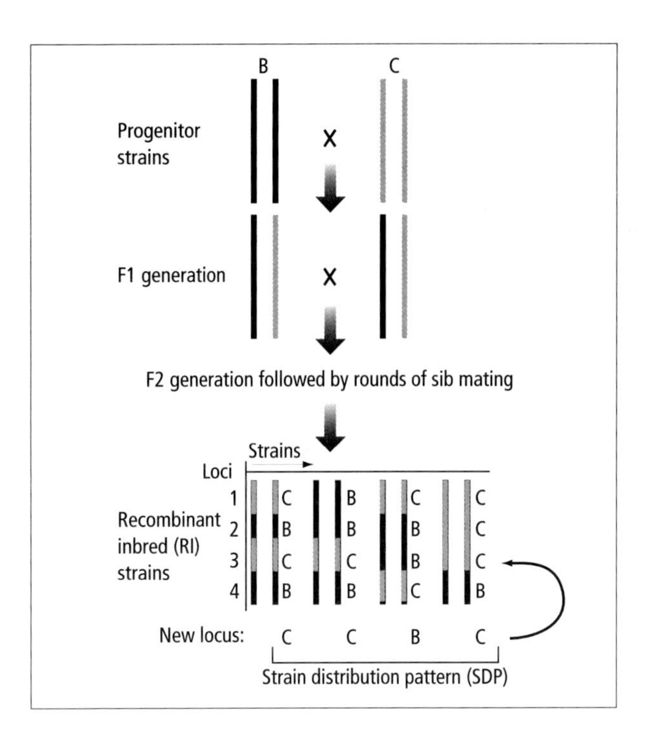

Fig. 26.2 The construction and analysis of recombinant inbred (RI) strains. The principle behind the mapping of new loci in RI strains is illustrated. A new locus is assigned by comparing the strain distribution pattern (SDP) observed with those of previously mapped loci. For the chromosome illustrated, the SDP has been determined at four loci (1–4) in four RI strains. For the new locus, the SDP was determined in each of the four RI strains. The SDP corresponds to that of locus 3, indicating that the new locus maps close to locus 3 on this chromosome.

26.3.2 Genetic mapping of the mouse genome: current status of the mouse genetic map

A recent report [4] describes an integrated map of over 7000 STSs, including genic probes and microsatellites, covering the entire mouse genome. Nearly 90% of the markers on this map were readily analysable SSLPs.

26.3.2.1 The microsatellite map of the mouse genome

The first microsatellites from the mouse genome were developed by Todd and colleagues and used to construct a preliminary map of the mouse genome using RI strains [2]. Subsequently, a map of 317 microsatellite markers was developed by Dietrich and co-workers [3]. $(CA)_n$ microsatellites were identified and sequenced from total genomic C57BL/6J M13 clone libraries. Each microsatellite was analysed through a variety of inbred strains as well as *M. spretus* and *M. castaneus* DNA. SSLPs were analysed through the 46 F2 progeny arising from an intercross of an inbred obese mouse strain and *M. castaneus*. This mapping panel provides 92 meioses, giving a genetic resolution of 1 crossover per 1.1 cM. In addition, the available SSLPs were analysed through RI lines to anchor the microsatellite map to the known genetic map.

The Massachusetts Institute of Technology (MIT) microsatellite map was constructed using the MAPMAKER linkage package. The 317 SSLPs mapped initially had an average spacing of 4.3 and covered 99% of the mouse genome. The latest update of the microsatellite map [4] contains 6580 SSLPs with an average spacing of around 0.2 between markers. This analysis clearly provides a microsatellite map of the mouse genome at high density, but of only intermediate resolution. Many microsatellites remain unresolved and unordered.

26.3.2.2 Further integration of the microsatellite map with the gene map of the mouse

Copeland and Jenkins and colleagues have in parallel developed a dense gene map of the mouse genome, principally by the analysis of cDNA RFLVs through an interspecific back-cross [10,11]. Around 800 gene loci have been ordered across all 20 chromosomes. Efforts have been made to integrate the gene and microsatellite maps by the analysis of over 1000 SSLPs through the interspecific back-cross used for mapping the genic markers [4]. The result provides an integrated map of the mouse genome but, nevertheless, again a map of only intermediate resolution, where many markers remain unresolved and unordered.

26.3.3 Towards completing the mouse genetic map: a high-resolution genetic map of the mouse genome

There are considerable advantages to establishing a high-resolution genetic map where the bulk STSs are resolved and ordered. Such a map provides a strong basis for the construction of high-integrity physical maps of the mouse genome (see below). Interspecific back-crosses can provide very fine genetic resolution. For example, 1000 back-cross progeny offers genetic resolution at the 0.3 level with 95% confidence. The DNA content of the mouse haploid genome is 3×10^9 base pairs. Thus, 0.3 represents $\approx$0.5 Mb of DNA. This is a level of resolution approachable by physical mapping techniques (see Section 26.4). An STS genetic map of the mouse genome approaching 0.5 Mb resolution will be completed by the end of 1996 and will provide a template for the global physical mapping of the mouse genome. The intermediate resolution map of over 6000 microsatellites that is now complete (see ref. 4 and above) is a critical resource from which to develop a high-resolution map.

A collaborative programme is under way in Europe to develop a high-resolution microsatellite map of the entire mouse genome [8]. This programme—the European Collaborative Interspecific Back-cross (EUCIB) programme—aims to complete a microsatellite map of the mouse to 0.3 resolution. Nine hundred and eighty-two interspecific back-cross progeny have been generated, of which 501 mice were from F1 females back-crossed to *M. spretus* and 481 mice were from F1 females back-crossed to C57BL/6. All 982 back-cross progeny were analysed for over 70 anchor loci distributed across all chromosomes. Completion of the anchor map has allowed the identification of pools of animals recombinant in individual chromosome regions and enables the high-resolution mapping of new markers in a chromosome region by their analysis through a limited collection of animals. Each SSLP that has already been mapped to intermediate resolution and to a particular chromosome region is analysed through the relevant recombinants in the EUCIB back-cross panel thus developing a high-resolution, ordered microsatellite map of the mouse.

As half of the EUCIB back-cross was generated by back-crossing to *M. spretus*, it is possible to use the EUCIB resource for the rapid, high-resolution genetic mapping of IRS-PCR products [6]. Many of the IRS-PCR products generated in laboratory strains of mice do not detect a product when hybridized to IRS-PCR products of *M. spretus*. In *M. spretus* the arrangement of repeat sequences can be different, thus leading to a presence/absence variation between the species. This allows for a relatively rapid system for scoring the segregation of IRS-PCR products through back-cross progeny. IRS-PCR products are hybridized to gridded arrays of IRS-PCR products from the progeny derived by back-crossing to *M. spretus*. Presence or absence of signal for each back-cross progeny is rapidly scored and as with microsatellite or genic markers, linkage and haplotype ordering can be computed. IRS-PCR markers are set to make an increasing contribution to the high-resolution maps of the mouse genome.

The EUCIB programme is supported by the MBx database [8]. The MBx database was constructed using distributed client/server database Sybase, Sybase application development tools (APT), and the C and XView programming languages. MBx stores all locus, probe and SSLP data. Allele data at each locus is presented as a scrollable matrix on screen. MBx can compute genetic linkage between loci and, in addition, can automatically perform the necessary haplotype analysis by minimizing the number of observed recombinants across any chromosome region in order to derive genetic order of loci. Furthermore, MBx can abstract all mice carrying recombinants in any chromosome region. Such information when downloaded to robots is important for the automated and error-free selection of the correct DNAs for subsequent high-resolution microsatellite mapping in a particular chromosome region. Map information is displayed through a modified front-end version of the ACeDB developed for *Caenorhabditis elegans* (see Chapter 29).

In Fig. 26.3, the latest high-resolution microsatellite map for proximal mouse X chromosome (A. Haynes, N. Quaderi & S.D.M. Brown, unpublished data) is presented in ACeDB format. The ACeDB display is a multimap format demonstrating both the MIT microsatellite map at low resolution and the

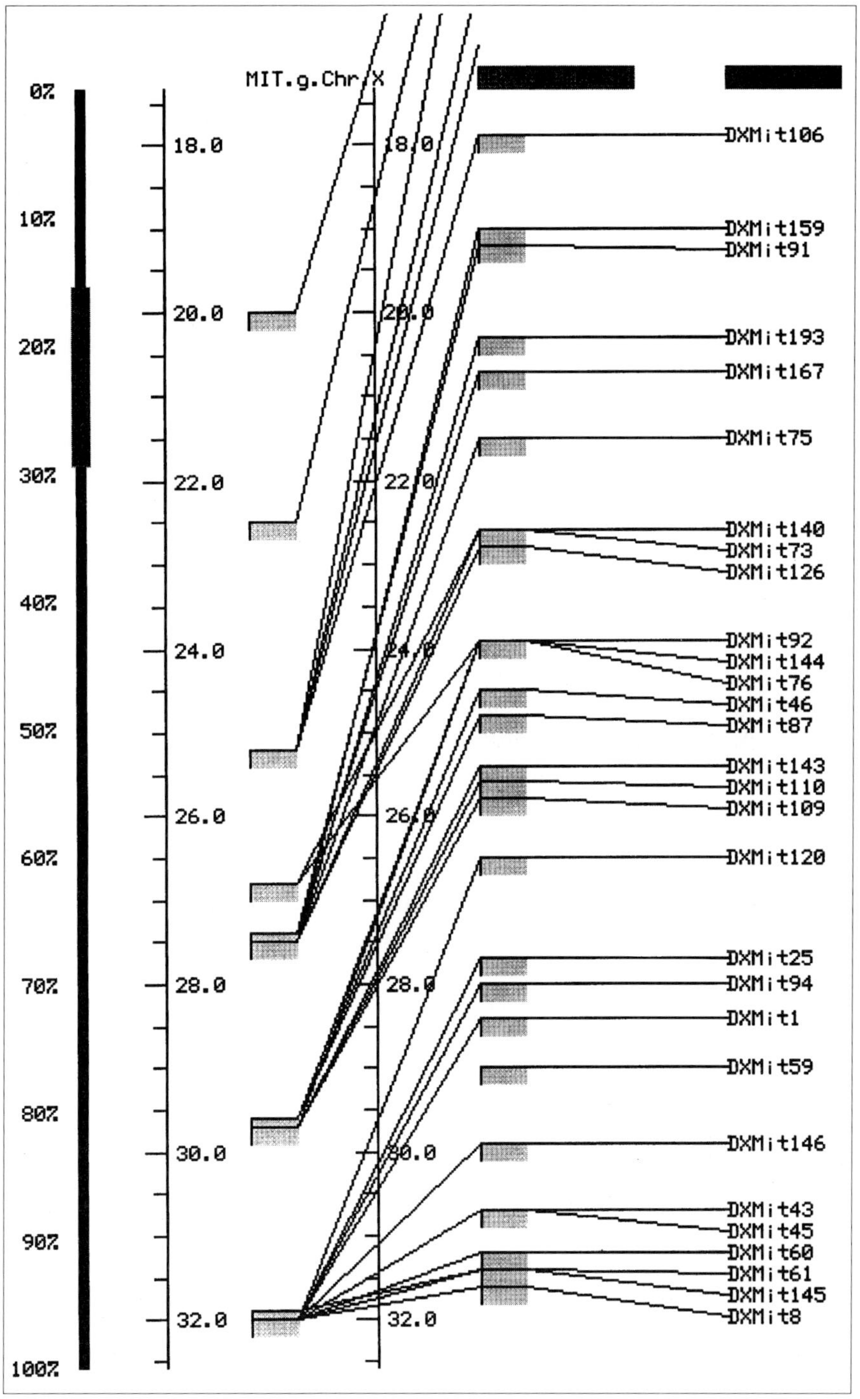

Fig. 26.3 High-resolution microsatellite map of the mouse X chromosome. An ACeDB database MultiMap display is shown for microsatellite maps of the mouse X chromosome. On the left is shown the low-resolution MIT map (MIT.g.Chr X) (see text). On the right is the high-resolution microsatellite map generated on the EUCIB back-cross (MBX.g.Chr.X). ACeDB Multimap displays connecting lines between identical loci mapped in different crosses allowing both comparisons of gene order as well as comparisons of resolution. The higher resolution of the EUCIB map is immediately apparent.

high-resolution map completed on the EUCIB back-cross. The increase in resolution is immediately apparent.

26.3.4 Accessing mouse genetic map information

26.3.4.1 The Mouse Genome Database and the Encyclopedia of the Mouse Genome

Mouse genetic mapping information from centre programs, collaborative programs and single laboratory efforts worldwide is regularly transferred to the Mouse Genome Database (MGD) at the Jackson Laboratory (Bar Harbor, ME, USA) and presented in the latest issue of the *Encyclopedia of the Mouse Genome*—a tool for the presentation of Mouse Genome and related information. MGD contains mouse locus information; genetic mapping data; mammalian homology data; probes, clones and PCR primers; genetic polymorphisms; the Mouse Locus Catalogue (gene descriptions) and characteristics of inbred strains. MGD is available over the World Wide Web as are various other services offered by the Jackson Lab. Information can be found on the MGD WWW home page at http://www.informatics.jax.org.

26.3.4.2 Updates on the Whitehead/MIT Centre for Genome Research microsatellite maps
Information on the latest releases of the MIT microsatellite map can be obtained via the World Wide Web: http://www-genome.wi.mit.edu.

26.3.4.3 The MBx database: accessing the high-resolution microsatellite map
The latest EUCIB maps and mapping data held on the MBx database are available the World Wide Web: http://www/hgmp.mrc.ac.uk/MBx/MBxHomepage.html.

26.3.4.4 Other sources of mouse genetic map information
There are a number of other routes to access current mouse genetic map information. The genetic maps are collated, examined and updated yearly by chromosome committees. yearly chromosome committee reports that include the current genetic maps are published in the journal *Mammalian Genome* (e.g. ref. 12). These reports form much of the basis for updates of MGD.

Mouse Genome [13] is a specialist mouse genetics publication with four issues per year that contains updates on gene names, chromosomal localization of genes and new loci, updates on nomenclature as well as short communications on genetic and physical maps. The first issue each year deals with linkage maps and maps of chromosomal anomalies. The second issue contains listings of gene symbols and chromosome anomalies. The third issue alternates between (a) information on the history and location of inbred strains, congenic strains and recombinant inbred strains and (b) lists of DNA clones and probes. The fourth issue has listings of RFLPs. Information about subscribing to *Mouse Genome* can be obtained via e-mail: j.peters@har.mrc.ac.uk.

Finally, at regular intervals, the publication *Genetic Variants and Strains of the Laboratory Mouse* [14] is updated and reissued. This contains extensive listings and discussions of all that is valuable for the mouse geneticist, including the mouse locus catalogue, linkage homologies between mouse and human, nomenclature rules, chromosomal anomalies, inbred strain listings amongst others. The third edition has just been published.

26.3.4.5 Nomenclature in the mouse
Refer to recent editions of *Mouse Genome* [13] for the latest revisions on the extensive rules that govern nomenclature in mouse genetics.

26.3.5 Using the mouse genetic map

A high-resolution, ordered genetic map of the mouse genome incorporating highly variable DNA markers has a number of important uses.

26.3.5.1 Detailed genetic mapping of mouse mutations as a prelude to positional cloning
The current genetic maps provide the necessary resources for the cloning of genes associated with the plethora of interesting mutations available in the mouse. To access the gene underlying a mutation it is necessary to carry out a specific cross that segregates the mutation of interest. An interspecific or intersubspecific back-cross carrying the mutation is produced as described in Section 26.3.1—often with 1000 or more back-cross progeny—allowing a high-resolution genetic analysis of the mutation's position. Back-cross progeny are analysed not only for the segregation of the mutation but also for sequence variation of DNA markers from the vicinity of the mutation. The genetic analysis allows us to determine our most closely flanking DNA markers; an important first stage in localizing and ultimately accessing the mutation through further physical mapping (see Section 26.4). In an interspecific back-cross of 1000 or more progeny, an STS nonrecombinant with the mutation would lie within 0.5 Mb of the mutated gene. This is the first stage in a positional cloning strategy based upon the powerful use of the high-resolution genetics available in the mouse.

26.3.5.2 Identification of candidate genes for mouse mutations
Once a new gene has been added to the genetic map, its position can be compared with the map position of known mutations. Depending upon what is known of the function of the gene and the phenotype of the mutation, the newly mapped gene may be a plausible candidate for the site of the mutation and this can be tested by direct sequence analysis. For example, the β-subunit glycine receptor subunit gene (*Glrb*) maps to mouse chromosome 3 in the vicinity of the spastic mutation (*spa*). It was found that *Glrb* mRNA was reduced throughout brains of *spa* mice, apparently as the result of a LINE-1 element insertion in intron 6 of the *Glrb* gene [15].

26.3.5.3 Genetic mapping of polygenic loci
The rapid expansion in the density of microsatellite markers on the mouse genetic map has been an

important advance in helping investigators to identify the map position of loci involved in polygenic traits in the mouse. It is important to emphasize that the mouse is a very important model organism for the analysis of traits that are determined polygenically. The analysis of polygenic traits is often more straightforward in the mouse or other laboratory organisms because defined genetic strains are available. For example, a back-cross between the non-obese diabetic (NOD) mouse that is predisposed to Type I insulin-dependent diabetes and normal laboratory strains provides back-cross progeny segregating the diabetes phenotype. Analysis of the back-cross progeny with microsatellites encompassing the whole of the mouse genome allows the investigator to identify linkage association between individual microsatellites and susceptibility loci. In the mouse, a number of susceptibility loci on a number of chromosomes have been mapped that appear to predispose to Type I diabetes [16].

26.3.5.4 Comparative genetic maps of the mouse and human genomes

Detailed mouse genetic maps are a powerful resource for comparison with the genome maps of other organisms, especially human. When the genetic maps of mouse and human are compared, it is found that there are many areas in the two genomes where gene content and gene order is conserved [17,18]. These regions are called conserved ordered segments. Given the density of genes mapped in common between mouse and human, it is believed that the bulk of conserved ordered segments between the two organisms have already been identified.

It is worth discussing three examples where conserved ordered segments between mouse and human can be of significant use.

1 *Identification of putative homologous mutations between the mouse and human genomes.* For example, human hereditary hyperefplexia, or startle disease (STHE), maps to distal chromosome 5q, and point mutations in exon 6 of *GLRA1*—the gene encoding the $\alpha1$ subunit of the glycine receptor (mouse homologue *Glra1*)—have been identified in patients from STHE families. STHE shows a similar phenotype to the mouse mutation spasmodic (*spd*) which maps on mouse chromosome 11 in a conserved linkage group with human 5q. The homology of *spd* and hyperefplexia was confirmed by the identification of mutants in the *Glra1* gene in *spd* mice [19].

2 *Identification of genes in the mouse genome that may be candidates for mutations in the human genome.* For example, in the mouse, the gene underlying the shaker-1 mouse deafness mutation was recently identified by a positional cloning route [20,21]. The *shaker-1* gene encodes an unconventional myosin molecule—myosin VII. This gene represented an attractive candidate for one form of the deaf–blind syndrome, Usher type 1b, in humans, which maps to human chromosome 11q13.5. The shaker-1 mouse mutant was known to map in a region of mouse chromosome 7 that forms part of a conserved ordered segment encompassing 11q13.5 in humans [22]. Indeed, it was shown that the myosin VII gene underlies Usher syndrome type 1b and a number of mutations have been identified in the myosin VII gene in affected families [23].

3 *Identification of genes mapped in the human genome that are potential candidates for mutations in the mouse genomes.* One recent example is the identification of a gene on the mouse X chromosome that carries the xid (X-linked immunodeficiency) mutation that causes defects in B-cell development. In humans, the X-linked agammaglobulinaemia mutation (XLA), which is also associated with disorders in B-cell development, lies in the same conserved linkage group as the mouse xid mutation. The gene affected in XLA encodes a cytoplasmic tyrosine kinase—Bruton's tyrosine kinase, Btk [24]. The mouse *Btk* gene maps close to the xid mutation on the mouse X chromosome and it has been shown that in xid mice the *Btk* gene carries a point mutation [25,26].

26.4 Physical mapping of the mouse genome

Genetic mapping provides an ordered array of STSs along the chromosome. A physical map is an ordered, overlapping array of DNA clones covering an entire chromosome or chromosome region. An overlapping set of DNA clones is known as a contig. The genetic map provides a framework upon which the physical map can be built by linking adjacent STSs on the genetic map into a set of overlapping DNA clones.

26.4.1 Converting the genetic map into a physical map

26.4.1.1 Clone resources for physical mapping

Yeast artificial chromosome (YAC) clones [27] have played a pivotal role in the construction of physical maps of mammalian genomes. YAC clones can contain large inserts up to 1 Mb or more, thus allowing considerable coverage of any chromosome region with a small number of clones. The use of YAC clones for coverage of large or entire chromosome regions is most advanced in human genomics studies [28] (see Chapter 15). There are, however, two drawbacks to YAC libraries. First, most YAC libraries, including mouse YAC libraries, demonstrate a high frequency of chimaeric clones—clones carrying insert material from noncontiguous regions of the genome [29]. Second, some clones are markedly unstable demonstrating deletions or other rearrangements of material. Nevertheless, these problems are outweighed by the advantages of very large insert size. Five mouse YAC libraries are currently available (Table 26.2). One of these libraries, the St Mary's library, was constructed in a recombination-deficient (*rad52*) strain of yeast—the first such YAC library to be constructed—in order to lower chimaerism rates and improve stability of YAC clones [30].

A number of other vector systems that have the capacity for large insert size have been developed, including P1 [31], PACs [32] and bacterial artificial chromosomes (BACs) [33]. The P1 cloning system can accommodate inserts up to around 100 kb (see Chapter 15), while PACs and BACs have a larger capacity, of around 130–150 kb and >300 kb, respectively. These cloning systems, although unable to accommodate the very large insert sizes of YAC clones, have considerable advantages in terms of clone stability and low chimaerism rates within libraries. Mouse P1 [34] and BAC libraries are available.

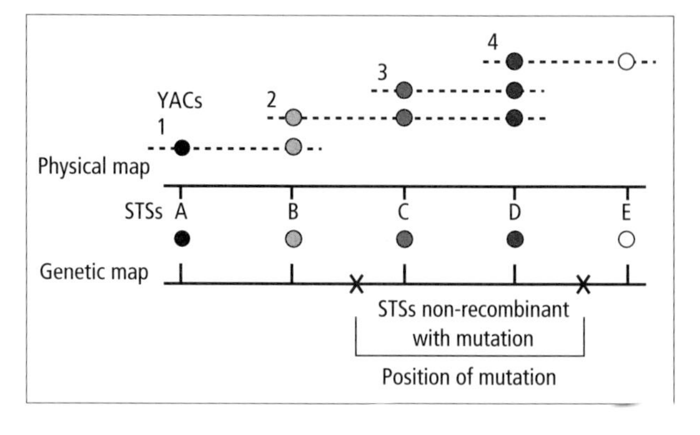

Fig. 26.4 STS YAC contigging to create and analyse physical maps across the mouse genome. Five STSs (A–E) which have been shown by genetic mapping to lie close to each other on a chromosome are used to screen for overlying YAC clones. STSs shared by the four YACs allow the assembly of a YAC contig (STS YAC content mapping). An overlapping series of YAC clones that provides access to all the DNA sequences is generated in that region. In addition, the STSs may be analysed in an interspecific back-cross segregating for a mutation in that region. Some STSs (A, B and E) may be recombinant with the mutation in back-cross progeny and represent STSs flanking the region of the mutation; other STSs may be non-recombinant (C and D). The genetic analysis defines a region on the overlying YAC physical map where the mutation is likely to lie. The YAC contig in this region provides the necessary cloned sequences to test for and identify the relevant gene sequences carrying the mutation.

26.4.1.2 Construction of physical maps by STS content

Screening of mouse YAC libraries using PCR or filter-based methods allows us to identify YAC clones covering adjacent STSs on the genetic map (Fig. 26.4). YAC clones carrying an adjacent series of STSs can be linked into an overlapping series of clones called a YAC contig, a procedure often referred to as *STS content mapping*. It can be expected that between 20 000 and 30 000 STSs might be needed to establish a robust physical map of the mouse genome with almost complete coverage.

PCR screening is the most rapid and often the most reliable method of identifying clones from a YAC library. But it requires a highly organized scheme for pooling clones to identify the relevant YACs from the large number of clones in a YAC

Table 26.2 Available mouse YAC libraries.

Library	Insert size (kb)	Yeast strain	Mouse strain	Genome equivalents	Screening method	Availability (see footnotes)
ICRF [43]	750	AB1380	C3H	4	HF* and PCR	†
St Mary's [30]	240	*rad52*	C57BL/10	3.5	PCR	‡
Princeton [44]	250	AB1380	C57BL/6J	4	PCR	§
MIT [45]	680	AB1380	C57BL/6J	4	PCR	II
MIT#2 [46]	820	J57D	C57BL/6J	10	PCR	¶

* HF, high-density hybridization filters.
† Genome Analysis Laboratory, ICRF, PO Box 123, Lincoln's Inn Fields, London WC2A 3PX, UK.
‡ MRC Mouse Genome Centre, Harwell, Oxfordshire OX11 0RD, UK.
§ No longer widely used.
II Research Genetics, 2130 Memorial Parkway SW, Huntsville, AL 35801, USA.
¶ Personal communication, Joyce Miller, Whitehead/MIT Center for Genome Research, 9 Cambridge Center, Cambridge MA 02412, USA.

library. For a programme of work directed towards the construction of a complete YAC physical map of the mouse X chromosome, we have rearrayed two mouse YAC libraries—the St Mary's and ICRF libraries (see Table 26.2)—into a three-dimensional (3-D) format for rapid PCR screening. The combined library—the 3-D library—represents seven genome equivalents with an overall average insert size of 500 kb and can be screened by two rounds of PCR screening to identify the relevant YAC coordinate. Briefly, microtitre plates, each carrying 96 (12×8) clones, are arranged into 3-D stacks with each stack containing 72 microtitre plates: 6 plates in each floor and 12 floors in each stack (Fig. 26.5). The 3-D library contains eight stacks in total. DNA pools are prepared from all the clones in each of the 12 floors, 24 rows and 24 columns in each stack (Fig. 26.5). In addition, DNA pools—superpools—representing all of the clones in each stack are prepared. In a first round of PCR screening using the superpools, those stacks containing positive clones are identified. This is followed by a second round of PCR screening of the relevant floor, column and row pools from the positive stacks providing a unique identifier for each positive YAC clone in the library.

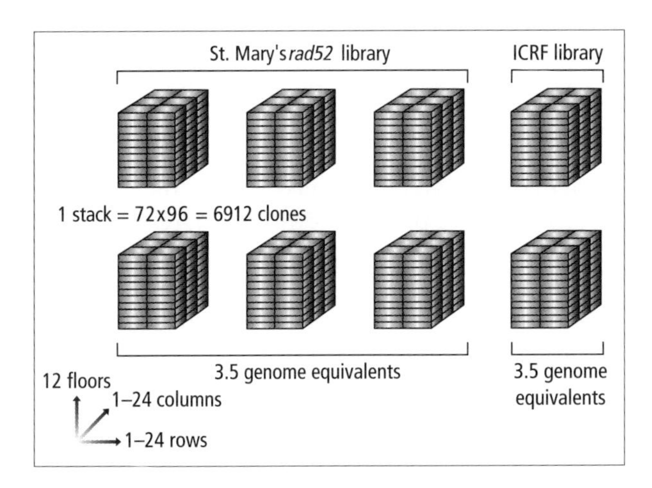

Fig. 26.5 Arraying mouse YAC libraries for rapid PCR screening. A schematic representation of the mouse 3-D yeast artificial chromosome (YAC) library that was prepared by combining the clones of the St Mary's and ICRF YAC libraries (see Table 26.2). Each of eight stacks in the library contains 72 microtitre plates, each plate holding 96 (8×12) clones. Each stack consists of 12 floors, with 6 plates in each floor. Combined clones from each floor, column and row in each stack (a total of 60—that is, 12 + 24 + 24—pools) were used to prepare DNA. In addition, DNA from the 60 pools was also combined to prepare a superpool for each stack. The library can be rapidly screened by PCR by first screening the superpools to identify positives in stacks followed by screening of the relevant set of 60 floor/rows/columns pools to identify a unique coordinate.

The alternative to PCR screening is hybridization screening using available STSs. High-density gridded filters of YAC clones can be used for hybridization screening of mouse YAC libraries [35]. High-density filters can be screened with available probes representing STSs from any region. It is not, however, possible to screen YAC libraries with microsatellite markers using hybridization techniques. As discussed below, YAC libraries can also be screened with IRS-PCR markers using filter hybridization techniques.

26.4.1.3 Construction of anchored YAC framework maps of mouse chromosomes

The first stage of our programme directed towards the coverage of the mouse X chromosome with YAC contigs is the establishment of an anchored YAC framework map. This involves screening the 3-D library with all available genetically mapped STSs, including the use of microsatellite primers and primers developed from known X chromosome gene sequences. To date, this work has allowed us to develop an anchored YAC framework map of the mouse X covering an estimated 50% of the chromosome: over 370 YAC coordinates have been identified to 139 STSs (N. Quaderi, G. Argyropoulos, A. Haynes & S.D.M. Brown, unpublished data). From the available data, 18 contigs have already been identified by common STS content.

26.4.1.4 Databases to hold physical mapping data

The MBx database has been modified and expanded to carry physical mapping data as well as genetic mapping data. For each STS, the relevant YAC coordinate information is stored, along with details of clone size, chimaerism, etc. Physical maps are generated by the use of an algorithm SAM2 [36]. This algorithm enables the user to automatically generate and subsequently modify, if required, contig tiling paths in any one chromosome region. Following assessment and manipulation of contig information in SAM2, contig information is then displayed on the Web (see MBx WWW address above). Both multimaps relating physical map information to genetic maps as well as more detailed displays of the STS content maps are available (Fig. 26.6).

26.4.1.5 High-resolution genetic maps as an aid to physical mapping and contig closure

The integrity of the growing YAC contigs is greatly assisted by the development in parallel of the high-resolution microsatellite map of the mouse X chromosome (see above). The high-resolution genetic map resolves and orders STSs to very fine resolution

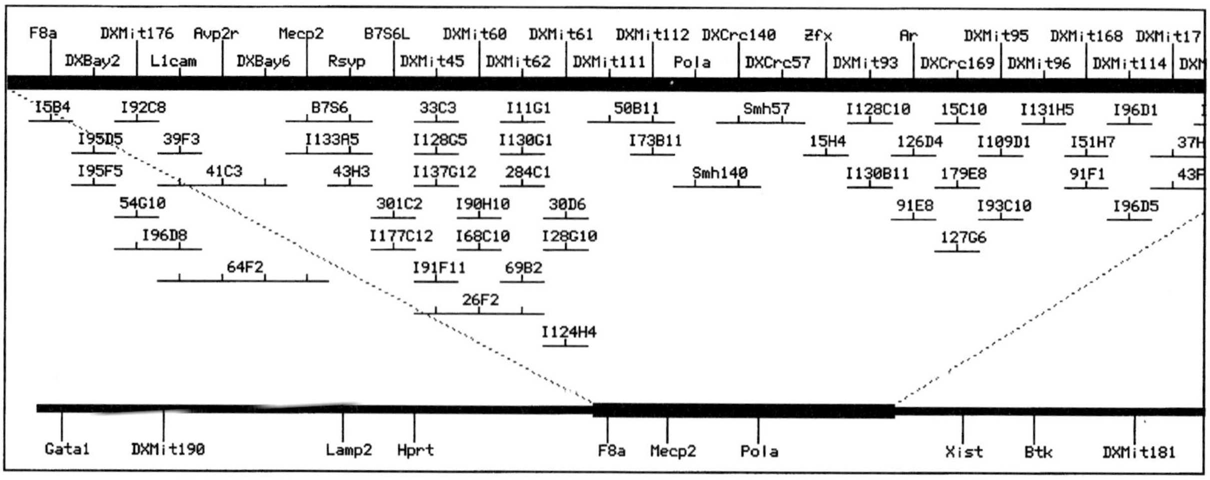

Fig. 26.6 The anchored YAC framework map of the mouse X chromosome. A display available through the WWW (http://www/hgmp.mrc.ac.uk/MBx/MBxHomepage.html) of a portion of the anchored YAC framework map of the mouse X chromosome. Above the upper bar, STSs are illustrated—the bulk of them anchored either at low or high resolution on the genetic map (see, for example, Fig. 26.3). Below the upper bar, the YACs detected by the anchored STSs are illustrated and a number of contigs can be identified. YAC coordinates for each anchored STS are given. The lower bar provides a schematic of the region of the X chromosome under examination.

and underpins the development of a high-integrity, overlying physical map. The genetic map provides confirmation on the integrity of growing contigs as well as eliminating false contig information. Furthermore, the genetic map allows us to orientate growing contigs and aids contig closure (Fig. 26.7). One route to contig closure across any chromosome region is the use of IRS-PCR to develop new markers as a tool to extend seed contigs.

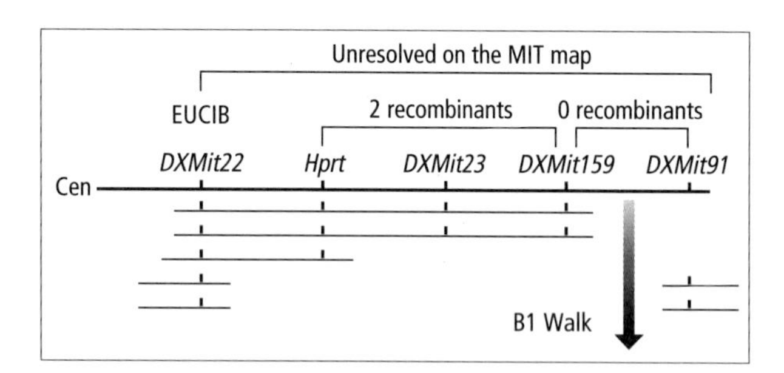

Fig. 26.7 High-resolution genetic maps allied to physical mapping for the efficient closure of chromosome YAC contigs. A YAC contig covering loci from *DXMit22*, *Hprt*, *DXMit23* and *DXMit159* is illustrated. These STSs along with *DXMit91*, for which YACs have also been isolated, are unresolved on the low-resolution MIT map. In the EUCIB back-cross, two recombinants have been detected between *Hprt* and *DXMit159/91*. No recombinants have been detected between *DXMit159* and *DXMit91*. This high-resolution map information orients the *DXMit22–DXMit159* contig with respect to *DXMit91* YACs and indicates the most efficient route to close the contig in this area by for example B1 walking.

26.4.2 Physical maps using interspersed repeat sequences

26.4.2.1 Recovering YACs to IRS-PCR markers

IRS-PCR markers such as the B1 and B2 repeats are being used for the genetic mapping of mouse chromosomes (see Section 26.3.1.1). Additionally, these markers can be screened against the available YAC libraries to provide additional YAC clones for generating contigs in any region. Screening of IRS-PCR products against YAC libraries must involve hybridization techniques and there are two principal approaches. First, an IRS-PCR marker can be hybridized directly to high-density gridded spots of IRS-PCR products from individual YAC clones. Alternatively, IRS-PCR markers can be hybridized to IRS-PCR products of DNA pools from three-dimensional arrays of YAC clones (see above and ref. 37). B1 IRS-PCR products are recovered from each of the three-dimensional DNA pools and robotically spotted onto filters. Following hybridization to the IRS-PCR marker, the positive signal in each pool (Fig. 26.8) indicates the correct YAC coordinate (floor, column and row) that contains the relevant IRS-PCR marker [37,38].

26.4.2.2 Extending seed contigs using IRS-PCR

One efficient way to extend the growing contigs on a particular chromosome is to apply IRS-PCR to recover new markers to seed contigs [37]. These new IRS-PCR markers can be used directly to extend the

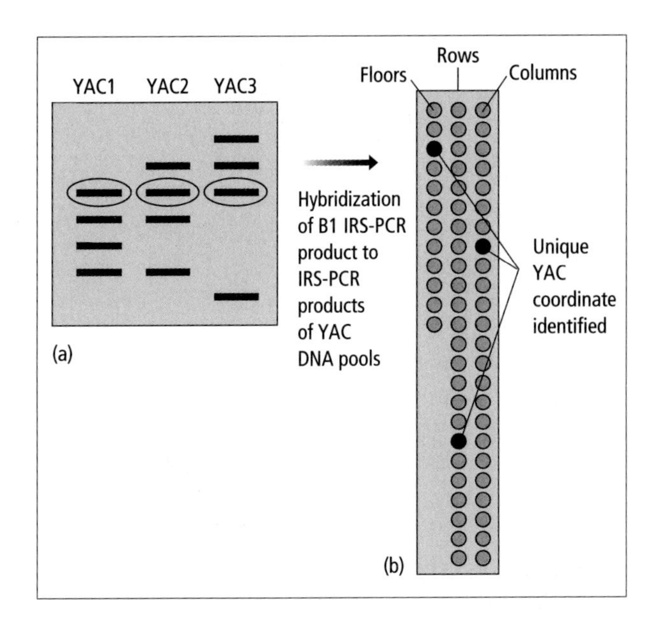

Fig. 26.8 The use of IRS-PCR products to isolate new YAC clones: an efficient route for extending YAC contigs. Schematic representation of the isolation of new YAC clones using B1 IRS-PCR in order to extend a growing YAC contig. PCR amplification using a B1 repeat primer of three YAC clones (YAC 1–3) from the end of a YAC contig generates a number of IRS-PCR products from each YAC when analysed by gel electrophoresis (a). Some bands appear to be held in common and may arise from a non-chimaeric portion of the YAC contig. Isolation of common B1 IRS-PCR products (circled) followed by hybridization to filters containing spotted arrays of B1 IRS-PCR products of YAC DNA pools from the 3-D library allows the identification of new YAC coordinates (positive floor, row and column) and the potential identification of new YAC clones that extend the growing YAC contig (b).

nascent contigs. IRS-PCR products can be recovered from the YACs towards the end of a contig and separated on an agarose gel (see Fig. 26.8). Identification of a common IRS-PCR product from two or more YACs at the end of a contig indicates that the product is unlikely to have originated from a chimaeric, noncontiguous portion of the YAC contig [37]. Subsequently, this IRS-PCR product can be cloned or used directly as a probe against filters containing IRS-PCR products of a YAC library (see above).

It is again important to emphasize that the strategy of extending seed contigs by the use of B1 IRS-PCR is greatly assisted by the availability of a high-resolution genetic map that aids contig orientation and closure. As indicated in Fig. 26.7, contig orientation often indicates the most efficient routes for contig closure by B1 walking or by other routes.

26.4.3 Uses of the physical map

The physical map gives us access to all of the underlying sequence in any particular region and, most importantly, access to the coding sequences. For most of the interesting mutations in both the mouse and human genomes, the underlying gene is not identified and there may be no suitable candidate gene to test. The only way of accessing the gene encoding the mutation is by a strategy known as positional cloning [1,39]. As indicated above, genetic crosses segregating the mutation can identify the most closely linked DNA markers. YAC contigs constructed using STSs that closely flank a mutation as identified by genetic analysis will contain the relevant gene and are an important start-point for identification of the relevant locus (see Fig. 26.4). YAC clones must be further analysed to identify the incumbent gene sequences and isolate potential candidates for the mutation. The techniques available for gene identification include techniques that are commonly used in a variety of organisms including:

1 exon trapping of YAC clones [40,41] (see also Chapter 17);
2 screening of YACs against libraries of cDNA clones [42];
3 identification of sequences conserved between species.

26.4.3.1 Confirming a candidate gene isolated by positional cloning

Once gene sequences are identified from YACs, it is necessary to assess their candidature for the mutation that is being positionally cloned. For many mutations, there is some understanding of the likely site that is affected by the defect and therefore some idea of the likely tissue within which the gene is normally expressed. Examination of tissue and developmental profiles of expression may eliminate certain candidates. This may have been taken into account when using YACs to screen cDNA libraries. The cDNA libraries used will, if possible, have been prepared from the tissue that is the likely site of expression. Finally, it will be necessary to determine by direct DNA sequencing of any candidate locus that the gene encodes the mutation. For many mouse loci, multiple mutations are available [14] and in many cases this will aid confirmation that the correct gene has been cloned.

26.5 Conclusions

The Mouse Genome Mapping Project aims to achieve comprehensive genetic and physical maps across all chromosomes, leading to further profound biological insights into mammalian gene function and organization. Conserved linkage groups between mouse and human chromosomes enable an extensive experimental interplay between these two pivotal mammalian genomes. Genes isolated and mapped in human can be examined as candidates for mouse mutations, while, equivalently, genes identified in the mouse can be analysed as candidates for mutations underlying human genetic diseases. The identification of mouse models for human genetic disease will have an increasing impact upon human biology and will be aided by the rapid expansion in mouse genomics. By the end of 1996, we can expect to see the completion of a high-resolution genetic map of the mouse genome, and by 1997, completion of comprehensive physical maps of a number of mouse chromosomes with the whole genome physical map following shortly thereafter. At the same time, we can expect the rapid development of high density transcript maps of the mouse genome that will position on the physical maps many more genes than currently mapped. Detailed physical and transcript maps can be expected to give us increasing and easier access to the various genes underlying the myriad mutations available in the mouse.

Finally, comparative sequencing of diverse genomes, including those of mouse and human, is expected to play an increasing role in our analysis of genome structure and function. A complete physical map of the mouse will provide the templates for the development of sequence-ready maps as a prelude to the acquisition of large tracts of sequence from the mouse genome.

Acknowledgements

We thank Nandita Quaderi and Andy Haynes for information in advance of publication. This work was supported by grant G-9201373 from the Medical Research Council, UK.

References

1 Brown, S.D.M. (1994) Integrating maps of the mouse genome. *Curr. Opinion Genet. Dev.* **4**, 389–394.
2 Love, J.M., Knight, A.M., McAleer, A.M. & Todd, J.A. (1990) Towards construction of a high-resolution map of the mouse genome using PCR-analysed microsatellites. *Nucleic Acids Res.* **18**, 4123–4130.
3 Dietrich, W., Katz, H., Lincoln, S.E. *et al.* (1992) A genetic map of the mouse suitable for typing intraspecific crosses. *Genetics* **131**, 423–447.
4 Dietrich, W., Miller, J.C., Steen, R.G. *et al.* (1996) A comprehensive genetic map of the mouse genome. *Nature* **380**, 149–152.
5 Hastie, N.D. (1996) Highly repeated DNA families in the genome of *Mus musculus*. In: *Genetic Variants and Strains of the Laboratory Mouse* (Lyon, M.F., Rastan, S. & Brown, S.D.M., eds), 1425–1442 (3rd edn, Oxford University Press, London/New York).
6 McCarthy, L., Hunter, K., Schalkwyk, L. *et al.* (1995) Efficient high-resolution genetic mapping of mouse interspersed repetitive sequence PCR products, toward integrated genetic and physical mapping of the mouse genome. *Proc. Natl Acad. Sci. USA* **92**, 5302–5306.
7 Avner, P., Amar, L., Dandolo, L. & Guenet, J.-L. (1988) Genetic analysis of the mouse using interspecific crosses. *Trends Genet.* **4**, 8–23.
8 Breen, M., Deakin, L., MacDonald, B. *et al.*, (The European Collaborative Group) (1994) Towards high-resolution maps of the mouse and human genomes—a facility for ordering markers to 0.1 cM resolution. *Hum. Mol. Genet.* **3**, 621–626.
9 Taylor, B.A. (1989) Recombinant inbred strains. In *Genetic Variants and Strains of the Laboratory Mouse* (Lyon, M.F. & Searle, A.G., eds), 773–796 (Oxford University Press, London/New York).
10 Copeland, N.G., Jenkins, N.A., Gilbert, D.J. *et al.* (1993) A genetic linkage map of the mouse: current applications and fututre prospects. *Science* **262**, 57–66.
11 Copeland, N.G. & Jenkins, N.A. (1991) Development and applications of a molecular genetic linkage map of the mouse genome. *Trends Genet.* **7**, 113–118.
12 *Encyclopedia of the Mouse Genome, I.V.* (1994) Chromosome Committee Reports. *Mammalian Genome* 5 (Special Issue), S1–S295.
13 Peters, J. (ed.). (1996) *Mouse Genome* **94**, 1–888.
14 Lyon, M.F., Rastan, S. & Brown, S.D.M. (1996) *Genetic Variants and Strains of the Laboratory Mouse* (3rd edn, Oxford University Press, Oxford).
15 Kingsmore, S.F., Giros, B. & Suh, D. (1994) Glycine receptor β-subunit gene mutation in spastic mouse associated with LINE-1 element insertion. *Nature Genet.* **7**, 136–141.
16 Todd, J.A., Aitman, T.J., Cornall, R.J. *et al.* (1991) Genetic analysis of autoimmune type I diabetes mellitus in mice. *Nature* **351**, 542–547.
17 Nadeau, J.H., Davisson, M.T., Doolittle, D.P. *et al.* (1992) Comparative map for mice and humans. *Mammalian Genome* **3**, 480–536.
18 Searle, A.G., Edwards, J.H. & Hall, J.G. (1994) Mouse homologues of human hereditary disease. *J. Med. Genet.* **31**, 1–19.
19 Ryan, S.G., Buckwalter, M.S., Lynch, J.W. *et al.* (1994) A missense mutation in the gene encoding the α1 subunit of the inhibitory glycine receptor in the spasmodic mouse. *Nature Genet.* **7**, 131–135.
20 Brown, K.A., Sutcliffe, M.J., Steel, K.P. & Brown, S.D.M. (1992) Close linkage of the olfactory marker protein gene to the mouse deafness mutation shaker-1. *Genomics* **13**, 189–193.
21 Gibson, F., Walsh, J., Mburu, P. *et al.* (1995) A type VII myosin encoded by the mouse deafness gene shaker-1. *Nature* **374**, 62-64.
22 Evans, K.L., Fantes, J., Simpson, C. *et al.* (1993) Human olfactory marker protein maps close to tyrosinase and is a candidate gene for Usher syndrome type I. *Hum. Mol. Genet.* **2**, 115–118.

23 Weil, D., Blanchard, S., Kaplan, J. *et al.* (1995) Defective myosin VIIA gene responsible for Usher syndrome type 1B. *Nature* **374**, 60–61.
24 Vetrie, D., Vorechovsky, I., Sideras, P. *et al.* (1993) The gene involved in X-linked agammaglobulinaemia is a member of the Src family of protein-tyrosine kinases. *Nature* **361**, 226–233.
25 Rawlings, D.J., Saffran, D.C., Tsukada, S. *et al.* (1993) Mutation of unique region of Bruton's tyrosine kinase in immunodeficient XID mice. *Science* **261**, 358–361.
26 Thomas, J.D., Sideras, P., Smith, C.I.E. *et al.* (1993) Colocalization of X-linked agammaglobulinemia and X-linked immunodeficiency genes. *Science* **261**, 355–358.
27 Burke, D.T., Carle, G.F. & Olson, M.V. (1987) Cloning of large segments of exogenous DNA into yeast by means of artificial chromosome vectors. *Science* **236**, 806–812.
28 Hudson, T., Stein., L.D., Gerety, S.S. *et al.* (1995) An STS-based map of the human genome. *Science* **270**, 1945–1954.
29 Green, E.D., Riethman, H.C., Dutchik, J.E. & Olson, M.V. (1991) Detection and characterization of chimeric yeast artificial-chromosome clones. *Genomics* **11**, 658–669.
30 Chartier, F.L., Keer, J.T., Sutcliffe, M.J. *et al.* (1992) Construction of a mouse yeast artificial chromosome library in a recombination-deficient strain of yeast. *Nature Genet.* **1**, 132–136.
31 Pierce, J.C. & Sternberg, N. (1992) Using the bacteriophage P1 system to clone high molecular weight (HMW) genomic DNA. *Meth. Enzymol.* **216**, 549–574.
32 Ioannou, P.A., Amemiya, C.T., Garnes, J. *et al.* (1994) A new bacteriophage P1-derived vector for the propagation of large human DNA fragments. *Nature Genet.* **6**, 84–89.
33 Shizuya, H., Birren, B., Kim, U.J. *et al.* (1994) Cloning and stable maintenance of 300 kilobase pair fragments of human DNA in *Escherichia coli* using an F-factor based vector. *Proc. Natl Acad. Sci. USA* **89**, 8794–8797.
34 Pierce, J., Sternberg, N. & Sauer, B. (1992) A mouse genomic library in the bacteriophage P1 cloning system: organization and characterization. *Mammalian Genome* **3**, 550–558.
35 Olsen, A.S., Combs, J., Garcia, E. *et al.* (1993) Automated production of high density cosmid and YAC colony filters using a robotic workstation. *BioTechniques* **14**, 116–123.
36 Soderlund, C.A. (1995) *SAM v 2.2. User Manual.* The Sanger Centre Technical Report, SC-95-01 (Hinxton, UK).
37 Hunter, K.W., Ontiveros, S.D., Stanton, V.P. Jr *et al.* (1994) Rapid and efficient construction of yeast artificial chromosome contigs in the mouse genome with interspersed repetitive sequence PCR (IRS-PCR): generation of a 5 cM, >5 megabase contig on mouse chromosome 1. *Mammalian Genome* **5**, 597–607.
38 Hunter, K.W., Riba, L., Schalkwyk, L. *et al.* (1996) Toward the construction of integrated physical and genetic maps of the mouse genome using interspersed repetitive sequence PCR (IRS-PCR) genomics. *Genome Res.* **6**, 290–299.
39 Collins, F. (1992) Positional cloning—Let's not call it reverse anymore. *Nature Genet.* **1**, 3–6.
40 Buckler, A.J., Chang, D.D., Graw, S.L. *et al.* (1991) Exon amplification: a strategy to isolate mammalian genes based on RNA splicing. *Proc. Natl Acad. Sci. USA* **88**, 4005–4009.
41 Gibson, F., Lehrach, H., Buckler, A.J. *et al.* (1994) Isolation of conserved sequences from yeast artificial chromosomes by exon amplification. *BioTechniques* **16**, 453–458.
42 Lovett, M., Kere, J. & Hinton, L.M. (1991) Direct selection: a method for the isolation of cDNAs encoded by large genomic regions. *Proc. Natl Acad. Sci. USA* **88**, 9628–9632.
43 Larin, Z., Monaco, A.P. & Lehrach, H. (1991) Yeast artificial libraries containing large inserts from mouse and human DNA. *Proc. Natl Acad. Sci. USA* **88**, 4123–4127.
44 Burke, D.T., Rossi, J.M., Leung, J. *et al.* (1991) A mouse genomic library of yeast artificial chromosome clones. *Mammalian Genome* **1**, 65.
45 Kusumi, K., Smith, J.S., Segre, J.A. *et al.* (1993) Construction of a large-insert yeast artificial chromosome library of the mouse genome. *Mammalian Genome* **4**, 391–392.
46 Haldi, M.L., Strickland, C., Lim, P. *et al.* (1996) A comprehensive large-insert yeast artificial chromosome library for physical mapping of the mouse genome. *Mammalian Genome* **7**, 767–769.

Chapter 27 Gene therapy

Richard G. Vile & Rosa Maria Diaz*

ICRF Laboratory of Cancer Gene Therapy, ICRF Oncology Unit, Hammersmith Hospital, DuCane Road, London W12 ONN, UK

**Richard Dimbleby Department of Cancer Research, Rayne Institute, St Thomas' Hospital, Lambeth Palace Road, London SE1 7EH, UK*

27.1 Introduction

As a direct result of the powerful techniques of genome analysis (see Sections 1–4), it has become possible to map, clone and sequence individual genes, mutations of which are responsible for the development of disease. Identification of such genes, and the disease-associated mutations, has raised the prospect that genetic disease may be treatable by direct correction of the underlying defect, that is, at the level of the genome itself.

Gene therapy was initially conceived as a way to treat diseases for which a (simple) genetic defect was known to be the cause. In its simplest form, gene therapy involves the delivery of a functionally correct copy of a mutated gene into the affected cells in order to obtain long-term correction of the physiological defect caused by the mutation. An example is the treatment of cystic fibrosis by delivery of the gene for the cystic fibrosis chloride ion transporter (CFTR) protein into the airway epithelial cells of patients with cystic fibrosis.

However, as the number of diseases that are known to have at least some genetic component increases, the definition of gene therapy has become much broader. Now gene therapy is routinely evoked to encompass the use of genetic material to alleviate the symptoms of a disease, even if the therapeutic genes are not strictly 'corrective' (in the sense of restoring a function known to be mutated in the affected cells). Hence, the delivery of cytotoxic genes to kill cancer cells (rather than to correct the oncogenic mutations within them) is also accepted as gene therapy. Therefore, in its broadest terms, gene therapy represents 'an opportunity for the treatment of genetic disorders in adults and children by genetic modification of human body cells' [1].

A further important classification is to distinguish the *heritable potential* of gene therapy. All of the gene therapy trials that are currently approved for use in human patients target those somatic cells that will live only as long as the patient. Barring inadvertent spread of the therapeutic genes to the gametes, the genetic treatment will only affect one generation and will not be able to alter the genetic make-up of any offspring. This is therefore known as *somatic gene therapy*. The purpose of somatic gene therapy is to alleviate disease in the treated individual, and that individual alone.

In contrast, it is also possible to target the gametes (sperm and ova) directly in order to modify the genetic profile, not of the current but of the subsequent generation, of unborn 'patients'. Gene transfer at an early stage of embryonic development may also have the same effects by achieving gene transfer to both somatic and germ line cells. This is *germline gene therapy*. The attraction of germline gene therapy for the treatment of disease is that, at least in theory, permanent genetic cures might be achieved by delivering a functional copy of a mutated gene to every cell of the resulting progeny. However, there is currently widely held apprehension about the development of germline gene therapy research programs. The ability to alter the genetic profile of subsequent generations rightly invokes many spectres. Apart from a complete inability to predict the long term sequelae of altering the germline by delivery of exogenous genetic material at the scientific level, there are many ethical issues raised by the prospect of treating 'patients' whose consent it is impossible to obtain.

In addition, although it is currently not possible to manipulate traits such as 'intelligence' or 'beauty' genetically, there is a perceived fear of such technology being abused in eugenic-type breeding programs in the future. As a result, the major ethical and regulatory bodies of gene therapy in both the USA and in Europe have placed a moratorium on the consideration of any germline gene therapy of human patients because of 'insufficient knowledge to evaluate the risks to future generations' [1]. However, it is important that such issues be addressed at a regulatory level sooner rather than later. Non-consideration of applications for germline trials in patients will in no way prevent continued research into the direct genetic modification of the germline and the relevant ethical and regulatory dilemmas will simply be deferred, rather than solved, by procrastination.

27.2 The perfect disease

Genetically, gene therapy is well advanced for many diseases: that is, the underlying genetic defect has been identified and the corrective version of the relevant mutated gene is available for delivery into and expression in target tissues. However, it is the imperfections of current *in vivo* gene delivery technologies which currently impose the most limiting restrictions upon the practical success of most proposed gene therapy protocols [2–5]. Therefore, when assessing candidate diseases for gene therapy, several considerations must be taken into account. The following checklist can be drawn up, against which a candidate disease can be compared when considering how it compares to the 'perfect' target disease for gene therapy (Fig. 27.1).

1 The pathology of the disease should be caused by a defect in just a single gene (monogenic disorders), the correction of which will restore normal physio-

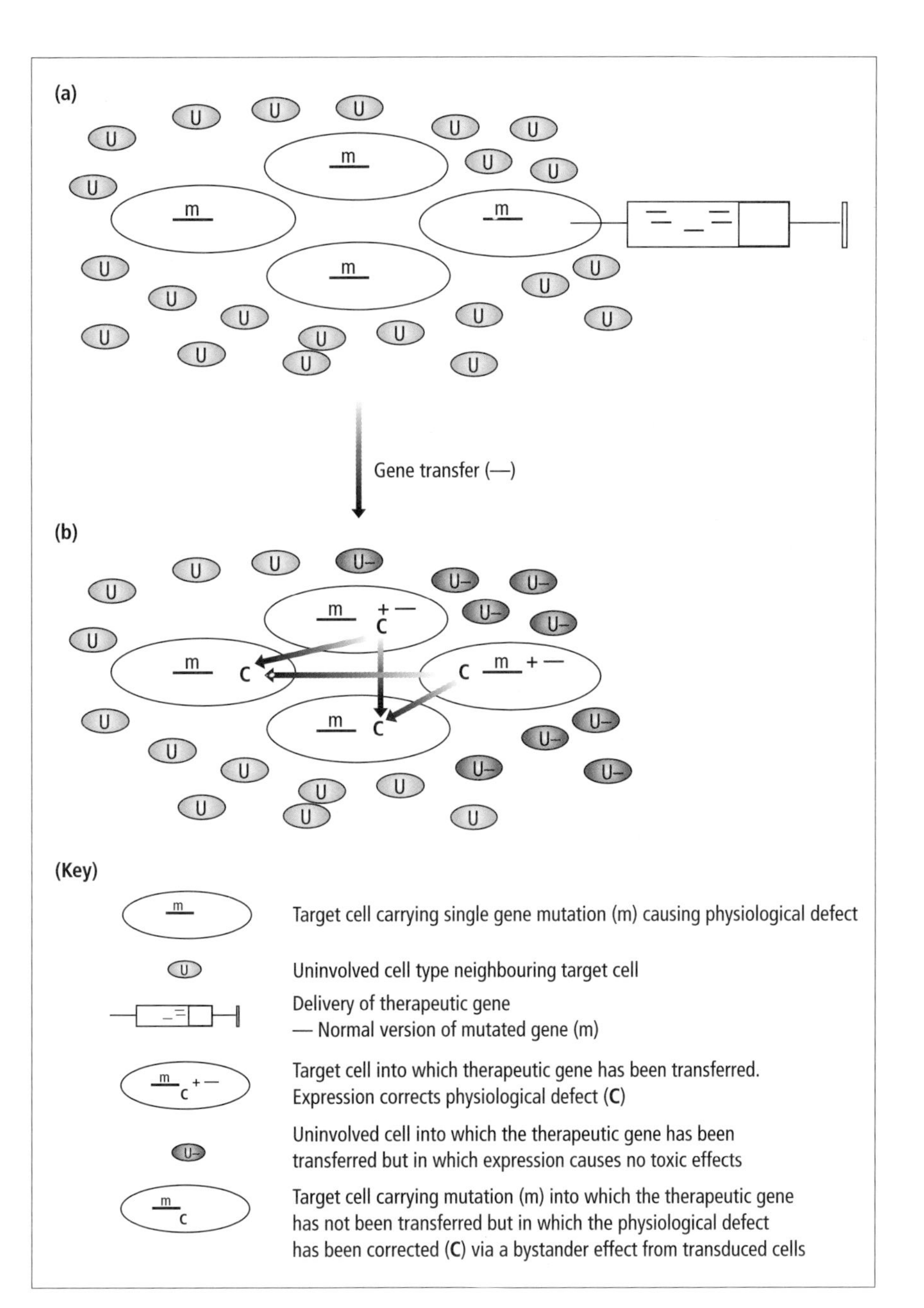

Fig. 27.1 An idealized protocol for the gene therapy of a simple monogenic disorder. (a) The pathology of the disease is caused by a mutation (m) in just a single gene in the target cells (large ovals), which are surrounded by uninvolved cells (small circles, U) which also carry the mutation but are not pathologically affected. The target cells should be in localized, anatomically accessible positions for direct *in vivo* gene delivery (for example, by direct injection by syringe). (b) Following gene transfer, the pathology can be reversed by simple, constitutive ON/OFF regulation of expression of the correct version of the gene (–) in the affected target cells. In addition, correction (c) of the overall physiological defect should be achievable by delivery of the corrective gene to only a proportion of the affected target cells. The biological properties of the gene delivery vehicle and expression of the therapeutic gene should be nontoxic to normal uninvolved cells so that perfectly targeted delivery only to the affected cell type is not required.

logical function to the affected cells and tissues. Hence, affected cells require only one gene to be delivered; the probability of delivering more than one gene to any given cell *in vivo* diminishes rapidly with increasing number.

2 The gene which is mutated in such a monogenic disorder should have been cloned and the mutations which cause disease should be well characterized.

3 Correction of the physiologcal defect caused by the mutation should be achievable by simple, constitutive ON/OFF regulation of expression of the correct version of the gene. Obtaining temporally regulated gene expression in target cells *in vivo* requires inclusion of regulatory elements which are still being characterized in most systems; in addition, quantitative regulation of exogenously introduced gene expression, relative to other endogenous genes *in vivo*, is also likely to be especially problematic.

4 The biological properties of the gene delivery vehicle and expression of the therapeutic gene should be nontoxic to normal cells so that perfectly targeted delivery only to the affected cell type is not required. This will permit relatively promiscuous gene delivery without widespread toxicity.

5 The target cells/tissue for gene correction should be in localized, anatomically accessible positions. Delivery of a single copy of any gene to every cell in the body is currently impossible, other than by germline or *in utero* gene therapy. This requirement will help to overcome the problems with efficiency of gene delivery, which is a major limitation to gene therapy for most diseases.

6 Indeed, delivery of a single copy of a gene to every cell even in a localized body compartment is highly improbable with current technologies. Therefore, correction of the physiological defect should be achievable by delivery of the corrective gene to only a proportion of the affected target cells.

7 Given the costs associated with the development of any new drugs for human use and especially considering the heightened safety concerns associated with the use of genetic treatments in human to justify the many regulatory hurdles which must be traversed for the use of gene therapy, there should be no effective currently existing treatment for the disease.

27.3 The real diseases

In contrast to the idealized situation described above, a wide variety of conditions have been proposed to be amenable to gene therapy, some more realistically than others [6]. These range from simple monogenic disorders (e.g. cystic fibrosis), which fulfil many of the criteria for the ideal candidate disease, through more complex monogenic and multifactorial genetic diseases (e.g. cancer), to diseases where the underlying genetic 'defect' is introduced into the patient in the form of pathogenic genomes of bacteria or viruses (e.g. HIV). Examples of the spectrum of diseases currently under active investigation with genetic therapies are given below; although it is not possible to describe each disease in great detail, examples are used from different classes to illustrate the potential, and pitfalls, of gene therapy.

27.3.1 Simple monogenic disorders

Not surprisingly, the diseases for which clinical trials are most advanced, and for which there is the most optimism for the clinical outcome, are those which conform the closest to the criteria 1–7 above. The best examples of such disorders are cystic fibrosis and severe combined immune deficiency (SCID).

27.3.1.1 Cystic fibrosis

Cystic fibrosis is a recessive disorder caused by mutation to a single gene encoding a chloride ion transporter protein, the CFTR [7]. When a patient inherits two mutated copies of the CFTR gene, ion transport across epithelial surfaces is disrupted. The most life-threatening pathology of CF presents as an accumulation of thick mucus in the airways accompanied by high risks of bacterial infection. This pathology is directly attributable to a defect in the chloride ion transport across the airway epithelial cells such that water is not secreted into the mucus-lined airway passages. However, CF patients also have other pathological consequences, especially in the gut and pancreas, but these conditions are usually managed effectively relative to the pulmonary symptoms [7].

The *CFTR* gene was cloned following extensive mapping studies and the range of mutations associated with the CF phenotype has been well documented [7]. *In vitro* and *in vivo* studies have shown that as few as 30% of cells in sheets of affected CF epithelial cells need to express the correct version of the CFTR gene for normal physiological levels of Cl^- ion transport to be restored to the entire cell layer [7]. Transgenic CF mice models have also been developed that show physiologically defective Cl^- ion transport across their airway epithelial cells. Unfortunately, these transgenic models do not necessarily develop CF-like disease so therapeutic gene therapy is difficult to demonstrate, although correction of the chloride ion transport defect has been conclusively shown [8,9].

Therefore, cystic fibrosis represents a near ideal candidate for classical gene therapy (Fig. 27.2). The pathology is caused by a single gene defect (Section 27.2, criterion 1) which can be corrected in *in vitro* and *in vivo* models by expression of the correct version of the gene (criterion 2) without the need for specific temporal or quantitative regulation of its expression (criterion 3). There is no evidence that expression of the *CFTR* gene in other tissues is toxic (criterion 4) and expression of the correct gene in only a proportion of affected epithelial cells is sufficient to restore normal function to epithelial cell layers (criterion 6). The target cell population for gene correction is relatively accessible to gene delivery by aerosols or even direct application (criterion 5) and although a range of conventional treatments can extend CF patients' lifespans into the mid-thirties, a lack of life-long treatments for CF more than justify the investment in gene therapy as a curative alternative (criterion 7).

Therefore, clinical trials of delivery of the *CFTR* gene to affected airway epithelial cells have now been approved and are underway in both the UK [10] and the USA [11]. In the first instance, these trials are aimed at assessing safety and are unlikely to show real therapeutic effects, not least because various technical hurdles still remain to be overcome. For example, although delivery of the CFTR expression vector to the airways is physically

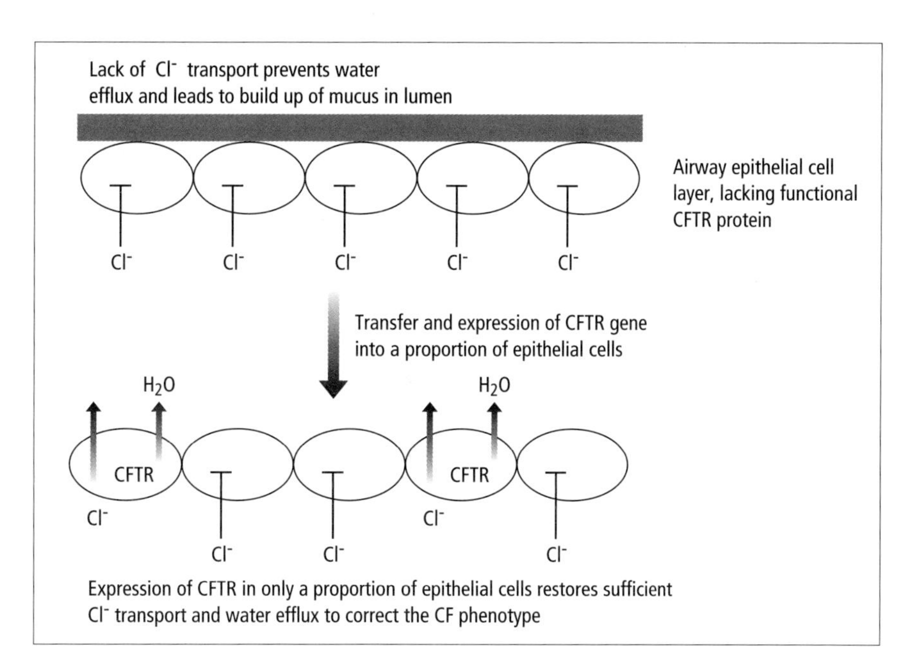

Fig. 27.2 Gene therapy for cystic fibrosis. The airway epithelial cells of a CF patient lack functional CFTR protein and cannot pump chloride ions across the cell layer. Consequently, water is not cotransported into the lumen and a life-threatening barrier of mucus accumulates. *In vivo* delivery of functional *CFTR* gene into at least some of the affected airway epithelial cells should generate sufficient Cl^- ion transport across the cell layer that enough water is now pumped into the lumen to clear the mucus barrier from the whole airway.

relatively simple (using either DNA complexed with cationic liposomes or high titre CFTR-adenoviral stocks), these vectors must penetrate the thick mucus to gain access to the epithelial cells before physiological correction can occur. It remains to be seen whether sufficient epithelial cells can be targeted in this way to generate clinical benefits to the patients.

In addition, other confounding factors associated with gene delivery make it unlikely that these early trials would be truly therapeutic. Since adenoviral vectors do not integrate into target cell chromosomes [3], any cells which become successfully transduced with the gene are most likely to express it only transiently. Hence, repeated administrations of viral vector would be required for chronic correction in the patient. However, development of immunity to the virus may well prevent such repeated administrations being effective [12]. In addition, inflammation in the lungs of animals treated with high-titre doses of recombinant adenovirus has been reported and one patient in a trial in the USA has already developed a life-threatening inflammatory reaction as a result of immune reactivity against very high dose adenoviral stock administered into the airway passage [13]. Alternative trials using CFTR expression vector plasmid DNA complexed with cationic liposomes seek to avoid such inflammatory problems by excluding the use of viral vectors. However, what such protocols seek to gain in terms of repeatability of dosing, they lose in terms of efficiency of gene transfer. Ideally, the corrective *CFTR* gene should penetrate the mucous barrier at sufficient levels for at least some of the stem cells of the continually self-renewing epithelial cell layer to become transduced. Only if stem cells can be stably transduced will the need for life-long administrations be avoided, a dogma which holds for many different forms of gene therapy.

Initial reports on the *in vivo* correction of Cl^- transport across small, treated areas (usually of the nasal lining of CF patients) are now appearing in the literature and look cautiously hopeful [14]. However, much technical work remains to be done before the inevitable compromises between efficiency and safety of gene delivery can be reconciled, and trials can proceed to protocols in which genuine clinical benefits are expected.

27.3.1.2 Severe combined immune deficiency

The second simple monogenic disorder which is at the forefront of human gene therapy trials is SCID. One form of SCID is caused by the absence of functional adenosine deaminase (ADA) in the patient's lymphocytes. However, animal models of SCID have shown that T-cell function can be corrected by removing affected lymphocytes *ex vivo* and expressing the cloned ADA gene in them. Return of 'corrected' lymphocytes to the animal can then provide sufficient enzyme levels systemically so that the immune system can function at normal levels [15]. Therefore, clinical trials are now well advanced in the USA in which a patient's lymphocytes are removed, transduced *ex vivo* with a retrovirus encoding the ADA gene, and returned *in vivo* to act as a source of ADA (Fig. 27.3) [16]. In this instance, many of the delivery problems associated with the cystic fibrosis trials are overcome by the *ex vivo* isolation of the target cells, their high level of transduction with viral vectors, and the potential to

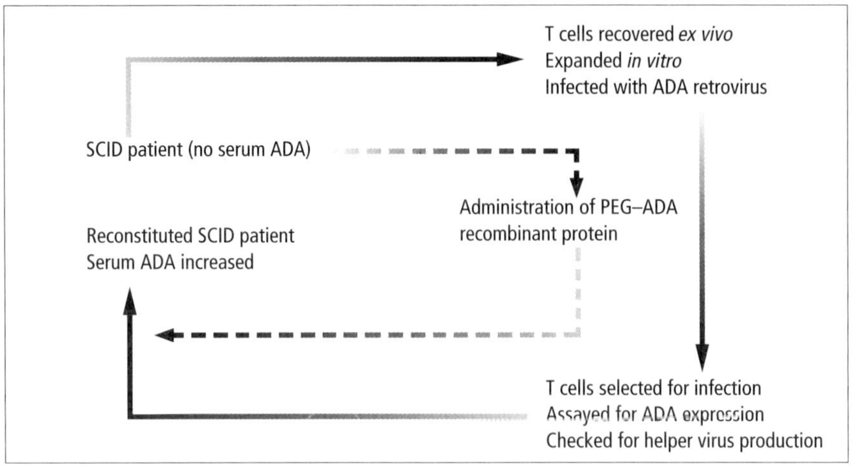

T cells recovered *ex vivo*
Expanded *in vitro*
Infected with ADA retrovirus
SCID patient (no serum ADA)
Administration of PEG–ADA recombinant protein
Reconstituted SCID patient
Serum ADA increased
T cells selected for infection
Assayed for ADA expression
Checked for helper virus production

Fig. 27.3 Gene therapy for ADA-deficient SCID patients. Patients' lymphocytes are removed, transduced *ex vivo* with a retrovirus encoding the *ADA* gene, and returned *in vivo* to act as a source of serum ADA. In these first trials, it was considered to be ethically unacceptable to withold the existing treatment of recombinant PEG-complexed ADA enzyme to patients treated with the genetically modified lymphocytes.

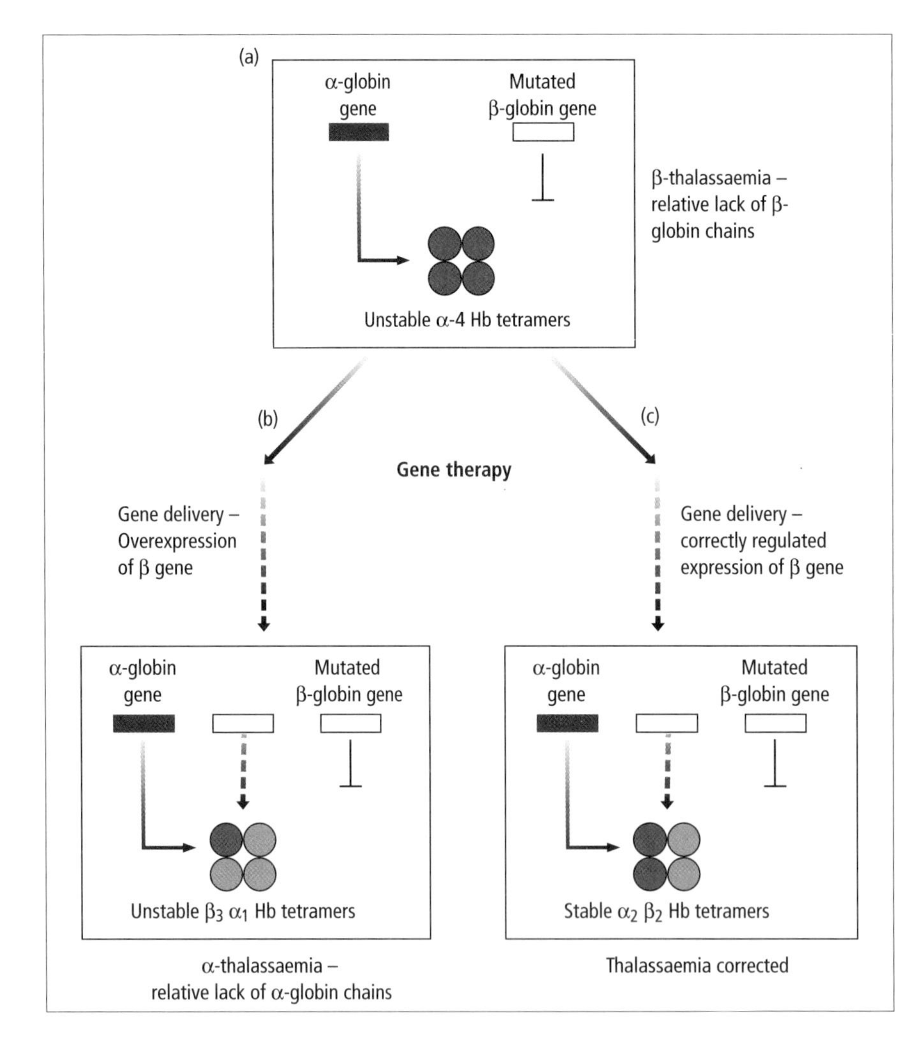

Fig. 27.4 Gene therapy for thalassaemia involves both effective gene delivery and appropriate regulation of gene expression. (a) β-thalassaemia is the result of a deficiency of β-globin chains, relative to α-chains, such that the resulting haemoglobin tetramers are unstable and defective in their normal oxygen carriage properties. (b) However, overexpressing the β-globin gene may simply deregulate haemoglobin synthesis in an equally detrimental way by converting a β-globin thalassaemia (relative lack of β-globin chains) into an α-thalassaemia (relative lack of alpha globin chains). (c) Correctly coordinated expression of the introduced β-globin gene, relative to the endogenous α-globin gene, in target cells could correct the β-thalassaemia (deficiency of β-globin chains).

gain long-term, stable expression of the therapeutic gene by using a retroviral vector which integrates into the genome [15].

The results of these trials (one of the first human gene therapy trials to be approved) are very encouraging, although in some respects they remain ambiguous. Since it was considered to be ethically unacceptable to withold the existing treatment of recombinant polyethylene glycol (PEG)–ADA complex to patients treated with the genetically modified lymphocytes (who now attend school and are apparently well), it has not been possible to attribute their continued immune function solely to the gene therapy rather than to the conventional treatment. None the less, lack of detectable treatment-related toxicity, detection of the introduced gene in circulating lymphocytes and continually elevated levels of ADA suggest that this form of gene therapy may eventually become standard in the treatment of the disease.

Although ADA deficiency is one of the flagships of human clinical gene therapy, it is actually a very rare disorder. Its adoption as a prototype disease is certainly not driven by consideration of the existence of large numbers of desperate patients (criterion 7 above). Rather, its amenability to the requirements of gene therapy (criteria 1–6) mean that it is the most likely to work [15]. Therefore, it is hoped that apparent success in this disease can be used as a justification to proceed with other similar monogenic disorders and even with other less ideal situations. Examples of such cases include a variety of metabolic disorders where the pathology is associated with the lack of a single identified enzyme [17]. Often, restoration of 5–25% of normal serum enzyme activity will protect from clinical disease in conditions such as haemophilia B, caused by a lack of the blood clotting factor IX. Therefore, the relevant gene can be delivered into ectopic tissues, or into fibroblasts *ex vivo* followed by implantation of the genetically modified cells to serve as a source of serum enzyme. Haemophiliac dogs have been 'cured' by implantation of cells modified by addition of the factor IX gene, or by direct gene modification of hepatocytes with retroviral vectors encoding the factor IX gene [18], and human trials based on these results have been proposed.

In summary, there are several disorders whose properties make them conceptually very attractive as candidates for gene therapy, as defined by the criteria listed above. However, even the most theoretically amenable diseases still present many technical difficulties which must be overcome before gene therapy becomes a routine tool in patient management.

27.3.2 Complex monogenic disorders

Treatment of certain other monogenic disorders will, however, be more complex from both the pragmatic and genetic standpoints. In these instances, simple replacement of the corrective gene into either the normal cells that produce the relevant gene product (e.g. airway epithelial cells in CF) or into more easily manipulated ectopic tissues (e.g. transplanted fibroblasts for secretion of factor IX), is not likely to be sufficient to alleviate disease symptoms.

For example, some monogenic metabolic disorders will require gene modification of specific tissues that provide cofactors for enzyme activity. Hence, correction of phenylketonuria requires delivery of the phenylalanine hydroxylase enzyme specifically to liver cells because of the cofactors produced in hepatic cells necessary for optimal enzyme activity. Similar requirements will be necessary for treatment of some disorders of glycogen metabolism or of the urea cycle where normal function of therapeutic genes requires additional hepatic enzymes [17].

Another example of a complex monogenic disorder is the haemoglobinopathies. Thalassaemias are the result of a deficiency of globin genes such that the resulting haemoglobin structures are unstable and/or defective in their normal oxygen-carrying properties [19]. As such, it is attractive to propose that simple delivery of the missing globin genes could be used to correct the relevant thalassaemic condition. Thus, expression of the β-globin gene in target cells could reverse β-thalassaemia (deficiency of β-globin chains). Unfortunately, synthesis of the haemoglobin tetramers involves very tight biochemical regulation, characterized by both temporal and quantitative controls on the production of several different globin species relative to each other. Therefore, simply overexpressing a particular globin molecule in the cell at any given time may simply deregulate haemoglobin synthesis in a different way—for instance, by converting a β-globin thalassaemia (relative lack of β-globin chains) into an α-thalassaemia (relative lack of α-globin chains).

Transcriptional control of the globin gene family is known to be highly regulated by tissue-specific enhancers and locus control regions (LCRs) [20,21], which determine the temporal switching of globin chain synthesis during development. Effective gene therapy aimed at control of globin synthesis will therefore have to incorporate such transcriptional regulation into the therapeutic constructs. Although retroviral vectors have been constructed that do appear to preserve the developmental regulation pattern of globin expression [22], there remains much to improve before gene therapy of thalassaemias can be confidently advanced into a clinical setting.

In summary, several diseases are caused by defects in just a single cloned gene (criteria 1 and 2); however, in many cases the therapeutic issue focuses not upon the gene itself but on achieving the correct levels and timing of its expression (criterion 3) relative to other proteins with which the gene product must interact in the relevant biochemical pathways *in vivo*. Identification of transcriptional control elements which can target and regulate gene expression promises to be one of the most important advances in the coming years in the field of gene therapy.

27.3.3 Multifactorial genetic disorders

Many diseases are now known that clearly have a genetic component but in which the genetic contribution is shared between several genetic loci and/or is also enhanced by epigenetic factors (see Chapter 2). For example, genetic linkages have been variously reported for several psychiatric disorders but the degree of genetic and environmental contributions remains unclear. Even when candidate genes for such diseases have been identified, as is potentially the case for Alzheimer's disease, the value of the genes for therapy remains unclear because of doubts as to the contributions of other genes and environmental influences [23].

However, an example of a disease with multiple genetic components that is widely cited as a target for gene therapy is cancer. However, if cystic fibrosis and ADA deficiency represent the conceptually easy end of the gene therapy spectrum, then cancer represents the other extreme [5]. It fulfils hardly any of the criteria set out earlier. The evolution of the malignant phenotype usually involves multiple genetic lesions within the same cell (see below) (criterion 1) and it is unlikely that the nature of every one of these oncogenic mutations is yet known (criterion 2); most cancer patients die because their primary cancers spread throughout the body to colonize essential tissues and organs as metastases. Hence, the target population for gene therapy is usually widely dispersed and often not very accessible (criterion 5); in addition, unlike the situation in CF or ADA deficiency, every tumour cell must, in theory, be 'corrected' to avoid the emergence of recurrent disease. Hence, every malignant tumour cell must be targeted by the therapy (criterion 6). Therefore, it would seem that cancer would not be a natural candidate for gene therapy, since the regulated delivery of just a single gene to localized areas of affected tissues remains highly problematic. None the less, the majority of human gene therapy trials currently under clinical assessment are targeted towards cancer. The rationalization of this almost certainly originates not in a common belief that cancer is particularly amenable to gene therapy, but rather in the fact that there is a large patient population lacking effective, tolerable treatments (criterion 7).

The conversion of a normal cell into a fully transformed malignant cell typically involves mutations in several genes of different classes (Fig. 27.5) [24]. Thus, so-called dominantly acting mutations convert proto-oncogenes into oncogenes and, within the same malignant clone, loss of function mutations abrogate the activity of tumour suppressor genes [25]. The genetic pathway of colorectal tumorigenesis is commonly believed to involve typically about five genetic mutations (or 'hits') to both proto-oncogenes (such as *RAS*) and tumour suppressor genes (such as *p53*, *DCC*, *APC*) [26] (see Appendix VII for details of genes). Therefore, it is far from obvious which of these genetic defects should be targeted for correction in a 'classical' gene therapy approach. It also seems improbable that, even if a single mutation could be corrected in every tumour cell, the malignant phenotype would necessarily be reversed, since the evolution of malignancy in human tumours is so multicomponent in nature.

None the less, several protocols have been proposed in which a mutation that is supposedly central to the continued maintenance of the transformed phenotype is targeted within the tumour cells, in the hope that its correction may reverse the malignant phenotype or induce apoptosis (Fig.

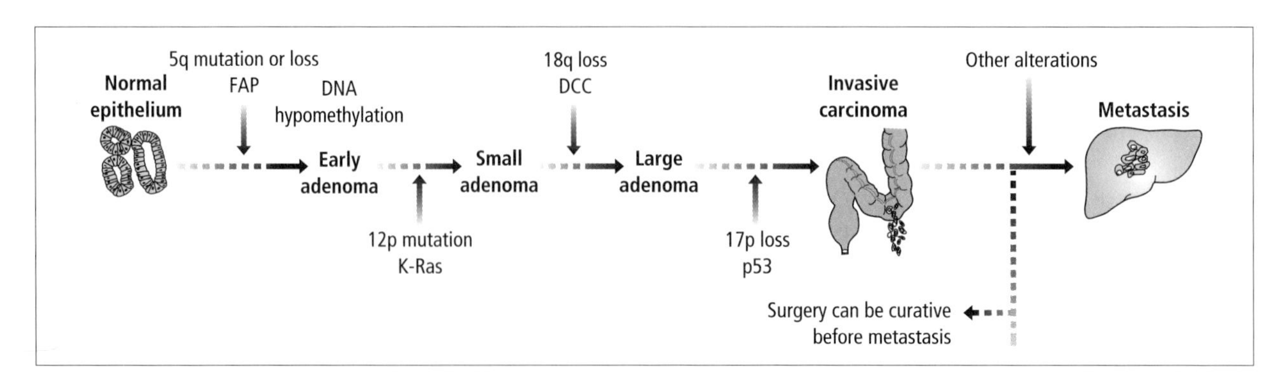

Fig. 27.5 The multifactorial basis of cancer. The evolution of the malignant phenotype of the cancer cell (in this case in the colon) involves multiple genetic mutations of different types, in both proto-oncogenes and tumour suppressor genes [24,25]. This makes it difficult to predict which, if any, of the many possible genetic targets, if corrected by gene transfer, would effectively reverse the malignant phenotype.

27.6a). Therefore, delivery of antisense constructs [27] targeted at abrogating the activity of activated oncogenes (such as *RAS*) have been proposed [28], as have protocols that seek to deliver a functional copy of tumour suppressor genes that are believed to be particularly important to maintenance of the malignant phenotype [29], such as p53 [30]. However, even if these genes are central enough to the tumourigenic process in some human cancers to be used as rational gene targets, the considerable problem remains of delivering at least a single copy of the therapeutic gene to every tumour cell carrying the mutation (although a bystander effect, of unknown origin, has been described which apparently leads to the killing of non-transduced tumour cells by an antisense construct to the *N-RAS* gene [28]). Such levels of gene delivery, even to all the cells in only a localized tumour mass, let alone to systemically dispersed metastatic deposits, is currently impossible [5], so such strategies remain more hopeful than realistic.

As a result of these considerations, genetic therapies for cancer have been proposed which necessarily have led to the creation of a broader definition of gene therapy. These strategies [31–35] represent a fundamental departure from the gene therapies already described, wherein the aim has been to preserve affected cells by correcting their basic genetic defects. Instead, the majority of gene therapy protocols for cancer seek to use noncorrective genes to enhance target (tumour) cell killing.

Gene therapy can be used to kill tumour cells either:

1 directly, by delivery of a *cytotoxic gene* to the tumour cells themselves; or

2 indirectly, by the delivery of an *immunomodulatory gene* which activates the immune system to recognize putative tumour antigens and leads to immune-mediated cell killing.

The delivery of cytotoxic genes to tumour cells has been used essentially for the treatment of localized tumour deposits, which are accessible for gene delivery but are inoperable (Fig. 27.6b). The most commonly used strategy involves delivering a gene encoding an enzyme that will activate a pro-drug to a toxic metabolite, leading to the death of the cell expressing the gene. An example of such a system currently in clinical trials is the herpes simplex virus

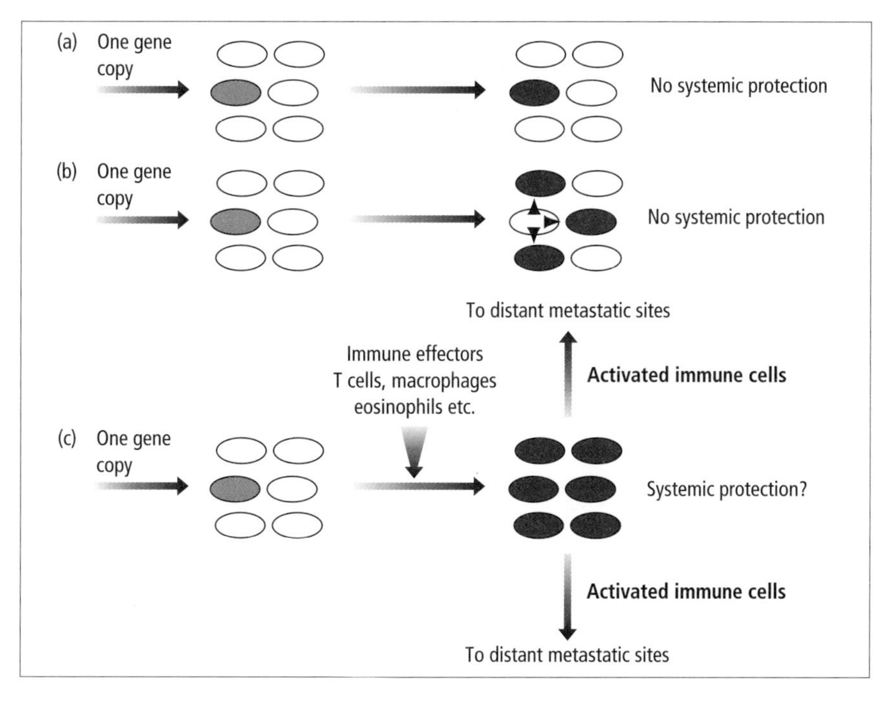

Fig. 27.6 Three possible approaches to the gene therapy of cancer [5]. (a) Corrective gene therapy requires the delivery of a single corrective gene (e.g. a tumour suppressor gene or antisense to an activated oncogene) to the tumour cell. Optimally, expression of the corrective gene reverses the malignant phenotype of the cell in which it is expressed. However, it is unlikely that this will have any effect on the continued growth of surrounding tumour cells (one hit, one kill). (b) Cytotoxic gene therapy leads to the death of the cell expressing the gene as well as its near neighbours by a local bystander killing effect (a single hit is amplified several-fold, see text for details). (c) Immunotherapy involves the expression of an immunomodulatory gene in a tumour cell. In theory, this 'reveals' putative tumour antigens, thereby recruiting immune effector cells to the deposit to kill similar antigen-expressing cells. The activated immune cells can also travel to distant sites of occult metastases to kill other tumour cells affording systemic protection against the cancer.

thymidine kinase gene (HSV*tk*) coupled with the antiherpetic drug ganciclovir [36]. This system has the added advantage that a local bystander killing effect leads to the killing of (non-transduced) cells neighbouring the cells expressing the HSV*tk* gene due to transfer of toxic metabolites between juxtaposed cells [37,38].

In a trial currently in progress at the National Institutes of Health in the USA, patients with inoperable gliomas receive retroviral vectors encoding HSV*tk* by stereotactic injection directly into the glioma followed by systemic ganciclovir [39]. In this design of trial, the chances of success have been maximized by reducing the clinical situation to as close a classical gene therapy approach as possible. Hence, a single gene (criterion 1) is delivered to a localized target tissue (criterion 5) in a manner requiring simple ON/OFF regulation of expression (criterion 3); the presence of the metabolic bystander effect means that the gene does not have to be delivered to every one of the target cell population (criterion 6). The problem of toxicity following inadvertent delivery of the toxic gene to surrounding normal brain tissue (criterion 4) has been partially overcome in this instance by the use of retroviral vectors for gene delivery; these viruses can only infect dividing (tumour) cells but cannot infect the neighbouring quiescent neural tissue [40]. Similar trials using cytotoxic genes delivered to tumour masses in other anatomical locations will require other forms of targeting to ensure minimal toxicity to surrounding tissues. Although in its infancy, the technology to provide this targeting will be provided by engineering specific tropisms into the delivery vectors, both at the level of the surface of the vector as well as by the use of transcriptional targeting [41,42].

An alternative approach to cancer gene therapy is to deliver genes that enhance the immunogenicity of tumour cells, thereby augmenting the immune response against them (Fig. 27.6c) [33,43,44]. Use of the immune system presents three major theoretical advantages for cancer gene therapy.

1 If it can be activated to recognize tumour-specific antigens on tumour cells, the specificity of the immune response should mean that systemic toxicity is reduced to a minimum, since only tumour cells expressing the antigens will be killed.

2 Once activated, immune responses have a natural response amplification mechanism so that only a small stimulus (low levels of gene transfer) is required to produce a large response. That response should, in theory, be body wide and protect against recurrence of disease.

3 Recruitment of the body's own immunity to recognize and destroy the tumour cells should be far less toxic than current treatments such as chemotherapy and radiotherapy.

In effect, if an immune response can be effectively activated against tumour cells, the burden of gene delivery efficiency, specificity and inadvertent toxicity should be transferred from the gene therapist onto the immune system.

There is now good evidence that at least some tumours express tumour antigens which can be recognized under certain circumstances by the immune system [45]. Therefore, it has been proposed that expression of various types of immunostimulatory molecules in tumour cells might enhance immune recognition, possibly by overcoming intrinsic defects in the pathways of antigen presentation by tumour cells [46]. The hope is that tumour cells engineered to express such molecules, either *ex vivo* as vaccines or directly by *in vivo* gene delivery, will generate long lasting immunity to unmodified tumour cells growing at distant sites in the body. Results from animal models have been encouraging, using tumour cells modified to express cytokines (e.g. interleukin-2 and -4 (IL-2, IL-4), granulocyte–macrophage colony-stimulating factor (GM-CSF), and interferons (IFNs)) [43,47], costimulatory molecules (e.g. members of the B7 family) [48,49], MHC molecules [50], allogeneic antigens [51] and syngeneic tumour antigens [52]. Human clinical trials are under way to see if these results translate into clinical gains in humans [43,47].

A modification of this approach has been to use immune cells recovered from excised tumours in adoptive transfer protocols [53]. Hence, immune cells infiltrating certain human tumours, principally melanoma, renal cell cancers and colorectal cancers, have been grown *ex vivo* to high numbers and reinfused into patients. These immune cells presumably have natural tumour recognition capabilities since they are originally isolated from growing tumours; when reinfused they should circulate through the body and concentrate in metastatic deposits, expressing whatever antigens they are primed to recognize (Fig. 27.7).

Initial patient trials using nonT/nonB cell tumour-infiltrating lymphokine-activated killer (LAK) cells in adoptive immunotherapy [54] were superseded by the use of a more specific T-cell population of IL-2 expanded tumour-infiltrating lymphocytes (TILs) [55,56]. Although these trials have reported only limited clinical success, TIL populations are now being used in gene therapy experiments. TILs recovered from patients will be engineered *ex vivo* to express either IL-2 or tumour necrosis factor (TNF) and will then be re-infused

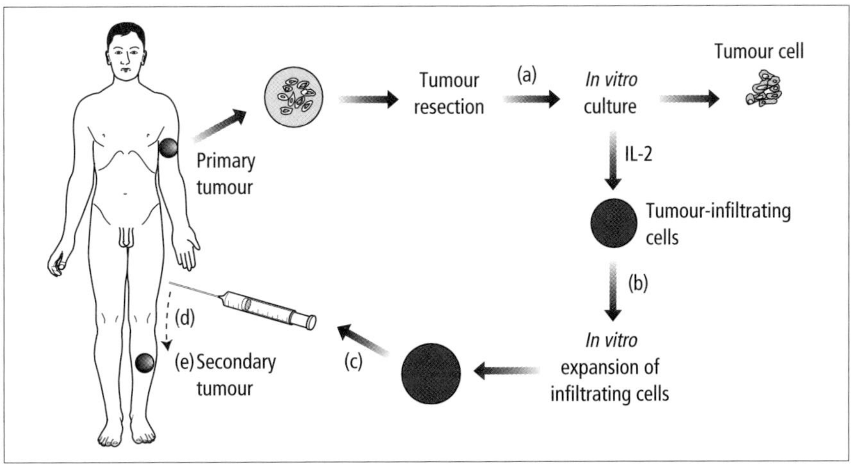

Fig. 27.7 Adoptive immunotherapy with tumour-infiltrating immune cells. The immune cells infiltrating a tumour can be recovered from the excised tumour (a), grown *ex vivo* to high numbers (b) and reinfused into patients (c). These immune cells presumably have natural tumour recognition capabilities since they are originally isolated from growing tumours; when re-infused they should circulate through the body (d) and concentrate in metastatic deposits expressing whatever antigens they are primed to recognize (e).

into the patient [33]. The TILs are effectively being used as tumour-specific delivery vehicles to express immune activating and/or tumoricidal cytokines at high concentrations within tumour deposits. It is not possible to reach therapeutically useful concentrations of such cytokines, especially TNF, by systemic administration of recombinant proteins because of the toxic effects associated with such treatments in humans. However, several technical difficulties have been encountered in achieving high levels of cytokine expression in patients' TILs. This combination of TIL and gene transfer is attractive if the TILs genuinely can localize to tumour deposits which the clinician cannot find/treat but the *in vivo* efficacy of TIL in most tumour types remains controversial.

Finally, gene therapy has been proposed as a means of improving the efficacy of conventional chemotherapeutic treatments. One of the most common causes of treatment failure is the emergence of drug-resistant tumour cells [57,58] which no longer respond to levels of chemotherapy that are acceptable to the patient. If chemotherapy doses could be increased, without the associated bone marrow toxicity, it may be that chemotherapy could be more effective against these resistant clones. Therefore, it has been proposed that transfer of the gene encoding the multidrug resistant protein (*MDR-1*) [57] into bone marrow cells may allow increased dosing with chemotherapeutic drugs [59]. Drug levels might be attainable which will now be toxic to tumour cells but will still be acceptable to the modified marrow because MDR protein actively pumps various chemotherapeutic drugs out of cells which express it. Chemoprotective gene therapy of bone marrow cells has been effective in animal models [60] and may prove clinically beneficial in dose escalation regimens in human patients.

In summary, gene therapy for diseases, such as cancer, that have a multifactorial genetic component, presents many more theoretical problems than for the simple monogenic disorders such as CF or ADA deficiency. For cancer, in particular, the scope of gene therapy has been expanded to include the use of cytotoxic and immunomodulatory genes, as well as the more conventional corrective approaches which are more analogous to CF or ADA deficiency. However, reduction of the clinical target to as close to the CF-type situation as possible may increase the chances of success for specific clinical situations (such as the treatment of gliomas with the HSV*tk*/ganciclovir system).

27.3.4 Infectious diseases

In theory, gene therapy for infectious diseases is attractive because the invading organism introduces pathogen-specific genetic material which is an ideal target for genetic intervention. For example, antisense oligonucleotides can be synthesized with high specificity for gene targets upon which replication of the pathogen is dependent, but which should not recognize any cellular genetic material [61]. Host target cells could then be transduced with such pathogen-protective constructs such that they become resistant to productive infection. Such approaches have been suggested to treat protozoan parasite infections for which drug therapy is currently inadequate [61].

Viral infections offer similar opportunities for

specific genetic interventions. Indeed, in cancers with a known viral aetiology, the presence of viral genes, upon which the evolution of the malignant phenotype depends, offers more cause for optimism than in the treatment of nonviral cancers, because of the presence of specific targets which are separate from cellular genes. Therefore, gene therapy designed to abrogate the expression of papilloma transforming proteins E6 and E7 might be effective in treatment of cervical cancer; similarly, hepatitis B (hepatocellular carcinoma), human T-cell lymphotropic virus types 1 and 2 (adult T-cell lymphoma/leukaemia) and Epstein–Barr virus (nasopharyngeal carcinoma and Burkitt's lymphoma) all offer virus-specific targets for gene therapy intervention in the infected target cells [62].

Similarly, gene therapy is becoming an increasingly attractive option for the treatment of AIDS in the continuing absence of an effective vaccine or drug treatment for the human immunodeficiency virus (HIV) [63]. HIV is a complex retrovirus whose genome expression is controlled by a series of regulatory proteins which control levels of viral protein production and the switch from latency to productive infection [64]. One of these proteins, TAT, is an obligatory transcriptional activator of the viral promoter in the long terminal repeat (LTR). It may be possible to use the complexity of the control of genome expression against the virus to protect the principal target of HIV infection, the $CD4^+$ T cells. For instance, T cells removed *ex vivo* can be transduced with constructs that use the HIV LTR to direct expression of a suicide gene such as the HSV*tk* gene (see earlier) (Fig. 27.8) [65]. When these T cells are returned *in vivo* the absence of TAT will prevent expression of the *tk* gene. However, if the modified T cells become infected with HIV, the wild type virus will provide TAT *in trans* and expression of the

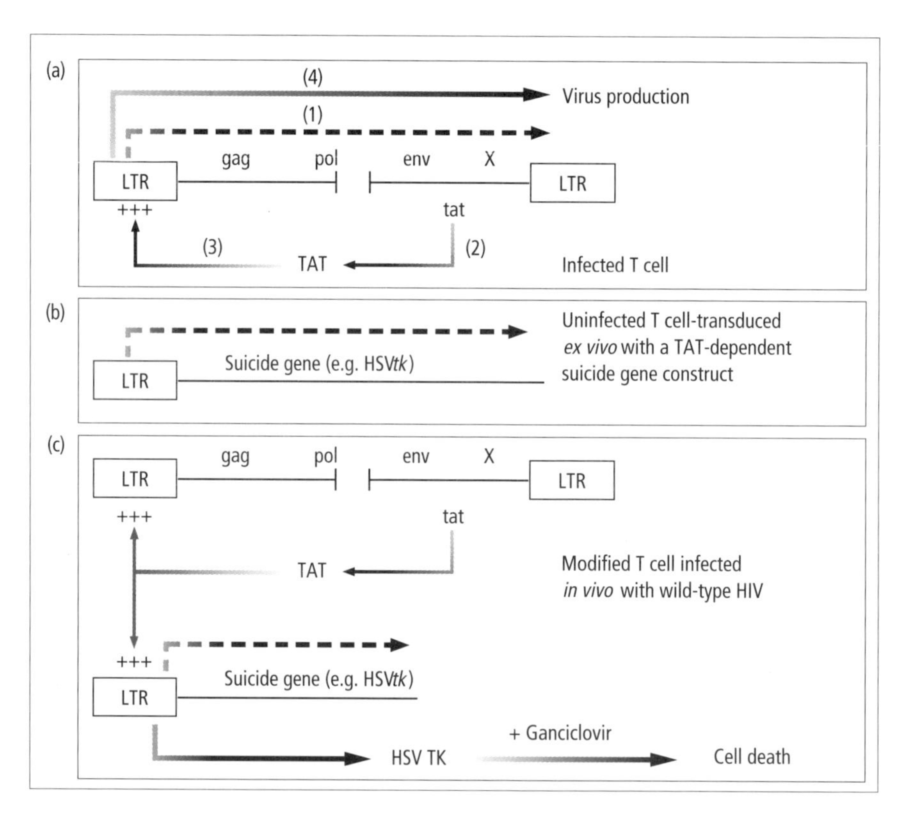

Fig. 27.8 A possible approach to T-cell protective gene therapy for infection with the human immunodeficiency virus (HIV). (a) In an HIV-infected T cell, the integrated proviral LTR promoter has only basal activity, which is insufficient to drive expression of viral structural proteins (1). However, this basal transcription is sufficient to lead to small levels of expression of TAT mRNA (2) and protein (3). TAT feedback on the viral LTR to activate transcription and expression of the viral structural proteins is greatly amplified (4) so that the infected cell becomes a source of virus production. (b) Uninfected T cells can be removed from the body and transduced with a TAT-dependent vector in which expression of the HSV*tk* suicide gene is dependent upon the HIV LTR and, hence, on the presence of TAT in the cell. These modified T cells are returned *in vivo*. (c) If such a modified T cell then becomes infected with wild-type HIV, production of TAT by the infecting HIV upregulates expression from the TAT-dependent LTR-HSVtk construct and the T cell becomes sensitive to ganciclovir. Therefore, infected T cells can be killed *in vivo* before they produce more infectious HIV.

transgene HSV*tk* will be activated. Treatment of the patient with ganciclovir would kill the infected T cells before they could serve as a reservoir of viral production, thereby limiting the ability of HIV to infect more cells.

However, such approaches would be unlikely to abolish infection and would, at best, only slow the progression of disease. Other gene therapy approaches have also been proposed which seek to interfere specifically with viral replication steps without killing the infected T cells [66]; these include the transduction of $CD4^+$ T cells with TAT-dependent HIV-specific antisense or ribozyme [67] constructs [68]. So far, *in vitro* experiments have shown promising results in that these constructs can protect tissue culture cells from infection with HIV. Applications are currently being approved for trials in HIV-infected patients.

27.4 Delivery systems for gene therapy

From the preceding discussions of the clinical situations in which gene therapy may have a role for the future, it is clear that the principal constraint is the ability to deliver the therapeutic gene effectively to the target cells. Vector systems must achieve gene transfer, depending upon the different clinical targets, with varying degress of efficiency, accuracy, stability and safety. These properties will be discussed in general terms below, but for detailed reviews of the properties of individual gene delivery systems the reader is referred to other reviews [3–5].

27.4.1 Efficiency of gene transfer

Physical, non-viral methods of gene transfer have been described for the transduction of cells both *in vitro* and *in vivo*.

Generally, the most efficient means of delivering genes to cells *in vivo* has been by complexing the DNA with cationic lipid and either injecting the complex into the target tissue directly (e.g. a tumour) [51], intravenously [69] or by direct application onto the target tissue [10]. However, these methods are usually much less efficient than virus-mediated vectors. Viruses are natural genetic vectors and have optimized their life cycles for the carriage of genes into target cells. The use of replication-defective, *recombinant viral vectors* has greatly increased the possible efficiencies of gene transfer *in vivo*. To date, only recombinant retroviral vectors and adenoviral vectors have been used in clinical trials [2]. Each has specific advantages and disadavantages which are reviewed elsewhere [3,4, 70]. With current vectors the order of decreasing efficiency of titres is: adenoviral vectors > retroviral vectors > plasmid vectors.

In order to improve existing efficiencies, novel liposome formulations are being developed for plasmid-based delivery [69,71] and improvements to viral titres have been achieved by various means [72]. None the less, currently available vectors often lack sufficient titres for the demands of the clinical situation and improvements in this area will be necessary especially where the target cell population is very large (such as tumours). These considerations have led to suggestions that the only way to achieve sufficient titres for certain disorders is to develop replication-competent vectors which can initiate spreading infections within the target cell population [73] but which have inbuilt safety features to prevent their spread to other cell types [70]. Currently, however, the use of such replicating vectors remains strictly a development for the future.

27.4.2 Accuracy of gene transfer

Ideally, the therapeutic gene should be delivered/expressed only in the target cells to prevent any treatment-related toxicities, although the importance of this requirement depends heavily upon the type of gene being used [5].

Accuracy of delivery of the vector can be achieved at several levels [41,74]. The vector can be delivered to the target area by physical means such as stereotactic injection into tumour deposits (HSV*tk*) [39] or topical application onto airway epithelial cells (CF) [75]. However, more sophisticated genetic means of gene targeting are required for vectors which encode potentially toxic genes and/or which are delivered systemically.

Vector-specific targeting has been used to target HSV*tk* encoding retroviral vectors to replicating glioma cells whilst avoiding infection of quiescent neural tissue around the tumour [40,76]. In addition, cytokine genes such as the IL-2 and TNF might be targeted to tumour deposits using the intrinsic tumour-homing properties of TILs [33]. Surface targeting of the delivery vehicle would be desirable, such that it only infects the appropriate cells. Incorporation of antibodies or ligands into liposomes [77] can target physical delivery of drugs and plasmids and engineering of (retro)viral envelopes may eventually allow cell-specific infection to occur via recognition of target cell-specific molecules (such as tumour antigens) [78–80]. To date, the most effective targeting has been achieved at the transcriptional level by inclusion of cell-type specific enhancer/locus control regions in to both plasmid and retroviral vectors [81–83], thereby restricting gene expression to target cell types even if delivery occurs to surrounding cells. Ultimately, the hope is that delivery vehicles will be developed which incorporate targeting at several levels including transcriptional and surface specificity [74].

27.4.3 Stability of gene transfer

To avoid the need for repeated administration of gene therapy, stable integration of a corrective gene into at least some of the self-renewing stem cells of the target cell population would be the ideal result of a single treatment dose. For diseases such CF or ADA deficiency, stability of expression is clearly of great importance. Plasmid DNA and retroviral vectors can integrate into host cell chromosomes (essentially at random sites), although retrovirus-mediated integration is much more efficient and precise. Adenoviral vectors, however, are maintained episomally in infected cells and are diluted out of the target cell population when the cells divide. Therefore, the order of decreasing efficiency of generating stable gene expression is: retroviral vectors > plasmids > adenoviral vectors.

27.4.4 Safety of gene transfer

A major concern about the advent of genetic therapies for patient treatment is the uncertainty of the consequences of introducing new genetic material into patients' cells. The use of plasmid DNA alone is perceived as carrying less threat than the use of viral vectors, partly because less DNA is usually transferred and partly because viral vectors usually retain viral regulatory sequences to improve efficiency of gene transfer. In the case of retroviral vectors, these regulatory sequences may cause activation of nearby cellular proto-oncogenes following viral integration, leading to transformation of the target cell [84–86], although the estimated risk of this occurring is low [87].

In addition, although *in vitro* tests for replication competent viruses are well developed, especially for retroviral vector stocks, there is a finite chance that contaminating, potentially pathogenic replicating viruses might be cotransferred into patients along with the recombinant stocks [88]. However, the amount of such replication-competent retrovirus that must be transferred to a patient to cause disease appears to be much greater than the quantities that can routinely be detected by current *in vitro* safety tests [89,90]. There is also a risk that naturally occurring, superinfecting viruses may rescue novel, pathogenic viruses by recombination between the wild-type virus and the vector genome. The risks associated with the generation of such new 'doomsday' viruses are difficult to quantify but probably represent more of a conceptual, than a real, risk.

Recombinant viral stocks are also naturally immunogenic by displaying viral antigens on their surfaces [12]. This may hinder the repeated use of such stocks if more than one treatment is required as immunity to the antigens would be expected after a single dose. Moreover, immune responses to even a single dose might be damaging to the patient, as seen in a potentially life-threatening inflammatory reaction of a CF patient treated with very high titre adenoviral stock (see earlier). Therefore, a ranking of currently used vectors for safety, in decreasing order, would be: plasmid vectors > retroviral vectors > adenoviral vectors.

27.4.5 Perspectives

Of those vectors that have currently been approved for use in human trials, no single vector system is likely to possess all the desired attributes for any given situation. The ranking of vectors for safety, efficiency and stability of gene expression does not give concordant results. Therefore, there is often likely to be conflict in the choice of the optimal vector system to use for any particular trial. For instance, where high efficiency of gene transfer should ideally be combined with long-term stable gene expression (such as in the treatment of CF), a compromise must be made between the high-titre adenoviral vectors, the stable integration of retroviral vectors and the safest option of plasmid transfer. In other situations, the dilemma as to which system to use may be less acute; for instance, transient expression of the HSV*tk* gene, or an immunomodulatory gene, in tumour cells would probably be sufficient for cytotoxic or immunological gene therapy of cancer, and in these cases the efficiency of adenovirus-mediated gene transfer may prove to be optimal.

Other viral vectors, not yet approved for clinical use, are currently in development, including herpes simplex virus, parvovirus and adeno-associated viruses [4]. As the number of vector systems that are well characterized enough to be used safely in patients increases, so the conflicts between the different requirements of each system should be easier to resolve. It may also soon be possible to synthesize custom-designed delivery vehicles by incorporating the best features of different vectors into hybrid constructs which have the specific, combined properties required for the gene therapy protocol of choice [74].

27.5 Prospects

When contemplating the 'perfect' disease for which intervention by gene therapy stands the greatest chance of success, several criteria can be proposed. This disease should be a simple monogenic disorder for which the gene has been cloned, and should require simple high-level gene expression for correction; the corrective gene should not be toxic to other cells if inadvertently expressed in them, but the affected cells should be accessible to gene delivery in a localized group, and correction of the target cells should be achievable even if only a fraction of the cells actually receive the gene. Finally, it would only be worthwhile developing gene therapy if the disease has no simple, safe and cheap treatment already.

In reality, gene therapy, in some form or other, has been proposed for a range of different diseases, and this list will continue to expand rapidly, although many do not conform at all closely to the above checklist. The most idealized of real diseases for which gene therapy has apparently been most successful, ADA deficiency, actually has fewer patients who suffer from the disease than researchers working on it. In contrast, cancer, the disease which least fits these criteria, is the one for which the majority of human trials currently exists, principally because many cancers have such poor prognoses that any novel therapeutic approach can be justified on the grounds of patient desperation.

Gene therapy for cancer and infectious diseases, such as HIV, has also led to a general broadening of the definition of gene therapy away from the original concept of the use of genes to correct genetic defects within target cells. However, the unrealistic expansion of the remit of gene therapy for treatment of disease also poses some serious threats to the credibility of the field for the future. Inflated claims regarding its clinical potential, in part as a justification to obtain dwindling research funding, will raise expectations so high that even moderate clinical success in a few limited disease situations will be unable to fulfil the over-hyped promise associated with gene therapy. It is important to define realistic and obtainable goals which gene therapy might actually be able to achieve. These goals will only be sensible if there is a clear knowledge of the capabilities, and limitations, of the gene delivery vectors which are currently available and these should be well understood.

We are currently in an exciting phase where the results of the first human clinical trials of gene therapy are beginning to be reported. The first priority is to ensure that the treatments administered to patients are safe and do not cause adverse reactions. It is unlikely that these early trials will show therapeutic effects, partly because of their inherent design and partly because it is generally end-stage patients who have been recruited. Provided no unforeseen toxicities are reported, the next decade should see gene therapies being administered to patients at earlier stages of disease, in circumstances where they may begin to have therapeutic effects. Eventually, it is to be hoped that, in certain well-designed clinical situations, gene therapy may emerge as effective adjuvant therapy for pre-existing treatment modalities and even, in some cases, as the treatment of choice in diseases as diverse as cystic fibrosis, cancer and HIV.

References

1 *Report of the United Kingdom Health Minister's Gene Therapy Advisory Committee* (1995) Guidance on making proposals to conduct gene therapy research on human subjects. *Human Gene Ther.* **6**, 335–346.
2 Hodgson, C. (1995) The vector void in gene therapy. *Biotechnology* **13**, 222–225.
3 Ali, M., Lemoine, N.R. & Ring, C.J.A. (1994) The use of DNA viruses as vectors for gene therapy. *Gene Ther.* **1**, 367–384.
4 Jolly, D. (1994) Viral vector systems for gene therapy. *Cancer Gene Ther.* **1**, 51–64.
5 Vile, R.G. & Russell, S.J. (1994) Gene transfer technologies for the gene therapy of cancer. *Gene Ther.* **1**, 88–98.
6 Anderson, W.F. (1992) Human gene therapy. *Science* **256**, 808–813.
7 Collins, F.S. (1992) Cystic fibrosis: molecular biology and therapeutic implications. *Science* **256**, 774–779.
8 Alton, E.W.F.W., Middleton, P.G., Caplen, N.J. *et al.* (1993) Non-invasive liposome-mediated gene delivery can correct the ion transport defect in cystic fibrosis mutant mice. *Nature Genet.* **5**, 135–142.
9 Hyde, S.C., Gill, D.R., Higgins, C.F. *et al.* (1993) Correction of the ion transport defect in cystic fibrosis transgenic mice by gene therapy. *Nature* **362**, 250–255.
10 Caplen, N.J., Gao, X., Hayes, P. *et al.* (1994) Gene therapy for cystic fibrosis in humans by liposome-mediated DNA transfer: the production of resources and the regulatory process. *Gene Ther.* **1**, 139–147.
11 Welsh, M.J., Smith, A.E., Zabner, J. *et al.* (1994) Clinical protocol: cystic fibrosis gene therapy using an adenovirus vector: in vivo safety and efficacy in nasal epithelium. *Hum. Gene Ther.* **5**, 209–219.
12 Yang, Y., Nunes, F.A., Berencsi, K. *et al.* (1994) Cellular immunity to viral antigens limits E1-deleted adenoviruses for gene therapy. *Proc. Natl Acad. Sci. USA* **91**, 4407–4411.
13 Yei, S., Mittereder, N., Wert, S. *et al.* (1994) *In vivo* evaluation of the safety of adenovirus-mediated

transfer of the human cystic fibrosis transmembrane conductance regulator cDNA to the lung. *Hum. Gene Ther.* **5**, 731–744.
14 Zabner, J., Couture, L.A., Gregory, R.J. & *et al.* (1993) Adenovirus-mediated gene transfer transiently corrects the chloride transport defect in nasal epithelia of patients with cystic fibrosis. *Cell* **75**, 1–20.
15 Vega, M.A. (1992) Adenosine deaminase deficiency: a model system for human somatic cell gene correction therapy. *Biochim. Biophys. Acta* **1138**, 253–260.
16 Blaese, R.M. *et al.* (1990) Clinical protocol: treatment of severe combined immune deficiency (SCID) due to adenosine deaminase deficiency with autologous lymphocytes transduced with a human ADA gene. *Hum. Gene Ther.* **1**, 327–362.
17 Kay, M.A. & Woo, S.L.C. (1994) Gene therapy for metabolic disorders. *Trends Genet.* **10**, 253–257.
18 Kay, M.A., Rothenberg, S., Landen, C.N. *et al.* (1993) In vivo gene therapy of hemophilia B: sustained partial correction in Factor IX-deficient dogs. *Science* **262**, 117–119.
19 Weatherall, D.J. (1985) *The New Genetics and Clinical Practice* (2nd edn, Oxford Medical Publications, Oxford).
20 Dillon, N. & Grosveld, F. (1993) Transcriptional regulation of multigene loci: multilevel control. *Trends Genet.* **9**, 134–137.
21 Dillon, N. (1993) Regulating gene expression in gene therapy. *Trends Biotechnol.* **11**, 167–173.
22 Dzierzak, E.A., Papayannopoulou, T. & Mulligan, R.C. (1988) Lineage-specific expression of a human β-globin gene in murine bone marrow transplant recipients reconstituted with retrovirus-transduced stem cells. *Nature* **331**, 35–41.
23 Friedmann, T. & Jinnah, H.A. (1993) Gene therapy for disorders of the nervous system. *Trends Biotechnol.* **11**, 192–197.
24 Vogelstein, B. & Kinzler, K.W. (1993) The multistep nature of cancer. *Trends Genet.* **9**, 138–141.
25 Vile, R.G. & Morris, A.G. (1992) The multiple molecular mechanisms of cancer: in search of unification. In *Introduction to the Molecular Genetics of Cancer* (Vile, R.G., ed.), 1–32 (John Wiley, Chichester).
26 Fearon, E.R. & Jones, P.A. (1992) Progressing toward a molecular description of colorectal cancer development. *FASEB J.* **6**, 2783–2790.
27 Mercola, D. & Cohen, J.S. (1995) Antisense approaches to cancer gene therapy. *Cancer Gene Ther.* **2**, 47–59.
28 Georges, R.N., Mukhopadhyay, T., Zhang, Y. *et al.* (1993) Prevention of orthotopic human lung cancer growth by intratracheal instillation of a retroviral antisense K-ras construct. *Cancer Res.* **53**, 1743–1746.
29 Friedmann, T. (1992) Gene therapy of cancer through restoration of tumor-suppressor functions? *Cancer* (Suppl.) **70**, 1810–1817.
30 Fujiwara, T., Grimm, E.A., Mukhopadhyay, T. *et al.* (1993) A retroviral wild-type p53 expression vector penetrates human lung cancer spheroids and inhibits growth by inducing apoptosis. *Cancer Res.* **53**, 4129–4133.
31 Russell, S.J. (1993) Gene therapy for cancer. *Cancer J.* **6**, 21–25.
32 Whartenby, K.A., Abboud, C.N., Marrogi, A.J. *et al.* (1995) The biology of cancer gene therapy. *Lab. Invest.* **72**, 131–145.
33 Rosenberg, S.A., French Anderson, W., Blaese, M. *et al.* (1993) The development of gene therapy for the treatment of cancer. *Annls Surg.* **218**, 455–464.
34 Culver, K.W. & Blaese, R.M. (1994) Gene therapy for cancer. *Trends Genet.* **10**, 174–178.
35 Dorudi, S., Northover, J.A. & Vile, R.G. (1993) Gene transfer therapy in cancer. *Br. J. Surg.* **80**, 566–572.
36 Culver, K.W., Ram, Z., Walbridge, S. *et al.* (1992) *In vivo* gene transfer with retroviral vector-producer cells for treatment of experimental brain tumors. *Science* **256**, 1550–1552.
37 Bi, W.L., Parysek, L.M., Warnick, R. & Stambrook, P.J. (1993) In vitro evidence that metabolic cooperation is responsible for the bystander effect observed with HSV tk retroviral gene therapy. *Hum. Gene Ther.* **4**, 725–731.
38 Freeman, S.M., S.M., Abboud, C.N., Whartenby, K.A. *et al.* (1993) The 'bystander effect': tumor regression when a fraction of the tumor mass is genetically modified. *Cancer Res.* **53**, 5274–5283.
39 Oldfield, E.H., Ram, Z., Culver, K.W. *et al.* (1993) Clinical protocol: gene therapy for the treatment of brain tumors using intra-tumoral transduction with the thymidine kinase gene and intravenous ganciclovir. *Hum. Gene Ther.* **4**, 39–69.
40 Ram, Z., Culver, K.W., Walbridge, S. *et al.* (1993) Toxicity studies of retroviral-mediated gene transfer for the treatment of brain tumors. *J. Neurosurg.* **79**, 400–407.
41 Vile, R.G. (1994) Tumour specific gene expression. *Semin. Cancer Biol.* **5**, 429–436.
42 Huber, B., Richards, C.A. & Austin, E.A. (1994) Virus directed enzyme/prodrug therapy (VDEPT). *Annls NY Acad. Sci.* **716**, 104–114.
43 Tepper, R.I. & Mule, J.J. (1994) Experimental and clinical studies of cytokine gene-modified tumor cells. *Hum. Gene Ther.* **5**, 153–164.
44 Nabel, G.J., Chang, A., Nabel, E.G. *et al.* (1992) Clinical protocol: immunotherapy of malignancy by in vivo gene transfer into tumors. *Human Gene Ther.* **3**, 399–410.
45 Boon, T., Cerottini, J.C., Van den Eynde, B. *et al.* (1994) Tumor antigens recognized by T lymphocytes. *Annu. Rev. Immunol.* **12**, 337–365.
46 Pardoll, D.M. (1993) Cancer vaccines. *Immunology Today* **14**, 310–316.
47 Gilboa, E. & Kim Lyerly, H. (1994) Specific active immunotherapy of cancer using genetically modified tumour vaccines. In *Biologic Therapy of Cancer* (De Vita, V.T., Hellman, S. & Rosenberg, S., eds), 1–16 (J.B. Lippincott, Philadelphia).
48 Townsend, S.E. & Allison, J.P. (1993) Tumour rejection after direct co-stimulation of CD8+ T cells by B7-transfected melanoma cells. *Science* **259**, 368–370.
49 Ramarathinam, L., Castle, M., Wu, Y. & Liu, Y. (1994) T cell co-stimulation by B7/BB1 induces CD8 T cell-dependent tumor rejection: an important role of B7/BB1 in the induction, recruitment, and effector function of antitumor T cells. *J. Exp. Med.* **179**, 1205–1214.
50 Browning, M.J. & Bodmer, W.F. (1992) MHC antigens and cancer: implications for T-cell surveillance. *Curr. Opinion Immunol.* **4**, 613–618.

51 Plautz, G.E., Yang, Z.-Y., Wu, B.-Y. *et al.* (1993) Immunotherapy of malignancy by in vivo gene transfer into tumors. *Proc. Natl Acad. Sci. USA* **90**, 4645–4649.
52 Hawkins, R.E., Winter, G., Hamblin, T.J. *et al.* (1993) A genetic approach to idiotypic vaccination. *J. Immunol.* **14**, 273–278.
53 Rosenberg, S.A. (1991) Immunotherapy and gene therapy of cancer. *Cancer Res.* (Suppl.) **51**, 5074s–5079s.
54 Rosenberg, S.A. (1984) Immunotherapy of cancer by systemic administration of lymphoid cells plus interleukin-2. *J. Biol. Response Modifiers* **3**, 501–511.
55 Rosenberg, S.A., Packard, B.S., Aebersold, P.M. *et al.* (1988) Use of tumor-infiltrating lymphocytes and interleukin-2 in the immunotherapy of patients with metastatic melanoma, special report. *New Engl. J. Med.* **319**, 1676–1680.
56 Ioannides, C.G. & Whiteside, T.L. (1993) T cell recognition of human tumors: implications for molecular immunotherapy of cancer. *Clin. Immunol. Immunopathol.* **66**, 91–106.
57 Endicott, J.A. (1995) The molecular basis of resistance of cancer cells to chemotherapy. In *Cancer Metastasis: From Mechanisms to Therapies*, 123–144 (John Wiley, Chichester).
58 Dalton, W.S., Grogan, T.M., Meltzer, P.S. *et al.* (1989) Drug resistance of multiple myeloma and non-Hodgkins lymphoma: detection of P-glycoprotein and potential circumvention by addition of verapamil to chemotherapy. *J. Clin. Oncol.* **7**, 415–424.
59 Gottesman, M.M. & Pastan, I. (1993) Biochemistry of multidrug resistance mediated by the mutidrug transporter. *Annu. Rev. Biochem.* **62**, 385–427.
60 Sorrentino, B.P., Brandt, S.J., Bodine, D. *et al.* (1992) Selection of drug resistant bone marrow cells *in vivo* after retroviral transfer of human MDR1. *Science* **257**, 99–103.
61 Miller, N. & Vile, R.G. (1994) Gene transfer and antisense nucleic acid techniques. *Parasitol. Today* **10**, 92–97.
62 Schulz, T.F. & Vile, R.G. (1992) Viruses in human cancer. In *Introduction to the Molecular Genetics of Cancer* (Vile, R.G., ed.), 137–176 (John Wiley, Chichester).
63 Gilboa, E. & Smith, C. (1994) Gene therapy for infectious diseases: the AIDS model. *Trends Genet.* **10**, 109–114.
64 Subbramanian, R.A., Cohen, E.A. *et al.* (1994) Molecular biology of the human immunodeficiency virus accessory proteins. *J. Virol.* **68**, 6831–6835.
65 Brady, H.J.M., Miles, C.G., Pennington, D.J. & Dzierzak, E.A. (1994) Specific ablation of human immunodeficiency virus Tat-expressing cells by conditionally toxic retroviruses. *Proc. Natl Acad. Sci. USA* **91**, 365–369.
66 Dropulic, B. & Jeang, K.T. (1994) Gene therapy for human immunodeficiency virus infection: genetic antiviral strategies and targets for intervention. *Hum. Gene Ther.* **5**, 927–939.
67 Altman, S. (1993) RNA enzyme-directed gene therapy. *Proc. Natl Acad. Sci. USA* **90**, 10 898–10 900.
68 Buchschacher, G.L. & Panganiban, A.T. (1992) Human immunodeficiency virus vectors for inducible expression of foreign genes. *J. Virol.* **66**, 2731–2739.
69 Zhu, N., Liggitt, D., Liu, Y. & Debs, R. (1993) Systemic gene expression after intravenous DNA delivery into adult mice. *Science* **261**, 209–211.
70 Vile, R.G. & Russell, S.J. (1995) Retroviruses as vectors. *Br. Med. Bull.* **51**, 12–30.
71 San, H., Yang, Z.Y., Pompili, V.J. *et al.* (1993) Safety and short-term toxicity of a novel cationic lipid formulation for human gene therapy. *Hum. Gene Ther.* **4**, 781–788.
72 Burns, J.C., Friedmann, T., Driever, W. *et al.* (1993) Vesicular stomatitis virus G glycoprotein pseudotyped retroviral vectors: concentration to very high titer and efficient gene transfer into mammalian and nonmammalian cells. *Proc. Natl Acad. Sci. USA* **90**, 8033–8037.
73 Russell, S.J. (1994) Replicating vectors for cancer therapy: a question of strategy. *Semin. Cancer Biol.* **5**, 437–443.
74 Miller, N. & Vile, R.G. (1995) Targeted vectors for gene therapy. *FASEB J.* **9**, 190–199.
75 Zabner, J., Couture, L.A., Gregory, R.J. *et al.* (1993) Adenovirus-mediated gene transfer transiently corrects the chloride transport defect in nasal epithelia of patients with cystic fibrosis. *Cell* **75**, 207–216.
76 Ram, Z., Culver, K.W., Walbridge, S. *et al.* (1993) In situ retroviral mediated gene transfer for the treatment of brain tumours in rats. *Cancer Res.* **53**, 83–88.
77 Ahmad, I., Longenecker, M., Samuel, J. & Allen, T.M. (1993) Antibody targeted delivery of doxorubicin entrapped in sterically stabilized liposomes can eradicate lung cancer in mice. *Cancer Res.* **53**, 1484–1488.
78 Russell, S.J., Hawkins, R.E. & Winter, G. (1993) Retroviral vectors displaying functional antibody fragments. *Nucleic Acids Res.* **21**, 1081–1085.
79 Salmons, B. & Gunzburg, W.H. (1993) Targeting of retroviral vectors for gene therapy. *Hum. Gene Ther.* **4**, 129–141.
80 Valsesia-Wittmann, S., Drynda, A., Deleange, G. *et al.* (1994) Modifications in the binding domain of avian retrovirus envelope protein to redirect the host range of retroviral vectors. *J. Virol.* **68**, 4609–4619.
81 Vile, R.G. & Hart, I.R. (1993) Use of tissue-specific expression of the Herpes Simplex Virus thymidine kinase gene to inhibit growth of established murine melanomas following direct intratumoral injection of DNA. *Cancer Res.* **53**, 3860–3864.
82 Harris, J.D., Gutierrez, A.A., Hurst, H.C. *et al.* (1994) Gene therapy for carcinoma using tumour-specific prodrug activation. *Gene Ther.* **1**, 170–175.
83 Huber, B.E., Richards, C.A. & Krenitsky, T.A. (1991) Retroviral-mediated gene therapy for the treatment of hepatocellular carcinoma: an innovative approach for cancer therapy. *Proc. Natl Acad. Sci. USA* **88**, 8039–8043.
84 Gunter, K.C., Khan, A.S. & Noguchi, P.D. (1993) The safety of retroviral vectors. *Hum. Gene Ther.* **4**, 643–645.
85 Cornetta, K. (1992) Safety aspects of gene therapy. *Br. J. Haematol.* **80**, 421–426.
86 Cornetta, K., Morgan, R.A. & Anderson, W.F. (1991) Safety issues related to retroviral-mediated gene transfer in humans. *Hum. Gene Ther.* **2**, 5–14.
87 Moolten, F.L. & Cupples, L.A. (1992) A model for predicting the risk of cancer consequent to retroviral gene therapy. *Hum. Gene Ther.* **3**, 479–486.

88 Miller, A.D. & Buttimore, C. (1986) Redesign of retrovirus packaging cell line to avoid recombination leading to helper virus formation. *Mol. Cell. Biol.* **6**, 2895–2902.
89 Donahue, R.E., Kessler, S.W., Bodine, D. *et al.* (1992) Helper virus induced T cell lymphoma in non human primates after retroviral mediated gene transfer. *J. Exp. Med.* **176**, 1125–1135.
90 Anderson, W.F. (1993) What about those monkeys that got T-cell lymphoma? *Hum. Gene Ther.* **4**, 1–2.

Chapter 28 Drosophila genome maps

Robert D.C. Saunders

Department of Anatomy and Physiology, University of Dundee, Dundee DD1 4HN, UK

28.1 Introduction

The fruit fly *Drosophila melanogaster* has been at the forefront of genetic research since it was adopted as an experimental organism by T.H. Morgan in the early years of this century. *Drosophila* has many features that make it an eminently suitable organism for laboratory research. *Drosophila* research has produced excellent cytological maps of the larval salivary gland polytene chromosomes, which have served their purpose well in the genetic analysis of this organism. Nevertheless, in the modern era of molecular genetics, the availability of a molecular map has become essential. It is often mistakenly assumed that the primary purpose of a molecular map is to facilitate whole genome sequencing. This is not the case for *D. melanogaster*: there is a wealth of existing genetic information that will be tied in with the molecular map. This includes a great many chromosome rearrangements whose breakpoints are known at the level of resolution afforded by the polytene chromosome map, several thousand characterized and mapped genes, many of which have been cloned, and a large collection of transposon insertions. In this chapter I shall review the current status of genome mapping projects in *Drosophila*, in the context of its use as an experimental model organism. Because *Drosophila* has immense value as a model for a very wide variety of biological and biomedical studies, I shall discuss the features which make it such a powerful experimental model.

28.2 Drosophila as a model system

D. melanogaster is highly amenable to laboratory study. Its requirements for culture are extremely modest, and large-scale genetic experiments are easily carried out. The life cycle of *Drosophila* is typical of holometabolous insects. Embryonic development is rapid, with larvae hatching about 22 h after fertilization. After hatching, the larva grows through three larval instars before pupation. During the pupal stage, the animal metamorphoses into the adult. At 25 °C, the life cycle takes ≈ 10 days. Several features of *Drosophila* development are important with respect to its use as a model organism. The maternally provided RNA and proteins fuel the embryo's development through the early syncytial stages of development, and through to cellularization. Indeed, this maternal provision can exert an effect on the progeny's development well into larval development (a phenomenon known as *perdurance*).

A consequence of this division of the life cycle is that different mutant alleles of a locus may have different phenotypic effects, and lethal phases. One example is the cell cycle gene *polo* [1], which is required for progression through mitosis. The stages at which cell division is required for viability in *Drosophila* are embryogenesis and metamorphosis, as most larval growth is by cell enlargement and endoreduplication of chromosomes during the three larval instars. Thus, weaker alleles of *polo* may yield homozygous adults, whose progeny die as embryos owing to insufficient maternal provision of the *polo* gene product (maternal-effect lethality), while nulls or strong hypomorphic alleles will not allow development of homozygotes through metamorphosis, with a consequent late larval lethal period. Null homozygotes can develop because their maternal provision of *polo* gene product is sufficient to permit the embryonic mitoses in the absence of functional zygotic *polo*. Further examples can be seen in the analysis of genes required to set up the segmental body plan. Such genes have been identified in screens of maternal-effect and zygotic lethals [2,3]. Some of the mutations isolated in this way have turned out to be alleles of viable mutants with a visible phenotype.

The relevance of *Drosophila* to modern medical and biological research stems from the conservation of basic biological processes. Examples are numerous. The differentiation of photoreceptor seven in the compound eye, and of the terminal structures in the embryo, have been shown to be mediated by receptor tyrosine kinases and signal transduction pathways very similar to those in vertebrates. Moreover, the biological function of genes identified by recessive lethal mutations can be directly studied *in vivo* by using mitotic recombination to generate homozygous mutant clones of cells within a viable background.

There are a number of books available which describe the biology of *Drosophila*. Perhaps the most useful in describing the general biology is that edited by Demerec [4]. Ashburner has published a single volume monograph dealing with all aspects of *Drosophila* genetics and biology [5], supplemented by a useful volume of methods [6]. An invaluable sourcebook on genetic loci and chromosome rearrangements, the 'Red Book', has recently been updated [7] and can be accessed electronically via FlyBase (see Section 28.3.5).

28.2.1 Genetic mapping

The first visible mutations of *D. melanogaster*, *speck* and *white*, were discovered in 1910 [8] and were rapidly followed by many others. A recombination map using six sex-linked mutants, the first in any organism, was conceived by Sturtevant and published in 1913 [9]. This map actually orders five loci, since two of the sex-linked factors turned out to be alleles of *white*. By 1925, the genetic map of *D. melanogaster* consisted of about 400 loci allocated to four linkage groups [10]. At that time, genetic research had also been carried out on some other *Drosophila* species, including *D. simulans*, a closely related sibling species [10]. Many mutants were found in *D. simulans* which had their homologous counterparts in *D. melanogaster*, as judged by phenotypic and mapping analysis. This work revealed the existence of a large inversion on chromosome arm 3R in *D. simulans* relative to *D. melanogaster*.

At the time of writing, over 7300 mutants have been characterized and mapped (FlyBase, personal communication). Many of these loci have been cloned and subjected to detailed molecular analysis.

28.2.2 Cytogenetics

28.2.2.1 Mitotic chromosomes

The genome of *D. melanogaster* comprises four pairs of chromosomes (Fig. 28.1). The sex chromosomes are heteromorphic: the X is subtelocentric, while the Y chromosome is entirely heterochromatic. In *Drosophila*, the primary signal for sex determination is provided by the ratio of X chromosomes to sets of autosomes: if the ratio is 1:2, the fly is male; if it is 1:1, the fly is female. Unlike mammalian systems, the Y chromosome is not male determining, but is required for male fertility, and XO individuals are fully viable, but sterile, males. The genetics of sex determination has been well characterized in *Drosophila*, revealing a cascade of genetic interactions which have illuminated many basic biological functions [11]. Again in contrast to mammalian systems, the process by which the level of expression of X-linked loci is adjusted to be the same in both males and females is not by inactivation of one of the X chromosomes in females, but by an increase in the transcriptional activity of the X chromosome in males (dosage compensation) [12].

The two major autosomes, the second and third chromosomes, are metacentric chromosomes in metaphase preparations, while the tiny fourth chromosome is dot-like. Each chromosome can be subdivided into euchromatin and heterochromatin, the heterochromatin being principally located around the centromeres. The definitions of heterochromatin and euchromatin are essentially morphological; heterochromatin remaining more condensed than euchromatin during the interphase of the cell cycle. The majority of the genetic loci are located in the euchromatic regions of the chromosomes.

The heterochromatic regions of the chromosomes have been the subject of intensive chromosome mapping, using DNA-intercalating fluorochromes (see, for example, ref. 13), although this provides a much lower resolution than is possible with polytene chromosome mapping. Nevertheless, these techniques have been fundamental to the genetic analysis of the Y chromosome, which contains no loci essential for viability, but several required for male fertility.

Euchromatin and heterochromatin Heterochromatin is defined on the basis of its condensation behaviour during the cell cycle, generally remaining in a condensed state during interphase, although with the correlation of satellite DNA with heterochromatin, the distinction between satellite DNA and heterochromatin has become a little blurred. Heterochromatin in *Drosophila* can be divided into two classes, α- and β-heterochromatin. In polytene nuclei, the α-heterochromatin is entirely unpolytenized, and appears as a small dot at the chromocentre, while β-heterochromatin is located at the bases of the chromosome arms at intermediate levels of polyteny, with a fuzzy, poorly banded appearance. Many transposable elements are known to be accumulated within the β-heterochromatin.

There are few genes located within heterochromatin, which in general appears to be in a transcriptionally inactive state. Chromosome rearrangements which bring euchromatic genes into close proximity to heterochromatic regions often display a phenomenon known as position effect variegation [14]. For example, when the *white* gene is relocated near toheterochromatin, a variegated or patchy distribution of *white*$^+$ activity can be seen in the ommatidia of the compound eye. The molecular explanation for this phenomenon is at present rather unclear, although many suppressors and enhancers of position effect variegation are known and have been characterized. These genes are implicated in the determination of chromatin structure. Interestingly, position effect variegation has been observed 'in reverse' for a heterochromatic gene, *light* [15], in which the expression of *light* is reduced when relocated by chromosome rearrangement to a euchromatic location.

28.2.2.2 Polytene chromosomes and cytogenetic mapping

Polytene chromosomes had been discovered in *Chironomus* in 1881 by Balbiani [16], though it was not until T.S. Painter published his work on the mapping of chromosomes in 1929 [17] that their importance to genetics was fully realized.

Polytene chromosomes are rather curious struc-

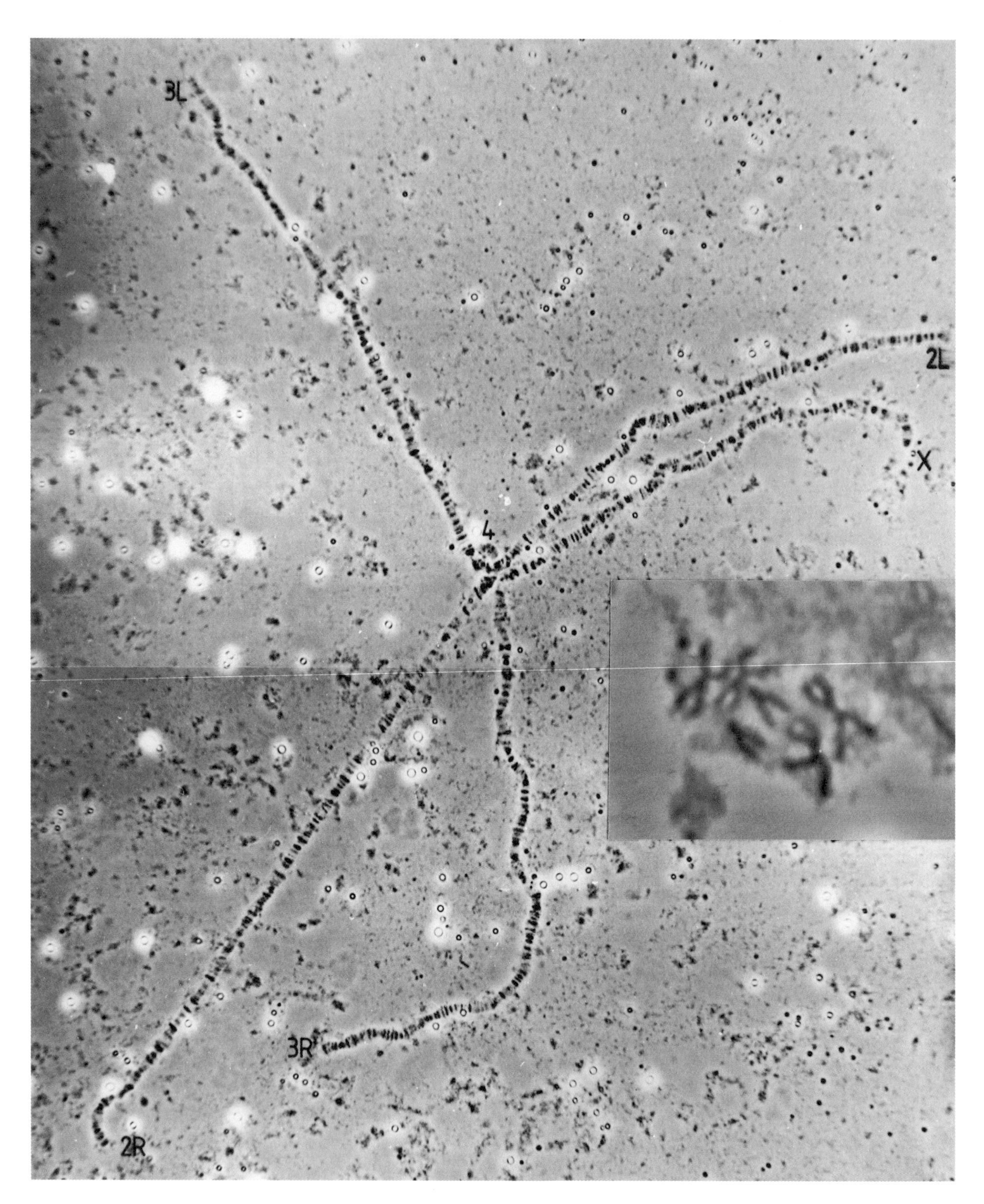

Fig. 28.1 Polytene and mitotic chromosomes of *Drosophila melanogaster*. The five major chromosome arms can be seen extending from the chromocentre. The chromocentre contains the centromeres, and the unpolytenized pericentromeric heterochromatin, with (in males) the heterochromatic Y chromosome. Note the characteristic transverse banding pattern, which is the basis of Bridges' map. The inset shows the mitotic complement of *D. melanogaster* at lower magnification. The two major autosomes, chromosomes 2 and 3, are clearly distinguishable from the other chromosomes of the complement. In favourable preparations, these chromosomes can be distinguished from each other by a secondary constriction of chromosome 2.

tures produced by the continued replication of chromosomal DNA in the absence of mitosis, and persistent synapsis of the homologues in a state of condensation resembling interphase. In mature third instar larval salivary gland nuclei of *Drosophila*, the polytene chromosomes contain ≈1000 tightly synapsed DNA molecules, yielding a characteristic and reproducible pattern of transverse bands. It is this banding pattern that is the real key to the utility of polytene chromosomes for mapping genetic loci and chromosome rearrangements. The polytene chromosome complement is shown in Fig. 28.1. The chromosome arms are joined at the chromocentre, which consists of the under-replicated pericentric heterochromatin. A diagrammatic representation of the structure of polytene chromosomes is shown in Fig. 28.2.

It should be emphasized at this point that the polytene chromosomes represent only the euchromatic fraction of the *Drosophila* genome: the heterochromatin, located close to the centromeres and throughout the Y chromosome is under-replicated in polytene nuclei. Heterochromatin, while representing 35% of the genome, is essentially absent from polytene chromosome maps, and indeed their molecular derivatives. This is not generally considered a problem since the great majority of genetic loci are found in the euchromatin.

The real breakthrough in the cytogenetic mapping of loci to the polytene chromosomes was the inspired mapping system devised by Bridges in 1935 [18]. Bridges' achievement was the adoption of a nomenclature system by which a particular band on the polytene chromosomes might be recognized virtually unambiguously by any other investigator. In so doing, he created what is effectively a usable physical map of the *Drosophila* genome capable of resolving loci as close as a few tens of kilobases in modern terminology. The importance of these chromosome maps for the course of genetics as a science cannot be overstated.

In Bridges' map, each major chromosome arm is divided into 20 sections, or divisions, and each division is subdivided into six subdivisions labelled A to F. In most cases, subdivisions begin with an easily recognized heavy band. This scheme allocates divisions 1–20 to the X chromosome, 21–40 and 41–60 to the left and right arms, respectively, of the second chromosome, and 61–80 and 81–100 to the left and right arms, respectively, of the third chromosome. The minute fourth chromosome, which appears as a dot in metaphase spreads, was allocated divisions 101 and 102. The divisions at the bases of the chromosome arms (divisions 20, 40, 41, 80, 81 and 101) have generally poorly defined band morphology, associated with the increasing quantities of β-heterochromatin found in these regions. Since its introduction 60 years ago, this map has been of central importance to genetic studies in *D. melanogaster* and its sibling species, and it has been used as a model for polytene chromosome maps in many other *Drosophila* species, other Diptera such as mosquitoes, and indeed for species of some of the few other insect orders that have polytene chromosomes (such as Collembola, the springtails).

Bridges' maps have been extensively improved, most notably by a partnership between Bridges and his son, P.N. Bridges [19–23], in which individual bands within each subdivision were given identifying numbers, a real *tour de force* of optical microscopy. The Bridges' revised map contains 5059 bands, which are all uniquely identifiable. This is in

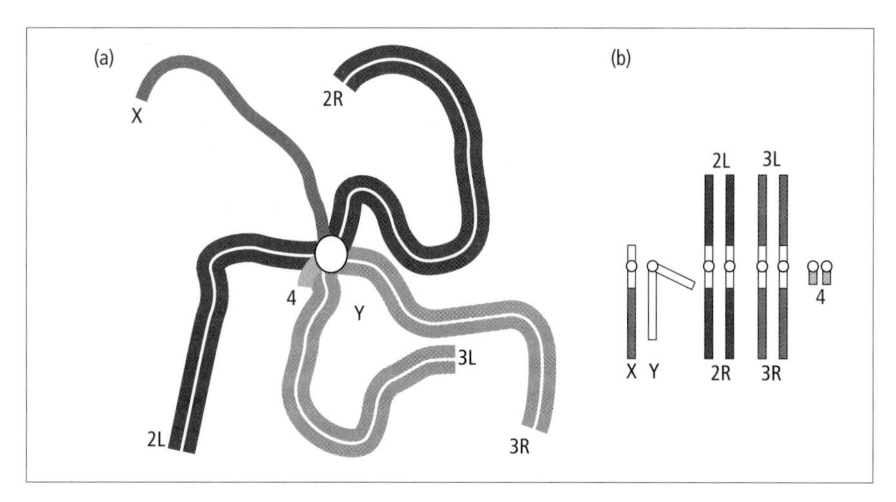

Fig. 29.2 Diagrammatic representation of the polytene chromosomes of *Drosophila melanogaster*. (a) The polytene chromosome complement. The polytene chromosome arms consist of the paired homologues replicated 1000-fold (in *Drosophila* larval salivary gland nuclei) and tightly synapsed in an interphase-like state. The central white circle represents the chromocentre in which heterochromatic regions are located. (b) Representations of individual chromosomes, showing heterochromatic regions as white blocks. Parts (a) and (b) are not to scale.

very good agreement with the total found by Sorsa [24], during electron microscopic studies of the polytene chromosomes of *D. melanogaster*. There are of course minor disagreements between the band counts of these two maps, although Bridges' map system can generally be relied upon to avoid confusion. A further set of maps, this time a photographic map aligned with Bridges' 1935 map, was published in 1976 [25]. Taken together, these three maps are an essential research tool in the genetic analysis of *D. melanogaster*.

An immense collection of chromosomal rearrangements is available to the *Drosophila* investigator: inversions, transpositions, translocations, and deficiencies (*Drosophila* terminology for deletions), many of which can be combined to yield a sometimes bewildering array of possible segmental aneuploids. Using these rearrangements, mutations may be mapped to a particularly high precision, often to within a few tens of kilobases.

28.2.2.3 Polytene chromosomes and evolutionary studies

Polytene chromosomes have facilitated studies of the phylogeny of *Drosophila* species. For example, within the *melanogaster* species group the phylogenetic relationship of the six species was worked out on the basis of fixed inversions of the polytene band sequence [26]. This is possible because the banding pattern of the chromosomes of these species, while extensively rearranged by inversions relative to one another, is essentially identical.

Muller [27] proposed an alternative nomenclature for chromosome arms, in which the standard designations X, 2L, 2R, 3L, 3R and 4 of *D. melanogaster* are replaced by elements A, B, C, D, E and F, respectively, reflecting evolutionarily conserved genetic elements [27,28]. This terminology was introduced since, for example, chromosome arm 2R of *D. melanogaster* may not correspond with the arm designated 2R in another species. Additionally, chromosome arms break and rejoin during evolution. For example, in *D. virilis*, each of Muller's elements are present as separate chromosomes. It appears that these elements have been maintained as units throughout the evolution of the genus *Drosophila*, as judged by the analysis of homologous mutations, chromosome banding patterns (where possible), and *in situ* hybridization. Indeed it has been suggested that Muller's elements are conserved in flies as distantly related to *Drosophila* as the blowfly *Lucilia* [29] and mosquitoes. As discussed in Section 28.3.6, there are many applications of the *D. melanogaster* physical map to the study of evolutionary relationships in the genus *Drosophila*, and further afield within the order Diptera.

28.2.3 Molecular genetics of Drosophila

Drosophila is in the vanguard of molecular genetic investigation of eukaryotic systems, and many important strategies and techniques were developed or perfected for use in *Drosophila*, such as *in situ* hybridization, genomic libraries and chromosome walking, to name but three. Consequently, it has remained the organism of choice for many researchers.

28.2.3.1 The structure of the Drosophila genome

The *D. melanogaster* haploid genome size is 0.18 pg, which corresponds to 170×10^6 bp [30]. In contrast to mammalian genomes, the DNA is unmethylated [31]. Table 28.1 describes the basic composition of the *D. melanogaster* genome.

Satellite DNA Twenty-one per cent of the haploid genome consists of satellite DNA of a number of families. Satellite DNA in *Drosophila* is principally located in the pericentromeric heterochromatin, and in the Y chromosome. In general, the satellite DNA consists of simple-sequence repeats arranged in large blocks, although one class, the 1.688 g ml^{-1} satellite located on the X chromosome, has a repeat unit of 359 bp [32]. There are several other classes of satellite DNA repeats, some of which display characteristic distribution patterns among the chromosomes.

Ribosomal RNA genes *D. melanogaster* rDNA is located in the nucleolar organizers, on the X and Y

Table 28.1 The composition of the *D. melanogaster* genome.

Category of DNA	% of genome	Size (bp)
Total genome		170×10^6
Single-copy sequences	61	103.7×10^6
Satellite DNA	21	35.7×10^6
Genes for rDNA, histones, etc.	3	5.1×10^6
Transposable elements	9	15.3×10^6
Foldback DNA	6	10.2×10^6

chromosomes. These include the genes for the 18S and 28S rRNAs [33]. *bobbed* (*bb*) is a mutation resulting from mutation of the rDNA arrays [34]. The *bb* phenotype is seen when the number of functional rDNA copies is reduced to 50% of normal wild-type levels. X-chromosomal *bb* alleles are complemented by the presence of rDNA arrays on the Y chromosome, Y*bb*⁻ chromosomes are known.

Histone genes There are ≈100 repeats of the histone genes per haploid genome [35], located on chromosome arm 2L, at map location 39D3–39E1.2 [36]. In addition, a number of histone gene variants appear to be present elsewhere in the genome [36].

Foldback DNA Because foldback DNA is implicated as transposable DNA, and in the function of the family of large transposable elements, the TEs, it is described in the section on transposable elements below [37–42].

Transposable elements Approximately 9% of the *D. melanogaster* genome is composed of middle repetitive transposable elements, being dispersed on average every 13 kb throughout the euchromatin. The families of *Drosophila* transposable elements and related elements are reviewed in refs 43 and 44. Some 50 families of transposable element have been identified in *D. melanogaster*, and they are typically present in copy numbers in the range 10–100. The most significant element with respect to the molecular genetic analysis of *Drosophila* has been the P element, and this is covered in greater detail in Section 28.2.3.2. Limitations of space prevent the listing of all transposable elements found in *D. melanogaster*. A reasonably complete listing can be found in ref. 5. Several of these elements are associated with hybrid dysgenesis. There is strong evidence for horizontal transmission of some transposable elements between species [45]. *Drosophila simulans*, a sibling species of *D. melanogaster*, appears to have a lower proportion of dispersed repetitive DNA, and different populations of transposable elements [46–48]. This difference can be exploited experimentally, as has been done, for example, by the European Consortium genome mapping project described in Section 28.3.2 below.

Copia-like elements Copia elements possess direct long-terminal repeats (LTRs) of several hundred base pairs, and within these LTRs there are short-terminal inverted repeats. The structure of these elements is similar to that of retroviruses. Other members of this class of transposable element include *gypsy*, *297*, *17.6*, *mdg-1* and *412*. Many *Drosophila* mutations are due to the insertion of *copia* or other members of this class of element.

Long-terminal inverted repeat elements A substantial proportion of the genome is composed of rapidly reannealing foldback DNA; it has been estimated that there are about 2000–4000 pairs of inverted repeats in the *D. melanogaster* genome. These structures have been shown to be transposable [40], and they are quite variable both in terms of the length and sequence of the repeats (which are themselves internally repetitious), and the length and sequence of the region between the repeats. The very large TE elements of Ising [49] are derived from foldback (FB) elements, and typically contain a section derived from the X chromosome spanning the *white* to *roughest* interval, sufficiently large to be seen cytologically in the polytene chromosomes in some cases.

Transposable elements with short inverted terminal repeats The most notable member of this class is the P element, which will be described in greater detail below. Another element in this class, *hobo*, is also implicated in a hybrid dysgenesis syndrome, and has been used as an insertional mutagen and as a germline transformation system in a similar way to the P element [50]. The activity of both P and *hobo* elements are associated with high levels of chromosome rearrangements [51].

Transposable elements without terminal repeats The I factor is the causative agent of IR hybrid dysgenesis in *D. melanogaster*. I factors are related to the mammalian LINE elements, and *Drosophila* F elements [52]. These elements appear to transpose via an RNA intermediate.

28.2.3.2 The P element and its uses

The most important transposable element from the point of view of *Drosophila* genetics is the P element. This element is the cause of P–M hybrid dysgenesis, a syndrome of multiple effects such as male recombination (which does not normally occur in *Drosophila*), high frequency of chromosome rearrangement, sterility, and high mutation rates [53]. The dysgenic effects are seen in the descendants of crosses of P strain males with M strain females, but not vice versa. P strains contain P elements and have the P cytotype, whereas M strains lack P elements, and have the M cytotype. The dysgenic effects appear to be due to the transposition of P elements within the genome of the P strain male, as it enters the permissive environment of M cytotype eggs at fertilization. The full-length P element is ≈2.9 kb, and encodes a single transcription unit of four exons, which are differentially spliced to encode either the transposase or a repressor of transposition [54]. Additionally, factors encoded in the host genome are required for P element transposition [55]. The elevated mutation rates typically seen in hybrid dysgenesis are due to the insertion of P elements within or in close proximity to transcription units.

Transposon tagging The mutagenic effect of P-element insertion has been of use in gene cloning strategies. Because the mutations due to P transposition generally retain a copy of the element either within or near the coding region, these mutant genes can be cloned by virtue of the inserted element [56]. This procedure is known as transposon tagging, and is of course not restricted to P elements. Transposon mutagenesis has been refined so that stable single insertions of P elements are recovered [57]. Collections of *Drosophila* stocks each bearing a single marked P insertion associated with a lethal or visible mutant phenotype are important components of the European and US *Drosophila* genome projects (see Sections 28.3.2.6 and 28.3.3.2)

P element-mediated germline transformation In the early 1980s, Rubin and Spradling developed a system by which the P element could be used to introduce DNA into the germline at high frequencies [58,59]. Essentially, P element-mediated germline transformation utilizes two components. The first component is the vector, a P element modified so it does not encode transposase, into which the DNA under investigation is inserted, along with a selectable marker. The second component is a helper P element, which encodes functional transposase but which cannot undergo transposition itself. The resulting transformant flies are genetically stable, since there is no active transposase present. Germline transformation can prove that a particular gene has been cloned, by rescue of a mutant phenotype with a candidate clone.

Germline transformation has enabled the development of a wide variety of additional techniques with applications in genetics, developmental and cell biology. Enhancer trapping was initially developed using a single element containing a promoterless bacterial *lacZ* gene to detect the transcriptional activity of neighbouring enhancer elements [60]; it is now available as two-component systems. In these systems, one element expresses the yeast transcription factor Gal4 in a temporal and spatial pattern determined by a neighbouring enhancer. The Gal4-containing element is then used to drive a reporter gene, borne on a second element, via an upstream activating sequence from the Gal4 promoter (UAS_{Gal4}), placed upstream of the reporter gene [61]. This is a highly versatile system: any Gal4 element insertion can be used with any responding gene cloned downstream of a UAS element. The responding gene can be a gene under experimental investigation, a reporter gene such as the bacterial chloramphenicol acetyl transferase (CAT) gene and *lacZ* [61], or a toxin used for cell ablation in a pattern corresponding to the Gal4 expression pattern [62].

P elements as mutagens *Drosophila* lacks a system analogous to homologous recombination in yeast which can be used to mutate a specific gene. In cases where genes have been identified solely from cloning experiments, corresponding mutants may not be available. P elements have been utilized in a sib-selection mutagenesis strategy [63,64], where potentially mutagenic insertions at a locus of interest are selected by PCR amplification of eggs laid by mutagenized females. A series of tenfold divisions of the pool of mutagenized flies can be screened, ultimately identifiying single flies containing a particular insertion. Of course, not all insertions will have a mutagenic effect.

28.3 Mapping projects

The *Drosophila* genome has been mapped at several levels of resolution, reflecting the insert size in yeast artificial chromosome (YAC), P1 phage, or cosmid vectors. Phage λ-vectors have too small a cloning capacity to be of use in large-scale mapping endeavours in *Drosophila*. There has been cooperation between the genome mapping groups in the exchange of materials and information, but in practice they have been run independently. I will describe these projects in a loose chronological order. The first to be discussed are those that have reached an acceptable degree of conclusion: the YAC-based maps of D. Hartl and colleagues, in which clones were ordered by *in situ* hybridization to polytene chromosomes. *In situ* hybridization of randomly selected clones has also been used to build a framework map of the *D. melanogaster* genome in P1 clones, with the same library used in the *Drosophila* Genome Center mapping project (see Section 28.3.3). A second project, which also makes extensive use of the polytene chromosomes, is a collaborative project funded by the European Community (EC) to map the genome in cosmids. While conventional fingerprinting techniques are used, chromosome microdissection (see ref. 65 and

Chapter 11) is used to subdivide the genome prior to mapping. Finally, a comprehensive effort to tie together a wealth of genetic and molecular information to a P1 map is being conducted by a large collaborative venture involving the laboratories of Rubin, Spradling, Hartl and Palazzolo, which together form the *Drosophila* Genome Center.

All these mapping endeavours are being carried out concurrently, and will provide a multilayered map which will be of immense utility. It should be pointed out that all of the above projects are dealing with the euchromatic portion of the genome. The heterochromatin, containing long arrays of simple repeat satellite DNA, is not easily amenable to physical mapping at this time.

28.3.1 The Drosophila YAC maps

Two YAC-based mapping projects have been carried out, principally in the laboratories of D.L. Hartl (Harvard, formerly at Washington University), and I. Duncan (Washington University). The procedures involved have been directly transferred to the analysis of P1 genomic libraries of *D. melanogaster* (Section 28.3.3) and *D. virilis* (Section 28.3.6.1).

28.3.1.1 Strategy

The strategy adopted by Hartl, Duncan and colleagues was that of mapping YAC clones directly to the polytene chromosomes by *in situ* hybridization. This is an approach appealing in its simplicity and straightforwardness. YACs were chosen in favour of other vector systems principally for the large size of their cloned DNA segments, with the consequence that a correspondingly small number of *in situ* hybridizations need be carried out. The band count of the revised polytene chromosome map of Bridges and Bridges is 5059 bands, and since the euchromatic fraction of the genome which this represents is 65–70% of the total genome of 165×10^6 bp, or 115×10^5 bp, the average band contains 22 kb DNA (the interbands contain very little DNA). One therefore expects to find that a typical YAC clone containing an insert of 200–220 kb would span about 10 bands, and this is indeed what was seen [66].

There are drawbacks to this mapping strategy, relating to the interpretation of hybridization signals on polytene chromosomes: the experimenter's interpretation of an *in situ* hybridization can be incorrect. This can be controlled relatively easily by including duplicate clones, and comparing the *in situ* readings. Cai *et al.* [67] describe the comparison of 38 YACs mapped by Ajioka *et al.* [68], and conclude that while the reported localizations are broadly correct, the precise end points defined by the two groups differ. In practice these differences are likely to be due to technical limitations on the resolution of *in situ* signals and chromosome bands. These include variable signal strength developed on the chromosomes, and variable chromosome morphology. The former can result in an overestimation or underestimation of the number of bands covered by a signal, and the second can prevent the accurate visualization of chromosome bands.

28.3.1.2 Libraries

Three YAC libraries were constructed, in the laboratories of D.L. Hartl and I.W. Duncan (Table 28.2) [66,67]. The first library consists of 768 clones, derived from randomly sheared DNA from the wild-type strain Oregon RC. The genomic DNA was prepared from embryos and was size selected for DNA fragments larger than 120 kb. The fragments were cloned into the vector pYACP-1. This vector has a number of features associated with its use as a *Drosophila* vector, including terminal fragments of the transposable P element that flank both the cloned DNA and a bacterial G418-resistance gene driven by the *Drosophila hsp70* promoter. In principle, these features should enable clones from this library to be reintroduced to the genome by P element-mediated germline transformation, though Garza *et al.* [69] do not report attempts to achieve this and, to date, successful germline transformation using YAC clones in this vector has not been reported.

The second library was derived from DNA partially digested with *Not*I, prepared by embedding cells from Oregon RC gastrulae in agarose plugs. Partial digestion with *Not*I was carried out in these plugs. The genomic DNA was

Table 28.2 *Drosophila* YAC libraries.

DNA source	Oregon RC	Oregon RC	*y; cn bw sp*
DNA preparation	random shear	*NotI*I partial digest	*Eco*RI partial digest
Size selection	> 120 kb	–	150–550 kb
Vector	pYACP-1	pYAC5	pYAC4
No. of clones	768	2688	4032
No. analysed	272	502	419

cloned in the vector pYAC5, resulting in 2688 clones.

The third library was prepared with DNA extracted from an isogenic *y; cn bw* sp. stock. The DNA was partially digested with *Eco*RI, and fragments of 150–550 kb selected for library construction. The vector used was pYAC4, and 4032 clones were obtained.

It was considered important that questions of the integrity and stability of YAC inserts be addressed for each library used. Repetitive sequences in particular might be expected to be susceptible to postcloning rearrangements. In the case of rRNA genes, about 50 YAC clones containing rDNA repeats were selected and analysed, indicating that there is no obstacle to the cloning of these regions in the YAC systems used. These clones hybridize *in situ* to the chromocentre of the polytene chromosome complement, as expected. However, some instability of clones was observed by comparing restriction patterns of different isolates of identical clones.

Long stretches of simple-sequence satellite DNA were not recovered from the YAC libraries examined. In the *Drosophila* genome, these sequences are located in heterochromatin, which is principally found near the centromeres of all chromosome arms, and throughout the Y chromosome. At these loci, satellite DNA is arranged in long blocks of repeated units. However, it is not immediately obvious which stage of the cloning process discriminates against the satellite sequences. Hartl [66] discusses a number of possibilities, including the hypothesis of Lohe and Brutlag [70] that this DNA is lost at the stage of DNA purification, due to the nature of DNA–chromatin protein interaction in heterochromatin.

The representation of single-copy sequences was assessed by hybridization of cloned genes to the library. In no case was there a failure to identify at least one clone in the library. In addition, examination of the organization of the inserts in comparison to the previously characterized clones revealed no detectable rearrangement during the YAC cloning. A further point is that since the libraries were made from DNA extracted from embryos of mixed sex, there is an expectation that there would be reduced representation of X-chromosomal regions relative to autosomal regions (assuming there to be an equal proportion of male and female, there will be only three X chromosomes and one Y chromosome for each four sets of autosomes).

A very similar approach is being taken with a library of P1 clones, which are being ordered by *in situ* hybridization to polytene chromosomes [71,72]. This P1 library is the library used by the *Drosophila* Genome Center for a large-scale mapping project, and will be described in Section 28.3.3.

28.3.1.3 Final status of the YAC maps

Estimation of the degree of YAC clone coverage of the genome is made by calculating the proportion of chromosome bands covered by YAC clones, as judged by *in situ* hybridization. These calculations are always approximate, due to technical considerations discussed in Section 28.3.1.1.

Without the resolution afforded by the polytene chromosomes, the linear relationships between YAC contigs and unattached YAC clones would be impossible to determine. However, since *in situ* hybridization to polytene chromosomes is the means by which these contigs were assembled, even those clones that remain unattached are a usable component of the final map. This mapping strategy results in a map in which clone overlaps are not molecularly characterized.

Hartl map The final YAC map is estimated to represent 90% of the euchromatic genome [66,73], as all but 550 of the 5157 bands of the Sorsa EM polytene chromosome map are covered by YAC clones in *in situ* hybridization experiments. It consists of 1193 YAC clones in 149 contigs. The estimate of the number of contigs is conservative: the number may well be lower, since YACs with *in situ* signal overlaps shorter than two bands are not scored as overlapping. The average insert size of the YAC clones is 200 kb.

Duncan map Cai *et al.* [67] estimate their map covers about 76% of the autosomal euchromatin, and 63% of the X chromosome euchromatin. The under-representation of the X chromosome in these libraries is expected, as described above. In addition, sequences derived from the fourth chromosome are under-represented, for reasons that are at present unclear.

28.3.1.4 Availability of clones

The complete list of mapped YAC clones is to be found in FlyBase (see Section 28.3.5). The listing contains clone names with the cytological map location on polytene chromosomes, and information on how to obtain clones.

28.3.2 European Consortium cosmid map

28.3.2.1 Strategy

The extensive use of polytene chromosomes is a feature of the cosmid-based physical map being constructed by a consortium of European laboratories, headed by F.C. Kafatos (EMBL, Heidelberg) [74]. The laboratories involved are those of C. Savakis and C. Louis (IMBB, Crete), M. Ashburner (Cambridge, UK), D.M. Glover and R.D.C. Saunders (Dundee, UK), and J. Modollel (Madrid, Spain). The strategy adopted is illustrated in Fig. 28.3. This approach uses fingerprinting techniques, as developed for the *Caenorhabditis elegans* cosmid-based mapping project (see Chapter 29) [65]. However, unlike the *C. elegans* methodology, cosmids for fingerprinting are not selected at random from the genomic library. Rather, the genome is separated into about 100 similarly sized segments, corresponding to each of the Bridges' map divisions, by chromosome microdissection and PCR amplification. DNA amplified in this way is used to screen the master cosmid library, in order to identify division-specific minilibraries of clones. Each minilibrary is treated as a separate small genome of 1.2 Mb for the purpose of assembling contigs of cosmids. This has a beneficial effect, as it permits the use of reduced stringency in the computer matching of clones, with no consequent increase in spurious overlap detection. Additionally, the map for each division approaches completion earlier than if the whole library were to be fingerprinted simultaneously.

All contigs are checked by *in situ* hybridization of some member clones to polytene chromosomes,

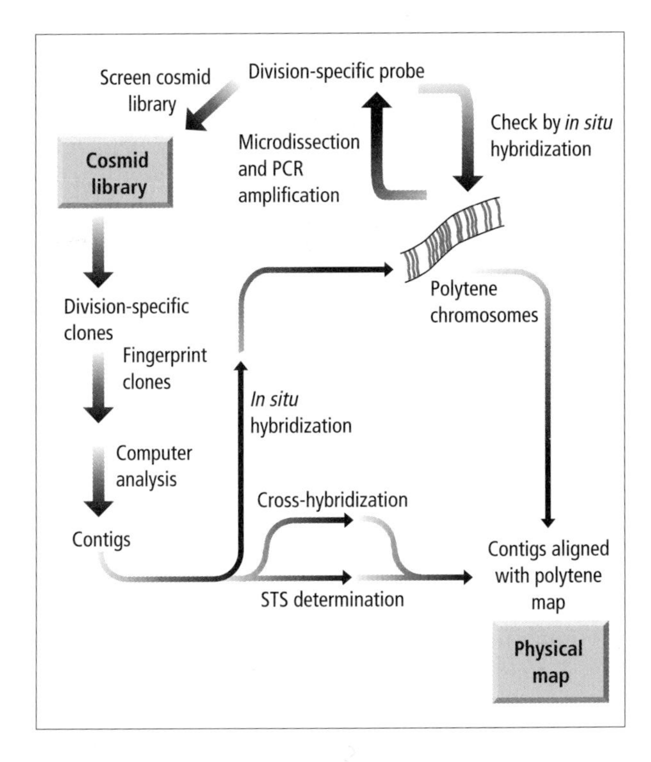

Fig. 28.3 Flow chart of the European Consortium genome mapping project.

and unattached cosmids are also mapped in this way. Any clones derived from unexpected locations are reassigned to their correct divisional map. Sequence-tagged sites (STSs) are determined from selected cosmids, which together with hybridization of cloned genes to the cosmid library permits alignment of the physical and genetic maps. A flow chart illustrating the strategy is presented in Fig. 28.3.

28.3.2.2 Library construction

The cosmid library was constructed in the cosmid vector Lorist 6 [75], using DNA extracted from the wild-type strain Oregon R. DNA was extracted from freshly eclosed adults, and was partially digested with *Sau*3 A, then ligated in the *Bam*HI site of Lorist 6. Nineteen thousand and two hundred independent colonies were transferred into wells of 200 microtitre plates, for long-term storage. For library screening, the clones were gridded manually on 25 filters of 768 clones (eight microtitre plates per filter). Subsequently, 192 of the microtitre plates were picked using a robotic device built and operated by H. Lehrach's laboratory at the ICRF in London. Robotically picked filters either consisted of two 22 cm^2 filters bearing 96 × 96 colonies on each, or one 22 cm^2 filter bearing 192 × 96 colonies, in a regular array.

28.3.2.3 Polytene chromosome microdissection

In the early stages of this project, conventional microcloning was used to prepare region-specific probes, a procedure in which DNA microdissected from polytene chromosomes is digested with restriction enzyme then cloned in a λ insertion vector [76]. However, this approach was not technically straightforward, and it proved impossible to routinely generate usable probes. Only one division was mapped using a microcloned probe. All subsequent division-specific probes were derived from PCR amplification of microdissected DNA (see Chapter 11) [77,78]. This procedure produces probes of much higher complexity than does microcloning. Microdissected DNA is cleaved to completion with *Sau*3A, and double-stranded adapter oligonucleotides are ligated to the cohesive termini, to provide priming sites for PCR amplification. Because the polytene chromosomes of *Drosophila* represent an initial amplification of 1000-fold because of polyteny, a single microdissection provides enough material to generate a representative probe.

A major problem in this approach was rapidly identified. The presence of dispersed middle repetitive DNA in the genome results in a high frequency of misidentification of clones using these probes. For example, a microdissection of division 1 will identify most or all of the cosmids containing inserts derived from that division. However, should there have been (for example) a *copia* element present in the initial microdissection, a number of clones will be identified solely because they contain further copies of *copia*. As described in Section 28.2.3.1, there is evidence that the genome of *D. simulans* contains a lower proportion of dispersed middle repetitive DNA [46–48]. Furthermore, only a proportion of that present in *D. simulans* is also present in *D. melanogaster*. For these reasons, the microdissections were carried out on the polytene chromosomes of *D. simulans*. These polytene chromosomes are essentially identical to those of *D. melanogaster*, differing principally by a large inversion in chromosome arm 3R [27].

It has been found empirically that this strategy alleviates the problem of transposable elements in microdissected probes, but that it does not eliminate it entirely. A proportion of clones in each screen are thus derived from an essentially random genomic location, though these clones are not lost from the analysis. The question of whether the nucleotide sequence conservation between the two species is sufficient for successful use of this strategy was addressed empirically [76], by probing a Southern blot of λ-phage clones from a walk in the *achaete-scute* region with an appropriate microdissected probe. Virtually all of the restriction fragments hybridized with the probe. It has been estimated that the sequence divergence between the majority of the single-copy sequences in these two species is only of the order of 3–4%. The probes prepared by microdissection consist of a pool of small restriction fragments, used to identify segments of DNA some 100 times their own size. In practice there appear to be no problems due to nucleotide sequence divergence.

28.3.2.4 Fingerprinting and contig assembly

Cosmids are fingerprinted and analysed essentially as described for the *C. elegans* genome mapping project [79] with the modification that the enzyme used for generating the fingerprints is *Hin*fI. Cosmids corresponding to each division are analysed together, and kept initially in an individual division-specific database.

Following contig assembly, a number of cosmids are selected from each contig and analysed by *in situ* hybridization to polytene chromosomes to verify their location on the cytogenetic map. Not all cosmids can be assigned to a single locus, however. In some cases a cosmid has, in addition to a primary site, a number of secondary sites with weaker signals, which probably correspond to dispersed transposable elements. In these cases, the primary site can still be determined. In other clones, a large number of sites of equal intensity are found, making it impossible to verify the cytogenetic location of the clone. Hybridization of these clones to *D. simulans* chromosomes can in many cases resolve the location of the clone, because of the differences between the population of repetitive DNA in *D. simulans* and *D. melanogaster*.

The cosmid-based physical map has been aligned with the recombination map in three ways. First, members of the research community have made existing cloned genes available for hybridization studies, and these clones have been used to screen the master cosmid library. In some cases contigs could be linked, or new cosmids mapped by this procedure. A simpler approach has been to synthesize oligonucleotides corresponding to genes with entries in the sequence databases, and to use these as hybridization probes. Oligonucleotides 20 or 25 bases long have been successfully utilized. To avoid problems with base composition affecting the T_m of the hybrids, the washing conditions of Wood *et al.* [80] were used. This eliminates the annealing strength differences of GC compared with AT base pairs, and means that all oligonucleotide probes may be washed at the same stringency. Finally, database searches conducted with STSs determined from the termini of cosmid clone inserts (see below) has also identified cloned genes which are therefore linked to the cosmid map [81].

28.3.2.5 Sequence-tagged sites

Sequence-tagged sites (STSs) [82] are being determined at intervals along the physical map, for two reasons. Firstly, they will make the map independent of the library with which the map is being constructed. Any investigator with access to a PCR machine and an oligonucleotide synthesis facility will be able to generate a locus-specific probe, by amplification from genomic DNA using primers deduced from the STS, with which to screen any genomic library available. Secondly, this part of the project has proved to be an efficient means by which the recombination map can be aligned with the physical map (and therefore also the cytogenetic map). STSs are determined by sequencing the termini of the cloned DNA segments within certain cosmid clones. These cosmid clones are selected on the basis of their location within contigs, and from the unattached cosmids with unambiguous cytological locations. A further use of the STSs is in the P1-based genome map of Rubin and colleagues (see

Section 28.3.3). All the STS sequences generated in this project are deposited in the EMBL sequence database, and listed in FlyBase (Section 28.3.5). Searches for homology between STSs and sequence databases reveals that a number of STSs correspond to known *Drosophila* genes, and others to *Drosophila* homologues of known genes from other organisms. For example, within 568 STSs determined from mapped cosmids derived from the X chromosome, 33 corresponded to known *Drosophila* genes, and nine represented homologues of genes cloned from other species.

28.3.2.6 Determination of sequence tags from single P element insertions

A more recent addition to the European Drosophila Genome Mapping project concerns the characterization of a collection of P element insertions consisting of ≈ 3000 *Drosophila* lines, each containing a single third chromosomal P element insertion associated with lethal or subvital mutations. These P element insertion lines are being analysed in order to provide further detail to the cosmid physical map. The P element constructs selected for the mutagenesis have a number of important features. First, the elements are not autonomous; they cannot transpose in the absence of transposase gene function, and *Drosophila* strains containing such elements are therefore stable. Second, they have either a *lacZ* or *Gal4* reporter gene acting as an enhancer trap. Third, they contain *Escherichia coli* plasmid sequences (an origin of replication, and an antibiotic resistance gene) enabling integrated elements to be cloned, together with flanking DNA, by plasmid rescue.

The initial round of analysis involves mapping the insertions by testing them against chromosome deficiencies. Subsequently, each group of lines is then further analysed by complementation testing. In addition to identifying duplicate loci, this step also enables the recognition of background lethals that may have been in the mutagenized fly strain prior to mutagenesis, and which will not be associated with P element insertions. The cytological position of each insertion is subsequently precisely determined by *in situ* hybridization to polytene chromosomes.

Because the integrated P element vector still contains a functional plasmid origin of replication and encodes ampicillin resistance, genomic DNA sequences flanking the site of insertion may be cloned by plasmid rescue. The DNA sequence immediately flanking the site of insertion will be determined as an STS. However, because each insertion is associated with a lethal phenotype, each of the STSs determined is tightly linked to a vital gene. Genomic sequences isolated in such plasmid rescues will be related to the cosmid-based physical map by hybridization or PCR, providing a link between genetic and physical maps, and would be expected to identify a number of novel genetic loci.

28.3.2.7 Current status of the map

Analysis of the X and second chromosomes is essentially complete. Some cytological divisions appear to be under-represented in this map [81], for reasons that are poorly understood. Estimates of coverage are calculated per cytological division range from 28% to over 100% and average 64%, and are based on the DNA content per band calculated by Sorsa [24], and an estimate of the size of the cosmid contigs. This takes into account those cosmids mapped to the division, but which are not members of contigs.

28.3.2.8 Availability of cosmid clones

All the cosmids mapped and analysed are freely available to the research community. The complete list of clones, with cytogenetic location and other information can be found in FlyBase. STS sequences are deposited in the EMBL sequence database.

28.3.3 The Drosophila Genome Center

A major initiative towards a physical map closely tied to the wealth of available genetic data is being undertaken by the *Drosophila* Genome Center (G.M. Rubin, personal communication). The principal investigator is G.M. Rubin (Berkeley), and the Center actually comprises a number of laboratories: those of C. Martin and M. Palazzolo (Lawrence Livermore, Berkeley), D.L. Hartl (Harvard), and A. Spradling (Carnegie Institution, Baltimore). Significantly, this project is working closely with the automation laboratories at Lawrence Livermore to develop software and hardware for the project. The Center can be accessed through FlyBase.

28.3.3.1 Construction of a complete P1 physical map

The *Drosophila* Genome Center is constructing a map using the P1 cloning technology developed by Sternberg [83]. The library used was constructed in D.L. Hartl's laboratory [71] and contains five genome equivalents. This library has been the subject of an *in situ*-based mapping strategy very similar to that carried out for the YAC library by Hartl's laboratory [71,72]. The *Drosophila* Genome Center is engaged in assembling these clones into contigs with molecularly defined overlaps, as determined by STS content mapping. Figure 28.4 shows a flow chart depicting the *Drosophila* Genome Center's mapping strategy.

In situ *hybridization mapping* A strategy analogous to that used in the YAC map project described in Section 28.3.1 has been used to create a 'framework' map of P1 clones aligned with the polytene chromosome cytogenetic map [72].

The P1 library consists of 9216 clones, of which ≈40% were made using the vector pNS582-tet14Ad10, and the remainder with pAd10sacBII. The genomic DNA was derived from nuclei isolated from adults of an isogenic *y*; *cn bw* sp. stock. The genomic DNA was partially digested with *Sau*3A before ligation into the *Bam*HI sites of the two cloning vectors. The two sets of clones have a very similar distribution of insert sizes, averaging slightly over 80 kb. Hartl *et al.* [72] describe the *in situ* analysis of 3104 clones. Of these clones, 388 hybridized to the chromocentre or to many euchromatic sites, and 191 clones were deliberate duplicates to assess the accuracy of the interpretation of *in situ* results. The presence of transposable elements was inferred for about 10% of the remaining 2461 clones, which gave dispersed multiple sites of hybridization. As with the cosmid *in situ* hybridizations described in Section 28.3.2, in general the primary signal is easily determined by its intensity. The mapped clones with unique or

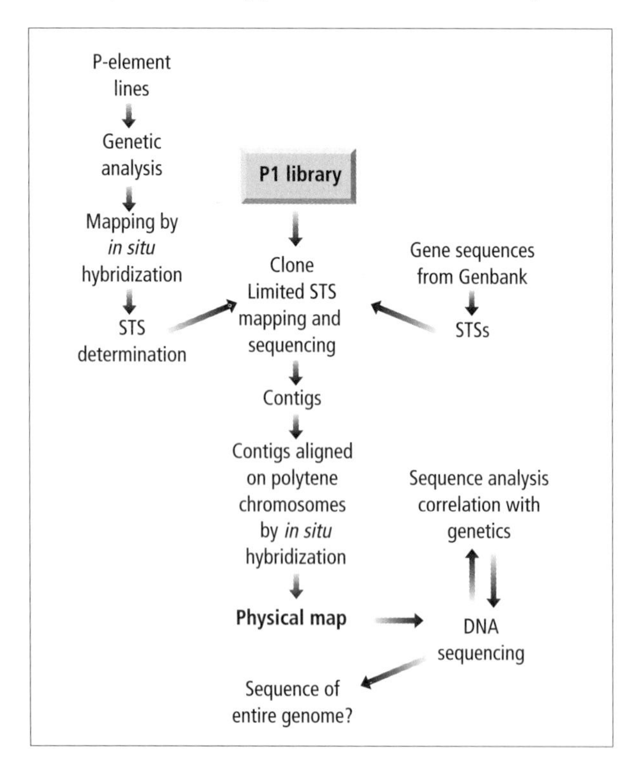

Fig. 28.4 Flow chart of the *Drosophila* Genome Center genome mapping project.

primary hybridization sites total 200 Mb of insert DNA (assuming an average insert size of 80 kb), equivalent to 1.8 copies of the euchromatic fraction of the haploid genome. The map represents an estimated 85% coverage of the genome. Hartl *et al.* [72] present a diagrammatic representation of the P1 clone map relative to the polytene map.

These *in situ* hybridization studies revealed 64 clones with two sites of hybridization. Although these clones were not investigated further, it is probable that they represent chimaeric clones.

STS content mapping STS content mapping is carried out using a two-stage PCR analysis of the entire P1 clone library. Primers are designed from the STS sequence itself. Positive signals are detected as bands on agarose gel electrophoresis. To streamline the PCR analysis, the P1 library, picked into 96 96-well microtitre dishes, has been divided into a collection of pools: there are 96 plate pools, each of which contains all 96 clones from a plate. In the first stage, PCR amplifications are performed on all 96 pools, by which the plates containing clones spanning a particular STS can be rapidly identified. In the second stage, row and column pools (containing 12 and 8 clones, respectively) from microtitre plates identified in the first round are screened to identify positive clones unambiguously.

Initially, STS content mapping was used to assemble contigs for the genomic regions containing the bithorax complex (300 kb), the Antennapedia complex (350 kb), and a 2000 kb region around the alcohol dehydrogenase locus (*Adh*). The latter region is a section of the genome that has been extensively characterized genetically and cytogenetically: many mutants and chromosome rearrangements are available.

STSs are determined by sequencing the termini of cloned inserts of the P1 library. Additionally, STSs are determined from the insertion sites of P elements (see Section 28.3.2.6) and from *Drosophila* gene sequences in the sequence databases.

28.3.3.2 The use of P elements as sequence-tagged sites

As described in Section 28.2.3.2, P element transformation vectors have been of great importance in modern *Drosophila* genetics, with some highly sophisticated systems in use. A.C. Spradling and his laboratory have accumulated a large number of *Drosophila* stocks, each bearing a single P element insertion and yielding a mutant phenotype as a consequence of that insertion. Each of these insertion stocks therefore is a marker for a gene. The P element used contains plasmid DNA positioned between the P element termini, and which is therefore inserted into the genome along with the element. The plasmid DNA enables the element to be 'rescued' from genomic DNA by restriction digestion followed by recircularization. These rescued plasmids also contain fragments of genomic DNA that originally flanked the P element. P elements are being retrieved from these stocks by plasmid rescue, and their sites of insertion sequenced to yield STSs.

Each P element line is analysed by *in situ* hybridization to determine the location of the insertion on the polytene chromosome map. Those insertions with similar map positions are checked by genetic complementation tests for allelism. This analysis also reveals those lines with more than one inserted element (about 7%), which are not suitable for further analysis.

In order to facilitate the dissemination of these materials, copies of the collection of stocks have been distributed to five laboratories, and to the Bloomington Stock Center. Stocks can be ordered by e-mail: through FlyBase. Several stock centres may be accessed in this way.

28.3.3.3 cDNA project

The original goals of this project were to accumulate 2000–3000 different cDNA sequences (expressed sequence tags, ESTs) for use as STSs. Analysis by database searches would be likely to identify many genes, and spot homologies. However, the emphasis has now shifted towards the characterization of all transcribed sequences in a defined region of the *Drosophila* genome, the 2 Mb region around the gene *Adh*. Accordingly, this project is now aimed at the large-scale sequencing of this genomic region, by using and developing novel nonrandom sequencing strategies, and the characterization of cDNAs encoded within the region.

28.3.3.4 Large-scale sequencing

Shotgun sequencing presents a number of problems when applied to large-scale DNA sequencing (see Chapter 20). In addition to the redundancy of repeatedly sequencing the same stretches of DNA, many projects end with the synthesis of many oligonucleotide primers with which gaps between sequence may be bridged. The *Drosophila* Genome Center has developed a directed sequencing strategy that overcomes many of the drawbacks of conventional shotgun sequencing. The overall strategy is based upon the transposon-insertion method of generating templates, developed by Palazzolo and colleagues [84]. A significant emphasis on the use of specifically designed software and hardware is being made by the *Drosophila* Genome Center.

The assembly of ordered arrays of subclones The strategy begins with the establishment of contiguous arrays of P1 clones. These clones have an average insert size of about 80 kb, which must be subdivided before sequencing. An ordered set of ≈960 subclones containing inserts of about 3 kb is made from each P1 clone. Clones are ordered by PCR amplification based upon limited sequence analysis of subclones. The subclone library is analysed by a pooling system similar to that used in the STS content mapping of the P1 library. However, in this case one of the primers is complementary to the plasmid vector flanking the site of insertion, while the second primer is complementary to sequence determined from one of the subclones. Thus each clone will yield a different sized DNA fragment upon PCR amplification, with the fragment size dependent upon the degree of overlap with the starting clone. In this way 30 DNA pools need be screened by PCR (10 plate pools, 8 row pools, and 12 column pools) to identify suitable clones. By determining sequence from the minimally overlapping clone, a further round of analysis can be undertaken. The end result of this stage of the analysis is an ordered array of small subclones corresponding to a contig of P1 clones.

Transposon-facilitated DNA sequencing Strathman *et al.* [84] describe the use of γδ transposon mobilization to provide priming sites for DNA sequencing. In brief, the *E. coli* host strain carrying an F factor with a γδ element is transformed with an ampicillin-resistant genomic subclone and allowed to conjugate with a kanamycin-resistant strain. By plating conjugants on medium containing both kanamycin and ampicillin, only cells receiving plasmids transferred during conjugation as γδ transposition cointegrates between the subclone plasmid and the F plasmid are recovered. Following resolution of the cointegrate, which yields plasmids with γδ insertions, the colonies are rapidly screened by PCR using one vector primer and one γδ-specific primer to establish the site of integration of the γδ element. A minimally overlapping set of clones is selected for sequencing.

Sequencing is carried out with chain termination protocols, using sequencing primers complementary to the γδ transposon. Contig assembly is performed using both commercially available software, and software developed by the Lawrence Berkeley Laboratory (LBL) Human Genome Center computing group (see Appendix V.1 for contact address).

28.3.3.5 Availability of clones

The computer database FlyBase should be consulted

for details of mapped P1 clones. Gridded copies of the P1 library have been distributed to a number of laboratories around the world, and clones can be obtained from these sources. This strategy was adopted in order that clones might be obtained by researchers quickly, efficiently and at low cost. Clone requests should be directed to the nearest laboratory holding a copy of the library. FlyBase contains details of these laboratories.

28.3.4 Sequencing the Drosophila genome

At present, genome sequencing of the *D. melanogaster* genome is being undertaken by two groups. The US *Drosophila* Genome Center is using the technology outlined in Section 28.3.3.4 to carry out large-scale sequencing, while the European Union is funding a consortium of laboratories to carry out a pilot project in which the sequence of the terminal three divisions of the X chromosome will be determined. In contrast to the US DGC strategy, the European project is using a shotgun cloning approach, in which individual mapped cosmids are fragmented by sonication, and the fragments subcloned to yield templates for sequencing.

28.3.5 FlyBase

FlyBase is a database containing information on the genetics and biology of *Drosophila*, being built by a collaboration between researchers funded by the National Institutes of Health in the USA, and the Medical Research Council in the UK. The laboratories involved in this collaboration are those of W. Gelbart (Harvard), M. Ashburner (Cambridge, UK), T. Kaufman and K. Matthews (Indiana University, Bloomington), and J. Merriam (UCLA). Access to FlyBase is most easily obtained by using a Web browser, such as Netscape. FlyBase is found at http: //morgan.harvard.edu:80/ where details of FlyBase mirrors in the UK, Australia and Japan may be found.

FlyBase contains a wealth of genetic and molecular information concerning *Drosophila*. The text of the Lindsley and Zimm 'Red Book' [7] is included (only in the Indiana copy, owing to copyright reasons), as are lists of chromosome aberrations (sorted by class and cytological breakpoints), molecular clones, the genetic map, and stock lists of the international *Drosophila* stock centres, amongst other material. These files can be interactively searched.

28.3.6 Genome mapping in other Diptera

The application of genome mapping to the genetic study of other insect groups can be by both direct usage of the *D. melanogaster* genome map, and transfer of techniques developed for mapping the *D. melanogaster* genome. As well as the obvious benefits such maps yield for the molecular genetic analysis of these species, they will have an impact on a variety of areas, such as evolutionary biology. For example, phylogenetic relationships of many *Drosophila* species have been deduced from the analysis of the banding sequence of their polytene chromosomes. Ashburner [85] has made a strong argument for the importance of genetic mapping in a variety of insect species, and in many cases, molecular physical genome maps will be important. I will discuss the application of large-scale genome analysis to some Diptera other than *D. melanogaster*.

28.3.6.1 Drosophila virilis

At 313 Mb, the genome of *Drosophila virilis* is approximately double the size of that of *D. melanogaster*. There are six chromosome pairs in the diploid complement: acrocentric sex chromosomes, four pairs of acrocentric autosomes, and a pair of tiny autosomes. In general terms, the distribution of genetic loci is in agreement with the conservation of Muller's elements (see Section 28.2.2.3), although the gene order within each element is scrambled relative to *D. melanogaster*. The X chromosome of *D. melanogaster* corresponds to the X chromosome of *D. virilis*, 2L to chromosome 4, 2R to chromosome 5, 3L to chromosome 3, 3R to chromosome 2, and 4 corresponds to *D. virilis* chromosome 6 [86].

Drosophila virilis is a member of the subgenus *Drosophila*, while *D. melanogaster* is a member of the subgenus *Sophophora*. The polytene chromosome banding pattern of *D. virilis* is not similar to that of *D. melanogaster*, and the degree of sequence divergence between the two species is sufficient that *D. melanogaster* clones cannot be directly used in hybridization studies with *D. virilis* DNA. A molecular genome map would be of use to those studying genome evolution in the genus *Drosophila*, and to those engaged in molecular and genetic research in *D. virilis*. The *D. melanogaster* P1 map described in Section 28.3.3.1 has been utilized in two ways in the *D. virilis* work. Firstly, the transfer of molecular biological technology has permitted the development of a useful P1 library and mapping strategies, and secondly, protocols have been developed that allow the polytene locations of *D. melanogaster* homologues to be determined.

A *D. virilis* P1 genomic library has been constructed by Lozovskaya *et al.* [87]. This library consists of more than 10 000 clones of average insert size 65.8 kb. These clones are being mapped by *in situ* hybridization to polytene chromosomes, as with the YAC- and P1-based maps. Because it has proved impossible to map *D. melanogaster* P1 clones to *D. virilis* polytene chromosomes reliably by *in situ* hybridization, owing to sequence divergence outside conserved regions, a scheme for identifying homologous regions was devised. In brief, DNA fragments were PCR amplified from *D. melanogaster* genomic DNA, and used to screen the *D. virilis* P1 library. These P1 clones were then used to map the *D. virilis* homologue by *in situ* hybridization.

28.3.6.2 Drosophila pseudoobscura

Extensive research has been carried out on the evolution of the obscura species group, and in particular *D. pseudoobscura*. Consequently, genome mapping of *D. pseudoobscura* has been initiated [72]. In contrast to the situation described above for *in situ* hybridization studies using *D. melanogaster* P1 clones to probe *D. virilis* polytene chromosomes, *D. pseudoobscura* is sufficiently closely related to *D. melanogaster* for the corresponding experiments to work with a reasonable degree of success. *D. pseudoobscura* belongs to the same subgenus as *D. melanogaster*. However, the polytene chromosome banding patterns are not comparable, although experiments confirming the existence of Muller's elements by *in situ* hybridization have been carried out.

28.3.6.3 Anopheles gambiae

The mosquito *Anopheles gambiae* is the major malarial vector in sub-Saharan Africa. As a consequence it has attracted a great deal of research interest. A low-resolution genome map based on microdissection of polytene chromosomes has been constructed [88]. This map consists of pools of DNA amplified by PCR following microdissection [65,78], and is intended to aid the molecular genetic characterization of this medically important insect.

One of the points of interest in the genomic analysis is the presence of six sibling species best distinguished by chromosomal inversions that can be visualized in the polytene chromosomes of larval salivary glands or the ovarian nurse cells. Genomic analysis of the species complex will be vital if attempts are to be made to genetically modify natural mosquito populations.

A detailed recombination map is being assembled for *A. gambiae* [89], using microsatellites as markers. Some of these markers were derived from the low-resolution microdissection genome map. All these microsatellite markers are effectively STSs tied to the recombination map, and by *in situ* hybridization to the polytene chromosome map.

28.4 Conclusions and prospects

At present, *Drosophila* researchers have access to three genome maps assembled by direct correlation of clones with the polytene chromosomes, using clone libraries constructed in YAC and P1 vectors. These maps are virtually complete, estimates of their coverage of the genome are currently 90% [66] and 76–63% [67] for the two YAC maps, and 85% for the P1 map [72]. The European Consortium cosmid-based map has also been essentially completed to an acceptable degree of coverage for many regions of the genome (the X, and many of the chromosome map divisions of the autosomes). In the case of the X chromosome, the overall coverage in cosmids is estimated at 65%, though this figure varies between divisions [81]. Furthermore, in the near future, the P1-based map will be refined by molecular analysis to yield molecularly demonstrable clone overlaps. *Drosophila* researchers already have excellent access to clones derived from specific regions of the genome.

The *Drosophila* genome is of similar size to a typical mammalian chromosome, so those mapping projects that are not tied into a specifically *Drosophila*-based approach should function well as models for mapping strategies for larger genomes. In particular, the approach taken by the *Drosophila* Genome Center is a good model. In these cases, the added benefit of the high-resolution cytological maps afforded by polytene chromosomes enable frequent checks on map quality to be made.

Particular issues of interest in physical mapping that may be approached in *Drosophila* include approaches to the long-range analysis of centromeres and heterochromatic regions of chromosomes. In particular, a powerful approach is to use deletion derivatives of chromosomes to narrow down regions absolutely required for accurate chromosome segregation.

Drosophila remains the model organism of choice in many areas of research, and the physical maps of the *D. melanogaster* genome now available represent a major asset in the future exploitation of this most important model organism.

Acknowledgements

I thank my colleagues in the *Drosophila* genome mapping field, in particular G.M. Rubin and D.L. Hartl, who supplied information, much of it prior to publication, my collaborators in the European *Drosophila* genome mapping consortium, and those members of the laboratory in Dundee who read drafts of the manuscript. I acknowledge the

Wellcome Trust and the UK Medical Research Council for financial support.

References

1 Sunkel, C.E. & Glover, D.M. (1988) *polo*, a mitotic mutant of *Drosophila* displaying abnormal spindle poles. *J. Cell Sci.* **89**, 25–38.
2 Nusslein-Volhard, C. (1979) Maternal effect mutations that alter the spatial coordinates of the embryo of *Drosophila melanogaster*. In *Determinants of Spatial Organization* (Subtelney, S. & Konigsberg, I.R., eds), 185–211 (Academic Press, New York).
3 Nusslein-Volhard, C. & Wieschaus, E. (1980) Mutations affecting segment number and polarity in *Drosophila*. *Nature* **287**, 795–801.
4 Demerec, M. (ed.) (1950) *The Biology of* Drosophila (John Wiley, New York). (Reprinted as a facsimile edn, 1994, Cold Spring Harbor Laboratory Press, Cold Spring Harbor, NY.)
5 Ashburner, M. (1989) Drosophila*: A Laboratory Handbook* (Cold Spring Harbor Laboratory Press, New York).
6 Ashburner, M. (1989) Drosophila*: A Laboratory Manual* (Cold Spring Harbor Laboratory Press, New York).
7 Lindsley, D.L. & Zimm, G.G. (1992) *The Genome of* Drosophila melanogaster (Academic Press, San Diego).
8 Morgan, T.H. (1910) Sex-limited inheritance in *Drosophila*. *Science* **32**, 120–122.
9 Sturtevant, A.H. (1913) The linear arrangement of six sex-linked factors in *Drosophila*, as shown by their mode of association. *J. Exp. Zool.* **14**, 43–59.
10 Morgan, T.H. (1925) *The Genetics of* Drosophila. (Reprinted 1988, Garland Publishing, New York).
11 Parkhurst, S.M. & Meneely, P.M. (1994) Sex determination and dosage compensation: lessons from flies and worms. *Science* **264**, 924–932.
12 Lucchesi, J.C. & Manning, J.E. (1987) Gene dosage compensation in *Drosophila melanogaster*. *Adv. Genet.* **24**, 371–429.
13 Gatti, M. & Pimpinelli, S. (1983) Cytological and genetic analysis of the Y chromosome of *Drosophila melanogaster*. I. Organization of the fertility factors. *Chromosoma* **88**, 349–373.
14 Karpen, G.H. (1994) Position effect variegation and the new biology of heterochromatin. *Curr. Opinion. Genet. Dev.* **4**, 281–291.
15 Wakimoto, B.T. & Hearn, M.G. (1990) The effects of chromosome rearrangements on the expression of heterochromatic genes in chromosome 2L of *Drosophila melanogaster*. *Genetics* **125**, 141–154.
16 Balbiani, E.G. (1881) Sur la structure du noyau des cellules salivaires chez les larves de Chironomus. *Zoologischer Anzeiger* **4**, 637–641, 662–666.
17 Painter, T.S. (1933) A new method for the study of chromosome rearrangements and the plotting of chromosome maps. *Science* **78**, 585–586.
18 Bridges, C.B. (1935) Salivary chromosome maps with a key to the banding of the chromosomes of *Drosophila melanogaster*. *J. Hered.* **26**, 60–64.
19 Bridges, C.B. (1938) A revised map of the salivary gland X chromosome of *Drosophila melanogaster*. *J. Hered.* **29**, 11–13.
20 Bridges, C.B. & Bridges, P.N. (1939) A new map of the second chromosome: a revised map of the right limb of the second chromosome of *Drosophila melanogaster*. *J. Hered.* **30**, 475–476.
21 Bridges, P.N. (1941) A revised map of the left limb of the third chromosome of *Drosophila melanogaster*. *J. Hered.* **32**, 64–65.
22 Bridges, P.N. (1941) A revision of the salivary gland 3R chromosome map of *Drosophila melanogaster*. *J. Hered.* **33**, 299–300.
23 Bridges, P.N. (1942) A new map of the salivary gland 2L chromosome of *Drosophila melanogaster*. *J. Hered.* **33**, 403–408.
24 Sorsa, V. (1988) Chromosome maps of *Drosophila* (2 vols, CRC Press, Boca Raton, FL).
25 Lefevre, G. Jr (1976) A photographic representation and interpretation of the polytene chromosomes of *Drosophila melanogaster* salivary glands. In *Genetics and Biology of* Drosophila. Vol. 1A (Ashburner, M. & Novitski, E., eds), 31–66 (Academic Press, London).
26 Lemeunier, F. & Ashburner, M. (1976) Relationships within the *melanogaster* species subgroup of the genus *Drosophila* (Sophophora). II. Phylogenetic relationships between six species based upon polytene chromosome banding sequences. *Proc. R. Soc. Lond.* **B193**, 275–294.
27 Muller, H.J. (1940) Bearings of the '*Drosophila*' work on systematics. In *The New Systematics* (Huxley, J., ed.), 185–268 (Clarendon Press, Oxford).
28 Sturtevant, A.H. & Novitski, E. (1941) The homologies of the chromosomal elements in the genus *Drosophila*. *Genetics* **26**, 517–541.
29 Foster, G.G., Whitten, M.J., Konovalov, C. *et al.* (1981) Autosomal genetic maps of the Australian sheep blowfly, *Lucilia cuprina dorsalis* R.-D. (Diptera, Calliphoridae), and possible correlations with the linkage maps of *Musca domestica* L. & *Drosophila melanogaster* (Mg.). *Genet. Res.* **37**, 55–69.
30 Rudkin, G.T. (1969) Non-replicating DNA in *Drosophila*. *Genetics* (Suppl. 1) **61**, 227–238.
31 Bird, A.P. & Taggart, M.H. (1980) Variable patterns of total DNA and rDNA methylation in animals. *Nucleic Acids Res.* **8**, 1485–1497.
32 Hilliker, A.J. & Appels, R. (1982) Pleiotropic effects associated with the deletion of heterochromatin surrounding rDNA on the X chromosome of *Drosophila*. *Chromosoma* **86**, 469–490.
33 Ritossa, F.M. & Spiegelman, S. (1965) Localization of DNA complementary to ribosomal RNA in the nucleolus organizer region of *Drosophila melanogaster*. *Proc. Natl Acad. Sci. USA* **53**, 737–745.
34 Ritossa, F.M., Atwood, K.C. & Spiegelman, S. (1966) A molecular explanation of the *bobbed* mutants as partial deficiencies of 'ribosomal' DNA. *Genetics* **54**, 819–834.
35 Lifton, R.P., Goldberg, M.L., Karp, R.W. & Hogness, D.S. (1978) The organization of the histone genes in *Drosophila melanogaster*, functional and evolutionary implications. *Cold Spring Harbor Symp. Quant. Biol.* **42**, 1047–1051.
36 Pardue, M.L., Kedes, L.H., Weinberg, E.S. & Birnstiel, M.L. (1977) Localization of sequences coding for

histone messenger RNA in the chromosomes of *Drosophila melanogaster*. *Chromosoma* **63**, 135–151.

37 Childs, G., Maxson, R., Cohn, R.H. & Kedes, L. (1981) Orphons, dispersed genetic elements derived from tandem repetitive genes of eucaryotes. *Cell* **23**, 651–663.

38 Wilson, D.A. & Thomas, C.A. Jr (1974) Palindromes in chromosomes. *J. Mol. Biol.* **84**, 115–144.

39 Schmid, C.W., Manning, J.E. & Davidson, N. (1975) Inverted repeat sequences in the *Drosophila* genome. *Cell* **5**, 169–172.

40 Potter, S., Truett, M., Phillips, M. & Maher, A. (1980) Eukaryotic transposable elements with inverted terminal repeats. *Cell* **20**, 639–647.

41 Potter, S. (1982) DNA sequence of a foldback transposable element in *Drosophila*. *Nature* **297**, 201–204.

42 Potter, S. (1982) DNA sequence analysis of a *Drosophila* foldback transposable element rearrangement. *Mol. Gen. Genet.* **188**, 107–110.

43 Finnegan, D.J. (1985) Transposable elements in eukaryotes. *Int. Rev. Cytol.* **93**, 281–326.

44 Finnegan, D.J. & Fawcett, D.H. (1986) Transposable elements in *Drosophila melanogaster*. In *Oxford Surveys on Eukaryotic Genes*. Vol. 3. (MacLean, N., ed.) (Oxford University Press, Oxford).

45 Kidwell, M.G. (1992) Horizontal transfer of P elements and other short inverted repeat transposons. *Genetica* **86**, 275–286.

46 Dowsett, A.P. (1983) Closely related species of *Drosophila* can contain different libraries of middle repetitive DNA sequences. *Chromosoma* **88**, 104–108.

47 Dowsett, A.P. & Young, M.W. (1982) Differing levels of dispersed repetitive DNA among closely related species of *Drosophila*. *Proc. Natl Acad. Sci. USA* **79**, 4570–4574.

48 Martin, G., Wiernasz, D. & Schedl, P. (1983) Evolution of *Drosophila* repetitive–dispersed DNA. *J. Mol. Evol.* **19**, 203–213.

49 Ising, G. & Ramel, C. (1976) The behaviour of a transposing element in *Drosophila melanogaster*. In *The Genetics and Biology of* Drosophila. Vol. 1B (Ashburner, M..& Novitski, E., eds), 947–954 (Academic Press, London).

50 Smith, D., Wohlgemuth, J., Calvi, B.R. *et al.* (1993) *hobo* enhancer trapping mutagenesis in *Drosophila* reveals an insertion specificity different from P elements. *Genetics* **135**, 1063–1076.

51 Lim, J.K. & Simmons, M.J. (1994) Gross chromosomal rearrangements mediated by transposable elements in *Drosophila melanogaster*. *BioEssays* **16**, 269–275.

52 Fawcett, D.H., Lister, C.K., Kellett, E. & Finnegan, D.J. (1986) Transposable elements controlling I–R hybrid dysgenesis in *D. melanogaster* are similar to mammalian LINEs. *Cell* **47**, 1007–1015.

53 Kidwell, M.G., Kidwell, J.F. & Sved, J.A. (1977) Hybrid dysgenesis in *Drosophila melanogaster*: a syndrome of aberrant traits including mutation, sterility and male recombination. *Genetics* **86**, 813–833.

54 O'Hare, K. & Rubin, G.M. (1983) Structures of P transposable elements and their sites of insertion and excision in the *Drosophila melanogaster* genome. *Cell* **34**, 25–35.

55 Rio, D.C. & Rubin, G.M. (1988) Identification and purification of a *Drosophila* protein that binds to the terminal 31-base-pair inverted region of the P transposable element. *Proc. Natl Acad. Sci. USA* **85**, 8929–8933.

56 Bingham, P.M., Kidwell, M.G. & Rubin, G.M. (1982) The molecular basis of P–M hybrid dysgenesis: the role of the P element, a P-strain-specific transposon family. *Cell* **29**, 995–1004.

57 Cooley, L., Kelley, R. & Spradling, A. (1988) Insertional mutagenesis of the *Drosophila* genome with single P elements. *Science* **239**, 1121–1128.

58 Rubin, G.M. & Spradling, A.C. (1982) Genetic transformation of *Drosophila* with transposable element vectors. *Science* **218**, 348–353.

59 Spradling, A.C. & Rubin, G.M. (1982) Transposition of cloned P elements into *Drosophila* germline chromosomes. *Science* **218**, 341–347.

60 O'Kane, C.J. & Gehring, W.J. (1987) Detection *in situ* of genomic regulatory elements in *Drosophila*. *Proc. Natl Acad. Sci. USA* **84**, 9123–9127.

61 Brand, A.H. & Perrimon, N. (1993) Targeted gene expression as a means of altering cell fates and generating dominant phenotypes. *Development* **118**, 401–415.

62 Moffat, K.G., Gould, J.H., Smith, H.K. & O'Kane, C.J. (1992) Inducible cell ablation in *Drosophila* by cold-sensitive ricin A chain. *Development* **114**, 681–687.

63 Ballinger, D.G. & Benzer, S. (1989) Targeted gene mutations in *Drosophila*. *Proc. Natl Acad. Sci. USA* **86**, 9402–9406.

64 Kaiser, K. & Goodwin, S.F. (1990) 'Site-selected' transposon mutagenesis of *Drosophila*. *Proc. Natl Acad. Sci. USA* **87**, 1686–1690.

65 Saunders, R.D.C., Glover, D.M., Ashburner, M. *et al.* (1989) PCR amplification of DNA microdissected from a single polytene chromosome band: a comparison with conventional microcloning. *Nucleic Acids Res.* **17**, 9027–9037.

66 Hartl, D.L. (1992) Genome map of *Drosophila melanogaster* based on yeast artificial chromosomes. In *Genome Analysis*. Vol. 4. *Strategies for Physical Mapping* (Davies, K. E..& Tilghman, S. M., eds), 39–69 (Cold Spring Harbor Laboratory Press, Cold Spring Harbor, NY).

67 Cai, H., Kiefel, P., Yee, J. & Duncan, I. (1994) A yeast artificial clone map of the *Drosophila* genome. *Genetics* **136**, 1385–1399.

68 Ajioka, J.W., Smoller, D.A., Jones, R.W. *et al.* (1991) *Drosophila* genome project: one-hit coverage in yeast artificial chromosomes. *Chromosoma* **100**, 495–509.

69 Garza, D., Ajioka, J.W., Burke, D.T. & Hartl, D.L. (1989) Mapping the *Drosophila* genome with yeast artificial chromosomes. *Science* **246**, 641–646.

70 Lohe, A.R. & Brutlag, D.L. (1986) Multiplicity of satellite DNA sequences in *Drosophila melanogaster*. *Proc. Natl Acad. Sci. USA* **83**, 696–700.

71 Smoller, D.A., Petrov, D. & Hartl, D.L. (1991) Characterization of bacteriophage P1 library containing inserts of *Drosophila* DNA of 75–100 kilobase pairs. *Chromosoma* **100**, 487–494.

72 Hartl, D.L., Nurminsky, D.I., Jones, R.W. & Lozovskaya, E.R. (1994) Genome structure and evolution in *Drosophila*: Applications of the framework P1 map. *Proc. Natl Acad. Sci. USA* **91**, 6824–6829.

73 Hartl, D.L., Ajioka, J.W., Cai, H. *et al.* (1992) Towards a *Drosophila* genome map. *Trends Genet.* **8**, 70–75.
74 Kafatos, F.C., Louis, C., Savakis, C. *et al.* (1991) Integrated maps of the *Drosophila* genome: progress and prospects. *Trends Genet.* **7**, 155–161.
75 Gibson, T.J., Rosenthal, A. & Waterston, R.H. (1987) Lorist 6, a cosmid vector with *Bam*HI, *Not*I, *Sca*I and *Hind*III cloning sites and altered neomycin phosphotransferase gene expression. *Gene* **53**, 283–286.
76 Scalenghe, F., Turco, E., Edström, J.E., Pirrotta, V. & Melli, M. (1981) Microdissection and cloning of DNA from a specific region of *Drosophila melanogaster* polytene chromosomes. *Chromosoma* **82**, 205–216.
77 Sidén-Kiamos, I., Saunders, R.D.C., Spanos, L. *et al.* (1990) Towards a physical map of the *Drosophila melanogaster* genome, mapping of cosmid clones within defined genomic divisions. *Nucleic Acids Res.* **18**, 6261–6270.
78 Johnson, D.H. (1990) Molecular cloning of DNA from specific chromosomal regions by microdissection and sequence-independent amplification. *Genomics* **6**, 243–251.
79 Coulson, A., Sulston, J., Brenner, S. & Karn, J. (1986) Towards a physical map of the genome of the nematode *Caenorhabditis elegans*. *Proc. Natl Acad. Sci. USA* **83**, 7821–7825.
80 Wood, W.I., Gitscher, J., Lasky, L.A. & Lawn, R.M. (1985) Base composition-independent hybridization in tetramethylammonium chloride: a new method for oligonucleotide screening of highly complex gene libraries. *Proc. Natl Acad. Sci. USA* **82**, 1585–1588.
81 Madueño, E., Papagiannakis, G., Rimmington, G. *et al.* (1995) A physical map of the X chromosome of *Drosophila melanogaster*: cosmid contigs and sequence tagged sites. *Genetics* **139**, 1631–1647.
82 Olson, M., Hood, L., Cantor, C. & Botstein, D. (1989) A common language for physical mapping of the human genome. *Science* **245**, 1434–1435.
83 Sternberg, N. (1990) A bacteriophage P1 cloning system for the isolation, amplification, and recovery of DNA fragments as large as 100 kbp. *Proc. Natl Acad. Sci. USA* **87**, 103–107.
84 Strathmann, M., Hamilton, B.A., Mayeda, C.A. *et al.* (1991) Transposon-facilitated DNA sequencing. *Proc. Natl Acad. Sci. USA* **88**, 1247–1250.
85 Ashburner, M. (1992) Mapping insect genomes. In *Insect Molecular Science* (Crampton, J.M. & Eggleston, P., eds), 51–75 (Academic Press, London).
86 Alexander, M.L. (1976) The genetics of *Drosophila virilis*. In *The Genetics and Biology of* Drosophila. Vol. 1C (Ashburner, M. & Novitski, E., eds), 1365–1427 (Academic Press, London).
87 Lozovskaya, E.R., Petrov, D.A. & Hartl, D.L. (1993) A combined molecular and cytogenetic approach to genome evolution in *Drosophila* using large-fragment DNA cloning. *Chromosoma* **102**, 253–266.
88 Zheng, L., Saunders, R.D.C., Fortini, D. *et al.* (1991) Low-resolution genome map of the malaria mosquito *Anopheles gambiae*. *Proc. Natl Acad. Sci. USA* **88**, 11187–11191.
89 Zheng, L., Collins, F.H., Kumar, V. & Kafatos, F.C. (1993) A detailed genetic map for the X chromosome of the malaria vector *Anopheles gambiae*. *Science* **261**, 605–608.

Chapter 29 **The genome of Caenorhabditis elegans**

John Sulston, Robert H. Waterston* and members of the *Caenorhabditis elegans* Genome Consortium

The Sanger Centre, Hinxton Hall, Cambridge CB10 1SA, UK

**Department of Genetics and Genome Sequencing Center, Washington University School of Medicine, St Louis, MO 63110, USA*

29.1 Introduction

Caenorhabditis elegans is a small, free-living soil nematode found in many parts of the world. It is a simple organism which is easily maintained and studied in the laboratory. The adult hermaphrodite has only 959 somatic nuclei, while the adult male has a total of 1031. The haploid genome is ≈100 Mb, which is about eight times larger than the yeast *Saccharomyces*, two thirds the size of the fruit fly *Drosophila* and 30 times smaller than the human genome.

About 80% of the *C. elegans* genome is composed of single-copy sequences, with the remainder being moderately repetitive sequences which occur in two to many copies per genome. Genes of *C. elegans* have been mapped into six linkage groups which correspond to the six haploid chromosomes. More than 1000 *C. elegans* genes, distributed over the six linkage groups, have been identified; study of these genes is leading to important new insights in neurobiology and developmental biology.

In support of this enterprise, the *C. elegans* genome project was begun in the early 1980s. Since its inception the genome project has pioneered approaches to physical mapping and genome sequencing. Today, the physical map of the genome is among the largest and most complete yet constructed for a multicellular organism. It consists of 17 000 cosmids and 2500 YACs, which have been positioned relative to each other by gel fingerprinting and cross-hybridization. More genomic sequence and more sequenced genes are now available from the worm than from any other multicellular organism, and we are on target for completion of the full genomic sequence by the end of 1998. The utility of these resources can be judged by the many *C. elegans* laboratories now using the map and sequence to study mutationally defined genes, and by the use of sequence homologies by many more laboratories. The genome sequence is critical to gaining a thorough understanding of this important model organism and will aid in studies of human disease.

In this chapter we describe the underlying philosophy and the general approaches that we feel have been important for the success of the project. These points are applicable not only to other small genome projects, but also to the much larger and more challenging Human Genome Project.

29.2 The genome map

The physical map of the *C. elegans* genome consists largely of overlapping cosmid and YAC clones [1–3]. Both components are essential: the YACs, by virtue of their large inserts and propagation in yeast, provide long-range continuity and can hold DNA that is unclonable in bacterial cosmid clones; the cosmids provide high resolution locally and a more convenient substrate for biochemistry.

The principal techniques for physical mapping have been:

- restriction enzyme-based fingerprinting for construction of cosmid contigs (Fig. 29.1);
- hybridization of individual YAC and cosmid clones to gridded arrays of cosmids and YACs, respectively;
- sequence-tagged site (STS) assays for direct detection of YAC/YAC overlaps;
- hybridization of YAC and cosmid clones to *C. elegans* chromosomes for long-range ordering.

The topological constraints imposed by the ordered cosmids were important for the interpretation of YAC/cosmid hybridizations, in that they helped to distinguish genuine matches from spurious matches due to repetitive sequences.

This physical array of cloned DNAs is made into a genome map by the wealth of genetic markers that have been attached to specific clones. This was facilitated by the early and unrestricted distribution of the clone resources and by the readiness of the community of *C. elegans* researchers to share information prior to publication. In fact, the genetic matches were also critical mapping tools, in that they, along with *in situ* hybridization, provide the longest-range linkage.

Unlike the physical mapping, which has been carried out mainly by the two laboratories above, the genetic linkage has been achieved by the cooperative effort of the entire *C. elegans* community. The communal approach is important in two ways. From the point of view of the map, it ensures that the specialized knowledge of all individuals and groups is brought to bear on the project. From the point of view of effort and funding, it means that everyone is involved and allows the central resources to be as lean and focused as possible.

The map as a whole gains from being a multilevel construct: no single technique is sufficient by itself to provide full linkage, and strength arises from partial redundancy between the levels. This is important, because all mapping information is to some degree stochastic.

The gel fingerprinting method is readily scalable, and is being applied increasingly to the human genome. Since the original nematode work, the procedure has been automated, fluorescent labels have been introduced instead of radioactive labels, and new generations of assembly software are appearing. In contrast to the worm, long-range

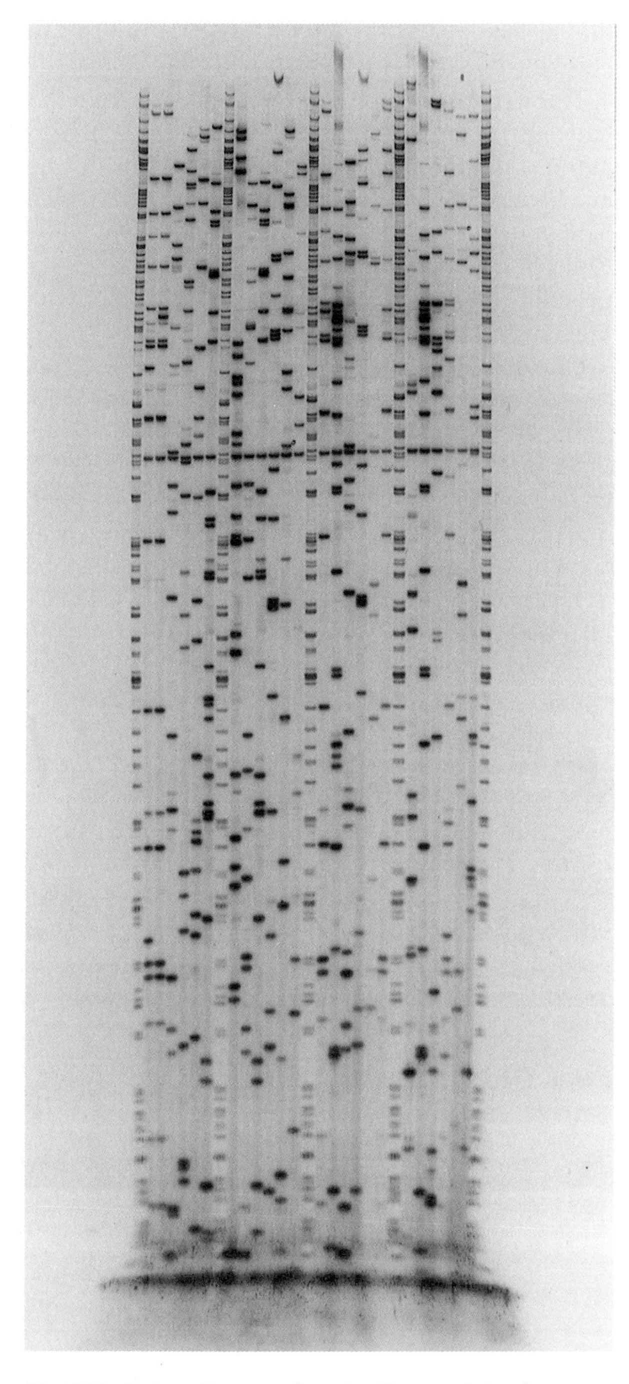

Fig. 29.1 Autoradiogram showing fingerprints of nematode cosmids. The lanes with closely spaced bands contain markers, and the rest contain samples. Many different fingerprinting methods can be, and have been, used for genome projects. The only requirement is that a pattern is generated from each clone, such that the partial identity of patterns from a pair of overlapping clones can be recognized. For the *C. elegans* project, a method based upon cutting with *Hin*dIII and *Sau*3aI was used; it was arranged that the *Hin*dIII sites, but not the *Sau*3aI sites, were labelled radioactively by end-filling and the resulting fragments were separated on a denaturing polyacrylamide gel of the type used for sequencing [1,19]. From [19] by permission of Oxford University Press.

order in the human genome is being achieved at an earlier stage, by STS analysis of YACs and radiation hybrids, and by *in situ* hybridization. However, just as in the worm, bacterial clones, whether cosmids, P1 clones (P1s), P1 artificial chromosomes (PACs), or bacterial artificial chromosomes (BACs) (see Chapter 15) will provide the preferred substrates for biochemistry.

29.3 The genome sequence

In sequencing the *C. elegans* genome, we have as far as possible adopted the same philosophy of collective endeavour. The task of the two central laboratories is restricted to collecting the data as efficiently as possible. They refrain strictly from exploiting the sequence data for their own research purposes before its release. As soon as the raw data has been assembled into contigs, it is available for screening by anyone by being placed on an anonymous ftp server. When each cosmid sequence is finished, it is analysed by computer to find possible genes, database similarities and other features; the annotated sequence is then immediately submitted to GenBank or the EMBL sequence database, as well as being placed in the *C. elegans* database ACEDB [4]. In this way, the expertise not only of the worm community but of the whole world is brought to bear at the earliest possible stage.

Sequencing concentrated at first on the central regions of the autosomes and the whole of the X chromosome, totalling roughly 60% of the genome, because genetic and cDNA mapping data indicated that these areas contain the majority of the genes (perhaps more than 80% of the total; refs 5 and 6 and Y. Kohara, personal communication). The focus of sequencing has now shifted to the autosomal arms; although these may be less rich in genes, there are many important aspects beyond the protein coding elements that will only be addressed by the sequence of the whole genome.

Starting in this way had the added benefit for the sequencing part of the project that we began by sequencing cosmids. Now, the 'YAC bridges'—the regions cloned in YACs but not in cosmids—are being dealt with. Starting from complete YACs is more difficult than from cosmids [7], because of the limited amounts and purity of material and their greater size. However, with improved analysis software, and with the whole yeast sequence available to identify and remove contaminating host sequences, this method is practical. Smaller bridges have been successfully rescued by recombination from YACs in yeast, and others by long-range PCR. Some regions are susceptible to cloning in fosmid vectors (Stephanie Chissoe, personal communication) and many to cloning in λ-vectors on permissive hosts.

Our principal sequencing strategy is an initial shotgun followed by directed finishing (described in detail in ref. 8, see also Chapter 20, Section 20.2). It is worth emphasizing here that shotgun sequences provide extremely detailed map information (in the form of the sequence itself), albeit only from a fraction of the subclone insert length. The proper assembly of these sequences allows the relative positions of the random subclones to be established, while at the same time producing the bulk of the final sequence. Like all mapping methods, however, it is vulnerable to repeated sequences. The density and accuracy of sequence information, however, compensates for the relatively short length of individual reads. Improved assembly programs (see below) are taking greater advantage of this information, and the sequence itself provides powerful means of evaluating alternative maps. In uncertain areas, additional sequence—for example, from the opposite end of the insert—can provide additional map information. In addition, restriction enzyme digests of the parent clone provide a simple and direct means of testing overall map accuracy.

Cost and accuracy are key considerations in evaluating effectiveness of any strategy. Current direct and indirect costs for the production of the final annotated sequence are below $0.45 a base, and total costs of all activities (including development and related research efforts) are below $0.70 per base. In general, the accuracy of the sequence appears to be better than 99.99%, on the basis of comparisons with previously sequenced genes, though these estimates are of limited reliability in the absence of a truly independent means of checking the sequence.

Throughout and the ability to scale-up the effort have also been important. As a result of increased automation/mechanization, improved software and better biochemistry, the combined production of nematode sequence exceeded 70 Mb in 1996. Collection of shotgun data for the remainder of the genome will be essentially completed in 1997, with finishing and gap-filling continuing in 1998.

Our objective is to extract the information from the genome rather than to exhaustively sequence every last base. For example, even now we sometimes describe long tandem repeats simply in terms of the consensus sequence and number of copies. The occurrence of such instances is more frequent as we move into repetitive regions, and so this method of reporting will increase. Conversely, we sequence all other regions as accurately as possible.

The shotgun/directed approach can be applied equally well to the human genome, provided that the extensive repeat families are allowed for in the assembly algorithm. At first, we adapted R. Staden's XGAP by screening the input so that Alu sequences were excluded from the initial assembly process. As for the nematode, we now begin with Phil Green's (P. Green, personal communication 1994) PHRAP which makes positive but selective use of repeat sequences in assembly, and then feed the results to XGAP or other editing programs.

29.4 Status of the sequencing project and its applications

The current status of the sequencing effort is summarized in Table 29.1. All of the sequence has been

Table 29.1 Current state of *C. elegans* 100-Mb genome sequencing project.

Physical map	17 500 cosmids
	3500 YACs
	Five autosomes: total of eight gaps (all in gene poor regions)
	X chromosome: single contig of ≈ 18 Mb
DNA sequence	65 Mb completed as of March 1997
	1300 putative protein coding genes (≈ 1 per each 5 kb)
	approximately 45 % have significant similarity to non-*C. elegans* genes

As well as having access to finished sequences in the GenBank, EMBL and DDBJ databases, investigators can search these sequence data and also more preliminary unfinished sequences at the two genome centers. Currently, the searchable database contains 85 Mb of sequence data (65 Mb finished, 20 Mb unfinished) which is estimated to contain about 90% of the genes in *C. elegans*. Searches with a nucleotide or protein query use the BLAST programs and are submitted via a World Wide Web interface (see URL http://www.sanger.ac.uk/ and http://genome.wustl.edu/). The Web pages provide additional information about the genome and include help addresses. All data pertaining to the genome (including genetic and physical maps and the sequence) are combined in the database ACeDB. The latest release can be obtained by anonymous ftp from the following: USA, ncbi.nlm.nih.gov (130.14.20.1) in repository/acedb; UK, ftp.sanger.ac.uk (193.60.84.11) in pub/acedb; or France, lirmm.lirmm.fr (193.49.104.10) in genome/acedb.

subjected to a series of programs to provide an initial interpretation of its features. Comparison with expressed sequence tags (ESTs) (ref. 6 and Y. Kohara, personal communication) allows the construction of a confirmed transcription map for a third of the predicted genes.

The *C. elegans* genome map is being strengthened by several other systematic studies. These projects are entirely independent, but their findings are united through the map and some of them draw on its resources. At the University of Leeds, UK, Ian Hope is collecting expression data using transgenic reporter constructs of the predicted genes (ref. 9 and I. Hope, personal communication 1996). Targeted gene disruption by transposon insertion was pioneered by Ronald Plasterk (Amsterdam, The Netherlands), and is now carried out in a number of laboratories: in this way, functionality for the predicted genes can be determined [10]. At the National Genetics Institute, Mishima, Japan, Yuji Kohara is

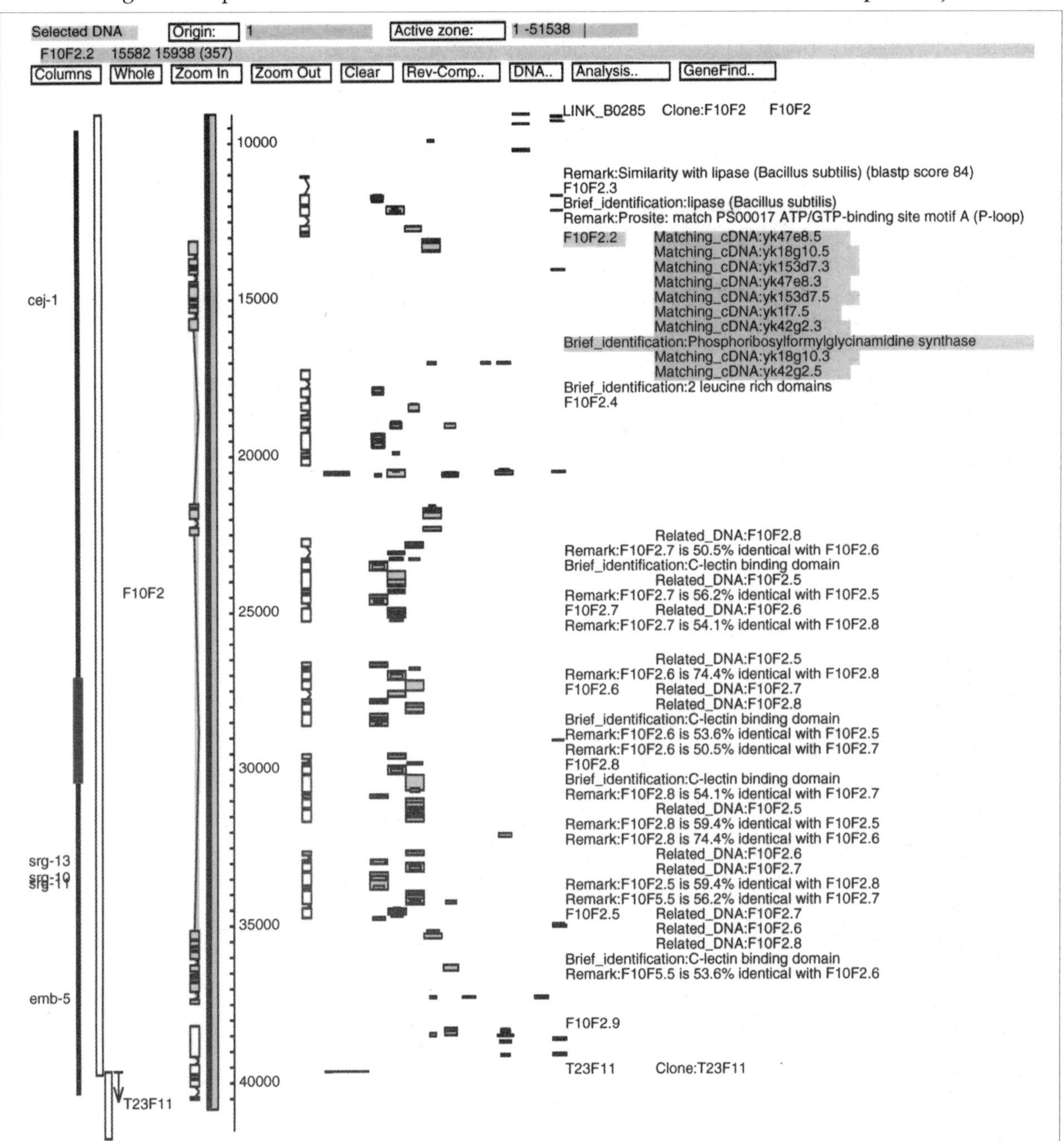

Fig. 29.2 The feature window of ACEDB showing a region of chromosome III cloned in the cosmid F10F2. To the left of the kilobase scale (i.e. on the negative DNA strand) a gene with two large introns is shown (F10F2.2; similarity to phosphoribosylformylglycinamidine synthase; highlighted). To the right of the scale (on the positive strand) a family of five genes (F10F2.4, F10F2.7, F10F2.6, F10F2.8 and F10F5.5) is seen to lie within the introns.

continually adding to his set of sequence-tagged cDNAs and is determining their expression patterns by *in situ* hybridization (Y. Kohara, personal communication 1996). At Vancouver, Canada, David Baillie (Simon Fraser University) and Ann Rose (University of British Columbia) have generated

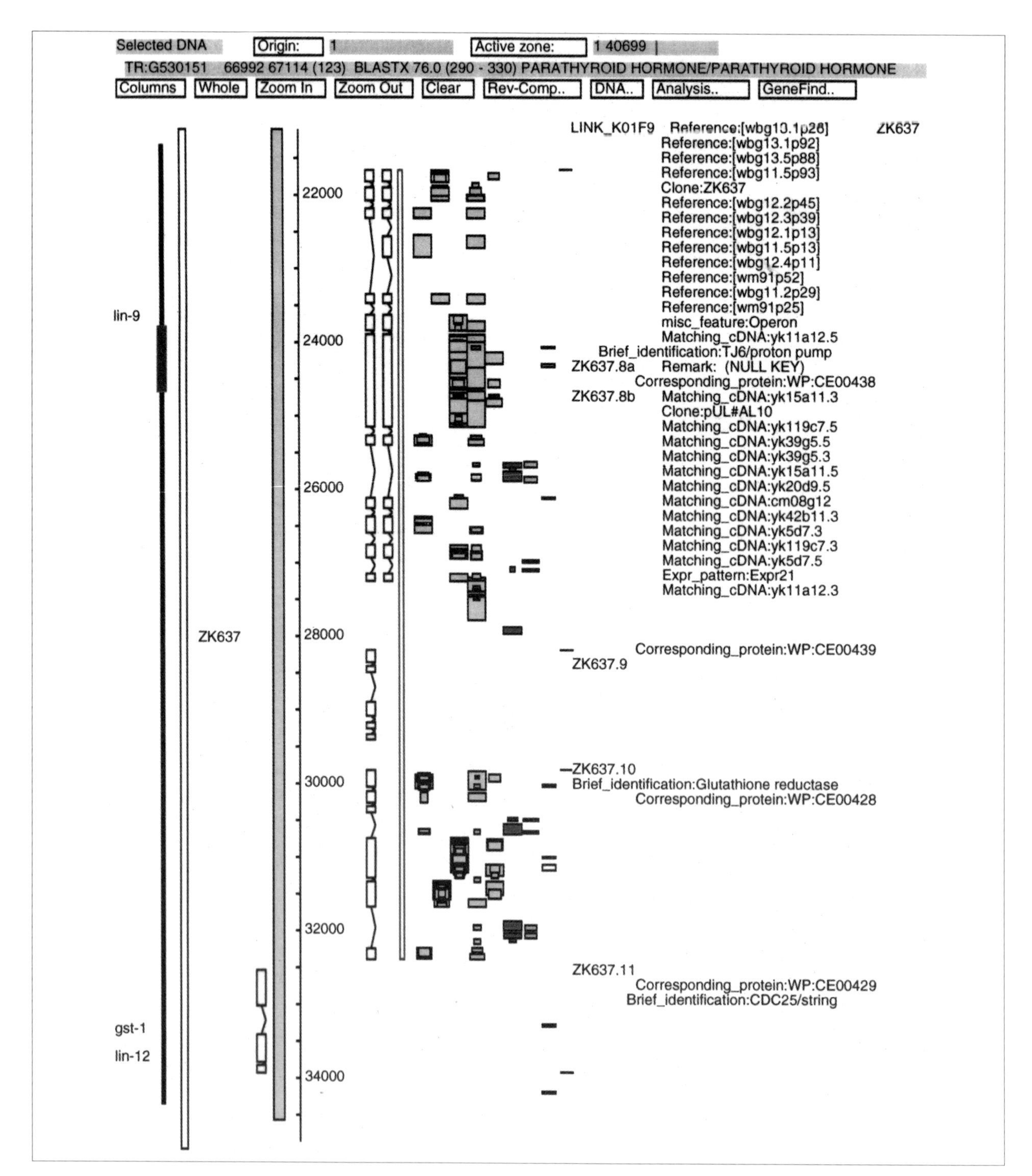

Fig. 29.3 The feature window of ACEDB showing a region of chromosome III cloned in the cosmid ZK637. The three genes ZK637.8, ZK637.9 and ZK637.10 lie head-to-tail on the positive strand. They have been shown to be transcribed as one operon [14]. Comparison with cDNA sequences (open boxes below the zoom-out button) shows that ZK637.8 has two alternative splicing patterns at position 23000.

Table 29.2 Features of interest in the *C. elegans* genome.

Direct analysis of the sequence yields many interesting features, with applications to gene function, evolution and medicine. A few examples are:
Genes within the introns of other genes. In one case five such genes were found within a single gene (Fig. 29.2)
Clusters of tRNAs, containing five members in one case and six in another
Gene families, some where the family members are dispersed, and others where they are close together in tandem arrays. We can begin to look at the evolution of the individual members
A relatively high incidence of inverted repeats within the introns of predicted genes. The functional significance of this, if any, is unknown
Head-to-tail patterns of genes, which have been shown to be indicative of operons (Fig. 29.3) [14]
Predictions of homologues to genes of medical importance, for example C50C3.7, the only known invertebrate gene similar to the human gene *OCRL*-1, implicated in Lowe's syndrome
The largest known *C. elegans* protein, predicted from a 45-kb gene in cosmid K07E12. It contains 13 000 amino acid residues and contains multiple copies of the cell adhesion molecule motif [16,17]
Many repeat families. Some are thought to be 'dead' transposons from an unknown and possibly extinct transposon type

transgenic strains incorporating sequenced cosmids; rescue of lethal and visible mutants by these strains will allow precise correlation of the genetic map with the sequence (refs 11–15 and A. Rose, personal communication 1996).

By far the largest use of the genome map and sequence is for the study of specific *C. elegans* genes. Virtually all *C. elegans* laboratories worldwide make use of the map and most of them request clones from it. Increasingly, laboratories working on other organisms are also using these resources. Genome and sequence data for *C. elegans* is held in the data base ACEDB (accessible at various sites, see Chapter 37). The database software (ACEDB) was designed as part of the *C. elegans* project to provide a data base tool specifically for use by biologists. ACEDB is the creation of Richard Durbin (The Sanger Centre, Cambridge, UK) and Jean Thierry-Mieg (CNRS, Montpellier, France), who also maintain the database itself.

Applications of the sequence are on several levels. At the most mundane level, the prior determination of the sequence by an efficient large-scale operation simply saves subsequent effort and resources. Importantly, the sequence provides ready-made tools, such as a restriction map and information for making primers, that facilitate experimental design.

More creatively, the sequence provides new entry points to the genome. Homologues of known genes or parts of genes can be sought by computer. Not only is this style of searching faster than physical probing, but also it is more thorough. Weak similarities, beyond the detection limit of hybridization, can be picked up and evaluated. At present, investigators can search a total of about 85 Mb (65 Mb finished, 20 Mb unfinished), containing perhaps 90% of these genes. Searches will become virtually complete by the end of 1997, and will be greatly enhanced as the emerging families of genes and domains are subjected to cluster analysis and grouped by similarity.

As the sequenced regions extend along the chromosomes, the large-scale structure of the genome starts to emerge. We are only just beginning to explore this level, but some of the early findings are illustrated in Figs 29.2 and 29.3 and Table 29.2. Figure 29.2 shows an annotated region of sequence of 30 000 bp from chromosome III. Figure 29.3 shows 10 000 bp from chromosome III. Apart from patterns of genes, we begin to see the matrix of duplicated, inverted and transposed pieces of which the genome is composed. Somewhere, there are elements that mediate replication, recombination and segregation of the chromosomes, and others that control sex determination, dosage compensation and global gene expression. The sequence will provide the framework in which hypotheses about such mechanisms can be developed and tested.

Finally, the sequence forms a permanent archive whose value we can only begin to tap at the first pass. The analysis, modification, and above all comparison of sequences from different organisms will provide a major route to a full understanding of biology.

Acknowledgements

This work was suported by the NIH National Human Genome Research Institute, the UK Medical Research Council and the Wellcome Trust. This chapter is an updated and extended version of an article that first appeared in *PNAS Genome Review* [16].

References

1 Coulson, A.R., Sulston, J.E., Brenner, S. & Karn, J. (1986) Towards a physical map of the genome of the nematode *Caenorhabditis elegans*. *Proc. Natl Acad. Sci. USA* **83**, 7821–7825.

2 Coulson, A.R., Waterston, R.H., Kiff., J.E. *et al.* (1988) Genome linking with yeast artificial chromosomes. *Nature* **335**, 184–186.

3 Coulson, A., Kozono, Y., Lutterbach, B. *et al.* (1991) YACs and the *C. elegans* genome. *BioEssays* **13**, 413–417.

4 Durbin, R. & Thierry-Mieg, J. (1994) The ACEDB genome database. In *Computational Methods in Genome Research* (Suhai, S., ed.), 45–55 (Plenum Press, New York).

5 Edgley, M.L. & Riddle, D.L. (1990) The nematode *Caenorhabditis elegans*. *Genetic Maps* **5**, 3.

6 Waterston, R., Martin, C. & Craxton, M. *et al.* (1992) A survey of expressed genes in *Caenorhabditis elegans*. *Nature Genet.* **1**, 114–123.

7 Vaudin, M., Roopra, A., Hillier, L. *et al.* (1995) The construction and analysis of M13 libraries prepared from YAC DNA. *Nucleic Acids Res.* **23**, 670–674.

8 Wilson, R., Ainscough, R., Anderson, K. *et al.* (1994) 2.2 Mb of contiguous nucleotide-sequence from chromosome III of *C. elegans*. *Nature* **368**, 32–38.

9 Hope, I.A. (1994) *pes*-1 is expressed during early embryogenesis in *Caenorhabditis elegans* and has homology to the fork head family of transcription factors. *Development* **120**, 505–514.

10 Zwaal, R.R., vanBroeks, A., Meurs, J. & Plasterk, R.H.A. (1993) Target-selected gene inactivation in *Caenorhabditis elegans* by using a frozen transposon insertion mutant bank. *Proc. Natl Acad. Sci. USA* **90**, 7431–7435.

11 Howell, A.M. & Rose, A.M. (1990) Essential genes in the HDF6 region of chromosome I in *Caenorhabditis elegans*. *Genetics* **126**, 583–592.

12 Rose, A.M., Edgley, M.L. & Baillie, D.L. (1994) In *Advances in Molecular Plant Nematology* 19–33 (Plenum Press, London).

13 Schein, J.E., Marra, M.A., Benian, G.M. *et al.* (1993) The use of deficiencies to determine essential gene content in the *let-56-unc* 22 region of *Caenorhabditis elegans*. *Genome* **36**, 1148–1156.

14 Zorio, D.A.R., Cheng, N.S.N., Blumenthal, T.E. & Spieth, J. (1994) Operons as a common form of chromosomal organization in *C. elegans*. *Nature* **372**, 270–272.

15 Leahey, A.M., Charnas, L.R. & Nussbaum, R.L. (1993) Nonsense mutations in the *OCRL*-1 gene in patients with the oculocerebrorenal syndrome of Lowe. *Hum. Mol. Genet.* **2**, 461–463.

16 Waterston, R. & Sulston, J. (1995) The genome of *Caenorhabditis elegans*. *Proc. Natl Acad. Sci. USA* **92**, 10836–10840.

17 Coulson, A. & Sulston, J. (1988) Genome mapping by restriction fingerprinting. In *Genome Analysis: A Practical Approach* (Davies, K.E., ed.), 19–39 (IRL Press, Oxford).

Chapter 30 The yeast genome project

Horst Feldmann

Adolf-Butenandt-Institut für Physiologische Chemie, Physikalische Biochemie und Zellbiologie der Universität, Schillerstrasse 44, D-80336 München, Germany

30.1 Introduction

The budding yeast *Saccharomyces cerevisiae* may be viewed as one of the most important fungi used in biotechnology. Used in making bread and alcoholic beverages, yeast has served mankind for several thousands of years. The use of yeast as an experimental system dates back to the mid-1930s [1] and it has since received increasing attention. The elegance of yeast genetics and the ease of manipulation of yeast, and finally the technical breakthrough of yeast transformation to enable it to be used in reverse genetics, have substantially contributed to the explosive growth in yeast molecular biology [2–4]. Its success is also due to the fact, which was not anticipated a dozen years ago, of the remarkable conservation of basic biological processes throughout the eukaryotes.

30.2 Yeast: an experimental system for molecular biology

The unique position of *S. cerevisiae* as a model eukaryote owes much to its intrinsic advantages as an experimental system, in which cell architecture and fundamental cellular mechanisms can be successfully investigated. It is a unicellular organism which, unlike more complex eukaryotes, can be grown on defined media, giving the investigator complete control over environmental parameters. Yeast is tractable to classical genetic techniques and functions in yeast have been studied in great detail by biochemical approaches [2–4]. In fact, a large variety of examples provides evidence that substantial cellular functions are highly conserved from yeast to mammals and that corresponding genes can often complement each other. No wonder then that yeast has again reached the forefront in experimental molecular biology by being the first eukaryotic organism for which the entire genome sequence is available [5,6]. The wealth of sequence information obtained in the yeast genome project has turned out to be extremely useful as a reference against which sequences of human, animal or plant genes may be compared. Moreover, the ease of genetic manipulation in yeast opens the possibility of functionally dissecting gene products from other eukaryotes in the yeast system.

30.2.1 The yeast genome

At 12.8 megabases (Mb), the yeast genome is about 200 times smaller than the human genome but less than four times bigger than that of *Escherichia coli*. At the outset of the sequencing project, knowledge of some 1200 genes encoding either RNA or protein products had accumulated [7]. The complete genome sequence now defines some 6000 open reading frames (ORFs) which are likely to encode specific proteins in the yeast cell. A protein-coding gene is found every 2 kb in the yeast genome, with nearly 70% of the total sequence consisting of ORFs [6]. In addition to the protein-coding genes, the yeast genome contains some 120 ribosomal RNA genes in a large tandem array on chromosome XII, 48 genes encoding small nuclear RNAs (snRNAs) and 275 tRNA genes (belonging to 43 families) which are scattered throughout the genome. Finally, the sequences of nonchromosomal elements, such as the 6 kb of the 2-μm plasmid DNA, the killer plasmids present in some strains, and the yeast mitochondrial genome (≈75 kb) have to be considered. None of the latter, however, has been included in the sequencing project; those sequences were largely determined in the 1980s.

The compact nature of the *S. cerevisiae* genome is apparent when compared to more complex eukaryotic systems. For example, the genome of *Caenorhabditis elegans* contains a potential protein-coding gene only every 6 kb [8] and, in the human genome, gene density might be as low as one gene in 30 kb [9]. Current data (obtainable from S. Bowman and B. Barrell at http://www.sanger.ac.uk/yeast/pombe. html) indicate that even the genome of the fission yeast, *Schizosaccharomyces pombe*, has a lower gene density (one gene per 2.3 kb) than *S. cerevisiae*. The difference between the two yeast genomes appears to be due to the fact that in the fission yeast ≈40% of the genes contain introns, whereas only 4% of the protein-coding genes in *S. cerevisiae* are interrupted by introns [6].

30.2.2 The chromosomes

The genome of *S. cerevisiae* is divided up into 16 chromosomes ranging in size from 250 to >2500 kb. Choosing appropriate conditions, it is feasible to separate all 16 chromosomes by pulsed field gel electrophoresis (PFGE). This provides definition of 'electrophoretic karyotypes' of strains by sizing chromosomes [10]. Laboratory strains possess different karyotypes, because of chromosome length polymorphisms and chromosomal rearrangements, but so do industrial strains. The gels can be utilized for Southern blotting followed by hybridization, or to isolate chromosome-specific DNA.

30.2.3 Genetic mapping

The first genetic map of *S. cerevisiae* was published

by Lindegren in 1949 [11]; many revisions and refinements have appeared since, and the latest version, mapping some 1200 genes, was edited in 1992 [7]. Both meiotic and mitotic approaches have been developed to map yeast genes. The life cycle of *S. cerevisiae* normally alternates between diplophase and haplophase. Both ploidies can exist as stable cultures. In heterothallic strains, haploid cells are of two mating types, **a** and α. Mating of **a** and α cells results in **a**/α diploids that are unable to mate but can undergo meiosis. The four haploid products resulting from meiosis of a diploid cell are contained within the wall of the mother cell (the ascus). Digestion of the ascus and separation of the spores by micromanipulation yield the four haploid meiotic products. Analysis of the segregation patterns of different heterozygous markers among the four spores constitutes tetrad analysis and reveals the linkage between two genes (or between a gene and its centromere) [12]. On the whole, genetic distance in yeast appears to be remarkably proportional to physical distance, with a global average of 3 kb cM^{-1}. Deviations from this rule and results from direct comparisons of the genetic and physical maps will be discussed below.

30.2.4 Manipulations in yeast

Yeast has a generation time of ≈80 min and mass production of cells is easy. Simple protocols are available for the isolation of high molecular weight DNA, rDNA, mRNA, and tRNA. It is possible to isolate intact nuclei or cell organelles such as intact mitochondria (maintaining respiratory competence).

High efficiency transformation of yeast cells is achieved, for example, by the lithium acetate procedure [13] or by electroporation. A large variety of vectors have been designed to introduce and to maintain or express recombinant DNA in yeast cells (see, for example, refs 4 and 14). Furthermore, a large number of yeast strains carrying auxotrophic markers, drug resistance markers or defined mutations are available. Culture collections are maintained, for example, at the Yeast Genetic Stock Center and the American Type Culture Collection (ATCC). In the near future, mutant strains with defined gene deletions together with clones carrying the corresponding gene cassettes will emerge from the EUROFAN project (see Section 30.7).

A comprehensive library of recombinant lambda clones constructed as part of an *S. cerevisiae* physical mapping project and grouped in contigs [15] is maintained and distributed by ATCC. Ordered cosmid libraries using different vectors were constructed during the yeast sequencing project (see, for example, refs 16–18).

The ease of gene disruptions and single step gene replacements is unique to *S. cerevisiae* and offers an outstanding advantage for experimentation. Yeast genes can functionally be expressed when fused to the green fluorescent protein thus allowing the localization of gene products in the living cell by fluorescence microscopy [19]. The yeast system has also proved invaluable to clone and to maintain large segments of foreign DNA in yeast artificial chromosomes (YACs), being extremely useful for other genome projects [20], and to search for protein–protein interactions using the two-hybrid approach [21].

30.3 The yeast genome sequencing project

30.3.1 Strategy

The yeast sequencing project was initiated in 1989 within the framework of the European Union biotechnology programs. It was based on a network approach into which initially 35 European laboratories became involved [22], and chromosome III—the first eukaryotic chromosome ever to be sequenced—was completed in 1992 [23]. In the following years and engaging many more laboratories, sequencing of further complete chromosomes was tackled by the European network. Soon after its beginning, laboratories in other parts of the world joined the project to sequence other chromosomes or parts thereof, ending up in a coordinated international enterprise [24]. Finally, more than 600 scientists in Europe, North America and Japan became involved in this effort. Figure 30.1 shows how the tasks were distributed. The sequence of the entire yeast genome was completed in early 1996 and released to public databanks in April 1996.

30.3.2 Cloning and mapping procedures

The sequencing of chromosome III started from a collection of overlapping plasmid or phage lambda clones that were distributed by the DNA coordinator to the contracting laboratories. In the following, cosmid libraries were constructed to aid large-scale sequencing (see, for example, refs 16–18 and 25). For yeast, cosmids turned out to be the most convenient tools in the construction and handling of genomic libraries, as 35–45 kb of DNA can be accommodated in a cosmid vector. Obvious advantages of cloning DNA segments in cosmids were:

1 larger genes could be obtained on a single recombinant clone;

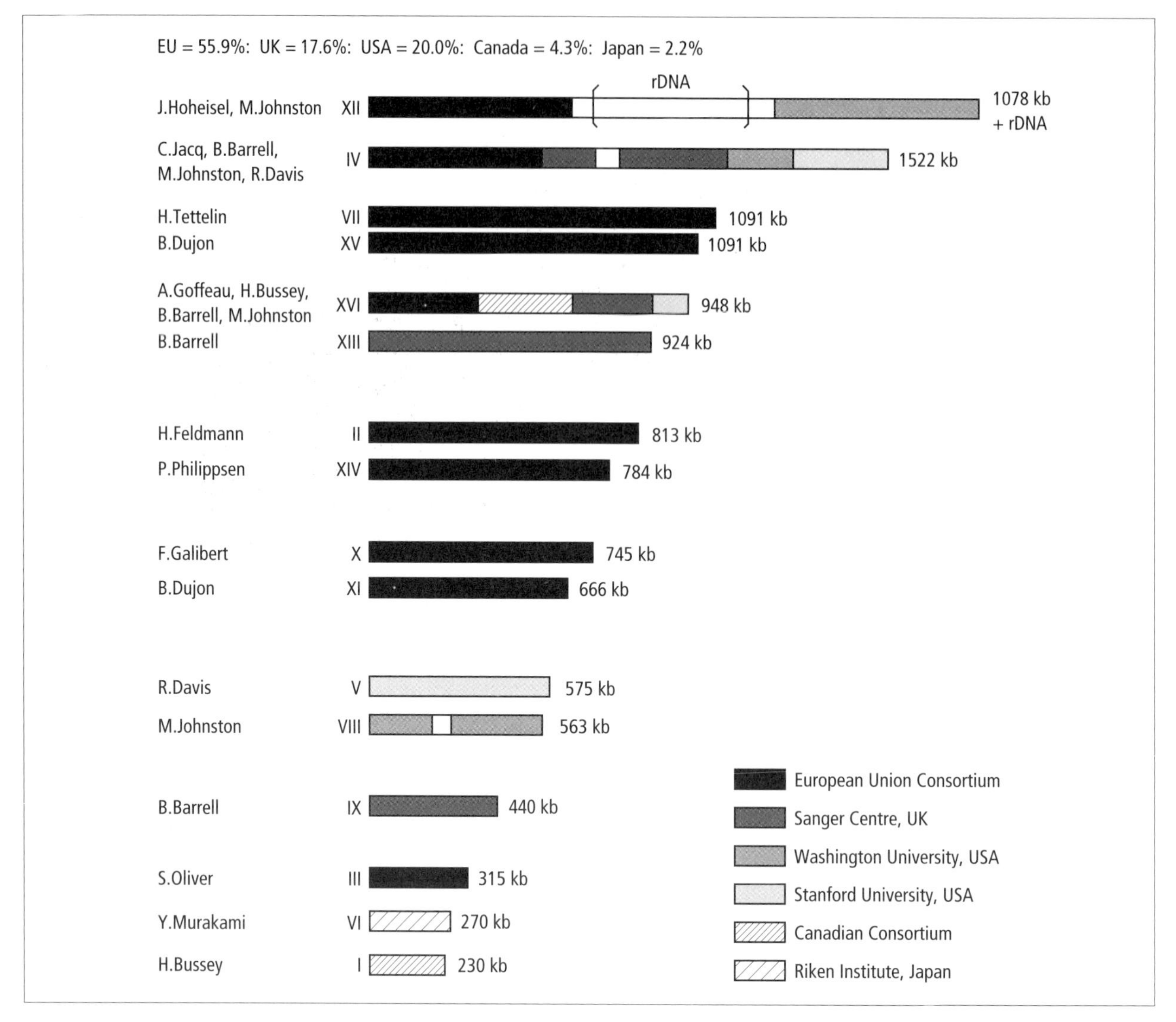

Fig. 30.1 The yeast genome project. The scheme represents the yeast chromosomes in that order as they would appear after separation by PFGE. The chromosomes are drawn to scale and coded according to the different collaborating programs (names are those of the coordinators) for determiming the sequences of particular chromosomes or regions thereof. Large tandem arrays such as those containing the ribosomal DNA repeats (chromosome XII), *CUP1* repeats (chromosome VIII), and *PMR2* repeats (chromosome IV) are each represented in the sequence by only two repeat units (white boxes). The complete sequence (12 052 kb) is available in annotated database entries; a compilation of useful computer addresses is presented in the annex. References to publications of the single chromosomes: I [52]; II [28]; III [23]; VI [100]; VIII [18]; X [101]; XI [71]. Publications to further chromosomes (IV [102], V [103], VII [104], IX [105], XII [106], XIII [107], XIV [108], XV [109], XVI [110]) and a general overview [111] will appear cumulatively [112].

2 several linked genes could be isolated together with their intergenic regions;
3 fewer colonies had to be maintained and screened to isolate a clone of interest;
4 cosmid clones turned out to be stable for many years under usual storage conditions.

Additionally, the isolation of sequentially overlapping cosmid clones has facilitated physical linkage over the entire yeast chromosomes.

To construct a library with as complete coverage as possible with as few clones as possible, the cloned DNA fragments should be randomly distributed on the DNA. Under these conditions, the number of clones (N) in a library representing each genomic segment with a given probability (P) is

$$N = \ln(1-P)/\ln(1-f)$$

where f is the insert length expressed as fraction of the genome size [26]. Assuming an average insert length of 35 kb, a cosmid library containing 4600 random clones would represent the yeast genome at $P = 99.99\%$, i.e. about 12 times the genome equivalent. The actual number of cosmid clones obtained by the usual procedures is very high (> 200 000 per microgram of DNA).

This small number of clones was advantageous in setting up ordered yeast cosmid libraries or sorting out and mapping chromosome specific sublibraries. For example, a chromosome XI specific sublibrary composed of 138 clones had been sorted out from an unordered cosmid library by colony hybridization using DNA from chromosome XI purified by PFGE as a probe. The *nested chromosomal fragmentation* method [27] was then applied to rapid sorting of these clones. Finally, a set of some 30 overlapping cosmids was sufficient to build a contig of chromosome XI. In the following, this approach has been successfully applied to most of the other chromosomes sequenced in the yeast genome project. We have selected 3500 independent clones from a total yeast cosmid library to prepare the DNA of each single cosmid from minilysates of 5 ml cultures [16]. These samples were numbered and the corresponding cultures kept as glycerol stocks at –70 °C. By chromosomal walking, 43 of these clones have been sorted out to cover yeast chromosome II [28]. For cosmid cloning of chromosome II DNA, we employed a vector that carries a yeast marker and therefore could be used in direct complementation experiments [16].

For sequencing chromosome VIII [18], partially overlapping phage lambda and cosmid clones were used, which were previously mapped for *Hin*dIII and *Eco*RI sites [29]. It may be noted that by convention of all laboratories engaged in sequencing the yeast genome, the strain αS288C or isogenic derivatives thereof were chosen as the source of DNA, as these strains have been fairly well characterized and employed in many genetic analyses.

High-resolution physical maps of the respective chromosomes were constructed by application of classical mapping methods (fingerprints, cross-hybridization) or by novel methods developed for this programme, such as site-specific chromosome fragmentation [27] or the high-resolution cross-hybridization matrix [30], to facilitate sequencing and assembly of the sequences. These techniques might be of interest for other genomes as well and, particularly, for mapping YAC inserts.

30.3.3 Sequencing strategies, sequence assembly and quality control

30.3.3.1 Sequencing strategies

In the European network, chromosome-specific clones were distributed to the collaborating laboratories according to a scheme worked out by the DNA coordinators. Each contracting laboratory was free to apply sequencing strategies and techniques of its own provided that the sequences were entirely determined on both strands and unambiguous readings were obtained. Two principal approaches were used to prepare subclones for sequencing:

1 generation of sublibraries by the use of a series of appropriate restriction enzymes or from nested deletions of appropriate subfragments made by exonuclease III;

2 generation of shotgun libraries from whole cosmids or subcloned fragments by random shearing of the DNA. Sequencing by the Sanger technique (see Chapter 22) was either done manually, labelling with [^{35}S]dATP being the preferred method of monitoring or by automated devices (on-line detection with fluorescence labelling or direct blotting electrophoresis system) following the various established protocols. Similar procedures were applied to the sequencing of the chromosomes contributed by the Sanger laboratory and the laboratories in North America, Canada and Japan. The American laboratories largely relied on machine-based sequencing.

30.3.3.2 Sequencing telomeres

The yeast chromosome telomeres presented a particular problem. Due to their repetitive substructure and the lack of appropriate restriction sites, conventional cloning procedures were successful only for a few exceptions. Largely, telomeres were physically mapped relative to the terminal-most cosmid inserts using the I-*Sce*I chromosome fragmentation procedure [27]. The sequences were then determined from specific plasmid clones obtained by *telomere trap cloning*, an elegant strategy developed by E. Louis [31,32].

30.3.3.3 Sequence assembly and quality control

Within the European network, all original sequences were submitted by the collaborating laboratories to the Martinsried Institute of Protein Sequences (MIPS) who acted as an informatics centre. They were kept in a data library, assembled into progressively growing contigs, and updated during the course of the project. In collaboration with the DNA coordinators, the final chromosome sequences were derived. Starting with chromosome XI, all sequences submitted by the collaborating laboratories were subjected to quality controls. 'Verifications' (amounting to a total of 450 kb) were achieved by anonymous resequencing of selected regions, either long fragments (total of 15–20% per chromosome) or short segments (total of 1–2% per chromosome) chosen from suspected or difficult zones which were resequenced directly from cosmids using designated pairs of oligonucleotides as primers.

Similarly, automated procedures were employed for sequence assembly in the other laboratories, based, for example, on the programme package developed at Cambridge (see, for example, ref. 33) or on the ACeDB programme developed for the *C. elegans* genome project [34]. In any case, correct assembly of the sequences was guaranteed by establishing that the order of restriction sites predicted from the sequence was consistent with the physical maps of these sites that had been determined independently and care was taken to perform quality controls that would result in a high accuracy [6].

In spite of all precautions, determination of a sequence cannot be error-free. From theoretical considerations [6] taking all types of errors together, it follows that with an average sequence accuracy of 99.9%, only a third of all yeast genes are properly described, whereas fidelity is brought to 85%, if a sequence accuracy of 99.99% is reached. The systematic sequencing programs with verifications resulted in a sequence accuracy of ≈99.97% corresponding to a gene accuracy of some 75%. In practice, care was taken to minimize frameshift errors, which represent about two thirds of all sequencing errors and will have the most deleterious effects on gene interpretation.

30.3.3.4 Sequence analysis

Along with data submission by the single laboratories, and finally when the complete sequences of the chromosomes were available, they were subjected to analysis by various algorithms. The sequences have been interpreted using the following principles:

1 all intron splice site/branch-point pairs detected by using specially defined patterns (ref. 35; Kleine, K. and Feldmann, H., unpublished) were listed;

2 all ORFs containing at least 100 contiguous sense codons and not contained entirely in a longer ORF on either DNA strand were listed (this includes partially overlapping ORFs);

3 the two lists were merged and all intron splice site/branch-point pairs occurring inside an ORF but in opposite orientation were disregarded;

4 centromere and telomere regions, as well as tRNA genes and Ty elements or remnants thereof, were sought by comparison with previously characterized datasets of these elements (Kleine, K. and Feldmann, H., unpublished data) including the data-base entries provided in a continuously updated library of tRNAs and tRNA genes [36] as well as the program tRNA Scan [37].

In the European network, special software developed for the VAX at MIPS was used to locate and translate open reading frames (ORFEX and FINDORF), to retrieve noncoding intergenic sequences (ANTIORFEX), and to display various features of the sequence(s) on graphic devices (XCHROMO; an interactive graphics display program).

Searches for similarity of proteins to entries in the databanks were performed by FASTA [38], BLAST [39], and FLASH [40], in combination with public protein sequence databases. Protein signatures were detected by using the PROSITE dictionary [41]. ORFs were considered to be homologues or to have probable functions when the alignments from these searches showed significant similarity and/or protein signatures were apparent. Compositional analyses of the chromosome (base composition; nucleotide pattern frequencies, GC profiles; ORF distribution profiles, etc.) were performed by using the X11 program package (C. Marck, unpublished). For calculations of CAI and GC content of ORFs the algorithm CODONS [42–44] was used. Comparisons of the chromosome sequences with databank entries at MIPS were based on a new algorithm developed there [45]. Furthermore, particular nucleotide patterns were searched for, which will be mentioned below. Basically, the same strategies were used by other laboratories to interprete their sequences, again combining well-established routines with special software developed in these laboratories.

30.4 Life with 6000 genes

30.4.1 The proteome: open reading frames and gene function

The term *proteome* has been coined to describe the complete set of proteins synthesized by a living cell [46]. With the completion of the yeast genome sequence, for the first time, we can now define the proteome of a eukaryotic cell.

The sizes of the majority of the ORFs in yeast vary between 100 to more than 4000 codons (Fig. 30.2). Less than 1% of the ORFs is estimated to be below 100 codons; the smallest mature peptides that have been characterized are the two mating pheromones.

Comparison of the final sequence with public databases revealed that some 28.11% of the yeast ORFs correspond either to previously known protein-coding genes or to genes whose functions have been determined previously or during the course of the project. An estimated 6% of the total remain questionable ORFs. Thus, 66% of the total ORFs represent novel putative yeast genes. As far as

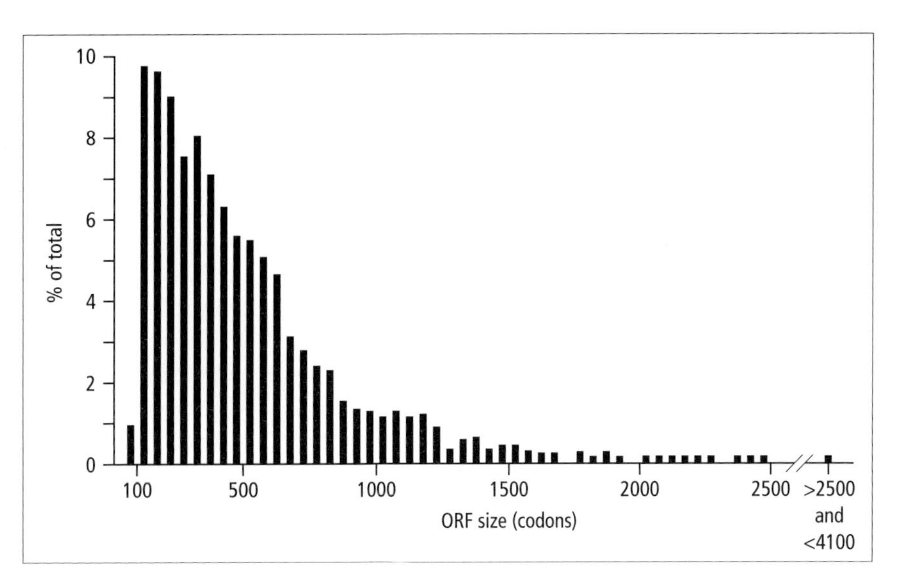

Fig. 30.2 Distribution of ORF length in the yeast genome. The panel indicates the relative fraction represented by each size class (increment by 50 codons) from the first nine chromosomes to be sequenced [6]. For systematic evaluation, the lower limit of ORF size was set to 100 codons. The fraction of shorter ORFs (< 100 codons) individually identified is included.

we can extrapolate from transcriptional mapping of particular chromosomes, the majority of these ORFs should represent 'real' genes, though many of those appear to be transcribed at an extremely low level [47,48]. Of the total ORFs, 14.8% have homologues among gene products from yeast or other organisms whose functions are known, whereas another 14.4% of the total have recognizable motifs or weak homologies to genes for experimentally characterized functions. The remaining 37.7% of the total ORFs have either homologues to ORFs of unknown function on other chromosomes (26.2% of the total) or no homologues in data libraries at all (10.5% of the total). Thus, ≈2200 of the yeast genes have to be catagorized as 'genes of unknown function', sometimes called 'orphans' [6].

It is noteworthy that the number of yeast genes to which functions could be attributed through comparisons with the database entries from all organisms has not substantially increased on completion of the sequence, although this information grew rather exponentially in the past few years. A similar observation was made when the complete genomes of small prokaryotes, such as *Haemophilus influenzae* (1.8 Mb) [49], *Mycoplasma genitalium* (0.6 Mb) [50], and *Methanococcus jannaschii* (1.7 Mb) [51] were determined: a large proportion of the genes have no counterparts in other organisms. Probably, therefore, we deal with a general phenomenon where it looks as if many of the novel functions only require transient or low-level transcription in an organism or are primarily phylum specific.

Now that the complete sequence of the yeast genome is available, it will be interesting to systematically compare all of the ORFs which can be classified according to the presence of known functional motifs (or protein signatures). A useful inventory list of the yeast proteins has been compiled by J.I. Garrells (http://quest7.proteome.com/YPDhome.html).

30.4.2 Overlapping ORFs, pseudogenes and introns

A few cases have been found where overlapping ORFs indeed exist and are expressed. In one particular case, it was even shown that expression of the two ORFs occurs at different stages of yeast growth. Another interesting question was, how many pseudogenes might be present in the yeast genome. From earlier studies, it was anticipated that this number in yeast should be low compared with that in mammalian genomes. Generally, this assumption seems to hold true for most of the yeast chromosomes, but chromosome I turned out to be an exception [52]. Chromosome I is the smallest naturally occurring functional eukaryotic nuclear chromosome so far characterized. The central 165 kb resemble other yeast chromosomes in both high density and distribution of genes. In contrast, the remaining sequences flanking this DNA (the two ends of the chromosome) have a much lower gene density, are largely not transcribed, contain no essential genes for vegetative growth, and contain four apparent pseudogenes and a 15-kb redundant sequence. These terminally repetitive regions consist of a telomeric repeat, flanked by DNA closely related to *FLO1*, a yeast gene involved in cell flocculation and encoding a large serine/threonine-rich cell wall protein with internal repeats. The pseudogenes are related to known yeast genes but have internal stop codons. Extreme care has been taken in such cases to reconfirm the sequences of the regions in question by independent laboratories.

Only a minor fraction of the yeast genes, around 4% of the total, are predicted (or already experimentally shown) to be interrupted by introns. To date, only two cases have been encountered where two introns are present: the *MAT* locus on chromosome III and a ribosomal protein gene, *RPL6 A*, on chromosome VII. In the latter, the second intron encodes a small RNA. Generally, the intron is located at the extreme 5′-end of each gene, sometimes even preceding the coding region. The predominant population of intron-containing genes is recruited by the ones encoding ribosomal proteins. The functional significance of the introns is by no means clear, and despite the low number of intron-containing genes yeast maintains a highly sophisticated and complex machinery for splicing.

30.4.3 Putative membrane proteins, mitochondrial proteins

The ALOM algorithm [53] can be applied to predict putative membrane spans. A first estimate for chromosome III revealed that some 38% of the 'real' genes may code for transmembrane proteins containing 1–14 potential membrane transversions [54]. A similarly high figure (142 ORFs out of 410) was found for chromosome II [28]. Data obtained from other systematically sequenced yeast chromosomes suggest that this appears to apply as a general rule in yeast [55,56]. Even though the algorithm may give a somewhat high estimate, possibly a third of yeast proteins have to be considered to be associated with membrane structures.

Examination of the ORFs for the occurrence of putative mitochondrial target signal sequences is difficult due to the complex character of these signatures [57] this can only be achieved by visual inspection. Since not all of the proteins participating in mitochondrial biogenesis are imported via particular signal sequences, the exact number of proteins involved in maintaining mitochondrial function in yeast remains unknown at present. A rough estimate is that some 8% of the yeast proteins may be involved in mitochondrial biogenesis.

30.4.4 Other genetic entities

In addition to the genes encoding proteins, we have obtained detailed information on the organization of the genes for tRNAs and other small RNAs, the yeast retrotransposons (termed Ty elements), as well as the telomeric and centromeric sequences. The genes for the ribosomal RNAs are clustered in some 100 copies on the right arm of chromosome XII, whereas the multiple copies of tRNA genes are found scattered throughout the genome.

Five different types of Ty elements, which exhibit substantial homology to retroviruses and retrotransposons from plants and animals, are present in the yeast genome: Ty1, Ty2, and Ty4 belong to the 'copia' class of retrotransposons, while Ty3 is a member of the 'gypsy' family (for a review, see ref. 58). αS288C contains 32 complete Ty1 and 13 Ty2 elements, whereas Ty4 is present at only three locations and Ty3 occurs in two copies. Ty5, found in chromosome III, appears to be a new class of yeast transposon. Like retroviruses, the Ty elements transpose through an RNA intermediate and by reverse transcription. Transposition rates are low, and the number of elements is kept fairly constant by balancing transposition and excision events [59]. This is manifest from the presence of 268 long terminal repeats (LTRs) or remnants thereof that are footprints of previous transposition events. Due to the vagabond life-style of the retrotransposons, yeast strains differ with respect to the sometimes rather complex 'patterns' formed by these elements resulting from multiple integrations and excisions. However, comparison of different yeast strains (see, for example, refs 60 and 61) and experimental data [62] revealed that spontaneous transposition events do not appear to occur randomly along the length of individual chromosomes but that the Ty elements are preferably integrated into the upstream regions of tRNA genes [63]. Since these regions do not contain any special DNA sequences, the region-specific integration of the Ty elements may be due to specific interactions of the Ty integrase(s) with the transcriptional complexes formed over the intragenic promoter elements of the tRNA genes or triggered by positioned nucleosomes in the 5′ flanking regions of the tRNA genes (see, for example, ref. 64). In any case, the Ty integration machinery can detect regions of the genome that may represent 'safe havens' for insertion, thus guaranteeing both survival of the host and the retroelement.

Analysis of the sequenced yeast genome clearly substantiates the earlier observations of considerable plasticity of the yeast genome around tRNA gene loci and the existence of 'transposition hotspots' (see, for example, refs 60, 61 and 65).

30.5 Genome architecture and gene organization

30.5.1 Gene density and gene arrangement

It is now well established that the gene density in all yeast chromosomes is rather similar. Excluding the ORFs contributed by the Ty elements, ORFs occupy, on average, 70% of the sequences. This leaves only limited space for the intergenic regions which can be thought to harbour the major regulatory elements involved in chromosome maintenance, DNA replication and transcription. Regarding transcription of protein-coding genes, a variety of elements have been identified and characterized that are operative in transcriptional initiation, regulation and termination. Not all of the yeast genes are preceded by a canonical TATA box, and it remains still open which type of AT-rich sequences or other elements can act as transcriptional initiation sites [66]. In some cases, terminator sequences have been defined, but no general consensus sequences can be deduced. The same holds true for polyadenylation sites and polyadenylation signal sequences. Where experimentally determined, it appears that there is a much larger variability to these sequences than in mammalian systems [67]. As in mammalian or plant systems, a number of regulatory *cis*-acting elements (upstream acting sequences; UAS) and the corresponding *trans*-activating factors have been experimentally characterized in yeast (for a review, see ref. 68). Also negative regulatory elements (upstream repressing sequences; URS) have been shown to control the expression of some genes. However, in a few instances, precise ideas on the intimate interplay of the various regulatory components mediating gene expression are beginning to evolve. The knowledge of the entire genome sequence, combined with the powerful genetic tools available for yeast, should now foster research along these lines.

Generally, ORFs appear to be rather evenly distributed among the two strands of the single chromosomes. In some chromosomes (e.g. I, II, VIII), there is a slight excess of coding capacity on one of the strands, the significance of which is not known. Figure 30.3 presents a scheme of how the single transcriptional units are organized along yeast chromosomes. Three principal arrangements are possible:

1 'head-to-tail' orientation of two adjacent genes, so that transcription occurs in the same direction and the intergenic regions should carry a terminator for one gene and a promoter for the next one to follow;

2 'head-to-head' orientation, in which transcription of two genes is divergent from a common 'promoter' region;

3 'tail-to-tail' orientation, by which two genes share a 'terminator' region.

There is no predominance of one or the other type of gene arrangement, although arrays longer than eight genes that are transcriptionally orientated in the same direction can be found on several chromosomes. The extreme seems to be a region from chromosome VIII, where 17 in a run of 18 ORFs are located on the 'top' strand.

In the 'head-to-tail' arrangements, the intergenic regions between two consecutive ORFs sometimes are extremely short, raising the question of whether they are maintained as separate units or coupled for transcription and translation. There are cases in which different functions have been combined in one genetic unit but, to the best of our knowledge,

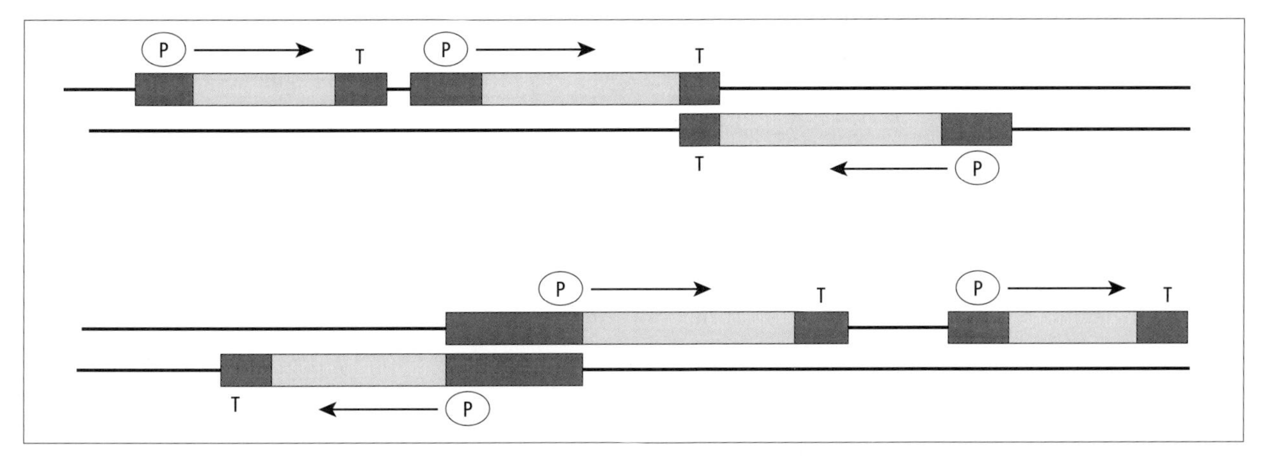

Fig. 30.3 Organization of yeast genes along chromosomes. The average base composition of yeast DNA is 38.4% (G + C). As expected, the protein-coding regions have a higher GC content, on average (40.2%), than the noncoding regions (35.1%). In sliding windows, coding regions may be discriminated from intergenic regions, since 'transitions' in GC content are rather sharp at their borders. An almost symmetrical distribution of dinucleotide frequencies over the entire chromosome is apparent, whereas the base composition of ORFs shows a significant excess of homopurine pairs on the coding strand. Normally, coding regions are evenly distributed between the two strands. The average ORF size is 1450 bp. The average sizes of interORF regions vary between 630 and 945 bp for different chromosomes, they are 618 bp on average for 'divergent promoters' (36.2% GC) and 326 bp for 'convergent terminators' (29.3% GC), while 'promoter-terminator combinations' (34.2% GC) are 517 bp in length on average.

polycistronic messages have not been observed in yeast to date.

Initially, the intervals between divergently transcribed genes might be interpreted to mean that their expression is regulated in a concerted fashion involving the common promoter region. This, however, seems not to hold for the majority of the genes and might be a principle reserved for a few cases, in which these genes belong to the same regulatory pathway (e.g. *GAL1/GAL10* [69]). By contrast, many examples are known in which a constitutively expressed gene shares its upstream sequences with that of a highly controlled gene. Regarding the fact that most of the intergenic regions are relatively short (cf. Fig. 30.3), an intriguing question becomes apparent: are regulatory elements confined to these sequences or could they also be present in coding sequences of neighbouring genes located upstream? Experimental data obtained for several genes involved in meiosis point to the latter possibility [70]. This would enable two different kinds of constraint to be superimposed on sequences during evolution, one for maintaining function of coding sequences and one for preserving regulatory sequences. By employing catalogues of consensus sequences of the known regulatory elements (H. Feldmann, unpublished data), one can detect many sites within the intergenic regions which could be thought to be functional, but at the same time, such sites are found scattered throughout the coding regions as well. Although the functional significance of these sequences might strictly depend on the regional context, it is difficult, at the moment, to discriminate between functional and nonfunctional elements merely by inspection of the sequence. Only experimental approaches will answer this problem.

30.5.2 Base composition and gene density

Average base composition has been found to be symmetrical over the entire genome (the symmetry being even more apparent with dinucleotide frequencies), but this only reflects the almost equal numbers of ORFs encoded on each DNA strand of most of the yeast chromosomes, the base composition of ORFs themselves showing a significant excess of homopurine pairs on the coding strand [71].

Regional variations of base composition with similar amplitudes were first noted along chromosome III [72], with major GC-rich peaks in the middle of each arm. Results from chromosome XI confirmed this finding, but owing to its larger size, revealed an almost regular periodicity of the GC content, with a succession of GC-rich and GC-poor segments. A most interesting observation was

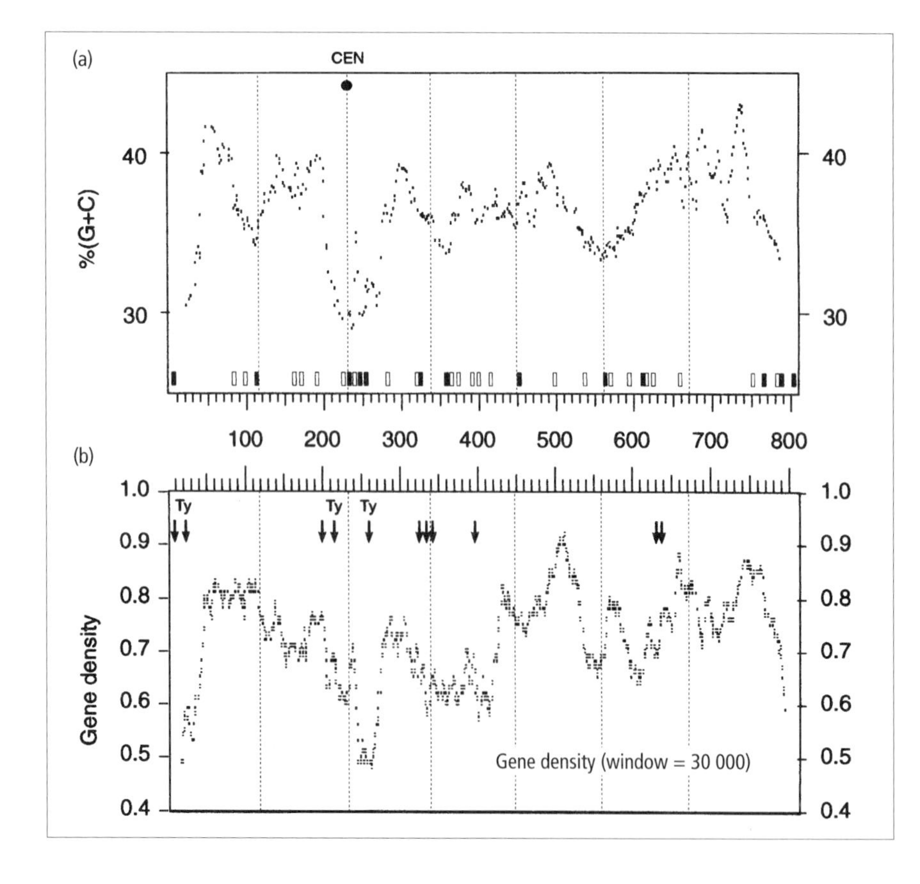

Fig. 30.4 Periodic variation of GC content and gene density along chromosome II. (a) Compositional variation was calculated according to ref. 71. Each point represents the average GC composition calculated from the silent positions only of the codons of 15 consecutive ORFs. Similar slopes were obtained when the GC composition was calculated from the entire ORFs or from the interORF regions, or when the averages of 13–30 elements were plotted. The location of perfect ARS consensus sequences is indicated by the rectangles, whereby filled boxes represent those ARS fulfilling the criteria attributed to functional replication origins [76, 77]. (b) Gene density was expressed as the fraction of nucleotides within ORFs vs. the total number of nucleotides in sliding windows of 30 kb. Arrows indicate location of tRNA genes and intact Ty elements, respectively.

that the compositional periodicity correlates with local gene density, which reaches more than 85% in GC-rich regions, followed by segments of comparably lower gene density (50–55%) in AT-rich regions [71]. Other chromosomes also show compositional variation of similar range along their arms, with pericentromeric and subtelomeric regions being AT-rich, though spacing between GC-rich peaks is not always regular. In most cases, however, there is a broad correlation between high GC content and high gene density as is the case in more complex genomes in which isochores of composition are naturally much larger [73].

The profiles obtained from the analyses of chromosomes II [28] are exemplified in Fig. 30.4. GC-poor peaks coinciding with relatively low gene densities are located at the centromere (around coordinate 230) and at both sides of the centromere, with a periodicity of ≈110 kb. These minima are more pronounced around coordinates 120, 340 and 560, while they are less so at coordinates 450 and 670. Remarkably, most of the tRNA genes reside in GC-poor 'valleys' and the Ty elements became eventually integrated into these regions. When analysing chromosome II for the occurrence of simple repeats, putative regulatory signals, and potential ARS elements, we noticed that the latter ones were not found randomly distributed. In Fig. 30.4, we have listed the location of 36 ARS elements which completely conform to the 11 bp degenerate consensus sequence [74,75]. Several of these were found associated at their 3′ extensions with imperfect (1–2 mismatches) parallel and/or antiparallel ARS sequences or putative ABF1-binding sites, reminiscent of the elements reported to be critical for replication origins [76,77]. Remarkably, these patterns are found within the GC valleys, suggesting that functional replication origins might preferably be located in AT-rich regions. This phenomenon was also apparent from an analysis of chromosome XI and, more convincingly, when the distribution of functional replication origins mapped in chromosome VI [78] or in 200 kb of chromosome III [79] were compared to the GC profiles of these chromosomes. Functional ARS elements have yet to be defined for the remainder of chromosome III and the other yeast chromosomes. In this context, it would be interesting to see whether the origins of replication reveal a regular spacing [80] and whether these and the chromosomal centromeres might maintain specific interactions with the yeast nuclear scaffolding [81]. In all yeast chromosomes analysed thus far, ARS elements located in the subtelomeric regions are closely associated with specific OBF-binding sites [82,83].

Although the fairly periodic variation of base composition is now evident for the yeast chromosomes, its significance remains unclear. Several explanations for the compositional distribution and the location-dependent organization of individual genes have been offered. For example, the compositional periodicity of a yeast chromosome could reflect the evolutionary history of the chromosome, together with the folding of that chromosome, its attachment to the nuclear matrix or structural elements involved in chromosome segregation, or in the 'homology search' that precedes synapsis in the early meiotic prophase. Other possibilities could be tested experimentally. For example, transcription mapping of a whole chromosome could give a clue as to whether such rules may influence the expression of genes. Furthermore, long-range determination of DNase I sensitive sites may be used to find a possible correlation between compositional periodicity and chromatin structure along a yeast chromosome.

30.5.3 Telomeres

The organization of the yeast telomeres (Fig. 30.5) has become clear from the work of E. Louis and his collaborators in conjunction with the chromosome sequences. All yeast chromosomes share characteristic telomeric and subtelomeric structures [32]. Telomeric (C1–3 A) repeats, some 300 nucleotides in length, are found at all telomere ends. Thirty-one out of 32 chromosome ends contain the X core subtelomeric elements (400 bp), and 21/32 of the chromosome ends carry an additional Y′ element. There are two Y′ classes, 5.2 kb and 6.7 kb in length, both of which include an ORF for a putative RNA helicase of yet unknown function. Y′ element show a high degree of conservation but vary among different strains [84]. Experiments with the *est1* (ever shortening telomeres) mutants, in which telomeric repeats are progressively lost, have shown that the senescence of these mutants can be rescued by a dramatic proliferation of Y′ elements [85]. Several additional functions have been suggested for these elements (for a review, see ref. 86), such as extension of telomere-induced heterochromatin or protection of nearby unique sequences from its effects; a role in the positioning of chromosomes within the nucleus.

Comparisons of the chromosome termini between each other revealed that, in addition to the common subtelomeric repeats, they share extended similarities in their subtelomeric regions: genetic redundancy is the rule at the ends of yeast chromosomes. The 'duplicated' regions contain copies of genes of known or predictable function as well as several

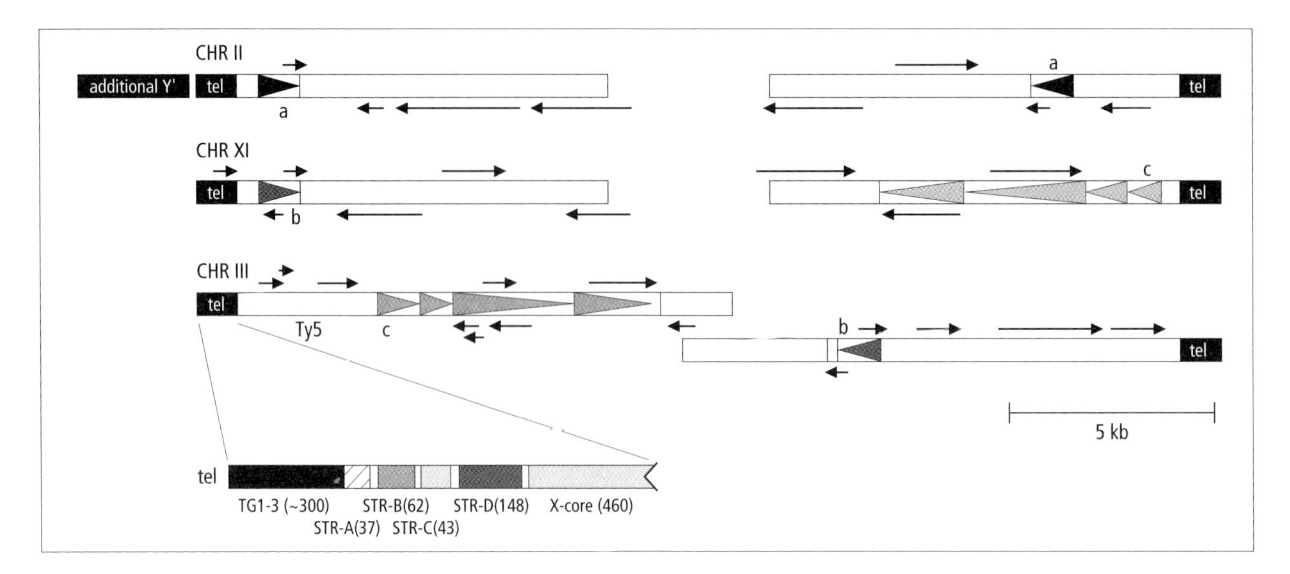

Fig. 30.5 Organization of yeast telomeres. The chromosome ends of the first three chromosomes to be completed are compared. Repetitious sequences of high similarity at the nucleotide level are indicated by the filled triangles. ORFs are represented by the arrows. The consensus telomere sequences are shown in black; their substructure [32] is indicated in the insert (not drawn to scale).

ORFs the putative products of which exhibit high similarity. The functions of the latter remain unclear as no homologues of known function have been found in the databases.

For instance, the two terminal domains of chromosome III show considerable sequence homology (up to 18 kb), both to one another and to the terminal domains of other chromosomes (II, V, and XI). In chromosome VIII, extensive duplications being present on other chromosomes have been observed: 30 kb near the right telomere is more than 90% identical to the similar region on the right arm of chromosome I. A smaller portion of this is also duplicated on the left arm of chromosome I. Shorter duplicated segments in the subtelomeric region of the left arm of chromosome VIII have been encountered in the subtelomeric regions of chromosomes III and XI (showing 54–94% identity).

30.5.4 Complex and simple repeats

Overall, the yeast genome is remarkably poor in repeated sequences. The unique constellation of repetitive sequences at the two ends of chromosome I has already been pointed out. Approximately 30 kb in each subtelomeric region carry similar (but non-essential) genes and a 15-kb repeat. These features are consistent with the idea that these terminal regions represent the yeast equivalent to heterochromatin and the occurrence of this type of DNA suggests that its presence gives this chromosome the critical length required for proper stability and function. The 30-kb region can be removed from each end without affecting vegetative growth, although chromosome stability is considerably reduced. Most likely, these repeated regions contribute to chromosome I size polymorphisms which have been observed [52]. Besides the Ty elements, it is the rDNA on chromosome XII that most significantly contributes to repetitiveness. A cluster of some 15 tandem repeats (2 kb each) containing the *CUP1* gene and contributing to polymorphic variation is found on chromosome VIII [18].

Repeated stretches of short oligonucleotides exist. These include poly(A) or poly(T) tracts, alternating poly(AT) or poly(TG) tracts, and direct or inverted long repeats. Even short stretches of the simple sequence repeat $(TG_{1-3})_n$ normally 'sealing' the chromosome ends have been encountered internal to some chromosomes. This type of internal repeats are probably relics of events during breakage and healing of chromosomes.

By applying the program PYTHIA [87] to search for simple repeats (Chapter 25), we detected at least 12 sets of regularly repeated trinucleotides along chromosome II representing repetitious codons for particular amino acids, thus forming homopeptide stretches. In some cases, even more complex amino acid patterns result (Table 30.1). A systematic study on the distribution and variability of trinucleotide repeats in the yeast genome is under way [88]. Perfect and imperfect repeats ranging from four to 130 triplets were recognized and the repartition of different triplet combinations was found to differ between ORFs and intergenic regions. Examination of various laboratory strains revealed polymorphic

Table 30.1 Simple repeat sequences on chromosome II corresponding to amino acid homopolymer stretches.

ORF	Gene	Amino acid repeat	Nucleotide repeat
YBL084c	CDC27	Asn (>24)	AAT
YBL081w		Asn	AAC
		Ser	complex
YBL029w		Asn	AAT
		Ser & Pro	complex
YBL011w		Glu	GAGGAA
YBL007c	SLA1	QQPQMMN	AC-rich
YBR016w		Gln	CAG/CAA
YBR040w		FLLI	CAAA
YBR067c	TIP1	Ser	complex
YBR112c	SSN6	Gln	CAA/CAG
		Gln-Ala	CAAGCA
YBR150c		Asn	TTA
		Asp (>21)	GAT
YBR289w	SNF5	Gln	CAACAG

size variations for all perfect repeats, compared to an absence of variation for the imperfect ones. These findings are particularly interesting in view of the fact that several human genetic disorders are caused by trinucleotide expansion. The yeast system may now provide an experimental approach to study the mechanisms of their expansion.

30.5.5 Comparison of genetic and physical maps

The genetic map of *S. cerevisiae* [7] was of considerable value to yeast molecular biologists before physical maps became available. In fact, we and others have used DNA probes from some known genes mapped to particular chromosomes for chromosomal walking. Finally, however, physical maps of all chromosomes have been constructed without reference to the genetic maps.

Beside local expansion or contraction of the genetic map, and the fact that the overall frequency of meiotic recombination increases with shortening chromosome size, the order of the genes positioned on the chromosomes by genetic and physical mapping grossly agree. Thus, the comparison of the physical and genetic maps show that most of the linkages give the correct gene order but that in many cases the relative distances derived from genetic mapping are imprecise. The obvious imprecisions of the genetic maps may be due to the fact that different yeast strains have been used in establishing the linkages. It is even possible that some strains used in genetic mapping experiments show inversions or translocations which then might contribute to discrepancies between physical and genetic maps. Clearly, the accuracy of genetic mapping will depend on the experimental approaches used. For example, a deviation between the genetic and the physical maps initially observed with chromosome XI [71] could be corrected by repeating the genetic mapping of a segment located next to the left telomere [89]. A more widespread phenomenon, however, that may lead to imprecisions of the genetic maps are strain polymorphisms caused by the extended repetitive sequences or subtelomeric duplicated genes mentioned above, and particularly by the Ty elements. Altogether, the experience gained from the yeast genome project shows that genetic maps provide valuable information but that independent physical mapping and determination of the complete sequences is needed to unambiguously delineate all genes along chromosomes. At the same time, the differences found between various yeast strains demonstrate the need to use one particular strain as a reference system.

30.6 Genome organization and evolutionary aspects

30.6.1 Genetic redundancy in yeast

A survey of previous sequence data and sequences obtained in the yeast sequencing project suggested that there is a considerable degree of internal genetic redundancy in the yeast genome. Although an estimate of sequence similarity (both at the nucleotide and the amino acid level) is now possible, it still remains difficult to correlate physical and functional redundancy, because even in yeast gene functions have been precisely defined only to a limited extent. Understanding the true nature of redundancy will help elucidate the biological role of every yeast gene.

30.6.1.1 Duplicated genes in subtelomeric regions
Classical examples of redundant genes in yeast are the *MEL, SUC, MGL* and *MAL* genes, which have been mentioned earlier. In fact, yeast strains differ by the presence or absence of particular sets of these genes. For example, three genes mapped on chromosome II of wild-type strains, *MEL1, SUC3*, and *MGL2*, are absent from the strain αS288C. A comparison at the molecular level of αS288C with brewer's yeast strain C836 clearly shows that the *SUC* genes are present on chromosome II of the latter strain [90]. Regarding the genes involved in carbohydrate metabolism, the presence of multiple gene copies could be attributed to selective pressure induced by human domestication, as it appears that they are largely dispensable in laboratory strains (such as αS288C) which are no longer used in fermentation processes. Nonhomologous recombination processes may account for the duplication of these and other genes residing in subtelomeric regions (see, for example ref. 91), reflecting the dynamic structure of yeast telomeres in general [79]. We have already mentioned the fact that the subtelomeric regions of several yeast chromosomes share highly conserved segments, in some instances up to 30 kb, which carry duplicated genes the functions of which are largely unknown.

30.6.1.2 Duplicated genes in internal chromosome regions
Analyses of individual chromosomes indicate a great variety of genes duplicated elsewhere in the chromosomes. Before complete chromosome sequences became available, a great variety of genes were known to occur in two or more identical, or nearly identical, copies located on different chromosomes, such as the histone genes, ribosomal protein genes, genes for ATP/ADP carriers, for enzymes of the glycolytic pathway, for sugar and amino acid transporters, and for many other proteins. Numerous examples can now be added when the completed chromosomes are searched for similarity at the nucleotide as well as at the protein level. These include dispersed families with related but nonidentical genes scattered singly over many chromosomes. The largest such family comprises the 23 *PAU* genes which specify the so-called seripauperines [92], a set of almost identical serine-poor proteins of unknown function. The *PAU* genes reside in the subtelomeric regions. Clustered gene families are less common, but a large family of this type occurs on chromosome I where six related genes encode a set of membrane proteins of unknown function [92]. Another 10 members of this family occur on five additional chromosomes; some are clustered, others are scattered singly, still others are located in subtelomeric regions.

30.6.1.3 Duplicated genes in clusters
Remarkably, duplicated genes have also been found in clusters. There are at least three examples of this kind in chromosome II [28]. Another case is a cluster of three hexose transporter genes on chromosome VIII [18], which appear to be the result of a less recent gene duplication. Rather unique cases of gene duplications are represented by a large clustered (tandem) gene family of membrane proteins on chromosome I, and a large cluster on chromosome VIII near *CUP1*. The *CUP1* gene-coding copper metallothionein is contained in a 2-kb repeat that also includes an ORF of unknown function. The repeated region has been estimated to span 30 kb in strain αS288C, which could encompass 15 repeats, but the number of repeats varies among yeast strains.

However, in these and other cases, the duplicated sequences are confined to nearly the entire coding region of these genes and do not extend into the intergenic regions. Thus, the corresponding gene products share high similarity in terms of amino acid sequence or sometimes are even identical and therefore may be functionally redundant. However, as suggested by sequence differences within the promoter regions, gene expression should vary according to the nature of the regulatory elements or other (regulatory) constraints. It may well be that one gene copy is highly expressed while another one is weakly expressed. Turning on or off expression of a particular copy within a gene family may depend on the differentiated status of the cell (such as mating type, sporulation, etc.). Biochemical studies also revealed that, in particular cases, 'redundant' proteins can substitute each other, thus accounting for the fact that a large portion of single gene disruptions in yeast do not impair growth or cause 'abnormal' phenotypes. This does not imply, however, that these 'redundant' genes were *a priori* dispensable. Rather they may have arisen through the need for yeast cells to adapt to particular environmental conditions. These notions are of practical importance when carrying out and interpreting gene disruption experiments.

30.6.1.4 Cluster homology regions
An even more surprising phenomenon became apparent when the sequences of complete chromosomes were compared to each other, revealing that there are large chromosome segments in which homologous genes are arranged in the same order, with the same relative transcriptional orientations, on two or more chromosomes. The occurrence of

such cluster homology regions (CHRs) is now manifest for a great deal of the yeast genome and might account for some 30–40% of total redundancy [5,93].

Chromosomes II and IV share the longest CHR, comprising a pair of pericentric regions of 170 and 120 kb, respectively, that share 18 pairs of homologous genes (13 ORFs and 5 tRNA genes). The genome has continued to evolve since this ancient duplication occurred: the insertion or deletioin of genes has occurred, Ty elements and introns have been lost and gained between the two sets of sequences. In all, at least 10 CHRs (shared with chromosomes II; V, VIII, XII, and XIII) can be recognized on chromosome IV. Remarkably, the entire chromosome XIV can be subdivided into several segments that are found duplicated on other chromosomes.

To analyse the extent and pattern of redundancy in the yeast genome, a potent data structure, the HPT, has been developed at MIPS, allowing an allagainst-all comparison of fixed size blocks of nucleotides, the results of which can be visualized by a graphical interface showing similarities both at the nucleotide and the protein level (data are obtainable from K. Heumann & W. Mewes at http://www.mips.biochem.mpg.de/mips/yeast).

30.6.1.5 Redundancy and gene organization

In all, we can imagine two ways in which duplications may have arisen. First, some of the duplicated genes could represent processed genes that were inserted into the genome relatively recently; a view which is consistent with the conservation of sequence only in the coding regions. However, all of these cases would appear to be created by integration of full-length complementary DNAs, because none appears to be a pseudogene, and this is unexpected in this model. In addition, some of the homologous gene pairs include introns in both genes, which suggest that these genes at least were not duplicated by this mechanism. Second, the clustering of duplicated genes and the occurrence of extended regions of similarity compel us to consider the idea that entire genomic regions were duplicated. Several of these duplication events would appear to be ancient, because the DNA sequence has clearly diverged outside the coding regions; moreover, such clusters even share a number of tRNA genes both in the same location and orientation. However, duplications may occur at any time during evolution (see reference to Heumann and Mewes, above).

An interesting problem intimately related to evolution is the origin of the present organizational pattern of genes. Could we, for example, find any criteria, be it structural or functional in nature, that govern the regional arrangement of particular genes? In other words, is there an 'ordered grouping' of genes along the yeast chromosomes or are we left with a random succession of most of the genes? Presently, we have few clues to answer these questions. In some respects, however, we do have indications for a particular location or grouping of genes, and some examples have already been mentioned above. In several cases, highly expressed genes are found associated with ARS elements, so that one could speculate that replication and efficient transcription is intimately coupled. In chromosome XI, it appears that highly expressed genes occur in 'clusters' within preferred regions (B. Dujon, personal communication). Clearly, the *MAL* and *SUC* loci, and the *GAL* locus represent examples, in which functionally related genes involved in a particular metabolic pathway are closely associated with each other.

30.6.2 Sequence variation among yeast strains

The question of to what extent yeast strains differ with respect to their genetic content has implicitly been touched already. We have discussed a number of features that contribute to polymorphisms in different yeast strains: (i) variable number of gene copies from repeated gene families; (ii) individual patterns caused by the presence or absence of particular Ty elements; and (iii) plasticity of the chromosome ends. In all these cases, polymorphism becomes also manifest through length differences between corresponding chromosomes. In addition, excisions or inversions of particular gene regions have been observed to give rise to polymorphisms. Chromosome breakage has been found to occur in yeast, resulting in karyotypes deviating from the 'normal' picture. However, sequence variations within the coding regions of individual genes seem to be rare, as far as we can tell from comparisons of the homologous sequences obtained from different strains.

30.6.3 Other genomes

30.6.3.1 The human–yeast connection

The availability of the complete yeast genome sequence not only provides further insight into genome organization and evolution in yeast but extends the catalogue of novel genes detected in this organism [93]. Many of these may be of particular value to yeast molecular biologists only, but of general interest may be those that are homologues to genes that perform differentiated functions in multicellular organisms or that might be of relevance to malignancy. Although the roles of these genes have still to be clarified, yeast may offer a useful experimental system to identify their function. On the other hand, the wealth of information to be expected clearly demands that new routes are explored to investigate the functions of novel genes.

By comparing the catalogue of human sequences available in the databases with the ORFs on the completed yeast chromosomes at the amino acid level it is estimated that more than 30% of the yeast genes have homologues among the human genes. As expected, most of the genes of known function catagorized in this way represent basic functions in both organisms. More similarities become apparent when expressed sequence tags (ESTs) are included in the analysis. Undoubtedly, the most compelling protagonists among these homologues are yeast genes that bear substantial similarity to human 'disease genes'. Recently, a comparative study along these lines has been published (obtainable from http://www.ncbi.nlm. gov/XREFdb/). Table 30.2 summarizes these findings.

30.6.3.2 Other model organisms

Prior to the release of the complete yeast genome sequence, two complete bacterial genomes had been published [49,50]; another prokaryotic genome was released recently [51]. The sequences of several further bacterial genomes have apparently been completed and the sequences of a number of bacteria, mostly extremophiles, are under way. The genome sequence of *E. coli* is now completed (see Chapter 31) and *Bacillus subtilis* will be completed soon (reviewed in ref. 5). The genome sequences of the next two eukaryotic genomes, those of *Schizosaccharomyces pombe* and *C. elegans* (see Chapter 29), are within our reach; the systematic sequencing of larger model genomes, most notably *Drosophila melanogaster* (see Chapter 28) and *Arabidopsis thaliana* (see Chapter 33), has now been tackled. Undoubt-

Table 30.2 Human disease genes with similarity to yeast genes. The positionally cloned genes are listed in order of decreasing statistical significance of the best match in the databanks [100].

Human disease	Human gene	Yeast gene	Yeast protein description
Hereditary non-polyposis colon cancer	*MSH2*	*MSH2*	DNA mismatch repair enzyme
Hereditary non-polyposis colon cancer	*MLH1*	*MLH1*	DNA mismatch repair enzyme
Cystic fibrosis	*CFTR*	*YCF1*	Cadmium resistance protein
Glycerol kinase deficiency	*GK*	*GUT1*	Glycerol kinase
Bloom syndrome	*BLM*	*SGS1*	Mismatch repair enzyme
Adrenoleukodystrophy, X-linked	*ALD*	*PAL1*	Phenyl ammonia lyase
Ataxia telangiectasia	*ATM*	*TEL1*	Telomere associated gene, chr. II
Pleiotrophic lateral sclerosis	*SOD1*	*SOD1*	Superoxide dismutase
Myotonic dystrophy	*DM*	*YPK1*	cAMP-dependent protein kinase
Lowe syndrome	*OCRL*	*YIL002*	Putative IPP-5-phosphatase
Neurofibromatosis, type 1	*NF1*	*IRA2*	Inhibitory regulator of ras-cAMP, chr. II
Choriodermia	*CHM*	*GDI1*	GDP dissociation inhibitor
Diastrophic dysplasia	*DTD*	*SUL1*	Sulphate transport protein
Lissencephaly	*LIS1*	*MET30*	Methionine pathway factor
Thomsen disease	*CLC1*	*GEF1*	Chloride channel protein
Wilms' tumour	*WT1*	*FZF1*	Sulphite resistance protein
Achondroplasia	*FGFR3*	*IPL1*	Protein kinase
Menkes' disease	*MNK*	*PCA1*	Copper-transporting ATPase, chr II
Multiple endocrine neoplasia 2A	*RET*	*CDC15*	Cell division control protein 15
Duchenne muscular dystrophy	*DMD*	*MLP1*	Myosin-like protein
Aniridia	*PAX6*	*PHO2*	Regulator in phosphate metabolism
Gonadal dysgenesis	*SRY*	*ROX1*	Hypoxic function transcription repressor
Breast cancer, early onset	*BCRA1*	*RAD18*	DNA repair protein
Epidermolytic palmoplantar keratoderma	*KRT9*	*MLP1*	Myosin-like protein
Wardenburg syndrome	*PAX3*	*RPB1*	RNA polymerase, subunit 10
Familial polyposis coli	*APC*	*AMYH*	Adenylate cyclase
Neurofibromatosis, Type 2	*NF2*	*YNL161*	Putative protein kinase
Retinoblastoma	*RB1*	*CYP1*	Regulator of O_2-dependent genes
Wiskott–Aldrich syndrome	*WASP*	*CLA4*	Protein kinase
Xerodermal pigmentosum	*RAD27*	*YKL113*	Nucleotide excision repair enzyme

edly, the information accumulating from these projects will provide an important foundation to learn more about the extent to which various processes are conserved among organisms in different lineages [95].

30.7 Perspectives and conclusions

30.7.1 Functional analysis

From the beginning, it was evident to anyone engaged in the project that the determination of the entire sequence of the yeast genome should only be considered a prerequisite for functional studies of the many novel genes to be detected [96,97]. Following these considerations, a European project called EUROFAN (for European Functional Analysis Network) has been established to undertake a systematic functional analysis of the functions of novel yeast genes [96]. Similar activities are under way in Germany, Canada and Japan, and in the USA, initiatives have been started by the NIH. For EUROFAN, a first goal will be to systematically investigate the phenotypes resulting from disruptions (and possibly overexpression) of some 1000 yeast genes of unknown function. A special set of yeast strains has been constructed for this purpose using a PCR-mediated gene replacement technique for the deletion of individual genes [98]. Concurrently, complete transcriptional maps of entire chromosomes will be constructed. Likewise, the development of refined *in silicio* analysis methods will be used to improve prediction of function (see, for example, ref. 99). These data are then used as a basis for intensified functional analyses: relevant genes or groups of genes that are suggested to be involved in particular functions are attributed to consortia of specialized laboratories for further exploitation.

30.7.2 Outlook

The wealth of fresh and biologically relevant information collected from the yeast sequences and the functional analyses have an impact on other large scale sequencing projects. The important contribution of genome projects in determining gene function has begun to emerge. Clearly, those genes that are homologues to genes that perform differentiated functions in multicellular organisms or that are of relevance to malignancy will remain as being of outstanding importance. Given the high evolutionary conservation of a multitude of basal functions from yeast to man and the experimental advantages of the yeast system, it will be of great benefit to combine these potentials to assist the human and other genome projects.

Acknowledgements

I wish to thank all colleagues who have actively contributed to the success of the yeast genome project, particularly those who participated in sequencing chromosome II [28]. I gratefully acknowledge help with information and computer analyses by B. Dujon, H.W. Mewes, K. Heumann, K. Kleine and J. Hani. The work carried out in this laboratory was supported by the European Union under the BAP, BRIDGE and BIOTECH II Programs, the Bundesminister für Forschung und Technologie, and the Fonds der Chemischen Industrie.

References

1 Roman, H. (1981) Development of yeast as an experimental organism. In *The Molecular and Cellular Biology of the Yeast Saccharomyces* (Strathern, J.N., Jones, E.W. & Broach, J.R., eds) 1–9 (Cold Spring Harbor Laboratory Press, Cold Spring Harbor, New York).

2 Strathern, J.N., Jones, E.W. & Broach, J.R. (1981) *The Molecular Biology of the Yeast Saccharomyces* (Cold Spring Harbor Laboratory Press, Cold Spring Harbor, New York).

3 Broach, J.R., Pringle, J.R. & Jones, E.W. (1991) *The Molecular and Cellular Biology of the Yeast Saccharomyces* (Cold Spring Harbor Laboratory Press, Cold Spring Harbor, New York).

4 Guthrie, G. & Fink, G.R. (1991) *Guide to Yeast Genetics and Molecular Biology: Methods in Enzymology*, Vol. 194 (Academic Press, San Diego).

5 Goffeau, A., Barell, B.G., Bussey, H. *et al.* (1996) Life with 6000 genes. *Science* **274**, 546–567.

6 Dujon, B. (1996) The yeast genome project: what did we learn? *Trends Genet.* **12**, 263–270.

7 Mortimer, R.K., Contopoulou, R. & King, J.S. (1992) Genetic and physical maps of *Saccharomyces cerevisial*, edn 11. *Yeast* **8**, 817–902.

8 Hodgkin, J., Plasterk, R.H.A. & Waterston, R.H. (1994) The nematode *Caenorhabditis elegans* and its genome. *Science* **270**, 410–440.

9 Olson, M.V. (1993) The human genome project. *Proc. Natl Acad. Sci. USA* **90**, 4338–4344.

10 Carle, G.F. & Olson, M.V. (1985) An electrophoretic karyotype for yeast. *Proc. Natl Acad. Sci. USA* **82**, 3756–3759.

11 Lindegren, C.C. *The Yeast Cell: Its Genetics and Cytology* (Educational Publishers, St Louis).

12 Mortimer, R.K. & Schild, D. (1981) Genetic mapping in *Saccharomyces cerevisial*. In *The Molecular Biology of the Yeast* Saccharomyces (Strathern, J.N., Jones, E.W. & Broach, J.R., eds) 11–26 (Cold Spring Harbor Laboratory Press, Cold Spring Harbor, New York).

13 Ito, H., Fukuda, Y., Murata, K. & Kimura, K. (1983)

Transformation of intact yeast cells treated with alkali cations. *J. Bacteriol.* **153**, 163–168.
14 Johnston, J.R. (1994) *Molecular Genetics of Yeast—A Practical Approach* (Oxford, Oxford University Press).
15 Olson, M.V., Dutchik, J.E., Graham, M.Y. *et al.* (1986) Random-clone strategy for genomic restriction mapping in yeast. *Proc. Natl Acad. Sci. USA* **83**, 7826–7830.
16 Stucka, R. & Feldmann, H. (1994) Cosmid cloning of yeast DNA. In *Molecular Genetics of Yeast—A Practical Approach* (Johnston, J., ed.), 49–64 (Oxford, Oxford University of Press).
17 Thierry, A., Gaillon, L, Galibert, F & Dujon, B. (1995) Construction of a complete genomic library of *Saccharomyces cerevisiae* and physical mapping of chromosome XI at 3.7 kb resolution. *Yeast* **11**, 121–135.
18 Johnston, M., Andrews, S., Brinkman, R. *et al.* (1994) Complete nucleotide sequence of *Saccharomyces cerevisiae* chromosome VIII. *Science* **265**, 2077–2082.
19 Niedenthal, R.K., Riles, L., Johnston, M. & Hegemann, J.H. (1996) Green fluorescent protein as a marker for gene expression and subcellular localization in budding yeast. *Yeast*, **12**, 773–786.
20 Burke, D.T., Carle, G.F. & Olson, M.V. (1987) Cloning of large segments of exogenous DNA into yeast by means of artificial chromosome vectors. *Science* **236**, 806–812.
21 Fields, S. & Song, O.K. (1989) A novel genetic system to detect protein–protein interactions. *Nature* **340**, 245–246.
22 Vassarotti, A. & Goffeau, A. (1992) Sequencing the yeast genome: the European effort. *Trends in Biotechnology* **10**, 15–18.
23 Oliver, S.G., van der Aart, Q.J., Agostoni-Carbone, M.L., *et al.* (1992) The complete nucleotide sequence of yeast chromosome III. *Nature* **357**, 38–46.
24 Levy, J. (1994) Sequencing the yeast genome: an international achievement. *Yeast* **10**, 1689–1706.
25 Huang, M.E., Chuat, J.C., Thierry, A., Dujon, B. & Galibert, F. (1994) Construction of a cosmid config and of an *Eco*RI restriction map of yeast chromosome X. *DNA Seqn* **4**, 293–300.
26 Clarke, L. & Carbon, J. (1976) A colony bank containing synthetic ColE hybrid plasmids representative of the entire *E. coli* genome. *Cell* **9**, 91–99.
27 Thierry, A. & Dujon, B. (1992) Nested chromosomal fragmentation in yeast using the meganuclease *I-Sce* I: a new method for physical mapping of eukaryotic genomes. *Nucleic Acids Res.* **20**, 5625–5631.
28 Feldmann, H., Aigle, M., Aljinovic, G., *et al.* (1994) Complete DNA sequence of yeast chromosome II. *EMBO J.* **13**, 5795–5809.
29 Riles, L., Dutchik, J.E., Baktha, A., McCauley, B.K., Thayer, E.C., Leckie, M.P., Braden, V.V., Depke, J.E. & Olson, M.V. (1993) Physical maps of the smallest chromosomes of *Saccharomyces cerevisiae* at a resolution of 2.6 kilobase pairs. *Genetics* **134**, 81–150.
30 Scholler, P., Schwarz, S. & Hoheisel, J.D. (1995) High-resolution cosmid mapping of the left arm of *Saccharomyces cerevisiae* chromosome XII; a first step towards an ordered sequencing approach. *Yeast* **11**, 659–666.
31 Louis, E.J. (1994) Corrected sequence for the right telomere of *S. cerevisiae* chromosome III. *Yeast* **10**, 271–274.
32 Louis, E.J. & Borts, R.H. (1995) A complete set of marked telomeres in *Saccharomyces cerevisiae* for physical mapping and cloning. *Genetics* **139**, 125–136.
33 Dear, S. & Staden, R. (1991) A sequence assembly and editing program for efficient management of large sequencing projects. *Nucleic Acids Res.* **19**, 3907–3911.
34 Thierry-Mieg, J. & Durbin, R. (1992) ACeDb, *C. elegans* database. *Cahiers IMA BIO* **5**, 15–24.
35 Kalegoropoulos, A. (1995) Automatic intron detection in nuclear DNA sequences of *Saccharomyces cerevisiae*. *Yeast* **11**, 555–565.
36 Sprinzl, M., Stegborn, C., Hübel, F. & Steinberg (1996) Compilation of tRNA sequences and sequences of tRNA genes. *Nucleic Acids Res.* **24**, 68–72.
37 Fishent, G.A.. & Burkes, C. (1991) Identifying potential tRNA genes in genomic DNA sequences. *J. Mol. Biol.* **220**, 659–671.
38 Pearson, W.R. & Lipman, D.J. (1988) Improved tools for biological sequence comparison. *Proc. Natl Acad. Sci. USA* **85**, 2444–2448.
39 Altschul, S.F., Gish, W., Miller, W., Myers, E.W. & Lipman, D.J. (1990) Basic local alignment search tool. *J. Mol. Biol.* **215**, 403–410.
40 Califano, A. & Rigoutsos, I. (1993) FLASH: A Fast Look-up Algorithm for String Homology. In *Proceedings of the 1st International Conference on Intelligent Systems for Molecular Biology*, 56–64 (Bethesda, MD).
41 Bairoch, A., Bucher, P. & Hofmann, K. (1995) The PROSITE database: its status in 1995. *Nucleic Acids Res.* **24**, 189–196.
42 Sharp, P.M. & Li, W.H. (1987) The codon adaptation index—a measure of directional synonymous codon usage bias, and its potential applications. *Nucleic Acids Res.* **15**, 1281–1295.
43 Lloyd, A.T. & Sharp, P.M. (1992) Codons: a microcomputer program for codon usage analysis. *J. Hered.* **83**, 239–240.
44 Lloyd, A.T. & Sharp, P.M. (1992) Evolution of codon usage patterns. *Nucleic Acids Res.* **20**, 5289–5295.
45 Mewes, H.W. & Heumann, K. (1995) Genome analysis: pattern search in biological macromolecules. *Proceedings of the 5th Annual Symposium on Combinatorial Pattern Matching. LNCS* **937**, 261–285.
46 Wilkins, M.R., Pasquali, C., Appel, R.D. *et al.* (1996) From proteins to proteome: large scale protein identification by two dimensional electrophoresis and amino acid analysis. *Bio/Technol.* **14**, 61–65.
47 Yoshikawa, A. & Isono, K. (1990) Chromosome III of *Saccharomyces cerevisiae*: an ordered clone bank, a detailed restriction map and analysis of transcripts suggest the presence of 160 genes. *Yeast* **6**, 383–401.
48 Fairhead, C. & Dujon, B. (1994) Transcript map of two regions of chromosome XI of *Saccharomyces cerevisiae* for interpretation of sympatic sequencing results. *Yeast* **10**, 1403–1413.
49 Fleischmann, R.D., Adams, M.D., White, O. *et al.* (1995) Whole-genome random sequencing and assembly of *Haemophilus influenzae* Rd. *Science* **269**, 496–512.
50 Fraser, C.M., Gocayne, J.D., White, O., *et al.*(1995) The minimal gene complement of *Mycoplasma genitalium*. *Science* **270**, 397–403.

51 Bult, C.J., White, O., Olsen, G.J. *et al.* (1996) Complete genome sequence of the methanogenic archaeon, *Methanococcus jannaschii. Science* **273**, 1058–1073.
52 Bussey, H., Kaback, D.B., Zhong, W. *et al.* (1995) The nucleotide sequence of chromosome I from *Saccharomyces cerevisiae. Proc. Natl Acad. Sci. USA* **92**, 3809–3813.
53 Klein, P., Kanehisa, M. & Delisi, C. (1985) The detection and classification of membrane spanning proteins. *Biochim. Biophys. Acta* **815**, 468–476.
54 Goffeau, A., Nakai, K., Slonimski, A. & Risler, J.L. (1993) The membrane proteins encoded by yeast chromosome III genes. *FEBS Letter* **325**, 112–117.
55 Goffeau, A., Slonimski, P.P., Nakai, K. & Risler, J.L. (1993) How many yeast genes code for membrane proteins? *Yeast* **9**, 691–702.
56 Nelissen, B., Mordant, P., Jonniaux, J.L., DeWachter, R. & Goffeau, A. (1995) Phylogenetic classification of the major superfamily of membrane transport facilitators, as deduced from the yeast genome sequencing. *FEBS Letters* **377**, 232–236.
57 Hartl, F.U., Pfanner, N., Nicholson, D.W. & Neupert, W. (1989) Mitochondrial protein import. *Biochim. Biophys. Acta* **988**, 1–45.
58 Sandmeyer, S. (1992) Yeast retrotransposons. *Curr. Op. Gen. Dev.* **2**, 705–711.
59 Fink, G.R., Boeke, J.D. & Garfinkel, D.J. (1986) The mechanism and consequences of retrotransposition. *Trends Genet.* **2**, 118–123.
60 Hauber, J., Stucka, R., Krieg, R. & Feldmann, H. (1988) Analysis of yeast chromosomal regions carrying members of the glutamate tRNA gene family: various transposable elements are associated with them. *Nucl. Acids Res.* **16**, 10623–10634.
61 Lochmüller, H., Stucka, R. & Feldmann, H. (1989) A hot-spot for transposition of various Ty elements on chromosome V in *Saccharomyces cerevisiae. Curr. Genet.* **16**, 247–252.
62 Ji, H., Moore, D.P., Blomberg, M.A., Braiterman, L.T., Voytas, D.F., Natsoulis, G. & Boeke, J.D. (1993) Hot spots for unselected Ty1 transposition events on yeast chromosome III are near tRNA genes and LTR sequences. *Cell* **73**, 1007–1018.
63 Eigel, A. & Feldmann, H. (1982) Ty1 and delta elements occur adjacent to several tRNA genes in yeast. *EMBO J.* **1**, 1245–1250.
64 Kirchner, J., Connolly, C.M. & Sandmeyer, S. (1995) Requirement of RNA polymerase III transcription factors for *in vitro* position-specific integration of a retrovirus-like element. *Science* **267**, 1488–1491.
65 Warmington, J.R., Anwar, R., Newlon, C.S. *et al.* (1986) A hot-spot for Ty transposition on the left arm of yeast chromosome III. *Nucleic Acids Res.* **14**, 3475–3485.
66 Struhl, K. (1987) Promoters, activator proteins, and the mechanism of transcriptional initiation in yeast. *Cell* **49**, 295–297.
67 Proudfoot, N. (1991) Poly (A) signals. *Cell* **64**, 671–674.
68 Svetlov, V.V. & Cooper, T.G. (1995) Review: compilation and characteristics of dedicated transcription factors in *Saccharomyces cerevisiae. Yeast* **11**, 1439–1484.
69 Bram, R., Lue, N.F. & Kornberg, R.D. (1986) A GAL family of upstream activating sequences in yeast: roles in both induction and repression of transcription. *EMBO J.* **5**, 603–608.
70 Smith, H.E., Yu, S.S.Y., Neigeborn, L., Driswell, S.E. & Mitchell, A.P. (1990) Role of *IME1* expression in regulation of meiosis in *Saccharomyces cerevisiae. Mol. Cell. Biol.* **10**, 6103–6113.
71 Dujon, B., Alexandraki, D., Andre, B., *et al.* (1994) Complete nucleotide sequence of yeast chromosome XI. *Nature* **369**, 371–378.
72 Sharp, P.M. & Lloyd, A.T. (1993) Regional base composition variation along yeast chromosome III—evolution of chromosome primary structure. *Nucleic Acids Res.* **21**, 179–183.
73 Bernardi, G. (1993) The isochore organisation of the human genome and its evolutionary history—a review. *Gene* **135**, 57–66.
74 Newlon, C.S. (1988) Yeast chromosome replication and segragation. *Microbiol. Rev.* **52**, 568–601.
75 Van Houten, J.V. & Newlon, C.S. (1990) Mutational analysisof the consensus sequence of a replication origin from yeast chromosome III. *Mol. Cell. Biol.* **10**, 3917–3925.
76 Marahrens, Y. & Stillman, B. (1992) A yeast chromosomal origin of DNA replication defined by multiple functional elements. *Science* **255**, 817–823.
77 Bell, S.P. & Stillman, B. (1992) ATP-dependent recognition of eukaryotic origins of DNA replication by a multiprotein complex. *Nature* **357**, 128–134.
78 Shirahige, K., Iwasaki, T., Rashid, M.B., Ogasawa, N. & Yoshikawa, H. (1993) Localisation and characterization of autonomously replicating sequences from chromosome VI of *Saccharomyces cerevisiae. Mol. Cell. Biol.* **13**, 5043–5056.
79 Dershowitz, A. & Newlon, C.S. (1993) The effect on chromosome stability of deleting replication origins. *Mol. Cell. Biol.* **13**, 391–398.
80 Fangman, W.L. & Brewer, B.J. (1992) A question of time: replication origins of 58 eukaryotic chromosomes. *Cell* **71**, 363–366.
81 Amati, B.B. & Gasser, S.M. (1988) Chromosomal ARS and CEN elements bind specifically to the yeast nuclear scaffold. *Cell* **54**, 967–978.
82 Eisenberg, S., Civalier, C. & Tye, B.K. (1988) Specific interaction between a *Saccharomyces cerevisiae* protein and a DNA element associated with certain autonomously replicating sequences. *Proc. Natl Acad. Sci. USA* **85**, 743–746.
83 Estes, H.G., Robinson, B.S. & Eisenberg, S. (1992) At least three distinct proteins are necessary for the reconstitution of a specific multiprotein complex at a eukaryotic chromosomal origin of replication. *Proc. Natl Acad. Sci. USA* **89**, 11156–11160.
84 Louis, E.J. & Haber, J.E. (1992) The structure and evolution of subtelomeric Y′ repeats in *Saccharomyces cerevisiae. Genetics* **119**, 303–315.
85 Lundblad, V. & Blackburn, E.H. (1993) An alternative pathway for yeast telomere maintenance rescues *est1* sequences. *Cell* **73**, 347–360.
86 Palladino, F. & Gasser, S.M. (1994) Telomere maintenance and gene repression, a common end? *Curr. Op. Cell Biol.* **6**, 373–379.
87 Milosavljevic, A. & Jurka, J. (1993) Discovering

simple DNA sequences by the algorithmic significance method. *CABIOS* **9**, 407–411.
88 Richard, G.-F. & Dujon, B. (1996) Distribution and variability of trinucleotide repeats in the genome of the yeast *Saccharomyces cerevisiae*. *Gene* **174**, 165–174.
89 Simchen, G., Chapman, K.B., Caputo, E., Nam, K., Riles, L., Levain, D.E. & Boeke, J.D. (1994) Correction of the genetic map from chromosome XI of *Saccharomyces cerevisiae*. *Genetics* **138**, 283–287.
90 Stucka, R. (1992) PhD Thesis, University of Munich.
91 Michels, C.A., Read, E., Nat, K. & Charron, M.J. (1992) The telomere-associated *MAL3* locus of S.cerevisiae is a tandem array of repeated genes. *Yeast* **8**, 655–665.
92 Viswanathan, M., Muthukumar, G., Cong, Y.S. *et al.* (1994) Seripauperins of *Saccharomyces cerevisiae*: a new multigene family encoding serine-poor relatives of serine-rich proteins. *Gene* **148**, 149–153.
93 Goffeau, A. (1994) Yeast genes in search of functions. *Nature* **369**, 101–102.
94 Bassett, D.E., Boguski, M.S. & Hieter, P. (1996) Yeast genes and human disease. *Nature* **379**, 589–590.
95 Miklos, G.L.G & Rubin, G.M. (1996) The role of the genome project in determining gene function: insights from model organisms. *Cell* **86**, 521–529.
96 Oliver, S. (1996) A network approach to the systematic analysis of yeast gene function. *Trends Genet.* **12**, 241–242.
97 Johnston, M. (1996) Towards a complete understanding of how a simple eukaryotic cell works. *Trends Genet.* **12**, 242–243.
98 Wach, A., Brachat, A., Pohlmann, R. & Philippsen, P. (1994) New heterologous modules for classical or PCR-based gene disruptions in *Saccharomyces cerevisiae*. *Yeast* **10**, 1793–1808.
99 Casari, G., de Daruvar, A., Sander, C. & Schneider, R. (1996) Bioinformatics and the discovery of gene function. *Trends Genet.* **12**, 244–245.
100 Murakami, Y., Naitou, M., Hagiwara, H., *et al.* (1995) Analysis of the nucleotide sequence of chromosome VI from *Saccharomyces cerevisiae*. *Nature Genet.* **10**, 261–268.
101 Galibert, F., Alexandraki, D., Baur, A *et al.* (1996) Complete nucleotide sequence of *Saccharomyces cerevisiae* chromosome X. *EMBO J.* **15**, 2031–2049.
102 Jacq, C., Alt-Mörbe, J., Andre, B. *et al.* (1997) The nucleotide sequence of *Saccharomyces cerevisiae* from chromosome IV. *Nature* (in press) .
103 Dietrich, F.S., Mulligan, J., Hennessy, M.A. *et al.* (1997). Complete nucleotide sequence of yeast chromosome V. *Nature*, accepted.
104 Tettelin, H., Agostoni Carbone, M.L., Albermann, K. *et al.* (1997) The nucleotide sequence of *Saccharomyces cerevisiae* chromosome VII. *Nature* (in press).
105 Churcher, C., Bowman, S., Badcock, K *et al.* (1997). The nucleotide sequence of *Saccharomyces cerevisiae* from chromosome IX. *Nature*, (in press).
106 Johnston, M., Hillier, L., Riles, L. *et al.* (1997) The nucleotide sequence of *Saccharomyces cerevisiae* chromosome XII. *Nature*, (in press).
107 Bowman, S., Churcher, C., Badcock, K. *et al.* (1997) The nucleotide sequence of *Saccharomyces cerevisiae* chromosome XIII. *Nature*, (in press).
108 Philippsen, P., Kleine, K., Pöhlmann, R. *et al.* (1997) The complete nucleotide sequence of *Saccharomyces cerevisiae* chromosome XIV. *Nature* (in press).
109 Dujon, B., Albermann, K., Alden, M. *et al.* (1997). The complete nucleotide sequence of *Saccharomyces cerevisiae* chromosome XV. *Nature* (in press).
110 Bussey, H., Storms, R.K., Ahmed, A. *et al.* (1997). The nucleotide sequence of *Saccharomyces cerevisiae* chromosome XVI. *Nature*, (in press).
111 Mewes, H.W., Albermann, K., Bähr, M. *et al.* (1997). Dictionary of the Yeast Genome. *Nature* (in press).
112 Walsh, S. and Barrell, B. (1996) The *Saccharomyces cerevisiae* genome on the World Wide Web. *Trends Genet.* **12**, 276–277.

Appendix

A number of informational resources on yeast are available in the Internet [103]. Valuable information coupled with data libraries, routines for searches and data retrieval can be found in several home pages on the World Wide Web. Sequence data can also be retrieved from various databases by ftp or e-mail; likewise, these resources offer information and special services via e-mail addresses.

Information on the yeast genome and related projects, search and query facilities

http://www.mips.biochem.mpg.de/yeast/
http://www.embl-ebi.ac.uk
http://www.sanger.ac.uk/yeast/home.html
http://genome-www.stanford.edu/saccharomyces/
http://www.nig.ac.jp
http://www.ncbi.nlm.nih.gov/
http://www.ncbi.nlm.nih.gov/XREFdb
http://quest7.proteome.com/YPDhome.html
http://expasy/hcuge.ch/cgi.bin/list?yeast.txt

E-mail addresses

mewes@mips.embnet.org (information on the European yeast project)
barrell@sanger.ac.uk (information on the yeast project at Cambridge)
linder@urz.unibas.ch (information on ListA, a yeast gene catalogue)
yeast-curator@genome.stanford.edu (informatio on the American yeast project)
NetServ@ebi.ac.uk (general information on data bases)
DataLib@ebi.ac.uk (general information on data bases)

Data retrieval by ftp

ftp://mips.embnet.org/yeast/
ftp://ftp.ebi.ec.uk/pub/databases/yeast
ftp://genome-ftp.stanford.edu/yeast/genome_seq

Chapter 31 Bacterial genomes: the Escherichia coli model

Millicent Masters & Guy Plunkett III*

Institute of Cell and Molecular Biology, The University of Edinburgh, Darwin Building, King's Buildings, Mayfield Road, Edinburgh EH9 3JR, UK

**University of Wisconsin-Madison, Laboratory of Genetics, 445 Henry Mall, Madison, WI 53706, USA*

31.1 Introduction

There can be few practising molecular biologists who have not made use of *Escherichia coli,* its plasmids or bacteriophages, even if only as tools to aid in the analysis of genes originating from other species. Yet today *E. coli,* for several heady decades itself the focus of genetic and molecular analysis, no longer occupies centre stage: most recently trained molecular biologists have interests focused on one or other eukaryote. In consequence, the elegance of *E. coli*'s molecular and genetic systems are a closed book to many who might profit from their use. We will try to partly redress this state of affairs here. We hope to do this by providing, in addition to information on the availability of sequences, both an overview of *E. coli* and its genome, and an up-to-date compendium of information which will serve to familiarize the readers of this book with, or remind them of, both the simplicities and the complexities of this versatile microorganism.

31.2 Escherichia coli as a model system

Escherichia coli offers its dedicated constituency the tantalizing possibility of fully defining the elements and processes that make up a simple free-living organism. The life cycle of *E. coli* is superficially uncomplex. The genome is replicated with a concomitant doubling in mass and length of the cell (fuelled by an intricate, but dissectable, intermediary metabolism). These events culminate in the division of the parent cell into two daughter cells. If this were a satisfactory paradigm, *E. coli* could be fully defined by dissecting its machinery of replication, protein synthesis, division and energy metabolism. It has become abundantly clear, however, that such a view is a gross oversimplification. Not only is *E. coli* capable of altering its metabolism and composition to cope with growth under a wide variety of conditions (free-living, life in the gut, aerobic and anaerobic states, conditions of plenty and starvation, etc.) but it is also supplied with a selection of inducible systems that enable it to respond to a panoply of possible stresses. This it does with alterations in composition, and with the production of groups of proteins whose function it is to repair and limit stress-induced damage. Two such proteins, active in DNA repair, have recently received attention because of their homology with a human protein associated with susceptibility to colon cancer [1,2], demonstrating, yet again, that *E. coli* continues to be a useful model for previously uncharacterized eukaryotic processes, including some linked to inherited human disease.

Gene nomenclature

Escherichia coli genes are identified by four-letter italicized names. The first three letters are a lower-case mnemonic describing gene function or mutant phenotype. The fourth distinguishes genes which share a mnemonic and is assigned alphabetically as genes are discovered (when the appropriate information was available, genes have been named to denote the order of their activity in a biochemical pathway). Early names relate to auxotrophies (*arg, thr* mutants require the amino acids arginine or threonine for growth), inability to ferment (*lac* mutants cannot use lactose as a sole carbon source), or drug-resistance (*amp* mutants are ampicillin resistant). Conditional mutants in essential processes such as macromolecular synthesis are named accordingly: thus *dnaA–dnaX* strains are defective in DNA synthesis and *fts* mutants form filaments because they have division defects. A new mutation with a particular defect is given an allele number; *dnaA52* will have been assigned to the *dnaA* gene. If a gene has not been assigned and only phenotypic information is available, a phenotypic designation is properly used. Dna-52 could be the phenotype of a strain with an otherwise uncharacterized defect in DNA synthesis while *dna-52* could designate a mutation in one of several *dna* genes which has not yet been assigned. By convention a mutant strain is described with the mnenomic (i.e. a *leu* strain and the nonmutant parent, if necessary, as *leu*$^+$).

Locations, such as transposon insertion sites, are assigned names with a similar form; these all start with *z.* The succeeding letters give positional information; *a-j* in the second position indicate the 10-min interval in which the site is located, the third position provides for subdivision into minutes. Thus *zbd* indicates 13–14 min, and *zej* 49–50 min. A similar system has come into use to indicate the positions of ORFs of unknown function identified during sequencing. These are assigned names starting with *y* with the following two letters giving positional information. For these putative genes the fourth letter is also assigned and here distinguishes multiple ORFs in the same region.

Nomenclature is discussed at length each year in the first (January) issue of the *Journal of Bacteriology.*

Although *E. coli* is less simple in its responses than once thought, it still leads the field as the first organism most likely to be fully understood. This remains true despite the fact that the smaller genomes of several other bacterial species were fully sequenced earlier (see Section 31.9 for a fuller discussion). The reasons for this are twofold. First, *E. coli* is genetically tractable. Its genome of 4.7 megabases (Mb) has at last been fully sequenced (completed in January 1997 [182]). Its well-defined genetic systems make easy the identification of genes and their products, and facilitate the analysis of the effects of specific mutations in, or deletions of, individual genes. New physiologic information can easily be collected using simple and well-tested methodologies. Second, there is an immense base of accumulated specific knowledge, its accessibility much increased by the publication, in 1987, of the encyclopaedic two-volume work, Escherichia coli *and* Salmonella typhimurium: *Cellular and Molecular Biology* [3]. This work has recently been revised and extended. The second edition includes not only genetic and physical maps, but a gene–protein and gene–function index to supplement the 155 information-dense chapters describing all aspects of these related organisms.

The earliest work with *E. coli* utilized a variety of different strains and isolates. However, the discovery of conjugation in the K-12 strain (isolated from a human patient in the 1920s) by Tatum and Lederberg in the 1940s [4] followed by the identification in the 1950s of derivatives able to transfer chromosomal markers with high frequency (Hfr donors) [5, 6] led to the adoption of *E. coli* K-12 as the strain of preference for subsequent genetic studies. The original K-12 isolate was both lysogenic for the bacteriophage λ and harboured the F-plasmid. In a series of efforts to isolate mutant derivatives the original strains were subjected to numerous genetic insults, including repeated irradiations with X-rays and ultraviolet, and treatment with assorted chemical mutagens. These treatments, coupled with continual passages and selections in many laboratories, have given rise to stocks of considerable diversity [7], even though all retain the specific modification and restriction system that defines K-12 strains [8]. Strains now described as 'wild-type' K-12 strains are usually prototrophic and cured of lambda and F. However, their precise histories differ [7] and considerable variations in restriction patterns are not uncommon [9–13]. It is to be anticipated that some of the substantial differences in genetic composition found amongst native *E. coli* isolates [14] will have been acquired secondarily by K-12 cultivars and may well account for at least some of the heterogeneities found in the sequence databases.

31.3 The genome of Escherichia coli

The basic genome of *E. coli* is a single circular chromosome (Fig. 31.1), 4.64 Mb in length (GenBank entry U00096, see ref. 15 for a brief overview, ref. 16 for a symposium volume, ref. 17 for a comprehensive handbook and ref. 170 for maps). The genome may be augmented, in particular strains, by integrated elements such as lysogenic bacteriophages (the best known of these, λ, is 48.5 kb in length, and was present in the original K-12 isolate, as described above) or plasmids, as in Hfr strains, in which the F-plasmid is integrated into the chromosome. Alternatively, or in addition, *E. coli* may harbour extrachromosomal genetic elements. Plasmids, autonomously replicating, circular DNA molecules (see refs 18–21 for reviews) range in size from about 5 kb (i.e. ColEl, the prototype of the group that has yielded multicopy cloning vectors) to about 100 kb (F). The larger plasmids are characteristically self-transmissible, encoding elaborate transfer systems which include hairlike cell-surface appendages, or pili, that promote conjugation. The smaller plasmids are adapted to parasitize the transfer systems of cohabiting larger plasmids and are thus also transmissible. Certain bacteriophages, notably P1 [22], adopt a plasmid form on lysogenization, but rely on lysis and reinfection as a method of spreading to new hosts. The larger plasmids are generally maintained at 1–5 copies per host chromosome, the smaller at 15–20 copies. Thus an *E. coli* cell which is host to several types of plasmid can easily have its DNA content significantly increased in consequence.

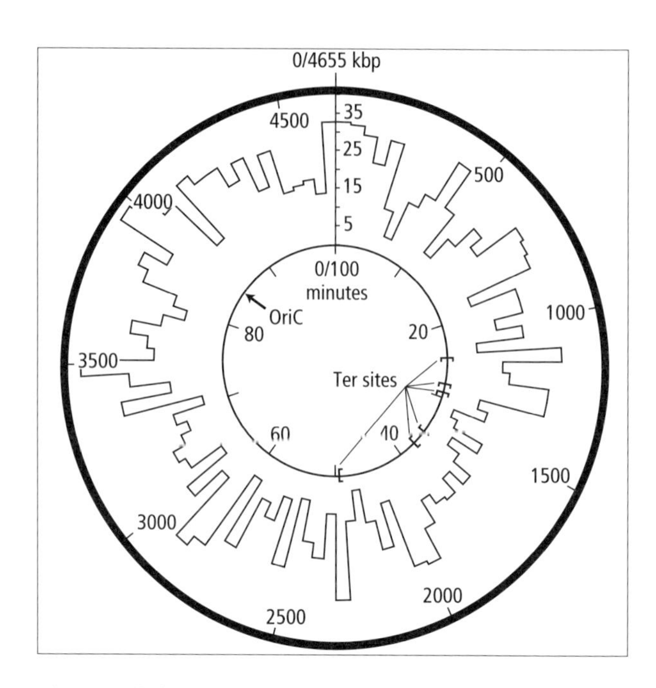

Fig. 31.1 Schematic representation of the *E. coli* circular chromosome. The outer ring shows chromosome lengths in kbp. The central ring shows the location of the replication origin and directional termination sites. The histogram shows the numbers of genes mapped to each minute of the chromosome on Edition 9 of the linkage map [170].

The smaller plasmids are single replicons, replicating from a particular origin, most often unidirectionally. Replication of these plasmids requires only host proteins. The larger plasmids tend to be chimaeras composed of more than one replicon; each intact replicon usually encodes a 'Rep' protein essential both to initiate replication, and to control replication frequency. Replication of plasmids may be uni- or bidirectional. In contrast, the chromosome itself is a single replicon, replicating bidirectionally [23,24] from an origin located at 84 map units; termination of replication occurs when the forks meet at a position approximately opposite to the terminus. Termination (reviewed in refs 25 and 26) is confined to the terminus region by the action of several orientated *ter* sites which prevent replication forks from proceeding through the terminus region back toward the origin. These sites act by binding a protein, Tus, which impedes the strand-separating action of the DnaB helicase. Their directionality arises from the fact that DnaB travels ahead of the replicating fork on one strand only. Replication is initiated once per cell division cycle when the initiation mass is attained; the details of replication and its control have been extensively reviewed [27–34].

The extended *E. coli* chromosome is 1000 times longer than the cell that contains it [35]. Prokaryotes lack a nuclear compartment; despite this the *E. coli* chromosome is not evenly dispersed throughout the cell but is centrally located within it [36]. It forms a discrete structure, termed the folded chromosome or nucleoid, which can be isolated intact from the cell (see refs 37–39 for further reviews). Nucleoid preparations lack the viscosity of unfolded DNA and have been observed to contain chromosomes with unstable beaded structures reminiscent of nucleosomes (see Fig. 31.2) [40]. Although *E. coli* does possess a variety of small, basic, histone-like DNA-binding proteins [41–43], none of these has been found to be essential for cell survival [42,44], nor has any been convincingly demonstrated to fulfil a histone-like role *in vivo*. The way in which *E. coli* DNA is compacted within the cell thus remains far from fully understood.

All circular DNAs in *E. coli* are maintained in an underwound state by the opposing actions of topoisomerases that introduce and remove turns in the helix [45]; thus circular DNAs are found to be supercoiled when examined *in vitro*. The intact chromosome (as recovered from the cell in folded form) has been estimated to be organized into 50–100 separate domains of supercoiling (see refs 37–39 for reviews). Nascent RNA and associated protein appear responsible for maintenance of these domains but no specific protein has been found to be essential. Supercoiling is likely to be responsible for some but not all of the compaction of the DNA. The chromosome contains several hundred similar, but not identical, intergenic sites (REP sequences, to be discussed in Section 31.6.4.2) of unknown function. These sites are able to bind DNA gyrase and DNA polymerase I (see ref. 46 for references) and, although it is attractive to suppose that REP sequences may have a role in maintaining chromosome structure, evidence is still lacking. The chromosome has also been reported to be associated with the membrane, again at 50–100 sites [47]. No specific chromosomal sequences (except possibly the replication origin [48]) have been identified as preferentially associated with the membrane [49], and it is thus thought that the interaction between chromosome and membrane is a dynamic one, and probably does not have a role in maintaining chromosome structure.

31.4 The physical map

The first steps toward constructing a physical map were taken in 1975 when Clarke and Carbon [50] prepared a hybrid plasmid library by AT-tailing mechanically sheared *coli* DNA and ligating it into ColEl. They collected 2200 independent clones to form a bank which, it was hoped, would contain the entire genome. These clones have been separately numbered, but, even today, not all have been characterized [51]. It soon became clear that some sequences were absent from the Clarke–Carbon bank. This has been attributed to the fact that certain genes are lethal when cloned in high copy number. However, the bank has, none the less, proved very useful as a source of complementing cloned DNA and as the starting material for deriving the gene–protein index [52]. More recently, other cosmid banks have been constructed and characterized [13,53] with genome coverages of about 70% and 95%, respectively. Table 31.1 lists some available clone banks of *E. coli* DNA.

In 1987, Kohara constructed a λ library which included almost all the DNA of the K-12 strain W3110 [54]. Because λ does not need to be maintained as a high copy number plasmid, the prospects for cloning the entire genome were good, although in fact, eight small regions proved to be missing from the original clone set. A later cosmid library, based on a low copy number plasmid origin, filled in these gaps [53]. This demonstrated that, although perhaps poorly maintained when cloned in high copy number, these were clonable regions. The Kohara library was used to construct a restriction endonuclease map, measured in kilo-bases (kb), of

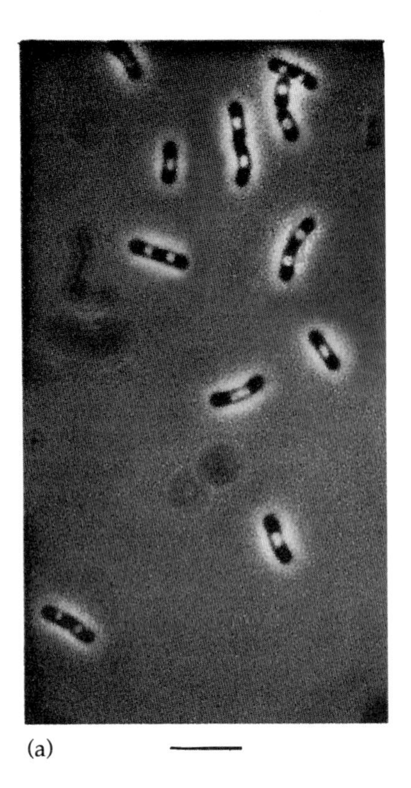

(a)

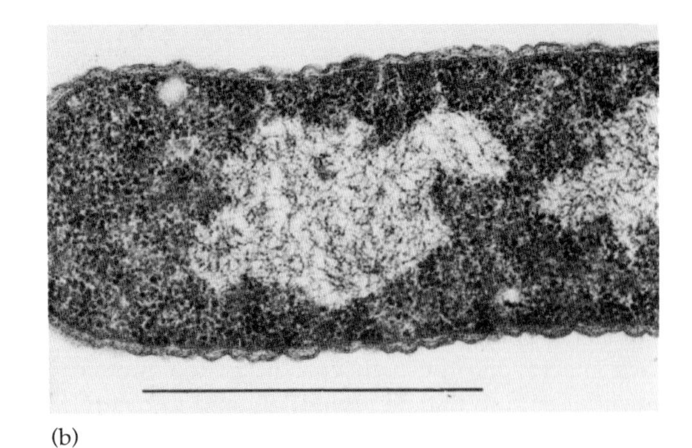

(b)

(c)

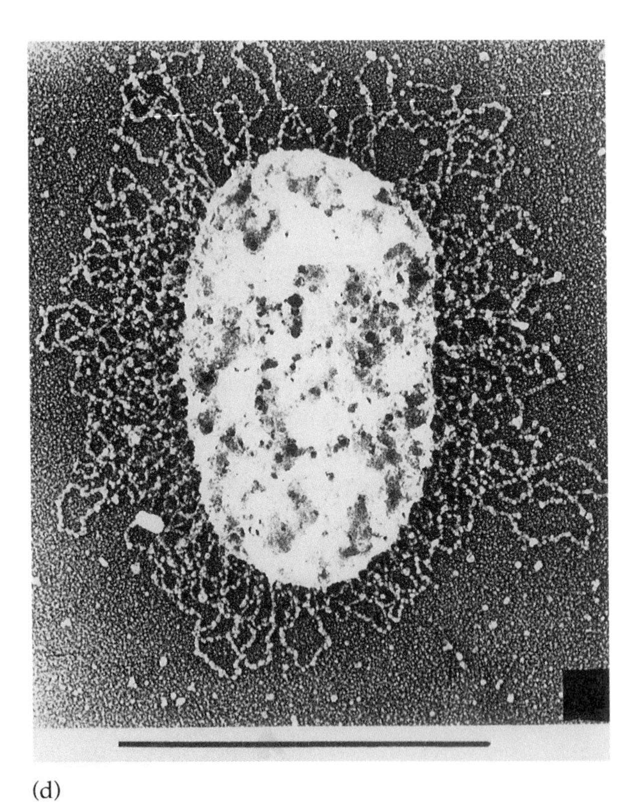

(d)

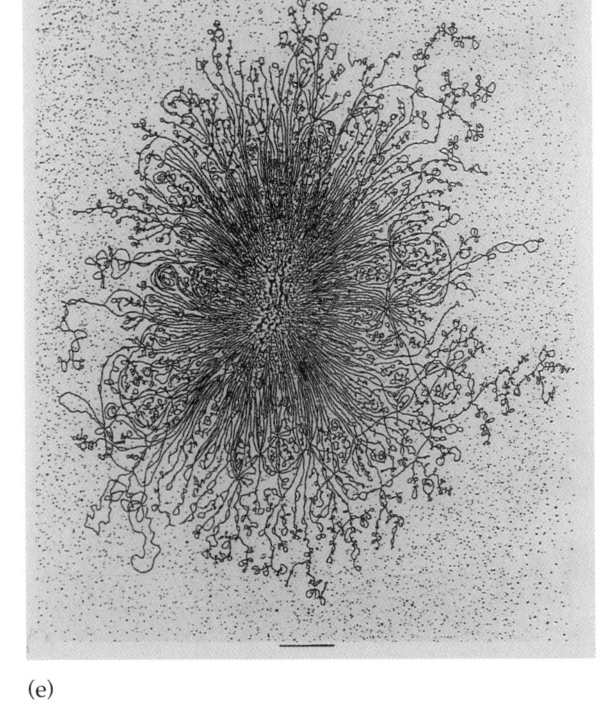

(e)

Fig. 31.2 The *E. coli* chromosome visualized. (a) *E. coli* cells, photographed with phase contrast and fluorescence microscopy. Nucleoids have been condensed by treatment with chloramphenicol (200 μg ml^{-1}, 5 min) and stained with DAPI. Photograph courtesy of K. Begg. (b, c) Electron micrographs of sectioned whole cells. The small spheres are ribosomes. Part b shows conventionally fixed cells with sharply demarcated condensed chromatin with distinct fibres. In part c, cells have been cryofixed and freeze-fractured; the DNA is more diffuse, occupies a greater volume, and has a finer fibrillar structure than seen in part b. Note that the cell envelope also seems free from distortion. Photographs courtesy of E. Kellenberger. For further discussion, see [36]. (d) DNA with an nucleosome-like appearance emerging from a cell which has been briefly lysed in 1% Triton X-100 directly onto the electron microscope grid. Reproduced, with permission, from [40]. (e) DNA derived from a single lysed cell, possibly held together by remaining membrane. The DNA is in supercoiled loops; each loop may represent a domain of supercoiling. This image is copyrighted as 'Bluegenes #1' 1983 with all rights reserved by DesignerGenes Posters Ltd, PO Box 100, Del Mar CA 92014, USA, from which posters and T-shirts are available. With permission of R. Kavenoff. Scale bars represent 5 μm in part a; 1 μm in parts b–e.

Table 31.1 Some available *E. coli* clone banks.

Type	Description	Source	Reference
λ (Kohara)	Ordered overlapping collection of 476 clones of W3110 DNA covering all but 7 small gaps	Kohara[a]	[54]
λ (Rose)	410 clones of MG1655 DNA; 7 recorded gaps, some abutments	*E. coli* genome project, University of Wisconsin[b]	Personal communication, see ref. 132
Phagemid library	W3110 DNA in SE6; NR1 origin allows maintenance as low copy number plasmid, unordered	ATCC (purchase)[c]	[169]
Clarke–Carbon plasmid collection	Sheared DNA in ColE1; 2200 individual clones (15 kb inserts), partial sets and characterized clones available	CGSC[d] (759 clones); GSRC[e] (518 characterized, 50% coverage, can supply whole set)	[50,51]
Cosmids (Tabata)	W3110 in pHC79; 40 kb inserts, 325 clones, 70% coverage	GSRC and 53	[53]
Cosmids (Birkenbihl)	Strain BHB2600; 95% coverage, 570 clones, 12 small gaps	R. Birkenbihl[f,g]	[13]
Cosmids(Knott)	W3110 DNA; a low copy number cosmid, pOU61cos, was needed to close 3/8 gaps in a conventional cosmid bank	Only 'gap-closing' cosmids available[g]	[55]

[a] Dr Y. Kohara, National Institute of Genetics, Mishima, Shizuoka-ken 1111, Japan. Fax: +81-559-81-6826.
[b] *E. coli* genome project, University of Wisconsin-Madison, Laboratory of Genetics, 445 Henry Mall, Madison WI 53706, USA. FAX: +608 263 745. E-mail: ecoli@genetics.wisc.edu.
[c] American Type Culture Collection, 12301 Parklawn Drive, Rockville, MD 20852, USA. WWW:http:/ /www.atcc.org/.
[d] *E. coli* Genetic Stock Center, Department of Biology, 3550ML, Yale University, PO Box 208104, New Haven, CT 06520-8104, USA. Fax: +203 432 3852. E-mail: berlyn@cgsc.biology.yale.edu. Also see 31.5.1.3.
[e] Dr A. Nishimura, Genetic Stock Research Center, address as note a. Fax: +81-559-81-6826.
[f] Dr R. Birkenbihl, Department of Genetics, University of Köln, Zülpicher Strasse 47, 50674 Köln, Germany.
[g] Please note that these collections are preserved by individual researchers unequipped for large-scale distributions.

the entire chromosome; the map shows the cutting sites for eight commonly used 6-bp cutters and the positions from which the insert of each of the clones in the Kohara clone library originates. The locations of certain genes which had been both genetically and physically mapped were used to align the physical with the genetic map. (At about the same time, a restriction map derived from pulsed field gel electrophoresis was published [9] showing sites for several rare cutters. Although they have proved of less use than the Kohara map in correlating the genetic and physical maps, maps of this sort have proved useful for the analysis of strain differences on a macro scale [9–13].)

The publication of the Kohara map and the subsequent availability of the ordered miniset of λ-clones revolutionized genetic mapping (see below) and provided a framework on which genes could be placed to construct a true physical gene map. Sequenced regions long enough to contain several of the mapped restriction sites could be assigned to unique locations on the physical map as could cloned genes for which a regional restriction map was available. Other genes were physically mapped by hybridization of cloned DNA to DNA prepared from miniset clones (see below). A section appeared in the *Journal of Bacteriology* from 1989 to 1993 dedicated entirely to short reports of physical map locations.

Several groups undertook the analysis of sequences in the databases or in the literature to compile maps, most notably those of Danchin [56] and Rudd [57]. Ecomap5, compiled by K. Rudd [58] is reproduced in a very useful two-volume laboratory manual and handbook of *coli* genetic data assembled by J. Miller [17]. Ecomap5 shows the locations of database entries on the physical map but does not attempt to delimit individual genes. A section from this map is included in Fig. 31.3. Later versions of the physical map have been compiled which do show the exact positions of individual genes; Ecomap6 is discussed, but not reproduced, in

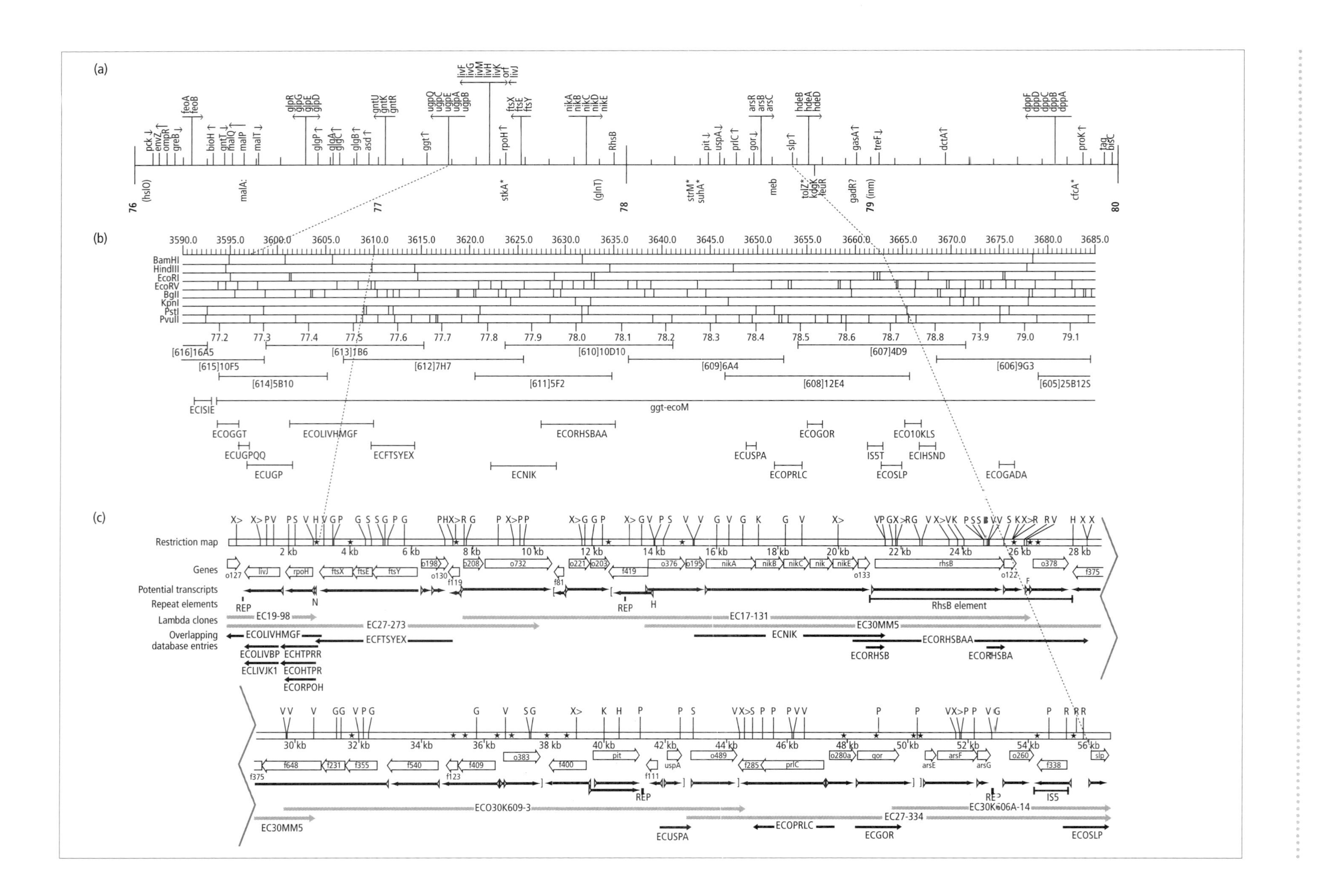
(a)
(b)
(c)
76
77
78
79 (min)
80
pck envZ ompR greB feoA feoB bioH gntT malQ malP malT
glpR glpG glpE glpD glpP glgA glgC glgB asd gntU gntK gntR
ggt ugpQ ugpC ugpE ugpA ugpB
livF livG livM livH livK orf livJ rpoH ftsX ftsE ftsY
nikA nikB nikC nikD nikE RhsB
pit uspA prlC gor arsR arsB arsC slp hdeB hdeA hdeD
gasA treF dctA dppF dppD dppC dppB dppA proK tag bisC
(hslO) malA: stkA* (glnT) strM* suhA* meb tolZ* kdgK feuR gadR? cfcA*
3590.0 3595.0 3600.0 3605.0 3610.0 3615.0 3620.0 3625.0 3630.0 3635.0 3640.0 3645.0 3650.0 3655.0 3660.0 3665.0 3670.0 3675.0 3680.0 3685.0
BamHI
HindIII
EcoRI
EcoRV
BglI
KpnI
PstI
PvuII
[616]16A5 [615]10F5 [614]5B10 [613]1B6 [612]7H7 [611]5F2 [610]10D10 [609]6A4 [608]12E4 [607]4D9 [606]9G3 [605]25B12S
ggt-ecoM
ECISIE ECOGGT ECUGPQQ ECUGP ECOLIVHMGF ECFTSYEX ECNIK ECORHSBAA ECUSPA ECOPRLC ECOGOR IS5T ECOSLP ECO10KLS ECIHSND ECOGADA
Restriction map
Genes
Potential transcripts
Repeat elements
Lambda clones
Overlapping database entries
2 kb 4 kb 6 kb 8 kb 10 kb 12 kb 14 kb 16 kb 18 kb 20 kb 22 kb 24 kb 26 kb 28 kb
o127 livJ rpoH ftsX ftsE ftsY o198 o130 f119 o208 o732 f81 o221 o203 f419 o376 o195 nikA nikB nikC nik nikE o133 rhsB o122 o378 f375
REP N H F
RhsB element
EC19-98 EC27-273 EC17-131 EC30MM5
ECOLIVHMGF ECFTSYEX ECNIK ECORHSBAA ECORHSB ECORHSBA
ECOLIVBP ECHTPRR ECLIVJK1 ECOHTPR ECORPOH
30 kb 32 kb 34 kb 36 kb 38 kb 40 kb 42 kb 44 kb 46 kb 48 kb 50 kb 52 kb 54 kb 56 kb
f375 f648 f231 f355 f540 f123 f409 o383 f400 pit f111 uspA o489 f285 prlC o280a qor arsE arsF arsG o260 f338 slp
IS5
ECO30K609-3 EC30K606A-14 EC27-334 EC30MM5
ECUSPA ECOPRLC ECGOR ECOSLP

ref. 59; Ecomap 7 appears in ref. 170. It is intended that the most recent Rudd version will be available electronically from the NCBI, as will be software for its use. A compilation of *E. coli* sequences extracted from the data bases by M. Kröger and colleagues has now been upgraded into a map which can be interactively accessed on the World Wide Web. This is discussed in more detail in Section 31.8.

Although this piecemeal placement of information on a developing physical map could be expected to result eventually in a complete map, the development of a complete map has been greatly accelerated by the systematic sequencing projects which will be described below. Analysis of the data from systematic sequencing allows all potential open reading frames to be identified, delimited, and, frequently, to be equated with previously identified and mapped genes which had either not been sequenced or not located on the physical map. An example of analysed sequence is shown in Fig. 31.3.

31.5 Genetic mapping in Escherichia coli

31.5.1 Classical methods: conjugation and transduction

Classical methods of mapping will be defined here as those that do not require the use of restriction enzyme technology. Classical mapping in *E. coli* relies on two methods of gene transfer, conjugation and transduction. A recent volume of *Methods in Enzymology* describes these techniques and others to be described below; it includes protocols [60].

31.5.1.1 Mapping by conjugation

Strains in which the F-plasmid is integrated into the chromosome are called Hfr strains. When mixed with strains that lack F (F^-, or female, strains), Hfr strains will initiate the formation of mating pairs and transfer a single-stranded copy of the chromosome, in an orientated fashion, beginning at the site of insertion of the F-factor; this process is called conjugation. Mating pairs usually separate before transfer is complete, but when complete transfer does occur, it requires about 100 min. For this reason the *coli* chromosome is divided into 100 map units termed minutes. (Conjugational crosses using a collection of donors differing in the site of F integration and orientation of transfer, supplied the initial evidence that the chromosome is circular [61].)

When conjugation is used for mapping [62], a mutant recipient is usually crossed with a non-mutant donor; selection of progeny is accomplished by selecting for transfer of the desired character and counter-selecting against the donor, usually with a drug (streptomycin and nalidixic acid are popular) to which the recipient has been made resistant. If the site of insertion of F on the donor chromosome is known, the time after mating at which the selected allele is first transferred indicates its position on the donor chromosome. Reasonably accurate estimates of position require Hfrs that transfer the desired allele early, and thus mapping of an unknown gene, which might be anywhere, requires the use of sets of Hfrs with different origins of transfer. Sets of Hfrs are described in refs 17 and 62 and are available from the *E. coli* Genetic Stock Center (CGSC, see Table 31.1 for address) and other sources.

More recently, sets of Hfrs have been developed in which a transposable element (Tn*10*) specifying tetracycline or kanamycin resistance, is transferred at about 20 min after the initiation of mating. When these Hfrs are used, drug resistance is selected after short mating periods. Transfer of the desired marker is scored amongst the drug-resistant progeny and, when transferred, can be presumed to be linked to the transposon or transferred earlier. Use of these sets allows approximate locations of new genes to be determined quickly. Such a set of Hfrs is available from the CGSC (Wanner set [63]) or from Dr C. Gross

Fig. 31.3 (*Opposite*) *Escherichia coli* linkage and physical maps. (a) A segment from the *E. coli* linkage map published in 1996 [171]. Each gene indicated here has been placed by genetic techniques, sometimes augmented, in the case of operons, by physical analysis. The numbers below the map represent map minutes. The arrows indicate direction of transcription. (b) Part of this region as derived from the physical map, based on Kohara [54] but updated to show regions that have been sequenced. The restriction map has been corrected, where necessary, with information derived from sequencing. The scale above the map is chromosomal length in kbp, with map minutes shown immediately below the map. The upper set of lines below the map indicate the inserts from the Kohara set of λ-clones and the lower set sequence contributions to GenBank. The continuous line (ggt-ecoM) indicates that this sequence is part of the large contig compiled by the Wisconsin mapping project. Contributed by K. Rudd. (c) Part of this segment based on information from the Wisconsin sequencing project which appeared in ref. 76. The λ-clones indicated are those isolated as part of this project. Restriction sites: B, *Bam*HI; G, *Bgl*I; R, *Eco*RI; V, *Eco*RV; H, *Hin*dIII; K, *Kpn*I; S,*Pst*I; P, *Pvu*II. X > marks the position of Chi sites; all are orientated similarly, * marks DNA bend sites. Potential promoters and terminators are shown with putative transcripts.

(Department of Stomatology, University of California, San Francisco, CA 941430, USA (Singer/Gross set [64]).

31.5.1.2 Mapping by transduction
In bacteriophage P1 lysates, about 1% of phage coats contain fragments of chromosomal DNA, 100 kb in length, which have been packaged in place of phage genomes [22]. These can be injected into recipient cells and will recombine with the recipient chromosome to replace homologous DNA, in a process termed *generalized transduction*. Once a gene has been located approximately (i.e. by conjugation) it can be located more precisely by scoring cotransduction with a possible neighbouring marker (which must be within 100 kb or 2 min to be cotransduced). The closer the pair of markers the greater is their frequency of cotransduction; cotransduction frequency can be used to calculate genetic distance by applying a function derived by Wu [65]. Sets of donors, each with a transposon located at a different known position, have simplified this form of mapping immensely. These donor sets are also available from C. Gross and the CGSC. The order of genes very close to one another, or of point mutations within genes, were classically determined by using three-point transductional crosses. Sequencing and other nonclassical approaches (to be discussed further below) have now supplanted these methods for fine-structure mapping.

31.5.1.3 The genetic map
An *E. coli* genetic map based on this type of study was first published in 1964 [66]. Subsequent editions of the map updated the coverage. Edition 9 includes about 1800 loci and is integrated with the physical map. The CGSC, now maintained by M. Berlyn at Yale University, has a collection of about 7000 strains which are supplied free of charge on request. Details of the strains in the collection can be browsed directly using the World Wide Web (http://cgsc.biology.yale.edu/top.html). Mutants for most genes that have been described are available, and strains with useful combinations of mutations can also be supplied. A less extensive collection is maintained in Japan; a catalogue of available strains can be obtained from Dr A. Nishimura (address in Table 31.1).

31.5.2 Methods dependent on the physical map: exploitation of clone banks

The Kohara restriction map, the availability of several ordered genome libraries, and the ease with which custom libraries can be constructed, has revolutionized gene mapping methodologies. Several commonly used approaches for mapping new genes are described here (see also ref. 60).

31.5.2.1 Conjugation followed by complementation
A commonly used approach is to roughly map the mutation of interest by conjugation and then to use a selected set from an ordered clone library to achieve complementation or recombination. In addition to the Kohara set of λ-phages there are now available a second independent λ-phage set and, as mentioned above, several plasmid and cosmid sets with full or partial coverage of the genome (Table 31.1). The complementing sequences can be pinpointed by subcloning from a complementing clone. It should be noted that high copy number suppression of mutations is common in *E. coli* (in such cases overproduction of a second protein reverses the effects of the deficiency of the first, see ref. 68 for an example) and it is necessary to confirm complementation attributed to DNA cloned in high copy number vectors with recombinational tests.

31.5.2.2 Direct identification of a complementing fragment
The second method dispenses with classical mapping entirely. A complementing clone can be sought in a genomic library made for the purpose (this need not be ordered) and the complementing fragment identified by comparing its physical map with that of the chromosome. This has been done by eye, but computer programs have now been devised to identify the chromosomal location of a particular restriction pattern [57,69]. Alternatively, DNA identified as complementing can be hybridized [70] to a commercially available filter (Takara Shuzo Co., Kyoto, fax: (+81 75) 241 5199) which contains an ordered array of DNA derived from the Kohara clone set. Now that the full *coli* sequence is available, it is to be anticipated that the chromosomal origin of cloned DNA will regularly be determined by sequence comparison.

31.5.2.3 Mapping an insertion mutation without cloning
A method has been described which exploits the fact that REP sequences (see Section 31.6.4.3) occur at frequent intervals [172]. Primers specific to the insertion are paired with generalized primers that will match most REP sequences and PCR is then used to amplify a fragment directly from chromosomal DNA. The resulting product can be cloned if desired or sequenced directly without cloning.

31.5.2.4 Reverse genetics
Finally, a method which starts with a protein pro-

duct rather than a mutant, and which is commonly used in eukaryotic systems, can be employed. To map by reverse genetics, the protein of interest is purified and N-terminal sequence obtained. The peptide sequence can be used to design a set of oligonucleotides, each of which encode it, and one of which must be the native sequence. In the past such sets have been hybridized to a library filter to identify the clone carrying the encoding DNA. Subsequent subcloning, hybridization and sequencing has been used to assign the gene to an exact physical map location (for an example, see ref. 71). Of course, now that the complete sequence is available, oligonucleotide 'probes' can be used to directly match a gene product associated with a phenotype with a previously identified coding unit.

31.6 The Escherichia coli chromosome

In this section we will summarize both information that has been available for some time and information derived from the analysis of data collected by the sequencing projects to be described in the final sections. The longest contig available in 1996, resulting from the combined systematic sequencing projects of the Madison, Wisconsin *E. coli* Genome Project (headed by F. Blattner) and a Japanese consortium, comprised a third of the genome. These groups have carefully analysed their sequence with regard not only to the identification and positioning of specific genes but with the goal of documenting global characteristics of gene and repetitive site arrangement. These analyses had been reported by 1994 in six primary publications from Wisconsin (refs 72–77) and in two from Japan [78,79]. Information in this section, where not specifically referenced, has been quoted or derived from these publications.

31.6.1 Arrangement of genes on the DNA

The chromosome of *E. coli*, like those of yeast, but unlike those of higher eukaryotes, is densely packed with genes. The contiguous sequence analysed to 1994 shows that about 85% of this DNA is likely to encode proteins, and another 4% to be transcribed to yield nonmessenger RNAs. Much of the remain-ing intergenic sequence is in very short stretches and is probably comprised mainly of sequences concerned with the control of transcription and translation. Although *E. coli* genes are commonly thought to be arranged in operons, in the regions where transcriptional units have been defined, only about two-thirds of the genes detected are likely to belong to transcriptional units containing two or more genes. The remainder have promoters of their own. Although there are numerous exceptions, cotranscribed genes tend to be functionally related.

Analysis undertaken as part of the sequencing projects listed above has allowed the assignment of definite functions to between 40 and 45% of the open reading frames (ORFs) identified. The majority of these correspond to previously defined or sequenced genes, but some functions could be assigned because of the near identity of new ORFs with those encoding better studied homologues in related organisms; these were corroborated, in most cases, with genetic evidence. A further 20–25% of translated ORFs were found to have significant amino acid sequence similarity to existing database entries; these similarities could be used to predict function. The remaining 30% of ORFS remained functionally uncharacterized. However, it now appears that closer analysis could well permit functions to be predicted for many of these as well. A recently published study [80], using highly sophisticated computer programs for protein sequence comparison, found that of 2300 known and proposed *coli* proteins (60% of the estimated total), more than 80% could be assigned at least a probable function. These authors describe 66% of *coli* proteins as having known functions and another 16% as sufficiently similar to already characterized proteins to permit an assignment of function.

Pairwise comparison of the sequences of about 1800 *coli* proteins reveals that about half have some sequence similarity over an extended region (100 or more amino acids) to one or more other *coli* proteins [81,82]. They can, on this basis, be assigned to groups which constitute small to large families. It is thought that each family is an evolutionary grouping whose members have diverged from an ancestral protein encoded by a single ancestral gene. A similar analysis of 2300 proteins is corroborative [83].

31.6.2 DNA sequence: base composition and codon usage

Escherichia coli DNA has an average G + C content of 51%, but there is considerable variation in base composition along the length of the DNA. The G + C content of overlapping 8-kb segments varies between 47 and 56%; the variation in base composition of 1-kb segments is even more extreme, spanning 30–65%. This variation is much greater than expected by chance, but currently remains unexplained. A similar degree of variation characterizes the contiguous sequence in yeast chromosome III (M. Masters, J. Collins & A. Coulson, unpublished data).

Analyses have also been made of the frequency of occurrence of short oligonucleotide sequences [83,84]. Although several such sequences are over- or under-represented to some degree, the sequence CTAG is strikingly deficient, occurring at only one-thirtieth the expected frequency within genes. Its component subsequences, CTA and TA, are also under-represented. As might therefore be expected, the Arg codons AGA and AGG, theLeu codon CUA, and the UAG stop signal are relatively rarely used in *E. coli* [85]. It has been suggested that CTAG has been eliminated by selection because it could promote undesirable DNA bending. Consistent with this idea is the fact that CTAG deficiency does not extend to eukaryotic organisms, in which the C would be modified. Karlin and coworkers have analysed and discussed these features and other sequence inhomogeneities [86,87]. The occurrence and possible role of CTAG has also been discussed in refs 79 and 88.

Sequences likely to promote static bending occur, at irregular intervals, about every 3 kb. Since they are more likely to be found in promoters than in coding sequences, a role for static bends in initiation of transcription can be inferred.

Codon usage varies considerably between species [89] and reflects the availability of the corresponding tRNAs [90,91]. *E. coli* genes were early divided into two classes differing in codon usage [92,93]. Genes that are constitutively expressed at high levels show a strong codon usage bias, with a preference for codons recognized by the commoner tRNAs. Included in this class are genes that encode abundant DNA-binding proteins, and those whose products are involved in protein synthesis. Genes expressed at moderate or low levels belong to the second class, with less biased codon usage. Recent work has extended these observations to a larger group of genes and, in addition, uncovered a third class [85], with still less biased codon usage. Many proteins encoded by this third class of genes are located on the cell surface; others are encoded by insertion sequences or phage remnants. Three-quarters of plasmid genes analysed belong to this third class. It has therefore been suggested that genes with this distribution of codon usage did not evolve in *E. coli*, but have been acquired relatively recently through horizontal transmission.

31.6.3 Arrangement of genes on the chromosome: some general principles

31.6.3.1 Genes concerned with transcription and translation tend to be close to the origin of replication
Because the *E. coli* chromosome replicates bidirectionally from a fixed origin to a terminus opposite [23,24], and because DNA replication requires the entire cell cycle for completion at moderate growth rates (and up to two cell cycles at high growth rates), the relative number of copies of any gene (its 'dosage') depends on its chromosomal position *vis-à-vis* the replication origin. In cells growing exponentially in broth, genes located close to the origin have four times the dosage of those at the terminus [23]. It might therefore be expected that genes whose products would be in particular demand in fast-growing cells, could satisfy that demand to some degree by 'choosing' to be located near the origin. The augmented gene dosage of genes proximal to the origin of replication in fast-growing cells could at least partly relieve the need for elaborate growth-rate dependent transcriptional and translational controls [94].

The most obvious class of gene products required at relatively higher concentrations in rapidly growing cells are the components of the transcriptional and translational apparatus: RNA polymerase subunits, the 50 or so ribosomal proteins, and 5S, 16S and 23S rRNAs. Ninety percentage of the genes for ribosomal subunits and all of those encoding RNA polymerase and its major σ (sigma)-factors are located in the third of the chromosome surrounding the origin, consistent with the gene dosage hypothesis presented above. Ribosomal RNAs cannot of course be translationally amplified; in order to obtain sufficient numbers of these molecules the genes encoding them are repeated. There are seven copies of the rRNA genes, five of them within 12 map units of the origin (see map, ref. 67).

Conversely, it might be expected that genes encoding inessential products, or those that are never needed in high quantity, would be located near the terminus. Consistent with this idea is the fact that the terminus region, although coding [95], is particularly variable in sequence when compared with related strains and species [96], suggesting that it contains inessential genes. This has been confirmed; almost all the DNA of the terminus region, with the exception of sequences likely to be concerned with the termination process itself, can be deleted with little ill-effect [97]. The terminus also appears to be a repository for a variety of elements of extrachromosomal origin (such as λ-related defective prophages [98–100]). Although the mechanism by which these elements were introduced to the terminus region is likely to involve enhanced, termination-related, recombination [101,102], their survival in the chromosome may be facilitated by the fact that their presence does not interfere with the gene dosage distribution of origin-proximal genes.

31.6.3.2 Are genes transcriptionally orientated with direction of replication?

In 1989, Brewer [103] noted that the transcriptional units specifying genes concerned with the synthesis of DNA, RNA and protein are orientated on the chromosome such that transcription from, and replication through, these genes occur in the same direction. She speculated that transcription proceeding in the opposing direction might interfere with replication, and predicted that genes with strong promoters (i.e. likely to be 'on' all the time) would be orientated so as to avoid potential conflict. Her analysis of 38 transcriptional units appeared to confirm this expectation. In an extended analysis of gene orientation [104], she found that of over 600 genes 70% appeared to be orientated in the same direction as replication; among 97 genes encoding RNAs and proteins active in translation, 92 were orientated with replication.

Burland *et al.* [73] tested the original Brewer hypothesis further, using the sequence from 81.5–86.5 min (which contains the origin). In this region two-thirds of the genes are orientated with replication, but no correlation between promoter strength and orientation could be demonstrated. Recent analysis of the much larger 1.6 Mb segment showed only a weak correlation between directions of transcription and replication overall, although there are some regions of strong correlation. However, genes likely to be highly expressible, as deduced from their biased codon usage, are found to be about five times more likely to be orientated with replication than against it.

31.6.3.3 Rearrangements

Is the order of genes on the chromosome important? This question arises because gene arrangement has, for the most part, been conserved between *E. coli* and *S. typhimurium* [105], organisms that have been separated, evolutionarily speaking, for millions of years. One approach to answering this question has been to determine whether inversions which radically alter gene order are in fact deleterious [106]. Inversions can occur by recombination between small chromosomal duplications placed at the desired end points of the inversion. Regeneration of a selectable gene from its separated halves was used for selection and required inversion of the DNA located between duplicated elements positioned at pairs of preselected points on the chromosome [107]. The results were striking. Strains with inversions, even very large ones, that included the region from 17 to 44 min, or that had occurred within a single replicating arm and did not include this region, grew normally, suggesting that, for the most part, gene order is not critical.

There were two major classes of exceptions. The first were inversions which resulted in the relocation of the 86–91 min region, which encodes RNA polymerase subunits and contains three of the seven rRNA cistrons, to a point distant from the origin. Strains with these rearrangements were found to be rich-medium sensitive, lending support to the idea that high gene dosage of these loci is required at rapid growth rates. The second group of inversions, with one or both ends between 17 and 44 min were either poorly tolerated or not obtainable at all. The reason for this is not understood. A plausible explanation, that inversions ending within the terminus region cause replication pausing at inverted Ter sites, has been excluded [108].

If inversions have little selective disadvantage, it might be anticipated that they would occur commonly. This does not appear to be the case. When K-12 strains are compared by restriction analysis using enzymes that cut at about 20 sites, they differ from one another by seven or eight insertions or deletions greater than 1 kb in length, but seldom exhibit inversions [11]. The genetic maps of *E. coli* and *S. typhimurium* are colinear, except for insertions and deletions, in all regions other than that near the replication terminus [109,110]. The two species are distinguished in this region by a single large inversion which spans the terminus. Thus, the order of genes on the chromosomes of enteric organisms appears to have been well conserved, suggesting that there is selective value in the existing arrangement. (Ironically, considering the rarity of inversions, the reference strain chosen by Kohara to construct the physical map harbours a large inversion relative to other K-12 strains [111]. The inverted DNA, generated by recombination between *rrnE*, at 90.5 min and *rrnD* at 72 min, is depicted in its more usual orientation on the standard physical map.)

31.6.4 Repeated sequences

Short tandemly repeated sequences, such as those of eukaryotic microsatellite DNA, are not characteristic of *E. coli* DNA, but there are several known families of interspersed repeats. Some of these have defined functions; others, present in relatively few copies, appear to have been horizontally transferred and are unlikely to have significant functional importance. A final group is present in many copies and may well have functional roles, although the exact nature of these roles remains obscure.

31.6.4.1 Interspersed repeats of known function
These include the seven *rrn* operons, each of which encodes 5S, 16S, and 23S rRNA in addition to several tRNAs (these differ between loci). Some of the equivalent RNA sequences are identical, others differ slightly. The loci contain some unusual oligonucleotide sequences: for instance they are rich in the rare CTAG tetranucleotide, and contain the only sites that can be cut by the enzyme I-Ceu I [112], which is encoded by an intron of chloroplast origin and is specific for a 26-bp site in 23S rRNA.

The other defined repeated sequence with a known functional role is the short, nonpalindromic sequence, 5′GCTGGTGG, termed Chi, which promotes homologous recombination in its vicinity (see ref. 113 for a review). Chi may occur either within or between genes. Chi-stimulated recombination is mediated by the major *E. coli* recombinational enzyme, the RecBCD complex, and Chi is thought to be a recognition site for the RecD subunit of this complex. An 8-bp sequence would be expected to occur by chance only about 70 times on the entire chromosome. Chi, in contrast, and consistent with its possessing a functional role, occurs over 20 times more frequently than this. Chi activity is directional, such that two Chi sequences, in opposing orientation, should be required to stimulate replacement of the DNA between them with homologous sequence. It is thus remarkable to find that over 90% of Chi sites are orientated relative to replication (discussed in ref. 73); Chi appears on one strand only, the leading strand of newly replicated DNA. This suggests that the primary role of Chi may be related to events at the replicating fork rather than to the stimulation of recombination between nonreplicating fragments; further information is required.

31.6.4.2 Sequences likely to have been acquired as a result of horizontal transmission
These include elements ranging from less than 1 kb long, to phage-sized sequences: all contain at least one ORF.

Insertion sequences Insertion sequences (IS elements) are transposable elements roughly 0.8–1.5 kb long; they are all similarly organized, with repeated sequences flanking a transposase gene (see ref. 114 for a collection of reviews). The *E. coli* chromosome contains about 10 different types of these elements, each present in different copy numbers. They transpose or are lost with frequencies well above that at which point mutations occur, resulting in extensive variation between strains as to numbers and positions of particular elements. Because IS elements transpose to positions within, as well as between, genes, they are often the agents of spontaneous mutation. Birkenbihl and Vielmetter [115] located the positions of five types of IS element in three K-12 strains. Copy number varied from 1 to 23; the three strains harboured, respectively, a total of 42, 46 and 57 IS sequences of the types studied. IS sequences are not genetically silent; they not only inactivate genes into which they are inserted, but can also activate genes either by the introduction of promoters, or by the alteration of DNA structure in their vicinity.

Transposons are related to IS elements and are probably derived from them. They are distinguished by the presence of genes for selectable characters which allows their genetic manipulation. Transposons are absent from wild-type K-12 cells but are transmitted in natural populations by plasmid vectors.

Longer elements Rhs elements (reviewed in ref. 116) are a family of long (7 kb is typical) complex sequences, with a tantalizing organizational similarity to the mammalian transposable sequence LINE–1. They were initially recognized as chromosomal hotspots for the initiation of duplication events. Rhs sequences are found in the chromosomes of many, but not all, naturally occurring *E. coli* strains. *E. coli* K-12 has five and part of a sixth; in total, they account for 0.8% of the chromosome. Each consists of a GC-rich core followed by an AT-rich extension, together defining an ORF which could encode a very large protein (up to 160 000 relative molecular mass) with some similarities to secreted or cell-surface proteins. It has not been possible to obtain evidence that the encoded proteins are expressed or possess active promoters. Because the base composition of Rhs elements is dissimilar to that of *E. coli*, and because they are present in only some strains, they are presumed to be of heterospecific origin. Consistent with this idea is the fact that some Rhs elements appear to contain a region with the characteristics of an IS sequence.

Lambdoid phages Escherichia coli strains can contain several defective prophages [98, 99, 117] with some sequence similarity to λ, together accounting for several percentage of the genome.

31.6.4.3 Short multicopy palindromic repeats
Two classes of repeat of this type have been well documented. Repetitive extragenic palindrome (REP) [118], also known as palindromic units (PU) sequences [119], and enterobacterial repetitive intergenic consensus (ERIC) [120] or intergenic repeat unit (IRU) sequences [121]. As can be deduced from

their names, these sequences occur between, rather than within, genes. They are located at the 3′-ends of transcripts in noncoding sequence; thus, they are always transcribed but never translated. Because they are palindromic, they can be expected to give rise to stem–loop structures in RNA and, possibly, also in DNA. These sequences, especially REP, are common, and, although short, are so numerous that they have been estimated to compose up to 1% of the genome. It is hard to imagine that sequences as numerous as these, and located only between genes, do not have a function; but, although several possible functions have been suggested, the role(s) of these repeated sequences remain elusive. A review describing REP and ERIC sequences appeared in 1992 and should be consulted for references not listed here [122].

REP sequences REP consists of a 30- to 40-bp consensus sequence, with dyad symmetry, which can form a stable stem–loop structure. REP sequences may occur singly, in pairs, or less often, in clusters containing several copies. Their locations have been identified both by hybridization (using a consensus probe), or by analysis of existing sequence. In the 1.6 Mb continuous sequence available in 1996, REP elements occur, on average, once per 13 kb. Because the REP palindrome is not perfect, left and right ends can be distinguished. Within a REP cluster, individual units alternate in orientation and are separated by one of a group of other conserved sequence motifs. The clusters can be divided into groups on the basis of the other sequences which they contain. Complex arrangements of this sort have been termed bacterial interspersed mosiac element (BIMES). The submotif structure of BIMES has been analysed [46].

Although the primary function of REP/PU elements is not known, at the RNA level REP sequences located between genes in an operon have been shown to be able to stabilize the upstream message, probably because they adopt a stem–loop configuration which limits degradation by exonucleases. There is also evidence that certain REP sequences, located between convergently transcribed genes, can act as transcription terminators. At the DNA level, REP sequences have been shown to bind DNA gyrase and DNA polymerase I, leading to the proposal that they may have a role in maintaining the domain structure of the nucleoid. A subclass of BIMES can bind IHF, a small DNA-binding protein which bends the DNA; this could facilitate gyrase action, or help to stabilize stem–loops [123].

REP sequences are not confined to *E. coli*. Hybridization analysis suggests that they are numerous in closely related species; surprisingly, they can even be found in quite distantly related phyla. PCR using REP-based primers has even been used to 'finger-print' bacterial strains of diverse eubacterial species; this method may be clinically useful for strain identification [124,125]. When the chromosomal locations of REP sequences in *E. coli* and the related species, *S. typhimurium*, are compared, it is evident that REP has been conserved in certain positions but not in others. Clearly REP is not an essential element for the control of expression of particular genes; perhaps the suggested role in chromosome structure maintenance is the more plausible.

Other repeated short palindromes The ERIC/IRU sequence is an imperfect palindrome 126 bp long. Like REP, it is located in transcribed, but nontranslated, portions of the genome. It too is widespread, at least among the enterobacteria. However, there are far fewer individual occurrences per genome; the initial 1.6 Mb of *coli* sequence includes only four IRUs. It appears to be commoner in *S. typhimurium* than in *E. coli*; several locations have been identified in the former where IRU does not occur in the latter. An attempt to find other short repeated sequences using a computational approach [126] identified additional groups of short palindromes; one of these is the sequence characteristic of rho-independent terminators. Two other short elements were defined, one within, the second between, genes but have not been further characterized.

31.7 Escherichia coli genome sequencing projects

Until a few years ago, all *E. coli* sequences had been contributed, piecemeal, to the databases by laboratories interested in particular genes and operons. Although each of these contributions was small (in this era of megabase sequencing efforts), by 1989 they had together provided about 20% of the total genomic sequence. At that time, several large-scale sequencing efforts were initiated. These had the goal of generating long contiguous sequences which would, ultimately, be merged to yield the complete sequence of the *E. coli* chromosome (derived, in at least one case, from a single K-12 strain). Publication of data from these large-scale efforts began in 1992. By December 1995, sequence data for ≈75% of the genome, originating from both large-scale and more limited efforts, was available in databases. The complete sequence became available in January 1997. An analysis of the sequence with annotations has been published [182].

31.7.1 Sequencing projects

31.7.1.1 The E. coli Genome Project at the University of Wisconsin, Madison

This group, led by F. Blattner, with US government funding, adopted a 'start-from-scratch' approach. A wild-type strain of *E. coli* K-12, MG1655, thought to be relatively undiverged from original wild-type isolates, was selected in consultation with B. Bachmann of the *E. coli* Genetic Stock Center at Yale University and used to make a new overlapping λ clone library [127]. Random sequencing, λ clone by λ clone, was then to be employed to sequence the entire chromosome, including previously sequenced segments. (There will be a more detailed description of the methodology used for sequencing and analysis/annotation in a following section.) The Wisconsin group chose to sequence the region counterclockwise from 100/0 minutes. Their publications to 1995 reported and analysed the following sequences:

ECOUW76 (U00039): 76.0–81.5 min [76]
ECOUW82 (L10328): 81.5–84.5 min [73]
ECOUW85 (M87049): 84.5–87.2 min [72]
ECOUW87 (L19201): 87.2–89.2 min [74]
ECOUW89 (U00006): 89.2–92.8 min [75]
ECOUW93 (U14003): 92.8–100 min [77]
ECOUW67 (U18997): 67.4–76.0 min

In the title of the first paper from this group [72] the map coordinates, taken from the version of the genetic map then in use, were indicated as 84.5–86.5 min. This sequence spans 84.5–87.2 min on the current map [67], and is so indicated above. There are no gaps between the sequences listed. The Wisconsin group, after overcoming some funding problems (see ref. 133 for a discussion of why the original estimates of the time required to sequence the *E. coli* genome proved overly optimistic) and altering certain techniques (see below) have now released the full sequence; ECOLI (U00096): *Escherichia coli* K-12, complete genome (4 638 858 bp) was deposited January 16, 1997 and released January 25, 1997 as a full annotated sequence. It is available via the Entrez Genomes division, GenBank, and the BLAST databases. In the Entrez Genomes division, the entire *E. coli* genome can be examined and explored at once. In GenBank itself, the 4.6 Mb *E. coli* sequence has been split into 400 records of approximately 11 500 bp each. These subsequences are also the entries used in the BLAST non-redundant databases for both peptide and nucleotide sequences.

One can also search *E. coli* separately in databases derived from the Wisconsin entry. In the Peptide Sequence Databases, 'E. coli' contains the *E. coli* genomic CDS translations, while in the Nucleotide Sequence Databases 'E. coli' contains the *E. coli* genomic nucleotide sequences. These databases can be searched using a BLAST client such as NCBI's web-based one at http://www.ncbi.nlm.nih.gov/BLAST/. See the Wisconsin WWW page: http://www.genetics.wisc.edu/.html for instructions on how to down-load the sequence. The Wisconsin group can be reached by e-mail at: ecoli@genetics.wisc.edu

31.7.1.2 The Japanese projects

An initial effort involved a consortium of laboratories in Japan. Using the λ clones from the Kohara set (prepared from strain W3110), they sequenced DNA to fill the gaps in already available sequence in order to merge the data into contiguous sequences. Working clockwise from 100/0 minutes, they have released ~285 kbp of this composite sequence so far:

ECO110K (D10483): 0–2.4 min [78]
ECO82K (D26562): 2.4–4.1 min [79]
ECOTSF (D83536): 4.0–6.0 min [unpublished]

Funding lapsed for the project as originally constituted, but sequencing was resumed following the assembly of the Japanese *Escherichia coli* genome project team (~36 scientists) coordinated by Professor T. Horiuchi (National Institute for Basic Biology in Okazaki, 444 Japan). This group has continued to sequence Kohara clones, proceeding from ~13 min in a clockwise direction (with a few gaps filled with other database entries) and have completed over half the genome. Their sequence is available in a sequene of database entries (D90699–D90892 with some numbers in the sequence not used) each corresponding to a Kohara phage. Their results have been reported in seven papers so far (others are in preparation): 12.7–28.0 min [131,132]; 28.0–40.1 min [128,173]; 40.1–50.0 min [174, 175]; 50.0–68.8 min [176].

They also merged their sequence with the earlier Wisconsin data to generate a complete sequence of the *E. coli* genome. Their data is available on World Wide Web servers at http://bsw3.aist-nara.ac.jp/ or http://mol.genes.nig.ac.jp/ecoli/. Information about numbered Kohara clones can be viewed at http://www.ddbj.nig.ac.jp/e-coli/ecoli_list.html

31.7.1.3 Harvard University

A group at Harvard University, headed by G. Church, has reported two large contiguous sequences which were determined as part of a program to establish new methods in automatic DNA sequence determination. They utilized a multiplex sequencing approach (ref. 129 and see Chapter 20) to sequence cosmid clones derived from the *E. coli* K-12 strain BHB2600, primarily as a test-bed for further technology development. Although their sequences, ECOHU47 (U00007) and ECOHU49 (U00008) from the 47–49 min region, are included in the databases, no paper has yet been published (P. Richterich, N. Lakey, G. Gryan *et al.*, unpublished data).

31.7.1.4 Stanford group

A group at Stanford, headed by R.W. Davis, prepared a sequencing library from pulsed field gel purified genomic *Avr*II fragments (minutes 4–25) of *E. coli* strain MG1655 (the same strain as used by the Blattner group). Four overlapping entries have been deposited in GenBank so far, together totalling 526 200 bp in length (September, 1996–January, 1997):

ECU70214 (U70214) *Escherichia coli* chromosome minutes 4–6.

ECU73857 (U73857) *Escherichia coli* chromosome minutes 6–8.

ECU82664 (U82664) *Escherichia coli* minutes 9–11 genomic sequence.

ECU82598 (U82598) *Escherichia coli* genomic sequence of minutes 11–12.

31.7.2 Analysis strategy employed by the Wisconsin group

As mentioned above, this group is completely sequencing the chromosome of a single strain of *E. coli* K-12, rather than targeting unsequenced intervals in a 'composite genome' constructed from various database entries. Since they have contributed more *coli* sequence than any other group, the strategy they have used will be described here in some detail.

Blattner's group divided the sequencing process into 10 steps, from strategy planning for initial DNA isolation to deposition of the final sequence in GenBank. These steps (see Table 31.2), or status levels, are used to define the degree of completion for a seqence segment. While this procedure is specific to Blattner's group, it provides a useful overview of the processes common to many genomic sequencing projects, and for that reason is described here.

31.7.2.1 Preparation of random clones for sequencing ('shotguns')

The first step in the process is the selection of a genomic region for sequencing, and choice of the source(s) of DNA for that sequencing. As noted in Section 31.9, several smaller genomes have subsequently been tackled as single units. However, when the *E. coli* project was initiated, the tools (especially sequence assembly software) necessary to succeed on such a scale were unavailable. For that reason, a set of clones from the MG1655 clone bank, chosen so as to minimize overlap and hence redundant effort, were selected for sequencing.

To prepare single-stranded sequencing templates, DNA from selected λ-clones is physically sheared and fragments of 0.7–2.0 kbp in length are cloned into the *Sma*I site of the M13 Janus vector [134]. Although physical shearing of DNA yields fragments which require end-repair treatment (by mung bean nuclease, or a combination of T4 and Klenow DNA polymerases) before cloning, it was adopted because enzymatic treatments (including DNAse) do not yield random-ended fragments of appropriate size. M13 shotgun clones containing inserts derived from the λ-vector arms, or containing DNA which had already been sequenced when analysing overlapping λ-clones, are identified by hybridization and discarded. The resulting library, from which unwanted clones have been removed, is archived and used for phage growth and DNA preparation [135].

31.7.2.2 Random data collection and assembly

The Wisconsin group collected its first 1.5 Mb of data using Sequenase and ^{35}S-label in Sanger dideoxy sequencing reactions (see Chapter 22). The reactions were performed by a sequencing reaction robot developed by the group, and resolved on large-format gels. Autoradiographs were digitized, or scanned photoelectrically with an experimental film scanner and base-calling software, to generate sequence files for assembly into contigs. Assembly software prescreened the individual sequences, removing any λ-sequence and trimming M13 vector sequences from the 5′ and/or 3′ ends.

After sufficient random data for five- to sevenfold coverage was collected, an initial assembly was attempted. Typically a number of 1–5 kbp contigs will have been generated for each 15–20 kbp λ-clone insert. The next step involves the systematic use of reverse-strand sequencing to extend the contigs and achieve better coverage of regions for which this is necessary. The Janus vector was developed to simplify reverse-strand sequencing [8]. Janus is an M13 derivative engineered so that a cloned insert may be sequenced from the opposite end, on the opposite strand, by inversion of the insert *in vivo* (flipping). The inversion is achieved by growing the phage on a host supplying phage λ Int recombinase, which acts at the *att* sites which flank the insert in the Janus vector. This permits all sequencing to be done from single-stranded template using standard primers and avoids the effort and expense of dealing with plasmids, PCR, or *in vitro* recloning to obtain reverse strand information.

Table 31.2 The 10-step sequencing process used in the Wisconsin *E. coli* Genome Project.

Step	Definition of the step, and summary of the processes involved
1	*Project strategy chosen* Select genomic region and sources of DNA for sequencing Plan handling of overlaps, gaps, and abutments between subsections Define strategy for assembly and coverage reduction
2	*Shotgun made* Prepare source DNA, and verify restriction map Subclone DNAs into M13 Janus vector Select M13 clones for sequencing with plaque hybridization Grow M13 clones and prepare template DNA
3	*Data gathering* Prepare sequencing reactions Load and run gels Collect sequence data
4	*Initial assembly* Remove vector sequences and attempt assembly Identify problems and remove bad data, if any Call for Janus flips using automated coverage analyser Use software to call for compression resolutions Repeat process with new data, as necessary
5	*Assembled* Automatically generate sequence with only a few gaps or trouble spots Update informatics system and transfer files to finishing team
6	*Edited* Edit alignments on screen Proofread primary sequence data (traces) in ambiguous regions Call for more data in thin or ambiguous areas Design primers for primer walking experiments where needed
7	*Provisional* Re-edit areas after any additional data has been added Generate provisional consensus sequence containing minimal ambiguities
8	*Bio-checked* Identify ORFs with software and examine codon-usage statistics Scan nucleic acid and protein databases for similarities Re-examine data in areas of disagreement, especially possible frameshift errors
9	*Annotated* Splice segments to create extended annotation unit, if appropriate Annotate a set of qualified ORFs Identify ORFs with gene or function where possible
10	*Finished* Splice to update single contig Deposit annotated sequence in Genbank

Choice of clones to flip was done automatically by a computer program, 'Spanner'. Working from an unedited initial assembly, the program identified clones with inserts bordering the ends of short contigs that could be flipped to achieve closure, inserts that could be flipped to provide second-strand sequence for regions sequenced on only one strand, and DNA originating from near the insert–λ junction which, if sequenced from the other end, could extend the total length of sequence determined. The program also could select a minimal set of M13 clones that fully covers the target segment; members of this set can be resequenced with dITP in order to resolve sequence compressions.

Addition of data from flipped clones generally serves to merge the smaller contigs into a single unit with sufficient coverage to constitute a sequence ready for alignment editing.

31.7.2.3 Editing and finishing

Contigs were then edited on screen: alignment errors were corrected, misaligned repeats realigned, and conflicts and ambiguities resolved by again consulting the sequence autoradiographs or fluorescent traces. Additional sequence data was obtained if necessary for resolution of ambiguities; this may involve a directed approach in which primers are designed to allow 'walking'. A minimum standard of completion required at least two determinations of every base, with at least one determination from each strand. The end-result of the editing process was a 'provisional' consensus sequence; discovery of the few remaining errors usually occurred in the course of analysing the sequence for its information content.

31.7.2.4 Identification of potential genes in provisional sequence

Sequencing accuracy was crucial: since an average *E. coli* gene is encoded by 1200 bp, even one base per kilobase, inserted or deleted as a result of sequencing error, could prevent correct identification of most genes. Once sequence that was as accurate as can be obtained, had been obtained, and alignment had been verified by the methods described above, the sequence was analysed for ORFs. Remaining errors that give rise to frameshifts could often be detected during this process.

Sequences were examined for potential genes using ORF searches and pattern of codon usage statistics. Geneplot, a commercially available (DNASTAR) codon usage based, gene-finding program with a graphical output, uses the methods of Staden [136], Gribskov [137], and Borodovsky [138]. It finds all potential ORFs in each of the six reading frames, and analyses their codon usages in comparison to a reference set (i.e. does the potential ORF contain the distribution of codons found in known *coli* genes?). This allows simple detection of possible frameshift errors, which appear as overlapping genes in different frames, each with codon usage scores falling to zero at the error point. The sequence for such regions was again rechecked and, if need be, resequenced.

Of the coding region detection programs, Borodovsky's Genemark [12], which identifies coding regions simultaneously on both strands, permitting overlapping and complementary reading frames to be assessed and compared, is the most effective, and the program has been used in several prokaryotic genome projects. Although the Wisconsin group uses an in-house implementation, the original Genemark is available, via an e-mail: server, to automatically analyse correctly formatted sequences included within an e-mail: message. For information, send e-mail containing the message 'help' to genemark@ford.gatech.edu or genemark@embl-ebi.ac.uk. Instructions for using Genemark can be found at: http://amber.biology.gatech.edu/william/genemark.html.

31.7.2.5 Annotation and analysis

Comparisons were made with any previously reported sequence from the same region and differences carefully evaluated; consultation with other scientists regarding nontrivial sequence differences was initiated. In some cases, errors have been detected in previous database entries. These take many forms, perhaps the most common being frameshifts found in the older entries for which no method of resolving compressed GC-rich sequences had been used. Other common errors include restriction fragments accidentally lost or included at subcloning, errors in assembling pieces of a sequence, and inclusion of unidentified vector sequences. Finally some differences cannot be explained (without further experimentation) except as strain variations. Thus great caution should be used in the interpretation of database matches and differences. While automated search routines have made such comparisons easy to do, a good deal of human judgement is still required to make sense of the results.

For previously unidentified potential genes, an attempt to identify possible homologues was made by comparing the new sequence to both the nucleotide and protein sequence databases. The ORFs were also checked for the occurrence of functional motifs, using the program MacPattern [139] to search the ProSite [140] and BLOCKS [141] databases. The results of these sequence searches and examination of the relevant literature have led to new gene discoveries. However, even in an organism as well studied as *E. coli*, about half of all newly sequenced ORFs have no certain homologues in the data bases and remain as hypothetical genes.

Other sequence features, including insertion elements, repetitive sequences, and prophages, can also be identified by database comparisons. Potential promoters were located in the sequence by matrix search programs, but it is clear that what constitutes a good promoter is not completely represented by a matrix score, since well characterized promoters do not always correspond to those found by the computer. The sequences were also searched using pattern-matching software to locate possible transcription terminators, potential static bends, Chi sites, and other features of global significance. Particular attention was paid to stretches of more than a few hundred bases in which no features are readily discovered. The term 'grey hole' has been coined for such regions [72]; they are especially intriguing in the otherwise densely packed genomes of prokaryotes (the final sequence contains ~7 of these).

The gene and feature identifications were correlated to generate a 'final' annotated sequence which was submitted to GenBank. In addition, using software developed to splice together such annotated sequences, each completed segment was added to the accumulating genomic contig.

31.7.2.6 Improvements to the process

In order to complete the *E. coli* genome sequence more rapidly the Wisconsin group made two major changes in methodology when sequencing was about half completed.

First, after a transitional period in which both were used, the group completely substituted fluorescent for radioactive sequencing. Fortunately, the ssDNA templates used for radioactive sequencing with Sequenase were found to be suitable, with no need for modification in preparation, for use in fluorescent sequencing with ABI automated sequencers. A potentially serious problem was that of GC compressions: since the ≈51% G+C content of *E. coli* leads to a compression per kilobase, a frameshift error could, in principle, occur in every gene. Comparison of the various fluorescent based sequencing chemistries demonstrated that dye-terminator chemistry with dITP substituting for dGTP would minimize this problem. DNA templates are sequenced, using *Taq* polymerase, by thermal cycling reactions with fluorescent-labelled dideoxy terminators. Reactions were set up in 96-well plates, and unincorporated dye-labelled dideoxynucleotides removed from the reactions by passage through Sephadex G50 size exclusion resin in a 96-well fritted plate, which can be centrifuged. These reactions yielded an average of 500–550 bases of usable sequence per template from a 3- to 4-h run on an ABI model 377.

A major problem faced by all groups sequencing *E. coli* DNA stems from the difficulty of subdividing the genome into appropriately sized fragments for sequencing. The theoretical advantages of sequencing from nonoverlapping (or minimally overlapping) fragments larger than those provided by λ-clones are obvious. These include reduction in the number of shotgun subclones that need to be prepared and processed, elimination of plaque hybridization or other steps to screen out λ-vector sequences from the subclones, and increasing over-all efficiency by avoiding excess depth of coverage in regions of overlap between adjacent (λ) clones.

A new approach by C. Bloch, a collaborator at the University of Michigan, seems to have solved the problem of segmenting the genome for sequencing without overlap redundancy. Transposons have long been useful as delivery vectors for *E. coli*, and have been used to introduce particular restriction sites to selected locations on the *E. coli* chromosome; recently engineered mini-Tn*10* derivatives make this easier [142,143]. I-*Sce*I is an intron-encoded site-specific DNA endonuclease originating from yeast [144, 145], for which no sites are present in the *E. coli* chromosome. Bloch's group has incorporated an I-*Sce*I cleavage site into these mini-Tn*10* derivatives, thus providing a system for introducing unique cutting sites into *E. coli* chromosomal DNA.

In order to use these sites to isolate DNA fragments, it is necessary to introduce them, pairwise, into the chromosome of MG1655, so that a specific fragment will be released by I-*Sce*I digestion. To facilitate this, Bloch has inserted I-*Sce*I elements into the MG1655 chromosome using three different mini-Tn*10* constructs, each with a different drug-resistance marker (spectinomycin, kanamycin, or chloramphenicol) [146,177]. Once the insert positions were mapped, a suitable set of *E. coli* lines can be constructed for the sequencing project by using P1 transduction to combine appropriately spaced pairs of sites linked to distinguishing antibiotic resistance markers.

Fragments were generated by digestion, with I-*Sce*I, of agarose-embedded genomic DNA originating from the appropriate double-insert strain. Preparative CHEF gel electrophoresis separates the fragment from the remainder of the genome, and the DNA is recovered by β-agarase digestion. The purified DNA fragment was then randomly fragmented by nebulization, size-fractionated, end-repaired, and ligated into the *Sma*I-cut M13 Janus vector to provide a library of source clones for DNA sequencing template preparation. Fifty-four percent of the genome was sequenced using I-*Sce*I fragments.

31.8 Escherichia coli databases

In contrast to other model organism genome projects, there is no single central database for *E. coli* genomic data. In part, this reflects the many years of pregenome era work and the diverse interests found within the *E. coli* community. Nevertheless, a number of databases are available which organize *E. coli* genetic, sequence, and biochemical data. Those of which we are aware are described below. A list is maintained by ECDC at http://susi.bio.uni-giessen.de/db_other.html.

31.8.1 EcoMap, EcoSeq, and EcoGene

EcoSeq is a nonoverlapping, that is nonredundant, *E. coli* DNA sequence collection which integrates information about genes, DNA and protein sequences. Vector sequences detected in GenBank/EMBL/DDJB entries have been removed, and adjacent or overlapping sequence entries melded to generate continuous sequences. EcoMap integrates EcoSeq with the genomic restriction map of Kohara *et al.* [54]. As additional DNA sequences are aligned with the restriction map, segments of the Kohara map are replaced with sequence-derived restriction maps. EcoGene contains information about identified and putative protein- and RNA-encoding genes, and translations of sequences thought to encode proteins. These data are correlated and cross-referenced with the SWISS-PROT protein sequence database.

A hard copy of version 5 has been published [58], and version 6 has been described [59]. An update of the dataset for version 7 is in preparation and will be made available, along with documentation and programs to access the datasets, via anonymous ftp (K. Rudd, personal communication): ftp://ncbi. nlm. nih.gov/repository/Eco/. For additional information, contact Kenn Rudd (rudd@ecogene.med. miami.edu).

31.8.2 The Escherichia coli database collection

This database contains information for the entire *E. coli* K-12 chromosome, and is organized like a genetic map. The database can be searched for gene names or map positions. Coding sequences (CDS) are indicated for each gene—whether putative ORF or untranslated RNA. Regulatory regions, promoters, terminators, and IS elements are also indicated. A hard copy of Release 20 has been published [147]. The complete ECDC dataset is available by anonymous ftp (susi.bio.uni-giessen.de) or together with a Windows application on the EMBL (EBI) CD-ROM. It can also be queried via the World Wide Web with a forms-capable client: http://susi.bio.uni-giessen.de/usr/local/www/html/ecdc.html. For additional information, contact Manfred Kröger (kroeger@embl-heidelberg.de).

31.8.3 Colibri

Colibri is a relational database dedicated to the analysis of the *E. coli* genome. It was developed as a part of a thesis project through a collaboration between the Unité de Regulation de l'Expression Genetique (Institut Pasteur–CNRS) and the Atelier de BioInformatique (Institut Curie). A complete description of the database and its organization has been published [148]. Colibri is a Macintosh application developed with the 4th Dimension database engine. Version 1.3 (30 October 1994) is available via anonymous ftp (ftp.pasteur.fr, in the directory pub/GenomeDB/Colibri); a new release will be available soon (A. Danchin, personal communication). For additional information, contact Ivan Moszer (moszer@pasteur.fr) or Antoine Danchin (adanchin@pasteur.fr).

31.8.4 The Escherichia coli Genetic Stock Center

The *E. coli* Genetic Stock Center (CGSC) at Yale University maintains a database of *E. coli* genetic information, including genotypes and reference information for the several thousand strains in the CGSC collection, a gene list with map and gene product information, and information on specific mutations. An electronic version of the *E. coli* linkage map is also under development.

A public version of the database includes the information of most interest to the community and is accessible via the WWW, with a forms interface for queries, at the URL: http://cgsc.biology.yale.edu/top.html

The CGSC Web server is experimental and does not include all information available through the Sybase data base. Address questions about the data base contents or requests for stocks, information, or a guest login to Mary Berlyn (mary@cgsc.biology. yale.edu).

31.8.5 The Encyclopedia of Escherichia coli Genes and Metabolism

The Encyclopedia of *E. coli* Genes and Metabolism (EcoCyc) is a database integrating information about *E. coli* genes and metabolism [149]. A graphical user interface creates drawings of metabolic pathways, of individual reactions, and of the *E. coli* genomic map. Users can call up objects through a variety of queries and then navigate to related objects shown in the display window. For example, a user could zoom in on a region of the genetic map, click on a gene to obtain detailed information about it, navigate to the enzyme product of the gene, and then view the metabolic pathway containing the enzyme.

For information on the EcoCyc, see the WWW URL: http://www.ai.sri.com/ecocyc/ecocyc.html

31.8.6 The Escherichia coli Gene Database

A compilation of *E. coli* genes and gene products, categorized by physiological function, the *E. coli* Gene Database (GenProTec) also includes homology information for proteins similar to at least one other *E. coli* protein [82]. The database is available by ftp://hoh.mbl.edu/pub/ecoli.zip or there is a WWW version at the URL: http://www.mbl.edu/~dspace/eco.html

31.8.7 The Escherichia coli Gene-protein Database

The *E. coli* Gene-protein Database (ECO2DBASE) is a database containing information about *E. coli* proteins obtained by the analysis of two-dimensional protein gels, and is maintained by F.C. Neidhardt. Full information and searching facilities are available at http://pcsf.med.umich.edu/eco2dbase.

Questions and comments can be sent to Ruth VanBogelen (vanbogr@aa.wl.com) or Fred Neidhardt (fcneid@umich.edu).

31.8.8 SWISS-PROT

SWISS-PROT is a carefully curated and highly reliable protein sequence database which strives to provide a high level of annotations (such as the description of the function of a protein, its domain structure, post-translational modifications, variants, etc.), a minimal level of redundancy and high level of integration with other databases [150]. SWISS-PROT is not an *E. coli*-specific database, but its cross-references to other databases make it an exceptionally useful source of information. A. Bairoch has coordinated the *E. coli* entries with K. Rudd, and an index of all *E. coli* K-12 entries is available from: http://expasy.hcuge.ch/cgi-bin/lists?ecoli.txt.

World Wide Web access through the ExPASy server allows forms-based searches of the entire data base by description or identification, accession number, author, or full text search: http://expasy.hcuge.ch/sprot/sprot-top.html

For additional information, contact Amos Bairoch (bairoch@cmu.unige.ch).

In addition to these electronic databases, sequence data have been used to produce compilations of a variety of functional sequence elements. These include promoters [151], terminators [152], and ribosomal binding sites [153].

31.9 Future prospects: what can we learn from sequencing prokaryotic genomes?

In the race to determine the first complete sequence of a free-living organism, *E. coli*—once a prime contender—has lost out to five prokaryotes with smaller genomes and lower G:C contents: first to *Haemophilus influenzae* [154] and then to *Mycoplasma genitalium* [155], followed by the archaebacterium *Methanococcus jannaschii* [178], the cyanobacterium *Synechocystis* sp. Strain PCC6803 [158,179], and a second mycoplasma *Mycoplasma pneumoniae* [180]. The 13 Mb genomic sequence of the yeast *Saccharomyces cerevisiae*, undertaken by a large consortium of laboratories, was also completed sooner [181].

The sequences of many more genomes are in progress with some virtually completed; web sites are maintained for many of these. Prokaryote genome projects currently near completion include those for the spore-forming soil bacterium *Bacillus subtilis* [156], the causative agent of leprosy, *Mycobacterium leprae* [157], *Mycobacterium tuberculosis* (two projects) and the archaeon *Methanobacterium thermoautotrophicum* (see web site at http://pandora.cric.com/htdocs/sequences/methanobacter/abstract.html). Under the auspices of its Microbial Genome Initiative, the United States Department of Energy (DOE) has launched a number of prokaryotic genome projects. An initial initiative in 1994 supported projects to sequence the genomes of several archaebacteria [159]. *M. jannaschii* and *M. thermoautotrophicum* have been completed; *Pyrococcus furiosus* is still underway. In a second round of support, projects to sequence two extremeophiles able to grow in boiling water, the archaebacterium *Archaeoglobus fulgidus* and the eubacterium *Thermotoga maritimae* were funded. Other possible candidate species for DOE genome-sequencing support include *Sulfolobus solfataricus*, which oxidizes sulfur, *Clostridium acetobutylicum*, of possible industrial use for alcohol production, and *Pseudomonas aeruginosa*, a particularly difficult-to-control hospital pathogen [160]. This is very much a partial list of species for which genome sequencing is in progress. For a tabulation of current prokaryotic/microbial genomes being sequenced, the reader is referred to the NCGR Microbial Genome Site (http://www.ncgr.org/microbe/) which attempts to keep track of current or completed eubacterial, archaeal and eukaryotic genome sequencing projects. Over 50 projects were listed at the time the site was last examined (May 1997)! The Institute for Genome Research also maintains this type of information at its web site.

In addition to genomes that are the subject of concerted sequencing efforts, physical maps of many other genomes have been constructed [161]. Such maps could provide the basis for efforts to survey the sequences of those genomes, without actually attempting to determine complete genomic sequences.

What can we look forward to as we enter a 'post-sequencing' era, with not just one but many versions of the so-called 'blueprint of life' available for examination (see ref. 183 for a discussion)?

31.9.1 The *E. coli* K-12 story

31.9.1.1 Identification of gene function

E. coli K-12 genes can be divided into three classes on the basis of our current information about them. Between one-half and two-third [83] of gene products can be assigned definite functions, either on the basis of existing genetic or biochemical knowledge or because strong similarity to a well-characterized gene product in another organism permits confident prediction of function. A further 16% of gene products show some sequence conservation, perhaps of a known motif, which provides a guidepost to function. For the remaining 20% of predicted gene products there are no functional clues. While some apparently coding DNA may not normally be transcribed, what we currently know about the organization of bacterial genomes suggests that 'junk DNA' is rare. How will functions for the ≈40% of genes which are still uncharacterized be deduced?

First, by application of existing and accumulating biochemical and genetic information: the community of *E. coli* researchers possesses a considerable body of knowledge which is continuing to be correlated with accumulating sequence. Identification of functional motifs in new sequences can help in their assignment to already described functions.

Second, by assigning new genes to existing regulons: for instance, a systematic approach to studying global regulatory mechanisms has been described for *E. coli*, in which mRNA levels expressed from various regions of the chromosome under different conditions are measured by hybridization to DNA derived from an ordered set of λ-clones [162]. The technique allows detection of both induced and repressed levels of gene expression, and is applicable to a variety of chemical, physical, or physiological treatments. For example, use of this method to examine gene expression under heat-shock conditions allowed the characterization of 26 new heat-shock loci in *E. coli* [163]. Sequence studies can also be used for regulon assignment: identification of DNA-binding sites for specific regulators upstream of new ORFs can permit their inclusion in existing regulons. Identification of a new motif upstream of a group of genes can allow them to be classed as a previously unrecognized regulon.

Third, knowledge of the positions and sizes of uncharacterized ORFs allows protein products to be sought (is the ORF transcribed/translated?) and the phenotypic consequences of insertional inactivation to be detailed (is the ORF dispensable? If not, what are the physiological consequences of inactivating it?). Such studies should also help to clarify what is actually regulated by genes classified as probable regulatory genes solely on the basis of sequence.

Finally, the continually accumulating sequence databases for all organisms will continue to provide strong homologues for uncharacterized ORFs that will serve as important pointers to function.

31.9.1.2 Comparisons with other E. coli strains: the power of genome scanning

Sequencing information and resources can be used to quickly identify major sequence variations between related strains which are associated with important phenotypic differences. Although *E. coli* K-12 is harmless, other *E. coli* strains cause disease. A 35-kbp locus has been identified in enteropathogenic and enterohaemorrhagic strains of *E. coli*, the presence of which is correlated with a specific histopathological effect on intestinal epithelial cells. Using mapping membranes of the Kohara *E. coli* K-12 clone set, and the sequence data from *E. coli* K-12, this locus was characterized as an insert relative to the K-12 genome. In uropathogenic *E. coli* strains, a different block of virulence genes is inserted at this same site, which is also the integration site for the *E. coli* retronphage phiR73 [164].

Similarly, a group of eight genes comprising the fimbrial gene cluster in pathogenic serotype b strains of *Haemophilus influenzae* was found to be an insertion relative to the genome of the nonpathogenic Rd strain [154].

Genome scanning—single-pass sequencing of the entire genomes of prokaryotic strains of interest, followed by comparison to the more fully characterized genomes of related strains/species—is likely to reveal other instances of such pathogenicity islands. Viewed in a more general manner as modules conferring new properties on a strain, the existence of these elements indicate that the old idea of a modular genome may well hold not only for bacteriophages [165], but for their hosts as well. One can predict that such modules will not be confined to pathogenicity determinants. Some of the gene clusters with atypical codon usage that have been described for *E. coli* K-12 and hypothesized to have been horizontally transmitted [85] may well owe their integration into the genome to mechanisms similar to those responsible for the acquisition of pathogenicity determinants.

31.9.2 Evolutionary insights and other biological questions

Prokaryotes (the Archaea and the Bacteria, the latter also referred to as Eubacteria) comprise two of the three superkingdoms of living organisms [166]. Full genomic sequences will soon be available for several representatives of each of these groups, and a full sequence is already available for the budding yeast *Saccharomyces cerevisiae*, a member of the Eucarya (eukaryotes), the third superkingdom. This will enable proteins to be divided into those that appear to have evolved relatively recently (conserved only in closely related genera) and those with much longer histories. Indeed, it has already been reported that 40% of *coli* proteins contain ancient conserved sequences, shared with Eucarya or Archaea [80], although it is not clear whether the extent of these similarities is sufficient to indicate evolutionary conservation of complete polypeptides as opposed to recruitment of useful motifs. To date, the construction of phylogenetic trees has mostly relied on comparison of rRNA sequences. The ability to construct trees based on sequence variation amongst a number of groups of homologous proteins should provide a powerful tool for phylogenetic analysis. In particular the relationship between archaebacteria and eubacteria and eukaryotes should become much clearer.

A second question that analysis of genome sequences should help to answer is what constitutes a 'minimal genome'. Is there a basic minimum set of genes required by any free-living organism and how big is it? Full analysis of the *M. genitalium* genome should help to answer this question. Possessing the smallest known genome of any free-living organism, this cell-wall deficient, fastidious prokaryote has a genetic complement of only 470 genes, perhaps 10–15% as many as *E. coli*. Although normally it obtains much of its nutrition directly from its host, *M. genitalium* can also grow independently. Thus, we may ask whether its small genome defines a basic set of genes that is shared by all free-living bacteria? The wealth of genome sequences soon to be available should enable us to answer this question.

If there prove to be core proteins common to all bacteria, and a population of horizontally transmissible gene clusters required by none, are there also unique genes which define a bacterial species? Or is a bacterial species simply a unique combination of genes, each of which is shared with some other species? It is to be hoped that comparative sequence studies will enable us to answer such questions.

31.9.3 Practical considerations

There are a number of clear practical benefits to be derived from functionally defining bacterial genes and we close with a short list of these.

First, gene functions are much easier to define in *E. coli* than in humans. Identification of the functions of homologous bacterial genes has helped to define the functions of genes involved in inherited human diseases such as colon cancer susceptibility [1,2] and cystic fibrosis [167,168].

Second, there are many potential economic benefits to be derived from using bacteria, or their enzymes, in industrial processes. It is no accident that several of the organisms for which sequencing projects are underway are thermophiles, able to carry out biological processes at high temperatures, or originate from deep sea vents, where pressures are extreme. It is to be expected that enzymes of industrial importance will be identified in these organisms. In addition to enzymes as potential chemical catalysts in heavy industry, there are certain to be new uses in the research market, where high-temperature DNA polymerases have already engendered PCR and greatly improved DNA sequencing. Contributions to energy production (methane generation) and toxic waste degradation are also likely.

Finally, let us not forget that bacteria cause disease, of both animals and plants. Even in the minimal *M. genitalium* genome, about 5% of the genes are devoted to evading the host's immune system. Comparison across many species of homologous proteins likely to form targets for antibacterial agents should help to identify conserved regions which could aid in targeted drug design. Comparison of virulent and avirulent strains should also identify surface proteins unique to the virulent organisms, providing a means of developing specific vaccines.

References

1 Fishel, R., Lescoe, M.K., Rao, M.R.S. *et al.* (1993) The human mutator gene homolog MSH2 and its association with hereditary nonpolyposis colon cancer. *Cell* **75**, 1027–1038.

2 Bronner, C.E., Baker, S.M., Morrison, P.T. *et al.* (1994) Mutation in the DNA mismatch repair gene homolog HMLH1 is associated with hereditary nonpolyposis colon-cancer. *Nature* **368**, 258–261.

3 Neidhardt, F.C. *et al.* (eds) (1987) Escherichia coli *and* Salmonella typhimurium*: Cellular and Molecular Biology* (American Society for Microbiology, Washington, DC).

4 Tatum, E.L. & Lederberg, J. (1947) Gene recombination in the bacterium *Escherichia coli*. *J. Bacteriol.* **53**, 673–684.

5 Cavalli-Sforza, L.L. (1950) La sessualita nei bateri. *Boll. Ist. Sieroter. Milan* **29**, 281–289.

6 Hayes, W. (1953) The mechanism of genetic recombination in *Escherichia coli*. *Cold Spring Harb. Symp. Quant. Biol.* **18**, 75–93.

7 Bachmann, B.J. (1987) Derivations and genotypes of some mutant derivatives of *Escherichia coli* K-12. In Escherichia coli *and* Salmonella typhimurium: *Cellular and Molecular Biology* (Neidhardt, F.C. *et al.*, eds), 1190–1219 (American Society for Microbiology, Washington, DC).

8 Bertani, J. & Weigle, J.J. (1953) Host controlled variation in bacterial viruses. *J. Bacteriol.* **65**, 113–121.

9 Smith, C.L., Econome, J.G., Shutt, A. *et al.* (1987) A physical map of the *Escherichia coli* K12 genome. *Science* **236**, 1448–1453.

10 Condemine, G. & Smith, C.L. (1990) Genetic mapping using large-DNA technology: alignment of *Sfi*I and *Avr*II sites with the *Not*I genomic restriction map of *Escherichia coli* K-12. In *The Bacterial Chromosome* (Drlica, K. & Riley, M., eds), 53–60 (American Society for Microbiology, Washington, DC).

11 Perkins, J.D., Heath, J.D., Sharma, B.R. & Weinstock, G.M. (1993) *Xba*I and *Bln*I genomic cleavage maps of *Escherichia coli* K-12 strain MG1655 and comparative analysis of other strains. *J. Mol. Biol.* **232**, 419–445.

12 Bloch, C.A., Rode, C.K., Obreque, V. & Russell, K.Y. (1994) Comparative genome mapping with mobile physical map landmarks. *J. Bacteriol.* **176**, 7121–7125.

13 Birkenbihl, R.P. & Vielmetter, W. (1989) Cosmid derived map of *Escherichia coli* strain BHB2600 in comparison to the map of strain W3110. *Nucleic Acids Res.* **17**, 5057–5069.

14 Milkman, R. & Bridges, M.K. (1993) Molecular evolution of the *Escherichia coli* chromosome. IV. Sequence comparisons. *Genetics* **133**, 455–468.

15 Masters, M. (1991) Prokaryotic chromosomes: damn fine progress but a long way to go. *Curr. Biol.* **1**, 63–64.

16 Drlica, K. & Riley, M. (eds) (1990) *The Bacterial Chromosome* (American Society for Microbiology, Washington, DC).

17 Miller, J.H. (1992) *A Short Course in Bacterial Genetics: A Laboratory Manual and Handbook for* Escherichia coli *and Related Bacteria* (Cold Spring Harbor Laboratory Press, NY).

18 Scott, J.R. (1984) Regulation of plasmid replication. *Microbiol. Rev.* **48**, 1–23.

19 Kües, U. & Stahl, U. (1989) Replication of plasmids in gram-negative bacteria. *Microbiol. Rev.* **53**, 491–516.

20 Cesareni, G., Helmer-Citterich, M. & Castagnoli, L. (1991) Control of ColE1 plasmid replication by antisense RNA. *Trends Genet.* **7**, 230–235.

21 Willetts, N. & Skurray, R. (1987) Structure and function of the F factor and mechanism of conjugation. In Escherichia coli *and* Salmonella typhimurium: *Cellular and Molecular Biology* (Neidhardt, F.C. *et al.*, eds), 1110–1133 (American Society for Microbiology, Washington, DC).

22 Sternberg, N.L. & Maurer, R. (1991) Bacteriophage-mediated generalized transduction in *Escherichia coli* and *Salmonella typhimurium*. *Meth. Enzymol.* **204**, 18–43.

23 Masters, M. & Broda, P. (1971) Evidence for the bidirectional replication of the *Escherichia coli* chromosome. *Nature New Biol.* **232**, 137–140.

24 Bird, R., Louarn, J., Martuscelli, J. & Caro, L. (1972) Origin and sequence of chromosome replication in *Escherichia coli*. *J. Mol. Biol.* **70**, 549–566.

25 Kuempel, P.L., Pelletier, A.J. & Hill, T.M. (1989) Tus and the terminators—the arrest of replication in prokaryotes. *Cell* **59**, 581–583.

26 Hill, T.M. (1992) Arrest of bacterial DNA replication. *Ann. Rev. Microbiol.* **46**, 603–633.

27 Masters, M. (1989) The *Escherichia coli* chromosome and its replication. *Curr. Opinion Cell Biol.* **1**, 241–249.

28 von Meyenburg, K. & Hansen, F.G. (1987) Regulation of chromosome replication. In Escherichia coli *and* Salmonella typhimurium: *Cellular and Molecular Biology* (Neidhardt, F. C. *et al.*, Eds) 1555–1577 (American Society for Microbiology, Washington, DC).

29 McMacken, R., Silver, L. & Georgopoulos, C. (1987) DNA replication. In Escherichia coli *and* Salmonella typhimurium: *Cellular and Molecular Biology* (Neidhardt, F.C. *et al.*, Eds) 564–612 (American Society for Microbiology, Washington, DC).

30 Bremer, H. & Churchward, G. (1991) Control of cyclic chromosome replication in *Escherichia coli*. *Microbiol. Rev.* **55**, 459–475.

31 Marians, K.J. (1992) Prokaryotic DNA replication. *Ann. Rev. Biochem.* **61**, 673–719.

32 Baker, T.A. & Wickner, S.H. (1992) Genetics and enzymology of DNA replication in *Escherichia coli*. *Ann. Rev. Genet.* **26**, 447–477.

33 Kornberg, A. & Baker, T.A. (1992) *DNA Replication* (W.H. Freeman and Company, New York).

34 Skarstad, K. & Boye, E. (1994) The initiator protein DnaA: evolution, properties and function. *Biochim. Biophys. Acta* **1217**, 111–130.

35 Cairns, J. (1963) The chromosome of *Escherichia coli* . *Cold Spring Harb. Symp. Quant. Biol.* **28**, 43–46.

36 Robinow, C. & Kellenberger, E. (1994) The bacterial nucleoid revisited. *Microbiol. Rev.* **58**, 211–232.

37 Pettijohn, D.E. & Carlson, J.O. (1979) Chemical, physical and genetic structure of prokaryotic chromosomes. In *Cell Biology*, Vol. 2, 1–57 (Academic Press, New York).

38 Drlica, K. (1987) The nucleoid. In Escherichia coli *and* Salmonella typhimurium: *Cellular and Molecular*

Biology (Neidhardt, F.C. *et al.*, eds), 91–103 (American Society for Microbiology, Washington, DC).
39 Schmid, M. (1988) Structure and function of the bacterial chromosome. *Trends Biochem. Sci.* **13**, 8131–8138.
40 Griffith, J.D. (1976) Visualization of prokaryotic DNA in a regularly condensed chromatin-like fiber. *Proc. Natl Acad. Sci. USA* **73**, 563–567.
41 Drlica, K. & Rouviere-Yaniv, J. (1987) Histone like proteins of bacteria. *Microbiol. Rev.* **51**, 301–319.
42 Pettijohn, D.E. (1988) Histone-like proteins and bacterial chromosome structure *J. Biol. Chem.* **26**, 12793–12796.
43 Schmid, M.B. (1990) More than just 'Histone-like' proteins. *Cell* **63**, 451–453.
44 Yamada, H., Yoshida, T., Tanaka, K. *et al.* (1991) Molecular analysis of the *Escherichia coli hns* gene encoding a DNA-binding protein, which preferentially recognizes curved DNA sequences. *Mol. Gen. Genet.* **230**, 332–336.
45 Drlica, K. (1992) Control of bacterial DNA supercoiling. *Mol. Microbiol.* **6**, 425–433.
46 Bachellier, S., Saurin, W., Perrin, D. *et al.* (1994) Structural and functional diversity among bacterial interspersed mosiac elements (BIMES). *Mol. Microbiol.* **12**, 61–70.
47 Ogden, G.B. & Schaechter, M. (1985) Chromosomes, plasmids, and the bacterial cell envelopes. In *Microbiology, 1985* (Leive, L., ed.), 282–286 (American Society for Microbiology, Washington, DC).
48 Kusano, T., Steinmetz, D., Henrickson, W.G. *et al.* (1984) Direct evidence for specific binding of the replicative origin of the *Escherichia coli* chromosome to the membrane. *J. Bacteriol.* **158**, 313–316.
49 Drlica, K., Burgi, E. & Worcel, A. (1978) Association of the folded chromosome with the cell envelope of *Escherichia coli*: nature of the membrane-associated DNA. *J. Bacteriol.* **134**, 1108–1116.
50 Clarke, L. & Carbon, J. (1976) A colony bank containing synthetic Col E1 hybrid plasmids representative of the entire *E. coli* genome. *Cell* **9**, 91–99.
51 Nishimura, A., Okiyama, K., Kohara, Y. & Horiuchi, K. (1992) Correlation of a subset of the pLC plasmids to the physical map of *Escherichia coli* K-12. *Microbiol. Rev.* **56**, 137–151.
52 Vanbogelen, R.A., Sankar, P., Clark, R.L. *et al.* (1992) The gene–protein database of *Escherichia coli*—edn 5. *Electrophoresis* **13**, 1014–1054.
53 Tabata, S., Higashitani, A., Takanami, M. *et al.* (1989) Construction of an ordered cosmid collection of the *Escherichia coli* K-12 W3110 chromosome. *J. Bacteriol.* **171**, 1214–1218.
54 Kohara, Y., Akiyama, K. & Isono, K. (1987) The physical map of the whole *E. coli* chromosome: application of a new strategy for rapid analysis and sorting of a large genomic library. *Cell* **50**, 495–508.
55 Knott, V., Blake, D.J. & Brownlee, G.G. (1989) Completion of the detailed restriction map of the *E. coli* genome by the isolation of overlapping cosmid clones. *Nucleic Acids Res.* **17**, 5901–5912.
56 Medigue, C., Viari, A., Henaut, A. & Danchin, A. (1991) *Escherichia coli* molecular genetic map (1500-kbp)—update 2. *Mol. Microbiol.* **5**, 2629–2640.
57 Rudd, K.E., Miller, W., Ostell, J. & Benson, D.A. (1990) Alignment of *Escherichia coli* K12 DNA sequences to a genomic restriction map. *Nucleic Acids Res.* **18**, 313–321.
58 Rudd, K.E. (1992) Alignment of *E.coli* DNA sequences to a revised, integrated, genomic restriction map. In *A Short Course in Bacterial Genetics* (Miller, J.H., ed.), 2.2–2.43 (Cold Spring Harbor Laboratory Press, NY).
59 Rudd, K.E. (1993) Maps, genes, sequences and computers: an *Escherichia coli* case study. *ASM News* **59**, 335–341.
60 Miller, J.H. (ed.) (1991) *Bacterial Genetic Systems. Meth. Enzymol.* **204.**
61 Jacob, F. & Wollman, E.L. (1958) Genetic and physical determinations of chromosomal segments in *E. coli*. *Symp. Soc. Exp. Biol.* **12**, 75–92.
62 Low, K.B. (1991) Conjugational methods for mapping with Hfr and F-prime strains. *Meth. Enzymol.* **204**, 43–62.
63 Wanner, B.L. (1986) Novel regulatory mutants of the phosphate regulon in *Escherichia coli* K-12. *J. Mol. Biol.* **191**, 39–58.
64 Singer, M., Baker, T.A., Schnitzler, G. *et al.* (1989) A collection of strains containing genetically linked alternating antibiotic resistance elements for genetic mapping of *Escherichia coli*. *Microbiol. Rev.* **53**, 1–24.
65 Wu, T.T. (1966) A model for three-point analysis of random general transduction. *Genetics* **54**, 405–410.
66 Taylor, A.L. & Thoman, M.S. (1964) The genetic map of *Escherichia coli* K-12. *Genetics* **50**, 659–677.
67 Bachmann, B.J. (1990) Linkage map of *Escherichia coli* K12, edn 8. *Microbiol. Rev.* **54**, 130–197.
68 Jenkins, A.J., March, J.B., Oliver, I.R. & Masters, M. (1986) A DNA fragment containing the *groE* genes can suppress mutations in the *E. coli dnaA* gene. *Mol. Gen. Genet.* **202**, 446–454.
69 Rudd, K.E., Miller, W., Werner, C. *et al.* (1991) Mapping sequenced *E. coli* genes by computer: software, strategies and examples. *Nucleic Acids Res.* **19**, 637–647.
70 Noda, A., Courtwright, J.B., Denor, P.F. *et al.* (1991) Rapid identification of specific genes in *E. coli* by hybridization to membranes containing the ordered set of phage clones. *BioTechniques* **10**, 474–477.
71 Cao, G.J. & Sarkar, N. (1992) Identification of the gene for an *Escherichia coli* poly (A) polymerase. *Proc. Natl Acad. Sci. USA* **89**, 10380–10384.
72 Daniels, D.L., Plunkett, G., Burland, V. & Blattner, F.R. (1992) Analysis of the *Escherichia coli* genome DNA sequence of the region from 84.5 to 86.5 minutes. *Science* **257**, 771–778.
73 Burland, V., Plunkett, G., Daniels, D.L. & Blattner, F.R. (1993) DNA sequence and analysis of 136 kilobases of the *Escherichia coli* genome—organizational symmetry around the origin of replication. *Genomics* **16**, 551–561.
74 Plunkett, G., Burland, V., Daniels, D.L. & Blattner, F.R. (1993) Analysis of the *Escherichia coli* genome. 3. DNA sequence of the region from 87.2 to 89.2 minutes. *Nucleic Acids Res.* **21**, 3391–3398.
75 Blattner, F.R., Burland, V., Plunkett, G. *et al.* (1993) Analysis of the *Escherichia coli* genome. 4. DNA

sequence of the region from 89.2 to 92.8 minutes. *Nucleic Acids Res.* **21**, 5408–5417.

76 Sofia, H.J., Burland, V., Daniels, D.L. *et al.* (1994) Analysis of the *Escherichia coli* genome. 5. DNA sequence of the region from 76.0 minutes to 81.5 minutes. *Nucleic Acids Res.* **22**, 2576–2586.

77 Burland, V., Plunkett, III., G., Sofia, H.J. *et al.* (1995) Analysis of the *Escherichia coli* genome. 6. DNA sequence of the region from 92.8 through 100 minutes. *Nucleic Acids Res.* **23**, 2105–2119.

78 Yura, T., Mori, H., Nagai, H. *et al.* (1992) Systematic sequencing of the *Escherichia coli* genome—analysis of the 0–2.4 min region. *Nucleic. Acids Res.* **20**, 3305–3308.

79 Fujita, N., Mori, H., Yura, T. & Ishihama, A. (1994) Systematic sequencing of the *Escherichia coli* genome—analysis of the 2.4–4.1 min (110, 917–193, 643 bp) region. *Nucleic Acids Res.* **22**, 1637–1639.

80 Koonin, E.V., Tatusov, R.L., Rudd, K.E. (1995) Sequence similarity analysis of *Escherichia coli* proteins: functional and evolutionary implication. *Proc. Natl Acad. Sci. USA*, **92**, 11921–11925.

81 Labedan, B. & Riley, M. (1995) Widespread protein sequence similarities: origins of *Escherichia coli* genes. *J. Bacteriol.* **177**, 1585–1588.

82 Labedan, B. & Riley, M. (1995) Gene products of *Escherichia coli*-sequence comparisons and common ancestries. *Mol. Biol. Evol.* **12**, 980–987.

83 Merkl, R., Kroger, M., Rice, P. & Fritz, H.J. (1992) Statistical evaluation and biological interpretation of nonrandom abundance in the *E. coli* K-12 genome of tetranucleotide and pentanucleotide sequences related to VSP DNA mismatch repair. *Nucleic Acids Res.* **20**, 1657–1662.

84 Burge, C., Campbell, A.M. & Karlin, S. (1992) Over-representation and under-representation of short oligonucleotides in DNA sequences. *Proc. Natl Acad. Sci USA* **89**, 1358–1362.

85 Médigue, C., Rouxel, T., Vigier, P. *et al.* (1991) Evidence for horizontal gene transfer in *Escherichia coli* speciation. *J. Mol. Biol.* **222**, 851–856.

86 Cardon, L.R., Burge, C., Schachtel, G.A. *et al.* (1993) Comparative DNA-sequence features in 2 long *Escherichia coli* contigs. *Nucleic Acids Res.* **21**, 3875–3884.

87 Karlin, S., Ladunga, I. & Blaisdell, B.E. (1994) Heterogeneity of genomes: measures and values. *Proc. Natl Acad. Sci. USA* **91**, 12837–12841.

88 Gutierrez, G., Casadesus, J., Oliver, J.L. & Marin, A. (1994) Compositional heterogeneity of the *Escherichia coli* genome: a role for VSP repair? *J. Mol. Evol.* **39**, 340–346.

89 Grantham, R., Gautier, C., Gouy, M. *et al.* (1980) Codon catalog usage and the genome hypothesis. *Nucleic Acids Res.* **8**, r49–r62.

90 Ikemura, T. (1981) Correlation between the abundance of *Escherichia coli* transfer RNAs and the occurrence of the respective codons in its protein genes. *J. Mol. Biol.* **146**, 1–21.

91 Bennetzen, J.L. & Hall, B.J. (1982) Codon selection in yeast. *J. Biol. Chem.* **257**, 3026–3031.

92 Blake, R.D. & Hinds, P. (1984) Analysis of the codons bias in *E. coli* sequences. *J. Biomol. Struct. Dynam.* **2**, 593–606.

93 Gouy, M. & Gautier, C. (1982) Codon usage in bacteria: correlation with gene expressivity. *Nucleic Acids Res.* **10**, 7055–7074.

94 Maaløe, O. & Kjeldgaard, N.O. (1966) *The Control of Macromolecular Synthesiz.* (W.A. Benjamin, New York & Amsterdam).

95 Moir, P.D., Spiegelberg, R., Oliver, I.R. *et al.* (1992) Proteins encoded by the *Escherichia coli* replication terminus region. *J. Bacteriol.* **174**, 2101–2110.

96 Masters, M. & Oliver, I.R. (1993) Lack of conservation of some TER region sequences. *J. Cell. Biochem.* **S17E**, 300.

97 Henson, J.M. & Kuempel, P.L. (1985) Deletion of the terminus region (340 kilobase pairs of DNA) from the chromosome of *Escherichia coli. Proc. Natl Acad. Sci. USA* **82**, 3766–3770.

98 Kaiser, K. & Murray, N.E. (1979) Physical characterization of the 'Rac prophage' in *Escherichia coli* K12. *Mol. Gen. Genet.* **175**, 159–174.

99 Espion, D., Kaiser, K. & Dambly-Chaudiere, C. (1983) A third defective lambdoid prophage of *Escherichia coli* K12 defined by the derivative, qin111. *J. Mol. Biol.* **170**, 611–633.

100 Bejar, S., Bouché, F. & Bouché, J.P. (1988) Cell division inhibition gene *dicB* is regulated by a locus similar to lambdoid bacteriophage immunity loci. *Mol. Gen. Genet.* **212**, 11–19.

101 Louarn, J.-M., Louarn, J., François, V. & Patte, J. (1991) Analysis and possible role of hyperrecombination in the termination region of the *Escherichia coli* chromosome. *J. Bacteriol.* **173**, 5097–5104.

102 Louarn, J., Cornet, F., François, V. *et al.* (1994) Hyperrecombination in the terminus region of the *Escherichia coli* chromosome—possible relation to nucleoid organization. *J. Bacteriol.* **176**, 7524–7531.

103 Brewer, B.J. (1988) When polymerases collide: replication and the transcriptional organization of the *E. coli* chromosome. *Cell* **53**, 679–686.

104 Brewer, B.J. (1990) Replication and the transcriptional organization of the *Escherichia coli* chromosome. In *The Bacterial Chromosome* (Drlica, K. & Riley, M., eds), 61–83 (American Society for Microbiology, Washington, DC).

105 Riley, M. & Krawiec, S. (1987) Genome organization. In Escherichia coli *and* Salmonella typhimurium: *Cellular and Molecular Biology* (Neidhardt, F.C. *et al.*, eds), 967–981 (American Society for Microbiology, Washington, DC).

106 Rebollo, J.E., François, V. & Louarn, J.M. (1988) Detection and possible role of two large nondivisible zones on the *Escherichia coli* chromosome. *Proc. Natl Acad. Sci. USA* **85**, 9391–9395.

107 François, V., Louarn, J., Patte, J. & Louarn, J.M. (1987) A system for in vivo selection of genomic rearrangements with predetermined endpoints in *Escherichia coli* using modified Tn*10* transposons. *Gene* **56**, 99–108.

108 François, V., Louarn, J., Patte, J. *et al.* (1990) Constraints in chromosomal inversions in *Escherichia coli* are not explained by replication pausing at inverted terminator-like sequences. *Mol. Microbiol.* **4**, 537–542.

109 Casse, F., Pascal, M.-C. & Chippaux, M. (1973)

Comparison between the chromosomal maps of *E. coli* and *S. typhimurium*: length of the inverted segment in the *trp* region. *Mol. Gen. Genet.* **124**, 91–97.
110 Riley, M. & Sanderson, K. (1990) Comparative genetics of *Escherichia coli* and *Salmonella typhimurium*. In *The Bacterial Chromosome* (Drlica, K. & Riley, M., eds), 85–95 (American Society for Microbiology, Washington, DC).
111 Hill, C.W. & Harnish, B.W. (1981) Inversions between ribosomal RNA genes of *Escherichia coli*. *Proc. Natl Acad. Sci. USA* **78**, 7069–7072.
112 Liu, S.L., Hessel, A. & Sanderson, K.E. (1993) Genomic mapping with I-Ceu I, an intron-encoded endonuclease specific for genes for ribosomal RNA, in *Salmonella* spp, *Escherichia coli*, and other bacteria. *Proc. Natl Acad. Sci. USA* **90**, 6874–6878.
113 Myers, R.S. & Stahl, F.W. (1994) Chi and the RecBCD enzyme of *Escherichia coli*. *Ann. Rev. Genet.* **28**, 49–70.
114 Berg, D.E. & Howe, M.M. (1989) *Mobile DNA* (American Society for Microbiology, Washington, DC).
115 Birkenbihl, R.P. & Vielmetter, W. (1989) Complete maps of IS1, IS2, IS3, IS4, IS5, IS30 and IS150 locations in *Escherichia coli* K12. *Mol. Gen. Genet.* **220**, 147–153.
116 Hill, C.W., Sandt, C.H. & Vlazny, D.A. (1994) RHS elements of *Escherichia coli*—a family of genetic composites each encoding a large mosaic protein. *Mol. Microbiol.* **12**, 865–871.
117 Anilioniz, A., Ostapchuk, P. & Riley, M. (1980) Identification of a second cryptic lambdoid prophage locus in the *E.coli* K12 chromosome. *Mol. Gen. Genet.* **180**, 479–481.
118 Higgins, C.F., Ames, G.F.-L., Barnes, W.M. *et al.* (1982) A novel intercistronic regulatory element of prokaryotic operons. *Nature* **298**, 760–762.
119 Gilson, E., Clément, J.-M., Brutlag, D. & Hofnung, M. (1984) A family of dispersed repetitive extragenic palindromic DNA sequences in *E. coli*. *EMBO J.* **3**, 1417–1421.
120 Hulton, C.S., Higgins, C.F. & Sharp, P.M. (1991) ERIC sequences: a novel family of repetitive elements in the genomes of *Escherichia coli*, *Salmonella typhimurium* and other enterobacteria. *Mol. Microbiol.* **5**, 825–834.
121 Sharples, G.J. & Lloyd, R.G. (1990) A novel repeated DNA sequence located in the intergenic regions of bacterial chromosomes. *Nucleic Acids Res.* **18**, 6503–6508.
122 Lupski, J.R. & Weinstock, G.M. (1992) Short, interspersed repetitive DNA sequences in prokaryotic genomes. *J. Bacteriol.* **174**, 4525–4529.
123 Oppenheim, A.B., Rudd, K.E., Mendelson, I. & Teff, D. (1993) Integration host factor binds to a unique class of complex repetitive extragenic DNA sequences in *Escherichia coli*. *Mol. Microbiol.* **10**, 113–122.
124 Versalovic, J., Koeuth, T. & Lupski, J.R. (1991) Distribution of repetitive DNA sequences in eubacteria and application to fingerprinting of bacterial genomes. *Nucleic Acids Res.* **19**, 6823–6831.
125 Woods, C.R., Versalovic, J., Koeuth, T. & Lupski, J.R. (1993) Whole cell repetitive element sequence based polymerase chain-reaction allows rapid assessment of clonal relationships of bacterial isolates. *J. Clin. Microbiol.* **31**, 1927–31.
126 Blaisdell, B.E., Rudd, K.E., Matin, A. & Karlin, S. (1993) Significant dispersed recurrent DNA-sequences in the *Escherichia coli* genome—several new groups. *J. Mol. Biol.* **229**, 833–848.
127 Daniels, D.L. & Blattner, F.R. (1987) Mapping using gene encyclopaedias. *Nature* **325**, 831–832.
128 Aiba, H., Baba, T., Hayashi, K. *et al.* (1996) A 570-kb DNA sequence of the *Escherichia coli* K-12 genome corresponding to the 28.0–40.7 min region on the linkage map. *DNA Res.* **3**, 363–377.
129 Church, G.M. & Kieffer-Higgins, S. (1988) Multiplex DNA sequencing. *Science* **240**, 185–188.
130 Swinbanks, D. (1994) Japan's *E. coli* genome project falling short. *Nature* **368**, 383.
131 Oshima, T., Aiba, H., Baba T. *et al.* (1996) A 718-kb DNA sequence of the *Escherichia coli* K-12 genome corresponding to the 12.7–28.0 min region on the linkage map. *DNA Res.* **3**, 137–155.
132 Oshima, T., Aiba, H., Baba T. *et al.* (1996) A 718-kb DNA sequence of the *Escherichia coli* K-12 genome corresponding to the 12.7–28.0 min region on the linkage map. *DNA Res.* **3**, (Suppl.), 211–123.
133 Danchin, A. (1995) Why sequence genomes? The *Escherichia coli* imbroglio. *Mol. Microbiol.* **18**, 371–376.
134 Burland, V., Daniels, D.L., Plunkett, III., G. & Blattner, F.R. (1993) Genome sequencing on both strands, the Janus Strategy. *Nucleic Acids Res.* **21**, 3385–3390.
135 Olson, C.H., Blattner, F.R. & Daniels, D.R. (1991) Simultaneous preparation of up to 768 single-stranded DNAs for use as templates in DNA sequencing. *DNA Methods* **3**, 27–32.
136 Staden, R. & McLachlan, A.D. (1982) Codon preference and its uses in identifying protein coding regions in long DNA sequences. *Nucleic Acids Res.* **10**, 141–156.
137 Gribskov, M., Devereux, J. & Burgess, R.R. (1984) The codon preference plot: graphic analysis of protein coding sequences and prediction of gene expression. *Nucleic Acids Res.* **12**, 539–549.
138 Borodovsky, M. & McIninch, J. (1993) GENMARK: Parallel gene recognition for both DNA strands. *Comput. Chem.* **17**, 123–134.
139 Fuchs, R. (1991) MacPattern: Protein pattern searching on the Apple Macintosh. *Comput. Applicat. Biosci.* **7**, 105–106.
140 Bairoch, A. (1992) PROSITE: a dictionary of sites and patterns in proteins. *Nucleic Acids Res.* **20**, 2013–2018.
141 Henikoff, S. & Henikoff, J.G. (1991) Automated assembly of protein blocks for database searching. *Nucleic Acids Res.* **19**, 6565–6572.
142 Kleckner, N., Bender, J. & Gottesman, S. (1991) Uses of transposons with emphasis on Tn10. *Meth. Enzymol.* **204**, 139–180.
143 Mahillon, J. & Kleckner, N. (1992) New IS10 transposition vectors based on a Gram-positive replication origin. *Gene* **116**, 69–74.
144 Colleaux, L., D'Auriol, L., Galibert, F. & Dujon, B. (1988) Recognition and cleavage site of the intron-encoded *omega* transposase. *Proc. Natl Acad. Sci. USA* **85**, 6022–6026.
145 Monteilhet, C., Perrin, A., Thierry, A. *et al.* (1990) Purification and characterization of the *in vitro*

activity of I-*SceI*, a novel and highly specific endonuclease encoded by a group I intron. *Nucleic Acids Res.* **18**, 1407–1413.
146 Rode, C.K., Obreque, V.H. & Bloch, C.A. (1995) New tools for integrated genetic and physical analyses of the *Escherichia coli* chromosome. *Gene* **166**, 1–9.
147 Wahl, R. & Kröger, M. (1995) ECDC—a totally integrated and interactively usuable genetic map of *Escherichia coli* K-12. *Microbiol. Res.* **150**, 7–61.
148 Medigue, C., Viari, A., Henaut, A. & Danchin, A. (1993) Colibri: a functional data base for the *Escherichia coli* genome. *Microbiol. Rev.* **57**, 623–654.
149 Karp, P. & Riley, M. (1993) Representations of metabolic knowledge. In *Proceedings of the First International Conference on Intelligent Systems in Molecular Biology* (Hunter, L., Searls, D. & Shavlik, J., eds), 207–215 (AAAI Press, Menlo Park, CA).
150 Bairoch, A. & Boeckmann, B. (1994) The SWISS-PROT protein sequence data bank: current status. *Nucleic Acids Res.* **22**, 3578–3580.
151 Lisser, S. & Margalit, M. (1993) Compilation of the *E. coli* mRNA promoter sequences. *Nucleic Acids Res.* **21**, 1507–1516.
152 d'Aubenton, C.Y., Brody, E. & Thermes, C. (1992) Prediction of Rho-independent *Escherichia coli* transcription terminators. *J. Mol. Biol.* **216**, 835–858.
153 Rudd, K.E. & Schneider, T. (1992) Compilation of *E. coli* DNA ribosome binding sites. In *A Short Course in Bacterial Genetics* (Miller, J.H., ed.), 17.19–17.45 (Cold Spring Harbor Laboratory Press, NY).
154 Fleischmann, R.D., Adams, M.D., White, O. *et al.* (1995) Whole-genome random sequencing and assembly of *Haemophilus influenzae* Rd. *Science* **269**, 496–512.
155 Fraser, C.M., Gocayne, J.D., White, O. *et al.* (1995) The minimal gene complement of *Mycoplasma genitalium*. *Science* **270**, 397–403.
156 Kunst, F., Vassarotti, A. & Danchin, A. (1995) Organization of the European *Bacillus subtilis* genome sequencing project. *Microbiology* **141**, 249–255.
157 Honore, N.T., Bergh, S., Chanteau, S. *et al.* (1993) Nucleotide sequence of the first cosmid from the *Mycobacterium leprae* genome project: structure and function of the Rif-Str regions. *Mol. Microbiol.* **7**, 207–214.
158 Kaneko, T., Tanaka, A., Sato, S. *et al.* (1995) Sequence analysis of the genome of the unicellular Cyanobacterium *Synechocystis* sp. strain PC68033. I. Sequence features in the 1 Mb region from map positions 64% to 92% of the genome. *DNA Res.* **2**, 153–166.
159 DOE launches bacterial genome analysis projects (1994) *ASM News* **60**, 530.
160 Holzman, D. (1996) DOE moves to second round of microbial genome sequencing. *ASM News* **62**, 8–9.
161 Cole, S.T. & Saint Girons, I. (1994) Bacterial genomics. *FEMS Microbiol. Rev.* **14**, 139–160.
162 Chaung, S.E., Daniels, D.L. & Blattner, F.R. (1993) Global regulation of gene expression in *Escherichia coli*. *J. Bacteriol.* **175**, 2026–2036.
163 Chaung, S.E. & Blattner, F.R. (1993) Characterization of 26 new heat shock genes of *Escherichia coli*. *J. Bacteriol.* **175**, 5242–5252.
164 McDaniel, T.K., Jarvis, K.G., Donnenberg, M.S. & Kaper, J.B. (1995) A genetic locus of enterocyte effacement conserved among diverse enterobacterial pathogens. *Proc. Natl Acad. Sci. USA* **92**, 1664–1668.
165 Campbell, A. & Botstein, D. (1983) Evolution of the lambdoid phages. In *Lambda II* (Hendrix, R.W., Roberts, J.W., Stahl, F.W. & Weisberg, R.A., eds), 365–380 (Cold Spring Harbor Laboratory, Cold Spring Harbor, NY).
166 Woese, C.R. (1994) There must be a prokaryote somewhere: microbiology's search for itself. *Microbiol. Rev.* **58**, 1–9.
167 Gibson, A.L., Wagner, L.M., Collins, F.S. & Oxender, D.L. (1991) A bacterial system for investigating transport effects of cystic fibrosis-associated mutations. *Science* **254**, 109–111.
168 Gibbs, T.W., Gill, D.R. & Salmond, G.P. (1992) Localized mutagenesis of the fts YEX operon: conditionally lethal missense substitutions in the FtsE cell division protein of *Escherichia coli* are similar to those found in the cystic fibrosis transmembrane conductance regulator protein (CFTR) of human patients. *Mol. Gen. Genet.* **234**, 121–128.
169 Elledge, S.J. & Walker, G.C. (1985) Phasmid vectors for identification of genes by complementation of *Escherichia coli* mutants. *J. Bacteriol.* **162**, 777–783.
170 Neidhardt, F.C. *et al* (Eds) (1996) Escherichia coli *and* Salmonella typhimimum: *Cellular and Molecular Biology* (American Society for Microbiology, Washington DC).
171 Berlyn, M., Low, K.B. & Rudd, K.E. (1996) Linkage Map of *Escherichia coli* K-12, Edition 9. In Escherichia coli *and* Salmonella typhimimum: *Cellular and Molecular Biology* (Neidhardt, F.C. *et al.*, eds) Chapter 109 (American Society for Microbiology, Washington DC).
172 Subramanian, P.S., Versalovic, J., McCabe, E.R.B. & Lupski, J.R. (1992) Rapid mapping of *Escherichia coli*: Tn5 insertion mutations by REP-Tn*5* PCR *PCR Meth. Appl.* **1**, 187–194.
173 Aiba, H., Baba, T., Hayashi, K. *et al.* (1996) A 570-kb DNA sequence of the *Escherichia coli* K-12 genome corresponding to the 28.0–40.7 min region on the linkage map. *DNA Res.* (Suppl.) **3**, 435–440.
174 Itoh, T., Aiba, H., Baba, T. *et al.*(1996) A 460-kb DNA sequence of the *Escherichia coli* K-12 genome corresponding to the 40.7–50.1 min region on the linkage map. *DNA Res.* **3**, 379–392.
175 Itoh, T., Aiba, H., Baba, T. *et al.* (1996) A 460-kb DNA sequence of the *Escherichia coli* K-12 genome corresponding to the 40.7–50.1 min region on the linkage map. *DNA Res.* (Suppl.) **3**, 441–445.
176 Yamamoto, Y., Aiba, H., Baba, T. *et al.* (1997) Construction of a contiguous 874 kb sequence of the Escherichia coli K-12 genome corresponding to 50.0–68.8 min region on the linkage map and analysis of its sequence features *DNA Res.* **4**, 91–113.
177 Bloch, C.A., Rode, C.K., Obreque, V.H. & Mahillon, J. (1996) Purification of *Escherichia coli* chromosomal segments without cloning. *Biochem. Biophys. Res. Commun.* **223**, 104–111.
178 Bult, C.J., White, O., Olsen, G.J. *et al.* (1996) Complete genome sequence of the methanogenic Archaeon, *Methanococcus jannaschii*. *Science* **273**, 1058–1073.

179 Kaneko, T., Sato, S., Kotani, H. *et al.* (1996) Sequence analysis of the genome of the unicellular cyanobacterium *Synechocystis* sp. Strain PCC6803. II. Sequence determination of the entire genome and assignment of potential protein-coding regions. *DNA Res.* **3**, 109–136.
180 Himmelreich, R., Hilbert, H., Plagens, H., Pirkl, E., Li, B.-C. & Herrmann, R. (1996) Complete sequence analysis of the genome of the bacterium *Mycoplasma pneumoniae*. *Nucleic Acids Res.* **24**, 4420–4449.
181 Goffeau, A., Barrell, B.G., Bussey, H. *et al.* (1996) Life with 6000 genes. *Science* **274**, 563–567.
182 Blattner, F.R., Plunkett III, G., Bloch, C.A. *et al.* (1997) The complete genome sequence of *Escherichia coli* K-12. *Science* **277**, 1453–1461.
183 Fox, J.L. (1997) Whole E. coli! Microbial sequencing in log-phase growth. *ASM News* **63**, 187–192.

Chapter 32 Problems in plant genome analysis

Prem P. Jauhar

USDA-ARS, Northern Crop Science Laboratory, Fargo, North Dakota 58105–5677, USA

32.1 Introduction

Knowledge of genome relationships within and between various plant taxa is of great use to many scientists, including cytogeneticists, plant breeders, taxonomists, evolutionists, molecular biologists and biotechnologists. Genome analysis provides useful information on chromosome pairing relationships and, hence, on the possibilities of transferring desirable traits between different species of higher plants. An understanding of genomic affinities helps to formulate effective breeding programmes designed to transfer desirable genes from wild relatives or primitive varieties of crop plants into otherwise superior cultivars. Many plants, and crop plants in particular, are polyploids, often with one or more sets of chromosomes derived from different sources; this means that the questions posed in plant genome analysis, and some of the problems encountered, are rather different from those pertaining to the analysis of animal genomes, especially those of humans and other mammals.

Before discussing problems encountered in plant genome analysis, we must define a genome and the specific sense in which it will be used in this chapter. In general biological terms, a genome refers to all genetic material or the complete gene complement contained in a set of chromosomes in eukaryotes, or in the equivalent in prokaryotes. Thus, a genome consists of a single chromosome in bacteria, or a DNA or RNA molecule in viruses. In higher organisms, a genome represents one complete haploid set of chromosomes. Eukaryotic genomes, and many plant genomes in particular, are characterized by the occurrence of highly repetitive noncoding sequences. The evolution of higher plant species has often been accompanied by quantitative changes in this noncoding DNA fraction [1].

For this chapter, a plant genome will be defined as a complete basic set of chromosomes (denoted by the small letter x) inherited as a unit from one parent. Thus, in diploid higher plants, a genome refers to only one complete haploid (monoploid) set of chromosomes, called the haplome [2]. A diploid species, for example, diploid wheat (*Triticum boeoticum* Boiss.) has a double dose of a single genome designated by the capital letter A (so the diploid has the constitution AA). A polyploid plant, on the other hand, has several different genomes; for example, hexaploid common wheat (*T. aestivum* L.) has three different genomes, called A, B and D, each present in two copies in the somatic cells of the hexaploid (AABBDD). Thus, wheat is a heterogenomic polyploid or allopolyploid. When a plant (e.g. alfalfa) has more than two doses of the same genome, it is called a homogenomic polyploid or an autopolyploid.

Traditionally, in genome analysis a genome refers to the nuclear genome, unless stated otherwise. However, mitochondrial and plastid DNA (e.g. chloroplast DNA) have been used in phylogenetic investigations [3–5]. Chloroplast DNA analysis showed that the chloroplast genome of the primitive wheat *T. timopheevii* is equivalent to that of the wild grass *Aegilops speltoides* [6].

Genome analysis attempts to determine the genomic constitution of polyploid species and elucidate genomic affinities among plant taxa. Various classical cytogenetic and biochemical techniques are used in studying genome relationships; these have more recently been supplemented with molecular methods. This chapter discusses some of the methods of genome analysis and the problems encountered in these studies, with the emphasis on the traditional cytogenetic approaches. I will discuss problems with particular reference to the cytogenetic analysis of relationships within the grasses, the large and important group of monocotyledons to which many of our staple crop plants belong. Progress in mapping and analysing the genome of rice (*Oryza sativa*) by molecular methods is described in Chapter 34, while the mapping of the genome of the model dicot *Arabidopsis thaliana* is described in Chapter 33.

Several techniques of genome analysis in plants have been employed over the years. These include:

- crossability: reproductive isolation between taxa;
- karyotypic analysis on conventionally stained somatic or pachytene chromosomes;
- karyotypic analysis on Giemsa-banded somatic chromosomes;
- chromosome pairing in hybrids at different ploidy levels;
- chromosome pairing in hybrids in the presence of the *Ph1* gene of wheat;
- discrimination of pairing between parental chromosomes;
- use of mathematical models on meiotic pairing data;
- protein electrophoresis;
- molecular tools of genome analysis.

Each of these techniques has certain inherent advantages and disadvantages. The relative usefulness of a technique depends upon the degree to which it measures, directly or indirectly, the similarity of the nuclear DNA of the related species. Genome analysis can be subjective at times and has certain limitations, as do most other biosystematic criteria. It is nevertheless one of the most useful methods for revealing phyletic relatedness. The merits and limitations of each of the criteria of genome analysis are discussed below.

32.2 Crossability: reproductive isolation between taxa

The phylogenetic relationship between two plant taxa is often indicated by the relative ease with which they can be hybridized. Crossability between two taxa and fertility of the resultant hybrid are useful criteria for elucidating genome relationships. These measures have long been used for estimating the degree of relationship between parental species [7]. When two taxa produce completely sterile hybrids and, hence, are reproductively isolated, it is probable that the taxa are genomically distinct. This technique has its own limitations, however.

1 Barriers to crossing may limit the production of hybrids. Crossability may be under genetic control. For example, crossability of wheat (*T. aestivum*) and rye (*Secale cereale*) is controlled by two recessive genes, *kr1* and *kr2*, in wheat [8]. Therefore, the success or failure of hybridization of two taxa may be influenced by the particular genotypes used. The crossability genes may act differently on different crosses; *kr1* and *kr2* genes prevent hybridization of wheat with its close relative rye but not with the remotely related maize [9]. The degree of fertility of a hybrid may also be influenced by the genotypes of the parents [10].

2 A cross may be successful only in one direction and its reciprocal may invariably fail, indicating the influence of cytoplasm on crossability, for example, in the legume genus *Glycine* [11]. Therefore, a hybrid studied only in one direction may produce erroneous results.

3 In some cases, two closely related taxa may cross very easily but the hybrids are completely sterile. For example, geographically isolated ecotypes of hexaploid tall fescue (*Festuca arundinacea* Schreb.) cross very readily but produce sterile hybrids. Interaction between parental genotypes results in the inactivation of the regulatory mechanism that controls diploid-like chromosome pairing, resulting in high homoeologous pairing and hence in sterility [12]. (Genetically and evolutionarily related chromosomes from different genomes within a heterogenomic polyploid or from related species are known as homoeologous chromosomes, which are capable of pairing among themselves.) This is a novel mechanism for the creation of reproductive isolation barriers between infraspecific categories within a species. Sterility in such intervarietal crosses obviously cannot be used for assessing genome relationships.

32.3 Karyotypic analysis on conventionally stained somatic or pachytene chromosomes

In simple terms, karyotype is defined as the morphology of chromosomes. In plants, the karyotype is generally studied at somatic metaphase in actively dividing root tips (or sometimes in shoot tips), in pollen mitosis, or at pachytene of meiosis. Chromosome number, total chromosome lengths, arm ratios, secondary constrictions and satellited regions or nucleolar organizer regions (NORs) constitute important parameters for karyotypic analysis. They have aided genome analysis and, hence, phylogenetic investigations.

32.3.1 Uses of karyotype analysis

Karyotype analysis has been used to study genome relationships in several plant groups, particularly the grasses. Avdulov [13] was among the first to use cytological features to establish evolutionary relationships among species and genera. Using cytological criteria, he attempted a phylogenetic subdivision of the grasses and his publication entitled *Cytotaxonomic Investigations in the Family Gramineae* marked the beginning of a new era in grass classification. Remarkably, Avdulov's classification was borne out by studies based on anatomy and geographical distribution.

The early studies on karyotype analysis used reconstructions prepared from serial sections of root tips and anthers and were therefore tedious and time consuming. However, the advent of squash techniques speeded up work on chromosome karyotyping, which has been carried out for numerous plant groups including the wheat group. Thus, based on the similarity of the satellite chromosomes of the diploid species *Aegilops speltoides* (Tausch) to those of polyploid wheats, Riley *et al.* [14] inferred that this diploid was the source of the B genome of polyploid wheats. This inference has since been challenged because karyotypic and pairing data show another diploid in the wheat group, *T. searsii* Feldman & Kislev, as a more likely source of the B genome [15]. This point remains controversial. Nevertheless, the B genome of polyploid wheats is very similar to the genome of *Ae. speltoides*.

32.3.2 Problems of using karyotypic features in genome analysis

Karyotypic data obtained from conventionally stained (e.g. acetocarmine, acetoorcein, or Feulgen-stained) condensed chromosomes have a limited usefulness for assessing genomic or phylogenetic relationships. There are inherent inaccuracies in measurements taken on condensed chromosomes at mitotic metaphase. Computer-aided karyotypic analyses [16] may facilitate precise measurements of chromosomes. Fukui [17] developed a chromosome image analysing system, CHIAS, especially for plant chromosomes. CHIAS offers two main advantages over conventional karyotyping techniques.

1 It enables the analysis of a large number of chromosome spreads in a relatively short time; it is possible, for example, to analyse more than 250 barley metaphase plates within a week by semiautomatic processing of the chromosome images [18].
2 Results obtained are accurate and reproducible.

The fact that somatic chromosomes from related species are similar in total size or arm ratio does not necessarily mean that they are similar in gene content, however. Similar genomes can certainly have dissimilar karyotypes and vice versa. Moreover, it is not generally easy to identify unequivocally different chromosomes or their homoeologous partners. Somatic karyotyping can therefore sometimes lead to erroneous conclusions [19–21].

The satellited (SAT) chromosomes are relatively easy to identify, and serve as useful karyotypic landmarks. Chennaveeraiah [22] found that chromosome identification in the wheat group, based on chromosome size and arm ratio, was difficult, although in many cases the SAT chromosomes could be identified unambiguously. However, in interspecific or intergeneric hybrids, amphiplasty can mask the visual expression of the NOR(s) of one of the parental species [23]. (Amphiplasty is the phenomenon of the masking of the NOR(s) of one species by those of the other in their hybrids.) The usefulness of karyotypic features in elucidating genomic relationships is therefore severely limited.

Mitotic karyotyping is of particularly limited value in species with small chromosomes, where it may be difficult even to sort out homoeologous members of the complement. Although this problem may be partly obviated by karyotyping at pollen mitosis (where the haploid complement is represented) or even at somatic metaphase in root tips of haploid plants, mitotic karyotyping has a limited usefulness. The pachytene stage of meiosis could prove relatively more useful for precise karyotypic analysis. In addition to accurate measurements of chromosomes, it permits a precise study of chromomere patterns and other finer details of the already paired chromosomes. Thus, one has to measure only the gametic (n) number of chromosomes. Generally, plant taxa with low chromosome number are suitable for pachytene karyotyping. Using this technique, Singh and Hymowitz [24] were nevertheless able to study genome relationships between species of *Glycine* with high chromosome number.

32.4 Karyotypic analysis on Giemsa-banded somatic chromosomes

Progress in different areas of cytogenetics depends on our ability to identify not only individual chromosomes but also parts of chromosomes. Some plants have similar chromosomes which cannot be distinguished by conventional staining. The advent of Giemsa banding techniques for mammalian chromosomes [25–27] provided cytogeneticists with a powerful tool for karyotypic analysis and chromosome mapping (see Chapter 7). These banding techniques differentially stain chromosome regions rich in constitutive heterochromatin that contains a large amount of highly repetitive DNA. This produces a unique pattern of dark and light bands along the length of a chromosome, which can be used in chromosome identification. Some of these banding techniques have been successfully used for karyotyping plant chromosomes [28–35].

32.4.1 Advantages of chromosome banding

Both C- and N-banding techniques have been used to identify individual chromosomes of numerous plant species. Giemsa N-banding was originally developed for differential staining of the NOR in both plant and animal chromosomes. However, this method was also found useful for detecting constitutive heterochromatin and hence for identification of specific chromosomes in cereals [32]. Giemsa C-banding gives the best resolution of chromosome-specific bands and has allowed the identification of all the 21 (A-, B- and D-genome) chromosome pairs and most chromosome arms in hexaploid wheat [35] and chromosomes of wild grass species in their hybrids with wheat (Fig. 32.1). The value of chromosome banding in genome analysis has been discussed by Friebe and Gill [36]. On the basis of the banding pattern, a particular genome in a polyploid crop plant may be traced back to a putative diploid progenitor. Thus, C-banding in *Ae. squarrosa* L. (= *T. tauschii* (Coss.) Schmalh) gives a pattern very similar to that of the D-genome chromosomes in hexaploid wheat [37]. To get good resolution of major and minor bands, C- and N-banding analysis should be done on relatively less condensed (prometaphase) chromosomes (see Chapter 7).

Chromosome banding has also proved useful in

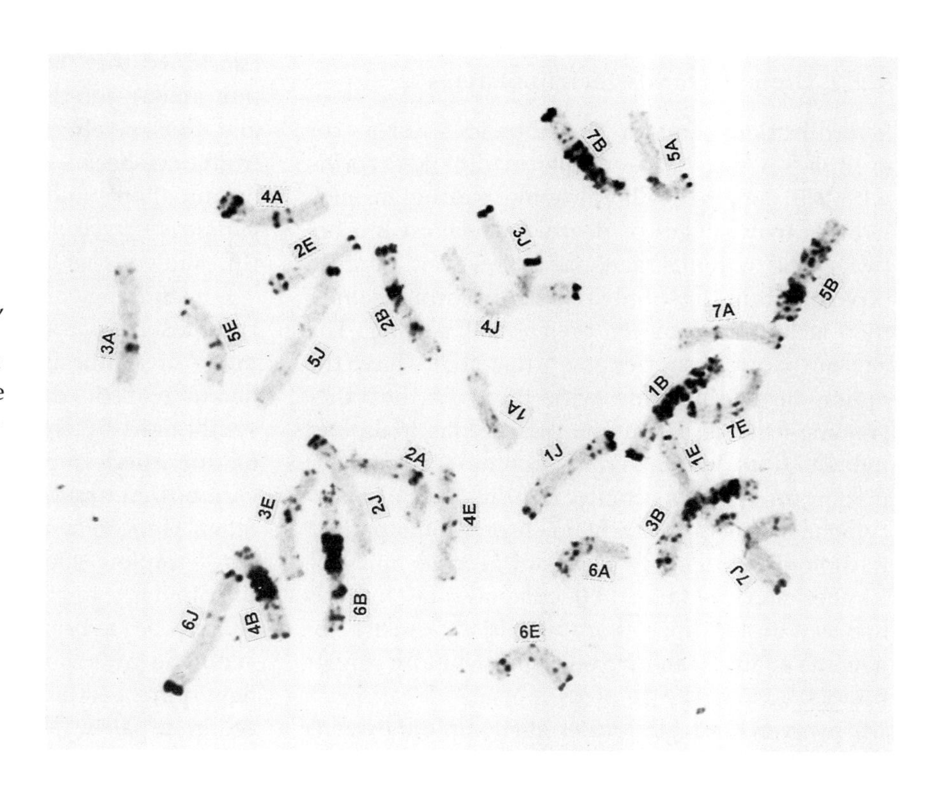

Fig. 32.1 C-banded somatic chromosomes of trigeneric hybrids ($2n = 4x = 28$; ABJE genomes). The hybrids were made between durum wheat (*T. turgidum*; $2n = 4x = 28$; AABB), *Thinopyrum bessarabicum* ($2n = 2x = 14$; JJ), and *Lophopyrum elongatum* ($2n = 2x = 14$; EE). Note that each of the 28 chromosomes can be identified on the basis of its diagnostic banding pattern. For the sake of convenience, chromosomes of each genome are numbered 1–7. The numbering of chromosomes of the J and E genomes does not reflect their homoeology with corresponding chromosomes of the A and B genomes of durum wheat. From ref. 64.

identifying chromosome segments involved in translocations. Using C-banding, Gill and Kimber [38] not only demonstrated translocations between two different wheat chromosomes but also between wheat and rye chromosomes. Similarly, reciprocal translocations involving chromosome 6A and 1G were revealed in *T. timopheevii* ($2n = 4x = 28$; AAGG) [39]. Thus, C-banding could also be useful in studying restructured genomes.

32.4.2 Problems in banding analysis

The usefulness of a banding technique depends upon the uniqueness of the banding patterns created. There are several variables in a C-banding protocol and it may not always be possible to obtain consistent and repeatable results. The interpretation of banding patterns is sometimes difficult, especially when dealing with small structural changes. Moreover, the banding pattern may vary among members of the same species [37,40,41]. Such a banding polymorphism could complicate banding and hence genome analysis in a particular plant group. Variation in C-banding patterns between homologous chromosomes within and between different plants and cultivars has also been reported in cross-pollinating rye (*S. cereale* L.) [42–44] and in barley [45]. Therefore, a 'chromosomal passport' based on a typical banded karyotype cannot always be prepared for a particular plant species. Such problems will tend to limit the utility of banding techniques.

Although barley (*Hordeum vulgare* L.) has a low chromosome number ($2n = 14$) and relatively large chromosomes compared to other plant species, problems have been encountered in its karyotypic analysis. It has a symmetrical karyotype, the morphology of the chromosomes being similar except for chromosomes 5, 6, and 7. It was therefore difficult to distinguish every chromosome until the C-banding technique [46] was used. However, even then, differences of opinion persisted among cytogeneticists. Linde-Laursen's [46] assignment of the short and long arm of chromosome 1 based on C-banding was revised by Noda and Kasha [47], and the revision accepted by Linde-Laursen [45]. However, when Singh and Tsuchiya [48] studied the chromosomes of barley by Giemsa N-banding they adopted Linde-Laursen's original short and long arm assignment for chromosome 1. This example well illustrates the difficulty encountered in plant chromosome studies.

32.5 Chromosome pairing in hybrids at different ploidy levels

The principal criterion for assessing genomic affinities between species has been and still is the study of chromosome pairing in their hybrids at different ploidy levels. The degree of chiasmate pairing between parental chromosomes is generally a reliable indicator of the degree of genomic relatedness. However, the situations under which chromosome pairing occurs must also be examined.

32.5.1 Diploid hybrids: their limitations

Hybridization between two diploid species—for example, AA and BB—results in diploid hybrids with AB genomes. Chromosome pairing in such hybrids may suggest a degree of relationship between the parental genomes. However, because diploid hybrids have only two sets of chromosomes, they lack conditions for preferential pairing. The chromosomes of one genome—that is, A—have the option of pairing only with those of the other genome—that is, B. Chromosome pairing in diploid hybrids does not therefore provide sound information on genomic affinity. For this reason, many cytogeneticists do not rely on chromosome pairing in diploid hybrids as a means of genome analysis. Kimber and Feldman [49] contend that hybrids between diploid species are 'essentially useless' for genome analysis, and we take an essentially similar position [21,50,51].

However, the degree of sterility in diploid hybrids may yield useful information on the relationship of the constituent genomes. If the hybrids are completely sterile, the two genomes in them may be considered to be differentiated.

32.5.2 Triploid hybrids: a realistic test of intergenomic affinity

To obviate the limitations of chromosome pairing encountered in diploid hybrids, autoallotriploid hybrids (e.g. AAB or ABB) derived by crossing a synthetic autotetraploid of one species (e.g. AAAA or BBBB) with a diploid of the other (e.g. BB or AA) should be studied. (An autoallotriploid hybrid has two doses of one genome and one of the other, e.g. AAB or ABB. Thus, AAB is auto with respect to genome A, and allo with reference to the two genomes.) The rationale of using such autoallotriploids is to create conditions for preferential pairing among chromosomes of different genomes to ascertain their relative affinities. If the chromosomes of the duplicated genome pair preferentially as bivalents and the chromosomes of the single genome remain largely unpaired, the two genomes may be considered to be distinct. However, if chromosomes of the single genome offer synaptic competition to the homologous chromosomes of the duplicated genome, resulting in a high frequency of trivalents, the two genomes could be considered to be closely related. For example, on the basis of chromosome pairing in various combinations of autoallotriploids among Italian ryegrass (*Lolium multiflorum* Lam.), perennial ryegrass (*L. perenne* L.), and meadow fescue (*Festuca pratensis* Huds.), we concluded that there is little structural differentiation among the chromosomes of the three species and, thus, no effective isolation barrier to gene flow from one species to another [52]. However, studies on autoallotriploids (JJE) between the grasses *Thinopyrum bessarabicum* (JJ) and *Lophopyrum elongatum* (EE) showed that J and E are distinct genomes [50].

For studying genome relationships on the basis of chromosome pairing in autoallotriploid hybrids, the mode of synthesis of these hybrids is critical. To obtain realistic data, autoallotriploids should be synthesized by crossing a synthetic autotetraploid of one species, produced through the use of a chromosome doubling agent, with the diploid of the other. Thus, to synthesize JJE hybrids, synthetic JJJJ autotetraploid should be crossed with the diploid EE. Similarly, to obtain JEE hybrids, synthetic EEEE should be hybridized with the diploid JJ. The synthetic autoallotriploids thus synthesized will have pure genomes, unadulterated by the homoeologous pairing and recombination that are characteristic of naturally occurring polyploids. Crosses of diploid species with naturally occurring tetraploids have been studied [53] but this approach may have problems since the naturally occurring tetraploids may have undergone homoeologous recombinations between the constituent genomes, thus altering the pattern of chromosome pairing in hybrids with their ancestral diploids. Naturally occurring tetraploids may have developed a genetic control of chromosome pairing [12,54,55] and therefore the pairing patterns in synthetic hybrids involving such tetraploids may be different from those in hybrids derived by crossing a diploid species with a synthetic autotetraploid.

32.5.3 Amphidiploids: opportunity for preferential pairing

Amphidiploids ($4x$) are obtained by chromosome doubling of diploid F1 interspecific or intergeneric hybrids ($2x$). The rationale for studying chromosome pairing in amphidiploids is the same as for autoallotriploids described above. If each of the two genomes is represented twice (for example, AABB), the genomes should maintain their meiotic integrity and pairing should be limited to homologous partners if the genomes are distinct. In other words, preferential pairing will lead to diploid-like meiosis and, hence, high fertility. However, if the genomes in question are closely related, the amphidiploids will form quadrivalents in addition to bivalents, and have varying degrees of sterility.

The fertility of an amphidiploid depends upon the degree of differentiation of the constituent genomes, which is evidenced by the degree of pairing in the parental diploid hybrid. There is generally a close inverse relationship between $2x$ pairing and $4x$ fertility [56], and between $2x$ chiasma frequency and $4x$ bivalent frequency [57]. In other words, the greater the divergence between the two genomes making up a hybrid, the greater the fertility in the derived amphidiploids. Thus, we have observed that a poor chiasma frequency in the diploid hybrid (JE) *Thinopyrum bessarabicum* × *Lophopyrum elongatum*, yielded a preponderance of bivalents in its amphidiploids (JJEE), which therefore were meiotically and reproductively stable and fertile [21,50]. These studies show the divergence of the J and E genomes.

32.5.4 Other limitations of chromosome pairing

In addition to the limitations of chromosome pairing discussed above, one may encounter some other problems.

32.5.4.1 Effect of environmental factors

The amount of pairing and chiasma frequency can be influenced by environmental factors such as temperature [58]. Therefore, hybrids being studied or compared should be grown under normal, uniform conditions. If the hybrids being compared are grown under very different environments, chromosome pairing data obtained from them may lead to erroneous conclusions on genomic affinities.

32.5.4.2 Effect of genotype

Chromosome pairing promoters and inhibitors [54] can influence the quantity of pairing and thus alter the pattern of affinity among genomes combined in hybrids. The amount of pairing in hybrids may be influenced by genotypes of the parents. Genotypes of some wild grasses when crossed with wheat suppress, in the hybrids, the activity of the homoeologous pairing suppressor gene, *Ph1*, and thus bring about unexpectedly high homoeologous pairing [59,60] (see also Section 32.6.2). Such hybrids do not offer sound conditions for interpreting genomic relationships.

32.5.4.3 Asynaptic or desynaptic phenomena

Asynapsis (total lack of chromosome pairing) and desynapsis (precocious separation of initially paired chromosomes), although rare, are generally under genetic control. Several major genes which, in the homozygous recessive condition, bring about failure or disruption of pairing are known [61]. These pairing variants may also be caused by environmental factors. They may prove to be impediments to genome analysis because lack of pairing in the asynaptic/desynaptic hybrids may be misinterpreted as being due to lack of pairing affinity between the parental chromosomes. An experienced cytogeneticist should, however, not be misled or biased by such rare situations because they can be recognized at diakinesis or early metaphase I of meiosis.

32.5.4.4 Possible cytoplasmic effects on chromosome pairing

There are some indications of cytoplasmic effects on chromosome pairing in hybrids (ref. 55 and G. Peterson and O. Seberg, personal communication 1994). In such cases, the amount of pairing in a hybrid would depend upon whether a particular genotype is used as a female or male parent. Failure to study reciprocal hybrids may produce erroneous results.

Despite all these limitations, genome analysis by studying chromosome pairing has yielded extremely useful information. In the tribe Triticeae, for example, this technique has been successfully employed for almost six decades and has revealed phylogenetic relationships that have been borne out by other criteria.

32.6 Chromosome pairing in hybrids in the presence of *Ph1*

Common wheat (*T. aestivum*) is an allohexaploid with three genomes, AA, BB and DD, derived from its three wild ancestors. Although the corresponding chromosomes of the three genomes are closely related, a gene called *Ph1* (located in the long arm of chromosome 5B) suppresses pairing between chromosomes of different genomes [54,62]. In other words, *Ph1* suppresses homoeologous pairing so that only homologous chromosomes can pair. This gene also suppresses pairing between homoeologous chromosomes of alien genomes introduced into wheat.

32.6.1 Advantages of *Ph1*-regulated chromosome pairing

Pairing or lack of pairing between chromosomes of two genomes in the presence of *Ph1* in the wheat background would provide a crucial test of their relationship. If chromosomes of two putatively related genomes pair with each other in the presence of *Ph1*, thereby passing its limits of discrimination, then they may be considered to be essentially homologous. Thus, the presence of *Ph1* provides a rigorous test of homology and may facilitate assessment of genome relationships. This approach has been applied to the study of the relationship between the J genome of *Thinopyrum bessarabicum* and the E genome of *Lophopyrum elongatum*. Forster and Miller [63] studied chromosome pairing in the AABBDDJE hybrids derived by crossing wheat–*Th. bessarabicum* amphidiploids (AABBDDJJ) with wheat–*L. elongatum* amphidiploids (AABBDDEE). The AABBDDJE hybrids showed mostly 21 bivalents (II) + 14 univalents (I). Clearly, the J- and E-genome chromosomes do not pair in the presence of *Ph1*. Essentially similar results were obtained in trigeneric hybrids (ABJE) involving durum wheat (*T. turgidum*, AABB), *Th. bessarabicum* (JJ), and *L. elongatum* (EE) [21,64]. It was therefore concluded that the chromosomes of J and E genomes are homoeologous at best [20,21].

32.6.2 Limitations of *Ph1*-regulated chromosome pairing

A limitation of the *Ph1*-regulated pairing is that sometimes, in the presence of more than one dose of *Ph1*, even homologous chromosomes present in more than two copies may pair as bivalents. This is exemplified by the formation of 22 bivalents in a large proportion of pollen mother cells (PMCs) in wheat tetrasomics ($2n=44$), which have four copies of one chromosome. Because of the genetic control of diploid-like pairing, hexaploid tall fescue ($2n=6x=42$; AABBCC) forms 21 II [65]. However, in the colchicine-induced dodecaploid tall fescue ($2n=12x=84$; AAAABBBBCCCC), which has four doses of each of the chromosomes, a preponderance of bivalents is observed instead of the expected quadrivalents [66]. It seems that the pairing control genes in four doses not only suppress homoeologous pairing, but also bring about bivalent pairing of the homologous sets. These are, nevertheless, exceptional situations and should not deter one from using the *Ph*-regulated pairing (e.g. in the wheat background) to decipher genome relationships.

Another limitation of the use of this tool is that the activity of *Ph1* can be suppressed in the hybrids by the genotype of the parental species. *Ae. speltoides*, for example, is known to suppress the activity of *Ph1* in its hybrids with wheat, resulting in high homoeologous pairing [67]. Similarly, certain genotypes of *Agropyron cristatum* (L.) Gaertner switch off or reduce the activity of *Ph1* in hybrids with wheat and thus bring about pairing among homoeologous chromosomes [59,60]. Chromosome pairing in such hybrids may not be used for deducing genome relationships because even relatively less related chromosomes pair in the absence of *Ph1*.

32.7 Discrimination of pairing between parental chromosomes

Chromosome pairing in hybrids may be due to autosyndesis, that is pairing within a parental complement, and/or allosyndesis, that is pairing between the parental chromosomes. Therefore, for assessing genome relationships, the nature of chromosome pairing should also be analysed because only the degree of pairing between parental chromosomes is indicative of the degree of relationship between parental species. The following criteria may help study specific pairing.

32.7.1 Size difference between parental chromosomes

A distinct size difference between chromosomes of parental diploid species can help estimate, in their diploid hybrids, the degree of autosyndetic (intragenomic) and allosyndetic (intergenomic) pairing. The intergenomic bivalents are generally heteromorphic, whereas intragenomic bivalents are relatively homomorphic, either large or small depending upon the size of parental chromosomes [68,69]. We [68] studied both auto- and allosyndetic pairing in interspecific hybrids between *Pennisetum typhoides* Stapf et Hubb. ($2n=14$ large chromosomes) and *P. purpureum* Schum. ($2n=28$ relatively smaller chromosomes) and deduced genomic relationships. The nature of pairing was also studied in hybrids between *P. typhoides* and *P. orientale* Rich. ($2n=18$ relatively smaller chromosomes) [69]. A certain degree of error may be involved in discriminating the parental chromo-somes in such hybrids.

32.7.2 Use of marked chromosomes

Cytologically marked and hence easily distinguishable chromosomes may be used to study pairing between specific chromosomes. Thus, a telocentric chromosome is readily recognizable in both somatic and meiotic plates and provides a ready marker to study pairing behaviour of particular chromosomes. Using double-double telocentric wheat, in which two homoeologous chromosomes were marked by their two separate telocentric arms, Alonso and Kimber [70] studied intergenomic chromosome pairing in wheat × *Aegilops* hybrids (ABDS). They found that the pairing frequencies of the B and S genomes were similar to those of the A and D.

This strategy of studying pairing between specific chromosomes may have certain limitations. Because a telocentric chromosome lacks one arm, the pairing potential of the entire chromosome will seldom be realized because recombination is strongly reduced near the centromere [71].

32.7.3 Meiotic chromosome banding

Distinct banding patterns of parental chromosomes are very helpful in studying pairing specificity between them in the hybrids of allopolyploid parental species. Meiotic C-banding of wheat and rye hybrids suggested pairing between chromosomes of the A and D genomes [72–74]. We have found that that N-banding analysis of meiosis in *Ph1*-deficient wheat euhaploids ($2n = 3x = 21$; ABD) offers an excellent means of elucidating pairing relationships among chromosomes of the A, B and D genomes without the competitive (or genetic) interference of an alien genome [62]. Chromosome arms involved in rod and ring bivalents were identified. Various chromosomes involved in bivalent and trivalent configurations were also identified. We showed that almost 80% of the metaphase I associations were between chromosomes of A and D genomes, indicating that these genomes are more closely related to each other than either one is to the B genome.

32.8 Application of mathematical models to meiotic pairing data

Giemsa banding of meiotic chromosomes may prove very useful in assessing the degree of pairing relationship between chromosomes of specific genomes. However, in the absence of banded or otherwise marked chromosomes, an acceptable—although relatively less reliable—substitute is the fitting of numerical models to the observed meiotic associations. The relative pairing affinities among chromosomes of the constituent genomes in polyploid hybrids can be assessed by applying appropriate mathematical models [75–80].

These theoretical models for deriving estimates of relative pairing affinities between chromosomes of different genomes are based on several assumptions. The models of Kimber and Alonso [77], for example, make a number of stringent assumptions about the chromosomes and the pattern of genomic affinity:

1 the long and short arms have equal chiasmate binding and have the same pattern of synapsis;

2 there are only two levels of genomic affinity, linked, respectively, to the proportioning constants x for the closest relative affinity and y for the remaining relative affinities.

Before applying a specific model, it is necessary to check if the assumptions of the model are met so that one does not arrive at erroneous conclusions. For example, pronounced differences in chiasma frequency between chromosome arms make the models of Kimber and associates less suitable.

32.9 Protein electrophoresis

Alcohol-soluble prolamins constitute a major storage protein fraction in cereals, such as the gliadins in wheat, avenins in oats, zeins in maize, and hordeins in barley. Electrophoretic similarity of seed gliadin proteins, for example, provides a reliable measure of species affinity [81–83], because the banding patterns of these proteins are not influenced by the environment [84], although their quantity is easily affected. Konarev *et al.* [85] identified two major groups of gliadin proteins, based on their relative mobilities, and these proteins have been used to assess species relationships. We have found, for example, that *Thinopyrum bessarabicum* and *Lophopyrum elongatum* can be identified on the basis of their distinct gliadin profiles. Seed-protein profiles have also been employed for assessing genome relationships between cultivated rice species and their wild progenitors [86].

The gliadins or other seed proteins may be limited in their ability to differentiate among genomes because only a small portion of the genome is analysed compared to the other techniques discussed above. Another limitation of this tool is the availability of seed. Moreover, the protocol for gliadin electrophoresis has several variables and may not always give consistent results.

Isoenzymes or isozymes—multiple molecular forms of enzymes—have been used to study phylogeny, genomic relationships, and synteny relationships among loci in related genomes in several plant groups [87–89]. They have, for example, played a valuable role in determining homoeologous relationships among the chromosomes of the tribe Triticeae. Although isozymes can serve as excellent markers for particular chromosomes, other molecular markers such as restriction fragment length polymorphisms (RFLPs), being essentially unlimited in number, offer more advantages as markers. Therefore, isozyme markers will probably have a limited role in comparative genome mapping in the future.

32.10 Molecular tools of genome analysis

In the past decade, major advances have been made in the molecular understanding of plant genomes. Molecular cytogenetic methods can help associate particular DNA sequences with particular sites on the chromosomes, thus helping to understand aspects of genome organization. Techniques are now available in which nucleic acid probes are used to locate specific nucleic acid sequences by *in situ* hybridization. Labelled DNA probes are hybridized to chromosomes after appropriate treatments, and the chromosomal regions that bind the probe are visualized (see, for example, Chapter 9).

In situ hybridization has proved valuable in studying genome relationships. In genomic *in situ* hybridization (GISH), total genomic DNA is used as a probe to chromosome spreads [90] or Southern blots [91]. This technique permits identification of the genomic origin of chromosomes from different species in allopolyploids, synthetic hybrids, and alien addition, alien substitution and translocation lines in various species. The method gives useful information about the similarities of DNA from related species. It is sometimes referred to as chromosome painting (see Chapter 10) because the probe is labelled with a fluorescent dye and the chromosomes detected by the probe thus become uniformly coloured. GISH has been successfully employed to study genomic relationships in common wheat, tobacco and other allopolyploid plant species and hybrids [91–93].

Fluorescence *in situ* hybridization (FISH) (Chapter 9) is another powerful tool for chromosome mapping and for analysing genome relationships. FISH has been used to detect parental genomes in hybrids [94] and allopolyploids [95], and for detecting alien chromosome segments in translocations [96–98]. Multicolour FISH has also been used to simultaneously discriminate several genomes in allohexaploids such as wheat [99]. Using a combination of Giemsa banding and FISH, Jiang and Gill [100] found a 4A-5A-7B cyclic translocation specific to *T. turgidum* ($2n = 4x = 28$; AABB) and a different cyclic translocation in *T. timopheevii* ($2n = 4x = 28$; AAGG), thereby supporting the diphyletic origin of tetraploid wheats.

In situ hybridization can help detect some chromosomal landmarks (see also Section 32.3.2), which, in turn, can be used for phylogenetic investigations. The diploid *T. monococcum* is known to have two pairs of satellite chromosomes with NOR loci [101, 102]. Recent *in situ* hybridization analysis using a 18S–26S rDNA probe detected a new NOR locus at the telomere of the long arm of chromosome 5A in *T. monococcum* ssp. *boeoticum* as well as in ssp. *urartu*, which is also present in chromosome 5A of *T. turgidum* and *T. aestivum*, and in $5A^t$ of *T. timopheevii* [103].

Molecular data generated from isozyme analyses and oligonucleotide fingerprinting have been used in genome analyses. Molecular marker-assisted genome analysis also offers certain advantages. Thus, RFLPs [104] and random amplified polymorphic DNAs (RAPDs) (refs 105 and 106; see also Chapter 5) have been successfully used for both genome analysis and mapping. The advent of molecular marker technology, using isozymes and RFLPs, has particularly revolutionized the linkage mapping strategies by providing easily mappable biochemical markers in addition to the classical morphological genes [107]. The molecular maps of several crop plants have been or are being constructed and are already proving useful in plant breeding programs in terms of locating genes of economic interest, and tagging and tracking genes to facilitate their transfer to desirable cultivars. Map-based technology is also useful in studying genome evolution and in revealing unusual synteny relationships among distantly related species and genera [108–111]. Thus, the RFLP map of the potato is almost identical in the order of markers with that of tomato [108]. One particular problem with RFLP analysis is that most current methods used for plants require the use of radioisotopes and may not be cost- and time-effective.

The RAPD-PCR technique (random amplification of polymorphic DNA by the polymerase chain reaction) provides a means of rapidly detecting polymorphisms for genetic mapping and strain identification [112,113]. The ability of RAPDs to survey numerous loci in the genome makes this technique particularly attractive for studying genetic distance and for phylogeny reconstruction. However, the method has several limitations [114]. The RAPD technique is useful in genetic analysis only if variation in banding patterns represents allelic segregation at independent loci. Polymorphism is detected as band presence vs. absence and may be caused either by failure to prime a site because of nucleotide sequence differences or by insertions or deletions in the fragments between two conserved primer sites. Intermittent PCR artefacts may sometimes be misread as true allelic segregation [115].

Other technical problems and some possible solutions have been outlined by Hadrys *et al.* [116]. A possibly serious problem with RAPD technology is its reproducibility, which would limit the comparison of RAPD analysis data from one laboratory to another. DNA fragments that were amplified by five primers and shown to be reproducibly polymorphic between two oat cultivars (at the Agriculture Canada, Ottawa laboratory) were tested in six other laboratories in North America. Four of these participating laboratories amplified very few or no fragments using the Ottawa protocol, although the same participants were able to generate a considerable number of amplified fragments using their own protocols [117]. Unless results obtained in one laboratory are reproducible both within and among laboratories, the potential benefits of RAPD technology will not be realized. This underscores the need for the use of uniform protocols when results are compared.

32.11 Conclusion and perspectives

Several techniques of classical cytogenetics and biochemistry have been successfully employed for studying genome relationships both within polyploid taxa and among various plant species. Each of these tools has its own merits and also certain problems associated with it. Nevertheless, major advances have been made in understanding genome relationships in numerous plant taxa, and this knowledge has helped in planning germplasm enhancement strategies in several crop plants and other species of economic value. Molecular maps of several crop plants are being constructed (see Chapters 33 and 34) and are already having an impact on plant breeding in respect of locating, tagging, and tracking genes of economic value.

The development of chromosome microdissection and microcloning techniques (see, for example, Chapter 11) has provided powerful tools for molecular analysis of the human genome [118,119]. Some plants, particularly those with small chromosome number but large chromosome size, such as rye (*S. cereale* L.) and barley (*H. vulgare* L.) may be amenable to microdissection. Like several other molecular genetic techniques that were first developed in human cytogenetic studies and have later been adopted for plants (e.g. flow cytometry, RFLPs, microsatellites, and FISH analysis), it is anticipated that chromosome microdissection and microcloning will soon find an application in plant genome analysis, particularly for fine structure physical mapping. This new technology may yet open an exciting new era in characterizing the molecular structure and organization of plant genomes.

In genome analysis by classical cytogenetics, a genome is considered to be a static entity. However, a genome of a plant species can be dynamic, that is, it can undergo structural changes spontaneously. Genome-restructuring genes have also been reported [120] and chromosome structural changes in one species can be induced by an alien chromosome introduced into it [121]. Chromosome mutations frequently occur in lines of common wheat with the addition of certain chromosomes, called gametocidal chromosomes, from its wild relative *Aegilops* [122,123]. The study of restructured genomes thus created, or of reconstituted genomes produced through plant breeding, cannot be carried out using traditional cytogenetic tools. However, during the past decade, molecular cytogenetic techniques have been developed that have added new dimensions to the study of genome relationships. Both GISH and FISH techniques can be successfully employed to study restructured genomes, for elucidating evolutionary changes in genomes, and even for identification of parts of chromosomes involved in translocations. Although most of the techniques outlined above have contributed useful information on genome relationships, a multidisciplinary approach to genome analysis is always preferable for obtaining a full picture of genome relationships within and between species.

Acknowledgements

I thank Drs James Anderson, Lynn Dahleen and Norm Williams for reading the manuscript and providing useful suggestions.

References

1 Rees, H., Jenkins, G., Seal, A.G. & Hutchinson, J. (1982) Assays of the phenotypic effects of changes in DNA amounts. In *Genome Evolution* (Dover, G.A. & Flavell, R.B., eds), 287–297 (Academic Press, London).

2 Löve, Å. (1984) Conspectus of the Triticeae. *Feddes Repertorium* **95**, 425–521.

3 Tsunewaki, K. (1996) Plasmon analysis as the counterpart of genome analysis. In *Methods of Genome Analysis in Plants:* (Jauhar, P.P., ed.), 271–299 (CRC Press, Boca Raton, FL).

4 Palmer, J.D., Jansen, R.K., Michaels, H.J. *et al.* (1988) Chloroplast DNA variation and plant phylogeny. *Ann. Missouri Bot. Gard.* **75**, 1180–1206.

5 Chase, M.W., Soltis, D.E., Olmstead, R.G. *et al.* (1993) Phylogenetics of seed plants: an analysis of nucleotide sequences from the plastid gene *rbcL*1. *Ann. Missouri Bot. Gard.* **8**, 528–580.

6 Ogihara, Y. & Tsunewaki, K. (1988) Diversity and evolution of chloroplast DNA in *Triticum* and *Aegilops* as revealed by restriction fragment analysis. *Theor. Appl. Genet.* **76**, 321–332.

7 Kihara, H. & Nishiyama, I. (1930) Genomanalyse bei *Triticum* und *Aegilops*. I. Genomaffinitäten in tri-, tetra-und pentaploiden Weizenbastarden. *Cytologia* **1**, 270–284.

8 Lein, A. (1943) The genetical basis of the crossability between wheat and rye. *Z. Indukt. Abstamm. Vererbungl.* **81**, 28–59 (in German).

9 Lauri, D.A. & Bennett, M.D. (1987) The effect of the crossability loci *Kr1* and *Kr2* on fertilization frequency in hexaploid wheat × maize crosses. *Theor. Appl. Genet.* **73**, 403–409.

10 Jauhar, P.P. (1995) Meiosis and fertility of F_1 hybrids between hexaploid bread wheat and decaploid tall wheatgrass (*Thinopyrum ponticum*). *Theor. Appl. Genet.* **90**, 865–871.
11 Singh, R.J., Kollipara, K.P. & Hymowitz, T. (1988) Further data on the genomic relationships among wild perennial species ($2n = 40$) of the genus *Glycine* Willd. *Genome* **30**, 166–176.
12 Jauhar, P.P. (1975) Genetic regulation of diploid-like chromosome pairing in the hexaploid species, *Festuca arundinacea* Schreb. & *F. rubra* L. (Gramineae). *Chromosoma* **52**, 363–382.
13 Avdulov, N.P. (1931) Karyo-systematic investigations in the family Gramineae. *Bull. Appl. Bot. Genet. Plant Breed. Suppl.* **43**, 428 pp. (in Russian).
14 Riley, R., Unrau, J. & Chapman, V. (1958) Evidence on the origin of the B genome of wheat. *J. Hered.* **49**, 91–98.
15 Feldman, M. (1978) New evidence on the origin of the B genome of wheat. In *Proceedings of the 5th International Wheat Genetics Symposium*, Vol. 1 (Ramanujam, S., ed.), 120–132 (Indian Society of Genetics and Plant Breeding, New Delhi, India).
16 Oud, J.L., Kakes, P. & DeJong, J.H. (1987) Computerized analysis of chromosomal parameters in karyotype studies. *Theor. Appl. Genet.* **73**, 630–634.
17 Fukui, K. (1986) Standardization of karyotyping plant chromosomes by a newly developed chromosome image analyzing system (CHIAS). *Theor. Appl. Genet.* **72**, 27–32.
18 Fukui, K. & Kakeda, K. (1990) Quantitative karyotyping of barley chromosomes by image analysis methods. *Genome* **33**, 450–458.
19 Wang, R.R.-C. (1985) Genome analysis of *Thinopyrum bessarabicum* and *T. elongatum*. *Can. J. Genet. Cytol.* **27**, 722–728.
20 Jauhar, P.P. (1990) Dilemma of genome relationship in the diploid species *Thinopyrum bessarabicum* and *Th. elongatum* (Triticeae: Poaceae). *Genome* **33**, 944–946.
21 Jauhar, P.P. (1990) Multidisciplinary approach to genome analysis in the diploid species, *Thinopyrum bessarabicum* and *Th. elongatum* (*Lophopyrum elongatum*), of the Triticeae. *Theor. Appl. Genet.* **80**, 523–536.
22 Chennaveeraiah, M.S. (1960) Karyomorphologic and cytotaxonomic studies in *Aegilops*. *Acta. Horti. Gotoburgensis* **23**, 85–178.
23 Lacadena, J.R., Cermeño, M.C., Orellana, J. & Santos, J.L. (1984) Evidence of wheat–rye nucleolar competition (amphiplasty) in triticale by silver-staining procedure. *Theor. Appl. Genet.* **67**, 207–213.
24 Singh, R.J. & Hymowitz, T. (1988) The genomic relationship between *Glycine max* (L.) Merr. & *G. soja* Sieb. & Zucc. as revealed by pachytene chromosome analysis. *Theor. Appl. Genet.* **76**, 705–711.
25 Sumner, A.T., Evans, H.J. & Buckland, R.A. (1971) New technique for distinguishing between human chromosomes. *Nature New Biol.* **232**, 31–32.
26 Arrighi, F.E. & Hsu, T.C. (1971) Localization of heterochromatin in human chromosomes. *Cytogenetics* **10**, 81–86.
27 Patil, S.R., Merrick, S. & Lubs, H.A. (1971) Identification of each human chromosome with a modified Giemsa stain. *Science* **173**, 821–822.
28 Vosa, C.G. (1970) Heterochromatin recognition with fluorochromes. *Chromosoma* **30**, 366–372.
29 Vosa, C.G. (1975) The use of Giemsa and other staining techniques in karyotype analysis. *Curr. Adv. Plant Sci.* **6**, 495–510.
30 Vosa, C.G. & Marchi, P. (1972) Quinacrine fluorescence and Giemsa staining in plants. *Nature New Biol.* **237**, 191–192.
31 Gill, B.S. & Kimber, G. (1974) Giemsa C-banding and the evolution of wheat. *Proc. Natl Acad. Sci. USA* **71**, 4086–4090.
32 Gill, B.S. (1987) Chromosome banding methods, standard chromosome band nomenclature, and applications in cytogenetic analysis. In *Wheat and Wheat Improvement* (Heyne, E.G., ed.), 243–254 (2nd edn, American Society of Agronomy, Madison, WI).
33 Lukaszewski, A.J. & Gustafson, J.P. (1983) Translocations and modifications of chromosomes in triticale × wheat hybrids. *Theor. Appl. Genet.* **64**, 239–248.
34 Friebe, B., Hatchett, J.H., Sears, R.G. & Gill, B.S. (1990) Transfer of Hessian fly resistance from 'Chaupon' rye to hexaploid wheat via a 2BS/2RL wheat-rye chromosome translocation. *Theor. Appl. Genet.* **79**, 385–389.
35 Gill, B.S., Friebe, B. & Endo, T.R. (1991) Standard karyotype and nomenclature system for description of chromosome bands and structural aberrations in wheat (*Triticum aestivum*). *Genome* **34**, 830–839.
36 Friebe, B. & Gill, B.S. (1996) Chromosome banding in genome analysis. In *Methods of Genome Analysis in Plants:* (Jauhar, P.P., ed.), 39–60 (CRC Press, Boca Raton, FL).
37 Friebe, B., Mukai, Y. & Gill, B.S. (1992) C-banding polymorphisms in several accessions of *Triticum tauschii* (*Aegilops squarrosa*). *Genome* **35**, 192–199.
38 Gill, B.S. & Kimber, G. (1977) Recognition of translocations and alien chromosome transfers in wheat by the Giemsa C-banding technique. *Crop Sci.* **17**, 264–266.
39 Gill, B.S. & Chen, P.D. (1987) Role of cytoplasm-specific introgression in the evolution of the polyploid wheats. *Proc. Natl Acad. Sci. USA* **84**, 6800–6804.
40 Endo, T.R. & Gill, B.S. (1984) The heterochromatin distribution and genome evolution in diploid species of *Elymus* and *Agropyron*. *Can. J. Genet. Cytol.* **26**, 669–678.
41 Friebe, B. & Gill, B.S. (1995) C-band polymorphism and structural rearrangements detected in common wheat (*Triticum aestivum*). *Euphytica* **78**, 1–5.
42 Singh, R.J. & Röbbelen, G. (1975) Comparison of somatic Giemsa banding pattern in several species of rye. *Z. Pflanzenzüchtg.* **75**, 270–285.
43 Weimarck, A. (1975) Heterochromatin polymorphism in the rye karyotype as detected by the Giemsa C-banding technique. *Hereditas* **79**, 293–300.
44 Giraldez, R., Cermeño, M.C. & Orellana, J. (1979) Comparison of C-banding pattern in the chromosomes of inbred and open pollinated varieties of rye. *Z. Pflanzenzüchtg.* **83**, 40–48.
45 Linde-Laursen, I. (1978) Giemsa C-banding of barley chromosomes. I. Banding pattern polymorphism. *Hereditas* **88**, 55–64.

46 Linde-Laursen, I. (1975) Giemsa C-banding of the chromosomes of 'Emir' barley. *Hereditas* **81**, 285–289.

47 Noda, K. & Kasha, K.J. (1978) A proposed barley karyotype revision based on C-band chromosome identification. *Crop Sci.* **18**, 925–930.

48 Singh, R.J. & Tsuchiya, T. (1982) Identification and designation of telocentric chromosomes in barley by means of Giemsa N-banding technique. *Theor. Appl. Genet.* **64**, 13–24.

49 Kimber, G. & Feldman, M. (1987) *Wild Wheat: An Introduction.* Special Report 353 (University of Missouri, Columbia, Missouri, USA).

50 Jauhar, P.P. (1988) A reassessment of genome relationships between *Thinopyrum bessarabicum* and *T. elongatum* of the Triticeae. *Genome* **30**, 903–914.

51 Jauhar, P.P. & Crane, C.F. (1989) An evaluation of Baum *et al.*'s assessment of the genomic system of classification in the Triticeae. *Am. J. Bot.* **76**, 571–576.

52 Jauhar, P.P. (1975) Chromosome relationships between *Lolium* and *Festuca* (Gramineae). *Chromosoma* **52**, 103–121.

53 Wang, R.R.-C. (1989) An assessment of genome analysis based on chromosome pairing in hybrids of perennial Triticeae. *Genome* **32**, 179–189.

54 Sears, E.R. (1976) Genetic control of chromosome pairing in wheat. *Ann. Rev. Genet.* **10**, 31–51.

55 Jauhar, P.P. (1993) *Cytogenetics of the Festuca-Lolium Complex: Relevance to Breeding,* 43–56 (Springer-Verlag, Berlin, Heidelberg, New York).

56 Sears, E.R. (1941) Chromosome pairing and fertility in hybrids and amphidiploids in the Triticinae. *Missouri Agric. Exp. Stn. Res. Bull.* **337**, 20 pp.

57 Jackson, R.C. (1982) Polyploidy and diploidy: new perspectives on chromosome pairing and its evolutionary implications. *Am. J. Bot.* **69**, 1512–1523.

58 Dowrick, G.J. (1957) The influence of temperature on meiosis. *Heredity* **11**, 37–49.

59 Chen, Q., Jahier, J. & Cauderon, Y. (1989) Production and cytogenetical studies of hybrids between *Triticum aestivum* L. Thell. and *Agropyron cristatum* (L.) Gaertn. *C.R. Acad. Sci., Paris* **308**, 425–430.

60 Jauhar, P.P. (1992) Chromosome pairing in hybrids between hexaploid bread wheat and tetraploid crested wheatgrass (*Agropyron cristatum*). *Hereditas* **116**, 107–109.

61 Kaul, M.L.H. & Murthy, T.G.K. (1985) Mutant genes affecting higher plant meiosis. *Theor. Appl. Genet.* **70**, 449–466.

62 Jauhar, P.P., Riera-Lizarazu, O., Dewey, W.G. *et al.* (1991) Chromosome pairing relationships among the A, B, and D genomes of bread wheat. *Theor. Appl. Genet.* **82**, 441–449.

63 Forster, B.P. & Miller, T.E. (1989) Genome relationship between *Thinopyrum bessarabicum* and *Th. elongatum. Genome* **32**, 930–931.

64 Jauhar, P.P. (1992) Synthesis and cytological characterization of trigeneric hybrids involving durum wheat, *Thinopyrum bessarabicum*, and *Lophopyrum elongatum. Theor. Appl. Genet.* **84**, 511–519.

65 Jauhar, P.P. (1975) Genetic control of diploid-like meiosis in hexaploid tall fescue. *Nature* **254**, 595–597.

66 Jauhar, P.P. (1975) Genetic control of chromosome pairing in polyploid fescues: its phylogenetic and breeding implications. *Rept Welsh Plant Breeding Stn.* **1974**, 114–127.

67 Dvořák, J. (1972) Genetic variability in *Aegilops speltoides* affecting homoeologous pairing in wheat. *Can. J. Genet. Cytol.* **14**, 371–380.

68 Jauhar, P.P. (1968) Inter- and intra-genomal chromosome pairing in an interspecific hybrid and its bearing on basic chromosome number in *Pennisetum. Genetica* **39**, 360–370.

69 Jauhar, P.P. (1981) *Cytogenetics and Breeding of Pearl Millet and Related Species,* 133–137 (A.R. Liss, New York).

70 Alonso, L.C. & Kimber, G. (1983) A study of genome relationships in wheat based on telocentric chromosome pairing. II. *Z. Pflanzenzüchtg.* **90**, 273–284.

71 Okamoto, M. & Sears, E.R. (1962) Chromosomes involved in translocations obtained from haploids of common wheat. *Can. J. Genet. Cytol.* **4**, 24–30.

72 Hutchinson, J., Miller, T.E. & Reader, S.M. (1983) C-banding at meiosis as a means of assessing chromosome affinities in the Triticeae. *Can. J. Genet. Cytol.* **25**, 319–323.

73 Naranjo, T., Roca, A., Goicoechea, P.G. & Giraldez, R. (1987) Arm homoeology of wheat and rye chromosomes. *Genome* **29**, 873–882.

74 Naranjo, T., Roca, A., Giraldez, R. & Goicoechea, P.G. (1988) Chromosome pairing in hybrids of *ph1b* mutant wheat with rye. *Genome* **30**, 639–646.

75 Driscoll, C.J., Bielig, L.M. & Darvey, N.L. (1979) An analysis of frequencies of chromosome configurations in wheat and wheat hybrids. *Genetics* **91**, 755–767.

76 Alonso, L.C. & Kimber, G. (1981) The analysis of meiosis in hybrids. II. Triploid hybrids. *Can. J. Genet. Cytol.* **23**, 221–234.

77 Kimber, G. & Alonso, L.C. (1981) The analysis of meiosis in hybrids. III. Tetraploid hybrids. *Can. J. Genet. Cytol.* **23**, 235–254.

78 Sybenga, J. (1988) Mathematical models for estimating preferential pairing and recombination in triploid hybrids. *Genome* **30**, 745–757.

79 Crane, C.F. & Sleper, D.A. (1989) A model of meiotic chromosome association in triploids. *Genome* **32**, 82–98.

80 Crane, C.F. & Sleper, D.A. (1989) A model of meiotic chromosome association in tetraploids. *Genome* **32**, 691–707.

81 Konarev, V.G. (1983) The nature and origin of wheat genomes on the data of grain protein immunochemistry and electrophoresis. In *Proceedings of the 6th International Wheat Genetics Symposium* (Sakamoto, S., ed.), 65–75 (Plant Germplasm Institute, Kyoto University, Kyoto, Japan).

82 Konarev, A.V. & Konarev, V.G. (1993) Use of genome-specific antigens and prolamin electrophoresis in the evaluation of wheat and its wild relatives. In *Biodiversity and Wheat Improvement* (Damania, A.B., ed.), 259–271 (Wiley, Chichester, UK).

83 Metakovsky, E.V., Kudryavstev, A.M., Iakobashvili, Z.A. & Novoselskaya, A.Y. (1989) Analysis of phylogenetic relations of durum, carthlicum and common wheats by means of comparison of alleles of gliadin-coding loci. *Theor. Appl. Genet.* **77**, 881–887.

84 Damania, A.B., Porceddu, E. & Jackson, M.T. (1983) A rapid method for the evaluation of variation of germplasm collections of cereals using polyacrylamide gel electrophoresis. *Euphytica* **32**, 877–883.

85 Konarev, V.G., Gavriljuk, I.P., Gubareva, N.K. & Choroshajlov, H.G. (1981) Electrophoretic and serological methods in seed testing. *Seed Sci. Technol.* **9**, 807–817.

86 Sarkar, R. & Raina, S.N. (1992) Assessment of genome relationships in the genus *Oryza* L. based on seed-protein profile analysis. *Theor. Appl. Genet.* **85**, 127–132.

87 Hart, G.E. (1987) Genetic and biochemical studies of enzymes. In *Wheat and Wheat Improvement* (Heyne, E.G., ed.), 199–214 (2nd edn, American Society of Agronomy, Madison, WI).

88 Crawford, D.J. (1989) Enzyme electrophoresis and plant systematics. In *Isozymes in Plant Biology* (Soltis, D.E. & Soltis, P.S., eds), 146–164 (Dioscorides Press, Portland, OR).

89 Crawford, D.J. (1990) *Plant Molecular Systematics* (Wiley, New York).

90 Schwarzacher, T., Leitch, A.R., Bennett, M.D. & Heslop-Harrison, J.S. (1989) *In situ* localization of parental genomes in a wide hybrid. *Ann. Bot.* **64**, 315–324.

91 Anamthawat-Jónsson, K., Schwarzacher, T., Leitch, A.R. *et al.* (1990) Discrimination between closely related Triticeae species using genomic DNA as a probe. *Theor. Appl. Genet.* **79**, 721–728.

92 Anamthawat-Jónsson, K. & Heslop-Harrison, J.S. (1993) Isolation and characterization of genome-specific DNA sequences in Triticeae species. *Mol. Gen. Genet.* **240**, 151–158.

93 Heslop-Harrison, J.S. & Schwarzacher, T. (1993) Molecular cytogenetics—biology and applications in plant breeding. *Chromosomes Today* **11**, 191–198.

94 Leitch, I.J., Leitch, A.R. & Heslop-Harrison, J.S. (1991) Physical mapping of plant DNA sequences by simultaneous *in situ* hybridization of two differently fluorescent probes. *Genome* **34**, 329–333.

95 Bennett, S.T., Kenton, A.Y. & Bennett, M.D. (1992) Genomic *in situ* hybridization reveals the allopolyploid nature of *Milium montianum* (Gramineae). *Chromosoma* **101**, 420–424.

96 Heslop-Harrison, J.S., Leitch, A.R., Schwarzacher, T. & Anamthawat-Jónsson, K. (1990) Detection and characterization of 1B/1R translocations in hexaploid wheat. *Heredity* **65**, 385–392.

97 Mukai, Y., Friebe, B., Hatchett, J. *et al.* (1993) Molecular cytogenetic analysis of radiation-induced wheat–rye terminal and intercalary chromosomal translocations and the detection of rye chromatin specifying resistance to Hessian fly. *Chromosoma* **102**, 88–95.

98 Jiang, J., Morris, K.L.D. & Gill, B.S. (1994) Introgression of *Elymus trachycaulus* chromatin into common wheat. *Chromosome Res.* **2**, 3–13.

99 Mukai, Y., Nakahara, Y. & Yamamoto, M. (1993) Simultaneous discrimination of the three genomes in hexaploid wheat by multicolor fluorescence *in situ* hybridization using total genomic and highly repeated DNA probes. *Genome* **36**, 489–494.

100 Jiang, J. & Gill, B.S. (1994) Different species-specific chromosome translocations in *Triticum timpoheevii* and *T. turgidum* support the diphyletic origin of polyploid wheats. *Chromosome Res.* **2**, 59–64.

101 Mukai, Y., Endo, T.R. & Gill, B.S. (1991) Physical mapping of the 18S–26S rRNA multigene family in common wheat: identification of a new locus. *Chromosoma* **100**, 71–78.

102 Kim, N.-S., Kuspira, J., Armstrong, K. & Bhambhani, R. (1993) Genetic and cytogenetic analysis of the A genome of *Triticum monococcum*. VIII. Localization of DNAs and characterization of 5S rRNA genes. *Genome* **36**, 77–86.

103 Jiang, J. & Gill, B.S. (1994) New 18S–26S ribosomal RNA gene loci: chromosomal landmarks for the evolution of polyploid wheats. *Chromosoma* **103**, 179–185.

104 Dvořák, J. & Zhang, H.-B. (1990) Variation in repeated nucleotide sequences sheds light on the phylogeny of the wheat B and G genomes. *Proc. Natl Acad. Sci. USA* **87**, 9640–9644.

105 González, J.M. & Ferrer, E. (1993) Random amplified polymorphic DNA analysis in *Hordeum* species. *Genome* **36**, 1029–1031.

106 Joshi, C.P. & Nguyen, H.T. (1993) Application of the random amplified polymorphic DNA technique for the detection of polymorphism among wild and cultivated tetraploid wheats. *Genome* **36**, 602–609.

107 Tanksley, S.D., Bernatzky, R., Lapitan, N.L. & Prince, J.P. (1988) Conservation of gene repertoire but not gene order in pepper and tomato. *Proc. Natl Acad. Sci. USA* **85**, 6419–6423.

108 Tanksley, S.D., Ganal, M.W., Prince, J.P. *et al.* (1992) High-density molecular maps of the tomato and potato genomes. *Genetics* **132**, 1141–1160.

109 Ahn, S., Anderson, J.A., Sorrells, M.E. & Tanksley, S.D. (1993) Homoeologous relationships of rice, wheat and maize chromosomes. *Mol. Gen. Genet.* **241**, 483–490.

110 Devos, K.M., Millan, T. & Gale, M.D. (1993) Comparative RFLP maps of the homoeologous group-2 chromosomes of wheat, rye and barley. *Theor. Appl. Genet.* **85**, 784–792.

111 Kurata, N., Moore, G., Nagamura, Y. *et al.* (1994) Conservation of genome structure between rice and wheat. *Bio/Technology* **12**, 276–278.

112 Welsh, J. & McClelland, M. (1990) Fingerprinting genomes using PCR with arbitrary primers. *Nucleic Acids Res.* **18**, 7213–7218.

113 Williams, J.G.K., Kubelik, A.R., Livak, K.J. *et al.* (1990) DNA polymorphisms amplified by arbitrary primers are useful as genetic markers. *Nucleic Acids Res.* **18**, 6531–6535.

114 Clark, A.G. & Lanigan, C.M.S. (1993) Prospects for estimating nucleotide divergence with RAPDs. *Mol. Biol. Evol.* **10**, 1096–1111.

115 Riedy, M.F., Hamilton, W.J. & Aquadro, C.F. (1992) Excess of non-parental bands in offspring from known primate pedigrees assayed using RAPD-PCR. *Nucleic Acids Res.* **20**, 918.

116 Hadrys, H., Balick, M. & Schierwater, B. (1992) Applications of random amplified polymorphic DNA (RAPD) in molecular ecology. *Mol. Ecol.* **1**, 55–63.

117 Penner, G.A., Bush, A., Wise, R. *et al.* (1993) Reproducibility of random amplified ploymorphic DNA (RAPD) analysis among laboratories. *PCR Meth. Appl.* **2**, 341–345.
118 Kao, F.T. (1993) Microdissection and microcloning of human chromosome regions in genome and genetic disease analysis. *BioEssays* **15**, 141–146.
119 Kao, F.T. (1996) Chromosome microdissection and microcloning: application to genome analysis. In *Methods of Genome Analysis in Plants:* (Jauhar, P.P., ed.) 329–343 (CRC Press, Boca Raton, FL).
120 Feldman, M. & Strauss, I. (1983) A genome restructuring gene in *Aegilops longissima*. In *Proceedings of the 6th International Wheat Genetics Symposium* (Sakamoto, S., ed.), 309–314 (Plant Germplasm Institute, Kyoto University, Kyoto, Japan).
121 Endo, T.R. (1988) Induction of chromosomal structural changes by a chromosome of *Aegilops cylindrica* L. in common wheat. *J. Hered.* **79**, 366–370.
122 Endo, T.R. (1990) Gametocidal chromosomes and their induction of chromosome mutations in wheat. *Jpn. J. Genet.* **65**, 135–151.
123 Endo, T.R., Yamamoto, M. & Mukai, Y. (1994) Structural changes of rye chromosome 1R induced by a gametocidal chromosome. *Jpn. J. Genet.* **69**, 13–19.

Chapter 33 The Arabidopsis nuclear genome

Michel Delseny & Richard Cooke

Laboratoire de Physiologie et Biologie Moléculaire des Plantes, URA 565 du CNRS, Université de Perpignan, 66860 Perpignan, France

33.1 Introduction

Plant species constitute a large proportion of living organisms, and all animals, including humans, are completely dependent on higher plants for their nutrition. As a consequence, improving crops has been a major activity for mankind since the very beginning of the human 'adventure'. In most countries, agriculture and plant breeding are among the principal economic resources and it is essential to improve our knowledge of plant genomes in order to progress in plant breeding. Towards this goal, many basic questions have to be addressed. How do we identify plant genes and determine how they are organized on the chromosomes? Which genes control agronomic traits? How can we improve classical breeding using the information obtained from in-depth studies of plant genomes? How is recombination capacity controlled? How can we reintroduce a new gene into a plant genome and also target its localization and expression?

Independently of these practical goals, higher plants have a number of specific properties and processes which suggest that novel genes and biological regulatory mechanisms could be uncovered by analysing their genomes. Examples of such processes are the structure and assembly of the cellulose cell wall, photosynthesis, nitrogen and carbohydrate assimilation, production of secondary metabolic products, perception of light, morphogenetic responses and response to environmental stress.

When trying to answer the questions posed above, plant breeders and plant physiologists have been faced with two problems. The first is the large number of crops that have to be studied for economic reasons. The second is the wide range of complexity of plant genomes, which range over two orders of magnitude [1–3] from a few hundred megabases up to 30 000 Mb (Table 33.1). To overcome these difficulties, plant scientists have focused their genome-sequencing projects on two species with reasonably sized genomes: a dicotyledon, *Arabidopsis thaliana*, and a monocotyledon, rice (*Oryza sativa*). These two plants have rather different habits and morphology. They also have very distinct codon usages. In addition to these two species, genomes from more than 40 economically important crops have been mapped by RFLP. This exceptionally favourable situation results essentially from the facility with which plants can be crossed and their progenies analysed.

Arabidopsis thaliana (or thale cress) is a small weed (see Plate 10) which is being used as a model plant species by a large part of the scientific community. It belongs to the family Cruciferae, which contains several major crops such as oilseed rape, mustard and cabbage. *Arabidopsis* has several key advantages over other plant species for genome analysis. The first is that its genome is one of the smallest in flowering plants. A second advantage is its short generation time (no more than two months in optimal conditions from seed to seeds). It is self-fertile and its progeny is abundant (more than 1000 seeds can be harvested from a single plant and a single fruit (silique) contains about 30 seeds). Many different ecotypes with distinct characters have been selected as sources of genetic variability. This set of features allows detailed genetic analysis, so that it is possible to combine genetic mapping of visible markers with molecular marker mapping. Because *Arabidopsis* seeds are rather small (1000 seeds weigh 20–22 mg), they are easy to mutagenize and a large number of mutants obtained by different methods have been described. Finally, *Arabidopsis* can be easily transformed using agrobacteria, opening the way to reverse genetic analysis for characterizing genes. Several reviews on *Arabidopsis* are available [4–9] and two textbooks of methods for *Arabidopsis* research have recently been published [10,11].

In this review, we shall describe the present status of our knowledge as well as the strategies which are being used to characterize the genome of this plant at increasingly greater resolution. Some of these strategies are specific to the analysis of plant genomes while others are common to other genomes. In addition, we shall discuss some of the implications of studies in *Arabidopsis* for finding human genes and analysing their function.

Table 33.1 Comparison of genome size in plant species with other genomes analysed in sequencing programmes.

Organisms	Size (Mbp)
Saccharomyces cerevisiae	15
Caenorhabditis elegans	100
Homo sapiens	3500
Arabidopsis thaliana	145
Oryza sativa	440
Brassica oleracea	600
Lycopersicon esculentum	1000
Zea mays	2500
Hordeum vulgare	4800
Triticum aestivum	16 000
Tulipa officinalis	30 000

33.2 Size of the genome and general organization

Arabidopsis thaliana, like all other higher plant

species, has three distinct genomes. The chloroplast genome is ≈150 kilobase pairs (kbp) and is very similar to those of the other plants in which it has been completely sequenced (tobacco, rice and pine) [12]. The mitochondrial genome is ≈370 kbp and is one of the simplest in plants. More than 75% of its sequence has now been determined [13]. In leaves, the chloroplast genome can account for as much as 27% of total DNA, whereas the mitochondrial genome usually represents less than 2% [12,14].

The *Arabidopsis* nuclear genome is organized in five chromosome pairs which are very small (on average, 2 μm) and are almost indistinguishable from each other by optical microscopy of metaphase plates [15–17]. However, preparation of synaptonemal complex complements for electron microscopy has recently allowed the analysis *Arabidopsis* chromosomes at high resolution [18]. Three types of data provide an estimation of the size of this genome. All these approaches have been used to estimate the size of other genomes and are not specific to plants. On the basis of DNA renaturation kinetics, the haploid genome size was initially estimated to be 70 megabase pairs (Mbp) [14]. Because the size of the *Escherichia coli* genome, which was used as a standard in these experiments, has been re-evaluated, the actual size is probably close to 100 Mbp. More recently, Goodman and coworkers fingerprinted ≈17 000 cosmids which could be grouped into 750 contigs covering ≈91 Mbp [19]. Finally, extensive cytometry measurements have been made on a large number of plant species using highly standardized conditions. The estimate for *Arabidopsis* was 145 Mbp [3]. Although this size might be overestimated, the actual genome size is probably slightly higher than 100 Mbp, which is comparable to that of other model organisms such as *Drosophila melanogaster* or *Caenorhabditis elegans* (see Table 33.1). This value has now been confirmed by physical mapping of individual chromosomes [20,21].

The initial DNA renaturation kinetics indicated that the *Arabidopsis* genome contained relatively few repeated sequences [14]. Highly repeated and foldback sequences comprise no more than 10% of the genome and another 10% is made of moderately repeated sequences. This gives *Arabidopsis* a considerable advantage as a model over other plants because these percentages are usually much higher: for instance, more than 80% of wheat DNA is made of repeated sequences [1,2]. Several tandemly repeated DNA families have been identified, cloned and sequenced. The genes encoding the 25S and 18S cytoplasmic rRNA are organized as 10 kbp tandem repeats. With ≈570 copies, they belong to the class of moderately repeated DNA. They exhibit some heterogeneity and were estimated to represent 7–8% of the DNA [14,22]. They have been located at two loci on the short arm of chromosome 4 and on the distal part of the shortest chromosome, chromosome 2 [16–18,23]. There are about 1000 copies of the 5S rRNA genes [24], with at least one locus on chromosome 4 [20]. Three other families of tandemly repeated DNA sequences have been described. They are members of the highly repetitive DNA sequences. The first comprises four to 6000 copies of a 180-bp unit which has been located essentially in the centromeric regions [25,26]. It accounts for about 1% of the genome. The two other types of repeats are made up of a 500-bp and a 160-bp unit, respectively [26]. Each constitutes around 0.3% of the genome. The 500-bp repeat is related to the 180-bp satellite DNA, but the 160-bp repeat corresponds to a completely different sequence. The last tandemly organized repeat to be described is the telomeric sequence which is made of around 350 copies of a 7-bp motif, CCCTAAA, as in many organisms. This block is present at each end of each individual chromosome [27]. Similar sequences have also been located in the chromosome 1 centromeric region [28]. In contrast to the satellite DNA sequences, which are species specific, the telomere sequences are highly conserved in all plant species studied so far.

The renaturation kinetics data also tell us that there is relatively little interspersion of repeated sequences with single-copy sequences [14], an observation which has been confirmed by physical mapping of chromosome 4 [20]. The most interesting type of dispersed repeated sequences are the transposable elements. Several subclasses of retrotransposable elements have been described by Ausubel's group, such as the *Ta1* element [29,30]. Another element, *Tat1*, with the properties of a transposon, has been isolated and detected as a linear extrachromosomal molecule [31]. Neither of them seem to be active and they are usually present in rather low copy number. Recently, an active element similar to the maize Ac transposon [32] as well as a new retrotransposon, *Athila* [33], have been discovered. *Athila*, which is 10.5 kbp long, is similar to the *Ulysses* element from *Drosophila* viroids. There are as many as 30 copies in the genome. The remainder of the sequence is essentially single-copy or low-copy number sequence. The next steps in the description of the *Arabidopsis* genome organization are the achievement of genetic and physical maps and the identification of the genes.

The *Arabidopsis* genome also contains simple repeat sequences such as microsatellites [34,35] (see Chapter 5). Minisatellites have been described recently [36]. In addition, it has recently been shown that the telomeric sequence is also dispersed along the chromosomes and statistically over-represented in the 5′ noncoding regions of the genes [37].

33.3 Arabidopsis genetic maps

The first *Arabidopsis* genetic map to be produced was based on morphological markers. It resulted from many crosses between lines carrying mutations, and from the analysis of trisomic lines [38]. Now, more than 800 mutants have been discovered following saturation mutagenesis with ethyl methanesulphonate (EMS), radiation and other mutagens [39]. For many mutants, several alleles have been isolated. These mutants include a wide variety of phenotypes corresponding to all aspects of development [39], responses to environmental biotic or abiotic factors, and metabolic pathways. Particularly impressive are the large number of mutants affected in embryo development [40,41,197], flower formation and setting [42], hormone sensitivity or deficiency and response to light [43,44] or in response to pathogens [45–47], most of these processes being largely specific to plants. More than 300 mutations have now been mapped and the most recent information can be obtained from AtDB (an *Arabidopsis thaliana* database) [48]. A list of these mutations has recently been published as a progress report from National Science Foundation [49]. Many of them are available from the Ohio State University and Nottingham University Stock Centres and can be ordered by electronic mail or WWW (see Section 33.8).

Several RFLP maps have been published [50,51] based on F2 segregation analysis and using essentially anonymous genomic probes (λ-phage or cosmid DNA). An attempt has recently been made to integrate them using common markers [52] and the JOINMAP program [53]. An example of the integrated map is shown in Fig. 33.1 as obtained from the AtDB. The major drawback of F2 mapping populations of annual species is that they are not permanent and cannot be distributed. More recently,

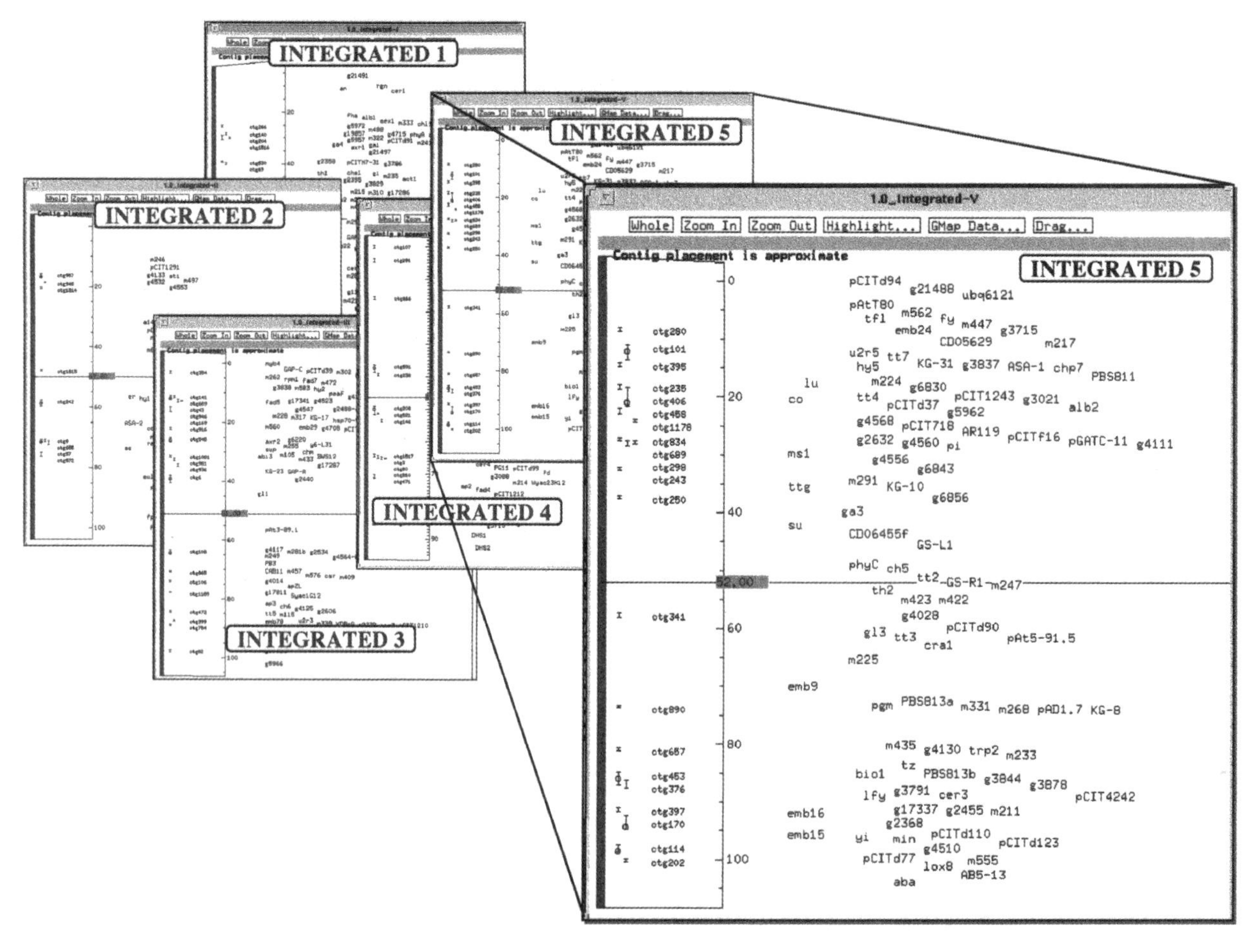

Fig. 33.1 A view of the integrated genetic map of *Arabidopsis*. The map of chromosome 5 has been enlarged to show the details. The right-hand part of each panel shows the approximate position of mutations (e.g. hy5, co, ms1, ttg, emb 9) and anonymous as well as identified RFLP markers ('g' series is from ref. 50 and 'm' from ref. 51). Symbols and figures to the left of the scale show contig placements.

new permanent mapping populations, made of recombinant inbred lines, have been produced by two groups [54,55]. This strategy is specific to the plant kingdom, and its first application to genome mapping was developed in maize [56].

Recombinant inbred lines are obtained by selfing individuals from an F2 progeny for several generations until the F7 or F8. Thus, the initial F2 segregation is fixed and each line is homogeneous and homozygous for the alleles that segregated in F2. Such populations can be used for ever, since each line is simply maintained by selfing. They can be distributed worldwide, thus allowing dispersed laboratories to contribute to the saturation of the maps. The population developed in the United States by the Dupont de Nemours company results from a cross between line W100F and Wassileskija ecotype and consists of 150 lines [54]. That developed in the UK was made by crossing ecotype Landsberg *erecta* (La-er) with ecotype Columbia (Col) [55]. Both populations are available from the stock centres and the mapping data are maintained and updated by the authors of the map. By now, more than 400 markers have been located on the La-er×Col map with 64 RFLP markers in common with the two original maps. Data concerning these markers are available from AtDB. All these molecular genetic maps gave similar estimates of the genetic complexity of the *Arabidopsis* genome of ≈600 cM. Therefore the average centiMorgan is only 170 kbp.

The trend for the future is to map expressed sequence tags (ESTs), which are coming out of the systematic and partial cDNA sequencing programmes in France [57,58] and in the USA [59] as well as new types of markers such as amplified fragment length polymorphisms (AFLPs) [60], microsatellites [35,36] or cleaved amplified polymorphic sequences (CAPS) [61,62]. AFLPs are obtained by digesting DNA with two restriction enzymes and adding adaptors; the next step consists in amplifying a subset of these fragments using primers overlapping the adaptors and extending two or three nucleotides (arbitrarily chosen) beyond the restriction site. The amplified fragments are then resolved on a sequencing gel and those which are polymorphic between the parental lines can be scored in the segregating population. The advantage of this fingerprinting technique is that several markers can be mapped at the same time.

More than 150 AFLP markers have already been mapped using La-er×Col recombinant inbred lines [21,60]. Thirty polymorphic microsatellite loci have been mapped by Ecker's group on the same lines [35] and corresponding primers can be purchased from Research Genetics Inc. (see Appendix III for address). A set of 18 CAPS has also been developed and is available from the same company [61]. Their map positions have been determined [62]. They are evenly dispersed on the genome and were derived from the sequences of mapped genes. This set of markers now allows one to map unambiguously any gene to one of the 10 chromosome arms in a single cross using a limited number of F2 progeny. Additional polymerase chain reaction (PCR)-based markers of this type should become available soon. The expectation is that within the next two years more than 1200 mapped molecular markers will be available with an average covering of one marker every 0.5 cM or every 85 kbp, which should transform tedious chromosome walking strategies into a straightforward 'chromosome landing' approach to isolate genes identified by a mutant phenotype.

33.4 Physical map

In order to be able to isolate genes corresponding to mapped mutations, it is necessary to establish physical maps. The first attempt to organize a physical map of the *Arabidopsis* genome was by Howard Goodman's group at Massachusetts General Hospital (MGH), Boston. They ordered a cosmid library according to the strategy used for *Caenorhabditis elegans* (see Chapter 29). Almost 20 000 cosmids were fingerprinted following labelling of *Hin*dIII sites, further digestion with *Sau*3 A and resolution of labelled DNA fragments by high-resolution polyacrylamide gel electrophoresis. Overlapping cosmids were organized into contigs by image and computer analysis, allowing the organization of 750 contigs [19]. Many of these data are also publicly available through AtDB, and a series of cosmid clones are distributed by the Ohio Stock Center. Filling the gaps between these contigs would have required enormous additional work and another strategy was preferred, based on ordering YAC libraries.

Three publicly available YAC libraries have been produced. They have relatively small inserts, in the range of 160 kbp [63–65]. They contain, respectively, 2100, 2700, and 2200 clones and collectively represent 10 genome equivalents. Two other libraries with larger inserts have been made [58,65] and several others are being developed. The library prepared by a collaboration between several French plant scientists and colleagues at the Centre d'Etude du Polymorphisme Humain, Paris (CEPH) shows an average insert of 420 kbp, which is much more convenient for chromosome walking [58]. All these libraries have been made using pYAC4 or derivatives as vectors. The major difficulty in preparing

YAC libraries from *Arabidopsis* is the DNA concentration: there is 35 times less DNA in nuclei from this species than in human cells, and nuclei have to be prepared from protoplasts, which constitutes a serious limiting factor. The three publicly available libraries have been partially screened with 125 RFLP probes by a small group of laboratories in England and the United States. As a result, a first set of 296 YACs covering ≈30% of the genome was described [66]. In addition to this large-scale coverage of the genome, several groups are organizing contigs around their favourite locus [67–71].

Most of the European effort has dealt with chromosomes 4 and 5. By the end of 1993 more than 85% of chromosome 4 was covered by 35 YAC contigs; due to the recent use of the new YAC library this number has now (early 1996) decreased to three and coverage outside of the centromere and ribosomal gene regions is now complete [20]. Chromosome 2 is covered by six overlapping contigs [72]. Two large contigs have been organized around the genes controlling late flowering—*fca* on chromosome 4 and *Co* on chromosome 5. The first is 2200 kbp long and covers ≈19 cM, whereas the second covers 2700 kbp [68,69]. The strategy for organizing these contigs has been to generate end-specific probes from YACs identified by colony hybridization with an RFLP probe, using either plasmid rescue or inverse PCR. However, this method is limited both by the presence of repeated sequences and by the occurrence of many chimaeric YACs. Therefore it is of utmost importance to control each step of the walk by generating a YAC-derived RFLP probe and demonstrating linkage with the initial RFLP marker. The YAC clones can also be used as probes for analysing genomic DNA separated by pulse field gel electrophoresis.

Several strategies are being developed to fill the gaps between the contigs. The first is to align the cosmid and YAC contigs. The second is to find more RFLP markers that identify new YACs. Finally, probes derived from the ends of the present YAC contigs are being used to identify overlapping YAC clones in the new libraries with larger inserts and less chimaerism. The combination of these strategies should lead in the very near future to a complete physical map of all five *Arabidopsis* chromosomes, which will considerably facilitate the sequencing of a complete plant chromosome and the isolation of any gene on this chromosome [21].

A library has also been made in a P1 phage vector [73] and, more recently, as bacterial artificial chromosomes (BACs) [74].

33.5 Strategies for gene identification

33.5.1 Classical strategies

The primary strategy to identify genes in plants has been the classical one, consisting of purifying the corresponding proteins, obtaining antibodies or amino acid sequence information and using these tools to isolate the corresponding cDNA and genomic clones. Many genes have also been picked up following differential screening of cDNA libraries, sequence determination and comparison with known genes in databases. Some examples illustrating this strategy are given in refs 75–79. This is the situation for many developmentally regulated genes or for genes responsive to environmental changes. In many cases, no well-defined function can be identified. The major limitation on this strategy is that the probability of isolating a gene that is not abundantly expressed is rather low. This is indeed the situation with most regulatory genes and most genes of agronomic interest, such as those controlling plant morphology, time of flowering or resistance to diseases. Many of these genes could not have been isolated by classical techniques.

Another relatively classic strategy consists in complementing *E. coli Saccharomyces cerevisiae*, mutants [80] or mammalian cell (see Table 33.2).

Complementation of a known mutant is also a useful tool for confirming the identification of a clone on the basis of sequence homology. Its usefulness is, however, limited by the availability of a mutant and by the divergence between yeast, bacteria and plant genes. Functional complementation of mammalian cells has also recently been used to isolate plant genes coding for plasma membrane proteins [101] and on apoptosis suppressor gene [102] (see Chapter 18 for general techniques in this area). Functional assays in *Xenopus* oocytes are also used to analyse transporter and ion channel genes. Finally, one can make use of genomic subtraction methods if deletion mutants are available. This strategy was illustrated with cloning of *GA1*, a gene involved in gibberellic acid biosynthesis [103].

33.5.2 Genetic approaches: map-based cloning, T-DNA and transposon tagging

Owing to the small size of the *Arabidopsis* genome and the ease with which it can be transformed, two major genetic strategies have been developed. They are partially specific to plants with small genomes.

The first is a map-based cloning approach. The gene of interest is identified by a mutation that can be located on the genetic map and flanked by several

Table 33.2 Examples of *Arabidopsis* genes cloned using a complementation strategy.

In *Escherichia coli*
RecA homologues [81]
UvrB, UvrC homologues [82,83]
5′-Phosphoribosyl-5-aminoimidazole synthetase [84,85]
Glycinamide ribonucleotide (GAR) synthetase [85]
GAR-transformylase [85]
3-Methyladenine DNA glycosylase [86]
cnx1 (Molybdenum cofactor biosynthesis) [162]
Phosphoribosylanthranilate isomerase [165]
In yeast
Potassium transporters [87, 88]
Glucose and sucrose transporters [89,90]
Amino acid and peptide transporters [91–93]
NTR1, nitrate transporter related [94]
NH_4^+ high-affinity transporter [95]
Cdc2-P34 protein kinase [96]
Chorismate mutase [97]
Cycloartenol synthase [98]
ATP-sulphurylase [99]
Aspartate transcarbamylase [100]
Orotate phosphoribosyltransferase / orotidine 5-phosphate decarboxylase [100]
CDC48 [166]
Mevalonate kinase [167]
Sar1, Sec12 [168]
Sec14 (M. Lepetit, personal communication 1995)
ERD2 [169]
AAT1 [170]
In mammalian cells
Plasma membrane integral proteins [101]

molecular markers. When the genetic distance between flanking markers is reasonably small a chromosome walk can be attempted. Once a contig spanning the mutation is organized, the region around the gene has to be narrowed down. This can be done in several ways: a new RFLP which cosegregates with the mutation can be found, the region covering the putatively identified gene can be transferred back to the mutant line by transformation, which should lead to complementation of the mutation in the progeny of the transgenic plants and finally, a cDNA detecting the expected expression pattern can be isolated and used to correct the mutation in transgenic plants. The first success of this strategy was the cloning of the *ABI3* locus [104] controlling sensitivity to abscisic acid, rapidly followed by that of the *FAD3* locus [105] which corresponds to a C18:2 fatty acid desaturase. Now more than thirty important genes have been isolated from *Arabidopsis* following this approach (Table 33.3). It is obvious that the more the genetic map is saturated and the physical map complete and reliable, the easier this approach becomes.

The second genetic approach is based on tagging the gene of interest with a piece of identified DNA. The plant biologist can use two tools: transposable elements as in *Drosophila*, or T-DNA derived from the Ti plasmid of *Agrobacterium tumefaciens*, the bacterium used for transferring foreign genes to plants. Most success has been achieved so far with the T-DNA approach pioneered by K. Feldmann and C. Koncz. The former developed a seed transformation protocol which avoids plant regeneration from tissue culture and the associated somaclonal variation. The latter used more classical techniques for transformation of leaf discs and root cells. Although the frequency of transformed plants was low, it was manageable, and several thousand individual lines were regenerated and screened for a variety of mutants [106–109]. Tagging should be established by selfing the mutant and analysing its progeny both for the selectable marker (usually resistance to kanamycin or bialaphos) and the presence of the mutation. To be sure that the mutation is tagged, no sensitive plant should be found amongst the mutants when 200–300 plants are analysed. Sequences flanking the T-DNA insertion can be recovered by plasmid rescue or inverse PCR and can be used as probes to isolate the wild-type allele of the gene that has been disrupted by the insertion. When the gene has been isolated, it should be demonstrated by transformation that it can correct the mutation.

During the last few years, several important *Arabidopsis* genes have been isolated using this strategy and several others will appear soon (Table 33.4). This technique, although very powerful, has several drawbacks: the seed transformation method is poorly reproducible; there are often several T-DNA insertions in the same plant: and, although somaclonal variation is reduced, many mutants are not tagged by the T-DNA. The more classical transformation method has the same limitations, and in addition is more time consuming. Recently, another transformation technique [110,111], in which the whole plant is vacuum-infiltrated with *Agrobacterium*, has allowed the production of several thousand additional transgenic plants transformed with an improved vector. It was calculated that 50 000 such independent lines would be enough to saturate the genome with well-defined insertions and pick up virtually any gene for which a mutant phenotype can be observed.

Such a collection of mutants will also be useful in determining the function of genes with no known function. In yeast, the strategy used to determine the

Gene	Phenotype or enzyme affected	Reference
Genes isolated by map-based cloning		
ABI1	Vegetative tissues, abscisic acid insensitivity	[70,71]
ABI3	Abscisic acid insensitivity in seed	[104]
AXR1	Auxin resistance	[171]
CONSTANS	Flowering gene	[172]
DET1 (FUSCA)	De-etiolated	[173]
ERT1	Ethylene resistance	[174]
FAD3	C18 : 2 desaturase	[105]
LEAFY	Inflorescence determination	[73]
PISTILLATA	Abnormal flower	[175]
RPM1	Disease resistance	[176]
RPS2	Disease resistance (*P. syringae* race avr Rpt2)	[47]
Other genes for which chromosome walking is in progress (cited in ref. 9)		
ABI2	Abscisic acid insensitivity	
ARA1	Arabinose sensitivity	
AXR2	Auxin resistance	
DET2	De-etiolated	
EIN2	Ethylene insensitivity	
EIN3	Ethylene insensitivity	
FCA	Late flowering	
FRI	Flowering gene	
FWA	Late flowering	
GA2	Gibberellic acid insensitivity	
GAI	Gibberellic acid insensitivity	
GI	Late flowering	
GNOM	Embryo pattern	
MS1	Male sterility	
PHOL	Phosphate translocator	
RPP5	Disease resistance (*Peronospora*)	
TEL1	Terminal flower	
TTG	Transparent testa glabrous	

Table 33.3 List of genes for which a map-based cloning strategy is used. Most of the genes in the lower part of this table have now been isolated, as well as several others.

function of a new gene is gene disruption via site-directed homologous recombination [112]. This method can also be used to some extent with animal cells but is not yet working with plants. An alternative would be to screen DNA pools of the collection of transgenic lines by PCR, to identify the line which is interrupted in the new gene and search for abnormal phenotypes. The first results using this strategy have been reported recently [113]. Most of the transgenic lines that have been reasonably well analysed are available from the stock centres for further screening.

The transposon tagging strategy is also very powerful, as demonstrated in maize, *Antirrhinum* and *Drosophila*. However, the problem with *Arabidopsis* was that, until recently [33], no active transposon had been identified in this species. This inconvenience could be overcome by introducing by T-DNA transformation a defective nonautonomous maize transposable element such as Ac/Ds or Enhancer/Inhibitor into one set of transgenic lines and an active transposase into another set of lines. When the two types of lines are crossed, the defective element is able to jump to another place. Because the transposase gene and the defective element are not linked, in most cases they would segregate in the progeny, and stable mutants, tagged with the defective nonautonomous element, should appear. The advantage of this method over the T-DNA tagging approach is that the defective element can be moved again by crossing with a line carrying the transposase gene, in which case revertants should be obtained. In addition, in the revertant, a footprint of the transposition event should be observed. Although several genes have been tagged in various laboratories, so far there have been very few reports of isolation of an *Arabidopsis* gene using this method [114–116]. The same proofs for tagging as in the case of a T-DNA tag should be obtained: mutation and insertion should cosegregate in the progeny when selfing the mutant; the mutation should be corrected by the wild type allele when

Table 33.4 List of genes isolated using T-DNA tagging.

Gene	Phenotype or enzyme affected	Reference
AGAMOUS1	Abnormal flower	[177]
APETALA1	Abnormal flower	[178]
APETALA2	Abnormal flower	[179]
AXI159	Auxin independent	[180]
CER1	Wax biosynthesis (aldehyde decarboxylase)	[181]
CER2	Wax biosynthesis	[182]
CHL1	Chlorate resistant (nitrate transporter)	[183]
CHLORATA42	Chlorophyll deficiency (protoporphyrin Mg^{2+} chelatase)	[184]
COP1 (FUSCA1)	Constitutive photomorphogenesis	[185]
COP9 (FUSCA7)	Constitutive photomorphogenesis	[186]
CTR1	Constitutive ethylene response	[187]
DWARF1	Gibberellin deficient	[188]
EMB30	Embryo pattern (GNOM)	[189]
FAD2	C18 : 1 desaturase	[190]
FAD3	C18 : 2 desaturase	[191]
FAE1	Deficient in fatty acid elongation	[192]
FUSCA6	Defective seed germination	[193]
GA4	GA3 β-hydroxylase	[194]
GLABRA2	Trichome absence	[195]
GLABROUS1	Trichome absence, Myb-like	[160]
HY4	Abnormal flowering (blue light receptor)	[196]
LUMINIDEPENDENS	Late flowering time	[197]
PALE CRESS	Abnormal plastid development	[198]
PFL	Abnormal leaf development	[199]
PISTILLATA	Abnormal flower	[175]
TOUSLED	Abnormal flower	[200]

Note: at least 20 additional genes have been isolated using this strategy.

isolated; and revertants should be obtained. Because transposons usually move only short distances on the same chromosome, they should be much more useful to generate mutations in a specific region of the genome provided that a collection of transgenic plants carrying a mapped defective element is available.

33.5.3 Promoter trapping

The use of T-DNA or transposon tags can only detect genes in which mutations cause a change in phenotype. Since many genes are usually present in more than one copy, it follows that many will not be detected using this strategy. In addition, some genes are not essential at all stages of development, and many mutations may be silent unless an appropriate screening procedure is designed. In order to overcome this problem, T-DNA vectors were constructed in which a promoter-less reporter gene is located close to one of the T-DNA borders. When such a construction is inserted near an active promoter it is then possible to detect its activity by looking for expression of the reporter gene. This strategy is being used by several groups to detect promoters that function specifically in one organ or one tissue [117,118,196]. A refinement of this strategy is the *enhancer trap*, using a transposable element rather than a T-DNA [120].

33.6 Sequencing cDNA: expressed sequence tags

With the double aim of obtaining a better understanding of gene expression in *Arabidopsis* and contributing to the genome mapping of this species, two groups of laboratories, one in France [57] and the other in the United States [59], have embarked on a project to partially sequence as many cDNAs as possible from various tissues. This will enable a gene to be identified by an expressed sequence tag (EST). The strategies and starting material differ to some extent between the two groups. The American group has been using the λ-YES vector which allows almost direct transfer of the cloned cDNA to yeast, whereas the French groups have essentially used λ-ZAP.

The American group at Michigan State University has made a single orientated library by pooling mRNA prepared from different parts of the plant.

The rationale for this strategy is to prepare a normalized library in which each clone should be equally represented, regardless of the enormous differences in gene expression. Up to now, sequencing has been from the 5′ end of each clone, using an automated sequencing apparatus. The French groups, which are dispersed in several places, have adopted a different approach and have made several orientated libraries corresponding to different tissues or culture conditions: young etiolated seedlings, young green plantlets, flower buds, immature siliques, dry seeds, wounded adult leaves, cell suspension cultures and cell suspensions elicited by a bacterial pathogen. After an initial period during which only one end of the cDNA was sequenced, they have sequenced each new cDNA from both ends using automated machines. This strategy has several advantages: complete sequences for many more cDNAs can be obtained; and sequencing of the 3′ untranslated region, which is usually gene-specific, allows the distinction between multigene family members for which the coding region is almost or completely identical. Studies on various libraries allow the expression patterns of a large number of genes to be determined from the frequency with which the genes are found in the different libraries.

The redundancy of clones selected at random is highly variable, depending on the libraries under analysis [57]: for instance, 35% of the sequenced clones in the immature silique library correspond to three multigene families, including the two major families of storage proteins (napins and cruciferins). There are five members in the napin family and three in the cruciferin. Most of the other clones have been observed only once or twice, although there is evidence for small multigene families. In contrast, in the cell suspension library, only 8% of the clones have been found to be redundant. Although this redundancy is becoming a hindrance in deriving a catalogue of *Arabidopsis* genes, it is a reliable source of information on the relative expression of gene family members. When more clones are isolated and sequenced it should be possible to determine an expression pattern simply from the frequency of a sequence in different libraries. Each group has now isolated the representatives of the most abundantly expressed genes in its own situation. The libraries are being cleaned by hybridization with characterized probes in order to overcome redundancy problems. An alternative way of finding new genes is to set up specific screening procedures which will replace random collection.

The analysis and editing of the sequences makes up a large part of the work. It was decided at the beginning of the French project to centralize information and to eliminate, as far as possible, sequences with too many uncertainties, as well as identical sequences, before submission to the EMBL data bank. Further, a minimum of sequence editing and correction is carried out, particularly for homologues to known proteins, with the aim of submitting sequences which are as accurate as possible so that they are useful for the scientific community. The submission protocol defined in collaboration with Rainer Fuchs at EMBL, includes citation of protein or nucleic acid homologues and definition of probable coding regions, providing valuable information when database searches are carried out.

Since September 1993 the program has been supported by the EC as part of the project European Scientists Sequencing *Arabidopsis* (ESSA) project. The ESSA project is funding only new sequences and an additional and independent verification for novelty is carried out at Martinsried by the Martinsried Institute for Protein Sequences (MIPS). Figure 33.2 shows the flow-chart for dealing with sequence data in use in our group and in most of

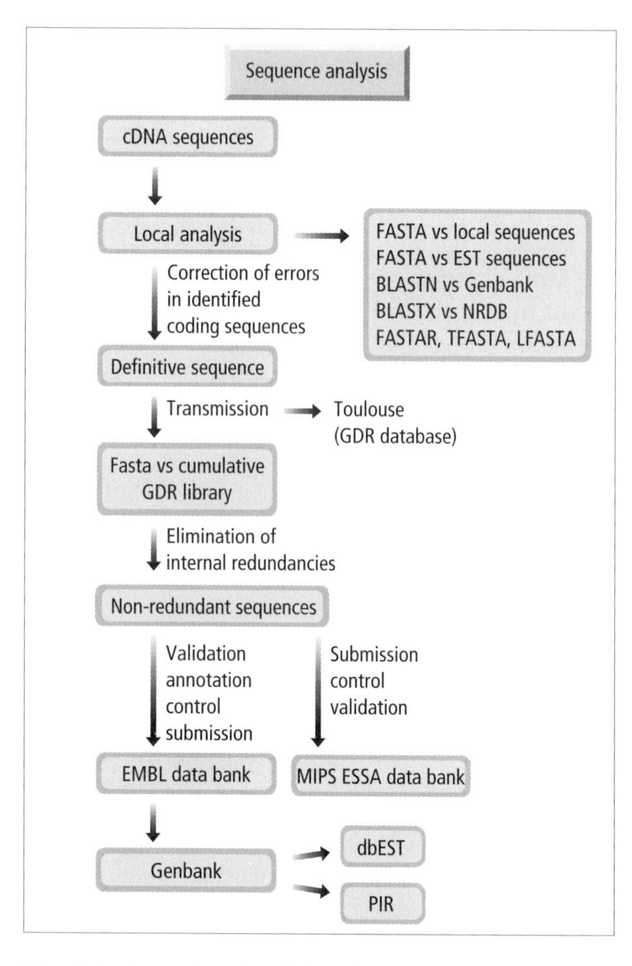

Fig. 33.2 Flow-chart for EST analysis and submission to databases.

the French groups. This protocol allows us to detect redundancies, rapidly resolve ambiguities in sequencing, and search for open reading frames with homologies in public databases. Thus, it is very often possible to extend useful sequence information well beyond the 300 nucleotide limit, after which the automatic analysis programme tends to insert spurious bases. Early alignment with the local database avoids repeatedly sequencing identical clones from both ends. Comparison with public databases allows one to detect significant similarity to known sequences from other organisms and thus to assign a putative function to the gene. When a disruption in an amino acid alignment but not in the nucleotide alignment is observed, this very often indicates a sequencing mistake and allows detection and correction of such errors. The most frequent errors are uncertainties (marked as N), misidentification of a base, and single base deletions or additions due to compression beyond 300–350 bases. These problems can be largely limited by using high-quality DNA (e.g. Qiagen-treated minipreps; see Chapter 21) in the reactions.

When a sequence has been corrected and edited, it is transferred to the consortium GDR database in Toulouse where it is compared with sequences produced by other groups in the consortium; redundant clones or remaining poor-quality sequences are identified at this stage. They are added to the local database for statistical analysis but are not necessarily submitted to public databases. The selected clones are then returned to each laboratory for validation and annotation before being automatically transferred to the public EMBL database. Although the European Union (EU) is paying only for new clones, we continue to deposit some sequences corresponding to a previously described cDNA if they significantly extend the already available data. In addition, because most sequences correspond to a single run of the machine and are not 100% accurate, it is certainly helpful to have some redundancy in the database to track sequencing errors. The plant origin of the sequences was routinely assessed using a quality control algorithm [121].

Altogether, we estimate that at the end of 1996, the French and American consortia had already produced more than 15 000 non-redundant ESTs, and probably nearly 35 000 ESTs if redundancy is not taken into account. By the end of February 1995 there were 24 352 entries for *Arabidopsis* ESTs in the centralized EST database at the National Institutes of Health, dbEST [122], comparing well with the rice and nematode programs (Table 33.5). During the first year of ESSA (September 1993–September 1994), 710 new and original cDNAs (not previously tagged by either the French or American programs) were partially sequenced from both ends. Out of the 5358 nonredundant ESTs contributed by the French group and released by the end of December 1995 in dbEST, 2144 showed a significant similarity at the protein level with other sequences in the databanks [123].

These similarities were generally established using the BLASTX program [124] and a score higher than 100 unless an obvious signature motif [125] was present, and were frequently improved using more sensitive programs such as TFASTA [126]. Another useful programme is the DOMAINER algorithm, which searches for homologous protein domains among the entries in SWISSPROT database [127]. Table 33.6 gives the statistics for the first 1152 published ESTs [57]. More than 7000 ESTs have now been analysed [20,21] but the general trends are not significantly changed.

It is striking that more than 60% of the genes do not yet have any homologues in the databases. With the effort made to identify genes in other organisms, this proportion is likely to decrease progressively in the future, as has been observed for yeast chromosome III [112]. Although it is difficult to make accurate comparisons because of the fact that some of the cDNAs have been sequenced only from one

Table 33.5 Evolution of ESTs entries in dbEST.

	Dates of release						
	July 1993	**Dec. 1993**	**Mar. 1994**	**Aug. 1994**	**Oct. 1994**	**Feb. 1996**	**Feb. 1997**
Homo sapiens	14556	16329	16943	22881	23945	349036	581794
Caenorhabditis elegans	4699	4699	4699	11590	12104	23438	30196
Arabidopsis thaliana	**1676**	**3432**	**4756**	**8010**	**8241**	**24352**	**29165**
Oryza sativa	1023	4221	4342	4342	4342	11301	12806
Zea mays		118	912	988	988	1183	1757
B. campestris			181	181	181	965	965
B. napus						1021	1425

Table 33.6 Analysis of the first 1152 published *Arabidopsis* ESTs.

Total number of ESTs	**1152**	
Number of different genes	895	
Number of genes with homologues in databases, of which:	286	(32%)
Already identified in *Arabidopsis*		35%
New member of an *Arabidopsis* multigene family		7%
Homologue to a previously described plant gene		33%
Not yet found in plants		25%

end and the sequences are not completely overlapping, preliminary calculations indicated that 35% of the putatively identified genes found by the French in the seed and silique libraries programme have been found in the American programme and that 30% are significantly similar to rice genes [128] (see Chapter 34). However, higher percentages might be found when other tissues, organs and physiological conditions are analysed. The percentage drops dramatically to only a few percentage when non-identified ESTs are searched for similarities. These preliminary data probably indicate that only the most abundantly expressed genes have been identified, and that there is still room for independent sequencing programmes.

Other relatively unexpected genes found in *Arabidopsis* include homologues of genes coding for the laminin receptor, an annexin, integrins, a *Schistosoma* haemoglobinase and a selenium-binding protein, as well as genes homologous to the *Drosophila* gene *abnormal wing disc* or to the mouse *unp* and *NEDD-6* genes. Approximately 25% of the newly identified genes in plants correspond to animal or microbial gene sequences, and it would have been very difficult or impossible to clone them using classical methods. They usually represent genes coding for highly conserved proteins such as those involved in protein synthesis (ribosomal proteins and translation factors), components of the cytoskeleton, such as actins and tubulins, or important enzymes from metabolic pathways (energy metabolism or amino acid biosynthesis). However, a number of sequences were completely unexpected, such as several tumour suppressor gene homologues, including Wilm's tumour suppressor, a few oncogenes such as *myb*, *raf* and *ras*, several supposedly brain-specific genes such as a 14-3-3-like protein and an acyl CoA-binding protein homologous to the benzodiazepam receptor [57]. On the other hand, several genes are clearly specific to plants, such as those for storage proteins, components of the photosynthetic apparatus, and components of the plant cell wall. This series of sequences, as well as those from rice [128], will help to determine the set of ancient conserved sequences which are represented in the plant genomes [129].

An interesting lesson from this systematic cDNA sequencing project is that it revealed that many proteins are encoded by multigene families (Table 33.7). This was confirmed when homologues were identified by PCR. Several of them are now being studied in great detail to unravel their pattern of expression and understand how they are regulated and how they evolved [130–144]. Although *Arabido-*

Table 33.7 Examples of multigene families in *Arabidopsis*.

Gene name	Minimum copy number
Cab	4
ACP	2
Rubisco	4
Cdc2	12
Calmodulin	6
Cruciferin 12S	3
Napin 2S	5
G-box binding factor	3
H1	2
Rab18	3
H3	3
Rib S18	3
Actin	10
Profilin	15
Thioredoxin	5
Agamous-like	6
β-1,3-Glucanase	3
Homeodomain protein (Leu zipper)	9
α-Tubulin	6
β-Tubulin	9
Ubiquitin	6
Ubiquitin-conjugating enzyme 2	7
Glucose transporter	3
Oleosin	4
LEA76	3
Major latex-like	2
Meri5	3
LEAD113	3
Em	2
Phytochrome	5
Anthanilate synthetase	2
H^+-ATPase	10
Kinesin	3
GF14 (14-3-3)	5
HSP70 and cognate	3
MAP kinase	7
HMG-CoA reductase	2

psis is considered to be a rather simple organism, some of these gene families have many more members than their animal counterparts.

Besides these two EST programmes, several cDNAs have been completely sequenced independently by several laboratories. From GenBank release 77 (June 1993), 220 full-length cDNA sequences were available. However, most of these sequences are also represented in the ESTs. Presently, ≈350 full-length cDNAs have been sequenced.

Another application of the EST programme has been to provide numerous probes to determine which genes are under the control of a specific regulatory protein—for instance, those whose activity is modified by the ABI3 gene product, which controls sensitivity to abscisic acid in the seed [145]. Due to increasing redundancy both the French and American EST programmes are almost completely stopped.

33.7 Genomic sequencing

Most genomic sequences that have been determined correspond to a few genes for which a cDNA could be characterized, and the effort towards extensive genomic sequencing has begun only recently. In GenBank release 77 there were fewer than 120 *Arabidopsis* protein-coding genes that had been sequenced as genomic clones. A more recent estimate, based on the EMBL database (March 1997), indicates that this figure is now around 600. However, this is almost certainly an underestimate because many genes have been characterized but not necessarily deposited in the database. This is the situation for the ESSA programme, which will soon release the sequence of 1.8 Mb of a region of chromosome 4 [146]. Nevertheless, we made a calculation on 200 available genes accounting for ≈531 000 bp of genomic DNA. The coding sequences represent 230 000 bp and the introns 86 500 bp. The remainder consists essentially of 5′- and 3′-noncoding flanking sequences. Figure 33.3 shows the distribution of the intron number per gene and Fig. 33.4 shows the size distribution of the 557 introns that have been scored in protein-coding genes. So far, 30% of the genes are found to be intronless and the vast majority have no more than four introns; an exception is the RNA polymerase II large subunit, which has 24 introns. In contrast to animal cells, introns are relatively short, 68% being smaller than 100 bp, while exons coding for no more than 12 amino acids have been identified [147]. Most of them have the canonical border sequences. However, these data concern only around one-hundredth of the genes and the situation should become much clearer when systematic genomic sequencing has accumulated several megabases of sequence.

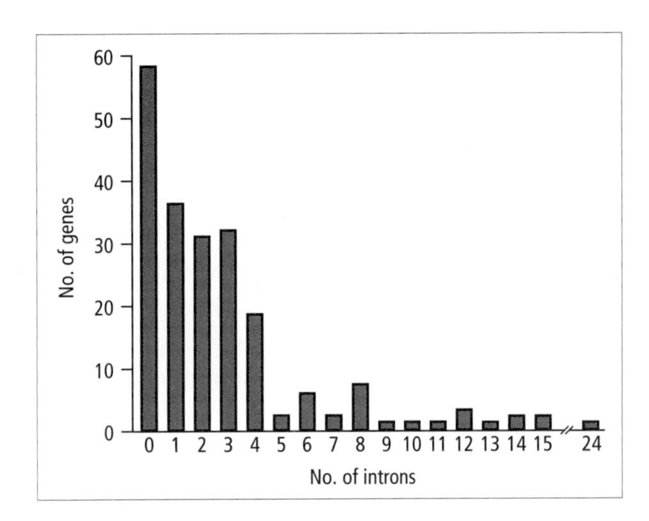

Fig. 33.3 Distribution of the intron number per gene in a sample of 205 sequenced *Arabidopsis* genes.

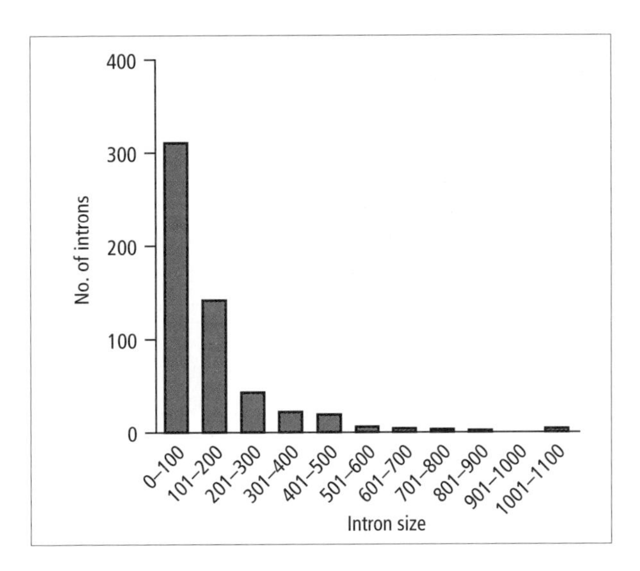

Fig. 33.4 Distribution of the intron size in a sample of 557 introns present in 205 sequenced genes. Size is in bp.

As observed from cDNA analysis, many genes belong to multigene families. In a few cases (napins, glycine-rich proteins, EF1-α proteins, kin1 protein) the genes are clustered at a single locus and we have information on the distance separating two adjacent genes. It is usually of the order of 1 kbp or less, indicating that genes are relatively densely organized. There is no accurate estimate of gene number. However, from genomic sequencing data from several laboratories, it seems that the gene density might be as high as four or five genes within 20 kbp. Assuming this is a general situation and that the genome size is 100 Mbp, including 15% repeated sequences, this gives a maximum gene number of the order of 17 000–21 500. On the other hand, if we estimate that the above value of 200 genes for 530 000 bp is more representative of the average gene density, then there would be room for as many as 30 000 genes. Obviously more data are needed, and a more accurate estimate will only be available when larger regions have been sequenced.

Toward this goal, the EU launched in September 1993 the three-year ESSA project as a model study to assess the feasibility of sequencing large portions of the *Arabidopsis* genome. This program concerns 14 groups who volunteered for either small-scale (25 kbp per year) or medium-scale (> 50 kbp per year) projects. Clones from two large contigs on chromosome 4 in the *FCA* and *APETALA2* regions (respectively 1500 kbp and 500 kbp) have been distributed to the participants. In addition, some of the participants focused on seven additional loci dispersed on several chromosomes. Each one has agreed to sequence 75 kbp of contiguous DNA around his favourite gene. This should provide an additional 525 kbp. As mentioned above, this program also includes some cDNA sequencing. All the sequence data are collected first and analysed by the MIPS in Martinsried, Germany, before being released to public databases. This group was already in charge of analysing yeast chromosomes for the EU programme (see Chapter 30) and this association guarantees careful assessment and comparison of the data derived from these two model organisms.

The goals for the first year were 370 kbp of genomic DNA and 384 kb of cDNA. After two years, more than 1 Mb of genomic DNA has been deposited at MIPS and considerably more was almost completed; nearly 1 Mb of cDNA sequence representing 1500 new cDNA clones has also been registered by MIPS. The programme was on time, and by the end of 1996 more than 2.5 Mbp of *Arabidopsis* DNA had been sequenced by EU scientists. The major results are the identification of about 500 genes, more than 80 % being completely new. Gene density was on average one gene every 4–5 kbp. Depending on the region analysed, between 37 % and 60 % of the predicted genes matched an EST, thus confirming that the EST programmes have tagged approximately half of the expressed *Arabidopsis* genes. From a functional point of view, it seems that there are relatively few examples of clustering of genes involved in a given metabolic or transduction pathway. A few genes with similar sequences are tandemly organized. Large-scale sequencing also revealed additional transposable elements.

By the end of 1996, a new EU Programme (ESSA 2) was initiated with the aim of sequencing between 5 and 7 Mbp on chromosome 4 long arm. Meanwhile three American consortia and one Japanese consortium have been organized with similar objectives so that by the end of the century approximately 40 % of the genome should be determined and the complete sequence of the nonrepeated part of the genome should be available around 2003–2004.

Because many genes are duplicated, by analysing the flanking regions of multigene family members it should be possible to learn how the duplication arose. Such a project will provide probes which are physically linked. Since homologues of some of the chosen loci exist in other plants and organisms, it will be interesting to determine how much of this local *synteny* has been conserved and how the intergenic regions have evolved in different species. Such a synteny is already obvious for the cereals, in which large portions of the different genomes are colinear [146,148–50]. How much of the information gained from analysis of the *Arabidopsis* genome can be transferred to other species remains to be determined and will be a goal for future programmes.

33.8 Arabidopsis stock centres and Arabidopsis-orientated software

Arabidopsis molecular geneticists are creating a huge amount of biological material in the form of seeds, mutants, individual cDNA and genomic clones, and a wide variety of libraries in different vectors. All these resources need to be preserved, stored and distributed in order to benefit the whole scientific community. At the same time, scientists are faced every day with more information concerning genetic resources, genetic and physical maps as well as sequence data. It is essential that all the information be available as easily as possible, not only to scientists interested in this plant, but also to other scientists and private companies. In addition to sending all the DNA sequences to be published to GenBank or EMBL, a number of additional sources of information have been set up. It should be emphasized, however, that new sites appear very frequently and that servers may move or be discontinued.

33.8.1 The Arabidopsis stock centres

There are two major stock centres functioning as an international network. They are in charge of distributing seeds, clones, and libraries. They are, respectively, located at Nottingham, UK (Nottingham *Arabidopsis* Stock Centre, NASC) and at Columbus, Ohio, USA (*Arabidopsis* Biological Resource Center, ABRC). In recent years, NASC and ABRC have collected and organized individual collections which were at the origin of genetic work on *Arabidopsis*, including those of Rédei, Koornneef, and Kranz. Altogether, these three collections represent more than 1000 accessions. In addition, the stock centres provide numerous mutants which have been deposited by individual researchers. They provide the recombinant inbred lines mapping populations, and a large collection of T-DNA-tagged transgenic lines. DNA stocks are preserved and distributed by ABRC. They include mapped RFLP phages [50] and cosmids [51], several YAC, cDNA and genomic libraries as well as individual clones. In particular, the cDNA clones partially sequenced by the American and French EST programmes are stored and distributed by ABRC. One of the goals of the stock centres is to encourage researchers to deposit seeds and clones as they are published. A US private company, Lehle Seeds, also provides mutagenized seeds as well as custom multiplication of F1 and F2 progenies.

33.8.2 AtDB: an Arabidopsis thaliana database

AtDB (formerly known as AAtDB) is a database [48] that can be accessed through a graphical interface using software developed for the *C. elegans* genome project (see Chapter 29). It was originally created by the US Department of Agriculture Plant Genome project through the National Agricultural Library, set up by Mike Cherry and Sam Cartinhour, who were in the Department of Molecular Biology at MGH, Boston, and curated by John Morris. The project is now under the auspices of the National Science Foundation (NSF) at the Department of Genetics, School of Medicine, Stanford University, where the second generation software is under development. Versions are available for all widely used computers (Sun, Digital, SG1, etc., as well as for the Macintosh) but it is also available through the graphical interface by remote login and by network communication tool clients such as Wais, Gopher and WWW hypertext clients such as Mosaic or Netscape. Information contained in AtDB was obtained directly from authors or from various public databases. It features a variety of information presented using graphical text and tabular format. A large number of interconnections allows the passage from one type of information to another, simply by clicking with the workstation mouse. As an example (Fig. 33.5), one can visualize a chromosome map on which RFLP markers are indicated, then zoom in on a specific region and obtain sequence and reference information on a gene or recombination frequencies for markers. Among the items stored in AtDB are:

- cosmid physical map from Goodman's lab;
- genetic maps, including the integrated RFLP map, the RAPD map and the visible markers map;
- recombination data for F2 and recombinant inbred line mapping populations;
- catalogues of lines and mutants available from the two stock centres in Nottingham (UK) and Columbus (USA);
- a list of *Arabidopsis* scientists with addresses;
- all *Arabidopsis* DNA sequences registered in GenBank and EMBL DNA sequence databases;
- bibliographic citations related to *Arabidopsis;*
- various software for DNA analysis.

AtDB is available without charge via Internet ftp transfer or on CD-ROM. A companion guide [48] is also provided. Data to be entered in the data base should be submitted to the curator of AtDB so that the integrity of the database is maintained.

33.8.3 AIMS: the Arabidopsis Information Management System

AIMS is being developed by S. Pramanik of Michigan State University in collaboration with R. Scholl of the ABRC. It is funded by the NSF. AIMS provides support for data management. Data items include stock centres, cloned genes available from ABRC, information on RFLP, RAPD and other markers, including annotations on enzymes to be used to detect polymorphisms, information on YAC libraries available in ABRC, and cross-homology between YAC and RFLP markers, genetic mapping data, sequence and homology search results for all EST cDNA clones, colour pictures of plant phenotypes for many of the stocks. AIMS features also include graphical display of genetic maps and ability to run linkage analysis programs. Contig information from the physical map has been added to the database. Seeds and clones can be ordered through on-line AIMS or EMAIL-AIMS from the stock centres (see Section 33.8.5). As is the case for AtDB, access is possible through the various network communication tools.

33.8.4 The Arabidopsis newsgroup

This is the privileged communication link within the *Arabidopsis* community. It allows both simple question-and-answer interactions in all fields of *Arabidopsis* research and immediate diffusion of important information on materials, meetings, etc. It

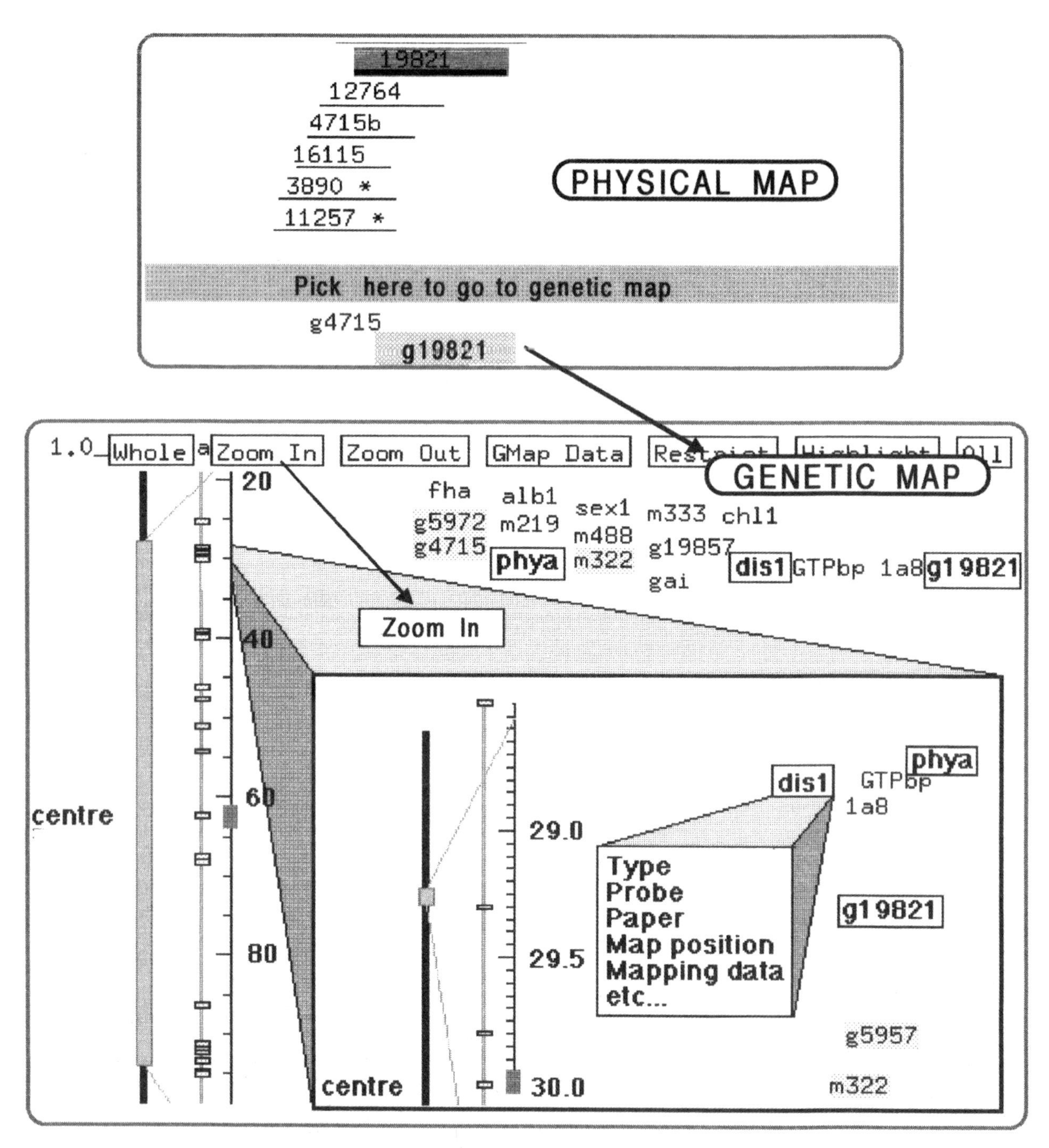

Fig. 33.5 An example of the use of AtDB. The bottom window shows the integrated genetic map as represented in Fig. 33.1. One can click on a specific DNA marker (e.g. g19821) to access the physical map and corresponding YACs and cosmids in this contig. A single click on any object highlights 'connected' items. Double-clicking on YAC and cosmid names brings up information on these clones, while a double click on a genetic symbol opens the genetic map. The 'buttons' at the top of the map allow opening of menus linked to other files associated with selected objects; for instance, it is possible to 'zoom in' on a particular region of the map, as shown. Double-clicking will bring up additional information on a marker, with links to literature references, sequences (if known), etc. The small rectangles on the second line from the left are linked to the physical map. (The original images used to prepare this figure were kindly provided by John Morris, AtDB curator. They have been slightly modified to emphasize details and facilitate presentation.)

is distributed world-wide through USENET news under the name of bionet.genome.*Arabidopsis* and through e-mail. It is one of the most active newsgroups, having a readership of ≈1300–1400 people and an average of three to four messages daily. *ARABIDOPSIS* postings are indexed in the general 'biosci.src' WAIS index and are accessible through all network clients. Previous postings can easily be queried by keyword searches by these methods. Many subjects of interest to *Arabidopsis* researchers are also discussed on the plant biology newsgroup.

33.8.5 Obtaining communication tools

Knowing how and where to obtain the tools necessary for using the Internet is possibly the hardest step in connecting to the different services available. An excellent starting point for those new to the Internet is *A Biologist's Guide to Internet Resources* by Una Smith. It can be obtained by sending an e-mail message to 'mail-server@rtfm.mit.edu' containing the line 'send pub/usenet/sci.answers/biology/guide/*'

Anonymous ftp A short guide to anonymous ftp can be obtained by sending the message 'help' to 'info@sunsite.unc.edu'. ftp archives can be searched using an Archie client, available from one of the many archie servers and other software sites.

Internet Gopher Gopher clients for most types of computers are available from the University of Minnesota, where Gopher was developed, by connecting to boombox.micro.umn.edu. Worldwide Gopher sites can be searched using the gopher tool Veronica.

World Wide Web This is another tool for retrieving information on the Internet and was developed at the European Particle Physics Laboratory (CERN). The ease of use of World Wide Web (WWW) clients simply by clicking with the mouse on highlighted words in the text has led to the creation of a very large number of server sites throughout the world. Mosaic software, developed at the University of Illinois, is available by anonymous ftp from ftp.ncsa.uiuc.edu or by Gopher from gopher.ncsa.uiuc.edu. The widely used Netscape software can be obtained from ftp.mcom.com.

33.8.6 Useful addresses

The following addresses are given to facilitate communication with the *Arabidopsis* community and to identify key persons or services. This is far from being an exhaustive list, but most sites have links to many others.

Stock centres
Mary Anderson: NASC (Nottingham *Arabidopsis* Stock Centre), Department of Life Science, University of Nottingham, University Park, Nottingham NG7 2RD, UK
Tel.: (+44 115) 979 1216
Fax: (+44 115) 951 3251
E-mail: *Arabidopsis*@nottingham.ac.uk
WWW: http://nasc.nott.ac.uk
Mary Anderson is the strain curator for AtDB.

Randy Scholl and Keith Davis: ABRC (Ohio State *Arabidopsis* Biological Resource Center), Ohio State University, 1735 Neil Avenue, Columbus, OH 43210, USA
E-mail: *Arabidopsis* +@osu.edu

Seed stocks
Tel.: (+1 614) 292-9371 for seed ordering information
Tel.: (+1 614) 292-1982, Randy Scholl (for general and seed-related questions)
Fax: (+1 614) 292-0603
E-mail: seeds@genesys.cps.msu.edu

DNA stocks
Tel.: (+1 614) 292-2115, Keith Davis (for DNA-related questions)
Fax: (+1 614) 292-0603
E-mail: dna@genesys.cps.msu.edu

AIMS
Contact Sakti Pramanik, Department of Computer Science, A729, Wells Hall, East Lansing, MI 48824, USA
For information on remote access, send an e-mail message with subject line 'help' to inquire-aims@aims.msu.edu
WWW to AIMS and ABRC: http://genesys.cps.msu.edu:3333/Help from: aims-manager@genesys.cps.msu.edu

AtDB
Located in the Department of Genetics, School of Medicine, Stanford University, USA
Contact atdb-curator@genome.stanford.edu
Anonymous ftp to ftp-genome.stanford.edu
WWW: http://genome www.stanford.edu/*Arabidopsis*/

***Arabidopsis* electronic newsgroup**
Information on all BIOSCI news groups and means of receiving messages can be obtained by anonymous ftp to net.bio.net in the folder pub/BIOSCI/doc or send the message help to biosci@daresbury.ac.uk (Europe, Africa and Central Asia) or biosci@net.bio.net (Americas and the Pacific rim). Do not send subscription messages to the list address.

Mutants curator
David Meinke, Oklahoma State University, Stillwater, OK 74078, USA
Fax: (+ 1 405) 744-7673
E-mail: btnydwm@mvs.ucc.okstate.edu
WWW: http://mutant.lse.okstate.edu/meinke.html

Physical and genetic mapping data
From the Ecker lab, University of Pennsylvania, USA
E-mail: atgc@atgenome.bio.upenn.edu
WWW: http://cbil.humgen.upenn.edu/~atgc/AT GCUP.html

***Arabidopsis* cDNA sequencing analysis project**
Contact Tom Newman, MSU, Michigan State University, East Lansing, MI 48824-1312, USA
A good description of the techniques to access information on ESTs is given in ref. 59.
Tel.: (+ 1 517) 353-0854
Fax: (+ 1 517) 353-9168
E-mail: tcn@msu.edu
WWW: http://lenti.med.umn.edu/*Arabidopsis*

French EST programme
Contact cooke@univ-perp.fr

Flower development
The Yanofsky Lab, San Diego, CA, USA
WWW: http://www-biology.ucsd.edu/others/yanofsky/home.html

Seeds
Lehle seeds, PO Box 2366, Round Rock, TX 78680-2366, USA
Tel.: (800) 881-3945 (USA only) or (+1 512) 388-3945 (outside USA)
Fax: (+ 1 512) 388-3974
WWW: http://www.*Arabidopsis*.com/

The Multinational Science Steering Committee
This committee is in charge of co-ordinating the Multinational Co-ordinated *Arabidopsis thaliana* Genome Research Project and is presently composed of: Chair: David Meinke, Oklahoma State University, Stillwater, Oklahoma, USA; Michel Caboche, Lab. Biol. Cellulaire, INRA, Versailles, France; Richard B. Flavell, John Innes Centre, Norwich, UK; Howard Goodman, Massachusetts General Hospital, Boston, Massachusetts, USA; Gerd Jurgens, University of Tübingen, Tübingen, Germany; Jose Martinez Zapater, DPTO de Proteccion Vegetal, Madrid, Spain; Bernard J. Mulligan, University of Nottingham, Nottingham, UK; Marc Van Montagu, University of Ghent, Ghent, Belgium; Robert Last, Boyce Thomson Institute, Ithaca New York, USA; Kiyotaka Okada, Kyoto University, Kyoto, Japan. Addresses available from AtDB.

Progress reports can be obtained from:
Fax: (+ 1 703) 644-4278
E-mail: pubs@nsf.gov
WWW: http://www.nsf.gov/bio/pubs/arabid/

EU ESSA Project
Mike Bevan, responsible for the EU ESSA project, John Innes Plant Science Centre, Cambridge Laboratory, Norwich, NR7 4UJ, UK
Tel.: (+ 44 1603) 452571
Fax: (+44 1603) 456844
E-mail: michael.bevan@bbsrc.ac.uk

General biological servers (excellent starting points)
Pedro's molecular biology research tools
WWW: http://www.public.iastate.edu/~pedro/research_tools.html
http://www.biophys.uniduesseldorf.de/bionet/research_tools.html
http://www.fmi.ch/biology/research_tools.html
http://www.peri.co.jp/Pedro/research_tools.html

Keith Robison's list of tools
WWW: http://golgi.harvard.edu/sequences.html

Atelier Bioinformatique de Marseille
WWW: http://www biol.univmrs.fr/biologie/logligne.html

Biological information servers, Stanford
WWW: gopher://genome-gopher.stanford.edu/11/bio
gopher: genome-gopher.stanford.edu/11/bio

Data banks: only WWW starting points are given
There is a wealth of information at all sites.

NCBI (GenBank, etc.)
http://www.ncbi.nlm.nih.gov

European Bioinformatics Institute (EMBL, etc.)
http://www.ebi.ac.uk

dbEST
http://www.ncbi.nlm.nih.gov/dbEST/index.html

Richard Cooke and Michel Delseny, the authors of this paper
cooke@univ-perp.fr and delseny@univ-perp.fr.

33.9 Arabidopsis as a model for other species

33.9.1 Arabidopsis as a model for plant species

The major advantage of deciphering the genome of a model plant species is certainly to progress in the understanding of plant biology and development, particularly of cultivated crops. Many genes regulating development are being genetically identified by mutations, but the corresponding gene has often still to be isolated and the biochemical function of its encoded protein elucidated. Such genes are likely to be conserved between all plant species. Owing to its small size, small genome and ease of genetic analysis, the model plant *Arabidopsis* allows most of these processes to be elucidated more easily and more rapidly than in any other plant species.

Arabidopsis will soon be the plant species for which we have the most detailed understanding of both its genome and physiology. Very close to it, at least for the cDNA and genome organization, is rice, for which more than 11 000 ESTs, a detailed RFLP map (at 1 cM resolution), and YAC, BAC and cosmid libraries are available (see Chapter 34). However, the rice genome is about four times bigger than that of *Arabidopsis* and therefore genomic sequence information should be available in the latter species much more rapidly. In addition, there are more known mutants in *Arabidopsis* and this species is still much easier to transform than rice.

Indeed, plant biologists need two models because higher plants are divided into two large groups, the monocotyledons and the dicotyledons, which, in addition to obvious morphological differences, differ in their codon usage. Monocotyledon codons are strongly biased toward C or G in the third position. Accordingly, few probes cross-hybridize between the two groups. *Arabidopsis* can be used as a model for dicots whereas rice is the model for monocots. Most *Arabidopsis* genes will cross-hybridize in relatively stringent conditions with genes from other crucifers or closely related families. Stringency will have to be reduced for more distantly related families.

Although it is not always possible to isolate genes from crops with *Arabidopsis* probes by cross-hybridization, the availability of large collections of ESTs from this species as well as from rice should allow identification of gene regions coding for conserved motifs. From these, it should be possible to derive PCR primers taking codon usage bias into account. Therefore, with the information from both model plants it should be possible to isolate most common genes without too great an effort [123]. In addition, the plant information, combined with that available from yeast, *C. elegans*, mouse and humans should help in designing PCR primers enabling homologous genes to be isolated from any living organism.

Many genes are not only conserved in function and sequence among plant species, but are also often in the same order on the chromosomes in different species. This phenomenon of synteny, which is most spectacular in cereals [150] (see Chapter 32), opens the way to comparative mapping and cloning. For example, it is now clear that rice chromosome 1 largely corresponds to wheat homeology group 3, and there are many examples of such synteny between rice, maize, sorghum, wheat and sugarcane [148,149]. Although such analysis is less advanced with *Arabidopsis*, which is a wild-type species, there is extensive information on the *Brassica* genomes, which belong to the same family [151–153]. The strategy for comparative cloning will be to map a gene of interest in a given crop, then move to the homologous region in the model species, isolate the functionally equivalent gene and then return with this probe to the original crop genomic library. Such a strategy should be extremely useful, for instance, in tracking disease-resistance genes since it has recently been demonstrated that the *RPG1* locus of soybean (conferring resistance to the pathovar *Pseudomonas syringae* pv *glycinea* avrB race) is functionally equivalent to the *Arabidopsis* locus *RPS3* [154]. Thanks to work on *Arabidopsis*, several important genes could be isolated in other species: a striking example is the *Arabidopsis AGAMOUS* gene, which controls flower development and for which homologues have been isolated in several species including maize and tomato [155,156].

A general strategy for identifying the function of an unknown gene is to antisense, overexpress or disrupt it and try to observe a phenotype in order to discover which process is altered. The first two strategies are not always successful, and gene disruption is not yet possible by homologous recombination in plants. However, the availability of a large collection of independent insertion mutants saturating the genome should allow the isolation of lines in which the desired gene is knocked out. Of course this strategy might not be successful either if, as in yeast, many gene disruptions do not confer a detectable phenotype. One of the major difficulties is certainly the presence of several copies of functionally similar genes.

The second problem, that of finding the gene corresponding to a mutant, can be solved without too much effort when fairly detailed genetic and physical maps are available. Because these two tools will soon be available for *Arabidopsis*, isolating a gene for which a mutant is known should become increasingly easier. Genes corresponding to specific mutations can also be isolated relatively easily if the mutation is tagged by a known insertion. Again, only *Arabidopsis* provides all these facilities in a plant, although some progress is also being made in rice. However, the final test to demonstrate that the correct gene has been isolated is to complement the mutant and this is still very difficult in most species.

33.9.2 Arabidopsis as a model for non-plant species

Arabidopsis can be used as a model for non-plant species, including animal genomes, in at least two ways. The first is to provide a source of homologous genes for comparison with those from other organisms. Comparative analysis enables identification of the conserved sequences and structures in a given protein that are really important for, say, its enzyme activity and its specific interactions with other proteins and substrates. The second is to use the ability to genetically manipulate the genome of this model plant in order to introduce genes from other genomes.

The major advantage of having several catalogues of ESTs from different species is in enabling the eventual determination of how many genes are common to all living kingdoms. Although we do not have any precise estimate, about a third of the non-redundant plant ESTs are homologous to something already in the databases. Many genes are indeed common to both animal and plant cells, and comparison of their sequences should tell us how basic conserved protein motifs have evolved and have become specialized for different functions. Several striking points have already emerged. Some well-conserved genes are represented in plants by multigene families that have many more members than their animal counterparts (Table 33.7). Each of these genes seems to have evolved a specific pattern of tissue- or organ-specific expression and differential response to environmental stress. Comparison of data from *Arabidopsis*, rice and other plant species suggests that most of these genes duplicated long ago, and now comprise several related subfamilies that have differentiated in different species.

One of the most striking examples of homology between plant and animal genes is in the homeobox genes that code for proteins with DNA-binding homeodomains [140]. These genes were initially described in *Drosophila*, where they are involved in pattern formation, segmentation and the specification of the various appendages. Similar genes have been found in vertebrates, where they also seem to be involved in segmentation. Recently, homologous homeobox-containing genes have been described in plants, where the homologous function does not exist [79]. Indeed, two such genes encode proteins of the phytochrome-controlled signal transduction cascade, and these genes are turned on by far-red light. However, *BELL1*, a member of the homeobox gene family, has recently been shown to be involved in pattern formation in the *Arabidopsis* ovule primordia [157]. The proteins encoded by these genes function as transcription factors, and the same motifs have most likely been re used to activate a different transduction pathway. Recently the *CURLY LEAF* gene (*CFL*) from Arabidopsis has been demonstrated to be an homologue of *Enhancer of Zeste* a member of the Polycomb gene family in *Drosophila* which control the activity of homeo domain genes. *CFL* is functionally similar in repressing the activity of *AGAMOUS*, a homeotic gene controlling flower formation, in vegetative tissues [158].

Several other genes coding for components of the light-signalling pathway (*COP1, COP9, COP11* and *DET1*) show striking similarities to *Drosophila*, human and *C. elegans* genes or ESTs, and might help in identifying developmental regulatory genes shared by plants and animals [159]. The same is true for the *myb* oncogenes, which code for a group of transcription regulatory proteins and which are represented by gene families in higher plants: at least five *myb*-related genes have been described in the snapdragon (*Antirrhinum*), and two from *Arabidopsis* have already been described and analysed in detail: one is responsible for the differentiation of trichomes [160] while the other is specifically induced by drought [77]. Another example of conservation is the 14-3-3-like protein gene [138], which was initially described as a brain-specific gene. In fact, it codes for a protein phosphatase inhibitor and this function has obviously been reused in several plant signal transduction pathways.

Some animal genes will repay study in the plant context, where they can be subject to fine genetic analysis and the effects of ectopic expression following transformation can be observed. For instance, the mode of action of some tumour suppressor genes and their relationship with the cell cycle might be much easier to analyse in *Arabidopsis* than in human cell cultures. This is certainly the case for the well-conserved Wilm's tumour suppressor, for which a plant homologue has been recently described in *Arabidopsis* and rice [161]. A further example is given by the recent demonstration that the *Arabidopsis* Cnx1 protein is capable of complementing the

Escherichia coli Moco (molybdenum cofactor) mutant *mogA* [162]. The plant protein shows homology to mammalian proteins whose function is uncertain, although the inborn loss of human Moco leads to impaired development of the brain and parts of the nervous system [163]. Further studies on Cnx1 could well shed light on the function of the mammalian homologues. Similarly, it might be of interest to learn that a gene which was initially supposed to be specific to the human brain but whose function is obscure is expressed in some plant tissues. Another similar example is the *dad1* gene (deficient in apoptotic death) which is also present in Arabidopsis [164]. This kind of link can provide hints concerning the function of some genes.

Acknowledgements

The authors wish to thank all their colleagues from the CNRS-GDR 1003 and from the EU-ESSA program, Tom Newman and Takuji Sasaki and their colleagues for fruitful and stimulating discussion. This work was supported by CNRS, ESSA and GREG (Groupe de Recherche et d'Etude des Génomes).

References

1 Flavell, R.B. (1980) The molecular characterization and organization of plant chromosomal DNA sequences. *Ann. Rev. Plant Physiol.* **31**, 569–596.

2 Vedel, F. & Delseny, M. (1987) Repetitivity and variability of higher plant genomes. *Plant Physiol. Biochem.* **25**, 191–210.

3 Arumuganathan, K. & Earle, E.D. (1991) Nuclear DNA content of some important plant species. *Plant Mol. Biol. Rpt.* **9**, 208–218.

4 Rédei, G.P. (1975) *Arabidopsis* as a genetic tool. *Ann. Rev. Genet.* **9**, 111–127.

5 Meyerowitz, E.M. (1987) *Arabidopsis thaliana*. *Ann. Rev. Genet.* **21**, 93–111.

6 Somerville, C.R. (1989) *Arabidopsis* blooms. *Plant Cell* **1**, 1131–1135.

7 Meyerowitz, E.M. (1989) *Arabidopsis*, a useful weed. *Cell* **56**, 263–270.

8 Sussman, M.R. (1992) Shaking *Arabidopsis thaliana*. *Science* **256**, 619.

9 Dean, C. (1993) Advantages of *Arabidopsis* for cloning plant genes. *Phil. Trans. R. Soc. Lond.* **B342**, 189–195.

10 Koncz, C., Chua, N.H. & Schell, J. (1992) *Methods in* Arabidopsis *research* (World Scientific, Singapore).

11 Meyerowitz, E. & Somerville, C. (1994) *Arabidopsis* (Cold Spring Harbor Laboratory Press, Cold Spring Harbor, NY).

12 Palmer, J.D., Downie, S.R., Nugent, J.M. *et al.* (1994) The chloroplast and mitochondrial DNAs of *Arabidopsis thaliana*: conventional genomes in an unconventional plant. In *Arabidopsis* (Meyerowitz, E. & Somerville, C., eds) (Cold Spring Harbor Laboratory Press, Cold Spring Harbor, NY).

13 Klein, M., Eckert-Ossenkopp, U., Schmiedeberg, I. *et al.* (1994) Physical mapping of the mitochondrial genome of *Arabidopsis thaliana* by cosmids and YAC clones. *Plant J.* **6**, 447–455.

14 Pruitt, R.E. & Meyerowitz, E.M. (1986) Characterization of the genome of *Arabidopsis thaliana*. *J. Mol. Biol.* **187**, 169–183.

15 Steinitz-Sears, L.M. (1963) Chromosome studies in *Arabidopsis thaliana*. *Genetics* **48**, 483–490.

16 Maluszynska, J. & Heslop-Harrison, J.S. (1991) Localization of tandemly repeated DNA sequences in *Arabidopsis thaliana*. *Plant J.* **1**, 159–166.

17 Maluszynska, J. & Heslop-Harrison, J.S. (1993) Molecular cytogenetics of the genus *Arabidopsis*: in situ localization of rDNA sites, chromosome numbers and diversity in centromeric heterochromatin. *Annls Bot.* **71**, 479–484.

18 Albini, S.M. (1994) A karyotype of the *Arabidopsis thaliana* genome derived from synaptonemal complex analysis at prophase I of meiosis. *Plant J.* **5**, 665–672.

19 Hauge, B.M. & Goodman, H.M. (1992) Genome mapping in *Arabidopsis*. In *Methods in* Arabidopsis *Research* (Koncz, C., Chua, N.H. & Schell, J., eds), 191–223 (World Scientific, Singapore).

20 Schmidt, R., West, J., Love, K. *et al.* (1995) Physical map and organization of *Arabidopsis thaliana* chromosome 4. *Science* **270**, 480–483.

21 Goodman, H.M., Ecker, J.R. & Dean, C. (1995) The genome of *Arabidopsis thaliana*. *Proc. Nat. Acad. Ssci. USA* **92**, 10831–10835.

22 Gruendler, P., Unfried, I., Pascher, K. & Schweizer, D. (1991) rDNA intergenic region from *Arabidopsis thaliana*, structural analysis intraspecific variation and functional implication. *J. Mol. Biol.* **221**, 1209–1222.

23 Copenhaver, G.P. & Pikaard, C.S. (1996) RFLP and physical mapping with an rDNA specific endonuclease reveals that nucleolus organizer regions of *Arabidopsis thaliana* adjoin the telomeres on chromosomes 2 and 4. *Plant J.* **9**, 259–272.

24 Campell, B.R., Song, Y., Posch, T.E. *et al.* (1992) Sequence and organization of 5S ribosomal RNA encoding genes of *Arabidopsis thaliana*. *Gene* **112**, 225–228.

25 Martinez-Zapater, J.M., Estelle, M.A. & Somerville, C.R. (1986) A highly repeated DNA sequence in *Arabidopsis thaliana*. *Mol. Gen. Genet.* **204**, 417–423.

26 Simoens, C.R., Gielen, J., Van Montagu, M. & Inzé, D. (1988) Characterization of highly repetitive sequences of *Arabidopsis thaliana*. *Nucleic Acids Res.* **16**, 6753–6766.

27 Richards, J. & Ausubel, F.M. (1988) Isolation of a higher eukaryotic telomere from *Arabidopsis thaliana*. *Cell* **53**, 127–136.

28 Richards, E.J., Goodman, H.M. & Ausubel, F.M. (1991) The centromere region of *Arabidopsis thaliana* chromosome 1 contains telomere similar sequences. *Nucleic Acids Res.* **19**, 3351–3358.

29 Voytas, D.F. & Ausubel, F.M. (1988) A copia-like transposable element family in *Arabidopsis thaliana*. *Nature* **336**, 242–244.

30 Konieczny, A., Voytas, D.F., Cumings, M.P. & Ausubel, F.M. (1991) A superfamily of *Arabidopsis thaliana* transposons. *Genetics* **127**, 801–809.

31 Peleman, J., Cottyn, B., Van Camp, W. *et al.* (1991) Transient occurence of extrachromosomal DNA of an *Arabidopsis thaliana* transposon-like element Tat 1. *Proc. Natl Acad. Sci. USA* **88**, 3618–3622.

32 Tsay, Y.F., Frank, M.J., Page, T. *et al.* (1993) Identification of a mobile endogenous transposon in *Arabidopsis thaliana*. *Science* **260**, 342–344.

33 Pélissier, T., Tutois, S., Derago, J.M. *et al.* (1995) *Athila*, a new retroelement from *Arabidopsis thaliana*. *Plant Mol. Biol.* **29**, 441–452.

34 Bell, C.J. & Ecker, J.R. (1994) Assignment of 30 microsatellite loci to the linkage map of *Arabidopsis*. *Genomics* **19**, 137–144.

35 Depeiges, A., Goubely, C., Lenoir, A. *et al.* (1995) Identification of the most represented repeated motifs in *Arabidopsis thaliana* microsatellite loci. *Theor. Appl. Genet.* **91**, 160–168.

36 Tourmente, S., Deragon, J.M., Lafleuriel, J. *et al.* (1994) Characterization of minisatellites in *Arabidopsis thaliana* with sequence similarly to the human minisatellite core sequence. *Nucleic Acids Res.* **22**, 3317–3321.

37 Regad, F., Lebas, M. & Lescure, B. (1994) Interstitial telomeric repeats within the *Arabidopsis thaliana* genome. *J. Mol. Biol.* **239**, 163–169.

38 Koornneef, M., Van Eden, J., Hanhart, C.J. *et al.* (1983) Linkage map of *Arabidopsis thaliana*. *J. Hered.* **74**, 265–272.

39 Rédei, G.P. & Koncz, C. (1992) Classical mutagenesis. In *Methods in* Arabidopsis *Research* (Koncz, C., Chua, N.H. & Schell, J. eds), 16–82 (World Scientific, Singapore).

40 Meinke, D.W. (1991) Embryonic mutants of *Arabidopsis thaliana*. *Dev. Genet.* **121**, 382–392.

41 Misera, S., Muller, A.J., Weiland-Heidecker. & Jürgens, G. (1994) The *FUSCA* genes of *Arabidopsis*: negative regulators of light responses. *Mol. Gen. Genet.* **244**, 245–252.

42 Ma, H. (1994) The unfolding drama of flower development: recent results from genetic and molecular analysis. *Genes Dev.* **8**, 745–756.

43 Jones, A.M. (1994) Surprising signals in plant cells. *Science* **263**, 183–184.

44 Deng, X.W. (1994) Fresh view of light signal transduction in plants. *Cell* **76**, 423–426.

45 Greenberg, J.T. & Ausubel, F.M. (1993) *Arabidopsis* mutants compromised for the control of cellular damage during pathogenesis and aging. *Plant J.* **4**, 327–341.

46 Parker, J.E., Szabo, V., Staskawicz, B.J. *et al.* (1993) Phenotypic characterization and molecular mapping of the *Arabidopsis thaliana* locus RPP5, determining disease resistance to *Peronospora parasitica*. *Plant J.* **4**, 821–831.

47 Mindrinos, M., Katagiri, F., Yu, G.L. & Ausubel, F.M. (1994) The *Arabidopsis thaliana* disease resistance RPS2 encodes a protein containing a nucleotide binding site and leucine-rich repeats. *Cell* **78**, 1089–1099.

48 Cherry, J.M., Cartinhour, S.W. & Goodman, H.M. (1992) AAtDB, an *Arabidopsis thaliana* database. *Plant Mol. Biol. Rpt.* **10**, 308–309.

49 Dennis, L., Dean, C., Flavell, R.B. *et al.* (1993) The multinational coordinated *Arabidopsis thaliana* genome research project. Progress report: year three. Publ. 94–4 (National Science Foundation, Washington DC).

50 Chang, C., Bowman, J.L., De John, A.W. *et al.* (1988) Restriction fragment length polymorphism linkage map for *Arabidopsis thaliana*. *Proc. Natl Acad. Sci. USA* **85**, 6856–6860.

51 Nam, H.G., Giraudat, J., Den Boer, B. *et al.* (1989) Restriction fragment length polymorphism linkage map of *Arabidopsis thaliana*. *Plant Cell* **1**, 699–705.

52 Hauge, B.M., Hanley, S.M., Cartinhour, S. *et al.* (1993) An integrated genetic/RFLP map of the *Arabidopsis thaliana* genome. *Plant J.* **3**, 745–754.

53 Stam, P. (1993) Construction of integrated genetic linkage maps by means of a new computer package JOINMAP. *Plant J.* **3**, 739–744.

54 Reiter, R.S., Williams, J.G.K., Feldmann, K.A. *et al.* (1992) Global and local genome mapping in *Arabidopsis thaliana* by using recombinant inbred lines and random amplified polymorphic DNA. *Proc. Natl Acad. Sci. USA* **89**, 1477–1481.

55 Lister, C. & Dean, C. (1993) Recombinant inbred lines for mapping RFLP and phenotypic markers in *Arabidopsis thaliana*. *Plant J.* **4**, 745–750.

56 Burr, M. & Burr, F.A. (1991) Recombinant inbreds for molecular mapping in maize: theoretical and practical considerations. *Trends Genet.* **7**, 55–60.

57 Hofte, H., Desprez, T., Amselem, J. (1993) An inventory of 1152 expressed sequence tags obtained by partial sequencing of cDNAs from *Arabidopsis thaliana*. *Plant J.* **4**, 1051–1061.

58 Creusot, F., Fouilloux, E., Dron, M. *et al.* (1995) The CIC library: a large insert YAC library for genome mapping in *Arabidopsis thaliana*. *Plant J.* **8**, 763–770.

59 Newman, T., De Bruijn, F., Green, P. *et al.* (1994) Genes galore: a summary of the methods for accessing the results of large scale partial sequencing of anonymous *Arabidopsis* cDNA clones. *Plant Physiol.* **106**, 1241–1255.

60 Vos, P., Hogers, R., Blecker, M. *et al.* (1995) AFLP: a new technique for DNA fingerprinting. *Nucleic Acids Res.* **23**, 4407–4414.

61 Konieczny, A. & Ausubel, F.M. (1993) A procedure for mapping *Arabidopsis* mutations using co-dominant ecotype-specific PCR based markers. *Plant J.* **4**, 403–410.

62 Jarvis, P., Lister, C., Szabo, V. & Dean, C. (1994) Integration of CAPS markers into the RFLP map generated using recombinant inbred lines of *Arabidopsis thaliana*. *Plant. Mol. Biol.* **24**, 685–687.

63 Ward, E.R. & Jen, G.C. (1990) Isolation of single copy sequence clones from a yeast artificial chromosome library of randomly-sheared *Arabidopsis thaliana* DNA. *Plant Mol. Biol.* **14**, 561–568.

64 Grill, E. & Somerville, C. (1991) Construction and characterization of a yeast artifical chromosome

library of *Arabidopsis* which is suitable for chromosome walking. *Mol. Gen. Genet.* **226**, 484–490.
65 Matallana, E., Bell, C.J., Dunn, P.J. *et al.* (1992) Genetic and physical linkage of the *Arabidopsis* genome: methods for anchoring yeast artificial chromosomes. In *Methods in* Arabidopsis *Research* (Koncz, C., Chua, N.H. & Schell, J. eds), 144–169 (World Scientific, Singapore).
66 Hwang, I., Kohchi, T., Hauge, B.M. *et al.* (1991) Identification and map position of YAC clones comprising one-third of the *Arabidopsis* genome. *Plant J.* **1**, 367–374.
67 Weigel, D., Alvarez, J., Smyth, D.R. *et al.* (1992) LEAFY controls floral meristem identity in *Arabidopsis. Cell* **69**, 843–859.
68 Schmidt, R. & Dean, C. (1993) Towards construction of an overlapping YAC library of the *Arabidopsis thaliana* genome. *BioEssays* **15**, 63–69.
69 Putterill, J., Robson, F., Lee, K. & Coupland, G. (1993) Chromosome walking with YAC clones in *Arabidopsis*: isolation of 1700 kb of contiguous DNA on chromosome 5 including a 300 kb region containing the flowering time gene Co. *Mol. Gen. Genet.* **239**, 145–157.
70 Leung, J., Bouvier-Durand, M., Morris, P.C. *et al.* (1994) *Arabidopsis* ABA response gene ABI 1: Features of a calcium-modulated protein phosphatase. *Science* **264**, 1448–1452.
71 Meyer, K., Leube, M.P. & Grill, E. (1994) A protein phosphatase 2C involved in ABA signal transduction in *Arabidopsis thaliana. Science* **264**, 1452–1455.
72 Zachgo, E.A., Wang, M.L., Dewdney, J. *et al.* (1996) A physical map of chromosome 2 of *Arabidopsis thaliana. Genome Res.* **6**, 19–25.
73 Liu, Y.G., Mitsukawa, N., Vazquez Tello, A. & Whittier, R.F. (1995) Generation of a high quality P1 library of *Arabidopsis* suitable for chromosome walking. *Plant J.* **7**, 351–358.
74 Choi, S.D., Creelman, R., Mullet, J. & Wing, R.A. (1995) Construction and characterization of a bacterial artificial chromosome library from *Arabidopsis thaliana. Weeds World* **2**, 17–20.
75 Braam, J. & Davis, R.W. (1990) Rain-, wind-, and touch-induced expression of calmodulin and calmodulin-related genes in *Arabidopsis. Cell* **60**, 357–364.
76 Yamaguchi-Shinozaki, K. & Shinozaki, K. (1993) Characterization of the expression of a dessication responsive rd29 gene of *Arabidopsis thaliana* and analysis of its promoter in transgenic plants. *Mol. Gen. Genet.* **236**, 331–340.
77 Urao, T., Yamaguchi-Shinozaki, K., Urao, S. & Shinozaki, K. (1993) An *Arabidopsis* myb homolog is induced by dehydration stress and its gene product binds to the conserved MYB recognition sequence. *Plant Cell* **5**, 1529–1539.
78 De Oliveira, D.E., Franco, L.O., Simoens, C. *et al.* (1993) Inflorescence specific genes from *Arabidopsis thaliana* encoding glycine-rich proteins. *Plant J.* **3**, 495–507.
79 Carabelli, M., Sessa, G., Baima, S. *et al.* (1993) The *Arabidopsis* Ath b2 and 4 genes are strongly induced by far-red-rich light. *Plant J.* **4**, 469–479.
80 Minet, M., Dufour, M.E. & Lacroute, F. (1992) Complementation of *Saccharomyces cerevisiae* auxotrophic mutants by *Arabidopsis thaliana* cDNAs. *Plant J.* **2**, 417–422.
81 Pang, Q., Hays, J.B. & Rajagopal, I. (1992) A plant cDNA that partially complements *E. coli* RecA mutation predicts a polypeptide not strongly homologous to RecA protein. *Proc. Natl Acad. Sci. USA* **89**, 8073–8070.
82 Pang, Q., Hays, J.B. & Rajagopal, I. (1993) Two cDNAs from the plant *Arabidopsis thaliana* that partially restore recombination proficiency and DNA damage resistance to *E. coli* mutants lacking recombination intermediate resolution activities. *Nucleic Acids Res.* **21**, 1647–1653.
83 Pang, Q., Hays, J.B., Rajagopal, I. & Schaefer, T.S. (1993) Selection of *Arabidopsis* cDNAs that partially correct phenotypes of E. coli DNA-damage sensitive mutants and analysis of two plant cDNAs that appear to express UV-specific dark repair activities. *Plant Mol. Biol.* **22**, 411–426.
84 Senecoff, J.F. & Meagher, R.B. (1993) Isolating the *Arabidopsis thaliana* genes for de novo purine synthesis by suppression of Escherichia coli mutants I.5′-phospho ribosyl-5-amino-imidazole-synthetase. *Plant Physiol.* **102**, 387–399.
85 Schnorr, K.M., Nygaard, P. & Laloue, M. (1994) Molecular characterization of *Arabidopsis thaliana* cDNAs encoding three purine biosynthetic enzymes. *Plant J.* **6**, 113–121.
86 Santerre, A. & Britt, A.B. (1994) Cloning of a 3-methyladenine-DNA glycosylase from *Arabidopsis thaliana. Proc. Natl Acad. Sci. USA* **91**, 2240–2244.
87 Sentenac, H., Bonneaud, N., Minet, M. *et al.* (1992) Cloning and expression in yeast of a plant potassium ion transport system. *Science* **256**, 663–665.
88 Anderson, J.A., Huprikar, S.S., Kochian, L.V. *et al.* (1992) Functional expression of a probable *Arabidopsis thaliana* potassium channel in Saccharomyces cerevisiae. *Proc. Natl Acad. Sci. USA* **89**, 3736–3740.
89 Sauer, N., Friedlander, K. & Grami-Wicke, U. (1990) Primary structure, genomic organization and heterologous expression of a glucose transporter from *Arabidopsis thaliana. EMBO J.* **9**, 3045–3050.
90 Sauer, N. & Stolz, J. (1994) Suc1 and Suc2: two sucrose transporters from *Arabidopsis thaliana*; expression and characterization in baker's yeast and identification of the histidine tagged protein. *Plant J.* **6**, 67–77.
91 Frommer, W.B., Hummel, S. & Riesmeier, W. (1993) Expression cloning in yeast of a cDNA encoding a broad specificity aminoacid permease from *Arabidopsis thaliana. Proc. Natl Acad. Sci. USA* **90**, 5944–5948.
92 Hsu, L., Chiou, T., Chen, L. & Bush, D.R. (1993) Cloning a plant aminoacid transporter by functional complementation of a yeast amino-acid transport mutant. *Proc. Natl Acad. Sci. USA* **90**, 7441–7445.
93 Steiner, H.Y., Song, W., Naider, F. *et al.* (1994) An *Arabidopsis* peptide transporter is a member of a new class of membrane transport proteins. *Plant Cell* **6**, 1289–1299.
94 Frommer, W.B., Hummel, S. & Rentsch, D. (1994) Cloning of an *Arabidopsis* histidine transporting protein related to nitrate and peptide transporters. *FEBS Lett.* **347**, 185–189.

95 Ninnemann, O., Jauniaux, J.C. & Frommer, W.B. (1994) Identification of a high affinity NH^+_4-transporter from plants. *EMBO J.* **13**, 3464–3471.
96 Ferreira, P.C.G., Hemerly, A.S., Villarroel, R. *et al.* (1991) The *Arabidopsis* functional homolog of the p34 cdc2 protein-kinase. *Plant Cell* **3**, 531–540.
97 Eberhard, J., Raesecke, H.R., Schmid, J. & Amrhein, N. (1993) Cloning and expression in yeast of a higher plant chorismate mutase: molecular cloning, sequencing of the cDNA and characterization of the *Arabidopsis thaliana* enzyme expressed in yeast. *FEBS Lett.* **334**, 233–236.
98 Corey, E.J., Matsuda, S.P.T. & Bartel, B. (1993) Isolation of an *Arabidopsis thaliana* gene encoding cycloartenol synthase by functional expression in a yeast mutant lacking lanosterol synthase by the use of a chromatographic screen. *Proc. Natl Acad. Sci. USA.* **90**, 11628–11632.
99 Leustek, T., Murillo, M. & Cervantes, M. (1994) Cloning of a cDNA encoding ATP-sulfurylase from *Arabidopsis thaliana* by functional expression in *Saccharomyces cerevisiae*. *Plant Physiol.* **105**, 897–902.
100 Nasr, F., Bertauche, N., Dufour, M.E. *et al.* (1994) Heterospecific cloning of *Arabidopsis thaliana* cDNA by direct complementation of pyrimidine auxotrophic mutants of *Saccharomyces cerevisiae*. I. Cloning and sequence analysis of two cDNAs catalysing the second, fifth and sixth steps of the de novo pyrimidine biosynthes is pathway. *Mol. Gen. Genet.* **244**, 23–32.
101 Kammerloher, W., Fisher, U., Piechottka, G.P. & Schüffuer, A.R. (1994) Water channels in the plant plasma membrane cloned by imonunoselection from a mammalian expression system. *Plant J.* **6**, 187–199.
102 Gallois *et al.* (1997) *Plant J.* in press.
103 Sun, T.P., Goodman, H.M. & Ausubel, F.M. (1992) Cloning the *Arabidopsis* GA1 locus by genomic subtraction. *Plant Cell* **4**, 119–128.
104 Giraudat, J., Hauge, B.M., Vallon, C. *et al.* (1992) Isolation of the *Arabidopsis* ABI 3 gene by positional cloning. *Plant Cell* **4**, 1251–1261.
105 Arondel, V., Lemieux, B., Hwang, I. *et al.* (1992) Map-based cloning of a gene controlling omega-3 fatty acid desaturation in *Arabidopsis*. *Science* **258**, 1353–1355.
106 Feldmann, K.A. (1991) T-DNA insertion mutagenesis in *Arabidopsis*: mutational spectrum. *Plant J.* **1**, 75–82.
107 Feldmann, K.A. (1992) T-DNA insertion mutagenesis in *Arabidopsis*: seed infection/transformation. In *Methods in* Arabidopsis *Research* (Koncz, C., Chua, N.H. & Schell, J., eds), 274–289 (World Scientific, Singapore).
108 Errampali, D., Patton, D., Castle, L. *et al.* (1991) Embryogenic lethals and T-DNA insertional mutagenesis in *Arabidopsis*. *Plant Cell* **3**, 149–157.
109 Koncz, C., Nemeth, K., Rédei, G.P. & Schell, J. (1991) T-DNA insertional mutagenesis in *Arabidopsis*. *Plant Mol. Biol.* **20**, 963–976.
110 Bechtold, N., Ellis, J. & Pelletier, G. (1993) In planta *Agrobacterium* mediated gene transfer by infiltration of adult *Arabidopsis thaliana* plants. *C.R. Acad. Sci. Paris Sci. Vie* **316**, 1194–1199.
111 Bouchez, D., Camilleri, C. & Caboche, M. (1993) A binary vector based on basta resistance for in planta transformation of *Arabidopsis thaliana*. *C.R. Acad. Sci. Paris, Sci. Vie* **316**, 1188–1193.
112 Koonin, E.V., Bork, P. & Sander, C. (1994) Yeast chromosome III: new gene functions. *EMBO J.* **23**, 493–503.
113 McKinney, E.C., Ali, N., Traut, A. *et al.* (1995) Sequence-based identification of T-DNA insertion mutations in *Arabidopsis*: actin mutants act 2-1 and act 4-1. *Plant J.* **8**, 613–622.
114 Aarts, M.G.M., Dirkse, W.G., Stiekema, W.J. & Peireira, A. (1993) Transposon tagging of a male sterility gene in *Arabidopsis*. *Nature* **363**, 715–717.
115 Long, D., Martin, M., Sundberg, E. *et al.* (1993) The maize transposable element Ac/Ds as a mutagen in *Arabidopsis*: identification of an albino mutation induced by DNA insertion. *Proc. Natl Acad. Sci. USA* **90**, 10370–10374.
116 Bancroft, I., Jones, J.D.G. & Dean, C. (1993) Heterologous transposon tagging of the DRL1 locus in *Arabidopsis*. *Plant Cell* **5**, 631–638.
117 Topping, J.F., Agyeman, F., Henricot, B. & Lindsey, K. (1994) Identification of molecular markers of embryogenesis in *Arabidopsis thaliana* by promoter trapping. *Plant J.* **5**, 895–903.
118 Devic, M., Hecht, V., Berger, C. *et al.* (1995) Assessment of promoter trap as a tool to study zygotic embryogenesis in *Arabidopsis thaliana*. *C. R. Acad. Sci. Paris, Vie* **318**, 121–128.
119 Topping, J.F., Wei, W. & Lindsey, K. (1991) Functional tagging of regulatory elements in the plant genome. *Development* **112**, 1009–1101.
120 Sundaresan, V., Springer, P., Volpe, T. *et al.* (1995) Patterns of gene action in plant development revealed by enhancer trap and gene trap transposable elements. *Genes Dev.* **9**, 1797–1810.
121 White, O., Dunning, T., Sutton, G. *et al.* (1993) A quality control algorithim for DNA sequencing projects. *Nucleic Acids Res.* **21**, 3829–3838.
122 Boguski, M.S., Lowe, T.M.J. & Tolstoshev, C. (1993) dbEST-database for 'expressed sequence tags'. *Nature Genet.* **4**, 332–333.
123 Cooke, R., Raynal, M. & Laudié, M. (1996) Further progress towards a catalogue of all *Arabidopsis* genes: analysis of a set of 5000 non-redundant ESTs. *Plant J.* **9**, 101–124.
124 Altschul, S.F., Gish, W., Miller, W. *et al.* (1990) Basic Local alignment search tool. *J. Mol. Biol.* **215**, 403–410.
125 Bairoch, A. (1992) 'PROSITE': a dictionary of sites and patterns in protein. *Nucleic Acids Res.* **20**, 2013–2018.
126 Pearson, W.R. & Lipman, D.U. (1988) Improved tools for biological sequence comparison. *Proc. Natl Acad. Sci. USA* **85**, 2444–2448.
127 Sonnhammer, E.L.I. & Kahn, D. (1994) Modular arrangement of proteins as inferred from analysis of homology. *Prot. Sci.* **3**, 482–492.
128 Sasaki, T., Song, J.Y. & Koga Ban, Y. *et al.* (1994) Toward cataloging all rice genes: large scale sequencing of randomly chosen rice cDNA from a callus cDNA library. *Plant J.* **6**, 615–624.
129 Green, P., Lipnau, D.L.H., Waterson, R., States, D. & Claverie, J.M. (1993) Ancient conserved regions in

new gene sequences and protein databases. *Science* **259**, 1711–1716.

130 Krebbers, E., Seurinck, J., Herdies, L. *et al.* (1988) Four genes in two diverged subfamilies encode the ribulose 1, 5 biphosphate carboxylase small subunit polypeptides of *Arabidopsis thaliana*. *Plant Mol. Biol.* **11**, 745–759.

131 Sharrock, R.A. & Quail, P.H. (1989) Novel phytochrome sequences in *Arabidopsis thaliana*: structure evolution and differential expression of a plant regulatory photoreceptor family. *Genes Dev.* **3**, 1745–1757.

132 Harper, J.F., Manney, L., Dewitt, N.D. *et al.* (1990) The *Arabidopsis thaliana* plasma membrane H+-ATPase multigene family. *J. Biol. Chem.* **265**, 13601–13608.

133 Ma, H., Yanofsky, M.F. & Meyerowitz, E.M. (1991) AGL 1–AGL 6, an *Arabidopsis* gene family with similarity to floral homeotic and transcription factor genes. *Genes Dev.* **5**, 484–495.

134 Kopczak, S.D., Haas, N.A., Hussey, P.J. *et al.* (1992) The small genome of *Arabidopsis* contain at least six expressed alpha tubulin genes. *Plant Cell* **4**, 539–547.

135 Snustad, D.P., Haas, N.A., Kopczak, S.D. & Silflow, C.D. (1992) The small genome of *Arabidopsis* contains at least nine expressed beta tubulin genes. *Plant Cell* **4**, 549–556.

136 Genschik, P., Durr, A. & Fleck, J. (1994) Differential expression of several E2 type ubiquitin carrier protein genes at different developmental stages in *Arabidopsis thaliana* and *Nicotiana sylvestris*. *Mol. Gen. Genet.* **244**, 548–556.

137 Mizoguchi, T., Hayashida, N., Yamaguchi-Shinozaki, K. *et al.* (1993) ATMPKS: a gene family of plant MAP kinases in *Arabidopsis thaliana*. *FEBS Lett.* **336**, 440–444.

138 Ferl, R.J., Lu, G.H. & Bowen, B.W. (1994) Evolutionary umplications of the family of 14-3-3 brain protein homologs in *Arabidopsis thaliana*. *Genetica* **92**, 129–138.

139 Villemur, R., Haas, N.A., Joyce, C.M. *et al.* (1994) Characterization of few new β-tubulin genes and their expression during male flower development in maize (*Zea mays* L). *Plant Mol. Biol.* **24**, 295–315.

140 Schena, M. & Davis, R. (1994) Structure of homeobox-leucine zipper genes suggests a model for the evolution of gene families. *Proc. Natl Acad. Sci. USA* **91**, 8393–8397.

141 Thümmler, F., Kirchner, M., Teuber, R. & Dittrich, P. (1995) Differential accumulation of the transcripts of 22 novel protein kinase genes in *Arabidopsis thaliana*. *Plant Mol. Biol.* **29**, 551–565.

142 Rounsley, S.D., Ditta, G.S. & Yanofsky, M.F. (1995) Diverse roles for MADS box genes in *Arabidopsis* development. *Plant Cell* **7**, 1259–1269.

143 Rivera-Madrid, R., Mestres, D., Marinho, P. *et al.* (1995) Evidence for five divergent thioredoxin h sequences in *Arabidopsis thaliana*. *Proc. Natl Acad. Sci. USA* **92**, 5620–5624.

144 Herzog, M., Dorne, A.M. & Grellet, F. (1995) GASA, a gibberellin regulated gene family from *Arabidopsis thaliana* related to the tomato GAST1 gene. *Plant Mol. Biol.* **27**, 743–752.

145 Parcy, F., Valon, C., Raynal, M. *et al.* (1994) Regulation of gene expression programs during *Arabidopsis* seed development: roles of the ABI 3 locus and endogenous abscisic acid. *Plant Cell* **6**, 1567–1582.

146 Bevan, M. (1996) Sequencing chromosome 4 of *Arabidopsis*. Plant Genome IV. San Diego, 14–18 January 1996. Abstract w21.

147 Gaubier, P., Wu, H.J., Laudié, M. *et al.* (1995) A chlorophyll synthetase gene from *Arabidopsis thaliana*. *Mol. Gen. Genet.* **249**, 58–64.

148 Anh, S. & Tanksley, S.D. (1993) Comparative linkage maps of the rice and maize genomes. *Proc. Natl Acad. Sci. USA* **90**, 7980–1984.

149 Kurata, N., Moore, G. & Nagamura, Y. *et al.* (1994) Conservation of genome structure between rice and wheat. *Biotechnology* **12**, 276–281.

150 Moore, G., Devos, K. & Gale, M. (1995) Grasses, line up and form a circle. *Curr. Biol.* **5**, 737–739.

151 Kowalski, S.P., Lan, T.H., Feldmann, K.A. & Paterson, A.H. (1994) Comparative mapping of *Arabidopsis thaliana* and *Brassica oleracea* chromosomes reveals islands of conserved organization. *Genetics* **138**, 499–510.

152 Lagercrantz, U., Putterill, J., Coupland, G. & Lydiate, D. (1996) Comparative mapping in *Arabidopsis* and *Brassica* fine scale genome collinearity and congruence of genes controlling flowering time. *Plant J.* **9**, 13–20.

153 Sadowski, J., Gaubier, P., Delseny, M. & Quiros, C.F. (1996) Genetic and physical mapping of a gene cluster from *Arabidopsis thaliana* in *Brassica* diploid species. *Mol. Gen. Genet.* **251**, 298–306.

154 Innes, R.W., Bisgrove, S.R., Smith, N.M. *et al.* (1993) Identification of a disease resistance locus in *Arabidopsis* that is functionally homologous to the RPG1 locus of Soy bean. *Plant J.* **4**, 813–820.

155 Schmidt, R.J., Veit, B., Mandel, M.A. *et al.* (1993) Identification and molecular characterization of ZAG 1 the maize homolog of the *Arabidopsis* floral homeotic gene AGAMOUS. *Plant Cell* **5**, 729–737.

156 Prueli, L., Hareven, D., Rounsley, S.D. *et al.* (1994) Isolation of the tomato AGAMOUS gene TAG1 and analysis of its homeotic role in transgenic plants. *Plant Cell* **6**, 163–173.

157 Reiser, L., Modrusan, Z., Margossian, L. *et al.* (1995) The BELL 1 gene encodes a homeodomain protein involved in pattern formation in the *Arabidopsis* ovule primordia. *Cell* **83**, 735–742.

158 Goodrich, J., Puangsomlee, P., Martin, M., Long. D., Meyerowitz, E. & Coupland, G. (1997) A Polycornb-group gene regulates homeotic gene expression in Arabidopsis. *Nature* **386**, 44–51.

159 Chamowitz, D.A. & Deng, X.W. (1995) The novel components of the *Arabidopsis* light signalling pathway may define a group of general developmental regulators shared by both animal and plant kingdoms. *Cell* **82**, 353–354.

160 Oppenheimer, D.G., Herman, P.L., Sivakumaran, S. *et al.* (1991) A myb gene required for leaf trichome differentiation in *Arabidopsis* is expressed in stipules. *Cell* **67**, 483–493.

161 Rivera-Madrid, R., Marinho, P., Chartier, Y. & Meyer, Y. (1993) Nucleotide sequence of an *Arabidopsis thaliana* cDNA clone encoding a homolog to a suppressor of Wilm's tumor. *Plant Physiol.* **102**, 329–330.

162 Stallmeyer, B., Nerlich, A., Schiemann, J. *et al.* (1995) Molybdenum co-factor biosynthesis: the *Arabidopsis thaliana* cDNA CN x 1 encodes a multifunctional two-domain protein homologous to a mammalian neuroprotein, the insect protein cinammon and three *Escherichia coli* proteins. *Plant J.* **8**, 751–762.
163 Johnson, J.L., Rajagopalan, K.V. & Wadman, S.K. (1993) Human molybdenum cofactor deficiency. *Adv. Exp. Med. Biol.* **338**, 373–378.
164 Gallois, P., Makishima, T., Hecht, V. *et al.* (1997) An Arabidopsis cDNA complementating a hamster apoptosis suppressor mutant. *Plant J.* (in press).
165 Li, J., Zhao, J. & Rose, A.B. (1995) *Arabidopsis* phosphoribosyl anthranilate isomerase: molecular genetic analysis of triplicate tryptophan pathway genes. *Plant Cell* **7**, 447–461.
166 Feiler, M.S., Desprez, T., Santoni, V. *et al.* (1995) The higher plant *Arabidopsis thaliana* encodes a functional CDC48 homologue which is highly expressed in dividing and expanding cells. *EMBO J.* **14**, 5626–5637.
167 Rion, C., Tourte, Y., Lacroute, F. & Karst, F. (1994) Isolation and characterization of a cDNA encoding *Arabidopsis thaliana* mevalonate kinase by genetic complementation in yeast. *Gene* **148**, 293–297.
168 D'Enfert, C., Gensse, M. & Gaillardin, C. (1992) Fission yeast and a plant have functional homologues of the Sar1 and Sec12 proteins involved in ER to Golgi traffic in budding yeast. *EMBO J.* **11**, 4205–4211.
169 Lee, H., Gal, S., Newman, T.C. & Raikhel, N.W. (1993) The *Arabidopsis* endoplasmic reticulum retention receptor functions in yeast. *Proc. Natl Acad. Sci. USA* **90**, 11433–11437.
170 Frommer, W.B., Hummel, S., Unseld, M. & Ninnemann, O. (1995) Seed and vascular expression of a high affinity transporter for cationic amino acids in *Arabidopsis*. *Proc. Natl Acad. Sci. USA* **92**, 12036–12040.
171 Leyser, H.M.O., Lincoln, C.A., Timpte, C. *et al.* (1993) *Arabidopsis* auxin-resistance gene axrl encodes a protein related to ubiquitin activating enzyme E1. *Nature* **364**, 161–164.
172 Putteril, J., Robson, F., Lee, K., Simon, R. & Coupland, G. (1995) The CONSTANS gene of *Arabidopsis* promoter flowering and encodes a protein showing similarities to zinc finger transcript factors. *Cell* **80**, 847–857.
173 Pepper, A., Delaney, T., Washburn, T. *et al.* (1994) DET1, a negative regulator of light mediated development and gene expression in *Arabidopsis*, encodes a novel nuclear localized protein. *Cell* **78**, 109–116.
174 Chang, C., Kwok, S.F., Bleeker, A.B. & Meyerowitz, E.M. (1993) *Arabidopsis* ethylene-response gene ETR1: similarity of product to two component regulators. *Science* **262**, 539.
175 Goto, K. & Meyerowitz, E.M. (1994) Function and regulation of the *Arabidopsis* floral homeotic gene PISTILLATA. *Genes Dev.* **8**, 1548–1560.
176 Grant, M.R., Godiard, L., Stranbe, E. *et al.* (1995) Structure of the *Arabidopsis* RPM1 gene enabling dual specificity disease resistance. *Science* **269**, 843–846.
177 Yanofsky, M.F., Ma, H., Bowman, J.L. *et al.* (1990) The protein encoded by the *Arabidopsis* homeotic gene AGAMOUS resembles a transcription factor. *Nature* **346**, 35–38.
178 Mandel, M.A., Gustafson-Brown, C., Savidge, B. & Yanofsky, M.F. (1992) Molecular characterization of the *Arabidopsis* floral homeotic gene APETALA 1. *Nature* **360**, 273–277.
179 Jofuku, K.D., Den Boer, B.G.W., Van Montagu, M. & Okamuro, J.K. (1994) Control of *Arabidopsis* flower and seed development by homeotic gene APETALA2. *Plant Cell* **6**, 1211–1225.
180 Hayashi, H., Czaja, I., Lubenow, H. *et al.* (1992) Activation of a plant gene by T-DNA tagging: auxin independent growth *in vitro*. *Science* **258**, 1350–1353
181 Aarts, M.G.M., Keijzer, C.J., Stiekema, W.J. & Pereira, A. (1995) Molecular characterization of the CER1 gene of *Arabidopsis* involved in epicuticular wax biosynthesis and pollen fertility. *Plant Cell* **7**, 2115–2127.
182 Negruk, V., Yang, P., Subramanian, M. *et al.* (1996) Molecular cloning and characterization of the CER2 gene of *Arabidopsis thaliana*. *Plant J.* **9**, 137–145.
183 Tsay, Y., Schroeder, J.I., Feldmann, K.A. & Crawford, N.M. (1993) The herbicide sensitivity gene CHL 1 of *Arabidopsis* encodes a nitrate-inducible nitrate transporter. *Cell* **72**, 705–713.
184 Koncz, C., Mayerhofer, R., Koncz-Kalman, Z. *et al.* (1990) Isolation of a gene encoding a novel chloroplast protein by T-DNA tagging in *Arabidopsis thaliana*. *EMBO J.* **9**, 1337–1346.
185 Deng, X.W., Matsui, M., Wei, N. *et al.* (1992) COP 1, an *Arabidopsis* regulatory gene encodes a protein with both a zinc binding motif and a Gb homologous domain. *Cell* **71**, 791–802.
186 Wei, N., Chamovitz, D.A. & Deng, W. (1994) *Arabidopsis* Cop 9 is a component of a novel signaling complex mediating light control of development. *Cell* **78**, 117–124.
187 Kieber, J.J., Rothenberg, M., Roman, G. *et al.* (1993) CTR1: a negative regulator of the ethylene response pathway in *Arabidopsis thaliana* encodes a member of the Raf family of protein kinase. *Cell* **72**, 427–441.
188 Feldmann, K.A., Marks, M.D., Cristianson, M.L. & Quatrano, R. (1989) A dwarf mutant of *Arabidopsis* generated by T-DNA insertion mutagenesis. *Science* **243**, 1351–1354.
189 Shevell, D.E., Len, W.M., Gillmor, C.S. *et al.* (1994) EMB 30 is essential for normal cell division, cell expension and cell adhesion in *Arabidopsis* and encodes protein that has similarity to Sec 7. *Cell* **77**, 1051–1062.
190 Okuley, J., Lightner, J., Feldmann, K. *et al.* (1994) *Arabidopsis* FAD2 gene encodes the enzyme that is essential for polyunsaturated lipid biosynthesis. *Plant Cell* **6**, 147–158.
191 Yadaw, N.S., Wiersbicki, A., Aergester, M. *et al.* (1993) Cloning of higher plant w-3 fatty acid desaturases. *Plant Physiol.* **103**, 467–476.
192 James, D.W., Lim, E., Keller, J. *et al.* (1995) Directed tagging of the *Arabidopsis* FATTY ACID ELONGATION (FAE1) gene with the maize transposon activator. *Plant Cell* **7**, 309–319.
193 Castle, L.A. & Meinke, D.W. (1994) A FUSCA gene of

Arabidopsis encodes a novel protein essential for plant development. *Plant Cell* **6**, 25–41.

194 Chiang, H.H., Hwang, I. & Goodman, H.M. (1995) Isolation of the *Arabidopsis* GA4 locus. *Plant Cell* **7**, 195–201.

195 Rerie, W.G., Feldmann, K.A. & Marks, M.D. (1994) The GLABRA 2 gene encodes a homeodomain protein required for normal trichome development in *Arabidopsis*. *Genes Dev.* **8**, 1388–1395.

196 Ahmad, M. & Cashmore, A.R. (1993) HY4 gene of *Arabidopsis thaliana* encodes a protein with characteristics of a blue light photoreceptor. *Nature* **366**, 162–166.

197 Lee, I., Auckerman, M.J., Gore, S.L. *et al.* (1994) Isolation of LUMINIDEPENDENS: a gene involved in the control of flowering time in *Arabidopsis*. *Plant Cell* **6**, 75–83.

198 Reiter, R.S., Coomber, S.A., Bourett, T.M. *et al.* (1994) Control of leaf and chloroplast development by the *Arabidopsis* gene pale cress. *Plant Cell* **6**, 1253–1264.

199 Van Lijsebettens, M., Vanderhaeghen, R., DeBlock, M. *et al.* (1994) An S18 ribosomal protein gene copy at the *Arabidopsis* PFL locus affects plant development by its specific expression in meristem. *EMBO J.* **13**, 3378–3388.

200 Roe, J.L., Rivin, C.J., Sessions, R.A. *et al.* (1993) The TOUSLED gene in *Arabidopsis thaliana* encodes a protein kinase homolog that is required for leaf and flower development. *Cell* **75**, 939–950.

201 Franzmann, L.H., Yoon, E.S. & Meinke, D.W. (1995) Saturating the genetic map of *Arabidopsis thaliana* with embryonic mutations. *Plant J.* **7**, 341–350.

Chapter 34 Rice as a model for genome analysis

Akio Miyao, Kimiko Yamamoto*, Ilkka Havukkala*, Masahiro Yano, Nori Kurata, Yuzo Minobe & Takuji Sasaki

Rice Genome Research Program, National Institute of Agrobiological Resources, 1–2 Kannondai 2-chome, Tsukuba, Ibaraki 305, Japan

** Rice Genome Research Program, Institute of Society for Techno-Innovation of Agriculture, Forestry and Fisheries, 446-1, Ippaizuka Kamiyokoba, Tsukuba, Ibaraki 305, Japan*

34.1 Introduction

Rice (*Oryza sativa* L.) is the main food in Asia, and for thousands of years varieties have been selected on the basis of the best agronomic characters. Rice breeding has been very successful in improving yields and crop reliability, but further improvements using biotechnology are necessary to keep up with the population growth of Asia.

During the past two decades the molecular biology of rice has advanced greatly, especially in Japan. Much information on its genetics is already available, with over 150 markers on the classical linkage map [1]. Rice has also recently emerged as a model plant for mapping all cereal genomes and isolating agronomically or scientifically important genes for plant breeding. This is due to its small genome size (430 megabases (Mb), the smallest in cereals), the availability of detailed linkage maps [2,3], and the large amount of synteny found between rice and other cereals [4]. Rice as a model monocot plant also complements the dicot model plant *Arabidopsis thaliana* (see Chapter 33), and together they will reveal the genome organization and gene function of most plants.

The Japanese Rice Genome Research Program (RGP) was launched in 1991 by the Ministry of Agriculture, Forestry and Fisheries to carry out comprehensive genome mapping and large-scale expressed gene sequencing. The aim is to promote biotechnological applications in rice and cereal breeding worldwide. For this purpose information on the research results is released as soon as possible.

Since 1992, RGP has published a semiannual newsletter, now being sent free of charge to some 2000 researchers in over 50 countries. Some 10000 restriction fragment length polymorphism (RFLP) markers and 1000 cDNA clones have been distributed to about 200 laboratories worldwide and over 10000 cDNA sequences made available in the international databanks. In December 1994, an Internet information service on rice genome research data was launched on the World Wide Web (address: http://www.staff.or.jp) as well as an ftp server (address: ftp.staff.or.jp) for distributing large data files to all plant genome researchers.

This chapter describes the current techniques being used and the results from the first three years of the programme, with technical details that should be helpful for people launching or expanding other plant genome projects.

34.2 Large-scale rice cDNA analysis

34.2.1 cDNA library construction and sequencing

In recent years, several large-scale cDNA projects have been in progress for various organisms, including humans [5], rice [6,7], *Arabidopsis* [8] and the nematode *Caenorhabditis elegans* [9,10]. Large-scale cDNA analyses have two great advantages for the investigation of genomes. First, isolated and partially characterized cDNA clones can be used as probes for genetic linkage and physical mapping. The cDNA clones have been used not only as expressed sequence tags (ESTs) on the RFLP linkage map [2] (Section 34.4) but also as good probes to screen YAC clones for the construction of the physical map of the chromosomes (Section 34.5). Second, isolated cDNA clones encode the amino acid sequence of expressed proteins. Therefore, if the cDNA library contains the cDNAs of all expressed proteins, the primary structure of any protein can be obtained from the library. A good-quality cDNA clone library will also be a powerful tool for isolation and characterization of useful genes for breeding and other applications.

In *Arabidopsis*, about 25000 genes are thought to be expressed [11,12]. In the case of rice, the size of genome is considered to be about three times larger than that of *Arabidopsis* [13], and the total number of the expressed genes in rice is estimated to be roughly 30000. The final aim of the large-scale cDNA analysis in rice is to catalogue the cDNAs of all expressed genes. Toward this goal, the RGP has been isolating and sequencing rice cDNAs. The isolation and characterization of several rice cDNA clones showing sequence similarities with known genes and proteins have previously been reported, such as the ATP/ADP translocator [14], the mitochondrial ATPase β-subunit [15], *cdc2* [16] and the NADP-dependent malic enzyme [17]. About 2200 callus cDNAs have recently been isolated and partially sequenced, and around 700 of these were found to have significant homologies with known proteins [6]. A *japonica* rice variety, 'Nipponbare', has been used for cDNA analysis.

In order to characterize the cDNAs of all expressed genes, it is necessary to construct various cDNA libraries prepared from different rice tissues, including callus, under different growing conditions, since many rice genes are expected to be expressed only in specific tissues and at specific growth stages, or only when the plant is exposed to specific environmental stresses. So far, we have prepared cDNA libraries from root, green shoot and etiolated shoot. We have also made libraries from calli grown in four different culture conditions: growth-phase callus grown in medium with 2,4-dichlorophenoxyacetic acid; 6-benzyladenine-treated callus grown in a medium with 6-benzyladenine (BA); redifferentiation callus grown in a medium with 6-benzyladenine and 1-naphthaleneacetic acid; and heat-treated callus grown at 37 °C (Figs 34.1 and 34.2).

Figure 34.3 shows the strategy for cDNA analysis used in the RGP. Total RNAs were isolated from the tissues by a single-step guanidinium thiocyanate method (see Chapter 18). The poly(A)$^+$ RNA was purified through Oligotex-dT 30 (Daiichi Kagaku, Japan). Then cDNAs were synthesized according to Superscript Plasmid System (BRL). Adaptors were ligated asymmetrically to cDNAs and cloned into pBluescript II SK+ having a *Sal*I site at the 5′ end and a *Not*I site at the 3′ end. Transformation was performed into NM522 competent cells (Stratagene). Clones were picked randomly for sequencing. The insert length of cDNAs in plasmid DNAs was checked by agarose gel electrophoresis after double digestion with *Sal*I and *Not*I. The plasmid DNAs were prepared by a robotic machine (Kurabo, Japan).

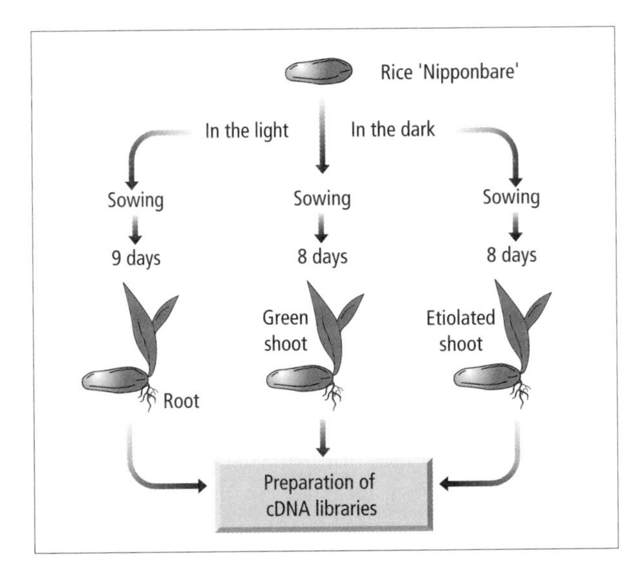

Fig. 34.1 Source materials for root and shoot cDNA libraries. Roots and green shoots were harvested from seedlings grown in the light at 9 and 8 days after sowing, respectively. Aetiolated shoots was harvested from seedlings grown in the dark at 8 days after sowing. Total mRNA was isolated from the harvested tissues and cDNA libraries were prepared.

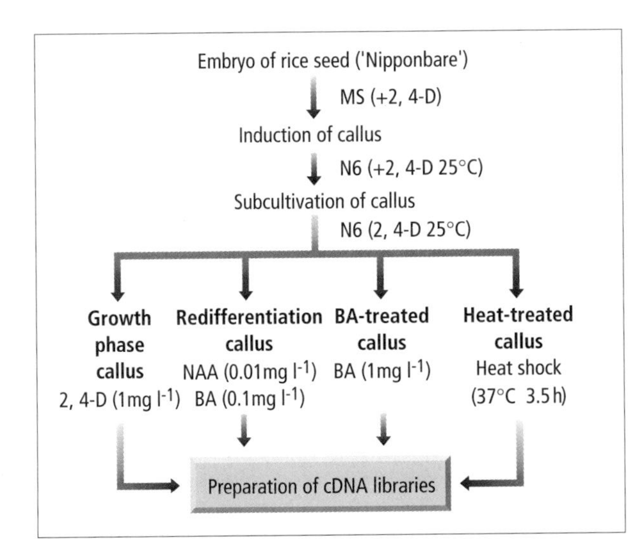

Fig. 34.2 Source materials for callus cDNA libraries. Induced callus was subcultivated in N6 medium containing 2,4-D and then grown in four different conditions as indicated. Callus was harvested after 12 days cultivation, and total mRNA isolated.

Template DNA was prepared as single-stranded (ss) DNA rescued by helper phage M13KO7 (see Chapter 21), and sequenced by the dideoxy method (see Chapter 22). Recently, several robotic workstations have been introduced to scale up this step. In the RGP, several manual and automated ssDNA preparation systems have been used in parallel as summarized below.

1 *Classical method* PEG-phenol method: a manual system requiring 3 h for preparation of 24 samples (see Chapter 21).

2 *Fast Magnetic Purification Kit (Amersham, UK)* Uses magnetic beads: a manual system requiring 1.5 h for preparation of 24 samples.

3 *EasyPrep M13 Prep Kit (Pharmacia Biotech, USA)* Uses glass-fibre filter: a manual system requiring 1.5 h for preparation of 24 samples.

Sequencing reaction:

1 *Cycle sequencing method* (see Chapter 22) *with a thermal cycler* Manual system requiring 3 h for processing of 24 samples.

2 *BcaBEST Dideoxy Sequencing Kit (TaKaRa, Japan)* Manual system requiring 2 h for processing of 24 samples.

3 *Catalyst Robotic Workstation (Perkin Elmer/Applied Biosystems, USA)* Automatic system requiring 5 h for processing of 24 samples.

We also use a DNA sequencing robot (Amersham) as a semiautomatic system for preparing ssDNA for sequencing reactions. It is a combination system with a Fast Magnetic Purification system and a Δ*Taq* cycle sequence. The system carries out these steps for 24 samples within 5 h. Many companies are now developing this type of automated system and their use will become mainstream in large-scale DNA template preparation technology. Sequencing is being carried out by automated DNA sequencers using chemical labelling (ABI model 373A, Perkin Elmer/Applied Biosystems). In the RGP at present, 11 DNA sequencers are being run in parallel. Currently, over 100 000 bp of cDNAs can be

Table 34.1 No. of analysed rice cDNAs as of February 1995.

cDNA library	Analysed clones	Hit clones[a]	%[b]	Submitted clones[c]
Callus				
BA[d] treatment	1915	472	24.6	608
NAA[e] + BA treatment	442	118	26.7	0
Heat shock treatment[f]	2673	581	21.7	0
Growth phase[g]	2492	740	29.7	2447
Root	1965	577	26.0	1849
Green shoot	4759	1280	26.9	3431
Etiolated shoot	4211	905	21.5	2654
Others	2511	567	22.6	0
Total	20968	5174	24.7	10989

[a] Clones with significant similarities to known proteins. The FASTA algorithm was used for the similarity search to the PIR database and an optimized score of at least 200 was required for putative assignment.
[b] (Hit clones/analysed clones) × 100.
[c] The sequence data of clones were submitted to the DDBJ database.
[d] BA, 6-benzyladenine.
[e] NAA, 1-naphthaleneacetic acid.
[f] This callus was treated at 37 °C for 3.5 h, after incubating at 25 °C for 12 days.
[g] Callus grown in medium with 2,4-dichlorophenoxyacetic acid (2,4-D).

sequenced per day. New types of sequencers that can analyse a larger number of samples and read longer sequences faster, are being developed. The rice cDNA analysis in the RGP was started in the autumn of 1991, and as of February 1995, we have isolated and partially sequenced over 20 000 cDNA clones from various cDNA libraries prepared from intact plant tissues and calli (Table 34.1). The sequence data from the cDNA clones have been stored in our in-house database RiceBase (see Section 34.6). Over 10 000 sequences have already been released and are available through DNA Data Bank of Japan (DDBJ), GenBank and EMBL.

34.2.2 cDNA identification, tissue-specificity and redundancy

Nucleotide sequences were transferred via a computer network to the main computer. The sequence data were translated to amino acid sequences for all three frames, and a similarity search of each sequence with known protein sequences in the PIR database was done using the FASTA algorithm [18]. The cDNA clones whose sequences showed an optimized similarity score over 200 were selected and considered to have homologies with the proteins in the database. We frequently found clones showing homologies with several different proteins. In such cases, we tentatively identified the cDNAs as encoding the protein showing the highest score among the candidate proteins.

The number of cDNA clones analysed in the RGP from October 1991 to February 1995 is summarized in Table 34.1. These sequenced clones were examined for similarities of predicted amino acid sequences to known proteins in the PIR database using the FASTA algorithm. Only 25% of the clones showed significant similarities with known proteins. Thus, most of the isolated cDNA clones are considered to encode unknown proteins. Table 34.2 shows some of the putatively identified genes from the root cDNA library and the growth-phase callus cDNA library.

The results of the similarity search showed that the cDNA libraries from various tissues were significantly different in their specificity of gene expression. As expected, the clones related to photosynthetic proteins were mainly obtained from the green shoot cDNA library. The cDNA clones from etiolated shoot complemented those from green shoot, but some proteins with unknown functions, such as viscotoxin, were also found. This suggests that some genes suppressed under light are stimulated and expressed in dark. In the root cDNA library, clones with significant homology to peroxidase, ribosomal proteins, and some metal-binding proteins were identified.

The characterization of the clones obtained from callus cDNA libraries also showed features of gene expression specific to the growth conditions (Fig. 34.2). Many ribosomal proteins and histone genes were found in growth-phase callus. For BA-treated callus, several chitinase genes, including an unknown plant chitinase class III, were identified. The cDNA library from redifferentiation callus included a higher percentage of clones encoding α-

Table 34.2 List of putatively identified proteins from rice callus and root cDNA.

Clone name	DDBJ accession no.	Length in aa	Match (%)	Initial score	Optimized score	Putatively identified protein name	Original species name
RA0578	D23922	92	89.8	233	234	(S)-tetrahydroberberine oxidase	*Coptis japonica*
RA1538	D24218	115	100	524	524	14-3-3 protein homologue	Rice
RA2648	D24845	112	91	415	415	2,3-bisphosphoglycerate-independent phosphoglycerate mutase, PGAM	Maize
RA0411	D23852	116	87.9	517	517	26S protease subunit 4	Human
RA1356	D39052	111	92.8	477	477	3-Oxoacyl-[acyl-carrier-protein] synthase precursor, chloroplast	Barley
CK0982	D15628	118	100	338	338	Actin 1	Rice
CK1555	D15925	120	84.9	271	275	Acyl carrier protein 3	Barley
RA3246	D25115	109	97.2	485	485	Adenosylhomocysteinase (EC 3.3.1.1)	Madagascar periwinkle
RA3368	D25150	90	100	478	478	ADP, ATP carrier protein	Rice
CK1681	D22874	120	97.8	476	476	Alcohol dehydrogenase, ADH1	Rice
RA1895	D24441	142	89.4	635	645	Aspartate aminotransferase	Proso millet
RA0210	D23806	124	89.1	321	330	Aspartic proteinase	Rice
RA1884	D24433	143	100	478	478	ATP/ADP translocator protein	Rice
CK0686	D15470	133	100	236	236	ATPase	Rice
CK2996	D23554	87	94	254	256	BBC1 protein	*Arabidopsis thaliana*
RA1753	D24337	119	98.9	396	396	Calmodulin	Rice
RA2856	D24965	146	86.3	614	614	Casein kinase II α-chain	Maize
RA2656	D24852	118	94.1	614	614	Catalase chain 1	Maize
RA2531	D24771	112	97.7	205	205	cdc2 protein kinase homolog 1	Rice
CK1328	D15815	131	97.5	500	504	Chaperonin 60	Cucurbit
RA2584	D24805	121	89.2	593	593	Cinnamyl-alcohol dehydrogenase	Kidney bean
CK0258	D15204	109	81.7	336	417	Cold-induced protein BnC24A	Rape
RA3384	D39350	131	86.2	512	524	Cyc07 protein, S-phase specific	Madagascar periwinkle
CK1525	D15917	116	93.5	273	278	Cytochrome b5	Rice
CK1991	D15999	109	97.3	325	327	Cytochrome c	Rice
CK1229	D15777	100	94.1	287	287	Elongation factor 1 β′-chain	Rice
RA2717	D24888	70	100	215	215	Elongation factor eEF-1 α-chain	Wheat
CK2630	D16057	130	80.2	456	456	Elongation factor eEF-1 β-A1 chain	*Arabidopsis thaliana*
RA2741	D24902	127	88.7	311	312	Elongation factor eEF-2	*Chlorella kessleri*
RA2075	D24506	122	95.1	546	546	Enolase	Maize
RA0111	D23770	140	83.6	649	649	Formate dehydrogenase precursor, mitochondrial	Potato
RA0876	D24022	91	100	246	246	Fructose-bisphosphate aldolase, cytosolic	Rice
CK1047	D15663	122	98.8	393	393	GF14-12 protein	Maize
CK1142	D15718	147	94.6	706	708	Glucose-1-phosphate adenylyltransferase	Barley

Continued on p. 794.

Table 34.2 *Continued.*

Clone name	DDBJ accession no.	Length in aa	Match (%)	Initial score	Optimized score	Putatively identified protein name	Original species name
CK2690	D16060	151	89.2	535	538	Glutamate synthase (NADH) (EC 1.4.1.14)	Alfalfa
CK2834	D16071	128	100	233	233	Glutamate–ammonia ligase(EC 6.3.1.2) 1 precursor, chloroplast	Maize
RA2455	D24733	118	89	481	492	Glycine hydroxymethyl-transferase (EC 2.1.2.1)	*Flaveria pringlei*
CK1818	D22933	110	97.7	272	392	Glycine-rich cell wall structural protein 2 precursor	Rice
CK0553	D38799	112	83.3	357	357	Glycine-rich protein	Maize
CK1467	D28232	136	100	382	382	GOS2 protein	Rice
CK0661	D15451	111	100	411	411	GTP-binding protein	Rice
CK2495	D16050	114	100	308	308	GTP-binding protein rab	Garden pea
CK1388	D15842	84	98.8	410	410	GTP-binding protein rgp2	Rice
CK0922	D22687	109	80.2	477	487	GTP-binding regulatory protein β-chain homologue	*Chlamydomonas reinhardtii*
RA2635	D24836	113	98.4	335	336	Guanine nucleotide regulatory protein	Fava bean
RA1512	D24194	119	100	326	326	H^+-transporting ATP synthase β chain, mitochondrial	Maize
RA1571	D24242	134	98.5	669	669	Heat shock protein 82, HSP82	Rice
CK1497	D15907	133	94	318	387	High mobility group protein	Maize
CK0893	D22681	87	100	247	247	Histone H2B.2	Wheat
CK1254	D22765	108	90	346	347	Histone H4	Maize
RA2248	D24610	95	95.1	272	272	Immunoglobulin-binding protein homolog b70	Maize
RA2031	D24486	92	100	479	479	Initiation factor 4A	Rice
CK1526	D22833	119	97.4	565	565	Initiation factor eIF-4A	Curled-leaved tobacco
RA2404	D24702	104	98.7	367	367	Initiation factor eIF-5A	Common tobacco
RA0195	D23800	108	92.5	479	498	Isocitrate dehydrogenase	Soybean
RA0078	D23757	169	84.2	430	430	KatA protein	*Arabidopsis thaliana*
CK2640	D23334	143	90.1	627	628	Ketol-acid reductoisomerase (EC 1.1.1.86)	*Arabidopsis thaliana*
RA2209	D24582	119	85.6	477	481	Lipoamide dehydrogenase, LDH, L subunit of glycine decarboxylase	Peas
RA0886	D24025	96	90.6	275	275	Malate dehydrogenase precursor, mitochondrial	Water-melon
RA3079	D39236	115	94.6	352	352	Methionine adenosyltransferase (EC 2.5.1.6)	Tomato
CK1912	D15997	126	97.1	469	469	Monoubiquitin-tail protein 2	Barley
CK0153	D28180	126	83.3	231	246	Nuclear antigen 21D7	Carrot

Continued.

Table 34.2 *Continued.*

Clone name	DDBJ accession no.	Length in aa	Match (%)	Initial score	Optimized score	Putatively identified protein name	Original species name
RA2290	D24637	127	82.9	493	503	Nucleoside diphosphate kinase I	Spinach
CK3114	D23625	65	92.2	269	269	Oleosin KD 18	Maize
RA0634	D23944	120	100	692	692	Oryzain α precursor	Rice
CK0180	D16134	135	87	483	492	Peptidylprolyl isomerase	Maize
RA1733	D24323	96	99	424	424	Phenylalanine ammonia-lyase	Rice
RA3349	D25146	106	92.5	524	524	Phospho-2-dehydro-3-deoxyheptonate aldolase (EC 4.1.2.15)	Tomato
CK3137	D23639	111	94.5	484	484	Phosphoglycerate kinase, cytosolic	Wheat
CK1074	D15678	217	90.1	385	385	Phospholipid transfer protein homologue	Rice
RA2304	D24644	115	94.8	607	607	Phosphoprotein Phosphatase (EC 3.1.3.16) type 2A	Alfalfa
RA0067	D38976	120	100	375	375	Polyubiquitin protein	*Arabidopsis thaliana*
CK2588	D16056	136	100	634	634	Proliferating cell nuclear antigen	Rice
CK0536	D15369	128	91.7	218	218	Pyruvate decarboxylase	Maize
CK1296	D38834	140	91.8	563	564	Pyruvate kinase, cytosolic	Potato
CK2680	D23355	87	86.2	258	260	Rab25 protein	Rice
RA0665	D23963	117	86	394	398	Rho1Ps = ras-related small GTP-binding protein	Garden pea
CK0385	D15270	108	96.4	411	411	Ribosomal 5S RNA-binding protein	Rice
CK2415	D23198	109	98.5	305	305	Ribosomal protein L2	Tomato (cv. Moneymaker)
CK2471	D38915	113	82.4	293	293	Ribosomal protein L23a	Rat
CK2111	D16034	135	100	660	660	Ribosomal protein L3	Rice
CK2450	D23222	100	83.6	321	335	Ribosomal protein L36a.e	Yeast (*Schwanniomyces occidentalis*)
CK0634	D22628	115	92.4	347	354	Ribosomal protein L37a	Turnip
CK2214	D23111	134	82.6	310	315	Ribosomal protein L38	Tomato (cv. Moneymaker)
CK2322	D16041	131	92.5	469	501	Ribosomal protein L7a	Rice
CK2268	D16039	95	89.6	352	352	Ribosomal protein S11	Maize
CK1804	D22928	96	86	208	208	Ribosomal protein S13	Maize
RA1809	D24377	133	100	488	488	Ribosomal protein S14 (clone MCH2)	Maize
CK1575	D22842	125	91.2	332	332	Ribosomal protein S16	Large-leaved lupine
CK1930	D22980	151	89.1	271	284	Ribosomal protein S20	Rice
CK1135	D15714	100	100	247	247	Ribosomal protein S21	Rice
RA0038	D38974	138	85	437	437	Ribosomal protein S25	Tomato
CK2303	D23139	152	81	355	357	Ribosomal protein S27	Rat
CK2977	D23541	66	100	207	207	Ribosomal protein S4	Potato
RA0528	D23895	125	88.9	205	210	Ribosomal protein S5	Rat
CK1004	D22713	101	90.5	218	218	RL5 ribosomal protein	Alfalfa
CK0904	D28208	134	98.1	506	506	Sa1T protein precursor	Rice
CK2540	D23280	109	82.4	509	511	Starch phosphorylase	Potato
RA1801	D24369	129	100	477	477	Sucrose synthase	Rice

Continued on p. 796

Table 34.2 *Continued.*

Clone name	DDBJ accession no.	Length in aa	Match (%)	Initial score	Optimized score	Putatively identified protein name	Original species name
RA1410	D24133	130	100	721	721	Sucrose-phosphate synthase	Rice
RA3035	D39209	140	100	401	401	Superoxide dismutase (Cu-Zn) (clone RSODA)	Rice
RA1625	D24279	119	91.1	479	483	T complex polypeptide 1	Oat
RA0821	D24003	104	86.9	375	377	Tat-binding protein-1	Human
RA0260	D38993	110	81.5	331	331	Tonoplast intrinsic protein gamma	*Arabidopsis thaliana*
RA2017	D24474	87	89.7	427	427	Trg-31 protein	Garden pea
CK2209	D23110	110	100	486	486	Triose-phosphate isomerase (EC 5.3.1.1)	Rice
CK2505	D23254	113	100	582	582	Tubulin α-1 chain	Rice
RA1623	D24277	119	95.5	569	575	Tubulin β-6 chain	*Arabidopsis thaliana*
CK0176	D22531	76	100	204	204	Ubiquitin	Rice
RA0552	D23906	120	100	346	346	Ubiquitin extension protein	White lupin
RA2710	D24887	115	93	516	521	UTP–glucose-1-phosphate uridylyltransferase	Potato
CK1761	D15975	119	84	496	496	Valosin-containing protein	Mouse
CK2490	D38921	135	87.3	303	308	Voltage dependent anion channel, VDAC	Wheat
RA0480	D23874	131	100	515	515	Ypt family	Maize

CK and RA are the cDNA libraries from growth phase callus in the presence of 2,4-D, and young root, respectively. The similarities of clones to the PIR database were examined using FASTA algorithm. The proteins that showed optimized scores greater than 200 were putatively assigned to the cDNA clones.

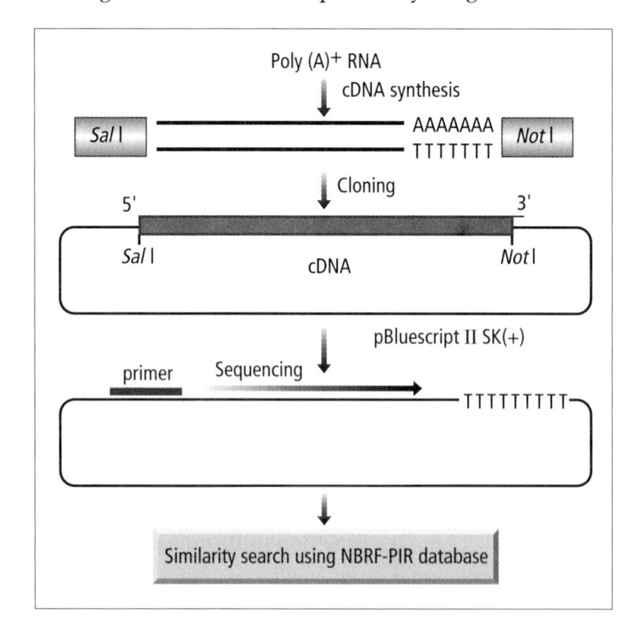

Fig. 34.3 Strategy for cDNA analysis. Poly(A)$^+$ RNA was isolated from each tissue and cDNAs were synthesized. Partial DNA sequences of randomly selected cDNA clones were determined and sequence similarities analysed. The details of each step are described in the text.

amylase when compared with other calli. It was not surprising that heat-shock proteins were prominent in the cDNA library derived from heat-treated callus. Some clones, such as the DNAs for ubiquitin and elongation factor, were isolated to some extent from all libraries.

These results clearly indicate that the composition of the clones in each cDNA library reflects the regulation of gene expression related to differentiation, growth conditions or environmental stress. Since each library includes a large number of uncharacterized cDNA clones, further investigation of the clones obtained from these cDNA libraries might provide new insights of the genes or proteins that play important roles in gene regulation in rice. Among the clones analysed, the percentage of unique clones was about 50% (Fig. 34.4). This indicates that many unique clones of expressed rice genes can be effectively isolated by large-scale cDNA analysis.

The redundancy analyses within each library were also performed using the BLAST algorithm [19]. After BLAST analysis, if the clone included a 50-bp region that showed more than 90% homology

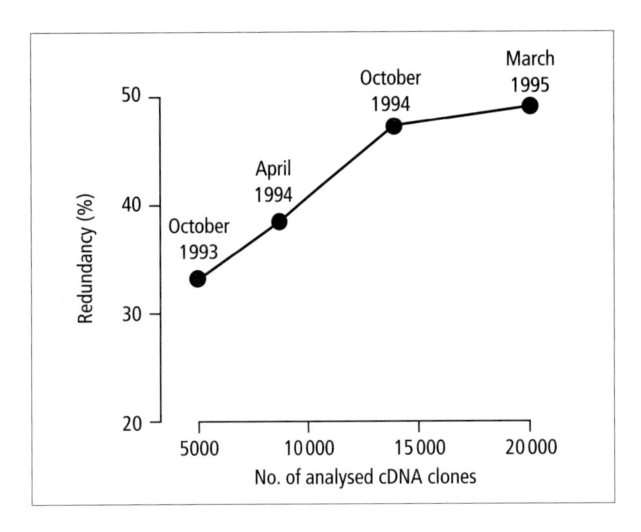

Fig. 34.4 Relationship between redundancy and the number of analysed cDNA clones. Each dot shows the redundancy of analysed cDNA clones and the number of clones at the times indicated.

or a 30-bp region that showed 100% homology with other clones, we regarded the clone as a redundant clone. Although the redundancy increased with the increase in the number of cDNA clones analysed, the increasing rate seemed to decline after the number of analysed clones became larger than 15000 (Fig. 34.4). The results are considered to be due to the properties of the cDNA libraries that were analysed. The ratio of unique clones in the libraries might be higher than in previously analysed libraries.

Many of the partially characterized cDNA clones are also effectively used as ESTs on our RFLP linkage map and for the construction of a physical map using YAC contigs (see Section 34.4).

34.2.3 Toward the complete catalogue of all rice genes

Large-scale cDNA analysis is a very useful method for the investigation of expressed genes in rice, and provides good probes for the construction of an RFLP or a physical map. In the RGP, we aim to catalogue all the cDNAs of the expressed genes, including tissue-specific, developmental stage-specific and stress-specific cDNA clones and the cDNAs from various tissues, including calli grown in different conditions. We have partially sequenced over 20000 cDNA clones and about 10000 clones (50%) were shown to be unique, as mentioned above. This means that we have already captured about one-third of the total expressed genes in rice.

All of the partially sequenced cDNA clones were analysed for similarities to PIR database entries. Some genes were found to be expressed differentially in specific tissues. These data will be useful in investigating the potential functions of such unknown proteins in plants and to widen the knowledge of protein families.

Our current method of cDNA analysis continues to be effective in identifying novel unique clones, although the ratio of unique clones has decreased with the increase in the number of clones analysed (around 50% at present). Furthermore, to identify other tissue-specific genes, cDNA analysis of clones from several stages of panicle development are now in progress.

As a new approach, we have been developing a high-density cDNA filter hybridization system which carries 768 (96×8 dot array) colonies per filter. By using these high-density filters, which include all isolated cDNA clones, we are now planning to screen all our cDNAs against each other. Simultaneous screening of all the isolated cDNAs in the RGP (about 20000 clones) will become possible by this system. Such screening will rapidly give a lot of useful genetic information, such as redundancy or tissue specificity of each cDNA clone. Furthermore, we also plan to construct a similar system using YAC or cosmid clones as probes. These systems are also expected to have large advantages in map-based cloning (Section 34.5).

34.3 PCR techniques for rice genome mapping

34.3.1 RAPD analysis

In the RGP, polymerase chain reaction (PCR) techniques are used extensively for detection of DNA polymorphism useful for linkage map construction [2,20,21], for screening YAC and cosmid clones, and for labelling probe DNAs, etc. For linkage analysis, we use both RFLP and PCR polymorphism techniques (see Chapter 5), although mainly we use sequenced cDNA clones, often characterized by similarity search. Using this strategy, we obtain map positions of genes with known functions, as indicated by similarity search. During the construction of our linkage map, however, some regions could not be mapped with cDNA clones. It seems that expressed genes are not equally distributed throughout rice chromosomes. For example, in the upper part of chromosome 9 in our map (Fig. 34.5), there was a largish region which could not be mapped with cDNA clones.

The random amplified polymorphic DNA (RAPD) method [21–23] (see Chapter 5, Section 5.4.2) can detect polymorphisms not only in coding regions, but also in noncoding regions, because RAPDs are amplified with genomic DNA templates.

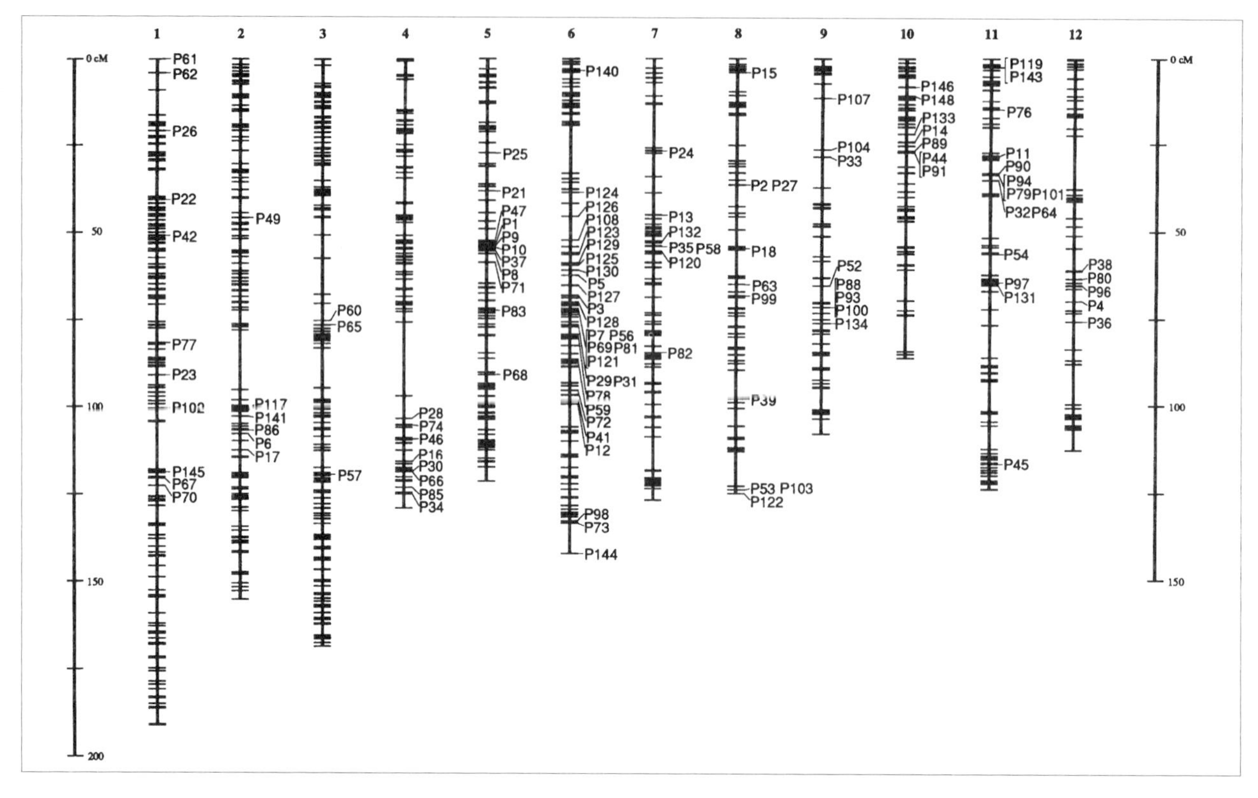

Fig. 34.5 Linkage map of PCR markers. RAPD markers are shown with locus names. Markers without names are RFLP markers.

For this reason we also used RAPD for linkage analysis. RAPD has also been used in other plants, such as *Arabidopsis* [24], soybean [25], black mustard [26], sugarcane [27] and alfalfa [28].

The RAPD method also has the following advantages.

1 It needs little space and simple equipment: only a PCR machine, an electrophoresis system and a system for detection of amplified DNA.

2 It needs a very small amount (nanograms for one reaction) of template DNA, much less than hybridization methods. This reduces the amount of sample plants needed, time for template DNA isolation, and the cost of handling materials.

3 It detects many bands per lane in gels—that is, many loci can be analysed at one time.

Normally, we use primers 20–25 nucleotides long in pairs for a simple PCR. Because these primers hybridize to a single locus of the genome, we can obtain specific DNA fragments from each locus. However, the best primer length is 9–10 nucleotides for the RAPD reaction. There are about 4^{10} times more homologous regions for a 10-nucleotide primer in the rice genome than for a 20-nucleotide primer. If we use such short primers for PCR amplification, many fragments are amplified at once. If mutations on primer annealing sites or large insertions/deletions between primer annealing sites exist in the genome, these loci cannot be amplified by PCR. If small insertions/deletions between annealing sites exist on the genome, the amplified fragments will show fragment length polymorphism. This way we can detect DNA polymorphisms that have become amplified in the genomes of different rice cultivars.

RAPD reactions are conventionally carried out with a single short primer, but in our laboratory, we routinely use pairs of 10-nucleotide primers. If we used a single primer for each RAPD reaction, 5000 different reactions would need 5000 different primers. However, if we use primer pair combinations, we only need 100 different primers for the same number of reactions ($100 \times 100/2$). There is also the additional advantage of detecting more amplified fragments. When we use primer A and primer B in one reaction, the amplified fragments can have three possible combinations of fragment ends:

1 both ends are primer A;

2 both ends are primer B; or

3 one end is primer A and the other end is primer B.

We can thus detect additional DNA fragments that have different primer sequences at opposite ends of the amplified fragment.

For linkage analysis, polymorphisms are screened

at first between parent plants. Using primer sets which showed polymorphism, genotypes of each F2 plant are obtained and then linkage calculated between markers from each genotype by linkage analysis software MAPMAKER [29]. Figure 34.5 shows the RAPD markers mapped on the RFLP linkage map of Kurata *et al.* [2]. Many of our RAPD markers have been useful in that they are located in areas with only few RFLP markers.

34.3.2 Determination of sequence-tagged sites from RAPD

Gel profiles of amplified fragments containing RAPD are sometimes complicated. When nonpolymorphic bands are observed near the RAPD band, it is difficult to determine the genotype from that RAPD band and to screen YAC and cosmid clones using RAPD fragments as probes. However, if RAPD fragments are cloned and sequenced, marker regions can be detected directly as sequence-tagged sites (STSs), which are also called sequence characterized amplified regions (SCARs) [30], by designing 20-mer primer pairs which amplify the target locus as a single band. We routinely clone RAPD fragments and design suitable STS primer pairs. If the amplified fragment with the STS primer pair [21,20] is used as a hybridization probe for RFLP analysis, it can detect codominant segregation for the linkage analysis. Moreover, because these STS primer pairs can amplify a single locus, these primer pairs are useful to isolate YAC and cosmid clones for making contigs.

34.3.3 Bulked segregant analysis using the RAPD method

The RAPD method can also be used to obtain markers that are closely linked to a specific target site. This method is called *bulked segregant analysis* [31,32]. Using this method, we were able to fill the gaps in linkage maps of chromosome 7 and 9. Bulked segregant analysis is an efficient method, not only for constructing many markers around a target locus, but also for tagging phenotypicly important loci with DNA markers. For gene tagging, we make two groups of F2 segregants according to the target phenotype. Genes for the phenotypes **A** and **a** are *A* and *a*, respectively. Group **A** has segregants whose phenotypes are **A** (genotypes are *AA* and *Aa*). Group **a** has segregants whose phenotypes are **a** (genotype is *aa*). Normally, we choose 10–15 segregants for making groups.

To make mixed templates of groups **A** and **a**, respectively, rice leaves of segregants are mixed and DNA extracted or each group of template DNAs mixed. The mixed template DNA of group **A** has the DNA segment which has the gene *A* and *a*. The mixed template DNA of group **a** has the DNA segment which has the gene *a*. Other loci of gene *A* and gene *a* are the same between the mixed templates of groups **A** and **a**. This means that polymorphisms are detected specifically around the genes *A* (and *a*) on the chromosome. When 10 segregants are mixed, the length of segments which have an *A* or an *a* gene is about 10 cM. These mixed (bulked) templates are then screened for RAPDs with many primers (or primer pairs). When a RAPD is detected in group **A**, the locus of the RAPD should be within 10 cM of the gene *A*.

Using near-isogenic lines [33], as well as bulked F2 segregants, this method can identify linked markers to important genes. For filling gaps and low-density marker regions in the RFLP map, one or two RFLP markers in the ends of these regions are chosen and groups of segregant DNA with genotypes of the two RFLPs are made. Using these templates, we can obtain RAPD markers in the target region. With this technique and 80 primers (by combination) we have set 18 markers in three regions in our RFLP map.

34.3.4 Single-strand conformation polymorphism analysis

It is difficult to detect polymorphisms between closely related rice cultivars, because there are few nucleotide differences between them. In such a case, more effective methods than RFLP and RAPD are necessary. Single-strand conformation polymorphism (SSCP) analysis [34,35] is a simple and sensitive PCR-based method for the detection of polymorphisms (see Chapter 5, Protocol 7). Single-stranded DNA in a nondenaturing polyacrylamide gel adopts a specific conformation depending on its nucleotide sequence. This affects its running in the gel and even one nucleotide difference between DNA fragments can be detected by the SSCP method. We mapped 17 SSCP markers in our RFLP linkage map [36]. Heat-denatured fragments amplified by PCR were electrophoresed in nondenaturing polyacrylamide gel. Separated fragments were visualized by silver staining. In our experiments it was difficult to detect SSCP with fragments over 300 bp. Because large fragments move slower than short fragments, they are closely spaced in the gel, so that we cannot detect the difference in migration between the fragments.

The SSCP method also is more costly in time and money compared with the RAPD method, because the SSCP method requires pairs of specific 20-mer primers, while the RAPD method requires randomly chosen 10-mer primers. However, the SSCP method is still effective for the detection of polymorphisms between genetically close lines, for which it is difficult to obtain polymorphic markers by the RFLP method.

34.3.5 Future of PCR analysis in plant genome research

We have already mapped about 2300 RFLP, RAPD and SSCP markers in the rice linkage map; of these about 7% are PCR markers. The advantage of PCR markers is that YAC clones can be screened directly. This enables us to link genetic loci to the physical map. In high-density marker regions, there are already enough markers to make YAC contig screening possible. For low-density marker regions the next step for producing markers will be screening markers in targeted loci by bulked segregant analysis. To accumulate more new markers, we need other methods for detection of polymorphism. Denaturing gradient gel electrophoresis (DGGE) [37] is a highly sensitive method of detecting polymorphism, which can detect even one point mutation. In plants, DGGE has been used to detect PCR fragments linked to the *S*-locus in rye [38]. However, the DGGE method costs more and takes longer than the RAPD method. The RAPD method has also been applied to detect differences of gene expression [39]. In the RGP, we have used cDNAs reverse transcribed from rice mRNAs from different tissues and growth stages. The transcribed genes were then amplified by RAPD using primers 10 bases long. Differences of expression level can be correlated with the amounts of amplified fragments. This method is very useful for differential screening of expressed genes.

In the RGP, much nucleotide sequence data on expressed rice genes has been accumulated in the past three years. From our sequence data, we have calculated decanucleotide frequencies and plan to adapt our construction of RAPD primers accordingly, so as to detect polymorphisms more easily and reliably. Moreover, using the decanucleotide frequency data, we can design primers flanked with di- or trinucleotide sequence repeats. These primers can detect fragment length polymorphism. These PCR-based techniques for detection of polymorphisms are simpler and more convenient than hybridization methods. Such methods could be adapted to detect polymorphisms in other plants.

34.4 Genetic linkage map of rice and its applications

Construction of a high-density linkage map with DNA markers is an important basis for genome analysis of rice as well as of other plant species. Here we describe the current status of a high-density linkage map of rice developed by the RGP using DNA markers [2]. General strategies and methods for linkage map construction and their application to gene isolation of rice will also be described.

34.4.1 Mapping populations and DNA probes

The initial choice of the segregation population is very important for linkage mapping of DNA markers. Several types of progenies, such as F2, BC1 (first back-cross generation) and recombinant inbred lines (RILs), can be used. Each of these populations has its their own advantages and disadvantages. F2 and BC1 populations can be easily obtained from a single cross of parental lines within a few years. However, leaf material for DNA extraction is limited. In this respect, RILs are very useful for linkage mapping with DNA markers. For example, as most chromosomal regions of these inbred lines are homozygous for one parental allele, self-pollinated seeds of these lines show no genotypic or phenotypic segregation, and so can be distributed as seeds to many researchers to share a mapping population. However, the main disadvantage of recombinant inbred lines is the time required for their construction. They are constructed by the single seed descent method from F2 individuals. It takes usually six or seven generations to establish inbred lines. Doubled haploid lines (DHLs) are another source of mapping populations. In rice, DHLs can be constructed by anther culture of F1 plants.

In general, wide crosses increase the probability of detecting polymorphism. In some plant species, crosses between a cultivar and a wild relative have been employed for linkage mapping. In rice, progenies derived from both intraspecific crosses [2,40–42] and interspecific crosses [3,43] have been used as mapping populations.

An efficient supply of single-copy or low-copy sequences is one requirement for linkage mapping of DNA markers. A genomic library is one of the sources for such DNA probes. In rice, many randomly selected genomic DNA clones, derived from a *Pst*I genomic library, were used as probes in the first linkage maps constructed with DNA markers [40,41]. cDNA libraries are another effective probe source for single-copy sequences in rice [2,3] and other plant species. Because of homology between related plant groups, DNA clones derived from other crop species, such as wheat [4], maize and oats [43], are also good probes for linkage mapping in rice.

34.4.2 Constructing the rice linkage map

The general procedure for linkage map construction in rice is shown in Fig. 34.6. To construct the RGP linkage map, we have used 186 F2 plants derived from an intraspecific cross between a *japonica* variety, 'Nipponbare', and an *indica* variety, 'Kasalath'. The total DNA of each individual has been extracted from frozen leaf tissue by the CTAB method [44]. In order to overcome a shortage of total DNA of F2 individuals for mapping analysis, bulked F3 seedlings derived by selfing each F2 individual have been used to restore F2 DNA. In our mapping strategy, we have used mainly cDNA clones as probes. As of May 1996, we have screened two parental lines with more than 5000 DNA clones, mainly cDNA clones from callus and root. After extensive RFLP screening, informative clones have been used to score genotypes of 186 F2 individuals by Southern hybridization analysis.

The first high-density and high-resolution rice genetic map was developed by RGP over three years [2]. In total, 1383 DNA markers have been mapped on the linkage map. The number of markers and their categories on the 12 chromosomes are shown in Table 34.3. The DNA markers are distributed along 1575 cM on 12 linkage groups and consist of 883 cDNAs, 265 genomic DNAs, 147 RAPDs, and 88 DNAs from other sources. cDNA markers were derived from rice callus and root libraries. Genomic markers were mainly randomly selected DNA clones, with some *Not*I linking, YAC end and

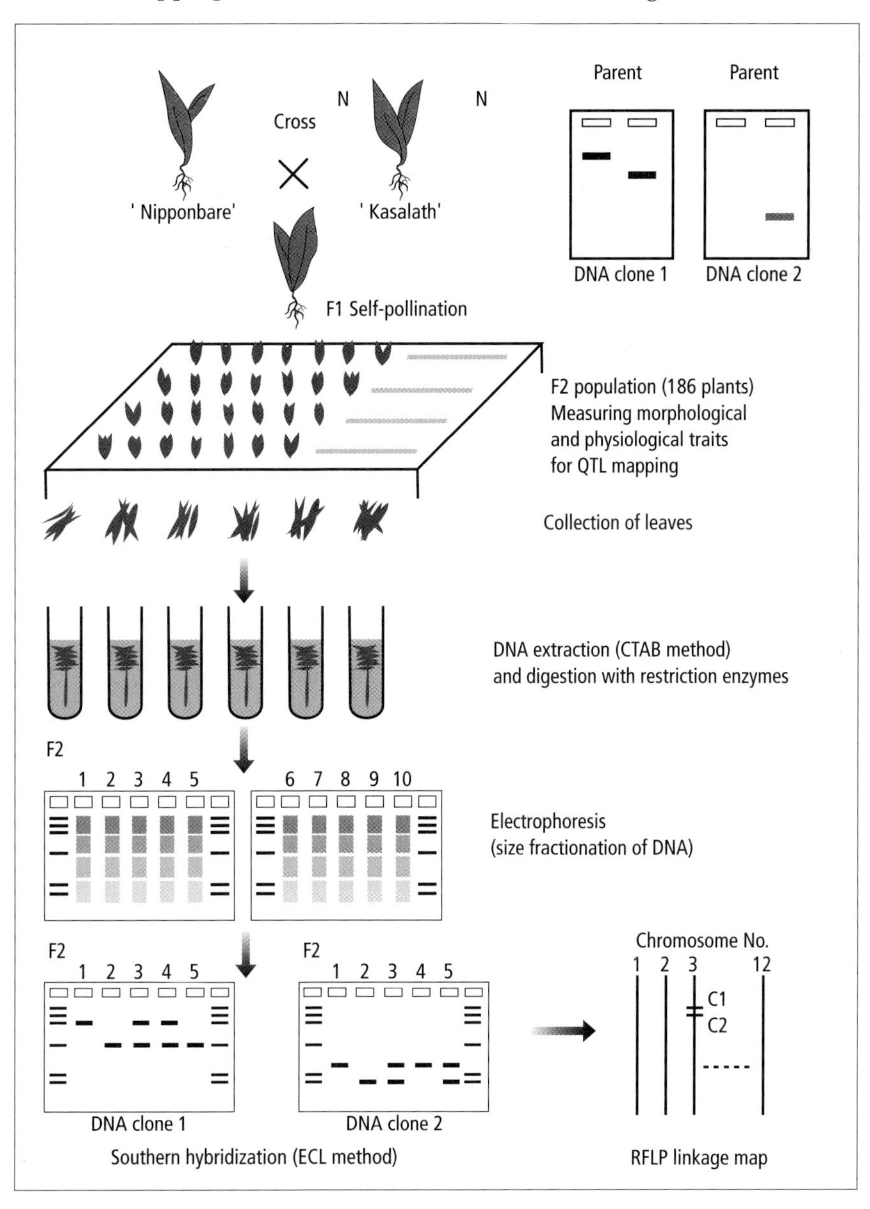

Fig. 34.6 Flow-chart of the construction of a high-density RFLP linkage map in the RGP.

Table 34.3 No. of markers and their categories in the high-density linkage map of rice [2].

	Chromosome												
	1	**2**	**3**	**4**	**5**	**6**	**7**	**8**	**9**	**10**	**11**	**12**	**Total**
cDNA from callus	67	51	59	38	45	41	41	25	25	28	28	19	467
cDNA from root	63	48	61	32	34	45	38	24	18	15	17	21	416
Random genomic	15	19	12	12	9	13	6	8	11	9	12	12	138
*Not*I linking	18	7	7	8	5	11	6	6	4	5	8	5	90
YAC end	4	3	3	2	4	6	3	0	1	2	3	2	33
Telomere	0	0	0	0	1	0	0	0	0	0	2	1	4
RAPD	10	5	3	8	11	25	7	10	8	5	13	5	110
STS	4	1	2	3	2	3	4	2	4	3	8	1	37
Wheat	9	4	7	4	7	7	4	4	5	0	3	4	58
Others	3	3	4	7	0	5	0	0	0	1	3	4	30
Total	193	141	158	114	118	156	109	79	76	68	97	74	1383
Total length (cM)	191.8	156.1	168.5	129	123.7	130.4	128.7	124.8	100.5	85.6	123.3	112.1	1575

telomere-associated clones (Table 34.3). RAPD markers were used for mapping to find out whether there is distribution bias for this type of marker and whether they can be used to fill marker-rare regions on the linkage map. A bulked segregant analysis [31] was used to develop markers in marker-rare regions (see Section 34.3). All the cDNA fragments and most of the genomic fragments were partially sequenced to convert them into STS (sequence-tagged site) or EST (expressed sequence tag) markers (see Section 34.2). The mapped markers have been characterized in detail by Kurata *et al.* [2]. As of May 1996, we have mapped about 2300 DNA markers on the linkage map. Green shoot and etiolated shoot cDNA libraries have been used as probe sources.

Analysis of molecular markers such as RFLPs is an effective way of revealing chromosomal rearrangements, such as translocations, deletions and duplications. Kishimoto *et al.* [45] indicated, by linkage mapping of cDNA clones, that sequences between some regions of chromosomes 1 and 5 have been conserved. In our linkage map of RGP, many clones showed more than one band by genomic Southern hybridization and were mapped at duplicate or triplicate loci among the 12 chromosomes. Extensive conservation in linkage alignment of 13 loci was observed between the lower distal regions of chromosomes 11 and 12 [46]. These conserved regions span 10 cM and 11.8 cM from the distal ends of the linkage maps of chromosomes 11 and 12, respectively. These results suggest that these conserved regions were generated by a duplication of chromosome segments.

In addition, 58 wheat genomic DNA fragments have been mapped on the high-density linkage map in collaboration with the Cambridge Laboratory at the John Innes Centre, UK. As a result, we have established that wheat chromosome groups 1, 2, 3, 4, 6 and 7 clearly correspond to rice chromosomes 5, 4 and 7, 1, 3, 2 and 6, respectively. Markers on wheat chromosome group 5 were mapped on rice chromosomes 1, 3, 9, 11 and 12. Surprisingly, in all chromosomes, most of the markers analysed showed the same linkage ordering between rice and wheat [4]. This preliminary synteny analysis of rice and wheat revealed high conservation for linkage ordering of DNA markers. Thus, the synteny map and the mapped rice probes will be very useful for the molecular analysis of the corresponding chromosomal regions in wheat. Rice will be a very useful model plant for genome analysis for several cereals, such as wheat, barley, maize and rye.

At present, four kinds of linkage map, constructed using independent probe sources, have already been constructed for rice [2,3,40,41]. Efforts to integrate these maps are in progress in collaboration with Cornell University. We are also developing a consensus framework linkage map using RILs, which have been constructed at Kyushu University [47]. As many research groups can share those RILs, the consensus framework map would facilitate an effective integration of all the information about linkage maps. Once a fully integrated genetic map with DNA markers is established, all the DNA markers will be very powerful tools, not only for the analysis of rice genome structure and function, but also for practical rice breeding [48].

34.4.3 Gene tagging for map-based cloning

It is often difficult to isolate genes conferring particular traits on a plant because of the lack of knowledge about gene functions and/or gene products. Map-based cloning is one method of isolating such genes, and has been used successfully in the tomato to isolate the disease-resistance gene *pto* and other genes [49]. Map-based cloning has also been employed in the RGP to isolate genes of biological and agronomic value. The first step of a map-based cloning strategy is the high-resolution linkage mapping of target genes. We have already mapped several genes for morphological and physiological traits, such as *Xa-1* (bacterial leaf blight resistance), *Se-1* (photoperiod sensitivity), *Gm-2* (gall midge resistance; in collaboration with Dr Mohan, ICGEB, India), e.g. (extra-glume), *Ph* (phenol staining), *Rc* (brown pericarp and seed coat) and *alk* (alkali degeneration) on our high-density linkage map.

Many rice geneticists and breeders have made linkage analyses using DNA markers. As a result, many DNA markers linked to useful genes have been identified. These include: genes for resistance to disease, such as rice blast [50,51] and bacterial leaf blight [52,53], genes for resistance to insects, such as brown plant hopper [54] and gall midge [55], a photoperiod sensitivity gene [56], photoperiod sensitive genic male sterility genes [57], semidwarf genes [58,59,60], scented kernel gene [61], a gene for accumulation of glucomannan in endosperm cell wall [62], etc. When we employ the map-based cloning method to isolate those genes, the linkage map of the target region will need to be of quite high resolution to select a single YAC or cosmid clone carrying the target gene (see Section 34.5).

To get high resolution, a large number of segregating individuals and a high degree of accuracy for genotyping of the target locus will be required. The segregating population derived from crosses between near-isogenic lines and their recurrent parents is usually the best material. The F3 line derived from F2 individuals should also be investigated in order to achieve a high reliability of genotyping for a target locus. However, this is time and labour consuming. The pooled-sampling approach is an alternative method for constructing a high-resolution map of the target region [63]. Application of this strategy requires several conditions, such as availability of a large segregating population and early and accurate classification of homozygous individuals for recessive or dominant alleles. When these conditions are met, it is possible to clarify the relative order of closely spaced markers, including the target locus of interest. In the RGP, we have made some progress in high-resolution linkage mapping for map-based cloning. A high-resolution linkage map of the *Xa-1* region, which is our first target gene to be isolated, has been constructed by a combination of the standard and pooled sampling methods.

34.4.4 Mapping quantitative trait loci

In contrast to disease and insect resistance, many important traits in breeding, such as yield, culm length, heading date and eating quality, show continuous variation in progenies. Inheritance of those traits are controlled by several genes. It is difficult to identify those genes, known as quantitative trait loci (QTL), because the individual effects of the genes on phenotype are relatively small. Recent progress in isolating DNA markers and their linkage maps now enables us to analyse these individual QTLs [64]. The strategy for detecting QTLs using linked major genes was developed many years ago [65], but was difficult to put into practice using conventional genetic markers. So far, many QTLs have been clarified using DNA markers in various crop plants, such as tomato [66,67] and maize [68,69]. In rice, QTL analysis with DNA markers has been employed to detect genomic regions conferring cooked-kernel elongation [70] and partial resistance to blast disease [51]. In these studies, putative genomic regions determining such complex traits could be identified.

From the beginning of the RGP, the feasibility of isolating genes at QTL with a map-based cloning system has been investigated. A large number of markers (857 loci) have been used to identify the QTLs affecting heading date, culm length, panicle length, etc. Many QTLs were detected, with a wide range of gene action on phenotypes, using the computer software MAPMAKER/QTL [71]. High-resolution QTL mapping also revealed evidence for the existence of multiple QTLs for the same trait in one chromosomal region and for specific gene interactions between identified individual QTLs, such as epistasis and suppression. Putative locations and DNA markers linked to QTLs will make marker-assisted selection feasible in rice breeding.

In general, it was difficult to determine the precise location and gene action of individual QTLs. The low accuracy of mapping of QTLs is bad for a map-based cloning system. In tomato, overlapping substitution lines have been used to map QTLs precisely [66]. To overcome these problems, we are constructing well-characterized genetic stocks, such as near-isogenic lines, carrying one or multiple chromosomal segments of the parental line, 'Kasalath', in the genetic background of the other parental line, 'Nipponbare' (Fig. 34.7). In this way, we will be able to combine/separate the desired/not desired chromosomal regions containing QTLs in selected

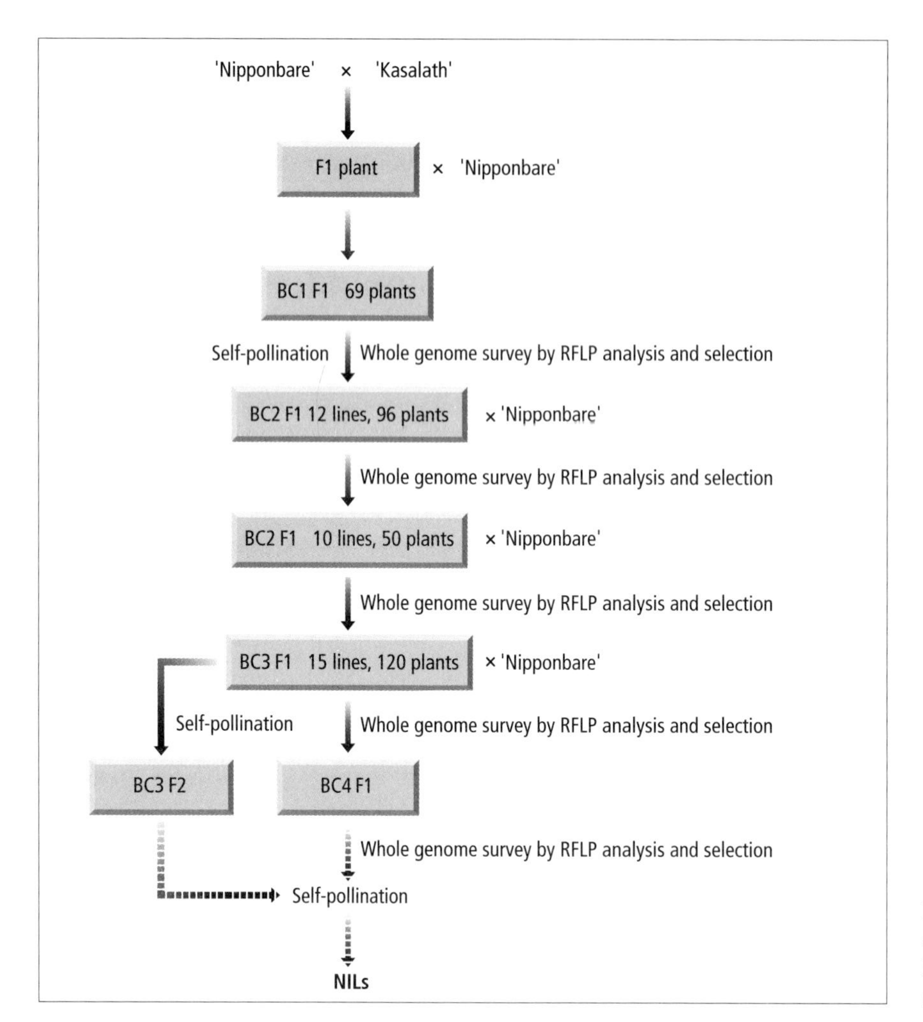

Fig. 34.7 Flow chart of the construction of near-isogenic lines (NILs) for the fine mapping of quantitative trait loci (QTLs).

individual plants. By using these substitution lines or near-isogenic lines for a given chromosomal segment, it will be possible to handle a given QTL as a single Mendelian factor. Thus we would be able to determine the accurate location of a given QTL on linkage map, to clarify the precise gene effects and to evaluate the genotype/environment interaction of individual QTLs. Once we succeed in mapping the genes of interest at high resolution, these genes will be isolated by map-based cloning.

34.4.5 Future prospects for linkage mapping

During the first three years of the RGP, a high-density linkage map using DNA markers has been constructed quickly. The linkage map and mapped DNA markers have already been used for physical map construction and for tagging genes with agronomic and biological interest. The number of mapped markers is already enough to embark on research into these topics. In order to progress further, it would be necessary to map DNA markers effectively in targeted chromosomal regions. We should also map cloned DNA fragments with already known function.

A detailed conventional rice linkage map has been compiled [1]. It is composed of many morphological and physiological marker genes and agronomically important genes, such as those for disease and insect resistance. Integration of the linkage map with conventional markers and with DNA markers is in progress [72,73]. In the RGP, our final objective is to construct a comprehensive genetic map including genes for morphological and physiological traits, genes for quantitative traits, as well as DNA markers. This genetic map will contribute greatly to rice genetics and breeding as well as to knowledge of the basic biology of the rice plant.

34.5 Making the physical map

34.5.1 Current state of the art in physical mapping

The cultivated rices, *O. sativa* and *O. glaberrima*, have 12 chromosomes ($2n = 24$), carrying about 430 million base pairs of DNA. The size of the rice genome is about 10-fold that of yeast and one-tenth of the human genome. Among plants, *Arabidopsis thaliana* has a genome three times smaller than rice, while rice has the smallest genome among cereal crops; maize has a genome eight times larger, and that of wheat is 40 times larger (see Chapter 32). The relatively small size of the rice genome facilitates construction of a physical map.

Two main results we want to obtain from a complete rice physical map are:

1 information on the basic structure of the rice genome as a model for monocot plants;

2 isolation of genes for agronomically or scientifically interesting traits by map-based cloning.

The first aim is based on the prospect that the establishment of a rice physical map, together with a high-density expression map, would be quite helpful not only for understanding the basis of monocot plant genome structure, but also for analysing genome evolution among various organisms. When utilizing a physical map, it is important to consider what kind of information one wants to extract from the map. From this point of view, the most useful information would be a map that has several hierarchical DNA contigs comprising long, medium sized and short DNA fragments, that is ordered YAC, cosmid and plasmid libraries overlapping each other. Furthermore, an expression genome map that has a complete array of genes (expressed sequences) on the ordered DNA fragments of the physical map should be the most informative comprehensive map for resolving genome organization.

The second aim in physical mapping is cloning of important genes that are often known only by phenotypic traits. Starting with the tagging of these trait genes by DNA markers located on the genetic linkage map, a detailed physical map of the target region will be needed for the next step. A large number of expressed sequences arrayed on the physical map, together with tagged DNA markers close to the target genes, make it possible to identify and clone the genes in a systematic manner.

In addition to the detailed physical map, maps of the synteny with related plants may help in map-based gene cloning. Recently, synteny relationships between several cereal crops have been reported [43,74,75]. The synteny analysis between rice and wheat in particular showed a strikingly high colinearity of gene order in all chromosomes [4]. Further work on microsynteny in limited regions, which in other cereal crops carry genes determining important traits, may make it possible to isolate such genes in other cereals using the rice physical map.

The cytogenetic map is also a kind of physical map. In the case of the human genome, such a map is being constructed by locating hundreds of DNA markers on chromosomes with fluorescence *in situ* hybridization (FISH) [76] (see Chapter 9). To visualize the location of genes directly on chromosomes or on isolated chromatin is an important way of generating a comprehensive genome map. In rice, several repetitive and single-copy sequences have been mapped on chromosomes by *in situ* hybridization [77–79]. However, detecting the exact location of single-copy sequences on the chromosomes is still not feasible. One reason is the similarity in size and shape of the metacentric or submetacentric rice chromosomes and their small size. Effective discrimination of rice chromosome regions by, for example, banding patterns or other cytogenetic characteristics is not yet possible, except by the use of a sophisticated densitometry-based imaging analysis system [80].

34.5.2 Use of YAC, BAC and cosmid libraries

The first step in the construction of a physical map is the preparation of genomic libraries. Several kinds of genomic libraries can be used for physical mapping, each having their own advantages and disadvantages both for library construction and in their utility for physical mapping. The three types of libraries most commonly used for physical mapping in other organisms are the libraries constructed with yeast artificial chromosome (YAC), bacterial artificial chromosome (BAC) and cosmid vectors. YACs can be used to clone very long genomic DNA fragments, from several hundred kilobases to over one megabase. The construction of a high-quality YAC library and its evaluation is, however, difficult and time consuming, because of the frequent occurrence of chimaeric clones. Such difficulties, however, can be overcome by combining several strategies in physical map construction, as discussed below.

After preliminary trials, we have constructed satisfactory YAC libraries [81] using protoplast cells of 'Nipponbare'. The two YAC libraries currently used comprise 7000 clones with an average insert size of 350 kb genomic DNA. These libraries cover the rice genome around 5.5 times and should cover over 80% of total chromosome length when all clones are aligned along the chromosomes. About 40% of the clones in the libraries are chimaeric. However, most of the chimaerism was observed in clones with inserts of over 400 kb, and was not frequent in the clones with smaller inserts. These libraries are now being successfully appplied to physical map construction [82].

In most cases, it is quite difficult to cover the whole genome with only one kind of genomic clone library. Therefore, the RGP also decided to make cosmid libraries. Cosmid libraries will be very useful not only to fill the gap regions of YAC contigs but also to divide the long YAC clones into several cosmid clones for further analysis, both to isolate target genes and to do detailed physical mapping. Several cosmid libraries have already been constructed [83], one with rice cultivar 'Nipponbare' DNA for physical mapping and others with different rice cultivars carrying interesting target genes for map-based cloning. A BAC library with inserts of about 150 kb DNA is also useful and easy to deal with for physical mapping. Wang *et al.* at the University of California at Davis have constructed a BAC library for map-based cloning of the rice bacterial blight resistance gene *Xa-21* [51]. Physical mapping will also be made easier by using complementary information from several different genomic libraries.

34.5.3 Map construction strategies and methods

There are several strategies applicable to physical map construction. A complete physical map should be a map which includes all DNA of all 12 chromosomes from one end to the other. For this purpose, it is best to obtain genomic DNA fragments as large as possible and order them along the 12 chromosomes. Isolation of the 12 chromosomes independently would be the best way for making chromosome-specific genomic DNA libraries as the source of starting materials for physical mapping. Rice chromosomes, however, cannot be sorted by laser beam chromosome sorting (flow sorting; see Chapter 12), because of the very small size and continuous variation in length (maybe also in DNA content) among the 12 chromosomes. However, because of the small genome size of rice, it is possible to construct a genomic library of large DNA fragments and to order them directly on the 12 chromosomes. In addition, the low proportion of repetitive DNA sequences (only about 50% of the rice genome) [84] makes it easier in rice to do cloning, selection and ordering of large DNA fragments to build up a physical map using a whole genomic library.

How to order all of the YAC clones in the libraries along chromosomes largely depends on the genome structure of the organisms in question. For instance, the human genome, which has highly dispersed Alu family sequences, can be reconstituted with a large number of YAC clones walked by Alu-sequence-based PCR methods [85] (see Chapter 15). In addition, human chromosome-specific libraries are available, so that formation of cosmid contigs by determining cosmid overlaps through fluorescence-based digested fragment mapping is also possible. For rice, we have not yet found any special genome features that could be utilized for efficient physical mapping, except for the small genome size and the low proportion of repetitive sequences. Therefore it seemed to us best to use a high-density and high-resolution rice genetic map to order YAC clones corresponding to the mapped DNA markers on it, as indicated in Fig. 34.8. Because we have already constructed a high-density 300-kb interval rice genetic map [2], it should be possible to cover the whole genome by selecting and ordering YAC clones with an average insert length of 350 kb. The practical methods for YAC contig formation used in RGP are as follows (see Fig. 34.9a, b).

1 We make high-density colony filters carrying 4×4×96 YAC clones on each filter. Our YAC libraries totalling 7000 clones can be spotted on five filters.
2 We screen the YAC libraries by colony hybridization with all individual RFLP markers of already mapped genomic and cDNA clones on our high-density rice genetic map.
3 DNAs of positive candidate YAC clones are further investigated by Southern hybridization analysis for detecting marker DNA sequences in them. The flow chart is presented in Fig. 34.9a.
4 Where we use STS markers (derived mainly from RAPD markers), all 7000 cloned DNAs are divided into several pool combinations in 96-well microtitre plates for PCR screening with site-specific primer sequences (see Fig. 34.9b).
5 After PCR screening of the first W pool, a second screening of X, Y and Z combination pools is performed to identify the position of the positive YAC clones. In Fig. 34.9b, as an example, PCR products are amplified in X5, Y7 and Z4 pools, identifying the YAC clone that has the STS marker sequence.
6 All the available data on YAC clones selected either by Southern hybridization or PCR screening is collected in the database for linking with other results useful in physical mapping.

By locating YAC clones on corresponding DNA markers using these methods, the RGP physical map so far (March 1996) covers about a half of the genome. With the aim of introducing more effective methods, trials in the use of fingerprinting to order YAC clones are also in progress. Isolation of high-copy DNA sequences in the rice genome has been carried out in microsatellite assays [78,86], genomic fingerprinting [87] and analysis for short nucleotide repeats. In the use of these repetitive DNAs for YAC fingerprinting, their specificity for rice, but not for yeast, is a very important factor. One of the microsatellites and one short repetitive nucleotide sequence have been found to produce specific

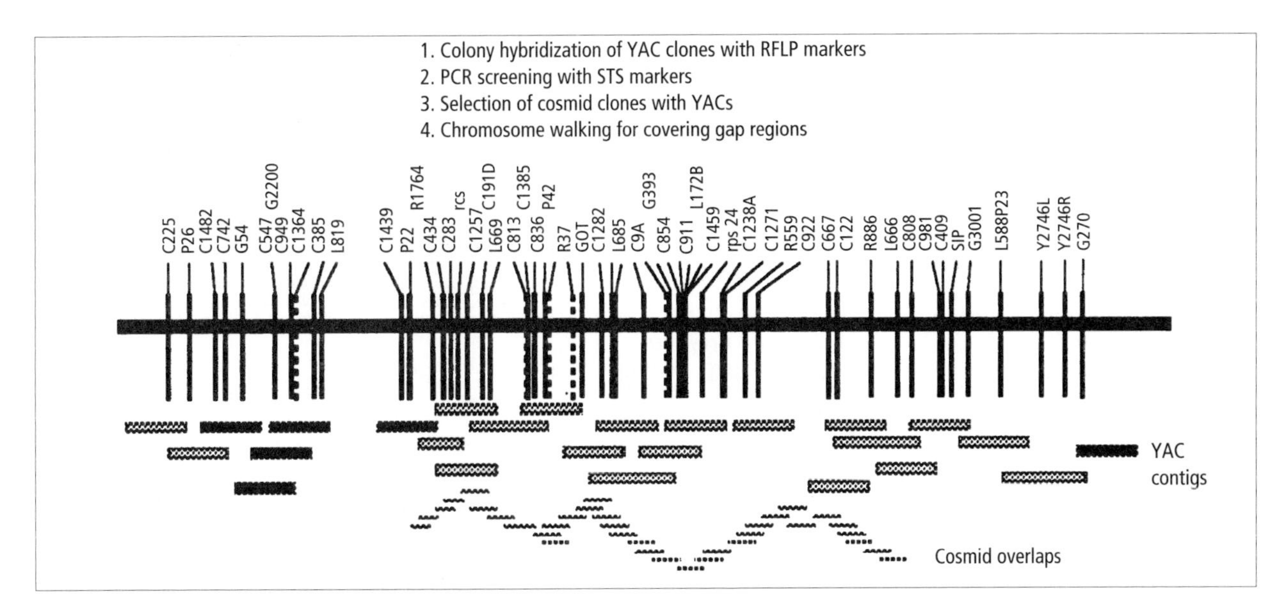

Fig. 34.8 Construction of the physical map by using DNA markers on the high-resolution genetic map. All DNA markers mapped on the high-density and high-resolution genetic linkage map are very useful to select and order YAC and/or cosmid clones. This should be the most reliable way to decide exact YAC overlaps on each chromosomes for physical mapping.

multiple banding in distinctive YACs. This should enable great progress in making YAC contigs.

34.5.4 Map-based cloning of target genes

Several examples of gene isolation through map-based cloning have been published recently in tomato [49], tobacco [88] and *Arabidopsis* [89,90]. Target genes for map-based cloning in rice are disease-resistance and insect-resistance genes, biotic and abiotic stress-tolerance genes, photoreactive genes and other genes of biological and/or biochemical importance (Sections 34.4.5 and 34.4.6). The strategy for map-based gene cloning in rice would be almost the same as that used in tomato and *Arabidopsis*. Easy focusing on the target gene in physical maps largely depends on how exact and near are the DNA markers tagging the gene.

In rice, the total length of the genetic map is about 1600 cM and that of the physical map is 430 Mb. This means that 1 cM corresponds to about 270 kb. A rough estimation of the number of expressed sequences in the rice genome tells us that 1 cM — that is, 270 kb — contains 30–60 genes on average, although naturally the gene density varies from region to region. At first, one should pick up a long DNA fragment of YAC, BAC or cosmid which carries the closely linked DNA marker sequence(s) to the target gene. Screening and identification of the target gene among the many expressed sequences on such DNA clones seems the most critical step. If one can use a plant population large enough for segregation analysis, it is possible to select out several candidate clones which show no segregation with the target gene from over several tens of genes on one YAC.

In the case of *Xa-1*, one of the bacterial leaf blight resistance genes in rice, we have been able to select as candidates about half of the expressed sequences from over 20 cDNA clones mapped on one 320-kb YAC using about 1000 F3 lines by bulking methods (see Section 34.4). This YAC clone was made from 'Nipponbare' DNA, and could be screened with overlapping cosmid clones made from the resistant near-isogenic lines [53]. Several cosmid clones covering all five to six candidate cDNA sequences could be selected from the cosmid library of the bacterial blight-resistant rice strain. The next step should include sequencing of these cDNA and cosmid clones to see sequence differences between resistant and susceptible strains and transformation of the susceptible strain by these cosmid clones to see whether any clone restores resistance to the susceptible rice strain.

In a similar way, other target genes are being tagged and then isolated as large DNA fragments in YACs. Although identification and characterization of target genes will take much more time, the isolated genes should be very useful for plant improvement through gene transfer into superior strains. Functional analysis of the isolated genes will also make it possible to compile biological databases for metabolic pathways.

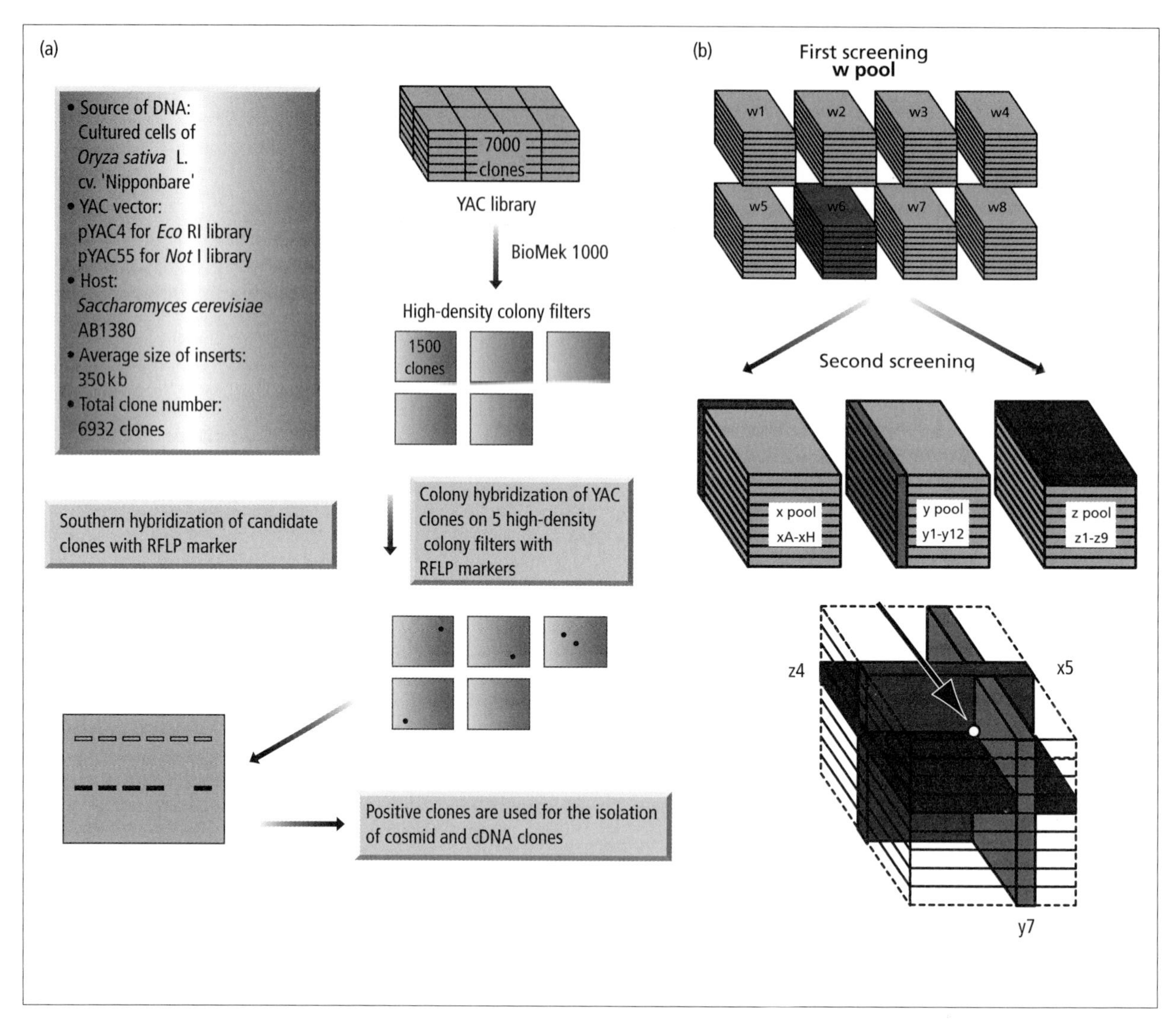

Fig. 34.9 Construction and screening of a rice YAC library. (a) Flow chart of the method of construction of an ordered YAC library in rice. Steps of colony hybridization on high-density colony filters and Southern hybridization needed for YAC selection are shown. (b) Schematic representation of the pooling and screening system of YAC clone DNAs using PCR methods. W is the pool for the first screening. If a clone which has a primer-specific sequence is contained in the w6 pool, one can screen a three-axis pool combination of xA – xH, y1 – y12, and z1 – z9 using the w6 pool as the second screening step.

34.5.5 Future directions for physical mapping

Completion of the physical map will need not only several kinds of DNA libraries but also methods to fill the gaps and to find chromosomal rearrangements and to reveal the degree and sites of structural complexity in the genome. The resulting physical map will be the main source for different types of analysis of rice genome structure. Many aspects of genome organization should be clarified by, for example, dissecting the whole genome into functional and nonfunctional segments, surveying for common functional domains and their functions both at the level of gene expression and chromatin/chromosome structure, and also by whole genome sequencing. The rice physical map should be the most valuable resource in monocot plants, especially in grass species, for further comparative analyses.

The physical map, the cytological map, the RFLP linkage map, the genetic linkage map of phenotypic traits and the expression map should be combined to generate a comprehensive genome map. Aiming to build up such a genome map, we expect to include the mapping of the over 20 000 cDNA clones which we will be isolating in our large-scale cDNA analysis (see Section 34.2). It may be difficult to locate all the cDNA clones on the linkage map, especially in the case of multiple-copy sequences such as those of isozymes and protein families. The difficulty comes from sequence similarities in the cDNA clones, lack of polymorphism and the limit of resolution in linkage analysis. The fine mapping of almost all genes, however, should be possible using ordered YAC libraries. In practice, we can locate almost all multicopy and gene family sequences on multiple YAC clones in our system for YAC contig formation. In the future therefore we wish to use all cDNA clones which have not yet been mapped on the linkage map. Such a full expression map will contribute greatly to unravelling genome organization for gene expression and gene evolution in rice and other plants.

34.6 Rice genome informatics

There are around 50 staff at the RGP. The 11 ABI sequencers can produce 100 000 bp of data daily, and every week hundreds of RFLP samples are prepared. The large amounts of data need to be moved around easily, and therefore a fast local network is important. Because the RGP was expanded rapidly to its current size, Macintosh computers were chosen for general use, as they are easy to learn to use. This section gives technical information on how the data processing and analysis have been arranged in our project. A simplifying factor for us is that we have the cDNA, PCR, RFLP and physical mapping laboratories in the same building, facilitating the planning and implementation of database activities from the beginning. This has made it possible to integrate the mapping and sequencing data for the rice genome analysis [91].

34.6.1 Data devices and raw data inputting and editing

We have a Hewlett Packard 9000/897S as the main computer (with 128 Mbyte main memory and a 23 Gbyte hard disk) several SUN computers and some 40 Macintoshes (mainly Macintosh Quadras for desktop analysis, Powerbooks as laboratory notebooks) in our system. They are linked with a 10 Mbyte s^{-1} local area network (LAN) with a star topology. The main host computer, HP9000, is linked to the LAN switching hub (LANplex 5012, Synernetics) at the speed of 100 Mbyte s^{-1}.

Macintoshes communicate using AppleTalk protocol to the hub at a speed of 10 Mbyte s^{-1}. Macintosh servers are easy to set up, and we have one fileserver on a Macintosh with a 4.5 Gbyte total memory for the database files and disk space allocated to researchers. Macintoshes communicate with the SUN computers, with the HP9000 and the Internet by EtherTalk, using the TCP/IP protocol. On the SUN computers we have set up our WWW server (address: http://www.staff.or.jp) and our ftp server (ftp.staff.or.jp), and use a POP server on the SUN computer to deliver electronic mail to the Macintoshes, which are equipped with Eudora mail reading software. All Macintoshes have been given an IP number, so that they can access the World Wide Web by Mosaic or Netscape client software. The computer network system is shown in Fig. 34.10

We have found the Microsoft Excel spreadsheet easy to use for almost all data input. After initial input and editing, the data files are imported to a relational database management system (RDBMS) called 4th DIMENSION running on Macintoshes. Often subsets of data are exported in text file format from 4th DIMENSION, imported again to Excel for manipulation, and graphs and statistics programs used via the Excel spreadsheet. The sequence data from the ABI 373 sequencers is likewise imported as text files from the network-linked Macintoshes controlling the sequencers.

34.6.2 Database management: 4th DIMENSION interface and SyBase processing

The 4th DIMENSION is a well-designed RDBMS that allows the designer to change the database structure easily and enables users to define desired display formats and to print and export files for many different purposes.

With the ease of use of a Macintosh, starting database activities is easy, and the database structure can be continuously expanded in a flexible way. Recently, server and client software has become available, so that Macintoshes on the network can share the common data without the need to have all the data in their own machine. In addition, local smaller subsets of data can be prepared for daily use by individual researchers with the same interface.

The current in-house database in the RGP, RiceBase2 combines the data from the cDNA, PCR, RFLP and physical mapping groups. In developing RiceBase2, special atention was given to designing a good way of storing experimental information, especially sets of consecutive experiments and protocols [92].

The limitation of 4th DIMENSION is in the memory space and processing speed of the local Macintoshes (though PowerMacs have performed quite well recently). As our database keeps growing, we are moving it to the HP9000 main computer into a SyBase RDBMS. We use a client–server database system, using 4th DIMENSION on Macintosh as a client and SyBase on HP9000 as a server. This way we keep the user-friendly 4th DIMENSION inter-

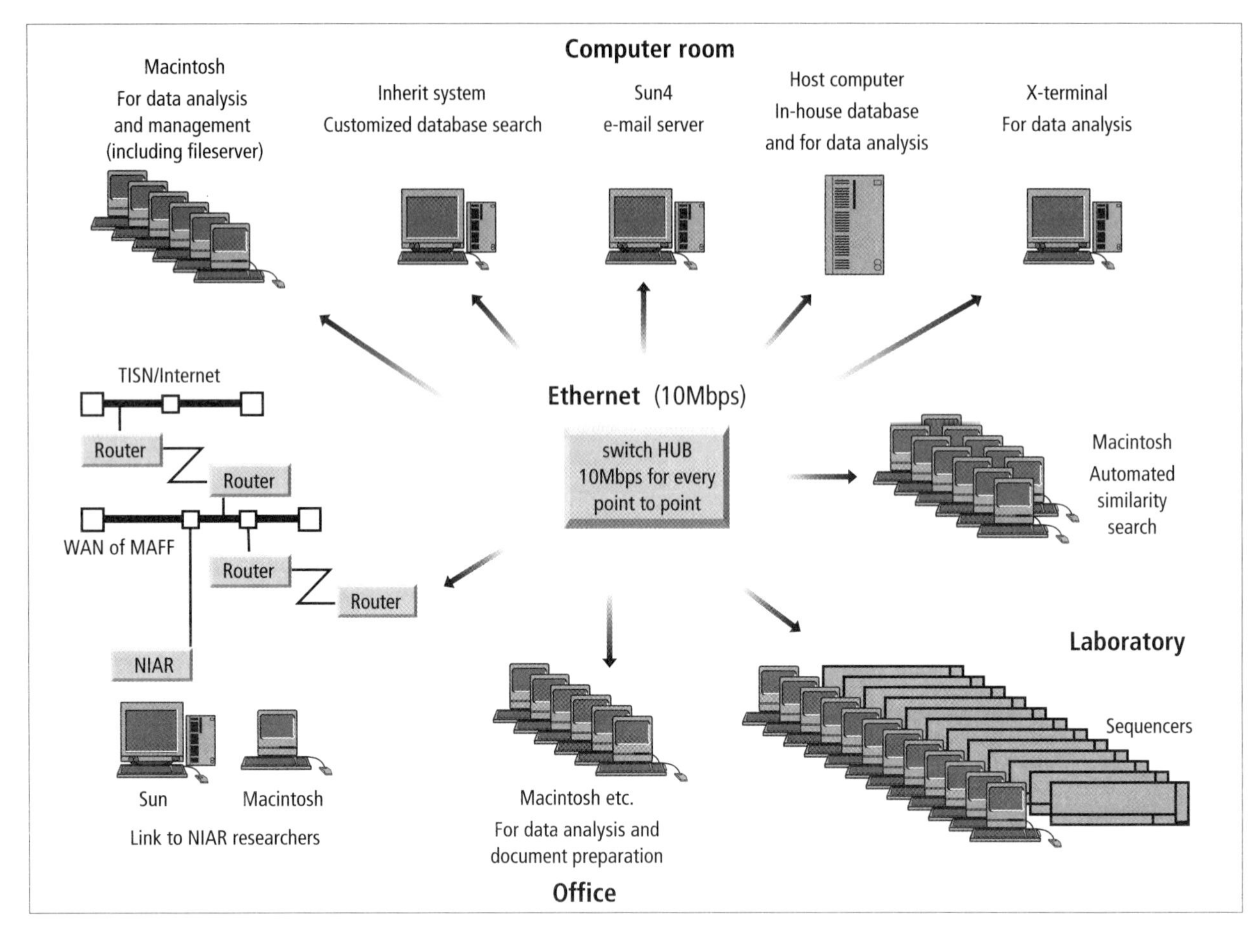

Fig. 34.10 Computer hardware and network system in the RGP.

face on the Macintoshes, which send the queries to the server where they are translated into the SyBase format SQL queries. This in-house database is called RiceBase3. It already needs 1 Gbyte memory for the SyBase tabulations of all data, with some 12 Mb of edited DNA sequence, information on some 30 000 clones and 320 000 individual plant genotypes used in linkage mapping, and thousands of gel images (Table 34.4). Many of the RFLP gel pictures for the linkage map [2] have been put already on our WWW server at http://www.staff.or.jp.

34.6.3 Toward a rice genome anatomy and international federated genome databases

Over 10 000 cDNA sequences are now available in the international database of expressed sequence tags (dbEST, a section of GenBank at the National Centre for Biotechnological Information, USDA, USA, see Chapter 37 and Appendix V for contact addresses), together with information on homologies to other released ESTs. In the release 86.0 (15 December 1994) rice is already the grass for which

Item	Amount
Edited DNA sequences, nucleotides	12 000 000
Number of cDNAs sequenced	20 968
Clones (cDNA)	35 000
Evaluated mapping genotypes (single plant individuals)	320 000
Mapped unique loci	1 600
Gel images	5 000
Total number of YACs	7 000

Table 34.4 Partial listing of information in the in-house database RiceBase3 of the Japanese Rice Genome Research Program, as at January 1995.

the most nucleotide sequence is available. All these sequences can now be linked to the large number of published *Arabidopsis* sequences and in fact to all gene products of conserved sequences.

Already, over 1100 rice cDNAs have been successfully converted to STSs and are used as standard landmarks in the RFLP map [2,20]. We also are progressing to integrate the YAC mapping data with the rest of the database. More modularity is needed in the database, and compatibility with the ACeDB software [93], which is the software adopted by many genome projects and also by the Plant Genome Database (PGD, at NCBI, USDA). It is possible that we will also add on an ACeDB interface into SyBase, as has been done in the Integrated Genomic Database (IGD, see ref. 94).

Whatever the interface to the researcher, international cooperation is needed to harmonize the semantic structures in the large number of genome databases being built in laboratories of very different sizes and resources. One step in that direction is the standardized plant gene nomenclature [95], promoted by the International Society of Plant Molecular Biology since 1991. The standardized names will later also be required information in publications and in the international sequence databases (DDBJ, EMBL, GenBank). Some of the sequenced rice genes have already been putatively assigned such standardized names [96].

As for the rapidly accumulating plant cDNA sequences, already some 14 000 *Arabidopsis* sequences [12] and over 10 000 rice sequences are available in the international databases. Of all the 30 000 or so rice genes, some 30% have already been partially sequenced in the RGP. Most of these and other cDNA sequences from big projects will be available only through the databases and will not be published in detail in the scientific journals. Even so, the release of large numbers of plant cDNA sequences makes this information available for research being done on structurally similar proteins in other organisms. Conversely, many released plant cDNA sequences show homology to proteins previously known only in organisms other than plants, giving valuable information on the function and evolution of conserved proteins and metabolic pathways.

In the future, the mere harmonization of a collection of genome map databases will not be enough, since the genomes of even related organisms have a large amount of noncoding (but not all nonfunctional!) sequences in their genomes. Relational information indicating the relative positions of gene transcription units and regulatory elements is essential to build up genome anatomies [97] that can be compared effectively between organisms. The role of rice genome anatomy in such comparisons will be very important for all plant researchers and especially for cereal genome researchers.

34.7 Rice as a model for cereal genome research

As high-density RFLP maps of several cereal crops, such as rice, wheat and maize are constructed with common DNA markers, the comparison of loci for the RFLP markers on these maps becomes possible. Already, extensive colinearity of single-copy DNA markers has been established between rice and wheat [4] and between rice and maize [43]. Comparative RFLP maps of the homoeologous group 2 chromosomes of wheat, rye and barley have also been constructed [75]. In the case of rice and wheat, clear colinearity is found within almost all chromosomes, in spite of the large differences in genome sizes and numbers of chromosomes (see Chapter 32).

The synteny between rice and wheat, based on our rice linkage map, is shown in Plate 11. Around 50 markers from wheat were used to detect polymorphism in a rice F2 population and their loci were determined on the rice RFLP map (see Section 34.4). About 50 markers from rice were also mapped on the wheat linkage map. Almost all the loci were identified as single-copy. This means that during evolution no duplications of these single-copy sequences have occurred in wheat or rice.

Clear synteny between rice and maize has also been recognized. In comparison to wheat, the chromosome structure of maize is rather complicated. It has duplicated chromosomes and the regions homoeologous to rice appear twice in the maize genome. Using rice as an anchor species, homoeology between wheat and maize could be elucidated easily. As for other cereal genomes, such as barley and millet, studies on synteny with the rice genome are currently in progress. The main advantages in using rice as an anchor species for elucidating synteny between various grasses are as follows.

1 The homologous alignment of nucleotide sequences also means the homologous alignment of genes corresponding to phenotypic traits. If some phenotypic trait is once tagged by DNA markers in one of the cereal crops linked by a synteny map, this trait is expected to be present at the corresponding genome location in other cereal crops. For example, the dwarf gene is located in wheat chromosome group 4 and maize chromosomes 1 and 5, and the respective chromosome segments have colinearity to each other. The corresponding segment in rice is located in chromosome 3, but unfortunately until now no such expressed gene has been located to this rice chromosome in the classical map. It may be that translocation of a short segment or a nucleotide replacement has suppressed the expression of the corresponding gene, or the gene has jumped to elsewhere in the genome.

2 The isolation of genes responsible for phenotypic traits in cereals other than rice is feasible by screening rice genomic libraries (such as YAC or cosmid libraries) with DNA probes linked with phenotypic traits in the target species. For species with large genomes, such as maize, wheat and barley, construction of YAC or cosmid libraries is not easy, and even if constructed, screening of target genes might be difficult because of the presence of a large amount of repetitive or noncoding sequence. If the targeted trait shows physiological characteristics or signs of resistance against pathogens quantitatively similar to those in rice, the comparison could be done directly for gene isolation. Even if that trait is not expressed in rice, the conservation of an ancestral sequence within the corresponding region is expected. This approach is currently being used in an effort to isolate and characterize the *Rpg1* gene (resistance gene to stem rust) in barley using rice YAC clones [98].

3 Information about synteny among the grasses is a powerful tool for studying evolution not only within grasses, but also between monocots and dicots. Rice has the smallest genome among the grasses and its chromosome structure is truly diploid. The high-density rice RFLP map has already been constructed as mentioned in Section 34.4. The nucleotide sequence analysis of randomly selected cDNA clones from rice callus library has enabled discovery of similar sequences in many other plant species and even many other eukaryotes [6]. Very similar sequences are found in cDNAs from other grasses, such as wheat, maize and barley. This suggests an extensive conservation, even in nucleotide sequence, among the grasses. For these reasons rice has been chosen as an anchor species for synteny mapping and will be a pivotal cereal crop for comparative genome research.

References

1 Kinoshita, T. (1994) Report of the committee on gene symbolization, nomenclature and linkage map. *Rice Genet. Newslett.* **10**, 7–39.

2 Kurata, N., Nagamura, Y., Yamamoto, K. *et al.* (1994) A 300 kilobase interval genetic map of rice including 883 expressed sequences. *Nature Genet.* **8**, 365–372.

3 Causse, M.A., Fulton, T.M., Cho, Y.G. *et al.* (1994) Saturated molecular map of the rice genome based on an interspecific backcross population. *Genetics* **138**, 1251–1274.

4 Kurata, N., Moore, G., Nagamura, Y. *et al.* (1994) Conservation of genome structure between rice and wheat. *Bio/Technology* **12**, 276–278.

5 Adams, M.D., Kerlavage, A.R., Fields, C. & Venter, J.C. (1993) 3,400 new expressed sequence tags identify diversity of transcripts in human brain. *Nature Genet.* **4**, 256–267.

6 Sasaki, T., Song, J., Koga-Ban, Y. *et al.* (1994) Toward cataloguing all rice genes: large-scale sequencing of randomly chosen rice cDNAs from a callus cDNA library. *Plant J.* **6**, 615–624.

7 Uchimiya, H., Kidou, S.-I., Shimazaki, T. *et al.* (1992) Random sequencing of cDNA libraries reveals a variety of expressed genes in cultured cells of rice (*Oryza sativa* L.). *Plant J.* **2**, 1005–1009.

8 Höfte, H., Desprez, T., Amselem, J. *et al.* (1993) An inventory of 1152 expressed sequence tags obtained by partial sequencing of cDNAs from *Arabidopsis thaliana*. *Plant J.* **4**, 1051–1061.

9 Waterston, R., Martin, C., Craxton, M. *et al.* (1992) A survey of expressed genes in *Caenorhabditis elegans*. *Nature Genet.* **1**, 114–123.

10 McCombie, W.R., Adams, M.D., Kelley, J.M. *et al.* (1992) *Caenorhabditis elegans* expressed sequence tags identify gene families and potential disease gene homologues. *Nature Genet.* **1**, 124–131.

11 Gibson, S. & Somerville, C.R. (1993) Isolating plant genes. *Trends Biotechnol.* **11**, 306–313.

12 deBruijn, F.J.T.N., Green, P., Keegstra, K. *et al.* (1994) Genes galore: a summary of methods for accessing results from large-scale partial sequencing of anonymous *Arabidopsis* cDNA clones. *Plant Physiol.* **106**, 1241–1255.

13 Arumuganathan, K. & Earle, E.D. (1991) Nuclear DNA content of some important plant species. *Plant Mol. Biol. Rpt.* **9**, 208–218.

14 Hashimoto, H., Nishi, R., Umeda, M. *et al.* (1993) Isolation and characterization of a rice cDNA clone encoding ATP/ADP translocator. *Plant Mol. Biol.* **22**, 163–164.

15 Sakamoto, M., Shimada, H. & Fujimura, T. (1992) Nucleotide sequence of a cDNA encoding a β subunit of the mitochondrial ATPase from rice (*Oryza sativa*). *Plant Mol. Biol.* **20**, 171–174.

16 Hashimoto, J., Hirabayashi, T., Hayano, Y. *et al.* (1992) Isolation and characterization of cDNA clones encoding *cdc2* homologues from *Oryza sativa*: a functional homologue and cognate variants. *Mol. Gen. Genet.* **233**, 10–16.

17 Fushimi, T., Umeda, M., Shimazaki, T. *et al.* (1994) Nucleotide sequence of a rice cDNA similar to a maize NADP-dependent malic enzyme. *Plant Mol. Biol.* **24**, 965–967.

18 Pearson, W.R. & Lipman, D.J. (1988) Improved tools for biological sequence comparison. *Proc. Natl Acad. Sci. USA* **85**, 2444–2448.

19 Altschul, S.F., Gish, W., Miller, W. *et al.* (1990) Basic local alignment search tool. *J. Mol. Biol.* **215**, 403–410.

20 Inoue, T., Zhong, H.S., Miyao, A. *et al.* (1994) Sequence-

tagged sites (STSs) as standard landmarkers in the rice genome. *Theor. Appl. Genet.* **89**, 728–734.

21 Monna, L., Miyao, A., Inoue, T. *et al.* (1994) Determination of RAPD markers in rice and their conversion into sequence tagged sites (STSs) and STS-specific primers. *DNA Res.* **1**, 139–148.

22 Welsh, J. & McClelland, M. (1990) Fingerprinting genomes using PCR with arbitrary primers. *Nucleic Acids Res.* **18**, 7213–7218.

23 Williams, J.G.K., Kubelik, A.R., Livak, K.J. *et al.* (1990) DNA polymorphisms amplified by arbitary primers are useful as genetic markers. *Nucleic Acids Res.* **18**, 6531–6535.

24 Reiter, R.S., Williams, J.G.K., Feldmann, K.A. *et al.* (1992) Global and local genome mapping in *Arabidopsis thaliana* by using recombinant inbred lines and random amplified polymorphic DNAs. *Proc. Natl Acad. Sci. USA* **89**, 1477–1481.

25 Wilde, J., Waugh, R. & Powell, W. (1992) Genetic fingerprinting of *Theobroma* clones using randomly amplified polymorphic DNA markers. *Theor. Appl. Genet.* **83**, 871–877.

26 Quiros, C.F., Hu, J., This, P. *et al.* (1991) Development and chromosomal localization of genome-specific markers by polymerase chain reaction in *Brassica*. *Theor. Appl. Genet.* **82**, 627–632.

27 Sobral, B.W.S. & Honeycutt, R.J. (1993) High output genetic mapping of polyploids using PCR-generated markers. *Theor. Appl. Genet.* **86**, 105–112.

28 Kiss, G.B., Csanadi, G., Kalman, K. *et al.* (1993) Construction of a basic genetic map for alfalfa using RFLP, RAPD, isozyme and morphological markers. *Mol. Gen. Genet.* **238**, 129–137.

29 Lander, E.S., Green, P., Abrahamson, J. *et al.* (1987) MAPMAKER: An interactive computer package for constructing primary genetic linkage maps of experimental and natural populations. *Genomics* **1**, 174–181.

30 Paran, I. & Michelmore, R.W. (1993) Development of reliable PCR-based markers linked to downy mildew resistance genes in lettuce. *Theor. Appl. Genet.* **85**, 985–993.

31 Michelmore, R.W., Paran, I. & Kesseli, R.V. (1991) Identification of markers linked to disease-resistance genes by bulked segregant analysis: a rapid method to detect markers in specific genomic regions by using segregating populations. *Proc. Natl Acad. Sci. USA* **88**, 9828–9832.

32 Giovannoni, J.J., Wing, R.A., Ganal, M.W. & Tanksley, S.D. (1991) Isolation of molecular markers from specific chromosomal intervals using DNA pools from existing mapping populations. *Nucleic Acids Res.* **19**, 6553–6558.

33 Martin, G.B., Williams, J.G.K. & Tanksley, S.D. (1991) Rapid identification of markers linked to a *Pseudomonas* resistance gene in tomato by using random primers and near-isogenic lines. *Proc. Natl Acad. Sci. USA* **88**, 2336–2340.

34 Orita, M., Iwahana, H., Kanazawa, H. *et al.* (1989) Detection of polymorphisms of human DNA by gel electrophoresis as single-strand conformation polymorphisms. *Proc. Natl Acad. Sci. USA* **86**, 2766–2770.

35 Orita, M., Suzuki, Y., Sekiya, T. & Hayashi, K. (1989) Rapid and sensitive detection of point mutations and DNA polymorphisms using the polymerase chain reaction. *Genomics* **5**, 874–879.

36 Fukuoka, S., Inoue, T., Miyao, A. *et al.* (1994) Mapping of sequence-tagged sites in rice by single strand conformation polymorphism. *DNA Res.* **1**, 271–277.

37 Cariello, N.F., Scott, J.K., Kat, A.G. *et al.* (1988) Resolution of a missense mutant in human genomic DNA by denaturing gradient gel electrophoresis and direct sequencing using *in vitro* DNA amplification: HPRT Munich. *Am. J. Hum. Genet.* **42**, 726–734.

38 Wehling, P., Hackauf, B. & Wricke, G. (1994) Identification of S-locus linked PCR fragments in rye (*Secale cereale* L.) by denaturing gradient gel electrophoresis. *Plant J.* **5**, 891–893.

39 Welsh, J., Chada, K., Dalal, S.S. *et al.* (1992) Arbitrarily primed PCR fingerprinting of RNA. *Nucleic Acids Res.* **20**, 4965–4970.

40 McCouch, S.R., Kochert, G., Yu, Z.H. *et al.* (1988) Molecular mapping of rice chromosomes. *Theor. Appl. Genet.* **76**, 815–829.

41 Saito, A., Yano, M., Kishimoto, N. *et al.* (1991) Linkage map of restriction fragment length polymorphism loci in rice. *Jpn.J. Breed.* **41**, 665–670.

42 Jena, K.K., Khush, G.S. & Kochert, G. (1994) Comparative RFLP mapping of a wild rice, *Oryza officinalis*, and cultivated rice, *Oryza sativa*. *Genome* **37**, 382–389.

43 Ahn, S.N. & Tanksley, S.D. (1993) Comparative linkage maps of the rice and maize genomes. *Proc. Natl Acad. Sci. USA* **90**, 7980–7984.

44 Murray, M.G. & Thompson, W.F. (1980) Rapid isolation of high molecular weight plant DNA. *Nucleic Acids Res.* **8**, 4321–4325.

45 Kishimoto, N., Higo, H., Abe, K. *et al.* (1994) Identification of the duplicated segments in rice chromosome 1 and 5 by linkage analysis of cDNA markers of known functions. *Theor. Appl. Genet.* **88**, 722–726.

46 Shimano, T., Inoue, T., Antonio, B.A. *et al.* (1995) Extensive conservation in linkage alignment of RFLP markers between rice chromosome 11 and 12. *Abstr. Plant Genome III* **3**, 75.

47 Tsunematsu, H., Hasegawa, H., Yoshimura, S. *et al.* (1993) Construction of an RFLP/RAPD linkage map by using recombinant inbred lines. *Rice Genet. Newslett.* **10**, 89–90.

48 Tanksley, S.D., Young, N.D., Paterson, A.H. & Bonierbale, M.W. (1989) RFLP mapping in plant breeding: new tools for an old science. *Bio/Technology* **7**, 257–264.

49 Martin, G.B., Brommonschenkel, S.H., Chunwongse, J. *et al.* (1993) Map-based cloning of a protein kinase gene conferring disease resistance in tomato. *Science* **262**, 1432–1436.

50 Yu, Z.H., Mackill, D.J., Bonman, J.M. & Tanksley, S.D. (1991) Tagging genes for blast resistance in rice via linkage to RFLP markers. *Theor. Appl. Genet.* **81**, 471–476.

51 Wang, G., Mackill, D.J., Bonman, J.M. *et al.* (1994) RFLP mapping of genes conferring complete and partial resistance to blast in a durably resistant rice cultivar. *Genetics* **136**, 1421–1434.

52 Ronald, P.C., Albano, B., Tabien, R. *et al.* (1992) Genetic and physical analysis of the rice bacterial blight disease resistance locus, *Xa-21*. *Mol. Gen. Genet.* **236**, 113–120.

53 Yoshimura, S., Yoshimura, A., Saito, A. *et al.* (1992) RFLP analysis of introgressed segments in three near-isogenic lines of rice for bacterial blight resistance genes, *Xa-1*, *Xa-3* and *Xa-4*. *Jpn. J. Genet.* **67**, 29–37.

54 Ishii, T., Brar, D.S., Multani, D.S. & Khush, G.S. (1994) Molecular tagging of genes for brown planthopper resistance and earliness introgressed from *Oryza australiensis* into cultivated rice, *Oryza sativa*. *Genome* **37**, 217–221.

55 Mohan, M., Nair, S., Bentur, J.S., Rao, U.P. & Bennet, J. (1994) RFLP and RAPD mapping of the rice Gm2 gene that confers resistance to biotype 1 of gall midge (*Orseolia oryzae*). *Theor. Appl. Genet.* **87**, 782–788.

56 Mackill, D.J., Salam, M.A., Wang, Z.Y. & Tanksley, S.D. (1993) A major photoperiod-sensitivity gene tagged with RFLP and isozyme markers in rice. *Theor. Appl. Genet.* **85**, 536–540.

57 Zhang, Q., Shen, B.Z., Dai, X.K. *et al.* (1994) Using bulked extremes and recessive class to map genes for photoperiod-sensitive genic male sterility in rice. *Proc. Natl Acad. Sci. USA* **91**, 8675–8679.

58 Ogi, Y., Kato, H., Maruyama, K. *et al.* (1993) Identification of RFLP markers closely linked to the semidwarfing gene at the *sd-1* locus in rice. *Jpn. J. Breed.* **43**, 141–146.

59 Cho, Y.G., Eun, M.Y., McCouch, S.R. & Chae, Y.A. (1994) The semidwarf gene, *sd-1*, of rice (*Oryza sativa* L.). III. Molecular mapping and markers-assisted selection. *Theor. Appl. Genet.* **89**, 54–59.

60 Liang, C.Z., Gu, M.H., Pan, X.B. *et al.* (1994) RFLP tagging of a new semidwarfing gene in rice. *Theor. Appl. Genet.* **88**, 898–900.

61 Ahn, S.N., Bollich, C.N. & Tanksley, S.D. (1992) RFLP tagging of a gene for aroma in rice. *Theor. Appl. Genet.* **84**, 825–828.

62 Yano, M., Zamorski, R., Saito, A. & Shibuya, N. (1994) A dominant gene controlling accumulation of glucomannan in the cell wall of rice endosperm. *Rice Genet. Newslett.* **10**, 107–108.

63 Churchill, G.A., Giovannoni, J.J. & Tanksley, S.D. (1993) Pooled-sampling makes high-resolution mapping practical with DNA markers. *Proc. Natl Acad. Sci. USA* **90**, 16–20.

64 Tanksley, S.D. (1993) Mapping polygenes. *Ann. Rev. Genet.* **27**, 205–233.

65 Sax, K. (1923) The association of size differences with seed-coat pattern and pigmentation in *Phaseolus vulgaris*. *Genetics* **8**, 552–560.

66 Paterson, A.H., DeVerma, J.W., Lanini, B. & Tanksley, S.D. (1991) Fine mapping of quantitative trait loci using selected overlapping recombinant chromosomes, in an interspecies cross of tomato. *Genetics* **124**, 735–742.

67 deVicente, M.C. & Tanksley, S.D. (1993) QTL analysis of transgressive segregation in an interspecific tomato cross. *Genetics* **134**, 585–596.

68 Edwards, M.D., Helentjaris, T., Wright, S. & Stuber, C.W. (1992) Molecular-marker-facilitated investigations of quantitative trait loci in maize. *Theor. Appl. Genet.* **83**, 765–774.

69 Stuber, C.W., Lincoln, S.E., Wolff, D.W. *et al.* (1992) Identification of genetic factors contributing to heterosis in a hybrid from two elite maize inbred lines using molecular markers. *Genetics* **125**, 823–839.

70 Ahn, S., Bollich, C., McClung, A. & Tanksley, S. (1994) RFLP analysis of genomic regions associated with cooked-kernel elongation in rice. *Theor. Appl. Genet.* **87**, 27–32.

71 Lander, E.S. & Botstein, D. (1989) Mapping mendelian factors underlying quantitative traits using RFLP linkage maps. *Genetics* **121**, 185–199.

72 Ideta, O., Yoshimura, A., Matsumoto, T. *et al.* (1992) Integration of conventional and RFLP linkage maps in rice. I. Chromosomes 1, 2, 3 and 4. *Rice Genet. Newslett.* **9**, 128–129.

73 Ideta, O., Yoshimura, A., Matsumoto, T. *et al.* (1993) Integration of conventional and RFLP linkage maps in rice. II. Chromosomes 6, 9, 10 and 11. *Rice Genet. Newslett.* **10**, 87–89.

74 Whitkus, R., Doebley, J. & Lee, M. (1992) Comparative genome mapping of sorghum and maize. *Genetics* **132**, 1119–1130.

75 Devos, K.M., Millan, T. & Gale, M.D. (1993) Comparative RFLP maps of homoeologous chromosomes of wheat, rye and barley. *Theor. Appl. Genet.* **85**, 784–792.

76 Stephens, J.C. (1993) Cytogenetic map of human (*Homo sapiens*, $2n = 46$) genes as of July 1992. *Human Maps* **5**, 5.1–5.45 (Cold Spring Harbor Laboratory Press, Cold Spring Harbor, NY).

77 Wu, H.K., Chung, M.C., Wu, T.Y. *et al.* (1991) Localization of specific repetitive DNA sequences in individual rice chromosomes. *Chromosoma* **100**, 330–338.

78 Gustafson, P.J. & Dille, J.E. (1992) Chromosome location of *Oryza sativa* recombination linkage groups. *Proc. Natl Acad. Sci. USA* **89**, 8646–8650.

79 Wang, Z.X., Kurata, N., Katayose, Y. & Minobe, Y. (1995) A chromosome 5-specific repetitive DNA sequence in rice (*Oryza sativa* L.). *Theor. Appl. Genet.* **90**, 907–913.

80 Fukui, K. & Iijima, K. (1993) Somatic chromosome map of rice by imaging methods. *Theor. Appl. Genet.* **81**, 589–596.

81 Umehara, Y., Inagaki, A., Tanoue, H. *et al.* (1995) Construction and characterization of rice YAC library for physical mapping. *Mol. Breed.* **1**, 79–89.

82 Umehara, Y., Tanoue, H., Wang, Z.-X. *et al.* (1995) Construction of a YAC ordered library for 12 rice chromosomes. *Abstr. Plant Genome III* **3**, 82.

83 Katayose, Y., Yoshino, K. & Kurata, N. (1993) Use of cosmid libraries for physical mapping. *Rice Genome* **2**, 6.

84 Deshpande, V.G. & Ranjekar, P.K. (1980) Repetitive DNA in three Graminae species with low DNA content. *Hoppe-Seyler's Z. Physiol. Chem.* **361**, 1223–1233.

85 Cohen, D., Chumakov, I. & Weissenbach, J. (1993) A first generation physical map of the human genome. *Nature* **366**, 698–701.

86 Zhao, X. & Kochert, G. (1993) Phylogenetic dis-

tribution and genetic mapping of a microsatellite from rice (*Oryza sativa* L.). *Plant. Mol. Biol.* **21**, 607–614.
87 Ramakrishnan, W., Gupta, V.S. & Ranjekar, P.K. (1994) DNA fingerprinting in rice microsatellite and minisatellites. *Proceedings of the 17th Meeting of the International Program on Rice Biotechnology* (Rockefeller meeting) **17**, 18.
88 Whitham, S., Dinesh-Kumar, S.P., Choi, D. *et al.* (1994) The product of the tobacco mosaic virus resistance gene: similarity to toll and interleukin-1 receptor. *Cell* **78**, 1101–1115.
89 Mindrinos, M., Katagiri, F., Yu, G.L. & Ausubel, F.M. (1994) The *A. thaliana* disease resistance gene RPS2 encodes a protein containing a nucleotide-binding site and leucine-rich repeats. *Cell* **78**, 1089–1099.
90 Bent, A.F., Kunkel, B.N., Dahlbeck, D. *et al.* (1994) RPS2 of *Arabidopsis thaliana*: leucine-rich repeat class of plant disease resistance genes. *Science* **265**, 1856–1860.
91 Havukkala, I., Ichimura, H., Nagamura, Y. & Sasaki, T. (1995) Rice genome analysis by integration of sequencing and mapping data. *J. Biotech.* **41**, 139–148.
92 Miyadera, N., Havukkala, I., Ohta, I. & Mukai, Y. (1993) Development of RiceBase, an electronic notebook database for genome research. *Rice Genome* **2**, 11–12.
93 Durbin, R. & Thierry-Mieg, J. (1994) The ACeDB genome database. In *Computational Methods in Genome Research* (Suhai, S., ed.) 45–55 (Plenum Press, New York).
94 Ritter, O. (1994) The integrated genomic database (IGD). In *Computational Methods in Genome Research* (Suhai, S., ed.) 57–73 (Plenum Press, New York).
95 C.P.G.N. (Commission on Plant Gene Nomenclature) (1993) Nomenclature of sequenced plant genes. *Plant Mol. Biol. Rpt.* Suppl. **12**, S1–S109.
96 Havukkala, I., Ichimura, H., Nagamura, Y. & Sasaki, T. (1995) Plant-wide names for partial cDNA sequences of rice using BLAST and 4th Dimension. *Plant Mol. Biol. Rpt.* **13**, 32–49.
97 Robbins, R.J. (1994) Representing genomic maps in a relational database. In *Computational Methods in Genome Research* (Suhai, S., ed.) 85–96 (Plenum Press, New York).
98 Kilian, A., Steffenson, B.J. & Kleinhofs, A. (1995) Towards the cloning of the Rgp1 locus in barley: via rice? *Abstr. Plant Genome III* **3**, 53–53.

Section 6 Internet resources

Section 6 Introduction

Stephen P. Bryant

Gemini Research Ltd, 162 Science Park, Milton Road, Cambridge CB4 4GH, UK

The anarchic Internet or World Wide Web (WWW), progenitor of the cyberpunk novels of the early 1980s, has achieved, after a period of relatively slow growth, a dominant position in the dissemination of information around the globe. As a medium, the WWW is likely to exceed the printed page in importance in the not-too-distant future, particularly for information that needs to be timely rather than long-lasting. The first electronic peer-reviewed journals, such as GENE/COMBIS, have begun to appear, and the construction of a Web site is a prerequisite of any collaborative project which requires a timely, low-cost delivery of results to the community or the general public.

As Martin Bishop points out in the opening to his chapter (Chapter 35), it is not possible to do molecular biology in any real depth without recourse to information technology and the Internet. For most researchers, it will be most useful as the gateway to large collections of dynamic data, characterizing the human genome with increasing resolution. Whatever the future of the Internet, it is vital that the researcher understands the basic principles behind it and the way it can add value to the work done. Like the telephone and the fax machine, it is an integral part of the way in which we now do science.

In an ideal world, scientists would not have to be engineers as well, but would be gently led through the database and analytical resources available on the Internet by interfaces that were completely intuitive and that did not require a degree in computer science to understand. However, reality is not like this at all, and researchers must be prepared to get their hands a little bit dirty, or else miss out. In Chapter 36, Jaime Prilusky has provided a guide that should help whenever it is necessary to dip into the ever-changing world of the computer operating system, in most cases UNIX. His is a first aid kit for the information age, the need for which will surely diminish as interfaces develop in their ease of use and utility.

Finding out what is on the information superhighway is the subject of the last part of the section. Here, a selective directory (Chapter 37) has been produced of important genomic resources. If the reader spends some time investigating these sites, they will find that such a directory represents a window of codependent sites that shifts around as the priorities of the Human Genome Project change. It is hoped that the window will retain a value which will persist long enough to be used as a jumping-off point for the increasing pool of relevant information resources. The Internet is clearly here to stay, and these contributions have the modest aim of making it just that little bit less arcane, and just a little bit more accessible.

Chapter 35 Internet resources and how to use them

Martin J. Bishop

HGMP Resource Centre, Hinxton, Cambridge CB10 1SB, UK

35.1 Introduction

Genome analysis is not possible without the aid of the computer as a tool. This applies both to the analysis of local data and to accessing information worldwide. Most people are now familiar with computers, so a detailed description of their components and usage will not be given here. This chapter will give a brief survey of the facilities available to access information and data for genome analysis. Addresses for particular sites may be found in the relevant chapters, in Chapter 37, and in Appendices III and V.

35.1.1 Computing hardware and operating system

The engine of the computer is the central processing unit (CPU), which is located with its interfaces on the mother board and controls the processing of information. Information is stored in memory, which may be volatile (the random access memory, or RAM), or the nonvolatile memory forming the file-store. The latter is usually a hard disk and a floppy disk drive using magnetic recording which can be read or written to, and an optical disk drive for reading large amounts of prerecorded information. Peripherals connected to the computer are the screen, the keyboard and the mouse. There may be a variety of other peripherals such as printers.

An *operating system* is a master control program which manages the function of the computer as a whole and the running of the *application programs*. The operating system runs continually while the computer is switched on and provides the means by which the user directs the operations performed by interaction from the keyboard and mouse. Interaction has evolved in recent years from the typing of a single command line, through full screen interaction at the level of individual characters, to the more sophisticated graphical user interfaces (GUIs) in use today.

The most commonly used operating systems relevant to genome analysis are:

- Microsoft Windows 3 and NT;
- Apple Macintosh OS;
- Unix X-windows (from a variety of vendors).

Windows and Macintosh GUIs must run on the local machine, whereas in X-windows the graphics can be generated on one computer and displayed over the network to another. Thus, in Unix it is possible to have an *X-terminal*, which is a computer dedicated to the GUI and user interaction. With suitable emulator software, a Windows (e.g. Vista eXceed program) or Macintosh (e.g. MacX program) machine can act as an X-terminal.

Historically, Windows and Macintosh machines were developed for office and home environments while Unix was developed for scientific and technical environments. Today, the hardware costs are similar for a similar configuration, irrespective of operating system. The important considerations are to have CPU and RAM giving adequate response times, sufficient disk space, and a screen of the highest quality, with a minimum diameter of 17 in and resolution of 1024×768 pixels. To interact with X-windows programs, a three-button mouse is desirable.

35.2 User interface

The quality of the user interface makes or breaks the success of a software application. A user interface requires some input devices to enter the necessary information. The common input devices are the keyboard and the mouse, but many others have been devised. Determinants of a good user interface are speed of learning, speed of use, elimination of error, rapid understanding and attractiveness to the user. In the genomic field, we can contrast the now obsolete Sybase APT forms interface to the Genome Data Base (GDB) with the World Wide Web (WWW) interface. The former involves learning arcane keystrokes, whereas the latter is driven by mouse clicks and form filling. Unfortunately, there are conflicts between interfaces being easy to learn and fast to use, and being able to exploit the full power of the program. There are also subjective factors: people differ in the sort of interface they prefer to use.

The *command line interface* is not encouraging to novices who have to wade through reams of documentation in order to understand the replies to instructions typed at a C:\ prompt. Incorrect typing may lead to serious errors such as deleting the wrong files.

Question and answer dialog, as, for example, in the text version of the Staden Sequence Analysis Package, assumes that the reader knows the answer to supply. If a mistake is made earlier in the sequence, it is not possible to go back.

Menus present multiple choices and act as an *aide-mémoire* to the options available. They appeal to novices but their attraction rapidly palls as the user becomes an expert.

Form filling with default answers allows the user to see and alter all the relevant items of data. There is a danger that inappropriate values may be provided if the user has failed to study the nuances of the analysis.

Talking to the computer in a natural language is not a practical proposition with today's technology.

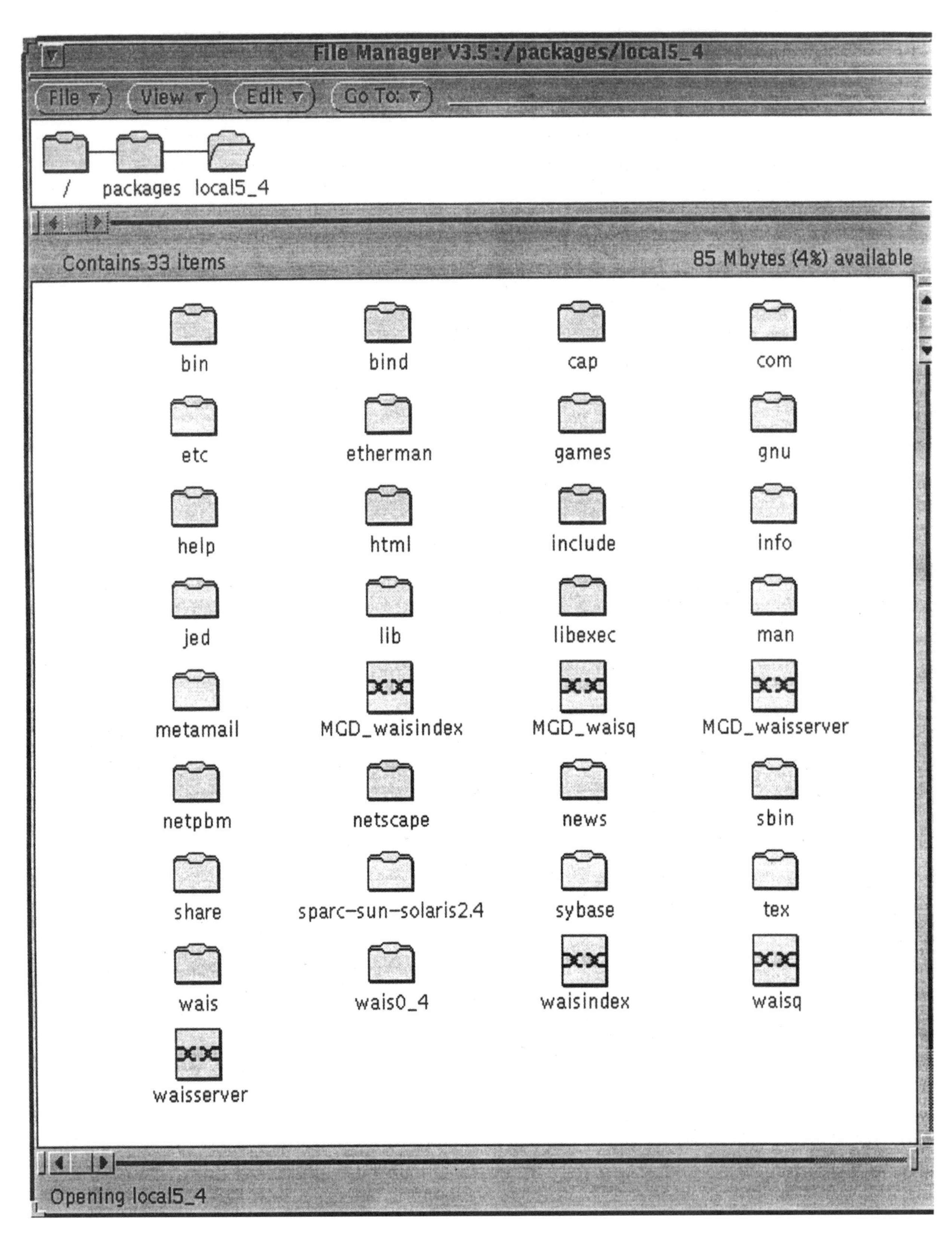

Fig. 35.1 A filing system represented by icons.

It may become an important user interface in the future.

The favoured interface of today is the GUI. Research at Xerox's Palo Alto Research Center was the start of the 'WIMPS' revolution. Windows, icons, mice, and pull-down menus are standard in all three of the common operating systems listed above. Operations are invoked by actions performed on visual representations—*icons*—of objects. The status of many applications running simultaneously can be displayed by the appearance of their icons (Fig. 35.1). For example, the arrival of new electronic mail can be signalled. Clicking on the icon with the mouse opens the mail folder and mail can be read in a window and a reply entered. Then the application can be reduced back to iconified form.

35.3 Network connectivity

A stand-alone computer can enable plenty of work to be done. However, it is not very interesting to receive electronic mail only from oneself. Increasingly, it is the interconnectivity of computers worldwide that is the major interest of their users.

A computer *network* consists of two or more intelligent communicating devices (personal computers, work stations, servers) linked in order to exchange information and share resources. A communicating device on a network is called a *node*. A *host* is a network node which provides individual network users with a variety of resources such as processing power, file stores, applications software and connections to other networks.

Computer networks now span the world, and the earlier classification into local area networks (LANs) and wide area networks (WANs) is becoming less useful. The only concern of the user is the slowest link in the connection from his or her work station to the host of interest. There are two main aspects of network operation: the physical medium which implements the network, and the method of data transfer or network protocol.

There are many transmission media in use today, including copper wire, fibre optic and microwave link. The most common media in the workplace are:

1 fibre optic cables made of plastics or glass which serve as a very high performance transmission medium unaffected by electrical interference; and

2 shielded or unshielded twisted pair (UTP) copper wire, which is also used in telephony for the less demanding applications.

There are a variety of ways in which the electronic signals can be placed on the LAN. Ethernet is most widely used; it operates on a bus configuration of a single strand of cable to which each node connects. Only one signal can travel on the cable at one time and the transmission speed is 10 megabits per second (Mbps) for standard ethernet or 100 Mbps for fast ethernet. For situations where heavy traffic is expected, direct node-to-node communication is possible using switched ethernet. More recent technologies are fibre distributed data interface (FDDI) using a token passing ring configuration and asynchronous transfer mode (ATM) using switching. These are capable of gigabit-per-second (Gbps) connection speeds.

Special cards (e.g. ethernet cards) need to be installed in the PC or work station to connect it to the LAN. Repeaters, bridges, switches and routers may be used to implement the LAN and its connection to the WAN.

It is possible to connect to analogue (voice) telephone channels using an external or internal device called a *modem* (modulator–demodulator) which converts digital data to analogue form. Speeds of about 20 kilobits per second (Kbps) are achievable.

Telephone: networks are being converted to digital circuits called the Integrated Services Digital Network (ISDN) which offers digital services for both voice and data. ISDN cards permit connectivity at 64 Kbps.

If a node is attempting to connect by a telephone circuit, the host needs to have a matching modem or ISDN card, and the network service provider will dictate the method used. Telephone: links are charged according to the time the connection remains open. It is also possible to install leased lines, where the connection remains open permanently and charging is according to a fixed monthly rental fee.

35.3.1 The Internet

The Internet is a collection of many national, regional, site, and individual network connections which all use the TCP/IP protocol suite. TCP/IP stands for Transmission Control Protocol/Internet Protocol and was originally developed for the US Defense Advanced Projects Agency (DARPA). The specifications of the protocol are publicly available so that any manufacturer can produce suitable equipment. This has led to TCP/IP becoming a *de facto* world networking standard.

To connect to the Internet you need to find an appropriate Internet service provider. The service provider will then inform you as to the equipment needed to connect to the service.

The service provider to the UK academic community is the United Kingdom Education and Research Networking Association, which runs a network called JANET (Fig. 35.2) and an IP service called JANET IP Service (JIPS). If you work at a university or Research Council institute this is likely to be the service available to you.

If you work on genome analysis in a hospital or at home you are likely to require a commercial Internet service provider. There are a number of these, including CompuServe, Demon, Pipex and Unit, who will specify the hardware and software which you need.

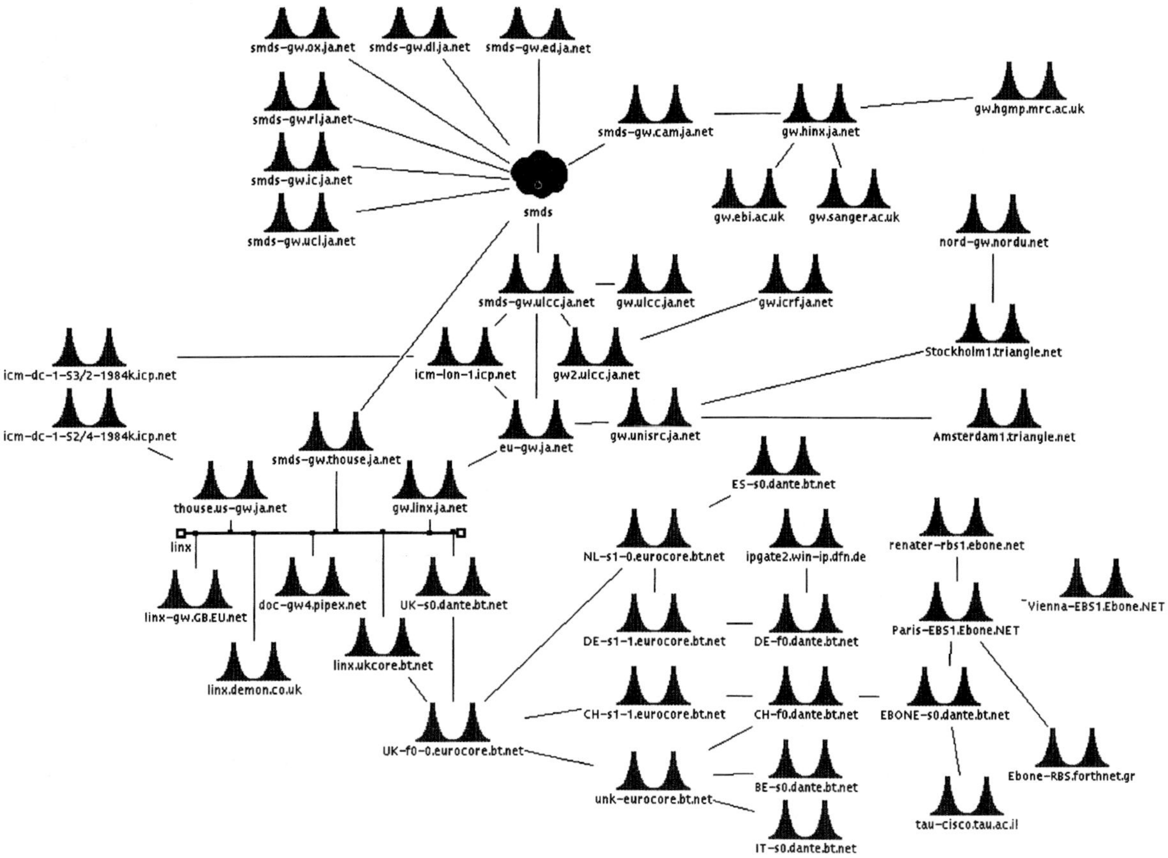

Fig. 35.2 Connection of the ICRF gateway (gw.icrf.ja.net) to the JANET network. SMDS means Synchronous Multimegabit Data Services and is used to implement the JANET backbone.

35.4 Client–server computing

The client–server model of computing is a means of distributing processing and graphical resources while sharing centralized resources such as file stores and data bases. Client computers issue requests and server computers respond to them over the network. These arrangements result in end-user operated facilities which can connect to remote systems throughout the world. The user is no longer communicating with a single computer, instead, 'the network is the computer'.

The components required for client–server computing are:

- the LAN and WAN technologies described in Section 35.3;
- desktop computing devices (personal computers, work stations, X-terminals or their emulations) described in Section 35.1;
- a GUI environment based on X-windows technology as described in Section 35.2. If a WWW client (such as Netscape) is used, it may not be necessary to have X-windows capability (depending on the services the user wishes to access);
- servers with the resources of interest.

35.5 Database technology

Databases are essential resources for genome analysis. The human genome contains perhaps 60 000 genes encoding genetic functions (dependent upon proteins and RNA molecules) and comprises some 3000 million base pairs (megabase pairs, Mbp) of DNA. Hundreds of differentiated cell types make up the human organism, and there are thousands of mechanisms for regulating gene expression. To record the basic information and to explain reproduction, development, function in life, and genetically programmed death is a major challenge in biological understanding and information technology which may be called *bioinformatics*.

Databases and the knowledge-based technologies required for bioinformatics are still an active area of computer science research. We do not yet possess all the tools needed.

A database project consists of three components:

1 developing the database structure that will permit storage and maintenance of the data;

2 entering and maintaining the data;

3 facilitating access by providing users with suitable analysis and display tools.

The three stages must not be confused (although they often are). The expense of development is minor in relation to on-going maintenance and support. A common experience is of moving goal posts: the database specification changes faster than the developers and maintainers can work, leading to projects running over budget, being late, or failing.

Genome analysis is at present in the data acquisition phase. Much of the data being collected is of ephemeral interest. (Will, for example, the contents of the St Louis YAC Library be of interest in 5 years' time?) The available data need to be analysed to discover what has to be represented in the database to produce good results. The problem for genome analysis is that there are many kinds of data from genetic mapping, physical mapping and sequencing which can only be linked if common markers are used. It is best to analyse the database requirement by working back from the desired end result: the human genome sequence, map position of phenotypes, and the nature of mutations of medical importance. This is merely the first step in understanding the organism, and further databases or knowledge bases will be required for development, the localization of gene expression, and cellular function.

Once the analysis is complete, the data are modelled to define how they will be internally represented in the database. This results in the conceptual schema of the database and is free of any assumptions of hardware and software. There will be a single conceptual view of the data.

In the database as implemented, the user is presented with an external view of the data and there may be many such views. Important databases for genome analysis are the GDB for human data and the Mouse Genome Database (MGD). Both are implemented in a relational database management system (RDBMS) which is a commercial product (Sybase).

Users of MGD do not have access to the database itself. Their view of the data is provided by a WWW browser system. In the case of GDB, users have a number of options. It is possible for them to perform Structured Query Language (SQL) queries which operate directly on the database to formulate questions of arbitrary complexity. In addition, there is a graphical user interface implemented in Galaxy and a WWW browser interface which provide views that are considered to be most frequently required.

The relational database model is well defined mathematically, with proven characteristics. However, it is slow and tedious to implement in practice. One approach to this difficulty is to build software tools to speed the implementation of the database and the development of the user interface. Such tools are becoming commercially available.

Another approach is to move away from the relational model towards an object-oriented model that considers the data as objects and classes that are more natural to the human perception of the problem. No commercial object oriented database management system is at present in popular use in genome analysis. However, a C language program written for the *Caenorhabditis elegans* Genome Project and called ACeDB operates along object-oriented lines. Acceptance of ACeDB by the genome analysis community relates to the appropriateness of the GUI rather than the robustness of the underlying storage and query method. The general user will be able to obtain all the information required by accessing the Web browser GUI of a genome dataset. Indeed, the power of such methods will increase with the introduction of executable code (aplets, that is small applications) running on the client, which is made possible by languages such as Java.

In my view, the accurate maintenance of the data and the possibility of arbitrary queries made possible by SQL as well as database integrity (that is, ensuring the database is accurate, correct, valid and consistent) give the edge to RDBMS systems. However, the general user should not need to know they exist.

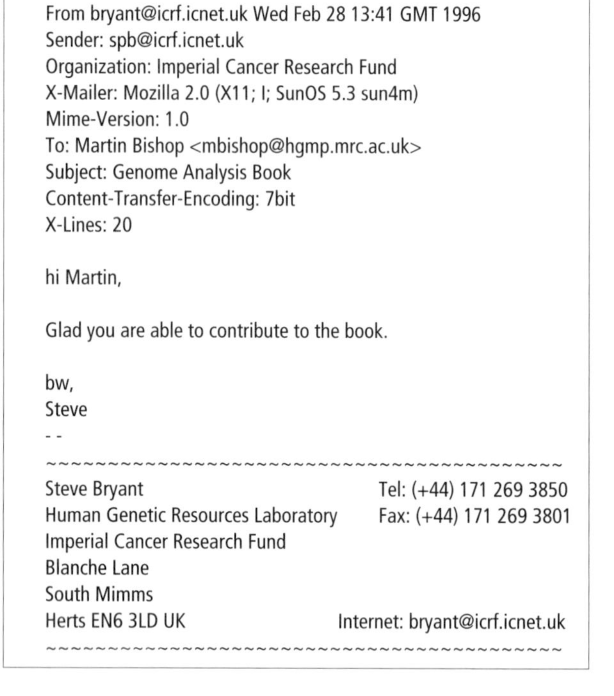

From bryant@icrf.icnet.uk Wed Feb 28 13:41 GMT 1996
Sender: spb@icrf.icnet.uk
Organization: Imperial Cancer Research Fund
X-Mailer: Mozilla 2.0 (X11; I; SunOS 5.3 sun4m)
Mime-Version: 1.0
To: Martin Bishop <mbishop@hgmp.mrc.ac.uk>
Subject: Genome Analysis Book
Content-Transfer-Encoding: 7bit
X-Lines: 20

hi Martin,

Glad you are able to contribute to the book.

bw,
Steve
--
~~~~~~
Steve Bryant Tel: (+44) 171 269 3850
Human Genetic Resources Laboratory Fax: (+44) 171 269 3801
Imperial Cancer Research Fund
Blanche Lane
South Mimms
Herts EN6 3LD UK Internet: bryant@icrf.icnet.uk
~~~~~~

Fig. 35.3 An e-mail message.

Table 35.1 List of bulletin boards relevant to biology.

bionet.agroforestry	bionet.molecules.peptides
bionet.announce	bionet.molecules.repertoires
bionet.audiology	bionet.mycology
bionet.biology.cardiovascular	bionet.neuroscience
bionet.biology.computational	bionet.neuroscience.amyloid
bionet.biology.grasses	bionet.organisms.pseudomonas
bionet.biology.n2-fixation	bionet.organisms.schistosoma
bionet.biology.symbiosis	bionet.organisms.urodeles
bionet.biology.tropical	bionet.organisms.zebrafish
bionet.biology.vectors	bionet.parasitology
bionet.biophysics	bionet.photosynthesis
bionet.celegans	bionet.plants
bionet.cellbiol	bionet.plants.education
bionet.cellbiol.cytonet	bionet.population-bio
bionet.cellbiol.insulin	bionet.prof-society.afcr
bionet.chlamydomonas	bionet.prof-society.ascb
bionet.diagnostics	bionet.prof-society.biophysics
bionet.diagnostics.prenatal	bionet.prof-society.cfbs
bionet.drosophila	bionet.prof-society.csm
bionet.ecology.physiology	bionet.prof-society.faseb
bionet.emf-bio	bionet.prof-society.navbo
bionet.general	bionet.protista
bionet.genome.arabidopsis	bionet.sci-resources
bionet.genome.chromosomes	bionet.software
bionet.glycosci	bionet.software.acedb
bionet.immunology	bionet.software.gcg
bionet.info-theory	bionet.software.sources
bionet.jobs.offered	bionet.software.srs
bionet.jobs.wanted	bionet.software.staden
bionet.journals.contents	bionet.software.www
bionet.journals.letters.biotechniques	bionet.software.x-plor
bionet.journals.letters.tibs	bionet.structural-nmr
bionet.journals.note	bionet.toxicology
bionet.metabolic-reg	bionet.users.addresses
bionet.microbiology	bionet.virology
bionet.molbio.ageing	bionet.women-in-bio
bionet.molbio.bio-matrix	bionet.xtallography
bionet.molbio.embldatabank	sci.med
bionet.molbio.evolution	sci.med.aids
bionet.molbio.gdb	sci.med.dentistry
bionet.molbio.genbank	sci.med.diseases.cancer
bionet.molbio.genbank.updates	sci.med.immunology
bionet.molbio.gene-linkage	sci.med.informatics
bionet.molbio.genome-program	sci.med.nursing
bionet.molbio.hiv	sci.med.nutrition
bionet.molbio.methds-reagnts	sci.med.occupational
bionet.molbio.molluscs	sci.med.pathology
bionet.molbio.proteins	sci.med.pharmacy
bionet.molbio.proteins.7tms_r	sci.med.physics
bionet.molbio.proteins.fluorescent	sci.med.psychobiology
bionet.molbio.rapd	sci.med.radiology
bionet.molbio.recombination	sci.med.telemedicine
bionet.molbio.yeast	sci.med.transcription
bionet.molec-model	sci.med.vision

35.6 Network services

Having described the context in which they operate, we now describe the services which are available on the Internet. These are evolving rapidly and one of the most popular, the World Wide Web, is only a few years old.

35.6.1 Electronic mail

E-mail is used for communicating messages to other people worldwide. It is very convenient because the recipient need not be contacted prior to delivery and messages are usually delivered within a few minutes (cf. fax, which can be connected to e-mail). E-mail has been adapted to other purposes such as the transfer of small files or delivery of the output of programs which take a while to run (e.g. FASTA).

A variety of e-mail programs are in common use, which vary considerably in their user interfaces. You will probably have a choice even on the same computer. It is unwise to run more than one mail program at a time, however—they will confuse each other! The WWW client Netscape 2.0 includes a mail facility. The common mail protocol on the Internet is called Simple Mail Transfer Protocol.

An e-mail message consists of an 'envelope' with address details and the message contents (Fig. 35.3). Important components of the envelope are the identity of the sender and recipient, date and subject matter. It may be possible to attach files to the message and these need not necessarily contain plain text. However, before sending formatted files—for example, word-processed text—make sure your recipient has the appropriate programs to read or convert the file.

One difficulty you may have with e-mail is finding the correct 'address' of the intended recipient. This information is harder to find than is a telephone number as there is no universal standard e-mail directory. In the first instance you may have to phone or fax to get the e-mail address. The form of the address is usually something like: persons_name @computer_address. The persons_name may be their computer user identifier, which is not necessarily anything meaningful. The computer_address may refer to a single machine or may be a domain. Many countries use a two-letter country code—for example, 'uk'. In the UK 'ac' means academic community and 'co' means commercial. So to contact user support at the UK Medical Research Council's Human Genome Mapping Project Resource Centre, the address is: support@ hgmp.mrc.ac.uk. In the US the country code is omitted (like the country name on British postage stamps). Main categories include 'edu' for education, 'gov' for government and 'com' for commercial. For example, to obtain information about the National Center for Biotechnology Information at the National Library of Medicine belonging to the National Institutes of Health, the address is: info@ncbi.nlm.nih.gov.

Mail lists are a form of public communication which enable people with common interests to exchange ideas and information. By subscribing to the list you are sent mail from every contributor to the list. This is useful for groups of people working closely together. It becomes unsatisfactory if you are not interested in the majority of the messages and you have to wade through masses of junk to find your urgent or important mails. The solution is the bulletin board.

35.6.2 Bulletin boards or newsgroups

Electronic bulletin boards are like the departmental notice board or the corner shop window. Often, anyone can post a message but sometimes the postings are moderated. There are boards for a huge variety of subjects but you must be careful to post to the correct board. It is suprising how unpleasant people can be when 'Flame' wars arise on mail lists and bulletin boards.

The advantage of bulletin boards over mail lists is that you can browse when you want and may be able to tell from the subject line what to avoid reading. There are usually a number of different

Subject: **Whitehead STS Mirror Site**
Date: 27 Feb 1996 06:19:54 -0800
From: pwoollar@hgmp.mrc.ac.uk (Peter M. Woollard x4523)
Reply-To: pwoollar@hgmp.mrc.ac.uk
Organization: UK MRC Human Genome Mapping Project
Newsgroups: bionet.announce

We are pleased to announce a European mirror to

The Whitehead Institute/MIT Center for Genome Research
Human Genomic Mapping Project:

"An STS-Based Map of the Human Genome"

Originating site: http://www-genome.wi.mit.edu/cgi-bin/contig/phys_map
Mirror site: http://www.hgmp.mrc.ac.uk/cgi-bin/contig/phys_map

Many thanks to Lincoln Stein for his help and patience in providing this.

The mirror still requires some work, but we hope that you find this useful in the mean time.

The WWW pages have an explanation of the data.

Best Regards,
Peter Woollard

Computing Services Section, Internet: p.woollard@hgmp.mrc.ac.uk
MRC Human Genome Mapping Project http://www.hgmp.mrc.ac.uk/
Resource Centre, Hinxton Hall,
Hinxton, Cambridge, CB10 IRQ, UK Tel: ++44 (0)1223 494 523

Fig. 35.4 A bulletin board entry.

```
mbishop@hydrogen% ftp ftp.hgmp.mrc.ac.uk
Connected to osmium.
220 osmium FTP server (Version wu-2.4 (2) Mon Apr 10 15:05:49 BST 1995) ready.
Name (ftp.hgmp.mrc.ac.uk:mbishop): anonymous
331 Guest login ok, send your complete e-mail address as password.
Password:
230-#####################################################################
230-
230-   Welcome to the UK HGMP Resource Centre anonymous ftp service
230-
230-          Please contact support@hgmp.mrc.ac.uk regarding
                    any problems with this service
230-#####################################################################
230-
230-Please read the file README
230- it was last modified on Tue Jul 5 13 : 41 : 56 1994 - 609 days ago
230 Guest login ok, access restrictions apply.
ftp> cd manuals/handbook
250-Please read the file README
250- it was last modified on Fri Apr 28 10 : 21 : 41 1995 - 312 days ago
250 CWD command successful.
ftp> dir
200 PORT command successful.
150 Opening ASCII mode data connection for /bin/ls.
total 4122
drwxrwxr-x    3 0   other       512 Sep 28  09:04 .
dr-xr-xr-x   19 0   other       512 Nov 23  16:28 . .
-rw-r--r--    1 0   other       306 Apr 28  1995 README
-rw-r--r--    1 0   other    169001 Apr 28  1995 man.asc
-rw-r--r--    1 0   other    558008 May 12  1995 man.ps
-rw-r--r--    1 0   other    177322 May 12  1995 man.ps.Z
-rw-r--r--    1 0   other    737792 May 12  1995 man.word
-rw-r--r--    1 0   other    393843 May 12  1995 man.word.Z
drwxr-xr-x    2 0   other      1024 Sep 23  15:50 manold
lrwxrwxrwx    1 0   other         6 Sep 28  09:04 manual -> manual
226 Transfer complete.
642 bytes received in 0.2 seconds (3.2 Kbytes/s)
ftp> get man.word.Z
200 PORT command successful.
150 Opening ASCII mode data connection for man.word.Z (393843 bytes).
226 Transfer complete.
local: man.word.Z remote: man.word.Z
395611 bytes received in 2.1 seconds (1.9e+02 Kbytes/s)
ftp> quit
221 Goodbye.
mbishop@hydrogen%
```

Fig. 35.5 An anonymous ftp session.

conversations ('threads') going on simultaneously. A threaded bulletin board reading program helps you to read about a single topic. Because of the mechanism by which bulletin board information is propagated around the world, the postings are quite likely to be in the wrong order.

On the Internet, bulletin boards are called 'Network News' or 'Usenet' and individual boards are known as 'newsgroups'. The protocol by which they are propagated is called Network News Transfer Protocol. The ratio of pearls of wisdom to dross should be monitored on a regular basis (otherwise you can waste a lot of time). The bulletin boards relevant to genome analysis are given in Table 35.1.

The structure of a bulletin is rather like a mail message, with an envelope and contents (Fig. 35.4). Network News contains little of lasting value, and articles are rapidly expired at most sites where you can read the bulletins, because of the large amounts of disk storage required to hold them. So if you find Network News useful, you should read it on a regular basis.

There is a wealth of news-reading programs (as with mail) and I have mentioned the threaded variety. Choose the most convenient reader available on your system. Netscape has the ability to read news.

35.6.3 File transfer

File transfer is a way of transferring larger files,

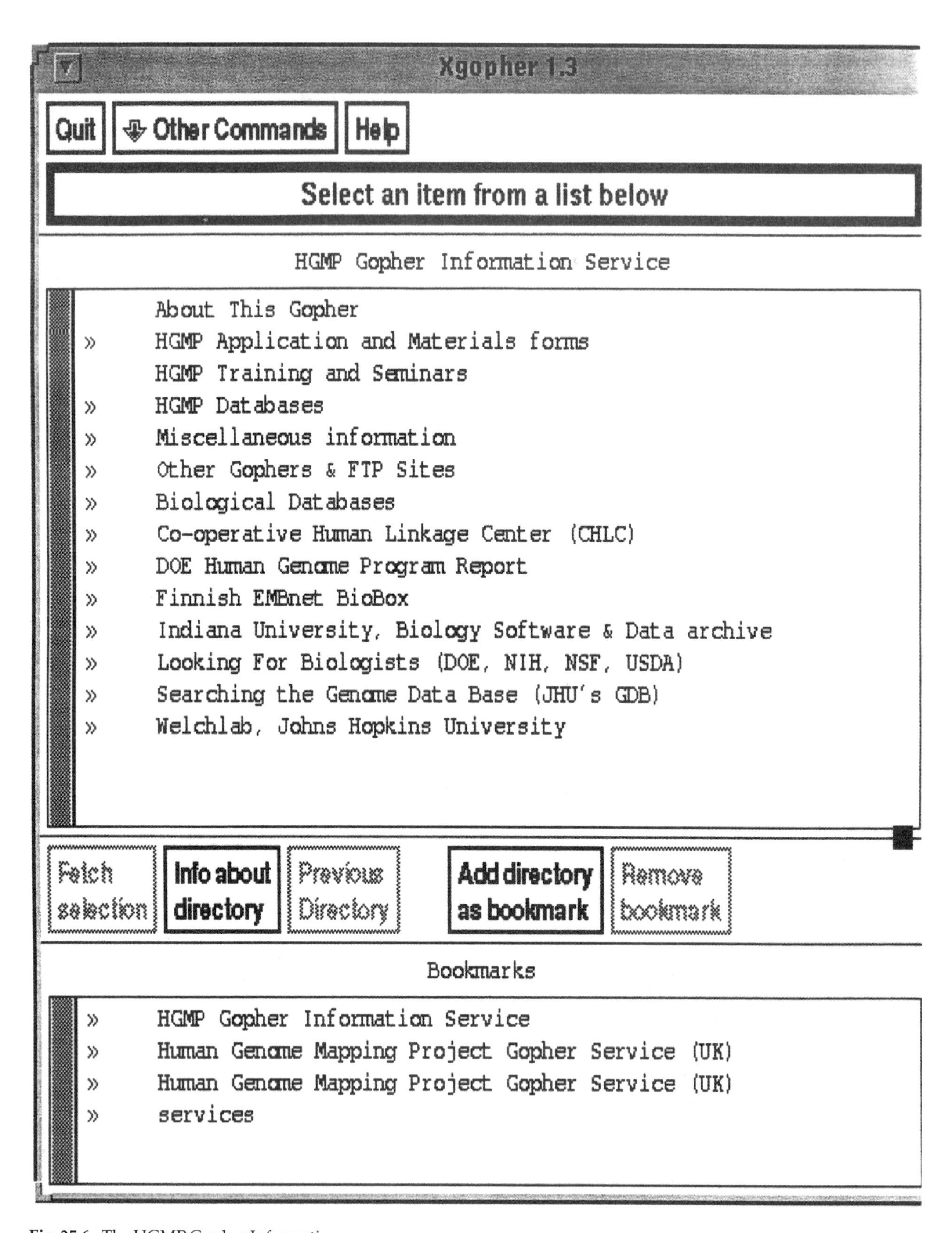

Fig. 35.6 The HGMP Gopher Information.

especially program files and large data files, from one computer to another. On the Internet, the file transfer protocol is called ftp. It includes a mechanism for identifying the user and giving a password, but the most useful form of ftp is *anonymous ftp*, by which you can access public files. These hold an enormous range of information, the most useful for our purposes being genome data

Fig. 35.7 The ICRF WWW server.

and computer programs for genome analysis. To find out what is available, users can access a database called Archie. To use Archie, telnet (see Section 35.6.6) to archie.doc.ic.ac.uk and when connected give login name 'archie' and command 'help'. Archie clients are also available from the many other archie servers and other software sites.

Once you have located files of interest you can use

Fig. 35.8 The Webcrawler front page.

the program ftp to download them to your system. An ftp session is shown in Fig. 35.5. You log in as 'anonymous' and give your e-mail address as the password. The command `dir` shows you the files available. If directories are listed (letter 'd' on the left of the permissions string) you can go into them with the `cd` command. When you have found the file or files you want, specify whether you want files transferred in text mode (ASCII mode), for text files, or binary mode, for programs. The command `get` will then transfer the file of interest to your computer (Fig. 35.5), or use `mget` if you wish to transfer multiple files. Gopher (Section 35.6.5) or a Web browser (Section 35.6.6) can also be used to ftp files, and they have a more user-friendly interface. A short guide to anonymous ftp can be obtained by sending the message 'help' to 'info@sunsite.unc.edu'.

35.6.4 Gopher

There is a huge amount of information on the Internet, much of which can be accessed by ftp. However, there are now more sophisticated methods of locating the information you require. Gopher was developed at the University of Minnesota to help find answers to questions on campus. It has been taken up worldwide to deliver documents to the user from a multitude of servers at centres which cooperate in providing this service. The client–server model is used and the gopher program presents a hierarchy of documents. The information around the world appears to the user as a single resource with a simple mechanism of browsing through it (Fig. 35.6). Key word searches can be made, as servers have full-text indexes for sets of gopher documents. This is often implemented using the Wide Area Information Server.

Gopher clients are available for all three of our named operating systems. Gopher clients for most types of computers are available from the University of Minnesota by connecting to boombox.micro.umn.edu. Worldwide Gopher sites can be searched using the Gopher tool Veronica. Gopher is a very valuable source of information, but has been replaced by the WWW.

35.6.5 World Wide Web

The WWW is an easily accessible set of hypertext images and information available around the world. There is a huge and growing amount of biological information, databases and analysis tools which may be accessed by WWW (see Chapter 37 and Appendix V for useful addresses). Normally, WWW is available to anyone, but it is also possible to have services with access limited by username and password and to charge for such services.

WWW is a client–server system. You run the client, called a browser, on your desktop computer. You open a connection to a server by specifying its universal relative locator (URL). For ICRF this is http://www.icnet.uk/. This will take you to the front page or 'home page' as shown in Fig. 35.7. A variety of browsers are available, such as Mosaic, Netscape, Lynx (character-based, no graphics), tkWWW and HotJava. Mosaic software, developed at the University of Illinois, is available by anonymous ftp from ftp.ncsa.uiuc.edu or by Gopher from gopher.ncsa.uiuc.edu. The widely used Netscape software can be obtained from ftp.mcom.com. When you access a server, a connection from the client is established, a page of information is delivered, and then the connection is closed.

```
mbishop@hydrogen% telnet menu.hgmp.mrc.ac.uk
Trying 193.62.192.50...
Connected to tin.
Escape character is '^]'.

UNIX(r) System V Release 4.0 (tin)

login: mbishop
Password:
Last login: Tue Mar  5 10 : 28 : 13 from hydrogen
Sun Microsystems Inc.   SunOS 5.4      Generic July 1994

** If you don't get the menu, please log out and try again later **
** If you still have problems, contact user support on 01223 494520 **

If you used the command 'telnet', and your terminal is running
X-Windows, you can start the using X-Windows versions of the
 programs at the HGMP.

You may also need to set the display permissions on your
machine by giving the local command 'xhost' followed by a list
of machines at the HGMP (see the Computing Handbook for details).

Do you wish to use X-Windows (y/n) > n
Setting up environment...
Starting menu....

      MOLECULAR BIOLOGY SOFTWARE FOR THE HGMP-RC
                      MAIN MENU
> > > > You Have NEW Mail. Choose Option 2
0)    Help
1)    Exit

2)    Electronic Mail
3)    BIOSCI/Network News (Biologist's Bulletin Boards)
4)    Information Services

5)    Analysis and Manipulation of Sequences
6)    Sequence Database Searching
7)    Genome Data
8)    Linkage Analysis
9)    Cell Lines, Clones & Probes Databases

10)   Other Molecular Data
11)   Utilities  (File Transfer & Management)
12)   UNIX Operating System
13)   Miscellaneous ('How to...'  etc)
14)   Queries, Suggestions and Comments to User Support

Enter a number, option-name or ? >
```

Fig. 35.9 A telnet session accessing the HGMP menu.

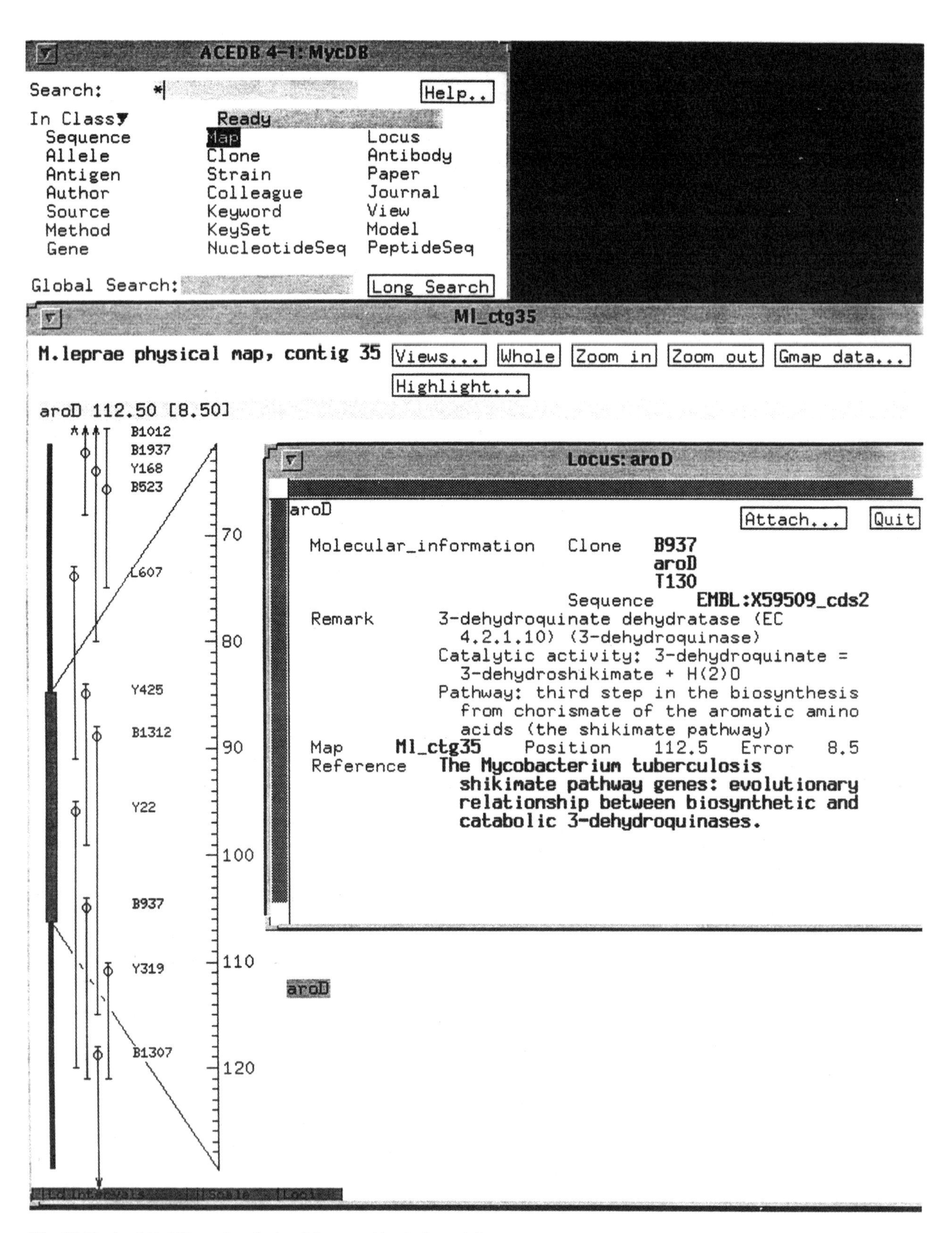

Fig. 35.10 An ACeDB session derived from an X-windows telnet access.

The pages are written in hypertext markup language (HTML) which is interpreted before being displayed on your screen. The pages may contain hot links to other pages or to different sites. By clicking on these links you are able to navigate the net. There are also searching facilities which enable you to look for sites of interest. Webcrawler (http://webcrawler.com/) is one example (Fig. 35.8).

In addition to delivering pages, the Java language enables *app lets* to be written, which are programs which run on the client. This is a relatively new development and we can expect to see many applications in the future, distributing the computational load to the clients, and reducing the need for telnet sessions.

35.6.6 Telnet

Telnet enables you to log in to a remote host from a terminal session and to use any of the programs available on the host. To log in you will need to be provided with a user name and password by the host administrator. The nature of the facilities available will depend on the terminal you are using. An example session accessing the HGMP menu is shown in Fig. 35.9.

In order to use graphical programs you need an X-terminal or emulation. The ACEDB software requires X-windows, and an example display from MycDB (the *Mycobacterium* database) is shown in Fig. 35.10.

Chapter 36 UNIX system survivors' guide

Jaime Prilusky

Bioinformatics Unit, Department of Biological Services, Weizmann Institute of Science 76100, Rehovot, Israel

36.1 Introduction

Programs used for genome analysis will often be run in a UNIX environment (see Chapter 3). This brief guide lists some of the basic commands you will need. Sources of further information are listed in Section 36.8. An example of a session using UNIX commands is given in Chapter 3.

36.1.1 Golden rules

There are two rules that you *must* remember when working in a UNIX environment.

- Case matters. You must provide the command or filename in exactly the same way as it exists, for the computer to recognize it. 'FILE' is different from 'File' and from 'file'.
- File overwritten/deleted is overwritten/deleted. There is no way back after deleting a file or saving a file using the same name of an existing one. *There is no way back.*

36.2 Basic UNIX commands

The list below presents some examples of basic UNIX commands. To use them:

- Type **boldface** text literally. Text that appears below in boldface are UNIX commands and should be typed as it appears in this text.
- Substitute actual value for *italicized* argument names.
- Optional arguments are in square brackets.
- Repeatable arguments are followed by ellipses (...).
- Text appearing within parentheses describes the action to be produced by the preceding command.

cat—Reads each filename in sequence and displays it on the standard output
% **cat** *file1 file2*—(displays the contents of file1 on the standard output)
% **cat** *file1 file2* > *file3*—(concatenates the contents of file1 and file2 and places the result into file3)
cd—Change working directory
% **cd**—(return to your login directory)
% **cd** *directory*—(change working directory to directory)
cp—File copy
% **cp** *file1 file2*—(make a copy of file1 and name it file2)
% **cp** *file1 ... dirname*—(copy one or more files into the specified directory)
(% cp gamma.seq temp.txt creates a new file—or writes over an existing one—called temp.txt that contains exactly the same data as the gamma.seq file)
du—Summarize disk usage
% **du** *[name]*—(disk usage for directory or file name)
grep—Pattern searching utility
% **grep** *pattern filename*—(looks through filename for the pattern character pattern and displays the lines that match the pattern)
jobs—Lists jobs
% **jobs**—(to see a list of all the jobs that are running in your login shell)
kill—Terminate a process
% **kill** *number...*—(terminate process number)
logout or **exit**—Ends your UNIX session. This command does not need extra arguments. Simply enter one of them in a line to close your current logiu shell
% **logout**—(ends your current UNIX session)
% **exit**—(ends your current UNIX session)
lpr—Send a file to printer
% **lpr -P***printername filename*—(send file filename to printer printername)
ls—List contents of directory
% **ls** *name ...* —(list contents of directory and information on files)
% **ls -l** *name ...* —(use long format)
man—Print On-Line manual
% **man** *command*—(displays on screen the pages of the manual corresponding to command. Can be used both to obtain information on the UNIX system itself and on the programs from the GCG package).
mkdir—Make a directory
% **mkdir** *dirname ...*—(create directory dirname in your current directory)
more—File displaying utility
% **more** *filename*—(displays the filename file on your terminal, one screen at a time)
mv—Move or rename files and directories
% **mv** *file1 file2*—(change name of file1 to file2)
% **mv** *dirname newdirname*—(change name of dirname to newdirname)
% **mv** *file1 dirname*—(move file1 from its position into directory dirname)
passwd—Change login password
% **passwd**—(will ask you for your current password and for the new one)
ps—Process status
% **ps**—(prints information about active processes)
pwd—Print working directory
% **pwd**—(displays the full pathname of your current directory)
rm—Remove files or directories
% **rm** *file1 ...* — (delete file1)
% **rm -r** *dirname*—(deletes dirname and all its included files)

tail—copies the last portion of a file to the standard output. If no other option is given it will show the last 10 lines of the file
% **tail** *file1*—(displays the last 10 lines of file1)
% **tail** *-30 file1* (displays the last 30 lines of file1)
% **tail** *-30c file1* (displays the last 30 characters of file1)
who—Who is on the system
% **who**—(list on-line user's login name, terminal and login time)
id—Displays your current user id. It does not require any other option or parameter

36.3 Basic UNIX activities

36.3.1 Your current directory and its files

Your current directory refers to whatever directory you are currently working in. To see the name of your current directory, use the % pwd (print working directory) command. To list the files in your current directory, use the % ls (list) command.

36.3.2 Entering file names

If you enter the filename gamma.seq, UNIX assumes that you are referring to a file in your current directory. You name files in other directories by including the directory path. A typical directory path has slashes (/) that separate each subdirectory. (You can think of a subdirectory as a hanging folder in a file drawer, or the directory.) To refer to files in other directories, include the directory path with the file name.

For example, typing:
% more /usr/users/maps/filename.txt
means that you want to 'view' the file filename.txt located in the directory /usr/users/maps, where maps is a subdirectory of users, and users is a subdirectory of usr.

To list all the files in another directory, use the % ls command with a directory path. For example, typing:
% ls /usr/users
lists all the files in directory /usr/users.

36.3.3 Working in other directories

In general, you should work in your current directory, just as you should work at your own lab bench. However, there may be files in other directories that you want to read or copy.

You can change from your current directory to another directory by using the % cd command. For example, typing:
% cd /usr/users/burgess
means you want to change your current directory to the directory /usr/users/burgess.
You can also use a relative path to refer to directories. For example, if your current directory is /usr/users/map and you want to change to the directory /usr/users/map/test_sequences, you would type
% cd test_sequences
to move down to that directory.

If your current directory is /usr/users/map/test_sequences and you want to change back to the directory /usr/user/map, you would type
% cd.
to move up to that directory. (Note that the '..' refers to the directory above your current directory.)

If your current directory is /usr/users/map and you want to change to the directory /usr/users/compare, you would type
% cd../compare
to move over to that directory.

36.3.4 Controlling program execution

With UNIX, you can run several jobs from the same terminal. A job is any program that you can start from the command line. You can run a job in either the foreground, which lets you control it with input from the keyboard as it is running, or in the background, which frees your terminal for other work. You can have many jobs in the background, but only one job in the foreground.

You may want to run a job in the background—for example, if you were processing a large file with illustrations for a PostScript printer which took a number of minutes. There are also some GCG (Genetics Computer Group) programs (Wisconsin Package™ programs) that require time to complete, so it makes sense to process the programs in the background while you do other work at your terminal.

36.3.5 Background and foreground processing

To run a job in the background, add an & (ampersand) to the end of the command line. For example,
% sort < unsorted.txt >> sorted.txt &

The shell assigns your background job a number and displays that number on your terminal screen, along with the process identifier (pid) of the job. For example,
[1]14999
shows the job number as [1] and the pid as 14999.

To see a list of all the jobs that are running in your login shell, use the % jobs command. On the list, each job is marked as either Running or Stopped. For example, this job is currently running:
% jobs

[1] Running sort < unsorted.txt >> sorted.txt
(Note that your login shell is never displayed in the jobs list.) A job will stop running if it needs information that you must enter at the keyboard, such as a filename or other parameter. When this happens to a background job, the shell displays a message like the following on your screen:
[1] + Stopped (tty input) sort
and the job stops running. To continue with this job, you need to enter information at the keyboard. However, before you can use the keyboard, you must bring the job into the foreground by using the % fg (foreground) command.

To do this, you would enter
% fg %1
where the %1 is the way to refer to job number [1].

After you enter the information, you can put a foreground job into the background again by using Ctrl-Z to stop the job, then entering the command % bg (background) to put it in the background for processing. Because there is only one foreground job at any given time, it is not meaningful, or possible, to specify a job number or process id when stopping a job with Ctrl-Z.

36.3.6 Wildchar characters

There are several characters that allow you to enter only portions of the directory or filenames. The three most frequently required are:

- A leading '~' expands to the home directory of a particular user.
- Each '*' is interpreted as a specification for zero or more of any character.
- Each '?' is interpreted as a specification for exactly one of any character.

For example, the pattern 'dog*' will find matches for, among others, files named 'dog', 'dogg', and 'doggy'. The pattern 'dog?' matches, among others, 'dogg' but not 'dog' or 'doggy'.

% cp *.seq /usr/burgess/mydirectory copies every file in the current directory that ends with the characters.seq to the directory called /usr/burgess/mydirectory.

36.4 Controlling output

Ctrl-C ends a program or an executing UNIX command.
Ctrl-Z suspends program execution.
Ctrl-Q starts screen output that has been stopped.
Ctrl-S stops screen output.
Ctrl-R refreshes the command line.
Ctrl-U deletes from the cursor to the beginning of a line.

Note: Whenever you see a key combination written as Ctrl-C in the documentation, it means you press the <Ctrl> key and hold it down while you press the letter key, which in this case is <C>.
To restart a suspended program, type
% fg %6
which means put job number 6 in the foreground;
% bg %2
means put job number 2 in the background.

If you cannot remember what programs you have suspended, type
% jobs
to list the jobs and job numbers

36.5 Comparison of VMS and UNIX commands

In both the VMS and UNIX versions of the Wisconsin Package™ for sequence analysis, you run a GCG program by entering information on the command line (after the VMS $ prompt and the UNIX % prompt). The command line can contain the name of a command, command qualifiers, qualified parameters (qualifiers with values), and unqualified parameters (usually file-names). The general syntax, or structure, of a VMS command is
Command /QUALifier /QUALifier=Parameter Parameter

As with all VMS commands, spaces on the command line are ignored, characters can be typed in upper case or lower case (case insensitivity), qualifiers are indicated by a '/' (slash), parameters are indicated by an '=' (equals sign), and the bold typeface indicates the fewest number of characters you enter.

An example of a GCG command using VMS syntax is
MapPlot /CIRcular /OUTfile=pBR322.MapPlot EMBL:pBR322

A UNIX command varies in several ways from a VMS command. The general syntax of a UNIX command is:
command -QUALifier -QUALifier=Parameter Parameter

The main difference between the VMS and UNIX versions is that command names must be typed in full (no shortcuts) and in lower case. In addition, qualifiers are indicated by a space and a '-' (hyphen), instead of a '/' (slash). In both versions, qualified parameters are indicated by an '=' (equals sign) and spaces are not accepted between a qualifier and its parameter(s). The case of unqualified parameters can vary, but if the unqualified parameter is a file name, you must enter the file name in the exact case

shown. An example of the mapplot GCG command using UNIX syntax is:
mapplot -CIRcular -OUTfile=pbr322.mapplot EMBL:pbr322

36.6 Control-key (^ key) differences

Control-key combinations that you use in VMS may produce different results when you use them in UNIX. (Note that in GCG documentation control-key combinations are written as Ctrl-C and not ^C.) The following table lists the control key combinations in UNIX that are different in VMS:

VMS	UNIX	Description
Ctrl-Y	Ctrl-C	ends a program
(None)	Ctrl-Z	suspends a program
Ctrl-Z	Ctrl-D	end of file

36.7 Program name changes in UNIX GCG

- Clear becomes ClearPlot
- Echo becomes EchoKey
- Extract becomes ExtractPeptide
- Find becomes FindPatterns
- Fold becomes FoldRNA
- Shift becomes ShiftOver
- Strings becomes StringSearch

36.8 URLs for live help on the Web

A Concise Guide to UNIX Books
http://rclsgi.eng.ohio-state.edu/Unix-book-list.html

All about Unix
http://ugrad-www.cs.colorado.edu/unix/ Home.html

Fundamentals of Unix
http://www.gl.umbc.edu/~banz/intro_unix.html

Introduction to Unix
http://musie.phlab.missouri.edu/IntroToUnix/unix-tutor/index.html

Top 10 Unix Questions at Dartmouth
http://coos.dartmouth.edu/~pete/top10.html

UNIXhelp for Users
http://www.ucs.ed.ac.uk/~unixhelp/servers.html

Unix Documentation
http://web.gmu.edu/bcox/Unix/00Unix.html

Unix Documentation
http://www.efs.mq.edu.au/unix/index.html

Unix FAQ
http://www.cis.ohio-state.edu/hypertext/faq/usenet/unix-faq/unix/intro/faq.html

Unix Resources
http://wwwhost.cc.utexas.edu/cc/services/unix/index.html

Unix for MS-DOS users
http://ugrad-www.cs.colorado.edu/unix/unix4dos.html

Unix is a Four Letter Word ...
http://tempest.ecn.purdue.edu:8001/~taylor/4ltrwrd/html/unixman.html

Unix on the Macintosh
http://www.astro.nwu.edu/lentz/mac/unix/

Acknowledgement

This chapter contains edited text from The Wisconsin Package User's Guide (Genetics Computer Group, 575 Science Drive, Madison, WI 53711 USA, http://www.gcg.com).

Chapter 37 Databases and genome resource centres

37.1 Subject indexes

37.1.1 Databases by organism

Aedes aegyptii
Mosquito Genomics WWW Server
Aedes aegypti Genome Data Base (AaeDB)

Anopheles gambiae
Mosquito Genomics WWW Server
AnoDB

Arabidopsis thaliana
Arabidopsis Biological Resource Centre
Arabidopsis Information Management System (AIMS)
Arabidopsis Stock Centre (Nottingham)
Arabidopsis thaliana Data Base (AtDB, formerly AAtDB)
USDA Agricultural Plant Genomes

Bacillus subtilis
NRSub
BSORF (*see* GenomeNet)
MICADO
National Centre for Biotechnology Information genomes database

Caenorhabditis elegans
Caenorhabditis elegans Database (AceDB)
Caenorhabditis elegans Genome Project
Caenorhabditis Genetics Center (CGC) (stock centre)
Canadian Genome Analysis and Technology Program (CGAT)
Moulon WWW server
Sanger Centre
Washington University School of Medicine Genome Sequencing Center
XREFdb

Candida albicans
University of Minnesota Medical School, Computational Biology Center
Virtual Genome Center

Cereals
USDA Agricultural Plant Genomes

Chicken
ChickMap
Japan Animal Genome Databases

Chlamydomonas
ChlamyDB

Comparative databases
OMIA
La Trobe University Comparative Genome Mapping Page
Vertebrate Comparative Database
XREFdb

Cotton
CottonDB

Cow
BovMap
Japan Animal Genome Databases
Meat Animal Research Center (MARC)

Cyanobacteria
GenomeNet, Japan

Dog
Dog Genome Project
DogMap

Drosophila
Drosophila database, Stanford
FlyBase
TBASE
University of California, Berkeley

Escherichia coli
Colibri
EcoCyc
ECO2DBASE: E. coli gene–protein database
EcoSeq, EcoMap, EcoGene
E. coli Genetic Stock Center
Escherichia coli database collection (ECDC)
GenProtEc: E. coli Gene Database

Horse
Japan Animal Genome Databases
University of California, Davis

Human (The entries listed here provide links to other sites; individual chromosomes are listed in numerical order in the main listing in Section 37.2.)
BodyMap
CEPH-Généthon Integrated map
Genetic Location Data Base (LDB)
Genome Data Base (GDB)
Harvard Biological Laboratories
Human Genome Project (HGP) at Oak Ridge National Laboratories
Human population genetics database (Geography)
Integrated Genomic Database (IGD)
National Center for Genome Resources

UK Human Genome Mapping Programme Resource Centre

Legumes
AlfaGenes
BeanGenes
CoolGenes

Maize
Maize Genome DataBase (MaizeDB)
University of Minnesota Medical School, Computational Biology Centers
USDA Agricultural Plant Genomes

Microbial genomes
DOE Microbial Genomes Initiative
Caltech Genome Research Laboratory
Canadian Genome Analysis and Technology Program (CGAT)
Microbial Advanced Database
TIGR

Mitochondrial genomes
Canadian Genome Analysis and Technology Program (CGAT)
MITOMAP
Organelle Genome Megasequencing Project (OGMP)

Mouse
Baylor College of Medicine Genome Center
Caltech Genome Research Laboratory
Dysmorphic Human–Mouse Homology Database (DHMHD)
European Collaborative Interspecific Backcross (EUCIB)
Gene Knockouts Database
Mouse Genome Database (MGD) and the *Encyclopedia of the Mouse Genome*
Mouse Locus Catalogue (MLC)
mousedb
TBASE
Whitehead Institute Mouse Genetic Map Information

Mycobacterium
Mycobacterium Database (MycDB)
Sanger Centre

Organelle genomes
Canadian Genome Analysis and Technology Program (CGAT)
MITOMAP
Organelle Genome Megasequencing Project (OGMP)

Pig
Meat Animal Research Center (MARC)
PigMap
TBASE

Plant genomes
USDA/ARS/NAL Plant Genome Data

Plant pathogens
PathoGenes

Plasmodium
Walter and Eliza Hall Institute of Medical Research

Rat
RATMAP
TBASE

Rice
Japanese Rice Genome Program (RGP)
RiceGenes
USDA Agricultural Plant Genomes
University of Minnesota Medical School, Computational Biology Centers

Saccharomyces cerevisiae
Saccharomyces Genome Project
Sanger Centre
Saccharomyces Genome Database
Yeast Protein Database

Schistosoma
Schistosoma Genome Project

Schizosaccharomyces pombe
Cold Spring Harbor Laboratory
Sanger Centre
Schizosaccharomyces pombe (NIH fission yeast information)

Sheep
Meat Animal Research Center (MARC)
SheepBase

Solanaceae (e.g. tomato, potato)
SolGenes

Trees
TreeGenes

Yeast
See *Saccharomyces cerevisiae* and *Schizosaccharomyces pombe*

Zebrafish
FishNet
Zebrafish Sequence Analysis Project

37.1.2 Other databases and resources

Cell lines
American Type Culture Collection (ATCC)
CBA-IST (Genova)
European Collection of Animal Cell Cultures
GDB
National Institute of General Medical Sciences (NIGMS) Human Genetic Mutant Cell Repository
Radiation Hybrid Database

DNA sequences and genes
CpG Island Database
dbEST (expressed sequence tags)
dbSTS (sequence tagged sites)
EGAD (Expressed Gene Anatomy Database)
EMBL Nucleotide Sequence Database
GenBank
Gene family database
Mendel (plant genes)
REPBASE (repetitive elements)
V BASE (Ig genes)

Proteins
Danish Centre for Human Genome Research (Human 2-D PAGE databases)
ENZYME (the Enzyme Nomenclature Database)
Kabat Database of Proteins of Immunological Interest
PIR (protein sequence database)
Prodom (protein domain database)
PROSITE (protein sites and patterns)
REBASE (restriction enzymes)
SWISS-PROT (protein sequence database)
SWISS-2DPAGE (two-dimensional polyacrylamide gel electrophoresis database)
SWISS-3DIMAGE (3D images of proteins and other biological macromolecules)

37.1.3 Genome resource centres and information

Australia
Australian Genomic Information Centre (AGIC)

Canada
Canadian Genome Analysis and Technology Program (CGAT)
InfoBiotech Canada

Denmark
Danish Centre for Human Genome Research

Europe
EMBL/EBI: the European Bioinformatics Institute

France
Centre d'Etudes du Polymorphism Humain (CEPH)
Genestream at EERIE
Généthon
InfoBioGen
Institut Pasteur
Moulon WWW server

Germany
Reference Library Database (RLDB) and Reference Library System

Israel
Weizmann Institute of Bio-informatics

Japan
GenomeNet, Japan

Spain
Centro Nacional de Biotecnologia (Madrid)

Switzerland
CBRG at ETHZ
Geneva University

United Kingdom
Sanger Centre
Roslin Institute, Edinburgh
UK Human Genome Mapping Project Resource Centre

United States
AGIS
Baylor College of Medicine Genome Center
Caltech Genome Research Laboratory
Cooperative Human Linkage Center (CHLC)
Harvard Biological Laboratories
Human Genome Program
I.M.A.G.E. Consortium
Lawrence Berkeley Laboratory
Lawrence Livermore National Laboratory
Los Alamos National Laboratory
Motif BioInformatics Server
National Center for Biotechnology Information (NCBI)
National Center for Genome Resources (NCGR)
National Center for Human Genome Research (NCHGR)

Stanford Human Genome Center
TIGR
University of Michigan Human Genome Center
USDA Agricultural Plant Genomes/ USDA/ARS/ NAL Plant Genome Data
Whitehead Institute/MIT Genome Center

37.1.4 Software: sources of information and programs

Computer-assisted design of oligonucleotide primers (e.g. OLIGO, OSP, PRIMER, PRIMEGEN)
Whitehead Institute
UK Human Genetic Mapping Project Resource Centre

Construction of integrated maps (e.g. SIGMA)
ICRF
UK Human Genome Mapping Project Resource Centre (Mapping Analysis Menu)
National Center for Genome Resources

DNA sequence assembly and analysis
George M. Church Laboratory
GCG (Wisconsin Package)
GenomeNet, Japan
NIH sequence analysis services
Oak Ridge National Laboratory
Philadelphia Genome Center
PYTHIA (*see* Chapter 25)
Sanger Centre
UK Human Genome Mapping Project Resource Centre
University of Minnesota

Linkage analysis and linkage mapping
Columbia University Linkage Analysis Web Server
Cooperative Human Linkage Center (CHLC)
UK Human Genome Mapping Project Resource Centre

Physical mapping
Sanger Centre
Imperial Cancer Research Fund (ICRF)
UK Human Genome Mapping Project Resource Centre
Généthon (QUICKMAP)
University of Michigan
Whitehead Institute

Radiation mapping
UK Human Genome Project Resource Centre
University of Michigan
Stanford Human Genome Center
Whitehead Institute

37.2 Alphabetical list of databases and genome resource centres accessible via the World Wide Web

Many of these sites can be accessed most conveniently through national or regional genome resource and information centres that have extensive links to other sites, such as the UK Human Genome Mapping Project Resource Centre, Harvard Biological Laboratories, the Motif Bioinformatics WWW Server, GenomeNet, Japan, AGIS, etc. You will need a WWW browser such as Netscape. Most services are publicly accessible with the appropriate software but some require registration or subscription to use fully. This list does not aim to be comprehensive but many other services and sites may be accessed through the sites listed here.

AaeDB *see Aedes aegypti* Genome Database

ACeDB *see Caenorhabditis elegans* DataBase

Aedes aegypti Genome Database (AaeDB)
http://klab.agsci.colostate.edu/acedb/AaeDB-acedb.html
Held at the Colorado State University, this database aims to collate both genetic and physical chromosome mapping data for the mosquito *Ae. aegypti*, the vector of yellow fever.

AGIC *see* Australian Genomic Information Center

AGIS *see* Agricultural Genome Information Server

Agricultural Biotechnology Information Center
http://www.nalusda.gov/bic

Agricultural Genome Databases *see* Agricultural Genome Information Server

Agricultural Genome Information Server (AGIS)
http://probe.nalusda.gov:8000
AGIS is a cooperative effort between the University of Maryland, Department of Plant Biology and the National Agricultural Library, and is sponsored by the US Department of Agriculture, Agricultural Research Service. You can browse and search a comprehensive collection of genome databases relevant to crop plants, pests and domesticated animals. Also links to molecular biology and informatics servers such as the European Bioinformatics Institute, EMBL, National Center for Biotechnology Information, National Center for Genome Resources, etc. AGIS also holds genome analysis tools and software. The 'how to' page has useful advice on

retrieving data for those new to genome research.

AIMS *see* *Arabidopsis* Information Management System

AlfaGenes
http://probe.nalusda.gov:8000/plant/
Database (still in the experimental stages) for alfalfa (*Medicago sativa*). Curators: Daniel Z. Skinner (e-mail: dzolek@ksu.ksu.edu) and Paul C. St. Armand (e-mail: pst@ksu.ksu.edu).

ANGIS (Australian National Genomic Information Service) *see* Australian Genomic Information Centre

American Type Culture Collection (ATCC)
http://www.atcc.org/

AnoDB
http://konops.imbb.forth.gr/AnoDB
Database for *Anopheles gambiae*, the vector of malaria.

Arabidopsis Biological Resource Center
Ohio State University, 1735 Neil Avenue, Columbus, OH 43210, USA (e-mail: *Arabidopsis*+@osu.edu).

Arabidopsis Stock Centre (Nottingham)
http://nasc.life.nott.ac.uk

Arabidopsis Information Management System (AIMS)
http://genesys.cps.msu.edu.3333/

Arabidopsis thaliana Database (AtDB, formerly AAtDB)
http://genome-www.stanford.edu/Arabidopsis
Located in the Department of Genetics, School of Medicine, Stanford University, USA. For queries contact atdb-curator@genome.stanford.edu
Anonymous ftp to ftp-genome.stanford.edu

AtDB *see* *Arabidopsis thaliana* Database

ATCC *see* American Type Culture Collection

Australian Genomic Information Centre (AGIC)
http://angis.su.oz.au/Agic/about.html
Objectives are to manage the Australian National Genomic Information Service (ANGIS) and to conduct collaborative bioinformatics research and development projects. Located in the Electrical Engineering Department of the University of Sydney. It also hosts ABNET, the Australian Bioinformatics Network. For further information contact Tim Littlejohn (e-mail: tim@angis.su.ox.au; tel./fax: (+61 2) 351 2948).

Bacillus subtilis Database (NRSub)
http://acnuc.univ-lyon1.fr/nrsub/nrsub.html
DNA sequences from *B. subtilis*. Additional data on gene mapping and codon usage are provided by links to other sites (e.g. InfoBioGen, EBI, HGMP). This database is mirrored in Japan at GenomeNet, Japan.

Baylor College of Medicine Genome Center
http://gc.bcm.tmc.edu.8088/home.html
Projects on human chromosomes 6, 15, 17, X and mouse X.

BeanGenes
http://probe.nalusda.gov:8000/plant/
Database for information on *Phaseolus* and *Vigna* species. Curator: Phil McClean, North Dakota State University, Fargo, North Dakota, USA (e-mail: cclean@beangenes.cws.ndsu.nodak.edu).

BodyMap
http://imcb.osaka-u.ac.jp/bodymap/welcome.html.
An anatomical expression database of human genes.

BovMaP
http://locus.jouy.inra.fr/~samson/bovmap/intro.html
Contains information on cattle loci, alleles, genetic and physical maps, polymorphisms, homologies, probes, primers and references.

Brookhaven National Laboratory (Biology)
http://bnlstb.bio.bnl.gov:8000/
Information on DNA sequencing project.

Caenorhabditis elegans Genome Project
http://www.sanger.ac.uk
http://genome.wustl.edu/gsc/gschmpg.html
Information on the sequencing work carried out at the Sanger Centre and Washington University, St Louis as part of the *C. elegans* genome project, including cosmid sequences, access to ACeDB, genome maps, analytical and assembly software, etc. Links to other sites.

Caenorhabditis elegans Database (ACeDB) Moulon server
http://moulon.inra.fr:8001/acedb/acedb.html

Caenorhabditis elegans WWW server (University of Texas Southwestern Medical Centre at Dallas)

http://eatworms.swmed.edu

Caenorhabditis Genetics Center (CGC)
http://elegans.cbs.umn.edu/
Based at the University of Minnesota, the Center maintains and distributes stocks of *Caenorhabditis elegans* mutant strains. Curator: Theresa L. Stiernagel, *Caenorhabditis* Genetics Center, 250 Biological Sciences Center, 1445 Gortner Avenue, University of Minnesota, St Paul, MN 55108-1095, USA (e-mail: stier@molbio.cbs.umn.edu).

Caltech Genome Research Laboratory
http://www.tree.caltech.edu/
Construction of Human Bacterial Artificial Chromosome (BAC) Library resource. Physical mapping of human chromosome 22 using BAC clones and YAC frameworks. Construction of Mouse Bacterial Artificial Chromosome (BAC) Library resource. Sequencing the 1.8 megabase genome of the archebacterium *Pyrobaculum aerophilum*.

Canadian Genome Analysis and Technology Program (CGAT)
http://cgat.bch.umontreal.c.
Contains information resources for genomics research in Canada, genomics data bases and information, data on chromosome X, *C. elegans* cosmid transgenics, the Organelle Genome Database (GOBASE) and Organelle Genome Megasequencing Program, the Rose Worm lab, and the *Sulfolobus solfataricus* Genome Project.

Candida albicans
http://alces.med.umn.edu/Candida.html
Contains information on the genetics, physical map, and sequence data of *C. albicans*.

CBA-IST (Genova)
http://www.ist.unige.it/
Contains the Cell Line Database listing 3000 human and animal cell lines from European culture collections.

CBRG at ETHZ the Computational Biochemistry Server at ETHZ (the Swiss Federal Institute of Technology)
http://cbrg.inf.ethz.ch/
Collection of programs for SWISS-PROT and other database searching and for constructing phylogenetic trees.

Cedars-Sinai Research Institute Molecular Genetics Laboratories
http://www.csmc.edu/genetics/korenberg/korenberg.html
Contains the integrated YAC/BAC/PAC resource for the human genome, the chromosome 21 phenotypic mapping project, gene mapping projects on chromosome 21, a BAC contig of the chromosome 21 congenital heart disease region.

Centro Biotecnologie Avanzate, Genova, Italy *see* CBA-IST

Centro Nacional de Biotecnologia (Madrid)
http://www.cnb.uam.es/

Centre d'Etudes du Polymorphism Humain (CEPH) (Fondation Jean Dausset)
http://www.cephb.fr/HomePage.html
CEPH is a research laboratory created in 1984 by Professor Jean Dausset, which constructs maps of the human genome. This site contains the CEPH-Généthon integrated maps (see below) and the CEPH genotype database (see Chapter 5) and the CEPH YAC library (see below).

CEPH-Généthon Integrated Map
http://www.cephb.fr/ceph-genethon-map.html
Contains information about the CEPH YAC library, contig maps, STS data, Alu-PCR hybridization data, fingerprint data, sizing data and FISH data. Primary copies of the CEPH YAC library of 33 000 clones are held at the following centres.
USA E.S. Lander or T. Hudson, Whitehead Institute/MIT Center for Genome Research, Cambridge, MA 02142, USA (e-mail: lander@genome.wi.mit.edu)
Europe
• D. LePaslier, Foundation Jean-Dausset-CEPH, 27 rue Juliette Dodu, 75010 Paris, France (e-mail: denis@ceph.cephb.fr)
• H. Lehrach, the Reference Library Database (RLDB), Max Planck Institute for Molecular Genetics, Ihnestrasse 73, 14195 Berlin-Dahlem, Germany (tel.: (+49 30) 8413 1627; fax: (+49 30) 8413 1395)
• D. Toniolo, GBE, CNR, via Abbiategrasso 207, 27100 Pavia, Italy (tel.: (+39 382) 546 340; fax: (+39 382) 422 286)
• G.J.B. van Ommen, YAC Screening Centre, Leiden University, Department of Human Genetics, Wassenaarseweg 72, 2333 A1 Leiden, the Netherlands (tel.: (+31 71) 276081; fax: (+31 71) 276075)
• K. Gibson, Human Genome Mapping Project Resource Centre (HGMP), Hinxton Hall, Hinxton, Cambridge CB10 1RQ, UK (tel.: (+44 1223) 494 500; fax: (+44 1223) 494 512)

Japan
• K. Yokoyama, 3–1-1 Koyadai, Tsukuba, Ibaraki 305, Japan (tel: (+81 298) 36 3612; fax: (+81 298) 36 9120)
• Y. Nakamura, Human Genome Centre, Institute of Medical Science, the University of Tokyo, 4–6-1 Shirokaneda, Minato, Tokyo 108, Japan (tel: (+81 3) 5449 5372; fax: (+81 3) 5449 5433)
China
Z. Chen, Shanghai Institute of Haematology, Rui-Jin Hospital, Shanghai Second Medical University, Shanghai 200025, China (tel.: (+86 21) 318 0300; fax: (+86 21) 474 3206)

CGC *see Caenorhabditis* Genetics Center

CGSC *see E. coli* Genetic Stock Center

ChickMap
http://www.ri.bbsrc.ac.uk/chickmap
Project to produce an integrated map of the chicken genome.

Children's Hospital of Philadelphia
http://www.cbil.upenn.edu/HGC22.html
Human chromosome 22.

ChlamyDB
http://probe.nalusda.gov:8000/plant
Database on *Chlamydomonas reinhardtii* including: genetic and molecular maps; information on genetic loci, mutant alleles, and sequenced genes; description of strains; contacts; bibliography; information on the Chlamydomonas Genetics Center. Curator: Elizabeth H. Harris (Duke University) (e-mail: chlamy@acpub.duke.edu)

CHLC *see* Cooperative Human Linkage Center

Chromosome 1 *see* Columbia Linkage Analysis Web Server for Chromosome 1 workshop

Chromosome 2 *see* Imperial Cancer Research Fund

Chromosome 3 A second-generation YAC contig map. All maps and supporting data tables are available by anonymous ftp from ftp://thor.hsc.colorado.edu.
See also Sanger Centre; University of Texas Health Science Center

Chromosome 4 *see* Sanger Centre; Stanford Human Genome Center

Chromosome 6 *see* Sanger Centre

Chromosome 9 *see* Galton Laboratory

Chromosome 10 *see* Genome Therapeutics Corporation

Chromosome 11 *see* Sanger Centre; University of Texas Southwestern Medical Center

Chromosome 12 *see* Yale Genome Center

Chromosome 13 *see* Columbia University Human Genome Project; Sanger Centre

Chromosome 15 *see* Baylor College of Medicine

Chromosome 16 *see* Los Alamos National Laboratory; Sanger Centre; TIGR

Chromosome 17 *see* Baylor College of Medicine

Chromosome 19 *see* Lawrence Livermore National Laboratory

Chromosome 21 *see* Cedars-Sinai Research Institute Molecular Genetics Laboratories; Lawrence Berkeley Laboratory

Chromosome 22 *see* Caltech Genome Research Laboratory; Computational Biology and Informatics Laboratory (CBIL) at Pennsylvania University; Sanger Centre; University of Oklahoma

Chromosome X *see* Baylor College of Medicine; Canadian Genome Analysis and Technology Program; Lawrence Livermore National Laboratory; Max Planck Institute for Molecular Genetics; Sanger Centre

Chromosome Y *see* Galton Laboratory

Cold Spring Harbor Laboratory
http://www.cshl.org
Schizosaccharomyces pombe sequencing and *Arabidopsis* sequencing.

Colibri
A relational database dedicated to the analysis of the *E. coli* genome. Macintosh application available via anonymous ftp (ftp.pasteur.fr, in the directory pub/GenomeDB/Colibri). For additional information, contact Ivan Moszer (moszer@pasteur.fr) or Antoine Danchin (adanchin@pasteur.fr).

Columbia Linkage Analysis Web Server
http://linkage.cpmc.columbia.edu/

Genetic linkage analysis software. Chromosome 1 Workshop. Links to other sites.

Columbia University Human Genome Project
http://genome1.ccc.columbia.edu/~genome/
Human chromosome 13 data (YACs, cosmids, STSs, markers).

CompDB *see* Vertebrate Comparative Database.

Computational Biology and Informatics Laboratory (CBIL) at Pennsylvania University
http://cbil.humgen.upenn.edu/
Human chromosome 22 data.

CoolGenes
http://probe.nalusda.gov:8000/plant/
Database (still being developed) for cool season food legumes (*Pisum, Lens, Cicer, Lathyrus, Vicia faba).* Curator: Fred Muehlbauer (e-mail: muehlbau@wsu.edu).

Cooperative Human Linkage Center (CHLC)
http://www.chlc.org/
Based at the University of Iowa, the goal of CHLC is to develop statistically rigorous, high heterozygosity genetic maps of the human genome that are greatly enriched for the presence of easy-to-use PCR-formatted microsatellite markers. Available at this site are: genetic maps showing the positions of genetic markers; integrated maps showing the position of genetic markers constructed using genotype data from the CEPH reference panel; CHLC marker maps showing the positions of CHLC generated markers in various reference maps; information on markers.

CottonDB
http://probe.nalusda.gov:8000/plant/
A database containing information on *Gossypium hirsutum* and related species. Curators: Gerard Lazo: (e-mail: lazo@tamu.edu) and Sridhar Madhavan: (e-mail: msridhar@tamu.edi)

CpG Island Database
http://biomaster.uio.no/cpgdb.html
The CpG island database is maintained at the Biotechnology Centre of Oslo. It deals with human genes appearing in major releases of the EMBL nucleotide sequence database but it is hoped that in the future it will include islands from other mammalian species.

Danish Centre for Human Genome Research
http://biobase.dk/cgi-bin/celis
Holds human 2-D PAGE data bases.

dbEST (Expressed Sequence Tags)
http://www.ncbi.nlm.nih.gov/dbEST/index.html
A division of GenBank that contains sequence data and other information on cDNA sequences characterized as single reads from DNA sequencing from a number of organisms.

dbSTS (Sequence Tagged Sites)
http://www.ncbi.nlm.nih.gov/dbSTS/index.html
An NCBI resource that contains sequence and mapping data on short genomic landmark sequences.

Dendrome Project
http://probe.nalusda.gov:8000/plant/
A genome database for forest trees.

DHMHD *see* Dysmorphic Human–Mouse Homology Database

DOE Human Genome Program
http://www.er.doe.gov/production/oher/hug_top.html
Information on the DOE projects; links to Los Alamos, Lawrence Berkeley, Lawrence Livermore and other laboratories involved in the Human Genome Project; a primer on the basic science of the Human Genome Project; project resources and meetings.

DOE Microbial Genomes Initiative
http://www.er.doe.gov/production/oher/mig_top.html
Projects to sequence a variety of microbial genomes. *See* TIGR.

Dog Genome Project
http://mendel.berkeley.edu/dog.html

DogMap
http://ubeclu.unibe.ch/itz/markma.html
A low-resolution map of the canine genome being constructed by an international collaboration under the auspices of the International Society for Animal Genetics. Information from Gaudenz Dolf, Institute of Animal Breeding, University of Berne, Bremgartenstrasse 109a, 3012 Berne, Switzerland (e-mail: gauden@itz.unibe.ch).

Drosophila database
http://www-leland.stanford.edu/~ger/drosphila.html

Dysmorphic Human-Mouse Homology Database (DHMHD)
http://www.hgmp.mrc.ac.uk/DHMHD/dysmorph.html
Three separate databases of human and mouse malformation syndromes together with a database of mouse/human syntenic regions. The mouse and human malformation databases are linked together through the chromosome synteny database. The purpose of the system is to allow retrieval of syndromes according to detailed phenotypic descriptions and to be able to carry out homology searches for the purpose of gene mapping.

EBI *see* European Bioinformatics Institute

E. coli clones
Dr Y Kohara, National Institute of Genetics, Mishima, Shizuoka-ken 1111, Japan (fax: +81 559) 81 6826.

E. coli Genetic Stock Center (CGSC)
http://cgsc.biology.yale.edu/top.html
The *E. coli* Genetic Stock Center (CGSC) at the Department of Biology, 3550ML, Yale University, PO Box 208104, New Haven CT 06520–8104, USA (fax: +1 203) 432-3852) maintains a database of *E. coli* genetic information, including genotypes and reference information for the several thousand strains in the CGSC collection, a gene list with map and gene product information, and information on specific mutations. For information contact Mary Berlyn (e-mail: mary@cgsc.biology.yale.edu).

ECACC *see* European Collection of Animal Cell Cultures

E. coli genome project
http://ecoliftp.genetics.wisc.edu

ECDC: *Escherichia coli* database collection
http://susi.bio.uni-giessen.de/usr/local/www/html/ecdc.html
Contains information for the entire *E.coli* K-12 chromosome, and is organized like a genetic map. the database can be searched for gene names or map positions. Coding sequences are indicated for each gene. Regulatory regions, promoters, terminators, and IS elements are also indicated. the complete ECDC dataset is available by anonymous ftp (susi.bio.uni-giessen.de) or together with a Windows application on the EMBL (EBI) CD-ROM. for information contact Manfred Kröger (e-mail: kroeger@embl-heidelberg.de).

ECO2DBASE: *E. coli* gene–protein data base
Contains information about *E. coli* proteins obtained by the analysis of two-dimensional protein gels, and is maintained by F.C. Neidhardt. 'Ed6.0195' is the database file (text format) for the sixth published version of the database. Updates will have a different extension after the decimal. Available by anonymous ftp by ftp://ncbi.nlm.nih.gov/repository/ECO2DBASE. Questions and comments can be sent to Ruth VanBogelen (e-mail: vanbogr@aa.wl.com) or Fred Neidhardt (e-mail: fcneid@umich.edu).

EcoCyc: Encyclopedia of *E. coli* Genes and Metabolism
http://www.ai.sri.com/ecocyc/ecocyc.html
A database integrating information about *E. coli* genes and metabolism. A graphical user interface creates drawings of metabolic pathways, of individual reactions, and of the *E. coli* genomic map. Users can call up objects through a variety of queries and then navigate to related objects shown in the display window.

EcoSeq, EcoMap, EcoGene
EcoSeq is a nonoverlapping *E. coli* DNA sequence collection which integrates information about genes, DNA and protein sequences. EcoMap integrates EcoSeq with a genomic restriction map. EcoGene contains information about identified and putative protein- and RNA-encoding genes, and translations of sequences thought to encode proteins. These data are correlated and cross-referenced with the SWISS-PROT protein sequence database. Available by anonymous ftp from ftp://ncbi.nlm.nih.gov/repository/Eco/
For additional information, contact Kenn Rudd (e-mail: rudd@ncbi.nlm.nih.gov).

EGAD (Expressed Gene Anatomy Database) *see* TIGR

EHCB (European Human Cell Bank) *see* European Collection of Animal Cell Cultures

EMBL *see* European Molecular Biology Laboratory

EMBL Nucleotide Sequence Database *see* European Bioinformatics Institute

ENZYME *see* Geneva University

European Bioinformatics Institute (EMBL/EBI)
http://www.ebi.ac.uk/
An outstation of the European Molecular Biology Laboratory. It holds the EMBL Nucleotide Sequence Database; SWISS-PROT protein sequence database;

dbEST; dbSTS; Radiation Hybrid Database; IMGT immunogenetics database; the PCR primers database; and Flybase. Also documentation and software.

European Collaborative Interspecific Backcross (EUCIB)
http://www.hgmp.mrc.ac.uk/MBx/MBxHomepage.html
The latest EUCIB high-resolution mouse microsatellite maps and mapping data.

European Collection of Animal Cell Cultures (ECACC)
http://www.gdb.org/annex/ecacc/HTML/geninf.html

European Human Cell Bank (EHCB) *see* European Collection of Animal Cell Cultures
Specialist collection of cell lines derived from patients with genetic disorders or chromosome abnormalities.

European Molecular Biology Laboratory (EMBL)
http://www.embl-heidelberg.de/

ExPASy *see* Geneva University

FishNet
http://zfish.oregon.edu
http://www-igbmc.u-strasbg.fr/index.html
Information on zebrafish genome projects and links to other sites.

FlyBase
http://morgan.harvard.edu:80/
http://flybase.bio.indiana.edu:82/
http://www.embl-ebi.ac.uk/flybase
http://www.angis.su.oz.au:7081/
http://shigen.lab.nig.ac.jp:7081/
A database containing information on the genetics and biology of *Drosophila* species. FlyBase contains the text of the Lindsley and Zimm 'Red Book' (only in the US copy, owing to copyright reasons), lists of chromosome aberrations (sorted by class and cytological breakpoints), molecular clones, the genetic map, and stock lists of the international *Drosophila* stock centres. Using Gopher, these files can be interactively searched.

Galton Laboratory
http://diamond.gene.ucl.ac.uk
Chromosome 9 Workshop reports, maps, contact addresses. Chromosome Y fingerprint data. Linkage software.

GCG *see* Genetics Computing Group

GDB *see* Genome Database

GenBank *see* National Centre for Biotechnology Information

Gene family database
http://gdbdoc.gdb.org/~avolz/home.html

Gene Knockouts Database
http://www.bayanet.com/bioscience/knockout/knochome.htm
Data on phenotypes obtained by the knockout of various molecules in mice.
See also Appendix VIII in this book.

Genestream at EERIE
http://genome.eerie.fr/Genome.html
The Southern France Human Genome Project Computing Resource Centre.

Généthon
http://www.genethon.fr/
See also CEPH.

Genetic Location Database (LDB)
http://cedar.genetics.soton.ac.uk/public_html/
An analytical database held at the University of Southampton, UK for constructing fully integrated genetic and physical maps (see Chapters 3 and 16). the ldb program generates an integrated map (known as the summary map) from partial maps of physical, genetic, regional, somatic hybrid, mouse homology and cytogenetic data. The summary maps and the data used to build up such maps are available from the WWW site. The files for each chromosome are stored in the same directory that includes the summary map, partial maps, lod files and the parameter files. Alternatively, the ldb program can be downloaded and used to create the user's own integrated maps. Submissions to arc@southampton.ac.uk

Genetics Computer Group (GCG)
http://www/gcg.com
Commercial suppliers of the Wisconsin Package™ for sequence analysis.

Geneva University (ExPASy)
http://expasy.hcuge.ch/
The molecular biology server of the Geneva University Hospital and the University of Geneva, which is dedicated to the analysis of protein and nucleic acid sequences and 2-D PAGE. You can

search the databases SWISS-PROT (annotated protein sequence database); PROSITE (a dictionary of protein sites and patterns); SWISS-2DPAGE (two-dimensional polyacrylamide gel electrophoresis database); SWISS-3DIMAGE (3D images of proteins and other biological macromolecules); ENZYME (the Enzyme Nomenclature Database); and SeqAnalRef (a sequence analysis bibliographic reference database). The site also contains tools and software packages.

GenoBase *see* NIH GenoBase server

Genome Database (GDB)
http://gdbwww.gdb.org/
The main repository for human genetic data. It contains entries for individual loci which are linked to other databases such as MGD, FlyBase and the Enzyme Nomenclature Database at SWISS-PROT so that homologies with other mammalian and *Drosophila* genes, and enzyme function data can also be obtained. The database also contains information on polymorphisms, some 70 maps (linkage, cytogenetic, radiation hybrid and the latest Généthon map), mutations, probes, clone libraries, cell lines, citations, and contact addresses.

Genome Therapeutics Corporation (GTC)
http://www.cric.com/
Commercial organization. Chromosome 10 physical mapping data.

GenomeNet, Japan
http://www.genome.ad.jp/
Provides access to other databases and sequence interpretation tools. Databases include BSORF, the *Bacillus subtilis* database held at the University of Tokyo; the *Escherichia coli* Databank; CyanoBase, a database for *Synechocystis* spp.; BodyMap, an expression database of human genes; SPAD, a signaling pathway database; and the Aberrant Splicing Database, and there are links to many of the main genomic and sequence databases (e.g. GDB, EMBL). It also holds the Kyoto Encyclopedia of Genes and Genomes, and software for sequence interpretation.

GenProtEc: E.coli Gene Database
http://www.mbl.edu/~dspace/eco.html
A compilation of *E. coli* genes and gene products, categorized by physiological function, this database also includes homology information for proteins similar to at least one other *E. coli* protein. Available by ftp://hoh.mbl.edu/pub/ecoli.zip

George M. Church Laboratory
http://twod.med.harvard.edu
Software for DNA sequence analysis.

GOBASE *see* Canadian Genome Analysis and Technology Program

GSDB *see* National Centre for Genome Resources (NCGR)

Harvard Biological Laboratories
http://golgi/harvard/edu
Extensive links to other sites. CGC software documentation. Genome databases for *Arabidopsis, C. elegans, Drosophila,* human, mouse, prokaryote, and yeast. Searches of sequence databases, Entrez, culture collections and REBASE, and information on internet resources and searching for information across the networks.

HGP: Human Genome Project at Oak Ridge National Laboratories
http://www.ornl.gov/TechResources/Human_Genome/home.hmtl
Links to all genome centres participating in the US Human Genome Project.

Human Genome Program
http://www/er/doe.gov/production/oher/hug_top.html
The human genome programme of the US Department of Energy.

Human population genetics database (Geography)
http://lotka.stanford.edu/genography.html.
Database and information on human population genetics.

ICRF *see* Imperial Cancer Research Fund

I.M.A.G.E. Consortium (Integrated Molecular Analysis of Genomes and their Expression)
http://www bio.llnl.gov/bbrp/image/image.html
Information on and availability of more than 200 000 arrayed human cDNA clones.

Imperial Cancer Research Fund (ICRF)
http://www.icnet.uk
Chromosome 2 mapping information. EUROGEM Project information. ICRF contig-building package from ftp.icnet.uk/icrf-public/GenomeAnalysis/icrf_contig_v2.tar.Z

Indiana University
http://ftp.bio.indiana.edu

InfoBioGen
http://www.infobiogen.fr/
Computing and information resource for the French molecular biology and genome projects. Mainly in French.

InfoBiotech Canada
http://www.ibc.nrc.c./ibc
Information on biotechnology in Canada and worldwide. Contains several databases and links to numerous other sites. Internet wide searches possible.

INRA (Institut National de la Recherche Agronomique) Biotechnology Laboratories
http://locus.jouy.inra.fr

Institut Pasteur
http://www.pasteur.fr/
Links to many databases and useful sites.

Integrated Genomic Database (IGD)
Moulon server http://moulon.inra.fr:8001/acedb/igd.html

Japan Animal Genome Databases
http://ws4.niai.affrc.go.jp/jgbase.html
Search genome databases on pig (cytogenetic map, USDA linkage map, PIGM linkage map); cattle (cytogenetic map, USDA linkage map); chicken (cytogenetic map; linkage map); horse (cytogenetic map).

Japanese Rice Genome Program (RGP)
http://www.staff.or.jp
Information service on rice genome research.

Johns Hopkins Bio-informatics WWW Server
http://www.gdb.org/hopkins.html
Holds protein databases including PIR (protein identification resource–protein sequence database) and REBASE (the restriction enzyme database). Holds TBASE and the DOE Human Subjects Database.

Johns Hopkins GDB WWW Server
http://gdbwww.gdb.org
Holds information about GDB and its future developments. Holds a GDB browser; OMIM; ideogram-based searching of GDB; maps of HUGO reference markers.

Kabat Database of Proteins of Immunological Interest
http://immuno.bme.nwu.edu/

La Trobe University Comparative Genome Mapping Page
http://www.latrobe.edu.au/www/genetics/compmap.html
CompMap—clickable human chromosome maps with corresponding loci from a wide range of species. Links to other sites.

LANL *see* Los Alamos National Laboratory

Lawrence Berkeley Laboratory (LBL) Human Genome Center
http://www-hgc.lbl.gov/GenomeHome.html
Human Chromosome 21 P1 and cDNA mapping data bases. *Drosophila* physical mapping. Instrumentation and informatics projects. Human chromosome and directed genome sequencing projects.

Lawrence Livermore National Laboratory (LLNL)
http://www_bio.llnl.gov.bbrp/genome.html
Biology and Biotechnology Research Program
http://wwwbio.llnl.gov/bbrp/bbrp.homepage.html
Human Genome Center. Physical maps of human Chromosome 19. Closure of the chromosome 19 map. Enhancement of the high resolution clone map of human chromosome X. DNA Sequencing. National Laboratory Gene Library Project. Alu Repeats: a novel source of genetic variation for mapping. Informatics and analytical genomics: Instrumentation for the Human Genome Project. I.M.A.G.E. Consortium home page.

LBL *see* Lawrence Berkeley Laboratory Human Genome Center

LDB *see* Genetic Location Database

LLNL *see* Lawrence Livermore National Laboratory

Los Alamos National Laboratory (LANL)
http://www-t10.lanl.gov/
Sigma chromosome maps. Chromosome 16 flat file. HIV databases.

MaizeDB: Maize Genome Database
http://www.agron.missouri.edu
Curated by the USDA Plant Genetics Unit located within the College of Agriculture of the University of Missouri-Columbia. Contains genetic maps, mapped loci, recombination and map score data,

probe data, genetic/cytogenetic stocks, locus variations, references, contact addresses of maize researchers.

Max Planck Institute for Molecular Genetics, Berlin
http://www.mpimg-berlin dahlem.mpg.de/~xteam
Chromosome X.

MBx database *see* European Collaborative Interspecific Backcross

Meat Animal Research Center (MARC)
http:/sol.marc.usda.gov
Pig and cattle genome data and maps.

Mendel
http://probe.nalusda.gov:8000/plant
A database of designations for plant-wide families of sequenced plant genes and designations for sequenced genes in individual plant species.

MGD *see* Microbial Database; Mouse Genome Database

MICADO *see* Microbial Advanced Database Organization

Microbial Advanced Database Organization (MICADO)
http://locus.jouy.inra.fr
Bacillus subtilis and *E. coli* databases.

Microbial Database (MGD)
http://www.tigr.org/
Sequence of the *Haemophilus influenzae, Mycoplasm genitalium*, and other bacterial genomes.

MITOMAP
http://www.gen.emory.edu/mitomap.html
A mitochondrial DNA database held at Emory University, Atlanta, which contains information on the human mitochondrial genome.

Mosquito Genomics WWW Server
http://klab.agsci.colostate.edu/
Holds the *Aedes aegypti* linkage and physical maps and AaeDB database. Holds MsqDB, a database of information across mosquito species. Holds FlyBase. Link to AnoDB.

Motif BioInformatics WWW Server
http://motif.stanford.edu
Extensive links to other sites and databases.

Moulon WWW server
http://moulon.inra.fr:8001
Holds the *C. elegans* database (ACeDB), the Integrated Genome Database (IGD), a metabolic database, and information on setting up ACeDB-style databases on the WWW.

Mouse Genome Database and the *Encyclopedia of the Mouse Genome*
http://www.informatics.jax.org/mgd.html
http://mgd.hgmp.mrc.ac.uk/mgd.html
Mouse genetic mapping information from centre programmes, collaborative programmes and single laboratory efforts worldwide is regularly transferred to the Mouse Genome Database (MGD) at the Jackson Laboratory (Bar Harbor, Maine, USA) and presented in the latest issue of the *Encyclopedia of the Mouse Genome*—a tool for the presentation of mouse genome and related information. MGD contains mouse locus information; genetic mapping data; mammalian homology data; probes, clones and PCR primers; genetic polymorphisms; the Mouse Locus Catalogue (gene descriptions) and characteristics of inbred strains.

Mouse Locus Catalog (MLC) *see* Mouse Genome Database (MGD)

mousedb
http://www.hgmp.mrc.ac.uk
The Harwell mouse database compiled from published information, includes man–mouse homologies, mouse gene list and data from the Mouse Chromosome Atlas maps.

MsqDB
http://klab.agsci.colostate.edu
Genetic and physical chromosome mapping data across mosquito species.

MycDB *see Mycobacterium* Database

Mycobacterium Database (MycDB)
http://www.biochem.kth.se/MycDB.html
An integrated mycobacterial database containing data on physical and genetic mapping, and nucleotide sequences of mycobacterial genomes. Curators: Staffan Bergh, Royal Institute of Technology, Stockholm (e-mail: staffan@biochem.kth.se) and Stewart Cole, Pasteur Institute, Paris (e-mail: stcole@pasteur.fr).

Mycobacterium tuberculosis
http://www.sanger.ac.uk/pathogens/
Cosmid sequences of the *M. tuberculosis* genome.

National Center for Biotechnology Information (NCBI)
http://www.ncbi.nlm.nih.gov/
Responsible for building, maintaining and distributing the DNA sequence database GenBank.

National Center for Genome Resources (NCGR)
http://www.ncgr.org/
Holds GSDB, a genome sequence database with links to GDB, SWISS-PROT, etc., the SIGMA system for integrated genome map assembly. Information on ethical, legal and social implications of biotechnology.

National Center for Human Genome Research (NCHGR)
http://www.nchgr.nih.gov/
The centre heading the Human Genome Project for the National Institutes of Health (NIH) in the United States. Information on projects underway, including clinical gene therapy; diagnostic development; education and training; genetic resources, Laboratory of Cancer Genetics; Laboratory of Gene Transfer; Laboratory of Genetic Disease Research; medical genetics; technology transfer.

National Institutes of Health
http://www.nih.gov/

National Library of Medicine
http://www.nlm.nih.gov/

NIGMS (National Institute of General Medical Sciences) Human Genetic Mutant Cell Repository
http://arginine.umdnj.edu/coriell/nigms.html
Human cell cultures are available in the following categories: inherited disorders with characterized mutation; well-characterized chromosomally aberrant cell cultures; CEPH Reference Families; a human diversity collection: and human–rodent somatic cell hybrid mapping panels.

NIH GenoBase Server
http://dcrt.nih.gov.8004
Molecular biology database which incorporates and links the contents of several large sets of data including the EMBL sequence database and SWISS-PROT. The server also holds data from the *Mycoplasma capricolum* genome project. Text and BLAST searches of GenBank possible.

NRSub *see Bacillus subtilis* Database

Oak Ridge National Laboratory
http://www.ornl.gov
Links to all genome centres participating in the US Human Genome Project.
http://avalon.epm.ornl.gov
Informatics Group. Software for sequence analysis.

Online Mendelian Inheritance in Animals (OMIA)
http://www.angis.su.oz.au/BIRX/omia/omia_form.html

Online Mendelian Inheritance in Man (OMIM)
http://gdbwww.gdb.org/omim/docs/omimtop.html
The on-line version of *Mendelian Inheritance in Man*. Selected tables of mapped human disease genes are reproduced with kind permission in Appendix VII of this book.

Organelle Genome Database (GOBASE) *see* Canadian Genome Analysis and Technology Program (CGAT)

Organelle Genome Megasequencing Program (OGMP)
http://megasun.bch.umontreal.c./ogmpproj.html
Organelle information and sequence databases. Protist image databases.

Oslo Biotechnology Centre
http://bioslave.uio.no:8001

PathoGenes
http://probe.nalusda.gov:8000/plant
A database about fungal pathogens of small-grain cereals.

Philadelphia Genome Center (University of Pennsylvania Computational Biology and Informatics Laboratory)
http://www.cbil.upenn.edu/HGC22.html
Human chromosome 22 sequencing and mapping information.
http://www.cbil.upenn.edu/~sdong/genlang_home.html
Sequence analysis software.

PigMap
http://www.ri.bbsrc.ac.uk/pigmap
Information on the collaborative project to construct a genetic linkage map of the pig genome, and database of pig genome information.

PIR (the Protein Identification Resource, a protein sequence database) *see* Johns Hopkins Bioinformatics WWW Server; Geneva University; UK Human Genome Mapping Project Resource Centre

PomBase (*Schizosaccharomyces pombe* database) *see* Sanger Centre

Prodom (Protein Domain Database) *see* Sanger Centre

PROSITE *see* Geneva University

Radiation Hybrid Database (Rhdb)
http://www.ebi.ac.uk/RHdb/
An archive of raw data with links to other related data bases.

RATMAP
http://ratmap.gen.gu.se/
Database covering genes physically mapped to chromosomes in the laboratory rat and kept by the Department of Genetics in Gothenburg. the service contains general information, rat genetic nomenclature, rat locus list, and literature references sorted by rat gene symbol.

REBASE
http://www.gdb.org/Dan/rebase/rebase.html
Database of restriction enzymes.

Reference Library Database (RLDB) and Reference Library System
http://rzpd.rz-berlin.mpg.de/RLDB/
Database of hybridization and PCR work on filters of cosmid, YAC, P1 and cDNA libraries. Lists of probes/clones freely available from the Reference Library.

REPBASE
Contains prototypical interspersed repetitive elements from primates, rodents, mammals, vertebrates, invertebrates, and plants, as well as a collection of prototypical simple DNA sequences in primates. the database also contains collections of occurrences of Alu, L1, MIR, and THE repetitive elements. Available via anonymous ftp to: ncbi.nlm.nih.gov. in the directory 'repository/repbase'.

Resourcen Zentrum Max-Planck-Institüt für Molekulare Genetik
http://rldb.rz-berlin.mpg.de/main_e.html
Arrayed cosmid, YAC, P1 and cDNA libraries of human chromosome-specific, *Drosophila*, and *Schizosaccharomyces pombe* clones.

Ribosomal DNA *see* Sanger Centre

RiceGenes
http://probe.nalusda.gov:8000/plant/
A database for the rice genome. Curator: Edie Paul, Cornell University (e-mail: epaul@nightshade.cit.cornell.edu)

Roslin Institute, Edinburgh
http://www.ri.bbsrc.ac.uk/homepage.html
Holds BovMaP, ChickMaP, PigMaP, SheepBase genome databases; also a wide range of sequence databases and tools, and lists of other genome databases. Will hold the comparative mapping database TCAGdb.

Saccharomyces Genome Database (SGD)
http://genome-www.stanford.edu

Saccharomyces Genome Project
http://www.mips.biochem.mpg.de/
http://www.embl-ebi.ac.uk
http://www.sanger.ac.uk/yeast/home.html
http://genome-www.stanford.edu (*Saccharomyces* Genome Database) http://www.nig.ac.jp
http://www.ncbi.nlm.nih.gov/
http://www.ncbi.nlm.nih.gov/XREFdb
http://quest7.proteome.com/YPDhome.html (the Yeast Protein Database)
http://expasy/hcuge.ch/cgi.bin/list?yeast.txt

Sanger Centre
http://www.sanger.ac.uk/
Information on the *C. elegans* genome sequencing project and access to ACeDB and its derivative data bases—Wormpep: predicted proteins from the *C. elegans* project; Prodom: Protein Domain Database. Information on the yeast genetics project at the centre and sequencing projects on *Mycobacterium tuberculosis*, *Saccharomyces cerevisiae*, *Schizosaccharomyces pombe* (PomBase), and human chromosomes. Software for DNA sequencing and sequence analysis.

Schistosoma Genome Project
http://www.nhm.ac.uk/schistosome

Schizosaccharomyces pombe (NIH fission yeast information) http://www.nih.gov/sigs/yeast/fission.html

SheepBase
http://dirk.invermay.cri.nz
An up-to-date compilation of published data from sheep genome mapping projects. Compiled by the New Zealand Sheep Genome Programme.

SolGenes
http://probe.nalusda.gov:8000/plant/

A genome database containing information about potatoes, tomatoes, and peppers. Curator: Clare Nelson, Cornell (e-mail: cnelson@nightshade.cit.cornell.edu).

SorghumDB
http://probe.nalusda.gov:8000/plant/
A genome database (still under development) for sorghum. Curator: Najeeb Siddiqi, Texas A&M University (e-mail: nus6389@tam2000.tamu.edu).

Southern France Human Genome Project Computing Resource Centre *see* Genestream at EERIE

Stanford Human Genome Center
http://shgc.stanford.edu
Chromosome 4. Generation of STSs throughout human genome. Construction of radiation maps of human genome.

Stanford University DNA Sequence and Technology Center
http://genome-www.stanford.edu/SDSATC/staff.html
Development of high throughput sequencing methodology.

Swiss Federal Institute of Technology, Zurich *see* CBRG at ETHZ

SWISS-2DPAGE *see* Geneva University

SWISS-3DPAGE *see* Geneva University

SWISS-PROT *see* Geneva University

TBASE: the Transgenic/Targeted Mutation Data Base
http://www.gdb.org/Dan/tbase/tbase.html
Information on transgenic animals and targeted mutations. Covers mouse, rat, pig, and *Drosophila*.

TIGR (the Institute for Genomic Research)
http://www.tigr.org
The Microbial Data Base (MDB) provides access to the genome sequences for the *Haemophilus influenzae*, *Mycoplasma genitalium* and *Methanococcus janaschii* genomes. The Human cDNA Database (HCD) provides researchers at non-profit institutions access to cDNA/EST sequence and related data. The Expressed Gene Anatomy Database (EGAD) links expression data, cellular roles, and alternative splicing information to a curated, non-redundant set of human transcript sequences and their function, cellular role, tissue distribution. The Sequences, Sources, Taxa database (SST) provides links between source, collection, taxonomy, and molecular sequence data. Also human chromosome 16 sequencing. Software tools available to academic researchers on request by e-mail to tools@tdb. tigr.org.

Tokyo University Insect Group
http://www.ab.a.u-tokyo.ac.jp/sericulture/shimada.html
Genetic maps of the silkmoth *Bombyx mori*.

TreeGenes
http://probe.nalusda.gov:8000/plant/
A genome database (still in development) for forest trees, part of the Dendrome Project.

Tumour Gene Database
http://condor.bcm.tmc.edu/oncogene.html

UK Human Genome Mapping Project Resource Centre (HGMP-RC)
http://www.hgmp.mrc.ac.uk
Access to many genomic databases and resources and extensive links to other sites. Software.

UniGene: Unique Human Gene Sequence Collection
http://www.ncbi.nlm.nih.gov/Schuler/UniGene
Database holding clusters of human EST sequences that represent the transcription products of distinct genes.

University of California, Berkeley
http://fruitfly.berkeley.edu
Drosophila Genome Center.

University of California, Davis
http://www.vgl.ucdavis.edu/~lvmillion
Horse genetics.

University of Michigan Human Genome Center
http://www.hgp.med.umich.edu/
DNA sequencing resources. Whitehead/MIT mouse map data. CEPH-Généthon physical map data. Links to many other sites.
http://www.sph.umich.edu/group/statgen/software
Radiation hybrid mapping software.

University of Minnesota Medical School, Computational Biology Center
http://www.cbc.med.umn.edu/
Info on sequence analysis projects on *Arabidopsis*, maize, rice, loblolly pine. Sequence analysis software. *Candida albicans* molecular biology.
http://lenti.med.umn.edu/zebrafish/zfish_top_page.html

Zebrafish sequence analysis project.
http://lenti.med.umn.edu/MolBio-man/
Unofficial guide to GCG ('Wisconsin') software package.

University of Texas Health Science Center
http://mars.uthscsa.edu/
Human chromosome 3.

University of Texas Southwestern Medical Center
http://mcdermott.swmed.edu
Human chromosome 11.
http://eatworms.swmed.edu
Caenorhabditis elegans WWW server.

University of Utah
http://www-genetics.med.utah.edu
Resources for genome sequencing.

University of Wisconsin *E. coli* Genome Center
http://ecoliftp.genetics.wisc.edu
Sequencing of the *E. coli* genome.

USDA Agricultural Plant Genomes/ USDA/ARS/ NAL Plant Genome Data
http://probe.nalusda.gov:8000/plant/index.html
Plant DNA libraries. Information on plant genome mapping projects. Contains molecular and phenotypic information about the genomes of *Arabidopsis*, alfalfa, *Phaseolus*, *Vigna*, *Chlamydomonas reinhardtii*, legumes, cotton, cereals, maize, fungal pathogens of small-grain cereals, rice, Solanaceae, sorghum, soybeans, forest trees.

US Human Genome Project: DOE and NIH Human Genome Research Sites
http://www.ornl.gov/TechResources/Human_Genome/CENTERS.HTML
Lists participating centres in the US Human Genome Project with contact addresses and information on projects.

V BASE
http://www.mrc-cpe.cam.ac.uk/imt-doc/vbase-home-page.html
A directory of human immunoglobulin V genes compiled from published sources.

Vertebrate Comparative Database (CompDB)
http://www.hgmp.mrc.ac.uk/Comparative/home.html
Contains homologue data between human genes and a range of other vertebrate species.

Virtual Genome Center
http://alces.med.umn.edu/VGC.html
Contains sequence analysis tools: query GenBank, SWISS-PROT databases. Useful databases and tables: codons; size of human chromosomes, human repeated DNA. Sequences of the *S. cerevisiae* chromosomes. *Candida albicans*: physical map, sequence data, strains, resources.

Walter and Eliza Hall Institute of Medical Research (WEHI)
http://www.wehi.edu.au
GCG programs and manuals. MHCPEP database. SRS malaria database. Graphics interface to the Brookhaven Protein Data Bank. GDB.

Washington University School of Medicine
http://genome.wustl.edu/
Human X chromosome.

Washington University School of Medicine Genome Sequencing Center
http://genome.wustl.edu/gsc/gschmpg.html
Caenorhabditis elegans sequencing. EST sequencing.

Washington University, Department of Pathology
http://www.pathology.washington.edu
Human and mouse chromosome ideograms. Horse idiographic karyotype. Cytogenetic gallery of scanned photomicrographs of abnormal human karyotypes. Scanned images of human and mouse chromosome spreads.

Weizmann Institute of Bioinformatics
http://dapsas1.weizmann.ac.il
Design and development of tools for Bioinformatics, especially in the areas of molecular biology and the human genome. Holds a list of genome and molecular biology sites.

Whitehead Institute Mouse Genetic Map Information
http://www-genome.wi.mit.edu/genome_data/mouse/mouse_index.html
Data representing the Whitehead Institute/MIT Center for Genome Research mouse genetic map.

Whitehead Institute for Medical Research/MIT Center for Genome Research
http://www-genome.wi.mit.edu/
Whitehead Institute STS/YAC Map. Links to other sites. Other resources.

XREFdb
http://www.ncbi.nlm.nih.gov/XREFdb/
Database cross-referencing the genetics of model organisms with mammalian phenotypes. Provides similarity search, mapping and relevant mammalian phenotype information, and also BLAST similarity search results that identify significant matches between sequences of model organism proteins and mammalian peptide sequences predicted by conceptual translation of ESTs. Originally funded by the NCHGR to cross-reference the *Saccharomyces cerevisiae* and mammalian genomes, XREFdb has recently been expanded to accept protein queries from other model organisms including *Caenorhabditis elegans, Drosophila melanogaster, Escherichia coli, Mus musculus, Rattus norvegicus, Schizosaccharomyces pombe* and *Xenopus laevis*. Information: info@qmail.bs.jhu.edu or basset@ncbi.nlm.nih.gov

Yale Genome Center
http://paella.med.yale.edu
Human chromosome 12.

Yeast Protein Database (YPD)
http://www.proteome.com/YPDhome.html
Database of *Saccharomyces cerevisiae* proteins.

Zebrafish Sequence Analysis Project
http://lenti.med.umn.edu/zebrafish/zfish_top_page.html
Zebrafish project at the University of Minnesota.

Appendix I Materials, media and solutions

acrylamide stock solution 40% (w/v) (acrylamide/*bis*-acrylamide, 37.5 : 1)
Dissolve 100 g acrylamide and 2.7 g *bis*-acrylamide in H_2O to a final volume of 250 ml. Store in dark glass bottles at 4 °C.
Acrylamide is toxic and is absorbed through the skin. Always wear gloves and work in the hood.

acrylamide solution 40% (w/v) for gradient gels for DNA sequencing (acrylamide/*bis*-acrylamide, 37.5 : 1)
Dissolve 380 g acrylamide and 20 g *bis*-acrylamide in H_2O to a final volume of 1 litre. Add 20 g mixed bed resin (e.g. Amberlite MB-1 or equivalent) and stir carefully for 20 min (this step removes metal ions and acrylic acid). Store in dark glass bottles at 4 °C.
Always wear gloves and work in the hood.

acrylamide stock solution 6% (w/v) (0% denaturant stock solution) in TAE buffer
For 500 ml: 75 ml acrylamide (40% stock), 25 ml TAE (20 × stock), and H_2O up to 500 ml.

alkaline lysis DNA miniprep solutions
Alkaline lysis I solution: 50 mM glucose, 25 mM Tris-HCl pH 8.0, 10 mM EDTA. Alkaline lysis II solution: 0.2 M sodium hydroxide, 1% SDS; Alkaline lysis III solution: 3 M potassium acetate, 2 M acetic acid.

ammonium persulphate stock solution (10% w/v)
Dissolve 10 g ammonium persulphate to 100 ml H_2O. This solution is usually freshly prepared but may also be stored in small aliquots (1 ml) at –20 °C.

beta-mercaptoethanol (β-ME)
The stock solution of β-ME is 14.4 M. To prepare a 1-M stock solution, for example, take 1 ml β-ME (14.4 M) and 13.4 ml double-distilled H_2O. For a final concentration of 5×10^{-3} M it may be preferable to make a 1 : 100 dilution and use this to make the final stock solution. Filter the 5×10^{-3} M β-ME, dispense into 5-ml aliquots and store at 4 °C.

buffers for reverse phase HPLC
Buffer A: 5% acetonitrile, 95% 100 mM TEA. Buffer B: 65% acetonitrile, 35% TEA.

Church buffer (for pre-hybridization and hybridization)
0.5 M sodium phosphate (pH 7.2), 7% SDS, 1 mM EDTA (pH 8.0).

denaturant stock solution for electrophoresis (80%) (6% acrylamide, 32% formamide, 5.6 M urea)
For 500 ml: 170 g electrophoresis-grade urea, 75 ml acrylamide (40% stock), 160 ml deionized formamide (100% stock), 25 ml TAE buffer (20 × stock), and H_2O to 500 ml. Store in dark glass bottles at 4 °C. To deionize formamide: add 2 g of mixed bed resin (J.T. Baker) to 100 ml formamide and stir for 30 in. Filter to remove resin and store in dark glass bottles.

denaturing solution for chromosomes
35 ml formamide, 5 ml 20 × SSC (pH 5.3), 10 ml sterile distilled H_2O.

Denhardt's solution
10 × : 0.2% bovine serum albumin/0.2% Ficoll 400/0.2% polyvinylpyrrolidone (MW ≈ 44 000). This can be prepared as a 100 × stock.

DNA
Herring or salmon sperm DNA: Dissolve 10 mg ml^{-1} DNA in sterile H_2O and sonicate to a fragment length of ~500 bp. Store frozen in aliquots of 100–200 µl.

ethidium bromide (10 mg ml^{-1})
Dissolve 1 g ethidium bromide in 100 ml H_2O.

fixative for chromosomes
Methanol/glacial acetic acid, 3 : 1.

formamide/SSC (50% formamide, 2 × SSC (pH 7.0) and 70% formamide, 2 × SSC (pH 7.0))
For 500 ml: 250 ml (= 50%) or 350 ml (=70%) formamide, 50 ml 20 × SSC. Adjust pH to 7.0 with HCl. Unlike formamide used in hybridization mixture, formamide for these two solutions does not need to be deionized. Both solutions can be stored at room temperature, but check that the pH = 7.0 prior to use.

GET buffer (for alkaline lysis DNA miniprep)
50 mM glucose, 25 mM Tris-HCl (pH 8.0), 10 mM EDTA.

HAT medium and supplements
Solution 1: methotrexate (alternatives are amethopterin or aminopterin). Add 0.045 g methotrexate to 10 ml distilled H_2O. Add 1 M NaOH until the methotrexate dissolves. Add 10 ml of distilled H_2O. Adjust the pH to between 7.5 and 7.8 with 1 M HCl. Make up to 100 ml. Filter-sterilize and store at –20 °C.
Solution 2: hypoxanthine and thymidine (HT). Add 0.14 g hypoxanthine to 30 ml distilled H_2O. Add 1 M NaOH until the hypoxanthine dissolves. Adjust the pH to 10 with 1 M HCl. Add 0.039 g thymidine to 35 ml distilled H_2O. Combine the hypoxanthine and thymidine solutions and adjust to 100 ml. Filter-sterilize and store at –20 °C. Add 1 ml of Solution 1 and 1 ml of Solution 2 to 98 ml of growth medium.
Supplements for HAT medium.
BUdR: 100 × = 0.3 g 5-bromo-2′-deoxyuridine per 100 ml H_2O (approx. 1×10^{-2} M, so that 1 × = 1×10^{-5} M). Store frozen. Light sensitive.
6-Thioguanine (2-amino-6-mercaptopurine). 50 × = 25 mg in 150 ml H_2O (so that 1 × = 2×10^{-5} M). Add 1 N NaOH to dissolve and adjust pH to 9.5 with 1 N acetic acid. Filter-sterilize and store at –20 °C.
8-Azaguanine: 100 × stock = 76 mg in 50 ml (so that 1 × = 1×10^{-4} M). Add 1 N NaOH to dissolve; heat to 37 °C if necessary and adjust pH to 9 with 1 N acetic acid.

Hirt squirt (for cell lysis)
0.8% SDS/10 mM EDTA.

hybridization buffer (for microFISH)
50% formamide, 10% dextran sulphate, 2 × SSC, 1% Triton X-100, sterile distilled H_2O.

hybridization mix (for FISH) (2 × SSC, 50% formamide, 10% dextran sulphate, 1% Tween 20)
For 10 ml: dissolve 1 g dextran sulphate in 1 ml 20 × SSC, 1 ml 10% Tween 20, and double-distilled H_2O to a total volume of 5 ml. Add 5 ml deionized formamide. Check that pH = 7.0 and store at –20 °C in aliquots of 12 µl for single use.

ionic extraction buffer (for detergent extraction of DNA from M13 phage)
Tris-HCl (pH 8.0), 1 mM EDTA, 125 mM potassium iodide, 0.16 mM potassium lauryl sulphate.

kinasing buffer
0.5 M Tris (pH 7.5), 10 mM ATP, 20 mM DTT, 10 mM spermidine, 1 mg ml^{-1} BSA, 100 mM $MgCl_2$.

Leishman's stain
3 g of Leishman's powder dissolved in 1 litre methanol.

ligation buffer (single-stranded, for use with T4 RNA ligase, for adding oligonucleotide to 5′ end of cDNA)
100 mM Tris-HCl (pH 8.0), 20 mM $MgCl_2$, 20 µg ml^{-1} BSA, 50% PEG, 2 mM hexamine cobalt chloride, 40 µM ATP.

ligation buffer 1× (for YAC ligation, for use with T4 DNA ligase)
50 mM Tris-HCl (pH 7.6), 30 mM NaCl, 10 mM $MgCl_2$, 1 × polyamines (0.75 mM spermidine, 0.30 mM spermine) (the polyamines may be omitted).

ligation buffer 10×
400 mM Tris-HCl (pH 7.6), 100 mM $MgCl_2$, 1 mM DTT, 5 mM ATP.

ligation buffer 5× (for use with DNA ligase, for ligating cosmid DNA into plasmids)
250 mM Tris-HCl (pH 7.6), 50 mM $MgCl_2$, 5 mM ATP, 5 mM DTT, 25% PEG 8000.

ligation buffer/mix 10×
Low salt buffer: 60 mM Tris (pH 7.5), 60 mM $MgCl_2$, 50 mM NaCl, 2.5 mg ml^{-1} BSA, 70 µM β-mercaptoethanol, with ligation additions: 1 mM ATP, 20 mM DTT, 10 mM spermidine, 1 mg ml^{-1} BSA, 100 mM $MgCl_2$.

linear polyacrylamide (LPA) carrier
A reliable and completely noninjurious inert carrier allowing efficient precipitation of picogram quantities of DNA. Prepare by polymerization of a 5% acrylamide solution with ammonium persulphate (0.1%) and TEMED (0.1%). This solution is 50 mg ml^{-1} and a working solution at 2 mg ml^{-1} is diluted from this. This is stored at –20 °C and may be frozen and thawed many times. Usually, 5–10 µg per precipitation reaction is sufficient.

loading buffer for RNA fractionation (poly(A) mRNA isolation on oligo-dT columns)
0.5 M lithium chloride, 50 mM Tris (pH 8.0), 5 mM EDTA, 1% SDS.

loading buffer for SSCP analysis
95% formamide, 20 mM EDTA (pH 8.0), 0.05% bromophenol blue, 0.05% xylene cyanol.

loading solution for DGGE 5×
0.25% (w/v) bromophenol blue, 0.25% (w/v) xylene

cyanol, 20% (w/v) Ficoll. Dissolve 20 g Ficoll, 250 mg bromophenol blue and 250 mg xylene cyanol in a final volume of 100 ml H_2O. Instead of Ficoll, glycerol can be used.

lysis buffer (for mammalian cells)
guanidinium thiocyanate (GuSCN) in 25% lithium chloride (LiCl).

MacIlvaine's buffer, pH 5.6
0.1 M anhydrous citric acid (solution A), 0.4 M anhydrous sodium phosphase dibasic (solution B). For buffer, use 92 ml solution A and 50 ml solution B.

middle wash buffer (MWB) (for poly(A) mRNA isolation on oligo-dT columns)
100 mM LiCl, 50 mM Tris (pH 8.0), 5 mM EDTA, 1% SDS.

mountant for chromosomes
Citifluor/PI: 1 ml citifluor mountant, 8 µl propidium iodide (50 µg ml^{-1}).

nick translation buffer 10×
500 mM Tris-HCl (pH 7.5), 50 mM $MgCl_2$, 500 µg ml^{-1} nuclease-free BSA.

nucleotide mix for nick translation with biotin 10×
500 µM dCTP, 500 µM dGTP, 500 µM dATP, 380 µM dTTP, 30 µM biotin-16-dUTP.

osmotic shock medium
PBS/10% dimethyl sulphoxide (DMSO)

panning buffer (for mammalian cells)
PBS, 2 mM EDTA, 0.1% sodium azide, 5% FCS.

PBS 1×
Solution A (1 litre): 8.20 g NaCl, 1.78 g $Na_2HPO_4 \cdot 2H_2O$.
Solution B: (500 ml) 4.14 g NaCl, 0.69 g $NaH_2PO_4 \cdot H_2O$.
Adjust the pH of solution A with solution B to 7.0.
Sterilize by autoclaving. Store at room temperature.

PCR buffer 1×
10 mM Tris-HCl (pH 8.9), 50 mM KCl, 1.5 mM $MgCl_2$, 0.1% (w/v) Triton X-100, 0.01% (w/v) gelatin.

PCR buffer 2×
20 mM Tris-HCl (pH 8.4), 100 mM KCl, 10 mM $MgCl_2$, 0.2 mg ml^{-1} gelatin.

PCR buffer 10×
100 mM Tris-HCl (pH 9.0) at 25 °C, 500 mM KCl, 15 mM $MgCl_2$, 1.0% Triton X-100.

PCR mix 1×
10 mM Tris-HCl (pH 8.3), 50 mM KCl, 1.5 mM $MgCl_2$, dNTPs (250 µM each).

PEG solution for transformation of yeast spheroplasts
20% polyethylene glycol 6000 MW, 10 mM Tris-HCl (pH 7.6), 10 mM $CaCl_2$.

PEG solution for DNA precipitation
26.2% PEG 8000, 6.6 mM $MgCl_2$, 0.6 M NaOAc (pH 5.2).

primer, Alu-BK33
5′CTGGGATTACAGGCGTGAGC3′.

primer, 6-MW
5′CCGACTCGAGNNNNNNATGTGG3′.

reverse transcription buffer 5×
250 mM Tris-HCl (pH 8.3), 400 mM KCl, 15 mM $MgCl_2$, 50 mM DTT.

RNase
Prepare a stock solution of 10 mg ml^{-1} RNase in sterile H_2O. Inactivate any contaminating DNase by boiling for 10 min. Cool to room temperature and store frozen.

SCE 1×
1 M sorbitol, 0.1 M sodium citrate (pH 5.8), 10 mM EDTA (pH 7.5).

Sørensen buffer
Solution A (100 ml): 0.946 g Na_2HPO_4 (anhyd.) (= 0.06 M).
Solution B (100 ml): 0.908 g KH_2PO_4 (= 0.06 M).
Adjust the pH of Solution A with Solution B to 6.8 (about 1 : 1). For 50% Sørensen buffer, mix 1 : 1 with double-distilled H_2O.

SOS solution
1 M sorbitol, 25% YPD, 6.5 mM $CaCl_2$, 10 µg ml^{-1} 1 µg ml^{-1} uracil.

SSC 1×
150 mM NaCl, 15 mM sodium citrate, pH 7.0.

SSC 20×
3 M NaCl, 0.3 M sodium citrate. For 1 litre: 175 g NaCl, 88 g sodium citrate. Adjust pH to 7.0 with HCl. Sterilize by autoclaving.

SSCT
4 × SSC, 0.05% Tween 20 (or Triton X-100) (pH 7.0). For 1 litre: 200 ml 20 × SSC, 5 ml 10% Tween 20 (or Triton X-100). Check pH = 7.0 and store at room temperature.

SSCT-BSA
3% (w/v) BSA in 4 × SSC, 0.05% (v/v), Triton X-100.

SSCTM
Prepare fresh. Dissolve 0.5 g 99% fat-free dried milk (Marvel) in 10 ml SSCT. Centrifuge at 1500 r.p.m. for 5 min in order to pellet undissolved particles. Soak off and discard the cloudy top layer and use the clear solution in the middle of the tube.

STC
1 M sorbitol, 10 mM Tris-HCl (pH 7.6), 10 mM $CaCl_2$.

TAE 50×
For 1 litre: 242 g Tris base, 100 ml 0.5 M EDTA, pH 8.0. Adjust pH to 7.2 with glacial acetic acid (about 57 ml).

TBE 0.5× 6% gel solution
460 g urea, 150 ml 40% acrylamide solution, 50 ml 10× TBE buffer. Make up to 1 litre with H_2O with dissolve. Filter through sintered glass funnel and store in the dark at 4 °C for up to several weeks.

TBE 5× 6% gel solution
115 g urea, 37.5 ml 40% acrylamide solution, 125 ml 10× TBE buffer, 10 mg bromophenol blue (optional). Make up to 250 ml with H_2O and dissolve. Filter through sintered glass funnel and store in the dark at 4 °C for up to several weeks.

TBE buffer 1×
89 mM Tris, 89 mM boric acid, 2.5 mM EDTA (pH 8.3).

TBE buffer 5×
For 1 litre: 54 g Tris base, 27.5 g boric acid, 20 ml 0.5 M EDTA (pH 8.0), make up to 1 litre with H_2O.

TBE buffer modified for DNA sequencing gels (does not precipitate) 1×
133 mM Tris, 44 mM boric acid, 2.5 mM EDTA (pH 8.8).

TBE buffer (pH 8.8) 10×
For 1 litre: 162 g Tris base, 27.5 g boric acid, 9.2 g Na_2EDTA, make up to 1 litre with H_2O.

TE
10 mM Tris, 1 mM EDTA (pH 8.0). For 1 litre: 1.21 g Tris base, 0.37 g EDTA. Adjust pH to 8.0 with HCl. Sterilize by autoclaving.

TE50
10 mM Tris-HCl (pH 7.6), 50 mM EDTA.

TEMED
N,N,N′,N′-tetramethylethylenediamine.

TEN9
50 mM Tris-HCl (pH 9.0), 100 mM EDTA (pH 8.0–9.0), 200 mM NaCl.

TENP buffer
10 mM Tris-HCl (pH 7.6), 20 mM EDTA, 30 mM NaCl + polyamines (0.75 mM spermidine, 0.30 mM spermine).

transformation buffer I (TFBI) (for bacterial transformation)
30 mM potassium acetate, 50 mM $MnCl_2$, 100 mM KCl, 10 mM $CaCl_2$, 15% glycerol (v/v).

transformation buffer II (TFBII) (for bacterial transformation)
10 mM Na-MOPS (pH 7.0), 75 mM $CaCl_2$, 10 mM KCl, 15% glycerol.

trisodium citrate (TSC)
For a 3.3% solution, and make up a solution of 33 g TSC with double-distilled H_2O to 1 litre. Dispense in 40-ml aliquots, autoclave and store at room temperature. Check volume and adjust with sterile double-distilled H_2O before use.

Triton-TE extraction buffer
0.5% Triton X-100, 10 mM Tris-HCl (pH 8.0), 1 mM EDTA (pH 8.0).

tRNA
E. coli tRNA: Dissolve 10 mg ml^{-1} *E. coli* tRNA in sterile H_2O and store frozen in aliquots of 100–200 µl.

Vectashield (Vector Laboratories)
A self-prepared mixture containing 22 mg 1,4-diazobicyclo (2.2.2) octane (DABCO) in 1 ml 20 mM $NaHCO_3$ (pH 8.0), 75% glycerol, or 10 mg ml^{-1} *p*-phenylenediamine in PBS mixed 1 : 9 with glycerol and adjusted to pH 8.0 with 0.5 M carbonate-bicarbonate buffer (pH 9.0).

Wright's stain stock solution
Dissolve 1.25 g Wright's stain in 500 ml methanol for around 1 h. Filter the solution through filter paper (Whatman no. 1) and store this Wright's stock solution at room temperature protected from light in a brown glass bottle. Older stock solutions usually give better results than fresh ones. Therefore, prepare the solution at least 2 weeks before use.

Media

chorionic villus sample transport medium
100 ml basal medium, e.g. Ham's F1, 10 ml FCS, 1 ml L-glutamine (200 mM), 3 ml penicillin or streptomycin (10 000 IU ml^{-1} or 10 000 µg ml^{-1}), 3 ml kanamycin (10 000 µg ml^{-1}), 0.3 ml mycostatin (1000 IU ml^{-1}), and 1 ml heparin (1000 IU ml^{-1}).

complete medium for culturing lymphocytes
100 ml Ham's F10 or RPMI 1640 medium, 10 ml FCS, 1.0 ml phytohaemagglutinin (purified), 1.0 ml penicillin (5000 IU ml^{-1}), 1.0 ml streptomycin (5000 µg ml^{-1}), 1.0 ml L-glutamine (200 mM).

double-selection growth medium
4% dextrose, 0.67% yeast nitrogen base (without amino acids), base and amino acid supplements (–uracil, –tryptophan).

freezing medium (for lymphocyte storage)
RPMI 1640 medium/FCS/DMSO, 2 : 2 : 1.

Hogness modified freezing medium (HMFM) 10×
63 g l^{-1} K_2HPO_4, 18 g l^{-1} KH_2PO_4, 4.5 g l^{-1} 1 sodium citrate, 9 g l^{-1} ammonium sulphate, 440 g l^{-1} glycerol, and 0.9 g l^{-1} $MgSO_4.7H_2O$.

LB broth/agar
Standard Luria broth/agar.

media for culturing solid tumours
see Chapter 8, Table 8.2.

RPMI/Hepes
RPMI 1640 medium containing 20 or 25 mM Hepes with (or without) L-glutamine.

SD broth
For 100 ml: 0.7 g yeast nitrogen base without amino acids, 2 g glucose, 5.5 mg adenine, 5.5 mg tyrosine. Adjust to pH 7.0 and autoclave (121 °C, 20 min). Add filter-sterilized solutions of: 7 ml casamino acids for double selection (–ura, –trp), and 2 ml 1% tryptophan for single selection (–ura).

thawing medium (for lymphocytes)
RPMI 1640/FCS, 9 : 1.

TYM agar
2% Bacto-Tryptone, 0.5% yeast extract, 0.1 M NaCl, 10 mM $MgSO_4$.

YAC regeneration medium (single-selection medium)
1 M sorbitol, 4% dextrose, 0.67% yeast nitrogen base (without amino acids), amino acid supplements (20× amino acid and adenine mixture—adenine, arginine, isoleucine, histidine, lysine, methionine: all at 400 mg l^{-1}, leucine 1200 mg l^{-1}, phenylalanine 1000 mg l^{-1}, valine 3000 mg l^{-1}, tyrosine 600 mg l^{-1}), 20 μg ml^{-1} tryptophan, 2% agar.

YPD medium
1% yeast extract, 2% bactopeptone, 2% dextrose. Make up with 2% agar for plates.

YT medium 2×
Per litre: 16 g yeast extract, 10 g tryptone and 5 g NaCl supplemented after autoclaving with the appropriate antibiotic.

Appendix II Preparation of blood bottles and processing of blood samples

Where sufficient sample is available, blood samples taken for genetic analysis are routinely placed into two separate blood bottles for transport to the laboratory. One part is stored in RPMI/Hepes to ensure cell survival. The separated lymphocytes (see II.5) from this sample can be used for cytogenetic analysis (see Chapter 7, Protocol 13), may be stored frozen, and may also be transformed by Epstein–Barr virus to produce immortalized cell lines (see later). The other part of the sample is placed into EDTA. This is used to prepare a batch of DNA for immediate use. Always wear gloves when dealing with blood and treat all human tissue samples as potentially infectious.

II.1 Preparation of blood bottles

Tube 1 Fifty-millilitre colour-coded (e.g. red-capped) flat-bottomed tube containing RPMI/Hepes to ensure cell survival. This will contain the blood sample to be used in the subsequent lymphocyte separations.
Prepare the following mixture:

- 200 ml RPMI/Hepes (e.g. from Gibco-BRL, Sigma);
- 40 ml 3.3% trisodium citrate;
- 2 ml 5×10^{-3} M β-mercaptoethanol (β-ME).

Add 20 ml per red-capped tube, seal top with a strip of parafilm, add label, date and store for up to 2 months at 4 °C.

Tube 2 Twenty-millitre sterile universal container containing EDTA. When filled with blood this will be frozen and thawed and together with the residues from the lymphocyte separation will be used to prepare a batch of DNA.
Add 4 ml 0.25 M EDTA to each 20-ml sterile universal. Seal with a strip of parafilm, add a label, date and store for up to 6 months at 4 °C.
Prepared blood bottles should be stored in the fridge.

II.2 Filling blood bottles with blood

Once taken, blood samples should be kept at room temperature; they will be stable up to a maximum of one week after collection.

Remove blood bottles from fridge several hours before using.

Blood bottles should be filled in the following order of priority for each blood sample:
1 25 ml blood into large tube (tube 1) with tissue culture medium;
2 15 ml blood into universal (tube 2) with clear medium (0.25 M EDTA).
Include a copy of pedigree if from a family, indicating the individual from which blood was taken.

II.3 Processing of blood tubes

Upon arrival at the laboratory, filled blood tubes should,

if possible, be processed immediately. All samples should be logged on a running list.

1 The red-capped tubes with growth medium and blood are used to prepare sterile separations of lymphocytes. They can be stored for up to 1 week (after collection of the blood sample) at room temperature.

2 The universal containing EDTA plus 10–15 ml of blood is stored at –80 °C. Before freezing, stand the tube upright to allow blood to separate from the plasma/EDTA. Take 2 × 500-μl aliquots of the plasma/EDTA and store at 80 °C. *After use, all glassware is decontaminated by overnight soaking in 2% chloros or by adding Weskodyne. All defibrinated clots plus glass beads, plastic disposables and paper tissue are double bagged, autoclaved and incinerated.*

Gloves should be worn throughout blood handling and centrifugation is carried out in sealed tubes with aerosol-preventing lids.

II.4 Lymphocyte sterile separations

Materials

- Blood sample (e.g. 50-ml Falcon tube containing 25 ml blood and 25 ml citrated medium as prepared in II.1 and II.2).
- RPMI/Hepes (e.g. Gibco-BRL). Store at 4 °C until needed. Prewarm to room temperature prior to use.
- 1 M calcium chloride (dihydrate). Prepare a 100-ml stock solution. Dissolve 14.7 g $CaCl_2$ in double-distilled water and make up to a final volume of 100 ml. Autoclave or filter stock and dispense into a 5-ml aliquots. Store stocks at 4 °C.
- glass beads 4-mm undrilled (500 g) (LIF). Put approximately 20 beads into a glass universal and autoclave.
- Lymphoprep (lymphocyte separation medium) (Nycomed).
- acetic acid: Use glacial acetic acid as 100% stock and dilute with double-distilled water, e.g. 50 ml stock at 4% (3 ml acetic acid (glacial) in 48 ml double-distilled water). Store at 4 °C. Acetic acid will oxidize with time, therefore the stock solution should be replaced each month.
- nigrosine (water soluble) (25 g) (BDH). Nigrosine is made up as a 1% solution in PBSA. Filter after preparation and store at room temperature.

Method

1 Pour contents of tube into a 250-ml flask labelled clearly with patient's name. Rinse out blood bottle with 4 ml RPMI/Hepes (the RPMI/Hepes *must* be at room temperature).

2 Add sterile beads (1 bead per ml blood).

3 Keep foil on the flask and add 0.6 ml sterile 1 M $CaCl_2$ through the foil using a 1-ml syringe and needle.

4 *Immediately* after adding the $CaCl_2$, defibrinate for 15 min at 260 r.p.m. on a gyratory shaker.

5 Add 20 ml RPMI/Hepes to flask.

6 Divide defibrinated blood between two 50-ml Falcon (type 2070) tubes each containing 15 ml Lymphoprep (diluted blood/Lymphoprep, 2 : 1), overlaying very carefully with a 25-ml pipette attached to a pipette aid (rinse out with RPMI/Hepes, add to tubes). The Lymphoprep *must* be at room temperature.

7 Spin at 1800 r.p.m. (700 *g*) for 20 min using a centrifuge with a swing-out rotor (e.g. Beckman TJ6), brining the speed up slowly.

8 Using a sterile pasteur pipette, remove the interface to a 50-ml Falcon (type 2070) tube. Dilute 1 : 1 with RPMI/Hepes and count the cells (e.g. in a Neubauer counting chamber) in 4% acetic acid. Nigrosine is used to check cell viability.

9 Spin at 2300 r.p.m. (1000 *g*) for 10 min.

10 Aspirate off the RPMI/Hepes and freeze cells in two freezing vials (labelled with patient's name, date and cell count), and three nonsterile straws. Vials should contain no less than 3×10^6 cells.

11 Keep residues for DNA extraction by aspirating off down to the Lymphoprep–RPMI/Hepes interface. Combine residues into one or two universals and freeze at –20 °C.

II.5 Freezing cells for storage

Cells for freezing must be viable, therefore cell lines (e.g. suspension cells or attached lines) should be growing rapidly and must be subconfluent 80–90% of maximum. Primary cells for future transformation (e.g. mixed lymphocyte populations) should be stored only after careful counting. Cell stocks can be stored in a freeze mix of fetal calf serum and dimethyl sulphoxide (FCS/DMSO) at a ratio of 90 : 10.

Preparation of FCS/DMSO freeze mix

1 Thaw a 500-ml bottle of FCS and aliquot extremely carefully into 90-ml lots (use twice autoclaved blue-capped Duran bottles).

2 Place serum into a waterbath preheated to 56 °C for 30 min. (This step eliminates complement and other components from the serum.)

3 Cool sample to room temperature and add 10 ml DMSO to the serum. Mix thoroughly. Aliquot into 1-ml or 5-ml lots. Store at –20 °C until needed. The mix can be thawed at 37 °C, though care should be taken in wiping off water and cleaning with industrial methylated spirits to avoid contamination.

Adding cells to freeze mix

1 After pelleting cells by centrifugation, aspirate all the medium from the cell pellet.

2 Label vials carefully with date, cell line, name and passage number if applicable.

3 Add 0.5–1 ml of freeze mix per vial of cells to be frozen.

4 It is essential that cells be quickly frozen once they are in freeze mix. If there are a large number of lines to be frozen, only prepare 4–6 vials at a time. Wrap vials in a couple of layers of tissue and place at –80 °C for at least 12 h.

5 Cells should be quickly transferred to liquid nitrogen. *Care should be taken in handling liquid nitrogen — wear gloves.* Take the liquid nitrogen tank to the freezer and add the vials to the appropriate space.

6 Carefully note the space in which the vials are placed, add the details of the cell line, date and space to a cell line card or an appropriate computer database. Also, add details of the computer databases. This must be done soon after storing the cells. Similarly, if cells are removed from storage the appropriate steps should be taken to edit the card and computer storage.

Note: DMSO comes in 500-ml bottles. Only a clean bottle should be used for tissue culture. *Handle with care*—it is taken up through the skin in seconds. It is used to apply drugs topically to skin lesions. DMSO is toxic and can kill cells if they are not frozen immediately.

II.6 Transformation of blood cells with Epstein–Barr virus (EBV) for long-term culture

Overview

(a) Preparation of EBV pool for transformation.
(b) Transformation of lymphocytes.

(a) Preparation of EBV pool for transformations

Materials

- Marmoset cell line B958 cells (tested to ensure they are free from mycoplasma infection)
- 10% FSC/RPMI 1640 medium

Method

1 Grow cells to 1×10^6 ml^{-1} in 10 % FSC/RPMI 1640 medium at 37 °C (e.g. 500-ml vols in large plastic TC flasks).
2 Dilute to 0.2×10^6 ml^{-1}.
3 Incubate at 33 °C for 2 weeks, mixing occasionally.
4 Allow cells to settle at 4 °C overnight.
5 Spin supernatant to clarify.
6 Filter supernatant to be sure all cells are removed.
7 Aliquot supernatant in 2-l vols. Store in liquid nitrogen tank.
8 Test by comparing transformation ability with previous batch. (Use duplicate vial of cells known to have been transformed successfully before.)

(b) Transformation of batches of frozen lymphocytes with varying numbers of cells

Materials

- frozen lymphocytes
- 20% FCS/RPMI 1640 medium
- 20% FCS/RPMI 1640 medium containing cyclosporin, prepared as follows.
  Dissolve 3 mg cyclosporin A powder in 200 μl absolute ethyl alcohol.
  Add 60 μl Tween 20.
  Add 740 μl *serum-free* RPMI 1640 dropwise using whirlmixer after every drop.
  Add 2 ml RPMI 1640 medium with 10% FCS dropwise with whirlmixing.
  Aliquot and store at –20 °C. This medium is stable for several months.
  Use diluted 1 : 1000 to give 1 μg ml^{-1} for use in transformations.

Method

1 Thaw lymphocytes quickly at 37 °C.
2 Wash with 10 ml 20% FCS/RPMI 1640 medium in a sterile conical container removing an aliquot for counting/viability testing before centrifuging at no more than 1000 r.p.m. for 5 min. Remove supernatant.
3 Add 0.2 ml EBV stock and incubate at 37 °C for 1 h.
4 Add 10 ml 20% FCS/RPMI 1640 medium containing 1 μg ml^{-1} cyclosporin A.
5 Distribute 2 ml to each of two tissue culture tubes containing 1×10^5 human fibroblasts as a feeder layer previously treated with mitomycin C or irradiation.
6 After 5 days incubation at 37 °C in an atmosphere of 5% CO_2 in air, add 1 ml 20% FCS/RPMI 1640 containing 1 μg ml^{-1} cyclosporin A to each tube.
7 Twice weekly thereafter remove 2 ml medium and add 2 ml fresh medium: 20% FCS/RPMI 1640 containing 1 μg ml^{-1} cyclosporin A. After 2 weeks' culture, the cyclosporin A should be omitted. To guard against loss of cultures from contamination (as the culturing will proceed, on occasion, for up to 3 months and normally for 2 months), use two different bottles of medium so that the two sets of tubes are fed from separate sources and the same pipettes etc. never touch the two tubes.
8 At 4 weeks, the tubes will normally transfer successfully to 25-cm^2 tissue culture flasks starting with the flask in the upright position and only 4 ml medium. Feeding with small volumes of medium regularly has been found to be more satisfactory than infrequent large amounts until the cultures are obviously well established.

Appendix III Commercial suppliers

Contact addresses are listed alphabetically at the end of the appendix. This list covers materials mentioned in protocols and is not intended to be comprehensive. Inclusion here does not imply any endorsement by the Imperial Cancer Research Fund. Suppliers' lists available on the World Wide Web include Anderson's Timesaving Comparative Guides (http://www.atcg.com) and Biosupplynet (http://www.biosupplynet.com).

Chromatography, filtration and separation media (e.g. treated columns, magnetic beads, DNA-binding resin, DNA purification kits)
Amicon
BIO 101
Bio-Rad
Collaborative Research
Dynal
Nycomed
Pharmacia
Promega
Qiagen

Cytogenetics, and fluorochromes, labelled nucleotides, labelled antibodies, chromosome paints and other materials for FISH
See Chapter 10, Table 10.1 for details for chromosome paints currently available commercially
Alpha Laboratories (UK)
Appligene Oncor
Boehringer Mannheim
Cambio
Citifluor Ltd. (UK)
Cytocell Ltd
Life Technologies (Gibco-BRL)
Sigma
Vector Laboratories
Vysis

Electrophoresis
Bio-Rad
FMC Bioproducts
Pharmacia (Hoefer)

Flow cytometry
Becton Dickinson
R&D systems

General laboratory products: chemicals, consumables, tissue culture products, media, etc.
BDH Laboratory Supplies (UK)
Becton Dickinson
Bibby Sterilin
Boehringer Mannheim
Eppendorf
Falcon
J.T. Baker
Life Technologies (Gibco-BRL)
Millipore
Pierce
Seromed
Sigma
Wellcome Diagnostics
Whatman

Microscopy and imaging (*see also* cytogenetics)
Alpha Laboratories
Carl Zeiss Jena GmbH
Chroma Technology Corp
Digital Scientific Instruments
Leica
Molecular Dynamics
Nikon
Olympus
Perceptive Scientific Instruments

Restriction enzymes, polymerases, ligases, plasmids, primers, clone libraries, kits etc. for cloning and PCR
Amersham
ATCC
BIO 101
Bio-Rad
Calbiochem-Novobiochem
Clontech
DuPont-Merck
Epicentre Technologies
Life Technologies (Gibco-BRL)
Invitrogen
New England Biolabs
Novagen
Perkin-Elmer
Research Genetics

Sigma
Stratagene
TaKaRa

Addresses

Advanced Biotechnologies
Unit 7, Mole Business Park, Randalls Road, Leatherhead, Surrey KT22 7BA, UK
Tel.: (+44 1372) 360 123
Fax: (+44 1372) 363 263

Alpha Laboratories Ltd
40 Parkham Drive, Eastleigh, Hants SO5 4NU, UK
Tel.: (+44 1703) 610 911
Fax: (+44 1703) 643 701

ATCC (American Type Culture Collection)
WWW.http://www.atcc.org/
E-mail: tech@atcc.org
12301 Parklawn Drive, Rockville, MD 20852-1776, USA
Tel.: (+1 301) 231 5585 and 881 2600
Fax: (+1 301) 231 5826 and 770 1848

Amersham International plc
WWW:
http://www.amersham.co.uk/life/
UK
Tel.: (+44 1494) 544 000
Fax: (+44 1494) 524 266
USA
Tel.: (+1) 800 323 9750
Fax: (+1) 800 228 8735
Europe Tel.: (+44 1494) 544 000
Japan Tel.: (+81 3) 38 16 1091

Amicon
Amicon Inc., 72 Cherry Hill Drive, Beverly, MA 01915, USA
Tel.: (+1) 800 343 1397
Europe (+49 2302) 960 600

Appligene Oncor
Pinetree Centre, Durham Road, Birtley, Chester-le-Street, Co Durham, DH3 2TD, UK
Tel: (+44 191) 429 0022

BDH Laboratory Supplies (Merck) (UK)
Tel.: 0800 223 344
Fax: (+44 1455) 558 586

Beckman
USA Tel.: (+1) 800 742 2345
UK Tel.: (+44 1494) 441 181
Fax: (+44 1494) 447 558
Germany Tel.: (+49 89) 38 871
France Tel.: (+33 1) 43 01 70 00
Australia Tel.: (+61 02) 816 5288
Japan Tel.: (+81 3) 3221 5831

Becton Dickinson
UK
Tel.: (+44 1865) 748844
Fax: (+44 1865) 781523
USA
Tel.: (+1) 800 223 8226/952 3222
Fax: (+1 498) 954 2009

Bibby Sterilin Ltd (UK)
Tilling Drive, Stone, Staffs ST15 0SA, UK
Tel.: (+44 1785) 812121
Fax: (+44 1785) 813748

BIO 101
WWW: http://www.bio101.com
USA
Tel.: (+1 619) 598 7299/800 424 6101
Fax: (+1 619) 598 0116
UK Tel.: (+44 1582) 456 666

Bio-Rad Laboratories
USA Tel.: (+1) 800 4BIORAD/(510) 741 100
UK Tel.: (0800) 181 134
Fax: (+44 1442) 259 118
France Tel.: (+33 1) 49 60 68 34
Germany Tel.: (+49 89) 318 840
Japan Tel.: (+81 3) 5811 6270
Australia Tel.: (+61 2) 805 5000

Boehringer Mannheim
WWW: http://biochem.boehringer.com
Boehringer Mannheim GmbH, D-68298 Mannheim, Germany
Tel.: (+49 621) 759 8545/0621 759 8568
UK Tel.: 0800 521 578
USA Tel.: (+1) 800 428 5433
France Tel.: (+33) 76 76 30 86
Australia Tel.: (+61 2) 899 7999
Japan Tel.: (+81 3) 3432 3155

Calbiochem-Novabiochem International
WWW: http://www.calbiochem.com
USA
Tel.: (+1) 800 854 3417/800 662 2616
Fax: (+1) 800 776 0999/617 577 8015
Germany:
Tel.: (+49 6196) 63955
Fax: (+49 6196) 62361
UK
Tel.: (+44 115) 943 0840
Fax: (+44 115) 943 0951
Japan
Tel.: (+81 3) 5443 0281
Fax: (+81 3) 5443 0271
Australia
Tel.: (+61 612) 318 0322
Fax: (+61 612) 319 2440

Cambio
E-mail: postmaster@cambio.demon.co.uk
34 Millington Road, Cambridge CB3 9HP, UK
Tel.: (+44 1223) 366 500
Fax: (+44 1223) 350 069

Cambridge Bioscience
24–25 Sigent Court, Newmarket Road, Cambridge CB5 8LA, UK
Tel.: (+44 1223) 316 855
Fax: (+44 1223) 60732

Carl Zeiss Jena GmbH
WWW: http://www.zeiss.com
E-mail: mikro@zeiss.de
Tel.: (+49 36 41) 64 29 36
Fax: (+49 36 41) 64 31 44
USA
Tel.: (+1 914) 747 1800/800 233 2343
Fax: (+1 914) 681 7446

Chroma Technology Corp
72 Cotton Mill Hill, Unit A-9, Brattleboro, VT 05301, USA
Tel.: (+1 802) 257 1800
Fax: (+1 802) 257 9400

Clontech
WWW: http://www.clontech.com
E-mail: tech@CLONTECH.com
USA
Tel.: (+1) 800 662-CLON/(415) 424 8222
Fax: (+1) 800 424 1350/(415) 424 1064
UK Distributed by Cambridge Bioscience
Germany Tel. (+49 6221) 303 907

Collaborative Research
Biomedical Products Division, Collaborative Research Inc., 2 Oak Park, Bedford, MA 01730, USA
Tel.: (+1 617) 275 0004
Fax: (+1 617) 275 0043

Cytocell Ltd
Somerville Court, Banbury Business Park, Adderbury; Oxfordshire OX17 3SN, UK
Tel: (+44 1295) 810 910

Difco Laboratories Ltd
PO Box 14B, Central Ave, East Molesey, Surrey KT8 0SE, UK
Tel.: (+44 181) 979 9951

Digital Equipment Corporation
WWW:
http://www.digital.com/info.html
E-mail: info@digital.com

Digital Scientific Instruments
36 Cambridge Place, Hills Road, Cambridge, CB2 1NS, UK
Tel.:
Fax:

Dupont (UK) Ltd
Nen Research Products, Wedgewood Way, Stevenage, Herts SG1 4QN, UK
Tel.: (+44 1438) 734026/28/31
Fax: (+44 1438) 734379

Dynal
Dynal AS, Norway
Tel.: (+47) 22 06 10 00
Fax: (+47) 22 50 70 15
UK
Tel.: (+44 151) 346 1234
Fax: (+44 151) 346 1223
USA
Tel.: (+1) 800 638 9416
Fax: (+1 516) 326 3298
Australia
Tel.: (+61 1) 800 623 435
Fax: (+61 3) 663 6660
Japan
Tel.: (+81 3) 3435 1558
Fax: (+81 3) 3435 1526

Epicentre Technologies
E-mail: techhelp@epicentre.com
1202 Ann St, Madison, WI 53713, USA
Tel.: (+1 608) 258 3080
Fax: (+1 608) 258 3088

Eppendorf
WWW: http://www.eppendorf.com/eppendorf
E-mail: eppendorf@eppendorf.com
Germany
Eppendorf-Netheler-Hinz GmbH
Tel.: (+49 40) 5 38 01 0
Fax: (+49 40) 5 38 01 556
USA Tel.: (+1) 800 645 3050

Falcon
Marathon Laboratory Supplies, Unit 6, 55–57 Park Royal Road, London NW10 7JJ, UK
Tel.: (+44 181) 965 6865/6886
Fax: (+44 181) 965 0989

Fluka
Fluka Chemie AG, Industriestrasse 25, CH-9470 Buchs, Switzerland
Tel.: (+41 81) 755 25 11
UK Tel.: (+44 1747) 822 211
USA Tel.: (+1 516) 467 0980

FMC BioProducts
WWW: http://www.bioproducts.com
FMC Corporation, Rockland, ME, USA
Tel.: (+1 207) 594 3400
UK
Distributed by Flowgen Instruments
Tel.: (+44 1795) 429 737
Germany Tel.: (+49 51) 52 2075
France Tel.: (+33 1) 34 84 6252
Australia Tel.: (+61 2) 520 2122
Japan Tel.: (+81 775) 43 7235

Genetics Computer Group (GCG)
(Wisconsin Sequence Analysis Package)
E-mail: info@gcg.com

Genetix
16 Riverside Park, Wimborne, Dorset BH21 1QU, UK
Tel.: (+44 1202) 881122
Fax: (+44 1202) 840577

Gibco-BRL ***see*** Life Technologies

Hybaid
WWW:/http://www.hybaid.co.uk
UK
Tel.: (+44 181) 614 1000
Fax: (+44 181) 977 0170
USA Tel.: (+1) 800 634 8886/516 244 2929

Imagenetics *see* Vysis

Invitrogen
WWW: http://www.invitrogen.com
USA
Tel.: (+1) 800 955 6288/619 597 6200
Fax: (+1 619) 597 6201
Europe
Tel.: (+31 594) 515 175
Fax: (+31 594) 515 312
E-mail: tech_service@invitrogen.nest.nl
Australia Tel.: (+61 3) 562 6888
Japan Tel.: (+81 3) 5684 1616

Jencons
Cherrycourt Way Industrial Estate, Leighton Buzzard, Beds LU7 8UA, UK
Tel.: (+44 1525) 372010
Fax: (+44 1525) 372010

J.T. Baker
Mallinckrodt Baker UK
Tel.: (+44 1908) 506 000
Fax: (+44 1908) 503 290
Germany
Tel.: (+49 6152) 90 33 72
Fax: (+49 6152) 90 33 99
France
Tel.: (+33 1) 48 44 65 44
Fax: (+33 1) 48 44 65 18
USA
Tel.: (+1 908) 859 2151
Fax: (+1 908) 854 9318

Leica
WWW: http://www.bodan.net/leica
PO Box 2040, D-35530 Wetzlar, Germany
Tel.: (+49 64) 41 29 0
Fax: (+49 64) 41 29 33 99
Switzerland
Tel.: (+41 71) 727 37 43
Fax: (+41 71) 727 46 67
USA
Tel.: (+1 708) 405 0123/800 248 0123
Fax: (+1 708) 405 0030

Li-Cor Biotechnology Division
4421 Superior St, PO Box 4000, Lincoln, NB 68504, USA
Tel.: (+1) 800 645 4267/402 467 0700
Fax: (+1 402) 467 0819
UK Tel.: (+44 181) 614 1000
Germany Tel. (+49 80) 92 82 890
Netherlands (+31 2946) 3119
Australia Tel.: (+61 2) 417 8877
Japan Tel.: (+81 422) 45 5111

Life Sciences International
WWW: http://www.lifesciences-intl.co.uk

Life Technologies (Gibco-BRL)
WWW: http://www.lifetech.com
WWW: http://www.lifetecheuro.co.uk (Europe)
8400 Helgerman Ct, PO Box 6009, Gaithersburg, MD 20884, USA
Tel.: (+1 301) 840 8000/800 828 6686
Fax: (+1 800) 331 2286/716 774 6783
UK
Tel.: 0 800 838 380/0800 838 380
Fax: (+44 141) 814 6260
Japan
Tel.: (+81 3) 3663 7974
Fax: (+81 3) 3663 8242

Molecular Dynamics
WWW: http://www.mdyn.com
USA Tel.: (+1) 800 333 5703
UK Tel.: (+44 1494) 793377
Australia Tel.: (+61 3) 9810 9572
Japan Tel.: (+81 3) 3976 9692

MWG-Biotech
WWW: http://www.mwgdna.com/biotech
E-mail: oligo@mwgdna.com
Tel.: (+49 80) 92 2 10 84
Fax: (+49 80) 92 82 89 77

NBS Biologicals
New Brunswick Scientific
Tel.: (+44 1707) 275 733
Fax: (+44 1707) 267 859

New England Biolabs
WWW: http://www.neb.com
USA
Tel.: (+1) 800 NEB LABS/508 927 5054
Fax: (+1 508) 921 1350
E-mail: info@neb.com
UK
Tel.: 800 31 84 86 (+44 1462) 420 616
Fax: (+44 1462) 421 057
E-mail: info@uk.neb.com
Germany
Tel.: (+49 130) 83 30 31/6196 3031
Fax: (+49 6196) 83639
E-mail: infro@de.neb.com
Australia Tel.: (+61 75) 94 0299
Japan Tel.: (+81 3) 3272 0671

Nikon (Electronic Imaging Division)
WWW: http:/www.klt.co.jp/Nikon
E-mail: nikonbio@aol.com
USA
Tel.: (+1 516) 547 8500
Fax: (+1 516) 547 0306
UK
Instruments Division, Nikon House, 380

Richmond Road, Kingston, Surrey KT2 5PR, UK

Novagen
WWW: http://www.novagen.com
E-mail: novatech@novagen.com
USA
Tel.: (+1) 800 526 7319
Fax: (+1 608) 238 1388
UK Tel. (+44 1993) 706 500/(+44 1670) 732 992

Nycomed (UK) Ltd
Nycomed House 2111 Coventry Road, Sheldon, Birmingham B26 3EA, UK

Olympus
Olympus Optical Co. (Europe)
Fax: (+49 40) 23 77 36 47

Perkin-Elmer
PE Applied Biosystems
WWW: www.perkin-elmer.com
E-mail: pebio@perkin-elmer.com
USA Tel.: (+1) 800 327 3002
UK Tel.: (+44 1925) 825 650
France Tel.: (+33 1) 49 90 18 00
Germany Tel. (+49 6150) 101 0
Other Tel.: (+49) 6103 708 301

Perceptive Scientific Instruments
2525 South Shore Boulevard, League City, TX 77573, USA
Tel.: (+1 713) 334 3207
Fax: (+1 713) 334 3116
International
Tel.: (+44 244) 682 288
Fax: (+44 244) 681555

Pharmacia Biotech
WWW: http://www.biotech.pharmacia.se
Tel.: (+46 18) 16 50 11)
UK
Tel.: (+44 1727) 814 000
Fax: (+44 1727) 814 001
USA
Tel.: (+1) 800 526 3593
Fax: (+1 908) 857 0557
Japan Tel.: (+81 3) 3492 6949

Pierce
WWW. http://www.piercenet.com
E-mail: PierceChem@mcimail.com
USA
PO Box 117, Rockford, IL 61105, USA
Tel.: (+1 815) 968 0747/800 874 3723
Fax: (+1 815) 968 8148/800 842 5007
UK Tel.: (+44 1244) 382525
Germany Tel.: (+49 22) 419 68 50
France Tel.: (+33) 70 03 88 55

Promega
WWW: http://www.promega.com
USA
Tel.: (+1) 800 356 9526/(+1 608) 274 4330
Fax: (+1) 800 356 1970/(+1) 608 277 2516
UK
Tel.: (+1) 800 378994
Australia Tel.: (+1) 800 225 123
France Tel.: (+33) 05 48 79 99
Japan Tel.: (+81 3) 3669 7981
Netherlands Tel.: (+31 71) 5324 244
Switzerland Tel.: (+41 1) 830 7037

QIAGEN
UK
Tel.: (+44 1306) 740 444/760 444
Fax: (+44 1306) 875885
USA
Tel.: (+1) 800 426 8157
Fax: (+1) 800 718 2056
Germany
Tel.: (+49 2103) 8920
Fax: (+49 2103) 892 222
Switzerland
Tel.: (+41 61) 317 9420
Fax: (+41 61) 317 9422

R&D Systems
WWW: http://www.rndsystems.com
USA
Tel.: (+1) 800 343 7475/(+1 612) 379 2958
Fax: (+1 612) 379 6580
Europe
Tel.: (+44 1235) 531 074
Fax: (+44 1235) 533 420
Australia Tel.: (+61 62) 008 25 1437
Japan
Tel.: (+81 3) 5684 1522
Fax: (+81 3) 5684 1633

Research Genetics
WWW. http://www.resgen.com/
Research Genetics Inc., 2130 Memorial Parkway, SW, Huntsville, AL 35801, USA
Tel.: (+1) 800 533 4363
UK
Tel.: 0 800 89 1393
Fax: (+44 205) 536 9016

Sigma-Aldrich
WWW: http://www.sigma.sial.com
E-mail: sigma-techserv@sial.com
USA Tel.: (+1 314) 771 5750 (collect/800 325 3010
UK
Tel.: (+44 1202) 733 114 (Sigma Chemical Co)
Tel.: (+44 1747) 822 211 (Aldrich Chemical Co)
Germany Tel.: (+49 130) 5155
France Tel.: (+33) 05 21 14 08

Stratagene
E-mail: tech_services@stratagene.com
UK
Tel.: 0800 585 370/(+44 1223) 420 955
Fax: (+44 1223) 420 234
USA
Tel.: (+1) 800 424 5444/(+1 619) 535 5400
Fax: (+1 619) 535 0045
Germany
Tel.: (+49 6221) 400634
Fax: (+49 6221) 400639
Switzerland
Tel.: (+1) 364 1106
Fax: (+41 1) 365 7707
Australia Tel.: 1800 252 204
Japan Tel.: (+81 3) 3660 4819/5684 1622

TaKaRa
Takara Shuzo Co. Ltd, Otsu, Shiga, Japan
Tel.: (+81 775) 43 7247
Fax: (+81 775) 43 9254
Europe
Tel.: (+33 1) 41 47 01 14
Fax: (+33 1) 47 92 18 80
UK
Distributed by Severn Biotech Ltd
Tel.: (+44 1562) 825 286
Fax: (+44 1562) 825 284

Vector Laboratories
UK
Tel.: (+44 1733) 265530
Fax: (+44 1733) 263048
USA
Tel.: (+1 415) 697 3600
Fax: (+1 415) 697 0339

Vysis (formerly Imagenetics)
USA
Vysis Inc.
Tel.: (+1) 800 553 7042
Europe (+49 711) 720 250
UK (+44 181) 332 6932

Wellcome Diagnostics
Temple Hill, Dartford, Kent DA1H 5AH, UK

Whatman International Ltd
WWW: http://www.Whatman.co.uk
E-mail: information@Whatman.co.uk
UK
Tel.: (+44 1622) 674 821/674 823
Fax: (+44 1622) 682 288

Appendix IV **Fluorochromes and filter sets for FISH**

Table IV.1 Fluorochromes commonly used for FISH, and fluorescent stains used for chromosome banding and identification.

Fluorochrome	Absorption wavelength (nm)	Emission wavelength (nm)
Aminomethyl coumarin acetic acid (AMCA)[a]	345	440
Cy5[a]	650	674
Diamidino-2-phenylindole-dihydrochloride (DAPI)[b]	359	461
Fluorescein isothiocyanate (FITC)[a]	495	525
Hoechst 33258[b]	365	480
Tetramethyl rhodamine isothiocyanate (TRITC)[a]	543	570
Texas red (TR)[a]	596	620

[a]Fluorochromes commonly used for FISH.
[b]Fluorescent stains.

Table IV.2 Fluorescence filter sets for the Nikon Optiphot microscope.

Filter set	Excitation (nm)	Dichroic (nm)	Barrier (nm)
B-2A (FITC)	450–490	DM 510	BA 520
G-2A (TR)	510–560	DM 580	BA 590
UV-2A (DAPI)	330–380	DM 400	BA 420

Table IV.3 Zeiss filter sets used for fluorescence detection and analysis.

Fluorochrome	Exciter filter	Dichroic reflector	Barrier filter	Filter set
FITC + propidium iodide	BP 450–490	510	LP515	09
Texas red/rhodamine	BP 546	580	LP590	15
AMCA/DAPI	G 365	395	LP420	02

BP, bandpass filter; LP, long-wave bandpass filter; G, solid glass filter.

Table IV.4 Fluorescence filter blocks on the MRC 600 confocal laser scanning microscope.

Filter block	Exciter filter (nm)	Dichroic reflector (nm)	Emission filter (nm)
Dual channel			
A1 (TR/rhodamine)	514 DF10	DR 527LP	
A2 (FITC)	540 DF30	DR 565LP	
Single channel			
BHS (FITC)	488 DF 10	510LP	OG 515LP
GHS (TR/rhodamine)	514 DF10	54LP	OG 550LP

Appendix V Useful addresses and Internet connections

(Specialist databases and other genome resource centres are listed in Chapter 37.)

This list does not aim to be complete, but many of the WWW addresses are home pages that will point to other sites of interest.

V.1 Useful addresses

American Society of Human Genetics
9650 Rockville Pike, Bethesda, MD 20814, USA
Tel.: (+1 301) 571 1825
Fax: (+1 301) 530 7079
WWW: http://www.faseb.org/genetics/ashg/ashgmenu.htm

American Type Culture Collection (ATCC)
12301 Parklawn Drive, Rockville, MD 20852-1776, USA
Tel.: (+1 301) 231 5585 and 881 2600
Fax: (+1 301) 231 5826 and 770 1848
E-mail: tech@atcc.org
WWW: http://www.atcc.org/

British Council
10 Spring Gardens, London SW1, UK
Tel.: (+44 171) 930 8466

Centre d'Etude du Polymorphisme Humain (CEPH)
27 rue Juliette Dodu, F-75010 Paris, France
Tel.: (+33 1) 4249 9862
Fax: (+33 1) 4018 0155

CIBA Foundation
41 Portland Place, London W1, UK
Tel.: (+44 171) 636 9456

Commission of the European Communities
Square de Meeus 8, B-1040 Brussels, Belgium

Cooperative Human Linkage Center (CHLC)
WWW: http://www.chlc.org/

US Department of Energy (DOE) Human Genome Program
WWW: http://www.er.doe.gov/production/oher/hug_top.html
Primer on Molecular Genetics
WWW: http://www.gdb.org/Dan/DOE/intro/html
White Paper on Bioinformatics
WWW: http://www.gdb.org/Dan/doe/whitepaper/contents.html

Deutsche Krebsforschungs Zenter (DKFZ) (Heidelberg, Germany)
WWW: http://genome.dkfz-heidelberg.de/

European Bioinformatics Institute
Hinxton Hall, Hinxton, Cambridge CB10 1RQ, UK
WWW: http://www.ebi.ac.uk/

European Collection of Animal Cell Cultures (ECACC)
Biologics Division, PHLS CAMR, Porton Down, Salisbury, Wilts SP4 0JG, UK
Tel.: (+44 1980) 610391
Fax: (+44 1980) 611315

European Molecular Biology Laboratory (EMBO)
Postfach 10.2209, Meyerhofstrasse 1, 6900 Heidelberg, Germany
Tel.: (+49 6221) 387258
Telex: 461613
Fax: (+49 6221) 387306

European Federation of Biotechnology
Cambridge Biomedical Consultants, Schuutstraat 12, NL 2517 XE Den Haag, The Netherlands
Tel.: (+31 70) 3653857
Fax: as phone number

Galton Laboratory (University of London)
WWW: http://diamond/gene.ucl.ac.uk

Généthon
1 rue de l'Internationale, 91000 Evry Cedex, France
Tel.: (+33 1) 6947 2965
Fax: (+33 1) 6077 1216
WWW:
http://www.genethon.fr/genethon_en.html

The Genetics Society of America
WWW: http/faseb/org/genetics/gsa/gsamenu.htm

Genome Data Base (GDB)
1830 E. Monument St., Baltimore, MD 21205, USA
Tel.: (+1 301) 955 9705
Fax: (+1 301) 955 0054
WWW: http://gdbwww.gdb.org/
Queries to GDB/OMIM: For e-mail Query Service put 'help' in the body of an e-mail message to mailserv@gdb.org

Howard Hughes Medical Institute/Human Gene Mapping Library
25 Science Park, Suite 457, New Haven, CT 06511, USA
Fax: (+1 203) 786 5534

Imperial Cancer Research Fund (ICRF)
PO Box 123, Lincoln's Inn Fields, London WC2A 3PX, UK
Tel.: (+44 171) 242 0200
Fax: (+44 171) 269 3469
WWW: http://www.icnet.uk

NIGMS Human Genetic Mutant Cell Repository
Coriell Cell Repositories, 401 Haddon Avenue, Camden, NJ 08103, USA
Tel.: 800 752 3805 (in USA); (+1 609) 757 4848 (elsewhere)
Fax: (609) 757 9737 (in USA); (+1 609) 964 0254 (elsewhere)
WWW:
http://arginine.umdnj.edu/ccr/ccr.html

Human Genome Management Information System (MGMIS)
WWW: http://ww.ornl.gov/Tech Resources/HumanGenome/home.html

The Human Genome Mapping Project Resource Centre (HGMP-RC) (UK)
Sanger Centre, Hinxton Hall, Hinxton, Cambridge CB10 1RQ, UK
WWW: http://www.hgmp.mrc.ac.uk

Human Genome Organization Europe (HUGO)
179 Great Portland Street, 5th Floor, London W1N 5TB, UK
Tel.: (+44 171) 436 7178
Fax: (+44 171) 436 1988

The Institute for Genomic Research (TIGR)
9712 Medical Center Drive, Rockville, MD 20850, USA
WWW. http://www.tigr.org/

Lawrence Berkeley Laboratory (Human Genome Center)
MS1-213 Lawrence Berkeley Laboratory, 1 Cyclotron Road, Berkeley, CA 94720, USA
Tel.: (+1 415) 486 6800
Fax: (+1 415) 486 5717
WWW: http://www-hgc.lbl.gov/GenomeHome.html
Resource for molecular cytogenetics
WWW: http://rmc-www.lbl.gov

Lawrence Livermore National Laboratory (Human Genome Project)
University of California, PO Box 5507, Livermore, CA 94550, USA
Tel.: (+1 415) 422 5698
Fax: (+1 415) 423 3608
Library information
WWW: http://www_bio.llnl.gov.bbrp/genome.html

Los Alamos National Laboratory (Center for Human Genome Studies)
Los Alamos National Laboratory, Los Alamos, NM 87545, USA
Tel.: (+1 505) 667 2746
WWW: http://www-t10.lanl.gov/

Medical Research Council (UK)
20 Park Crescent, London W1N 4AL, UK
Tel: (+44 171) 636 5422
Fax: (+44 171) 436 6179
WWW: http://www.mrc.ac.uk

MRC Mouse Genome Centre
Harwell, Oxfordshire OX11 0RD, UK

National Center for Biotechnology Information (NCBI) (USA)
WWW: http://www.ncbi.nlm.nih.gov/

National Center for Human Genome Research (NCHGR) (USA)
National Institutes of Health, Building 38A, Room 605, Bethesda, MD 20892, USA
Tel.: (+1 301) 496 0844
Fax: (+1 301) 402 0837
WWW: http://www.nchgr.nih.gov/

National Center for Genome Resources (NCGR) (USA)
WWW: http://www.ncgr.org/

National Institutes of Health (NIH) (USA)
Bethesda, MD 20892, USA
WWW: http://www.nih.gov/

NIH molecular biology
WWW: http://www.nih.gov/molbio

The National Library of Medicine (NLM) (USA)
WWW: http://www.nlm.nih.gov/

National Institute for Medical Research (UK)
The Ridgeway, Mill Hill, London NW7 1AA, UK
Tel.: (+44 181) 959 3666
Fax: (+44 181) 906 4477

Online Mendelian Inheritance in Man (OMIM)
WWW: http://www3.ncbi.nlm.nih.gov/Omim
The on-line version of *Mendelian Inheritance in Man*. Selected tables of mapped human disease genes are reproduced with kind permission in Appendix VII of this book.

Pasteur Institute
WWW: http://www.pasteur.fr/welcome-uk.html

Reference Library Database (RLDB)
Max-Planck-Institüt für Molekulare Genetik, Ihnestrasse 73, 14195 Berlin-Dahlem, Germany
Tel.: (+49 30) 8413 1627
Fax: (+49 30) 8413 1395
WWW: http://rldb.rz-berlin.mpg.de/main_e.html

Sanger Centre
Hinxton Hall, Hinxton, Cambridge CB10 1RQ, UK
Tel.: (+44 1223) 834244
Fax: (+44 1223) 1494919
WWW: http://www.sanger.ac.uk/

Wellcome Trust Centre for Human Genetics
Nuffield Department of Clinical Medicine, Windmill Road, Headington, Oxford OX3 7BN, UK
Tel.: (+44 1865) 740015
Fax: (+44 1865) 742187

Whitehead Institute for Biomedical Research/MIT Center for Genome Research
Cambridge, MA 02142, USA
WWW: http://www-genome.wi.mit.edu

Unité de Genetique Moleculaire Murine
Institut Pasteur, 28 rue du Dr Roux, 75724 Paris Cedex 15, France
Tel.: (+33 1) 4568 8000
Fax: (+33 1) 4568 8656

V.2 HUGO chromosome committees

The role of the editors on the chromosome committees is to approve new genes using names approved by the Nomenclature Committee. They are also charged with maintaining the consensus maps of each chromosome and sorting out disputes over marker order, etc. They are expected to maintain the quality and integrity of the Genome Data Base.

Chromosome 1
Gail A.P. Bruns, Associate Professor, Children's Hospital, Medical Center, Genetics Division, 300 Longwood Ave., Boston, MA 02115, USA
Tel.: (+1 617) 355 7575
Fax: (+1 617) 355 7588
E-mail: Bruns@rascal.med.harvard.edu

Tara Cox Matise, Columbia University, Department of Psychiatry, Unit 58, 722 West 168th St., New York, NY 10032, USA
Tel.: (+1 212) 960 2428
Fax: (+1 212) 568 2750
E-mail: tara@linkage.rockefeller.edu

Peter S. White, Philadelphia, PA, USA
Tel.: (+1 215) 590 4856
E-mail: white@kermit.oncol.chop.edu

Jeffrey M. Vance
Durham, NC, USA
Tel.: (+1 919) 684 6274
Fax.: (+1 919) 684 6514
E-mail: jett@dnadoc.mc.duke.edu

Andreas Weith, Research Institute for Molecular Pathology, Dr. Bohr-Gasse 7, 1030 Vienna, Austria
Tel.: (+43 1) 7973 0625
Fax: (+43 1) 798 7153
E-mail: weith@aimp.una.ac.at

Chromosome 2
Friedhelm Hildebrandt, Freiburg, Germany
Tel.: (+49 761) 270 4301
Fax: (+49 761) 270 4481
E-mail: hildebra@kk1200.ukl.uni-freiburg.de

Mansoor Sarfarazi, Farmington, CT, USA
Tel.: (+1 860) 679 3629
Fax: (+1 860) 679 2451
E-mail: msarfara@cortex.uchc.edu

Erwin A. Schurr, Montreal, Quebec, Canada
Tel.: (+1 514 937 6011)
Fax: (+1 514) 933 7146
E-mail: erwin@igloo.epi.mcgill.ca

Constantine Stratakis, Bethesda MD, USA
Tel.: (+1 301) 496 0610
Fax: (+1 301) 496 4686
E-mail: stratakC@ccl.nichd.nih.gov

Chromosome 3
Benjamin Carritt, MRC Human Biochemical Genetics Unit, University College of London, Wolfson House, 4 Stephenson Way, London NW1 2HE, UK
Tel.: (+44 171) 380 7415
Fax: (+44 171) 387 3496
E-mail: b.carritt@mrc-hbgu.ucl.ac.uk

Andreas Gal, Institüt für Humangenetik, MUL, Ratzenburger Allee 160, 23538 Lubeck, Germany
Tel.: (+49 451) 500 2622
Fax: (+49 451) 500 4187

Robert M. Gemmill, Eleanor Roosevelt Institute, 1899 Gaylord, Denver, CO 80206, USA
Tel.: (+1 303) 333 4515
Fax: (+1 303) 333 8423
E-mail: gemmill@loki.uchsc

Susan L. Naylor, The University of Texas Health Science Center at San Antonio, Dept. of Cellular and Structural Biology, 7703 Floyd Curl Drive, San Antonio, TX 78284-7762, USA
Tel.: (+1 210) 567 3842
Fax: (+1 210) 567 6781
E-mail: Naylor@uthscsa.edu

Chromosome 4
Michael Robert Altherr, Los Alamos, NM, USA
Tel.: (+1 505) 665 6144
Fax: (+1 505) 665 3024
E-mail: altherr@telomere.lanl.gov

Kenneth H. Buetow, Fox Chase Cancer Center, Division of Population Science, 7701 Burholme Ave., Philadelphia, PA 19111, USA
Tel.: (+1 215) 728 3152
Fax: (+1 215) 728 3574
E-mail: buetow@rudkin.rm.fcc.edu
E-mail: jekarl@morgan.popgen.fccc.edu

Jeffrey C. Murray, University of Iowa, Cooperative Human Linkage Center, 431 EMRB, Iowa City, IA 52242, USA
Tel.: (+1 319) 335 6946
Fax: (+1 319) 335 6970
E-mail: murray@uiowablue.weeg.edu

Olaf, Riess, Bochum, Germany
Tel.: (+49 234) 700 3831
Fax: (+49 234) 700 4196
olaf.riess@rz.ruhr-uni-bochum.de

Gert-Jan B. van Ommen, University of Leiden, Department of Human Genetics, Sylvius Laboratory, PO Box 9503, Wassenaarseweg 72, 2300 RA Leiden, The Netherlands
Tel.: (+31 71) 276 293
Fax: (+31 71) 276 075
E-mail: gvanomme@ruly46.Leidenuniv.nl

Chromosome 5
Michelle Le Beau, University of Chicago, Dept. of Medicine, Section of Hematology/Oncology, 5841 S. Maryland Ave., Chicago, IL 60637, USA
Tel.: (+1 312) 702 0795
Fax: (+1 312) 702 3163
E-mail:
mmlebeau@mcis.bsd.uchicago.edu

John D. McPherson, Genome Sequencing Center/Genetics, Washington University School of Medicine, 4444 Forest Park Blvd., 4th Floor, St. Louis, MO 63108, USA
Tel.: (+1 314) 286 1841
Fax: (+1 314) 286 1810
Tel.: (+1 714) 824 7447 (Lab) and 824 6792 (Lab)
E-mail: jmcphers@watson.wustl.edu

Chromosome 6
Stephan Beck, Hinxton, Cambs, UK
Tel.: (+44 1223) 834 244
Fax: (+44 1223) 494 919
E-mail: beck@sanger.ac.uk

R. Duncan Campbell, University of Oxford, MRC Immunochemistry Unit, Dept. of Biochemistry, South Parks Road, Oxford OX1 3QU, UK
Tel.: (+44 1865) 275 349
Fax: (+44 1865) 275 729
E-mail: rdcampbell@molbiol.ox.ac.uk

Howard M. Cann, Centre d'Etude du Polymorphisme Humain (CEPH), 27 Rue Juliette Dodu, 75010 Paris, France
Tel.: (+33 1) 4249 9862
Fax: (+33 1) 4018 0155
E-mail: howard@cephb.fr

Elizabeth Jazwinska, Brisbane, Queensland, Australia
Tel.: (+61 7) 3362 0179
Fax: (+61 7) 3362 0191
E-mail: lizJ@qimr.edu.au

Jiannis Ragoussis, London, UK
Tel.: (+44 171) 955 4438
Fax: (+44 171) 955 4444
E-mail: i.ragoussis@umds.ac.uk

Andreas Ziegler, Freie Universität Berlin, Institute for Experimental Oncology and Transplantation Medicine, Spandauer Damm 130, 14050 Berlin, Germany

Tel.: (+49 30) 3035 2617
Fax: (+49 30) 3035 3778
E-mail: aziegler@ukrv.de

Chromosome 7
Helen R. Donis-Keller, St. Louis, MO, USA
Tel.: (+1 314) 362 8629
Fax: (+1 314) 362 8630

Karl-Heinz Grzeschik, Med. Zentrum fur Humangenetik, Bahnhofstrasse 7, Abt. I—Allgemeine Humangenetik, 35037 Marburg, Germany
Tel.: (+49 642) 286232
Fax: (+49 6421) 288920
E-mail: grzeschi@mailer.uni-marburg.de

Lap-Chee Tsui, Hospital for Sick Children, Dept. of Genetics, 555 University Ave., Toronto, Ontario M5G 1X8, Canada
Tel.: (+1 416) 813 6015
Fax: (+1 416) 813 4931
E-mail: cfdata@sickkids.on.ca

Chromosome 8
Robin J. Leach, University of Texas Health Center at San Antonio, Dept. of Cellular and Structural Biology, 7703 Floyd Curl Drive, San Antonio, TX 78284-7762, USA
Tel.: (+1 210) 567 6947
Fax: (+1 210) 567 3803
E-mail: Leach@UTHSCSA.edu

Dan Wells, Houston, TX, USA
Tel.: (+1 713) 743 2671
Fax: (+1 713) 743 2636
E-mail: dwells@uh.edu

Stephen Wood, University of British Columbia, Department of Medical Genetics, 216 Wesbrook Building, 6174 University Blvd., Vancouver BC V6T 1Z3, Canada
Tel.: (+1 604) 822 6830
Fax: (+1 604) 822 5348
E-mail: swood@unixg.ubc.ca

Chromosome 9
Jonathan L. Haines, Massachusetts General Hospital, Neurogenetics Laboratory, Bldg. 149, 6th Floor, 13th St., Charlestown, MA 02129, USA
Tel.: (+1 617) 724 9571
Fax: (+1 617) 726 5736
E-mail: haines@helix.mgh.harvard.edu

Margaret Susan Povey, University College London, MRC Human Biochemical Genetics Unit, Wolfson House, 4 Stephenson Way, London NW1 2HE, UK
Tel.: (+44 171) 380 7410
Fax: (44 171) 387 3496
E-mail: sue@gallon.ucl.ac.uk

Brandon Wainwright, Centre for Molecular Biology and Biotechnology, University of Queensland, Brisbane QLD 4072, Australia
Tel: (+61 7) 3654542
Fax: (+61 7) 3717588
E-mail: B. Wainwright@cmcb.uq.edu

Jonathan Wolfe, University College London, Department of Genetics and Biometry, The Galton Laboratory, Wolfson House, 4 Stephenson Way, London NW1 2HE, UK
Tel.: (+44 171) 387 7050
Fax: (+44 171) 387 3496
E-mail: jwolfe@genetics.ucl.ac.uk

Chromosome 10
Jen-i Mao, Collaborative Research Division, Genome Therapeutics Corp., Genome Sequencing Center, 100 Beaver St., Waltham, MA 02154, USA
Tel.: (+1 617) 893 5007
Fax: (+1 617) 642 0310
E-mail: mao@genomecorp.com

Nicholas Moschonas, IMBB-FORTH, P.O. Box 1527, 71110 Heraklion, Crete, Greece
Tel.: (+30 81) 212 469
Fax: (+30 81) 230 469
E-mail: moschon@victor.imbb.forth.gr

Nigel K. Spurr, Harlow, Essex, UK
Tel.: (+44 1279) 622639
Fax: (+44 1279) 622 500
E-mail: Nigel_K_Spurr@sbphrd.com

Adrian R.N. Tivey, GDB Editorial Assistant, UK Human Genome Mapping Project, Resource Centre, Hinxton Hall, Hinxton, Cambs. CB10 1RQ, UK
Tel.: (+44 1223) 494528
Fax: (+44 1223) 494512
E-mail: A. Tivey@hgmp.mrc.ac.uk

Chromosome 11
Patrick Gaudray, LGMCH, CNRS URA 1462, Avenue de Valombrose, 06107 Nice, Cedex 2, France
Tel.: (+33) 93 37 77 95
Fax: (+33) 93 53 30 71
E-mail: gaudray@hermes.unice.fr

Daniela S. Gerhard, Washington University School of Medicine, Dept. of Genetics 4566 Scott Ave., Box 8232, St. Louis, MO 63110, USA
Tel.: (+1 314) 362 2736
Fax: (+1 314) 362 7855
E-mail: gerhard@sequencer.wustl.edu

Charles W. Richard, WPIC, 3811 O'Hara St., Room 1445, Pittsburgh, PA 15253, USA
Tel.: (+1 412) 624 1730
Fax: (+1 412) 624 1754
E-mail: richard+@pitt.edu

Veronica van Heyningen, Medical Research Council Human Genetics Unit, Western General Hospital, Edinburgh, Scotland EH4 2XU, UK
Tel.: (+44 131) 467 8405
Fax: (+44 131) 343 2620
E-mail: vervan@hgu.mrc.ac.uk

Chromosome 12
Ian W. Craig, University of Oxford, Genetics Laboratory, Department of Genetics, South Parks Road, Oxford OX1 3QU, UK
Tel.: (+44 1865) 275 327
Fax: (+44 1865) 275 318
E-mail: craig@bioch.ox.ac.uk
E-mail: icraig@hgmp.mrc.ac.uk

Raju S. Kucherlapati, Dept. of Molecular Genetics, Albert Einstein College of Medicine, 1300 Morris Park Ave., Bronx, NY 10461, USA
Tel.: (+1 718) 430 2069
Fax: (+1 718) 430 8778
E-mail: kucherla@aecom.yu.edu

Peter Marynen, Center for Human Genetics, University of Leuven, Campus Gasthuisberg, Herestraat 49, 3000 Leuven, Belgium
Tel.: (+32 16) 34 5891
Fax: (+32 16) 34 5997
E-mail:
Peter.Marynen@med.kuleuven.ac.be

Chromosome 13
Sarah Shaw, La Jolla, CA, USA
Tel.: (+1 619) 646 8281
Fax: (+1 619) 452 6653
E-mail: sarah@sequana.com

Dorothy Warburton, Columbia-Presbyterian Medical Center, Babies Hospital, Room BHS-B7, 3959 Broadway, New York, NY 10032, USA
Tel.: (+1 212) 305 7143
Fax: (+1 212) 305 7436
E-mail: cuh@cuccfa.ccc.columbia.edu

Chromosome 14
Diane W. Cox, Edmonton, Alberta, Canada
Tel.: (+1 403) 492 0874
Fax: (+1 403) 492 1998
E-mail: diane.cox@ualberta.ca

Torbjoern Nygaard, Columbia University, Dept. of Neurology, DB 3-330, 650 W. 168th St., New York, NY 10032, USA
Tel.: (+1 212) 305 1553
E-mail: tgn1@columbia.edu

Chromosome 15
Timothy A. Donlon, Chief, Molecular and Clinical Cytogenetics, Kapiolani Medical Center, Suite 400, 1946 Young St., Honolulu, HI 96826, USA
Tel.: (+1 808) 973 8349
Fax: (+1 808) 973 8053
E-mail:
Donlon@uhunix.uhcc.hawaii.edu

Susan Malcolm, London, UK
Tel.: (+44 171) 242 9789
Fax: (+44 171) 404 6191
E-mail: smalcolm@hgmp.mrc.ac.uk

Cynthia C. Morton, Brigham and Women's Hospital, Dept. of Pathology, 75 Francis St., Boston, MA 02115, USA
Tel.: (+1 617) 732 7980
Fax: (+1 617) 732 6996
E-mail:
CCMORTON@BICS.BWH.HARVARD.EDU

Chromosome 16
Anne-Marie Cleton-Jansen, Leiden, The Netherlands
Tel.: (+31 71) 526 6625
Fax: (+31 71) 524 8158
E-mail:
clet@pathology.medfac.leidenuniv.nl

David Frederick Callen, Head, Cytogenetics Unit, The Adelaide Children's Hospital, Dept. of Cytogenetics and Molecular Genetics, 72 King William Road, North Adelaide, SA 5006, Australia
Tel.: (+61 8) 204 6715
Fax: (+61 8) 204 7342
E-mail:
dcallen@dcallen.mad.adelaide.edu.au

Norman A. Doggett, Los Alamos National Laboratory, Life Sciences Division and Center for Human Genome Studies, Mail Stop: M888, Los Alamos, NM 87545, USA
Tel.: (+1 505) 665 4007
Fax: (+1 505) 667 7105
E-mail: doggett@lanl.gov

Chromosome 17
Doron Lancet, Weizmann Institute of Science, Dept. of Membrane Research and Biophysics, 76100 Rehovot, Israel
Tel.: (+972 8) 344 112
Fax: (+972 8) 343 683
E-mail:
bmlancet@weizmann.weizmann.ac.il

Jaime Prilusky, Bioinformatics Unit, Israel National Node—INN, Weizmann Institute of Science, PO Box 26, 76100 Rehovot, Israel
Tel.: (+972 8) 343 456
Fax: (+972 8) 344 113
E-mail:
lsprilus@weizmann.weizmann.ac.il

Ellen Solomon, Imperial Cancer Research Fund, Somatic Cell Genetics, PO Box 123, 44 Lincoln's Inn Fields, London WC2A 3PX, UK
Tel.: (+44 171) 269 3332
Fax: (+44 171) 269 3469
E-mail: e_solomon@icrf.ac.uk

Chromosome 18
Joan Overhauser, Thomas Jefferson University, Dept. of Biochemistry and Molecular Biology, Thomas Jefferson University, 233 South 10th St., Philadelphia, PA 19107, USA
Tel.: (+1 215) 955 5188
Fax: (+1 215) 923 9162
E-mail: J_Overhauser@lac.jci.tju.edu

Gary A. Silverman, Harvard Medical School of Pediatrics, Joint Program in Neonatology, 300 Longwood Ave., Enders-970, Boston, MA 02115, USA
Tel.: (+1 617) 355 6416
Fax: (+1 617) 355 7677
E-mail: silverman_g@al.tch.harvard.edu

Ad H.M. Geurts van Kessel, Catholic University of Nijmegen, Dept. of Human Genetics, Geert Grooteplein Zuid 20, 6500 HB Nijmegen, The Netherlands
Tel.: (+31 24) 361 4105
Fax: (+31 24) 361 4107
E-mail: A.GeurtsVankessel@antrg.azn.nl

Chromosome 19
Harvey Mohrenweiser, Lawrence Livermore National Laboratory, Biology and Biotechnology Research Program, 7000 East Ave., Livermore, CA 94550, USA
Tel.: (+1 510) 423 0534
Fax: (+1 510) 422 2282
E-mail: harvey@cea.llnl.gov

Anne Olsen, Livermore, CA, USA
Tel: (+1 510) 423 4927
Fax: (+1 510) 422 2282
E-mail: olsen2@llnl.gov

Chromosome 20
Ingo Hansmann, Halle, Germany
Tel.: (+49 345) 557 4291
Fax: (+49 345) 557 4293

Tim P. Keith, Collaborative Research, Inc., Dept. of Human Genetics and Molecular Biology, 1365 Main St., Waltham, MA 02154, USA
Tel.: (+1 617) 893 5007
Fax: (+1 617) 891 5062
E-mail: tim.keith@genomecorp.com

Chromosome 21
Stylianos E. Antonarakis, University of Geneva School of Medicine, Medical Genetics CMU-9, 9 Avenue de Champel, 1211 Geneva 4, Switzerland
Tel.: (+41 22) 702 5707
Fax: (+41 22) 702 5706
E-mail: sea@medsun.unige.ch

Jean Delabar, Paris, France
Tel.: (+33 140) 61 56 95
Fax: (+33 140) 61 56 90
E-mail: delabar@necker.fr

Kathleen Gardiner, Denver, CO, USA
Tel.: (+1 303) 333 4515
Fax: (+1 303) 333 8423
E-mail: gardiner@eri.uchsc.edu

Julie Ruth Korenberg, Los Angeles, CA, USA
Tel.: (+1 310) 855 7627
Fax: (+1 310) 652 8010
E-mail: jkorenberg@mailgate.csmc.edu

David Patterson, Eleanor Roosevelt Institute, 1899 Gaylord St., Denver, CO 80206-1210, USA
Tel.: (+1 303) 333 4515
Fax: (+1 303) 333 8423
E-mail: Davepatt@eri.uchsc.edu

Roger H. Reeves, Baltimore, MD, USA
Tel.: (+1 410) 955 6621
Fax: (+1 410) 955 0461
E-mail: rreeves@welchlink.welch.jhu.edu

Nobuyoshi Shimizu, Keio University School of Medicine, Dept. of Molecular Biology, 35 Shinanomachi, Shinjuku-ku, Tokyo, 160, Japan
Tel.: (+81 3) 3353 2370
Fax: (+81 3) 3351 2370
E-mail: shimizu@dmb.med.keio.ac.jp

Bruno Urbero, Service de Bioinformatique, UMS 825 CNRS-SC 13 INSERM, 7 rue Guy Moquet-BP 8, 94801 Villejuif, Cedex, France
Tel.: (+33 1) 4559 5252
Fax: (+33 1) 4559 5250
E-mail: bruno@infobiogen.fr

Christine Van Broeckhoven, Neurogenetics Lab, Born Bunje Foundation, University of Antwerp (UIA), Building T, room 5.35, Universiteits Plein 1, 2610 Antwerp, Belgium
Tel.: (+32 3) 820 2601
Fax: (+32 3) 820 2541
E-mail: cvbroeck@reks.uia.ac.be
E-mail: neurogen@reks.uia.ac.be

Chromosome 22

Kenneth H. Buetow, Fox Chase Cancer Center, Division of Population Science, 7701 Burholme Ave., Philadelphia, PA 19111, USA
Tel.: (+1 215) 728 3152
Fax: (+1 215) 728 3574
E-mail: buetow@rudkin.rm.fccc.edu

Jan Dumanski, Dept. of Clinical Genetics, Karolinska Hospital L-6, 104 01 Stockholm, Sweden
Tel.: (+46 8) 729 3922
Fax: (+46 8) 327 734
E-mail: Jan.Dumanski@molmed.ki.se

Beverly S. Emanuel, The Children's Hospital of Philadelphia, Dept. of Human Genetics and Molecular Biology, 10th Floor, Abramson Center, 34th St. and Civic Center Blvd., Philadelphia, PA 19104, USA
Tel.: (+1 215) 590 3856
Fax: (+1 215) 590 3764
E-mail: beverly@mail.med.upenn.edu

Chromosome X

Andrea Ballabio, Telethon Institute of Genetics and Medicine, Via Olgettina 58, 20132 Milano, Italy
Tel.: (+39 2) 21560 206
Fax: (+39 2) 21560 220
E-mail: ballabio@tigem.it

Anthony P. Monaco, Wellcome Trust Centre for Human Genetics, Windmill Road, Headington, Oxford OX3 7BN, UK
Tel.: (+44 1865) 740 019
Fax: (+44 1865) 742 186
E-mail: anthony.monaco@well.ox.ac.uk

David L. Nelson, Baylor College of Medicine, Institute for Molecular Genetics, One Baylor Plaza 902 E, Houston, TX 77030, USA
Tel.: (+1 713) 798 3122
Fax: (+1 713) 798 8854
E-mail: nelson@bcm.tmc.edu

Bruno Urbero, Service de Bioinformatique, UMS 825 CNRS—SC 13 INSERM, 7 rue Guy Moquet—BP 8, 94801 Villejuif, Cedex, France
Tel.: (+33 1) 4559 5252
Fax: (+33 1) 4559 5250
E-mail: bruno@infobiogen.fr.

Chromosome Y

Nabeel Affara, Cambridge University Dept. of Pathology, Tennis Court Road, Cambridge CB2 1QB, UK
Tel.: (+44 1223) 333 700
Fax: (+44 1223) 333 346
E-mail: na@mole.bio.cam.ac.uk

Michele Ramsay, South African Institute for Medical Research, Dept. of Human Genetics, PO Box 1038, Johannesburg 2000, Republic of South Africa
Tel.: (+27 11) 489 9214
Fax: (+27 11) 489 9226
E-mail: 058mrams@chiron.wits.ac.za

Mitochondrial DNA

Marie T. Lott, Emory University School of Medicine, 1462 Clifton Road, Room 403C, Atlanta, GA 30322, USA
Tel.: (+1 404) 727 3337
Fax: (+1 404) 727 3949
E-mail: mtlott@gmm.gen.emory.edu

Douglas C. Wallace, Chairman, Emory University Medical School, Genetics and Molecular Medicine, 1462 Clifton Road, Room 446, Atlanta, GA 30322, USA
Tel.: (+1 404) 727 5624
Fax: (+1 404) 727 3949
E-mail: dwallace@gmm.gen.emory.edu

Nomenclature Committee

Claude Boucheix, Hospital Paul Brousse, INSERM U-268, Avenue Paul Vaillant Couturier, 94800 Villejuif, France
Tel.: (+33 49) 581 068
Fax: (+33 49) 581 085
E-mail: boucheix@genome.vjf.inserm.fr
E-mail: jasmin@arthur.citi2.fr

Phyllis J. McAlpine, University of Manitoba, Dept. of Human Genetics, 250 Old Basic Sciences Building, 770 Bannatyne Ave., Winnipeg, Manitoba R3E 0W3, Canada
Tel.: (+1 204) 789 3393
Fax: (+1 204) 786 8712
E-mail:
mcal@genmap.hgen.umanitoba.ca

Joseph Nahmias, London, UK
Tel.: (+44 171) 380 7777
Fax: (+44 171) 387 3496
E-mail: j.nahmias@galton.ucl.ac.uk

Margaret Susan Povey, University College London, MRC Human Biochemical Genetics Unit, Wolfson House, 4 Stephenson Way, London NW1 2HE, UK
Tel.: (+44 171) 380 7410
Fax: (+44 171) 387 3496
E-mail: sue@galton.ucl.ac.uk

Thomas B. Shows, Roswell Park Cancer Institute, Dept. of Human Genetics, Elm and Carlton Streets, Buffalo, NY 14263, USA
Tel.: (+1 716) 845 3108
Fax: (+1 716) 845 8449
E-mail: tbs@shows.med.buffalo.edu

Hester M. Wain, London, UK
Tel.: (+44 171) 387 3496
Fax: (+44 171) 387 5096
E-mail: h.wain@galton.ucl.ac.uk

Julia A. White, University College of London, MRC Human Biochemical Genetics Unit, Wolfson House, 4 Stephenson Way, London NW1 2HE, UK
Tel.: (+44 171) 387 7050
Fax: (+44 171) 387 3496
E-mail: nome@galton.ucl.ac.uk

V.3 Scientific journals, bulletin boards and other information

The American Journal of Human Genetics
Editor: Peter H. Byers
The American Journal of Human Genetics, Dept. of Pathology, Box 357470, University of Washington, Seattle, WA 98195-7470, USA
Tel.: (+1 206) 685 9683
E-mail: ajhd@u.washington.edu

Annals of Human Genetics
Editor (UK): David Hopkinson
Department of Human Genetics, University College London, Wolfson House, 4 Stephenson Way, London NW1 2HE, UK

ARABIDOPSIS electronic newsgroup
Information on all BIOSCI news groups and means of receiving messages can be obtained by anonymous ftp to net.bio. net in the folder pub/BIOSCI/ doc or by sending the message 'help' to biosci@daresbury.ac.uk (Europe, Africa and Central Asia) or biosci@net.bio.net (Americas and the Pacific rim). Do not send subscription messages to the list address.

BIOSCI/BIONET-electronic news
anonymous ftp to: net.bio.net
gopher http://gopher.bio.net/

BIOSUPPLYNET (online directory of 15 000 products and 1400 suppliers)
WWW: http://www.biosupplynet.com

Cell
Editor: Ben Lewin
Cell, 1050 Massachusetts Avenue, Cambridge, MA 02138, USA
WWW: http://www.cell.com
The tables of contents and abstracts for *Cell*, *Immunity*, and *Neuron*.

Cell biology laboratory manual
http://www.gac.edu/cgi-bin/user/~cellab/phpl?index-1.html

Current Opinion in Genetics & Development
Editors: Ron Laskey and Matthew P. Scott

34–42 Cleveland Street, London W1P 6LB, UK
Tel.: (+44 171) 580 8377
Fax: (+44 171) 580 8428

Cytogenetics Cell Genetics
Editor: Harold Klinger
Department of Medical Genetics, Albert Einstein College of Medicine, 1300 Morris Park Avenue, Bronx, New York NY 10641-1602, USA

Genetic Linkage Bulletin Board
gopher http://gopher.bio.net/11/GENETIC-LINKAGE

Genomics
Editor: Victor McKusick
Editorial Office, 525 B St., Suite 1900, San Diego, CA 92101-4495, USA
Tel.: (+1 619) 699 6469
Fax: (+1 619) 699 6859

Human Molecular Genetics
Editors: Kay Davies and Willard Hunt
Dept. of Genetics, BRD 731, Case Western Reserve University, 2109 Adelbert Rd., Cleveland, OH 44106-4955, USA
Tel.: (+1 216) 368 0199
Fax: (+1 216) 368 3030
E-mail: HMGJournal@po.CWRU.edu

Hum-Molgen (news in Bioscience and Medicine)
WWW: http://www.informatik.uni-rostock.de/HUM-MOLGEN/NewsGen/

Nature Genetics
Editor: Kevin Davies
1234, National Press Building, Washington, DC 20045, USA
Tel.: (+1 202) 628 2513
Fax: (+1 202) 628 1609
E-mail: natgen@naturedc.com

Nature
Editor: Phil Campbell
UK
Porter's South, 4 Crinan Street, London N1 9XW, UK
Tel.: (+44 171) 833 4000
Fax: (+44 171) 843 45696/7
E-mail: nature@nature.com
WWW: http://www.nature.com
USA
1234 National Press Building, Washington, DC 20045, USA
Tel.: (+1 202) 737 2355
Fax: (+1 202) 628 1609
E-mail: nature@naturedc.com

Science
Editor-in-Chief: Floyd E. Bloom
1200 New York Avenue, NW
Washington, DC 20005, USA
WWW: http.//www.sciencemag.org

Trends in Genetics
Elsevier Trends Journals, 68 Hills Road, Cambridge CB2 1LA, UK
Tel.: (+44 1223) 315961
Fax: (+44 1223) 464430

WWW Virtual Library: Biochemistry and Molecular Biology
WWW:
http://golgi.harvard.edu/sequences.html

WWW Virtual Library: Biosciences
WWW:
http://golgi.harvard.edu/biopages.html

Appendix VI Basic data on human and other genomes

Table VI.1 Chromosome numbers of common species.

Species	Chromosome number
Bacteria	
Escherichia coli	1
Yeast	
Saccharomyces cerevisiae	16
Schizosaccharomyces pombe	3
Nematode	
Caenorhabditis elegans	6
Insects	
Drosophila melanogaster	4
Mammals	
Pig	19
Cat	19
Rabbit	22
Human	23
Sheep	27
Goat	30
Donkey	31
Horse	32
Dog	39
Birds	
Chicken	39[a]
Duck	40[a]
Turkey	40[a]
Plants	
Arabidopsis	5
Rice	12
Wheat	21 (A, B, D genomes)

All mammalian and bird chromosome numbers are haploid.
[a] Including microchromosomes.

Table VI.2 Estimated sizes of the human chromosomes.

Chromosome	Length (Mb)
1	263
2	255
3	214
4	203
5	194
6	183
7	171
8	155
9	145
10	144
11	144
12	143
13	114
14	109
15	106
16	98
17	92
18	85
19	67
20	72
21	50
22	56
X	164
Y	59

Table VI.3 Sizes of the *Saccharomyces cerevisiae* chromosomes.

Chromosome	Size (kb)
I	250
II	835
III	360
IV	1600
V	580
VI	280
VII	1125
VIII	580
IX	450
X	780
XI	690
XII	1090 + rDNA (~2000 kb)
XIII	950
XIV	810
XV	1125
XVI	970

Table VI.4 Genome size and physical data.

Organism	Estimated genome size (bp)
Bacteria	
Escherichia coli	4.45×10^6
Bacillus megaterium	3×10^6
Haemophilus influenzae	1.2×10^6
Viruses	
SV40	5243
Adenovirus	36×10^3
Polyoma virus	5292
Bacteriophage	
Lambda	48.5×10^3
Fungi	
Saccharomyces cerevisiae	15×10^6
Schizosaccharomyces pombe	14×10^6
Invertebrates	
Caenorhabditis elegans	100×10^6
Drosophila melanogaster	165×10^6
Vertebrates	
Amphibians	
Xenopus laevis	2.9×10^9
Reptiles	$1.6–5.1 \times 10^9$
Birds	
Chicken	1.125×10^9
Mammals	
Mouse (*Mus musculus*)	3.3×10^9
Rat (*Rattus norvegicus*)	3.0×10^9
Human	3.5×10^9
Plants	
Arabidopsis thaliana	1×10^8
Nicotiana tabacum	4.8×10^9
Oryza sativa (rice)	4.3×10^8

Fig. VI.I Cytogenetic divisions of human chromosomes.

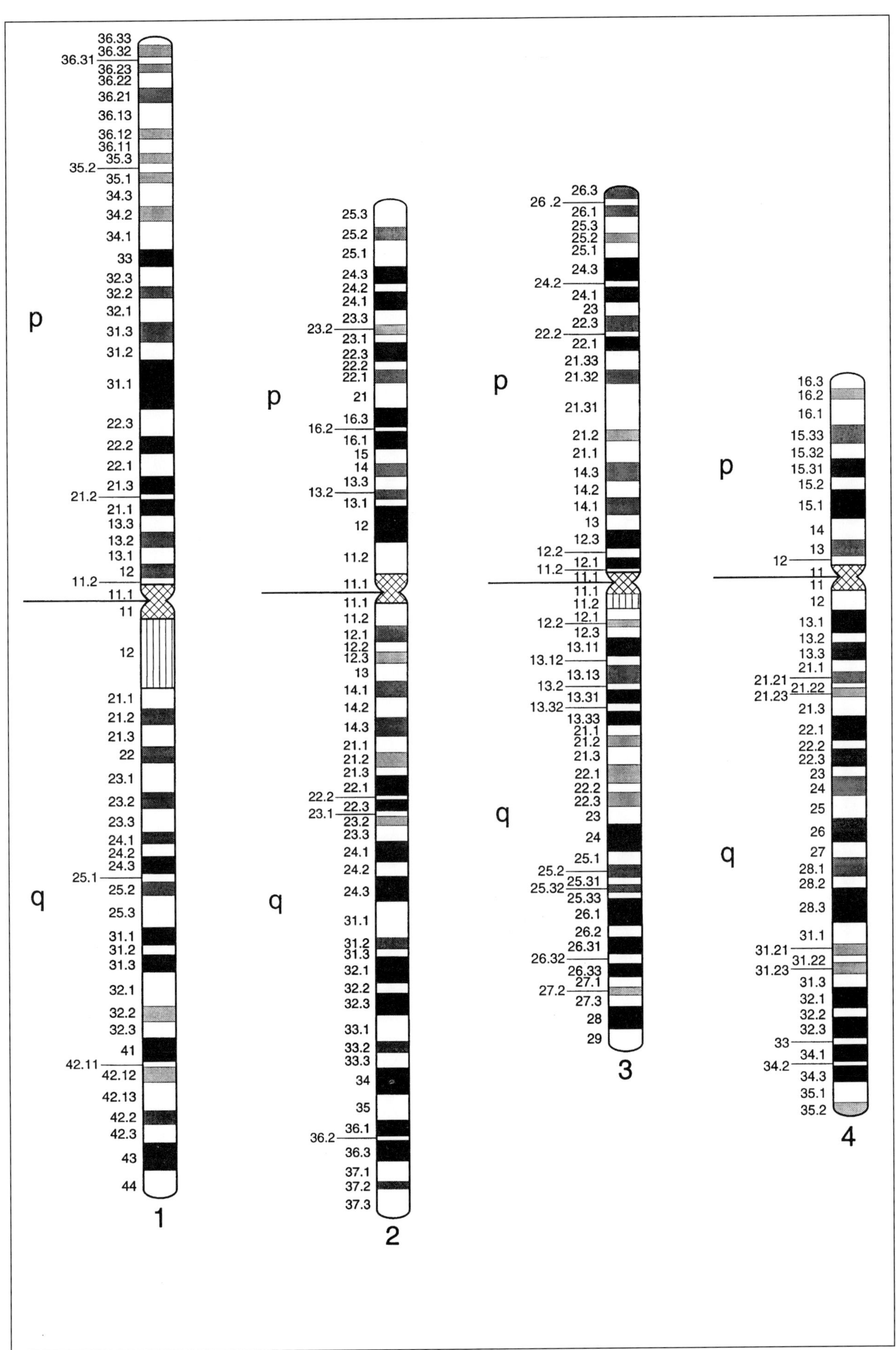

Continued on p. 890.

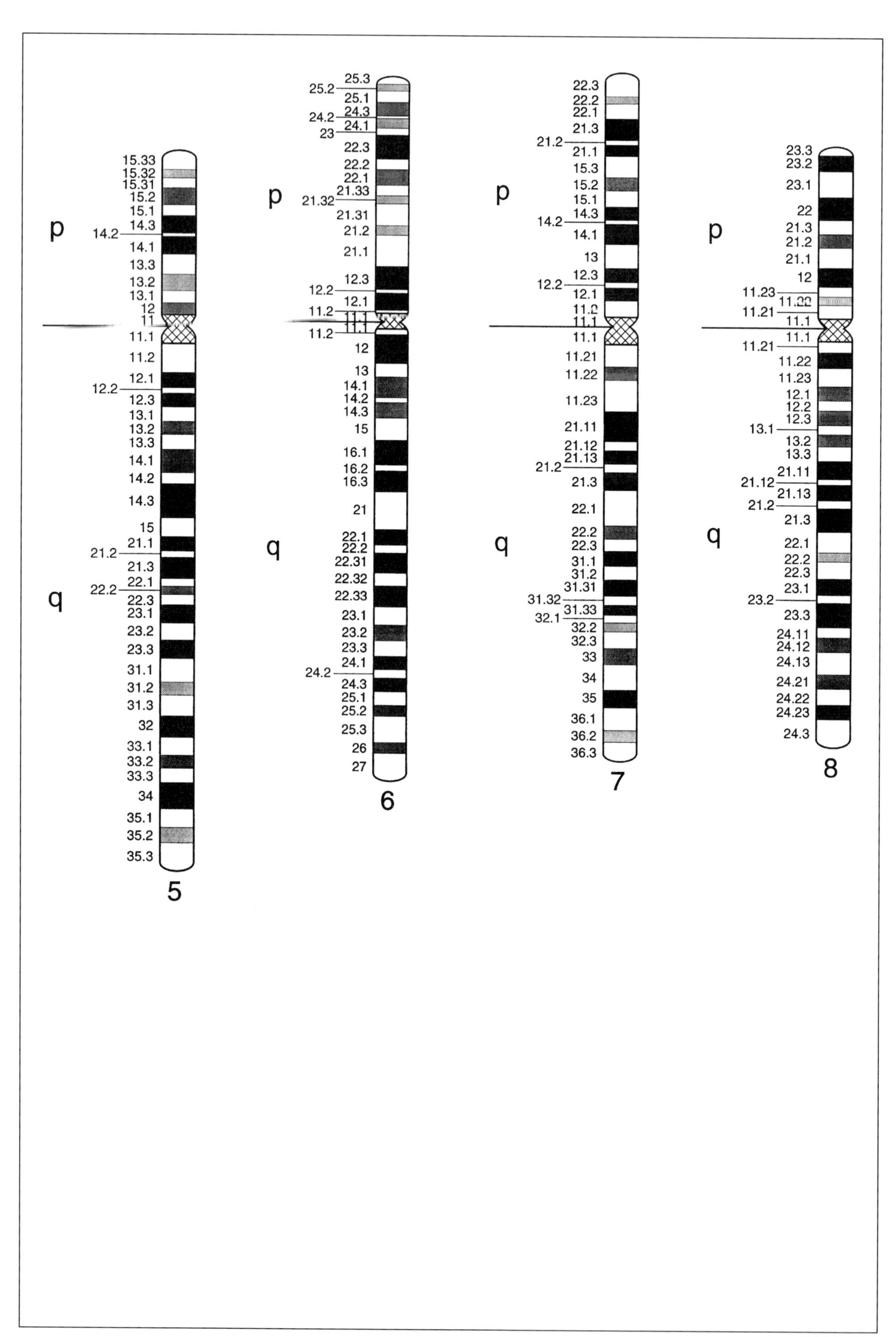
p
15.33
15.32
15.31
15.2
15.1
14.3
14.2
14.1
13.3
13.2
13.1
12
11
11.1
11.2
12.1
12.2
12.3
13.1
13.2
13.3
14.1
14.2
14.3
15
21.1
21.2
21.3
22.1
22.2
22.3
23.1
23.2
23.3
31.1
31.2
31.3
32
33.1
33.2
33.3
34
35.1
35.2
35.3
q
5
25.3
25.2
25.1
24.3
24.2
24.1
23
22.3
22.2
22.1
21.33
21.32
21.31
21.2
21.1
12.3
12.2
12.1
11.2
11.1
11.1
11.2
12
13
14.1
14.2
14.3
15
16.1
16.2
16.3
21
22.1
22.2
22.31
22.32
22.33
23.1
23.2
23.3
24.1
24.2
24.3
25.1
25.2
25.3
26
27
p
q
6
22.3
22.2
22.1
21.3
21.2
21.1
15.3
15.2
15.1
14.3
14.2
14.1
13
12.3
12.2
12.1
11.2
11.1
11.1
11.21
11.22
11.23
21.11
21.12
21.13
21.2
21.3
22.1
22.2
22.3
31.1
31.2
31.31
31.32
31.33
32.1
32.2
32.3
33
34
35
36.1
36.2
36.3
p
q
7
23.3
23.2
23.1
22
21.3
21.2
21.1
12
11.23
11.22
11.21
11.1
11.1
11.21
11.22
11.23
12.1
12.2
12.3
13.1
13.2
13.3
21.11
21.12
21.13
21.2
21.3
22.1
22.2
22.3
23.1
23.2
23.3
24.11
24.12
24.13
24.21
24.22
24.23
24.3
p
q
8

Continued.

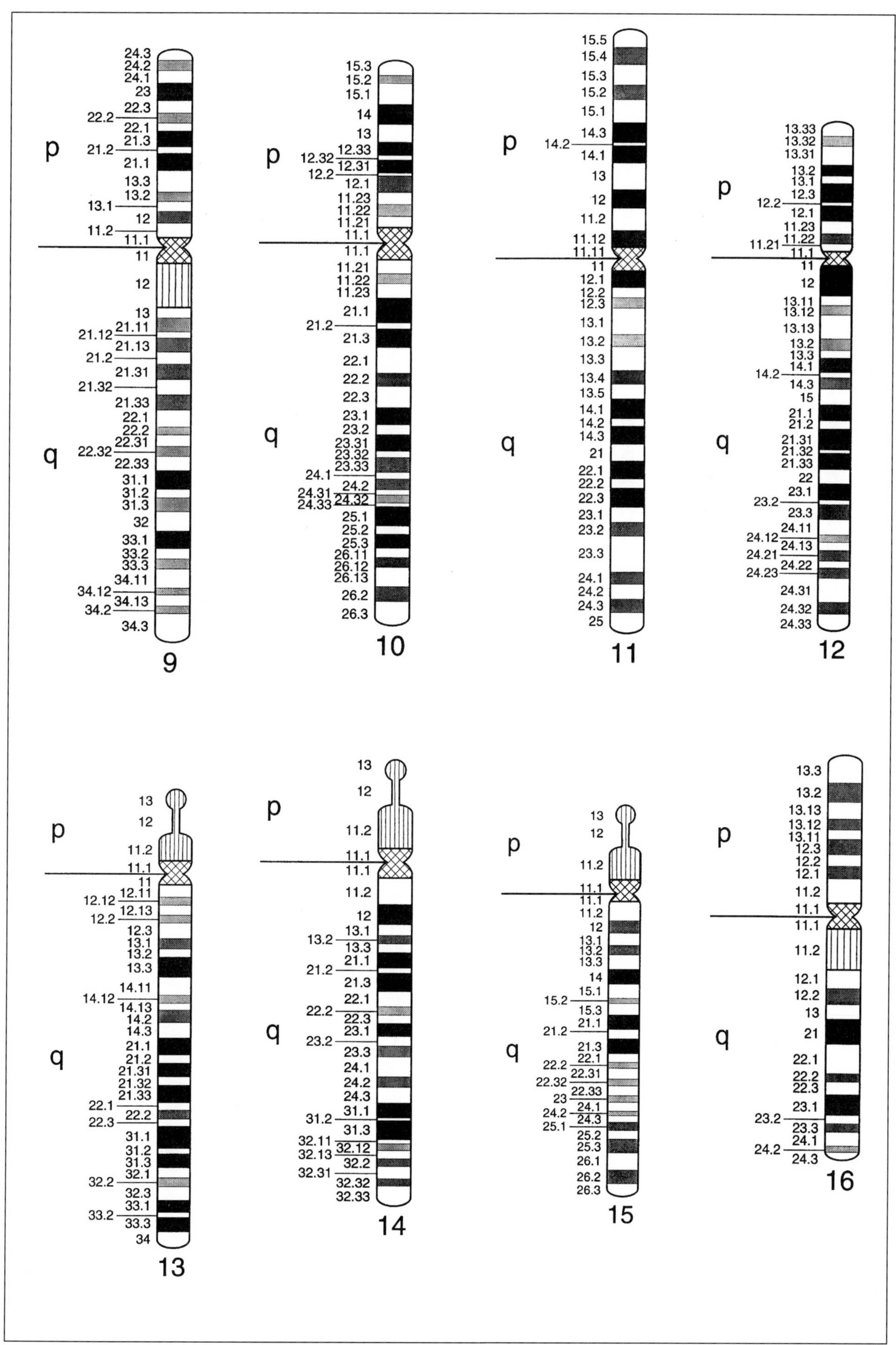

Continued on p. 892.

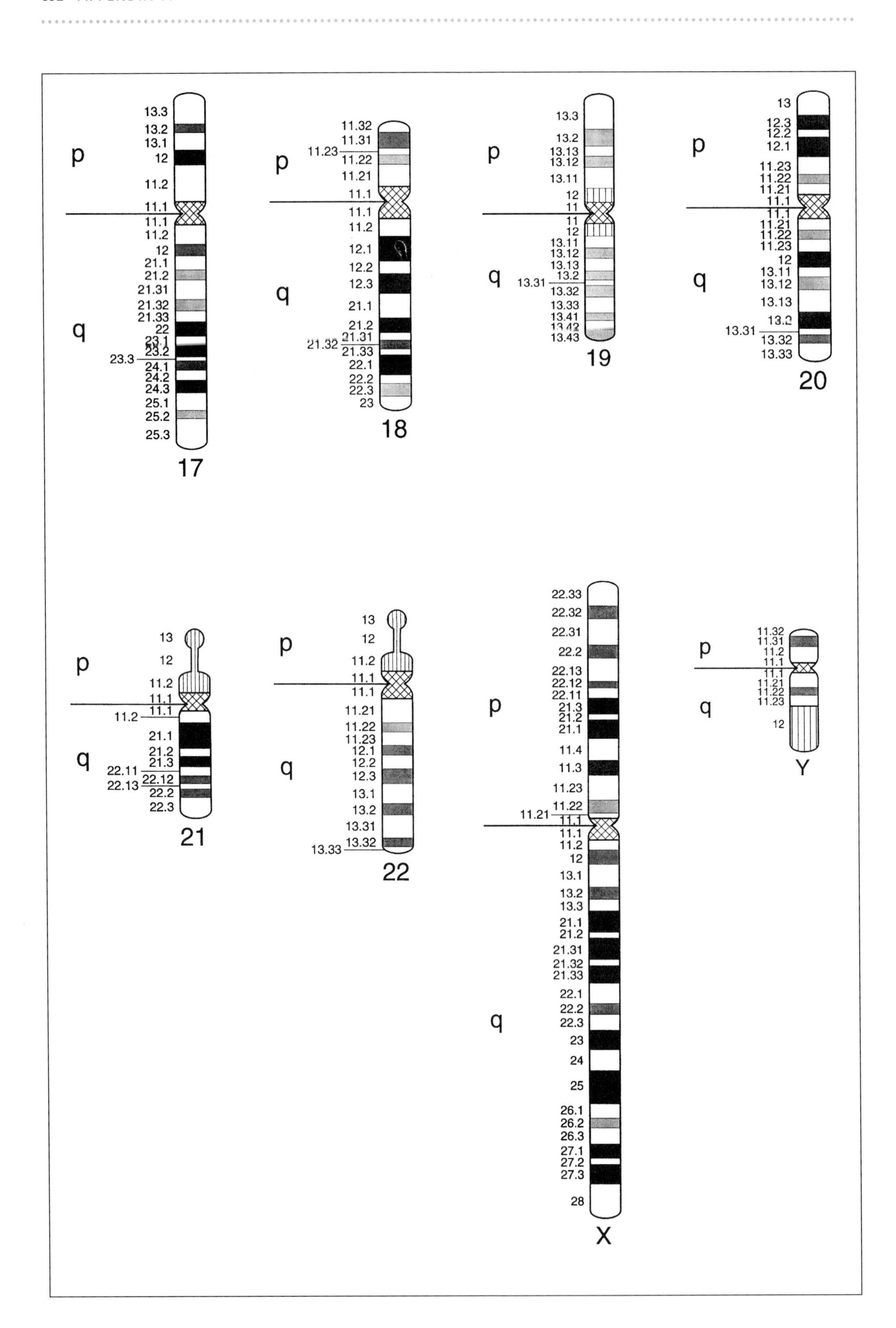
p
q
13.3
13.2
13.1
12
11.2
11.1
11.1
11.2
12
21.1
21.2
21.31
21.32
21.33
22
23.1
23.2
23.3
24.1
24.2
24.3
25.1
25.2
25.3
17
p
q
11.32
11.31
11.23
11.22
11.21
11.1
11.1
11.2
12.1
12.2
12.3
21.1
21.2
21.31
21.32
21.33
22.1
22.2
22.3
23
18
p
q
13.3
13.2
13.13
13.12
13.11
12
11
11
12
13.11
13.12
13.13
13.2
13.31
13.32
13.33
13.41
13.42
13.43
19
p
q
13
12.3
12.2
12.1
11.23
11.22
11.21
11.1
11.1
11.21
11.22
11.23
12
13.11
13.12
13.13
13.2
13.31
13.32
13.33
20
p
q
13
12
11.2
11.1
11.1
11.2
21.1
21.2
21.3
22.11
22.12
22.13
22.2
22.3
21
p
q
13
12
11.2
11.1
11.1
11.21
11.22
11.23
12.1
12.2
12.3
13.1
13.2
13.31
13.32
13.33
22
p
q
22.33
22.32
22.31
22.2
22.13
22.12
22.11
21.3
21.2
21.1
11.4
11.3
11.23
11.22
11.21
11.1
11.1
11.2
12
13.1
13.2
13.3
21.1
21.2
21.31
21.32
21.33
22.1
22.2
22.3
23
24
25
26.1
26.2
26.3
27.1
27.2
27.3
28
X
p
q
11.32
11.31
11.2
11.1
11.1
11.21
11.22
11.23
12
Y

Appendix VII Catalogue of mapped human disease genes

Status[a] C, confirmed; P, provisional; I, inconsistent (results of different laboratories disagree); L, in limbo (e.g. inferred by homology, correction of defect).

MIM#[b] Each entry is given a six-digit number whose first digit indicates the mode of inheritance of the gene involved: 1, dominant; 2, recessive; 3, X-linked; 4, Y-linked; 5, mitochondrial.

Method[c] Methods used for mapping the locus: A, *in situ* DNA–RNA or DNA–DNA hybridization; AAS, deduced from amino acid sequence of protein; C, chromosome-mediated gene transfer; Ch, chromosomal change associated with given phenotype; D, deletion or dosage mapping, trisomy mapping or gene dosage effects; EM, exclusion mapping; F, linkage studies in families; Fc, one trait is a chromosomal heteromorphism or rearrangement; Fd, one or both linked loci are identified by a DNA polymorphism; H, presumed homology; HS, solution hybridization; L, linkage to X-chromosome; LD, linkage disequilibrium; M, microcell-mediated gene transfer; OT, ovarian teratoma (centromere mapping); Pcm, PCR of microdissected chromosome segments; Psh, PCR of somatic cell hybrid DNA; R, irradiation of cells followed by rescue through fusion with nonirradiated (nonhuman) cells; RE, restriction endonuclease techniques (e.g. fine structure mapping); Rea, combined with somatic cell hybridization; Reb, combined with chromosome sorting; Rec, hybridization of cDNA to genomic fragment; Ref, isolation of gene from genomic DNA; Rel, isolation of gene from chromosome-specific genomic library; Ren, neighbour analysis in restriction fragments; S, segregation of human cellular traits and human chromosomes (or segments of chromosomes) in particular clones from interspecies somatic cell hybrids; T, telomere-associated chromosome fragmentation; V, induction of microscopically evident chromosomal change by a virus; e.g. adenovirus 12 changes on chromosomes 1 and 17; X/A, X-autosome translocation in female with X-linked recessive disorder.

Tables by kind permission of Dr V. A. Mc Kusick.

Table VII.1 The morbid anatomy of the human genome (by chromosome).

Location	Locus symbol	Status[a]	Title	MIM#[b]	Method[c]	Disorder(s)	Mouse locus
1pter-p36.13	ENO1, PPH	C	Enolase-1, α	172430	S, F, R, R, REa	Enolase deficiency (1)	4(Enol)
1pter-p33	HMGCL	P	3-hydroxy-3-methyl-glutaryl Coenzyme A lyase	246450	REa, A	HMG-CoA lyase deficiency (3)	
1p36.3	MTHFR	P	Methylenetetrahydrofolate reductase	236250 A		Homocystinuria due to MTHFR deficiency (3)	
1p36.3-p36.2	PLOD	P	Procollagen-lysine, 2-oxoglutarate 5-dioxygenase (lysine hydroxylase)	153454	REa, A	Ehlers–Danlos syndrome, type VI, 225400 (3)	4(Plod)
1p36.3-p34.1	C1QA	C	Complement component-1, q subcomponent, α-polypeptide	120550	REa, REb	?C1q deficiency (1)	
1p36.3-p34.1	C1QB	C	Complement component-1, q subcomponent, β-polypeptide	120570	REa, REb	?C1q deficiency (1)	(C1qb)

Continued on p. 894.

Table VII.1 *Continued.*

Location	Locus symbol	Status[a]	Title	MIM#[b]	Method[c]	Disorder(s)	Mouse locus
1p36.2-p36.1	NB, NBS	C	Neuroblastoma (neuroblastoma suppressor)	256700	Ch, D	Neuroblastoma (2)	
1p36.2-p34	EKV	C	Erythrokeratodermia variabilis	133200	F	Erythrokeratodermia variabilis (2)	
1p36.2-p34	EPB41, EL1	C	Erythrocyte surface protein band 4.1	130500	F, Reb	Elliptocytosis-1 (3)	4(Elp1)
1-36.2-p34	RH@	C	Rhesus blood group cluster	111700	F, D, Fd, A	Erythroblastosis fetalis (1); ?Rh-null hemolytic anemia (1)	
1p36.1-p34	ALPL, HOPS	C	Alkaline phosphatase, liver/bone/kidney	171760	S, H, Fd, F, A	Hypophosphatasia, infantile, 241500 (3): ?hypophosphatasia adult, 146300 (1)	4(Akp2)
1p36	BRCD2	P	Breast cancer, ductal	211420	Ch, F, D	Breast cancer, ductal (2)	
1p36	CMM, MLM, DNS	P	Cutaneous malignant melanoma/dysplastic naevus	155600	F, Fd, D	Malignant melanoma, cutaneous (2)	
1p36-p35	CMT2	P	Charcot–Marie–Tooth neuropathy-2 (hereditary motor sensory neuropathy II)	118210	Fd	Charcot–Marie–Tooth disease, type II (2)	
1p36-p35	GALE	C	UDP galactose-4-epimerase	230350	S, LD	Galactose epimerase deficiency (1)	
1p35-p34.3	CSF3R	C	Colony-stimulating factor-3 receptor (granulocyte)	138971	A, REb, Psh, Rea	Kostmann neutropenia, 202700 (3)	
1p34	FUCA1	C	Fucosidase, α-L- 1, tissue	230000	S, F, R, A, REa	Fucosidosis (3)	4(Fuca)
1p34	HUD, PNEM	P	HU-antigen D (a paraneoplastic encephalomyelitis antigen)	168360	A	Paraneoplastic sensory neuropathy (1)	
1p34	UROD	C	Uroporphyrinogen decarboxylase	176100	S, A, REa	Porphyria cutanea tarda (3); porphyria, hepatoerythropoietic (3)	4(Urod)
1p32	C8A	C	Complement component-8, α-polypeptide	120950	F, A, Ch, Fd	C8 deficiency, type I (2)	
1p32	C8B	C	Complement component-8, β-polypeptide	120960	F, A, Ch, H, Fd	C8 deficiency, type II (3)	4(C8b)
1p32	CLN1	C	Ceroid lipofuscinosis, neuronal-1, infantile	256730	Fd, LD, REn	Ceroid lipofuscinosis, neuronal-1, infantile (2)	
1p32	CPT2	C	Carnitine palmitoyltransferase II	255120	REa, A	Carnitine-palmitoyltransferase II deficiency (3)	
1p32	DFNA2	P	Deafness, autosomal non-syndromic sensorineural, 2	600101	Fd	Deafness, autosomal non-syndromic sensorineural, 2 (2)	
1p32	EDM2	P	Epiphyseal dysplasia, multiple 2	600204	Fd	Epiphyseal dysplasia, multiple 2 (2)	
1p32	TAL1, TCL5, SCL	C	T-cell acute lymphocytic leukaemia-1	187040	Ch, RE	Leukaemia-1, T-cell acute lymphoblastic (3)	4(Scl)
1p31	ACADM, MCAD	P	Acyl-Coenzyme A dehydrogenase, C-4 to C-12 straight chain	201450	REa, A	Acyl-CoA dehydrogenase, medium chain, deficiency of (3)	8(Acadm)
1p31	DBT, BCATE2	C	Dihydrolipoamide branched chain *trans*-acylase (E2 component of branched chain keto acid dehydrogenase complex)	248610	REa, A	Maple syrup urine disease, type II (3)	
1p22.1-qter	SDH	P	Succinate dehydrogenase	185470	S	?Myopathy due to succinate dehydrogenase deficiency (1)	
1p22	UOX	P	Urate oxidase	191540	REa, A	Urate oxidase deficiency (1)	1p22-q21
	DPYD, DPD	P	Dihydropyrimidine dehydrogenase	274270	Rea	Thymine-uraciluria (1); {fluorouracil toxicity sensitivity to} (1)	
1p22-p21	PXMP1, PMP70	P	Peroxisomal membrane protein-1 (70 kD)	170995	REa, A	Zellweger syndrome-2 (3)	3(Pmp70)
1p21	AGL, GDE	P	Amylo-1,6-glucosidase, 4-α-glucanotransferase (glycogen debranching enzyme)	232400	REc, A	Glycogen storage disease III (1)	
1p21-p12	STGD, FFM	P	Stargardt macular dystrophy	248200	Fd	Stargardt macular dystrophy (2); fundus flavimaculatus with macular dystrophy (2)	
1p21-p13	AMPD1	P	Adenosine monophosphate deaminase-1 (muscle)	102770	REa, A	Myoadenylate deaminase deficiency (3)	3(Ampd1)
1p21-p13	CSF1, MCSF	C	Colony-stimulating factor-1 (macrophage)	120420	A, REa, H	?Osteopetrosis, 259700 (1)	3(Csfm)

Continued.

Table VII.1 *Continued.*

Location	Locus symbol	Status[a]	Title	MIM#[b]	Method[c]	Disorder(s)	Mouse locus
1p13.1	HSD3B2	C	Hydroxy-δ-5-steroid dehydrogenase, 3 β- and steroid δ-isomerase, type 2 (adrenal, gonadal)	201810	A	3-β-hydroxysteroid dehydrogenase, type II, deficiency (3)	3(Hsd3B2)
1p13	TSHB	C	Thyroid-stimulating hormone, β-polypeptide	188540	REa, RE, Fd	Hypothyroidism, non-goitrous (3)	3(Tshb)
1p11-q11	CDCD	P	Cardiomyopathy, familial dilated, with conduction defect	115200	Fd	Cardiomyopathy, familial dilated, with conduction defect (2)	
1p11-qter	EPHX1	P	Epoxide hydroxylase 1, microsomal (xenobiotic)	132810	Rea	?Fetal hydantoin syndrome (1); diphenylhydantoin toxicity (1)	1(Eph1)
1p	PCHC	P	Phaeochromocytoma	171300	D	Phaeochromocytoma (2)	1cen-q32
	PFKM	P	Phosphofructokinase, muscle type	232800	S	Glycogen storage disease VII (3)	
1q	FMO4, FMO2	P	Flavin-containing monooxygenase 2 (adult liver)	136131	Psh	[Fish-odor syndrome] (1)	
1q2	CAE1	C	Cataract, zonular pulverulent-1 (FY-linked)	116200	F	Cataract, zonular pulverulent-1 (2)	
1q21	FLG	C	Filaggrin	135940	REa, A, REn	?Ichthyosis vulgaris, 146700 (1)	3(flg)
1q21	GBA	C	Glucosidase, β; acid	230800	S, A, D	Gaucher disease (3)	3(Gba)
1q21	PKLR, PK1	C	Pyruvate kinase, liver and RBC type	266200	REa, A	PK deficiency haemolytic anaemia (3)	
1q21	RCCP1	L	Renal cell carcinoma, papillary, 1	179755	Ch	?Renal cell carcinoma, papillary, 1 (2)	
1q21	SPTA1	C	Spectrin, α, erythrocytic-1	182860	REa, A, Fd	Elliptocytosis-2 (3); pyropoikilocytosis (3); spherocytosis, recessive (3)	1(Spna1)
1q21-q22	FY, GPD	C	Duffy blood group	110700	F, Fc, Fd, A	{Vivax malaria, susceptibility to} (1)	
1q21-q23	APCS, SAP	C	Amyloid P component, serum	104770	REa, A, Fd	{?Amyloidosis, secondary, susceptibility to} (1)	1(Sap)
1q21-q31	GLC1A, POAG, GPOA	C	Glaucoma 1, open angle	137760	Fd	Glaucoma, primary open angle, juvenile-onset (2)	
1q22	MPZ, CMT1B	C	Myelin protein zero	159440	REb, A, F, Fd, D	Charcot–Marie–Tooth neuropathy slow, nerve conduction type Ib, 118200 (3); Dejerine–Sottas disease, myelin P(0)-related, 145900 (3)	1(Mpp)
1q23-q25	CD3Z, TCRZ	C	CD3Z antigen, ξ-polypeptide (TiT3 complex)	186780	REa, A, REn	CD3, ξ-chain, deficiency (1)	1(T3z, Cd3z)
1q22-q23	TPM3, NEM1	C	Tropomyosin 3 (non-muscle)	191030	REa, A	Nemaline myopathy-1, 161800 (3)	1(Tpm3)
1q23	F5	C	Coagulation factor V (proaccelerin, labile factor)	227400	REa, A, Fd, Ren	Factor V deficiency (1); protein C cofactor deficiency (3)	1(Cf5)
1q23	FCGR3A, CD16, IGFR3	C	Fc fragment of IgG, low affinity III, receptor for (CD16)	146740	REb, REn	Lupus erythematosus, systemic, 152700 (1); neutropenia, immune (2)	
1q23	PBX1	C	Pre-B cell leukaemia transcription factor-1	176310	Ch, A	Leukaemia, acute pre-B-cell (2)	1(Pbx)
1q23-q25	AT3	C	Antithrombin III	107300	F, D, A, REa Fd,	Antithrombin III deficiency (3)	1(At3)
1q23-q25	SELE, ELAM1	C	Selectin E (endothelial leukocyte adhesion molecule-1)	131210	REn	{Atherosclerosis, susceptibility to} (2)	1(Elam)
1q23-q25	SELP, GRMP	C	Selectin P (granulocyte membrane protein, 140 kD; antigen CD62)	173610	REn, A	Platelet α/δ storage pool deficiency (1)	1(Grmp)
1q25	NCF2	C	Neutrophil cytosolic factor-2 (65 kD) due to deficiency of NCF-2 (1)	233710	REa, A	Chronic granulomatous disease	1(Ncf2)
1q25-q31	LAMC2, LAMNB2, LAMB2T	C	Laminin, γ2 (nicein (100 kD), (kalinin (105 kD), BM600 (100 kD))	150292	A, Fd	Epidermolysis bullosa, Herlitz junctional type, 226700 (3)	
1q3	TNNT2, CMH2	C	Troponin T2, cardiac	191045	REa, Fd	Cardiomyopathy, familial hypertrophic, 2, 115195 (3)	
1q31	EBR2A	P	Epidermolysis bullosa 2A, junctional Herlitz	226450	Fd, LD	Junctional epidermolysis bullosa inversa (2)	
1q31-q32.1	F13B	C	Coagulation factor XIII, B polypeptide	134580	Fd, A, RE	Factor XIIIB deficiency (3)	1(F13b)

Continued on p. 896.

Table VII.1 *Continued.*

Location	Locus symbol	Status[a]	Title	MIM#[b]	Method[c]	Disorder(s)	Mouse locus
1q31-q32.1	RP12	P	Retinitis pigmentosa-12 (autosomal recessive)	600105	Fd	Retinitis pigmentosa-12, autosomal recessive (2)	
1q32	CACNL1A3, CCHL1A3	C	Calcium channel, L type, α-1 polypeptide, isoform-3 (skeletal muscle)	114208	H, REa, A, Fd	Hypokalaemic periodic paralysis, 170400 (3)	1(Cchl1a3, mdg)
1q32	CR1, C3BR	C	Complement component (3b/4b) receptor-1	120620	F, REa, A, RE	CR1 deficiency (1); ?SLE (1)	
1q32	HF1, CFH	C	H factor-1 (complement)	134370	F, REa, RE, H	Factor H deficiency (1); membroproliferative glomerulonephritis (1)	1(Cfh)
1q32	MCP, CD46	C	Membrane cofactor protein (CD46, trophoblast lymphocyte cross-reactive antigen)	120920	REa, A, REn	{Susceptibility to measles} (1)	
1q32	REN	C	Renin	179820	REa, A, D, Fd, Ch	[Hyperproreninaemia] (3)	1(Ren1)
1q32	VWS, LPS, PIT	C	van der Woude syndrome (lip pit syndrome)	119300	Ch, Fd	van der Woude syndrome (2)	
1q41	RMD1	P	Rippling muscle disease 1	600332	Fd	Rippling muscle disease-1 (2)	
1q32	USH2A	C	Usher syndrome 2A (autosomal recessive, mild)	276901	Fd	Usher syndrome, type 2 (2)	
1q42	ADPRT, PPOL	C	ADP-ribosyltransferase NAD(+)	173870	REa, A	?Fanconi anaemia (1); ?Xeroderma pigmentosum (1)	
1q42-q43	AGT	C	Angiotensinogen	106150	A, Rea	{Hypertension, essential, susceptibility to} (3); {Pre-eclampsia, susceptibility to} (3)	8(Agt)
1q42.1	FH	C	Fumarate hydratase	136850	S, R, D, Psh	Fumarase deficiency (3)	
2p25.3	D2S448, MG50	P	Melanoma associated gene	600134	A	?Melanoma (1)	
2p25	POMC	C	Proopiomelanocortin (adrenocorticotropin β-lipotropin)	176830	REa	ACTH deficiency (1)	12(Pomc1)
2p24	APOB	C	Apolipoprotein B (including Ag(x) antigen)	107730	REa, A	Hypobetalipoproteinaemia (3); abetalipoproteinaemia (3); hyperbetalipoproteinaemia (3); apolipoprotein B-100, ligand-defective (3)	12(Apob)
2p24-p21	SPG4	C	Spastic paraplegia-4 (autosomal dominant)	182601	Fd	Spastic paraplegia-4 (2)	
2p23-p22	XDH	C	Xanthine dehydrogenase (xanthine oxidase)	278300	REb, A	Xanthinuria (1)	17(Xd)
2p21	HPE2, HPC	L	Holoprosencephaly-2, alobar or semilobar	157170	Ch	?Holoprosencephaly-2 (2)	
2p21	LHCGR	P	Luteinizing hormone/ chorionogonadotropin	152790	A	Precocious puberty, male, 176410 (3) receptor	
2p21	SLC3A1, ATR1, D2H, NBAT	C	Solute carrier family 3 (cystine, dibasic and neutral amino acid transporters), member 1	104614	REa, Fd, A	Cystinuria, 220100 (3)	
2p16-p15	COCA1, FCC1, MSH2	P	Colon cancer, familial, non-polyposis type 1	120435	Fd	Colon cancer, familial non-polyposis, type 1 (3)	
2p16-p13	LGMD2B	P	Limb-girdle muscular dystrophy 2B (autosomal recessive)	253601	Fd	Muscular dystrophy, limb-girdle, type 2B (2)	
2p13	TPO, TPX	C	Thyroid peroxidase	274500	REa, A, Fd	Thyroid iodine peroxidase deficiency (1); goitre, congenital (3); hypothyroidism, congenital (3)	12(Tpo)
2p12	IGKC	C	Immunoglobulin kappa constant region	147200	REa, A	[Kappa light chain deficiency] (3)	6(Igkc)
2p12-p11.2	SFTP3	C	Pulmonary surfactant-associated protein-3, 18 kD	178640	REa, A	Pulmonary alveolar proteinosis, congenital, 265120 (3)	6(Sftp3)
2q	TBS, BCG	L	Mycobacterial infections, susceptibility to	209950	H, Fd	{?Tuberculosis, susceptibility to} (2)	
2q12	ZAP70	C	Protein tyrosine kinase ZAP-70 (ξ-associated protein 70 kD)	176947	A	Selective T-cell defect (3)	
2q13	NPH1	C	Nephronophthisis-1 (juvenile)	256100	Fd	Nephronophthisis, juvenile (2)	
2q13-q14	PROC	C	Protein C (inactivator of coagulation factors Va and VIIIa)	176860	REa, A	Thrombophilia due to protein C. deficiency (3); purpura fulminans, neonatal (1)	

Continued.

Table VII.1 *Continued.*

Location	Locus symbol	Status[a]	Title	MIM#[b]	Method[c]	Disorder(s)	Mouse locus
2q14-q21	LCO	P	Liver cancer oncogene	165320	REa, REb, A	?Hepatocellular carcinoma (1)	
2q21	ERCC3, XPB	C	Excision-repair cross-complementing rodent repair deficiency, complementation group 3	133510	S, A	Xeroderma pigmentosum, group B (3)	
2q21	LCT, LAC	C	Lactase	223000	REa, Fd, A, Psh	?Lactase deficiency, congenital (1); ?lactase deficiency, adult, 223100 (1)	
2q31	COL3A1	C	Collagen, type III, α-1 polypeptide	120180	REa, A	Ehlers–Danlos syndrome, type IV, 1 (Col3a1) 130050 (3); aneurysm, familial, 100070 (3); fibromuscular dysplasia of arteries, 135580 (3); Ehlers–Danlos syndrome type III (3)	
2q31	GAD1	C	Glutamate decarboxylase-1, brain (67 kD)	266100	REa, H, A, Psh	?Pyridoxine dependency with seizures (1)	2(Gad1)
2q31	IDDM7	L	Insulin-dependent diabetes mellitus 7	600321	H	?Diabetes mellitus, insulin-dependent, 7 (2)	
2q32	WSS	P	Wrinkly skin syndrome	278250	Ch	Wrinkly skin syndrome (2)	
2q33-q34	NDUFS1	P	NADH dehydrogenase (ubiquinone), Fe-S protein-1 (75 kD)	157655	A	Lactic acidosis due to defect in iron–sulphur cluster of complex I (1)	
2q33-q35	ALS2	P	Amyotrophic lateral sclerosis-2 (juvenile)	205100	Fd	Amyotrophic lateral sclerosis, juvenile (2)	
2q33-q35	CRYGA, CRYG1	C	Crystallin, gamma A	123660	REa, A	Cataract, Coppock-like (3)	1(Cryg1)
2q33-q36	CPS1	P	Carbamoyl-phosphate synthetase 1, mitochondrial	237300	REa	Carbamoylphosphate synthetase I deficiency (3)	
2q33-qter	CYP27, CTX	P	Cytochrome P450, subfamily XXVII (sterol 27-hydroxylase)	213700	Rea	Cerebrotendinous xanthomatosis (3)	1(Cyp27)
2q34	FN1	C	Fibronectin-1	135600	S, REa, A	?Ehlers–Danlos syndrome, type X (1)	1(Fn1)
2q34	TCL4	P	T-cell leukaemia/lymphoma-4	186860	Ch, RE	Leukaemia/lymphoma, T-cell (2)	
2q34-q35	ACADL, LCAD	P	Acyl-Coenzyme A dehydrogenase, long chain	201460	A	Acyl-CoA dehydrogenase, long chain, deficiency of (3)	
2q35	DES	P	Desmin	125660	REa, A	?Cardiomyopathy (1); ? myopathy, desminopathic (1)	1(Des)
2q35	PAX3, WS1, HUP2	C	Paired box homeotic gene-3	193500	Ch, Fd, H, A, Psh	Waardenburg syndrome, type I (3); Waardenburg syndrome, type III, 148820 (3); rhabdomyosarcoma, alveolar, 268220 (3)	1(Sp)
2q35	NRAMP	P	Natural resistance-associated macrophage protein (might include Leishmaniasis)	600266	Ren	?Resistance/susceptibility to TB, etc. (1)	1(Nramp)
2q36	COL4A3	C	Collagen IV, α-3 polypeptide (Goodpasture antigen)	120070	REa, A, RE	Alport syndrome, autosomal recessive, 203780 (3)	
2q36-q37	AGXT, SPAT	P	Alanine-glyoxylate aminotransferase, liver-specific peroxisomal	259900	A, REa	Hyperoxaluria, primary, type 1 (3)	
2q36-q37	COL4A4	C	Collagen IV, α-4 polypeptide	120131	REa, A	Alport syndrome, autosomal recessive, 203780 (3)	
2q36-q37	GCG	C	Glucagon	138030	REa, A	[?Hyperproglucagonaemia] (1)	2(Gcg)
2q37	BDE	L	Brachydactyly type E	113300	Ch, D	?Brachydactyly type E (2)	
2q37	BMDR	P	Brachydactyly-mental retardation syndrome	600430	Ch	Brachydactyly-mental retardation syndrome (2)	
Chr.2	SRD5A2	P	Steroid-5-α-reductase, α-polypeptide-2 (3-oxo-5 α-steroid δ 4-dehydrogenase α-2)	264600	REa	Pseudovaginal perineoscrotal hypospadias (3)	
Ch.2	UGT1A1, GNT1	P	UDP-glucuronosyltransferase-1 family, member 1	191740	REa	Crigler–Najjar syndrome, type I, 218800 (3); ?Gilbert syndrome, 143500 (1)	1(Ugt1)
3p26-p25	VHL	C	von Hippel–Lindau syndrome	193300	Fd, D, RE	von Hippel–Lindau syndrome (3); renal cell carcinoma (3)	
3p25.3	PHS	L	Pallister–Hall syndrome	146510	Ch	?Pallister–Hall syndrome (2)	
3p25	BTD	P	Biotinidase	253260	A	Biotinidase deficiency (1)	

Continued on p. 898.

Table VII.1 *Continued.*

Location	Locus symbol	Status[a]	Title	MIM#[b]	Method[c]	Disorder(s)	Mouse locus
3p25	XPC, XPCC	C	Xeroderma pigmentosum, complementation group C	278720	REa, A, RE	Xeroderma pigmentosum, complementation group C (3)	
3p24.3	THRB, ERBA2, THR1	C	Thyroid hormone receptor, β (avian erythroblastic leukaemia viral (v-erb-a) oncogene homologue-2)	190160	REa, A, RE, Fd	Thyroid hormone resistance, 274300, 188550 (3)	
3p24-p21	SCN5A, LQT3	C	Sodium channel, voltage-gated, type V, α polypeptide	600163	Fd, A	Long QT syndrome-3 (3)	
3p23-p22	ACAA	P	Acetyl-Coenzyme A acyltransferase (peroxisomal 3-oxoacyl-Coenzyme A thiolase)	261510	REa, A	Pseudo-Zellweger syndrome (1)	
3p23-p21	SCLC1	C	Small-cell cancer of lung	182280	Ch, D	Small-cell cancer of lung (2)	
3p22-p21.1	PTHR	C	Parathyroid hormone receptor	168468	REa, A, Psh	Metaphyseal chondrodysplasia, Murk Jansen type, 156400 (3)	
3p21.33	GLB1	C	Galactosidase, β-1	230500	S, EM, A	GM1-gangliosidosis (3); mucopolysaccharidosis IVB (3)	9(Bgl)
3p21.3	COL7A1	C	Collagen VII, α-1 polypeptide	120120	REa, A	Epidermolysis bullosa dystrophica, dominant, 131750 (3); epidermolysis bullosa dystrophica, recessive, 226600 (3)	
3p21.3	MLH1, COCA2	P	mutL (*E. coli*) homologue 1	120436	Fd, A	Colorectal cancer, familial non-polyposis type 2 (3); Turcot syndrome with glioblastoma, 276300 (3)	
3p21.2-p21.1	AMT	P	Aminomethyltransferase (glycine cleavage system protein T)	238310	REa	Hyperglycinaemia, non-ketotic, type II (1)	
3p21.1-p12	SCA7, OPCA3	C	Spinocerebellar ataxia 7 (olivopontocerebellar atrophy with retinal degeneration)	164500	Fd	Cerebellar ataxia with retinal degeneration (2)	
3p14.3	TKT	P	Transketolase	277730	REa, A	{Wernicke–Korsakoff syndrome, susceptibility to} (1)	
3p14.2	RCA1, HRCA1	C	Renal carcinoma, familial, associated 1	144700	Fc, Ch	Renal cell carcinoma (2)	
3p14.1-p12.3	MITF, WS2A	C	Microphthalmia-associated transcription factor	156845	REa, A, Fd	Waardenburg syndrome, type 2A, 193510 (3)	6(mi)
3p13-p12	BBS3	P	Bardet–Biedl syndrome 3	600151	Fd	Bardet–Biedl syndrome 3 (2)	
3p12	GBE1	P	Glycogen branching enzyme	232500	REa	Glycogen storage disease IV (1)	
3p11.1-q11.2	PROS1	C	Protein S, α	176880	REa	Protein S deficiency (3)	
3p11	PIT1	C	Pituitary-specific transcription factor Pit-1	173110	Fd, A	Pituitary hormone deficiency, combined (3)	16(Pit1,dw)
3cen-q22	MER6, RHN	P	Antigen identified by monoclonal antibody 1D8 (Rh-null, regulator type)	268150	S	Rh-null disease (1)	
3q11-q12	GPX1	C	Glutathione peroxidase-1	138320	S, Rea	Haemolytic anaemia due to glutathione peroxidase deficiency (1)	
3q12	CPO	P	Coproporphyrinogen oxidase	121300	REa, A	Coproporphyria (3); harderoporphyrinuria (3)	
3q13	FIH	L	Hypoparathyroidism	164200	Fd	Hypoparathyroidism, familial (2)	
3q13	UMPS, OPRT	C	Uridine monophosphate synthetase (orotate phosphoribosyl transferase and orotidine-5 (fm-decarboxylase)	258900	S, A	oroticaciduria (1)	
3q13.3	DRD3	P	Dopamine receptor D3	126451	REb, A	{?Schizophrenia, susceptibility to} (2)	
3q2	AKU	C	Alkaptonuria	203500	Fd, H	Alkaptonuria (2)	16(aku)
3q21	TF	C	Transferrin	190000	S, H, Rea, D, A	Atransferrinaemia (1)	9(Trf)
3q21-q22	PCCB	C	Propionyl Coenzyme A carboxylase, β-polypeptide	232050	REa, A, D	Propionicacidaemia, type II or pccB type (3)	
3q21-q23	LTF	C	Lactotransferrin	150210	REa, A	?Lactoferrin-deficient neutrophils, 245480 (1)	9(Ltf)
3q21-q24	CP	C	Ceruloplasmin	117700	F, H, REa, A	[Hypoceruloplasminaemia, hereditary] (1)	9(Cp)

Continued.

Table VII.1 *Continued.*

Location	Locus symbol	Status[a]	Title	MIM#[b]	Method[c]	Disorder(s)	Mouse locus
3q21-q24	HHC1, FHH, PCAR1	P	Hypocalciuric hypercalcaemia-1 (parathyroid Ca(2+)-sensing receptor)	145980	Fd, Rea	Hypercalcaemia, hypocalciuric, familial (3); neonatal hyperparathyroidism, 239200 (3); hypocalcaemia, autosomal dominant (3)	
3q21-q24	RHO, RP4	C	Rhodopsin	180380	REa, A, Fd	Retinitis pigmentosa-4, autosomal dominant (3); retinitis pigmentosa, autosomal recessive (3); night blindness, congenital stationary, rhodopsin-related (3)	6(Rho)
3q21-q25	USH3	P	Usher syndrome-3	276902	Fd	Usher syndrome, type 3 (2)	
3q22-q23	BPES	C	Blepharophimosis, epicanthus inversus and ptosis	110100	Ch, Fd	Blepharophimosis, epicanthus inversus and ptosis (2)	
3q25-q26	SI	P	Sucrase-isomaltase	222900	REa, A, Fd	Sucrose intolerance (1)	3(Sis)
3q26	MDS1	P	Myelodysplasia syndrome-1	600049	Ch, REc, A	Myelodysplasia syndrome-1 (3)	
3q26-qter	KNG	C	Kininogen	228960	Psh, A	[Kininogen deficiency] (3)	
3q26.1-q26.2	BCHE, CHE1	C	Butyrylcholinesterase	177400	F, D, A	Apnoea, postanaesthetic (3)	
3q26.3	CDL	L	Cornelia de Lange syndrome	122470	Ch	?Cornelia de Lange syndrome (20	
3q26.3-q28	EHHADH, PBFE	P	Enoyl-Coenzyme A, hydratase/3-hydroxyacyl Coenzyme A dehydrogenase	261515	A	Peroxisomal bifunctional enzyme deficiency (1)	
3q27	BCL6	P	B-cell CLL/lymphoma-6	109565	Ch, A	Lymphoma, B-cell (2); lymphoma, diffuse large cell (3)	
3q28-q29	HRG	C	Histidine-rich glycoprotein	142640	REa, A, Fd	?Thrombophilia due to elevated HRG (1)	
3q28-qter	OPA1	P	Optic atrophy 1 (autosomal dominant)	165500	Fd	Optic atrophy 1 (2)	
Chr.3	TRH	P	Thyrotropin-releasing hormone	275120	REa	Thyrotropin-releasing hormone deficiency (1)	
4p16.3	FGFR3, ACH	C	Fibroblast growth factor receptor-3	134934	REn, Fd	Achondroplasia, 100800 (3); hypochondroplasia, 146000 (3)	5(Fgfr3)
4p16.3	HD, IT15	C	Huntingtin	143100	Fd	Huntington's disease (3)	5(Hdh)
4p16.3	IDUA, IDA	P	Iduronidase, α-L-	252800	REa, A, S	Mucopolysaccharidosis Ih (3); mucopolysaccharidosis Is (3); mucopolysaccharidosis Ih/s (3)	5(Idua)
4p16.3	PDEB, CSNB3	C	Phosphodiesterase, cyclic GMP (rod receptor) β-polypeptide	180072	REa, A, Fd	Night blindness, congenital stationary, type 3, 163500 (3)	5(Pdeb, rd)
4p16.3	WHCR	C	Wolf–Hirschhorn syndrome chromosome region	194190	Ch	Wolf–Hirschhorn syndrome (2)	
4p16.1	MSX1, HOX7	P	msh (Drosophila) homeobox homolog 1 (formerly homeobox 7)	142983	REa, A, D, Fd	?Wolf–Hirschhorn syndrome, 194190 (3)	5(Hox7)
4p16-p14	CDPR	L	Chondrodysplasia punctata, rhizomelic	215100	Ch	?Chondrodysplasia punctata, rhizomelic (2)	
4p15.31	QDPR, DHPR	C	Quinoid dihydropteridine reductase	261630	S, A, REa, D	Phenylketonuria due to dihydropteridine reductase deficiency (3)	5(Qdpr)
4p13-q12	TAPVR1	P	Total anomalous pulmonary venous return	106700	Fd	Total anomalous pulmonary venous return (2)	
4p	WFRS	P	Wolfram syndrome	222300	Fd	Wolfram syndrome (2)	
4q11-q13	AFP, HPAFP	C	α-fetoprotein	104150	H, A, Fd, F	[AFP deficiency congenital] (1); [hereditary persistence of α-fetoprotein] (3)	5(Afp)
4q11-q13	ALB	C	Albumin	103600	F, A, REa	Analbuminaemia (3); [dysalbuminaemic hyperthyroxinemia] (3); [dysalbuminaemic hyperzincaemia] (3)	5(Alb1)
4q11-q13	JPD	P	Periodontitis, juvenile	170650	F	Periodontitis, juvenile (2)	
4q12	KIT, PBT	C	Hardy–Zuckerman 4 feline sarcoma (v-kit) oncogene	164920	REa, A, H, Ch, H, Ren	Piebaldism (3); Mast cell leukaemia (3)	5(Kit; W)
4q13-q21	DGI1	C	Dentinogenesis imperfecta-1	125490	F, Fd	Dentinogenesis imperfecta-1 (2)	

Continued on p. 900.

Table VII.1 *Continued.*

Location	Locus symbol	Status[a]	Title	MIM#[b]	Method[c]	Disorder(s)	Mouse locus
4q21	IGJ	P	Immunoglobulin J polypeptide, linker protein for	147790	REa, A	?Leukaemia, acute lymphocytic, with 4/11 translocation (3)	5(Igj)
4q21-q23	GNPTA	P	UDP-*N*-acetylglucosamine-lysosomal-enzyme *N*-acetylglucosamine phosphotransferase	252500	F, S, D	Mucolipidosis II (1); mucolipidosis III (1)	
4q21-q23	PKD2, PKD4	C	Polycystic kidney disease-2 (autosomal dominant)	173910	Fd	Polycystic kidney disease, adult, type II (2)	
4q25	IF	C	I factor (complement)	217030	REa, Fd, A, RE	C3b inactivator deficiency (1)	
4q25-q27	RGS	C	Rieger syndrome	180500	Ch, Fd	Rieger syndrome (2)	
4q26-q27	IL2	C	Interleukin-2	147680	REa, A, F	Severe combined immunodeficiency due to IL2 deficiency (1)	3(Il2)
4q28	FGA	C	Fibrinogen, α-polypeptide	134820	RE, REa, H, D, LD, A	Dysfibrinogenaemia, α-types (3); amyloidosis, hereditary renal, 105200 (3)	
4q28	FGB	C	Fibrinogen, β-polypeptide	134830	RE, REa, D, LD, A	Dysfibrinogenaemia,α-types (3)	
4q28	FGG	C	Fibrinogen, γ-polypeptide	134850	F, REa, H, RE, D, LD, A	Dysfibrinogenaemia, gamma types (3); hypofibrinogenaemia, gamma types (3)	3(Fgg)
4q28-q31	ASMD	P	Anterior segment mesenchymal dysgenesis	107250	F	Anterior segment mesenchymal dysgenesis (2)	
4q28-q31	TYS	C	Sclerotylosis	181600	F	Sclerotylosis (2)	
4q31.1	MLR, MCR	C	Mineralocorticoid receptor (aldosterone receptor)	264350	REa, M, A	Pseudohypoaldosteronism (1)	
4q32-q33	AGA	C	Aspartylglucosaminidase	208400	S, F, D, A	Aspartylglucosaminuria (3)	
4q32-qter	ETFDH	P	Electron transfer flavoprotein: ubiquinone oxidoreductase	231675	REa, A	Glutaricacidaemia, type IIC (3)	
4q32.1	HVBS6	P	Hepatitis B virus integration site-6	142380	REa, A, D	Hepatocellular carcinoma (3)	
4q35	F11	C	Coagulation factor XI (plasma thromboplastin antecedent)	264900	A, H, Fd	Factor XI deficiency (3)	8(cf11)
4q35	FSHMD1A, FSHD	C	Facioscapulohumeral muscular dystrophy 1A	158900	Fd	Facioscapulohumeral muscular dystrophy 1A (2)	?8(myd)
4q35	KLK3	P	Kallikrein, plasma (Fletcher factor)	229000	A	Fletcher factor deficiency (1)	8(Kal3)
Chr.4	LAG5	P	Leukocyte antigen group 5	151450	S	Neutropenia, neonatal alloimmune (1)	
5p13	C6	C	Complement component-6	217050	A, H, RE, Fd, LD	C6 deficiency (1); Combined C6/C7 deficiency (1)	15(C6)
5p13	C7	C	Complement component-7	217070	A, H, RE, Fd, LD	C7 deficiency (1)	15(C7)
5p13	C9	C	Complement component-9	120940	REa, A, Fd, LD	C9 deficiency (1)	
5p13-p12	GHR	C	Growth hormone receptor	262500	REa, A	Laron dwarfism (3)	15(Ghr)
5p13-q12	BBBG	L	Hypospadias-dysphagia syndrome (Opitz BBBG syndrome)	145410	Ch	?Hypospadias-dysphagia syndrome (2)	
5q11-q13	ARSB	C	Arylsulphatase B	253200	S	Maroteaux–Lamy syndrome, several forms (3)	13(As1)
5q11.2	KFS	L	Klippel–Feil syndrome	214300	Ch	?Klippel–Feil syndrome (2)	
5q11.2-q13.2	DHFR	C	Dihydrofolate reductase	126060	S, REa, H, D	?Anaemia, megaloblastic, due to DHFR deficiency (1)	13(Dhfr)
5q11.2-q13.3	SCZD1	L	Schizophrenia disorder-1	181510	Ch, Fd	?Schizophrenia (2)	
5q12-q32	MAR	P	Macrocytic anaemia, refractory	153550	Ch	Macrocytic anaemia of 5q-syndrome, refractory (2)	
5q12.2-q13.3	SMA	C	Spinal muscular atrophy	253300	Fd	Werdnig–Hoffmann disease (2); spinal muscular atrophy II (2); spinal muscular atrophy III (2)	
5q13	HEXB	C	Hexosaminidase B (β-polypeptide)	268800	S, Ch, D	Sandhoff disease (3)	13(Hex2)
5q13.3	RASA, GAP	C	RAS p21 protein activator (GTPase activating protein)	139150	REa, A	Basal cell carcinoma (3)	13(Gap)
5q21	MCC	C	Mutated in colorectal cancers	159350	REn, D	Colorectal cancer (3)	18(Mcc)
5q21-q22	APC, GS, FPC	C	Adenomatous polyposis coli	175100	D, Fd, REn	Gardner syndrome (3); polyposis coli, familial (3); colorectal cancer (3)	18(Min, Apc)
5q22-q33.3	CDGG1	P	Corneal dystrophy, Groenouw type I	121900	Fd	Corneal dystrophy, Groenouw type I (2); corneal dystrophy, lattice type I, 122200(2); corneal dystrophy, combined granular/lattice type (2)	

Continued.

Table VII.1 *Continued.*

Location	Locus symbol	Status[a]	Title	MIM#[b]	Method[c]	Disorder(s)	Mouse locus
5q22.3-q31.3	LGMD1	P	Limb-girdle muscular dystrophy, autosomal dominant	159000	Fd	Muscular dystrophy, limb-girdle, autosomal dominant (2)	
5q23	DTS, HBEGF	C	Diphtheria toxin sensitivity (heparin-binding EGF-like growth factor)	126150	S, M	{Diphtheria, susceptibility to} (1)	
5q23-q31	FBN2, CCA	P	Fibrillin-2	121050	Fd, A	Contractural arachnodactyly, congenital (3)	18(Fbn2)
5q23-q31	ITGA2, CD49B, BR	P	Integrin, α-2 (CD49B; α-2 subunit of VLA-2 receptor; platelet antigen Br)	192974	S, Psh, A	Neonatal alloimmunethrombocytopenia (2); ?glycoprotein Ia deficiency (2)	
5q31	GRL	C	Glucocorticoid receptor, lymphocyte	138040	S, REa, Fd, H, A, D, REn	Cortisol resistance (3)	18(Grl1)
5q31-q33	DFNA1, LFHL1	P	Deafness, autosomal dominant-1	124900	Fd	Deafness, low-tone (2)	
5q31-q34	DTD	P	Diastrophic dysplasia	222600	Fd	Diastrophic dysplasia (3)	
5q31.1	IRF1	C	Interferon regulatory factor-1	147575	Fd, REa, A, Ren, D	Macrocytic anaemia refractory, of 5q-syndrome, 153550 (3); myelodysplastic syndrome, preleukaemic (3); myelogenous leukaemia, acute (3)	11(Irf1)
5q31.3-q33.1	GM2A	C	GM2 ganglioside activator protein	272750	S, REa, Psh, A	GM2-gangliosidosis, AB variant (3)	
5q32	GLRA1, STHE	C	Glycine receptor, alpha 1	138491	Fd, R, A	Startle disease/hyperefplexia, autosomal dominant, 149400 (3); startle disease, autosomal recessive (3)	11(spd)
5q32-q33.1	TCOF1, MFD1	C	Treacher Collins–Franceschetti syndrome-1	154500	Ch, Fd	Treacher Collins mandibulofacial dysostosis (2)	
5q33-qter	F12, HAF	C	Coagulation factor XII (Hageman factor)	234000	REa, A	Factor XII deficiency (3)	
5q34-q35	MSX2, CRS2, HOX8	C	msh (Drosphila) homeobox homologue 2	123101	Fd, REa, A	Craniosynostosis, type 2 (3)	
6pter-p22	SCZD3	P	Schizophrenia disorder 3	600511	Fd	Schizophrenia-3 (2)	
6p24.3	OFC1, CL	C	Orofacial cleft-1 (cleft lip with or without cleft palate; isolated cleft palate)	119530	Fd, Ch	Orofacial cleft (2)	
6p25-p24	F13A1, F13A	C	Coagulation factor XIII, A polypeptide	134570	F, Fd, A, D	Factor XIIIA deficiency (3)	
6p23	D6S231E, DEK	P	DEK gene	125264	Ch	Leukaemia, acute non-lymphocytic (2)	
6p23	SCA1	C	Spinocerebellar ataxia 1 (olivopontocerebellar ataxia 1, autosomal dominant)	164400	F, Fd, A	Spinocerebellar ataxia-1 (3)	
6p22-p21.3	STL2	P	Stickler syndrome, type 2	184840	Fd	Stickler syndrome, type 2 (2)	
6p22-p21	BCKDHB, E1B	C	Branched chain keto acid dehydrogenase E1, β-polypeptide	248611	REa, A	Maple syrup urine disease, type 3 (3)	
6p21.3	AS, ANS	P	Ankylosing spondylitis	106300	F, Fd	Ankylosing spondylitis (2)	
6p21.3	ASD2	P	Atrial septal defect, secundum type	108800	F	Atrial septal defect, secundum type (2)	
6p21.3	C2	C	Complement component-2	217000	F, LD, RE	C2 deficiency (3)	17(C2)
6p21.3	C4A, C4S	C	Complement component-4A	120810	F, H, RE, Fd	C4 deficiency (3)	17(C4)
6p21.3	C4B, C4F	C	Complement component-4B	120820	F, H, RE, Fd	C4 deficiency (3)	17(C4)
6p21.3	COL11A2	C	Collagen XI, α-2 polypeptide	120290	REa, A, REn, Fd	Stickler syndrome, type II, 184840 (3); OSMED syndrome, 215150 (3)	17 (Col11a2)
6p21.3	CYP21, CA21H	C	Cytochrome P450, subfamily XXI; steroid 21-hydroxylase	201910	F, RE	Adrenal hyperplasia congenital, due to 21-hydroxylase deficiency (3)	17(Cyp21)
6p21.3	DYLX2, DLX2	P	Dyslexia, specific, 2	600202	Fd	Dyslexia, specific, 2 (2)	
6p21.3	EJM1, JME	P	Epilepsy, juvenile myoclonic-1	254770	F, Fd	Epilepsy, juvenile myoclonic (2)	
6p21.3	GLYS1	P	Renal glucosuria-1	233100	F	[Renal glucosuria] (2)	
6p21.3	HFE	C	Haemochromatosis	235200	LD, F	Haemochromatosis (2)	
6p21.3	HLA-DPB1	C	Major histocompatibility complex, class II, DP β-1	142858	F, RE	{Beryllium disease, chronic, susceptibility to} (3)	
6p21.3	IDDM1	L	Insulin-dependent diabetes mellitus-1	222100	F, LD	?Diabetes mellitus, insulin-dependent-1 (2)	
6p21.3	NEU	I	Neuraminidase	256550	H, F	?Sialidosis (2)	17(Neu1)
6p21.3	PDB	L	Paget disease of bone	167250	F	?Paget disease of bone (2)	

Continued on p. 902.

Table VII.1 *Continued.*

Location	Locus symbol	Status[a]	Title	MIM#[b]	Method[c]	Disorder(s)	Mouse locus
6p21.3	RP14	P	Retinitis pigmentosa-14 (autosomal recessive)	600132	Fd	Retinitis pigmentosa-14 (2)	
6p21.3	RWS	L	Ragweed sensitivity	179450	F	?Ragweed sensitivity (2)	
6p21.3	TAP2, RING11, PSF2	C	Transporter-2, ABC (ATP-binding cassette)	170261	REn	Bare lymphocyte syndrome, type I, due to TAP2 deficiency (1)	17(Ham2)
6p21.3-p21.2	LAP	L	Laryngeal adductor paralysis	150270	F	?Laryngeal adductor paralysis (2)	
6p21.1-p12	PKHD1 ARPKD	C	Polycystic kidney and hepatic disease-1 (autosomal recessive)	263200	Fd	Polycystic kidney disease, autosomal recessive (2)	
6p21.1-pcen	RDS, RP7	C	Retinal degeneration, slow (peripherin)	179605	REa, A	Retinitis pigmentosa, peripherin related (3); retinitis punctata albescens (3); macular dystrophy (3); retinitis pigmentosa, digenic (3); butterfly dystrophy, retinal (3)	17(rds)
6p21	CCD	C	Cleidocranial dysplasia	119600	Ch, Fd, D	Cleidocranial dysplasia (2)	
6p21	MUT, MCM	C	Methylmalonyl Coenzyme A mutase	251000	REa, A, F, D	Methylmalonicaciduria, mutase deficiency type (3)	17(Mut)
6p	ICS1	L	Immotile cilia syndrome	1242650	F	?Immotile cilia syndrome (2)	6p
	PUJO	P	Pelviureteric junction obstruction	143400	F	Pelviureteric junction obstruction (2)	
6cen-q14	STGD3	P	Macular dystrophy with flecks, type 3	600110	Fd	Stargardt disease 3 (2)	
6q	SIASD, SLD	P	Sialic acid storage disease	269920	Fd	Salla disease (2)	
6q13-q15	OA3, OAR	L	Ocular albinism, autosomal recessive	203310	Ch	?Ocular albinism, autosomal recessive (2)	
6q14-q16.2	MCDR1	P	Macular dystrophy, retinal, 1 (North Carolina type)	136550	Fd	Macular dystrophy, North Carolina type (2)	
6q21-q22.3	COL10A1	C	Collagen, type X, α-1 polypeptide	120110	REa, A	Metaphyseal chondrodysplasia, Schmid type (3)	10(Col10a1)
6q22-q23	LAMA2, LAMM	C	Laminin, α-2 (merosin)	156225	REa, A, Fd	Muscular dystrophy, congenital, merosin-negative (2)	10(dy, Lamm)
6q23	ARG1	P	Arginase, liver	207800	Rea	Argininemia (3)	
6q25-q26	RCD1	L	Retinal cone dystrophy-1	180020	Ch	?Retinal cone dystrophy-1 (2)	
6q25.1	ESR	C	Estrogen receptor	133430	REa, A	Breast cancer (1); oestrogen resistance (3)	10(Esr)
6q26	PLG	C	Plasminogen	173350	REa, A, LD, F	Plasminogen Tochigi disease (3); dysplasminogenaemic thrombophilia (1); plasminogen deficiency, types I and II (1)	17(Plg)
6q26-q27	OVCS	P	Ovarian cancer, serous	167000	D	Ovarian cancer, serous (2)	
6q27	LPA	C	Apolipoprotein Lp(a)	152200	REa, A, F, Fd	{Coronary artery disease, susceptibility to} (1)	
Chr.6	PBCA	P	Pancreatic β-cell, agenesis of	600089	D	?Diabetes mellitus, insulin-dependent, neonatal (2)	
7p21.3-p21.2	CRS, CSO	C	Craniosynostosis, type I	123100	Ch	Craniosynostosis, type 1 (2)	
7p21	ACS3, SCS	C	Acrocephalosyndactyly-3 (Saethre–Chotzen syndrome)	101400	Fd, Ch	Saethre–Chotzen syndrome (2)	
7p21-p15	MDDC	P	Macular dystrophy, dominant cystoid (2)	153880	Fd	Macular dystrophy, dominant	
7p15.1-p13	RP9	P	Retinitis pigmentosa-9	180104	Fd	Retinitis pigmentosa-9 (2)	
7p15-p14	GHRHR	C	Growth hormone releasing hormone receptor	139191	REa, A	?Growth hormone deficient dwarfism (1)	6(Lit, Ghrhr)
7p15-p13	GCK	P	Glucokinase (hexokinase-4)	138079	Psh, Fd	MODY, type II, 125851 (3)	
7p14	AQP1, CHIP28, CO	C	Aquaporin 1 (channel-forming integral protein, 28 kD)	107776	REa, A, Fd	Colton blood group (3)	
7p13	GLI3	C	GLI-Kruppel family member GLI3 (oncogene GLI3)	165240	REa, A	Greig cephalopolysyndactyly syndrome, 175700 (3)	13(Xt)
7p13-p12.3	PGAM2, PGAMM	C	Phosphoglycerate mutase, muscle form	261670	REa, A	Myopathy due to phosphoglycerate mutase deficiency (3)	
7p13-p11.2	OGDH	P	Oxoglutarate dehydrogenase (lipoamide)	203740	Rea	α-ketoglutarate dehydrogenase deficiency (1)	
7p	GHS	L	Goldenhar syndrome	141400	Ch	?Goldenhar syndrome (2)	7cen-q11.2
	ASL	C	Argininosuccinate lyase	207900	S, REa, A	Argininosuccinicaciduria (3)	5(Asl)
7q11-q22	CAM	P	Cavernous angiomatous malformations	116860	Fd	Cavernous angiomatous malformations (2)	

Continued.

Table VII.1 *Continued.*

Location	Locus symbol	Status[a]	Title	MIM#[b]	Method[c]	Disorder(s)	Mouse locus
7q11.2	CD36	P	CD36 antigen (collagen type I)	173510	A	[Macrothrombocytopenia] (1); platelet glycoprotein IV deficiency (3)	
7q11.2	ELN	C	Elastin	130160	REa, A, F, Fd	Supravalvar aortic stenosis, 185500 (3); Williams–Beuren syndrome, 194050 (3)	5(Eln)
7q11.2-q21.3	EEC	L	Ectrodactyly, ectodermal dysplasia, cleft lip/palate	129900	Ch	?EEC syndrome (2)	
7q11.23	NCF1	P	Neutrophil cytosolic factor-1 (47 kD)	233700	REa, A	Chronic granulomatous disease due to deficiency of NCF-1 (3)	
7q11.23	ZWS1	C	Zellweger syndrome-1	214100	Ch	Zellweger syndrome-1 (2)	
7q21	EPO	C	Erythropoietin	133170	REa, A, REb, Fd	?Erythraemia (1)	5(Epo)
7q21-q22	CACNL2A	C	Calcium channel, L type, α-2 polypeptide	114204	Psh, Fd, A	Malignant hyperthermia susceptibility-3, 154276 (3)	
7q21.11	GUSB	C	Glucuronidase, β-	253220	S, D, EM	Mucopolysaccharidosis VII (3)	5(Gus)
7q21.2-q21.3	SHFM1, SHFD1, SHSF1	C	Split hand/foot malformation, type 1	183600	Ch	Split-hand/split-foot malformation type 1 (2)	
7q21.3-q22	PLANH1, PAI1	C	Plasminogen activator inhibitor, type I	173360	REa, REb, Fd, A, D	Thrombophilia due to excessive plasminogen activator inhibitor (1); haemorrhagic diathesis due to PAI1 deficiency (1)	
7q22-q31.1	DRA	P	Down-regulated in adenoma	126650	A	?Colon cancer (1)	
7q22.1	COL1A2	C	Collagen, type I, α-2 polypeptide	120160	S, REa, D, A	Osteogenesis imperfecta, 4 clinical forms, 166200, 166210, 259420, 166220 (3); Ehlers–Danlos syndrome, type VIIA2, 130060 (3)	6(Cola2)
7q31	CLD	P	Chloride diarrhoea, congenital	214700	Fd	Chloride diarrhoea, congenital (2)	
7q31	OBS	P	Obesity	164160	H, REa	?Obesity (2)	6(ob)
7q31-q32	DLD, LAD, PHE3	C	Dihydrolipoamide dehydrogenase (E3 component of pyruvate dehydrogenase complex, 2-oxo-glutarate complex)	246900	REa	Lipoamide dehydrogenase deficiency (3)	
7q31-q34	BPGM	P	2,3-bisphosphoglycerate mutase	222800	A	Haemolytic anaemia due to bisphosphoglycerate mutase deficiency (1)	
7q31-q35	RP10	C	Retinitis pigmentosa-10 (autosomal dominant)	180105	Fd	Retinitis pigmentosa-10 (2)	
7q31.1-q31.3	LAMB1	C	Laminin, β-1	150240	REa, A, Ch	?Cutis laxa, marfanoid neonatal type (1)	1(Lamb1)
7q31.2	CFTR, CF	C	Cystic fibrosis transmembrane conductance regulator	219700	F, Fd	Cystic fibrosis (3) congenital bilateral absence of vas deferens (3)	6(Cftr)
7q31.3-q32	BCP, CBT	C	Blue cone pigment	190900	REa, A	Colour blindness, tritan (3)	6(Bcp)
7q32-qter	TRY1	P	Trypsin-1	276000	REa	Trypsinogen deficiency (1)	6(Try1)
7q34	TBXAS1	C	Thromboxane A synthase 1 (platelet)	274180	A	Thromboxane synthase deficiency (2)	
7q34-qter	SLO	L	Smith–Lemli–Opitz syndrome	270400	Ch	?Smith–Lemli–Opitz syndrome (2)	
7q35	CLCN1	P	Chloride channel-1, skeletal muscle	118425	H, REa, Fd	Myotonia congenita, recessive, 255700 (3); myotonia congenita, dominant, 160800 (3)	6(adr, Clc1)
7q35-q36	HERG, LQT2	C	Long (electrocardiographic) QT syndrome-2	152427	Fd, REn, A	Long QT syndrome-2 (3)	
7q36	HPE3, HLP3	C	Holoprosencephaly-3	142945	Ch, Fd	Holoprosencephaly, type 3 (2)	
7q36	HPFH2	L	Hereditary persistence of fetal haemoglobin, heterocellular, Indian type	142335	Fd	?Hereditary persistence of fetal haemoglobin, heterocellular, Indian type (2)	
7q36	TPT1	C	Triphalangeal thumb-polysyndactyly syndrome	190605	Fd	Triphalangeal thumb-polysyndactyly syndrome (2)	

Continued on p. 904.

Table VII.1 *Continued.*

Location	Locus symbol	Status[a]	Title	MIM#[b]	Method[c]	Disorder(s)	Mouse locus
Chr.7	HADHB	P	Hydroxyacyl-Coenzyme A dehydrogenase/3-ketoacyl-Coenzyme A thiolase/enoyl-Coenzyme A hydratase (trifunctional protein), β-subunit	143450	S	3-hydroxyacyl-CoA dehydrogenase deficiency (1)	
8pter-p22	EPMR	P	Epilepsy, progressive, with mental retardation	600143	Fd	Epilepsy, progressive, with mental retardation (2)	
8p22	LPL, LIPD	C	Lipoprotein lipase	238600	REa, A, Fd	Hyperlipoproteinaemia I (1); lipoprotein lipase deficiency (3); hyperchylomicronaemia syndrome, familial (3)	8(Lpl)
8p21.1	GSR	C	Glutathione reductase	138300	S, D	Haemolytic anaemia due to glutathione reductase deficiency (1)	8(Gr1)
8p21-p12	CLU, CLI, SGP2, TRPM2	C	Clusterin (complement lysis inhibitor, SP-40,40; sulphated glycoprotein 2; testosterone-repressed prostate message-2; apolipoprotein J)	185430	REa, REb, A, RE	?{Atherosclerosis, susceptibility to} (3)	14(Sgp2)
8p21-p11.2	LHRH, GNRH	P	Luteinizing hormone releasing hormone (gonadotropin releasing hormone)	152760	REa, A	?Hypogonadotropic hypogonadism due to GNRH deficiency, 227200 (1)	14(Gnrh)
8p12	PLAT, TPA	C	Plasminogen activator, tissue type	173370	REa, A, REb	Plasminogen activator deficiency (1)	8(Plat)
8p12-p11.2	FGFR1, FLT2	C	Fibroblast growth factor receptor-1 (fms-related tyrosine kinase-2)	136350	REa, A	Pfeiffer syndrome, 101600 (3)	
8p12-p11	WRN	C	Werner syndrome	277700	Fd	Werner syndrome (2)	
8p12-q13	SPG5A	P	Spastic paraplegia 5A (autosomal recessive)	270800	Fd	Spastic paraplegia 5A (2)	
8p11.2	ANK1, SPH2		Ankyrin-1, erythrocytic	182900	F, Ch, D, REa, A, Fd, REb	Spherocytosis-2 (3)	8(nb)
8p11-q21	RP1	P	Retinitis pigmentosa-1	180100	Fd	Retinitis pigmentosa-1 (2)	
8q	EBN2	P	Epilepsy, benign neonatal-2 (benign familial neonatal convulsions)	121201	Fd	Epilepsy, benign neonatal, type 2 (2)	
8q	TTP1, AVED	C	Tocopherol transfer protein, α	600415	Fd, LD, REc	Ataxia with isolated vitamin E deficiency, 277460 (3)	
8q11	HYRC1, DNPK1	C	Hyperradiosensitivity of murine SCID mutation, complementing-1	202500	C, A	?Severe combined immunodeficiency, type I (1)	16(scid)
8q12	SGPA, PSA	P	Salivary gland pleomorphic adenoma	181030	Ch	Salivary gland pleomorphic adenoma (2)	
8q13-q21.1	CMT4A	P	Charcot–Marie–Tooth neuropathy-4A (autosomal recessive)	214400	Fd	Charcot–Marie–Tooth disease,type IV A (2)	
8q13.3	BOR	C	Branchio–otorenal syndrome	113650	Ch, Fd	Branchio–otorenal dysplasia (2)	
8q21	CYP11B1, P450C11	C	Cytochrome P450, subfamily XIB, polypeptide-1; 11-β-hydroxylase; corticosteroid methyl-oxidase II (CMO II)	202010	REa, A, Ch	Adrenal hyperplasia, congenital, due to 11-β-hydroxylase deficiency (3); Aldosteronism, glucocorticoid-remediable (3)	
8q21	CYP11B2	C	Cytochrome P450, subfamily XIB, polypeptide-2	124080	REa	CMO II deficiency (3)	
8q21.1	PXMP3, PAF1, PMP35	C	Peroxisomal membrane protein-3 (35 kD)	170993	RE	Zellweger syndrome-3 (3)	
8q22	CA2	C	Carbonic anhydrase II	259730	REa, H	Renal tubular acidosis-osteopetrosis syndrome (3)	3(Car2)
8q22-q23	CSH1	P	Cohen syndrome 1	216550	Fd	Cohen syndrome (2)	
8q24	EBS1	C	Epidermolysis bullosa simplex-1 (Ogna)	131950	F	Epidermolysis bullosa, Ogna type (2)	
8q24	PDS	L	Pendred syndrome	274600	Ch	?Pendred syndrome (2)	
8q24	VMD1	C	Macular dystrophy, atypical vitelliform	153840	F	Macular dystrophy, atypical vitelliform (2)	
8q24.11-q24.13	EXT1	C	Exostoses (multiple) 1	133700	Ch, Fd	Exostoses, multiple, type 1 (2)	
8q24.11-q24.13	LGCR, LGS, TRPS2	C	Langer–Giedion syndrome chromosome region	150230	Ch	Langer–Giedion syndrome (2)	
8q24.12	TRPS1	P	Trichorhinophalangeal syndrome, type I type I (2)	190350	Ch	Trichorhinophalangeal syndrome,	
8q24.12-q24.13	MYC	C	Avian myelocytomatosis viral (v-myc) oncogene homologue	190080	REa, A	Burkitt lymphoma (3)	15(Myc)

Continued.

Table VII.1 *Continued.*

Location	Locus symbol	Status[a]	Title	MIM#[b]	Method[c]	Disorder(s)	Mouse locus
8q24.2-q24.3	TG	C	Thyroglobulin	188450	A, REa, REb	Hypothyroidism, hereditary congenital (3); goitre, adolescent multinodular (1); goitre, non-endemic. simple (3)	15(Tgn; cog)
Chr.8	RTS	L	Rothmund–Thomson syndrome	268400	Ch	?Rothmund–Thomson syndrome (2)	
9p24	EAAC1	P	High-affinity glutamate transporter EAAC1	133550	REa, A	?Dicarboxylicaminoaciduria, 222730 (1)	
9p24	OVC	P	Oncogene OVC (ovarian adeno-carcinoma oncogene)	164759	Ch	Ovarian carcinoma (2)	
9p23	TYRP, CAS2	C	Tyrosinase-related protein 1	115501	Psh, REa, A	Albinism, brown, 203290 (1)	4(b;trp1)
9p22	GLDC, HYGN1, GCSP	C	Glycine dehydrogenase (decarboxylating; glycine decarboxylase, glycine cleavage system protein P)	238300	Ch, A	Hyperglycinemia, isolated non-ketotic, type I (3)	
9p22-p21	LALL	P	Lymphomatous acute lymphoblastic leukaemia	247640	Ch	Leukaemia, acute lym-phoblastic (2)	
9p21	CDKN2, MTS1, P16	P	Cyclin-dependent kinase inhibitor 2 (p16, inhibits CDK4)	600160	RE, D	Melanoma (1)	
9p21	MLM, CMM2, MLM2	C	Melanoma	155601	D, Fd	Melanoma, cutaneous malignant (2)	
9p21	IFN1α, IFNA	C	Interferon, type 1, cluster	147660	REa, A, RE	Interferon, α, deficiency (1)	4(Ifa)
9p21-q21	AMCD1, DA1	P	Arthrogryposis multiplex congenita, distal, type 1	108120	Fd	Distal arthrogryposis-1 (2)	
9p13	GALT	C	Galactose-1-phosphate uridyl-transferase	230400	S, D, F	Galactosaemia (3)	4(Galt)
9q21	GCNT2	P	Glucosaminyl (*N*-acetyl) transferase 2, I-branching enzyme	600429	A	[Ii blood group, 110800] (1)	
9p13-q11	CHH	P	Cartilage-hair hypoplasia	250250	Fd	Cartilage-hair hypoplasia (2)	
9p11	MROS	L	Melkersson–Rosenthal syndrome	155900	Ch	?Melkersson–Rosenthal syndrome (2)	
9p	VMCM	C	Venous malformations, multiple cutaneous and mucosal	600195	Fd	Venous malformations, musltiple cutaneous and mucosal (2)	
9q13-q21.1	FRDA	C	Friedreich ataxia	229300	Fd	Friedreich ataxia (2)	
9q22	ALDOB	C	Aldolase B, fructose-bisphosphatase	229600	REb, REa, A, D	Fructose intolerance (3)	
9q22	HSD17B3, EDH17B3	P	Hydroxysteroid (17-β) dehydrogenase 3	264300	A	Pseudohermaphroditism, male, with gynaecomastia (3)	
9q31	ESS1	P	Epithelioma, self-healing, squamous 1, Ferguson–Smith type	132800	Fd	Epithelioma, self-healing, squamous 1, Ferguson–Smith type (2); ?Basal cell carcinoma (2)	
9q31	NBCCS, BCNS	C	Naevoid basal cell carcinoma syndrome	109400	Fd, D	Basal cell naevus syndrome (2)	
9q31	TAL2	P	T-cell acute lymphocytic leukaemia-2	186855	REa, A, RE, Ch	Leukaemia-2, T-cell acute lymphoblastic (3)	4(Tal2)
9q31-q33	DYS	P	Dysautonomia (Riley–Day syndrome, hereditary sensory autonomic neuropathy type III)	223900	Fd, LD	Dysautonomia, familial (2)	
9q31-q33	FCMD	P	Fukuyama type congenital muscular dystrophy	253800	Fd, LD	Fukuyama type congenital muscular dystrophy (2); ?Walker–Warburg syn-drome, 236670 (2)	
9q32	AFDN	L	Acrofacial dysostosis, Nager type	154400	Ch	?Acrofacial dysostosis, Nager type (2)	
9q32-q34	DYT1	C	Dystonia-1, torsion (autosomal dominant)	128100	Fd	Torsion dystonia (2)	
9q33-qter	ITO	I	Hypomelanosis of Ito	146150	X/A	?Hypomelanosis of Ito (2)	
9q34	ALAD	C	Aminolaevulinate, δ-, dehydratase	125270	F, S, A, REa	Porphyria, acute hepatic (3); {lead poisoning, sus-ceptibility to} (3)	4(Lv)
9q34	ASS	C	Argininosuccinate synthetase	215700	S, D, REa, Fd	Citrullinaemia (3)	2(Ass1)
9q34	DBH	C	Dopamine-β-hydroxylase	223360	F, A	Dopamine-β-hydroxylase deficiency (1)	2(Dbh)
9q34	GSN	P	Gelsolin	137350	A, REa, RE	Amyloidosis, Finnish type, 105120 (3)	2(Gsn)
9q34	TSC1	C	Tuberous sclerosis-1	191100	F, Fd	Tuberous sclerosis-1 (2)	

Continued on p. 906.

Table VII.1 *Continued.*

Location	Locus symbol	Status[a]	Title	MIM#[b]	Method[c]	Disorder(s)	Mouse locus
9q34.1	ABL1	C	Abelson murine leukaemia viral (v-abl) oncogene homolog 1	189980	REa, Ch, A	Leukaemia, chronic myeloid (3)	2(Abl)
9q34.1	AK1	C	Adenylate kinase-1	103000	F, S, D, Fc	Hemolytic anemia due to adenylate kinase deficiency (1)	2(Ak1)
9q34.1	C5	C	Complement component-5	120900	REa, A	C5 deficiency (1)	2(Hc)
9q34.1	CRAT, CAT1	P	Carnitine acetyltransferase	600184	REa	?Carnitine acetyltransferase deficiency (1)	
9q34.1	D9S46E, CAN	P	CAIN gene	114350	Ch	Leukaemia, acute myeloid (2)	2(Can)
9q34.1	ENG, END, HHT, ORW	C	Endoglin	131195	A, H, Fd	Hereditary haemorrhagic telangiectasia,187300 (3)	2(Eng)
9q34.1	EPB72	C	Erythrocyte membrane protein band 7.2 (stomatin)	185000	REa, Ch, A	?Stomatocytosis I (1)	2(Epb7.2)
9q34.1	NPS1	C	Nail-patella syndrome	161200	F, Fd	Nail-patella syndrome (2)	
9q34.1	XPA	C	Xeroderma pigmentosum, A complementation group	278700	S, A, M	Xeroderma pigmentosum, type A (3)	4(Xpa)
9q34.2-q34.3	COL5A1	C	Collagen V, α-1 polypeptide	120215	REa, A	Ehlers–Danlos syndrome, type unspecified (3)	2(Col5a1)
9q34.2	NOTCH1, TAN1	C	Notch (Drosophila) homologue 1 (translocation-associated)	190198	Ch, H, A	Leukaemia, T-cell acute lymphoblastic (2)	2(Notch1)
10p12-q23.2	GBM	C	Glioblastoma multiforme	137800	D	Glioblastoma multiforme (2)	
10q	EPT	P	Epilepsy, partial	600512	Fd	Epilepsy, partial (2)	
10q	PEO	P	Progressive external ophthalmoplegia, autosomal dominant, with multiple mitochondrial DNA deletions	157640	Fd	PEO with mitochondrial DNA deletions (2)	
10q11.2	RET, MEN2A	C	RET transforming sequence; oncogene RET	164761	A, REn, Fd, Ch, D	Multiple endocrine neoplasia IIA, 171400 (3); medullary thyroid carcinoma, 155240 (3); multiple endocrine neoplasia IIB, 162300 (3); Hirschsprung disease, 142623 (3)	
10q11.2-q21	MBL	C	Mannose-binding lectin, soluble (opsonic defect)	154545	REa, A, Fd	{Chronic infections, due to opsonin defect} (3)	14(Mbl1)
10q11	ERCC6, CKN2	P	Excision repair cross complementing rodent repair deficiency, complementation group 6	133540	A	Cockayne syndrome-2, late onset, 216410 (2)	
10q11-q12	D10S170, TST1, PTC, TPC	C	DNA segment, single copy, probe pH4 (transforming sequence, thyroid-1, from papillary thyroid carcinoma)	188550	REa, A	Thyroid papillary carcinoma (1)	
10q21-q22	PSAP, SAP1	C	Prosaposin (sphingolipid activator protein-1)	176801	S, REa, A, D	Metachromatic leukodystrophy due to deficiency of SAP-1 (3); Gaucher disease, variant form (3)	10(Psap)
10q22	DCOH	C	Dimerization cofactor of hepatic nuclear factor 1α (TCF1)	126090	REa, H, A	Hyperphenylalaninemia due to pterin-4a-carbinolamine dehydratase deficiency, 264070 (3)	10(Dcoh)
10q22	HK1	C	Hexokinase-1	142600	S, D, A, REa	Haemolytic anaemia due to hexokinase deficiency (1)	10(Hk1)
10q23-q24	RBP4	C	Retinol-binding protein-4, interstitial	180250	REa, A	?Retinol binding protein, deficiency of (1)	19(Rbp4)
10q24	HOX11, TCL3	P	Homeobox-11 (T-cell leukaemia-3 associated breakpoint, homologous to Drosophila Notch)	186770	Ch	Leukaemia, T-cell acute lymphocytic (2)	
10q24-q25	LIPA	C	Lipase A, lysosomal acid, cholesterol esterase	278000	S, H	Wolman disease (3); cholesteryl ester storage disease (3)	19(Lip1)
10q24.1-q24.3	CYP2C, CYP2C19	C	Cytochrome P450, subfamily IIC; mephenytoin 4\(fm-hydroxylase)	124020	REa, A	Mephenytoin poor metabolizer (3)	19(P4502c)
10q24.3	BPAG2	C	Bullous pemphigoid antigen-2	113811	A, H	Generalized atrophic benign epidermolysis bullosa, 226650 (1)	19(Bpag2)
10q24.3	CYP17, P450C17	C	Cytochrome P450, subfamily XVII; steroid 17-α-hydroxylase	202110	REa, H, A	Adrenal hyperplasia, congenital, due to 17-α-hydroxylase deficiency (3)	19(Cyp17)
10q25	PAX2	C	Paired box homeotic gene-2	167409	REa, A	Optic nerve coloboma with renal anomalies, 120330 (3)	19(Pax2)
10q25.2-q26.3	UROS	P	Uroporphyrinogen III synthase	263700	REa, Psh	Porphyria, congenital erythropoietic (3)	7(Uros)

Continued.

Table VII.1 *Continued.*

Location	Locus symbol	Status[a]	Title	MIM#[b]	Method[c]	Disorder(s)	Mouse locus
10q26	FGFR2, BEK, CFD1, JWS	C	Fibroblast growth factor receptor-2 (bacteria-expressed kinase)	176943	A, Psh, Fd	Crouzon craniofacial dysostosis, 123500 (3); Jackson–Weiss syndrome, 123150 (3); Apert syndrome, 101200 (3); Pfeiffer syndrome, 101600 (3)	7(Fgfr2)
10q26	OAT	C	Ornithine aminotransferase	258870	S, REa, A, Fd	Gyrate atrophy of choroid and retina with ornithinemia, B6 responsive or unresponsive (3)	7(Oat)
10q26.1	PNLIP	P	Pancreatic lipase	246600	REa, A	Pancreatic lipase deficiency (1)	
11pter-p15.4	BWS, WBS	C	Beckwith–Wiedemann syndrome	130650	Ch, Fd	Beckwith–Wiedemann syndrome (2)	
11pter-p13	AMPD3	P	Adenosine monophosphate deaminase 3 (isoform E)	102772	REa	[AMP deaminase deficiency, erythrocytic] (3)	
11p15.5	IDDM2	P	Insulin-dependent diabetes mellitus-2	125852	Fd	Insulin-dependent diabetes mellitus-2 (2)	
11p15.5	HBB	C	Haemoglobin β	141900	LD, AAS, F, Fd	Sickle cell anaemia (3); thalassaemias, β- (3); methoglobinaemias, β- (3); erythraemias, β- (3); Heinz body anaemias, β- (3); HPFH, deletion type (3)	7(Hbb)
11p15.5	HBGR	C	Haemoglobin, γ, regulator of	142270	RE	?Hereditary persistence of fetal haemoglobin (3)	
11p15.5	HBG1	C	Haemoglobin, γA	142200	RE	HPFH, non-deletion type A (3)	
11p15.5	HBG2	C	Haemoglobin, γG	142250	RE	HPFH, non-deletion type G (3)	
11p15.5	INS	C	Insulin	176730	HS, A, REb, Fd, D	Diabetes mellitus, rare form (1); MODY one form (3); hyperproinsulinaemia, familial (3)	6(Ins1); 7(Ins2)
11p15.5	LQT1	P	Long (electrocardiographic) QT syndrome-1; Ward–Romano syndrome	192500	Fd	Long QT syndrome-1 (2)	
11p15.5	MTACR1, WT2	C	Multiple tumour associated chromosome region-1	194071	D	Wilms' tumour, type 2 (2); adrenocortical carcinoma, hereditary, 202300 (2)	
11p15.5	RMS1	P	Rhabdomyosarcoma, embryonal	268210	D	Rhabdomyosarcoma (2)	
11p15.5	TH, TYH	C	Tyrosine hydroxylase	191290	REa, A, Fd, RE	Segawa syndrome, recessive (3)	7(Th)
11p15.4	LDHA, LDH1	C	Lactate dehydrogenase A	150000	S, D, REb, C, A	Exertional myoglobinuria due to deficiency of LDH-A (3)	7(Ldh1)
11p15.4-p15.1	SMPD1, NPD	P	Sphingomyelin phosphodiesterase-1, acid lysosomal	257200	REa, A	Niemann–Pick disease, type A (3); Niemann–Pick disease, type B (3)	7(Smpd1)
11p15.3-p15.1	PTH	C	Parathyroid hormone	168450	REa, REb, A, Fd	Hypoparathyroidism, autosomal dominant (3); hypoparathyroidism, autosomal recessive (3)	7(Pth)
11p15.1	USH1C	C	Usher syndrome-1C (autosomal recessive, severe)	276904	Fd	Usher syndrome, type 1 C(2)	
11p15.1-p14	PHHI	C	Persistent hyperinsulinaemic hypoglycaemia of infancy (nesidioblastosis)	256450	Fd, LD	Persistent hyperinsulinaemic hypoglycaemia of infancy (2)	
11p15	RBTN1, RHOM1	C	Rhombotin-1	186921	Ch, D	Leukaemia, T-cell acute lymphoblastic (2)	7(Ttg1)
11p14-p13	HVBS1	C	Hepatitis B virus integration site-1	114550	REa, A, Ch	Hepatocellular carcinoma (1)	
11p13	CAT	C	Catalase	115500	S, D, Fd	Acatalasaemia (3)	2(Cas1)
11p13	CD59	C	CD59 antigen (p18-20)	107271	REa, A, D	CD59 deficiency (3)	15(Ly6)
11p13	FSHB	C	Follicle-stimulating hormone, β polypeptide	136530	D, REa	?Male infertility, familial (1)	2(Fshb)
11p13	FSHB	C	Follicle-stimulating hormone, β polypeptide	136530	D, REa	?Male infertility, familial (1)	2(Fshb)
11p13	PAX6, AN2	C	Paired box homeotic gene-6	106210	Ch, Fd	Aniridia (3); Peters anomaly (3); cataract, congenital, with late-onset corneal dystrophy (3)	2(Sey)

Continued on p. 908.

Table VII.1 *Continued.*

Location	Locus symbol	Status[a]	Title	MIM#[b]	Method[c]	Disorder(s)	Mouse locus
11p13	RBTNL1, RHOM2, TTG2	P	Rhombotin-like 1	180385	REa, REc	Leukaemia, acute, T-cell (2)	
11p13	TCL2	P	T-cell leukaemia/lymphoma-2	151390	Ch, RE, A, REa	Leukaemia, acute T-cell (2)	
11p13	WT1	C	Wilms' tumour 1	194070	Ch	Wilms' tumour (3); Denys–Drash syndrome (3)	2(Wt1)
11p13-q13	CMH4	P	Cardiomyopathy, hypertrophic, 4	115197	Fd	Cardiomyopathy, familial hypertrophic, 4 (2)	
11p12-p11.12	PFM	L	Parietal foramina	168500	Ch	?Parietal foramina (2)	
11p12-p11	ACP2	C	Acid phosphatase 2, lysosomal	171650	S, REa	?Lysosomal acid phosphatase deficiency (1)	2(Acp2)
11p11-q11	EXT2	I	Exostoses (multiple) 2	133701	Fd	?Exostoses, multiple, type 2 (2)	
11p11-q11	SCA5	P	Spinocerebellar ataxia 5	600224	Fd	Spinocerebellar ataxia, type 5 (2)	
11p11-q12	F2	C	Coagulation factor II (thrombin)	176930	REa, A	Hypoprothrombinaemia (3); Dysprothrombinaemia (3)	2(Cf2)
11q	CPT1	P	Carnitine palmitoyltransferase I	255110	Psh, A	?Carnitine palmitoyltransferase I deficiency (2)	
11q	JBS	L	Jacobsen syndrome	147791	Ch	?Jacobsen syndrome (2)	
11q	PC	P	Pyruvate carboxylase	266150	REa, H	Pyruvate carboxylase deficiency (1)	19(Pc)
11q11-q13.1	C1NH	C	Complement component-1 inhibitor	106100	REa, A	Angio-oedema, hereditary (3)	
11q12-q13	IGER, APY	C	IgE responsiveness (atopic)	147050	Fd	Atopy (2)	
11q13	BBS1	P	Bardet–Biedl syndrome 1	209901	Fd	Bardet–Biedl syndrome 1 (2)	
11q13	CCND1, PRAD1	C	Cyclin D1	168461	REn, R, REa, A	Parathyroid adenomatosis 1 (2); centrocytic lymphoma (2)	
11q13	IDDM4	P	Insulin-dependent diabetes mellitus 4	600319	Fd	Diabetes mellitus, insulin-dependent, 4 (2)	
11q13	MEN1	C	Multiple endocrine neoplasia, type I	131100	Fd, D	Multiple endocrine neoplasia I (1); prolactinoma, hyperparathyroidism, carcinoid syndrome (2)	
11q13	NDUFV1, UQOR1	P	NADH dehydrogenase (ubiquinone) flavoprotein 1 (51 kD)	161015	REa, A	?Mitochondrial complex I deficiency, 252010 (1)	
11q13	PYGM	C	Phosphorylase, glycogen, muscle	232600	REb, Fd, REn	McArdle disease (3)	19(Pygm)
11q13	ROM1, ROSP1	P	Rod outer segment membrane protein-1	180721	REa, A	Retinitis pigmentosa, digenic (3)	19(Rosp1)
11q13	RT6	P	RT6 antigen (rat) homologue	180840	REa, A	?{Susceptibility to IDDM} (1)	7(Rt6)
11q13	SMTN	P	Somatotrophinoma	102200	D	Somatotrophinoma (2)	
11q13	ST3	C	Suppression of tumorigenicity-3 (tumour-suppressor gene, HELA cell type)	191181	S, D	Cervical carcinoma (2)	
11q13	VMD2	C	Vitelliform macular dystrophy (Best disease)	153700	Fd, Psh	Macular dystrophy, vitelliform type (2)	
11q13	VRNI	P	Vitreoretinopathy, neovascular inflammatory	193235	Fd	Vitreoretinopathy, neovascular inflammatory (2)	
11q13-q23	EVR1, FEVR	C	Exudative vitreoretinopathy-1 (autosomal dominant; Criswick–Schepens syndrome)	133780	Fd	Vitreoretinopathy, exudative, familial (2)	
11q13.3	BCL1	C	B-cell CLL/lymphoma-1	151400	RE, Ch	Leukaemia/lymphoma, B-cell, 1 (2)	
11q13.5	DFNB2, NSRD2	P	Deafness, autosomal recessive	600060	Fd	Deafness, non-syndromic, recessive, 2 (2)	7(sh1)
11q13.5	USH1B	P	Usher syndrome-1B (autosomal recessive, severe)	276903	Fd	Usher syndrome, type 1B (2)	
11q14-q21	TYR	C	Tyrosinase	203100	REa, A, H, F	Albinism, oculocutaneous, type IA (3)	7(Tyr)
11q22-qter	ANC	L	Anal canal carcinoma	105580	Ch	?Anal canal carcinoma (2)	
11q22.3	ATA, AT1	C	Ataxia-telangiectasia (complementation) groups A, C, D)	208900	Fd, C, M	Ataxia-telangiectasia (2)	
11q22.3-q23.1	ACAT	C	Acetyl-Coenzyme A acetyltransferase (acetoacetyl Coenzyme A thiolase)	203750	A	3-ketothiolase deficiency (3)	
11q22.3-q23.2	PGL, CBT1	C	Paraganglioma (carotid body tumours)	168000	Fd	Paraganglioma (2)	
11q22.3-q23.3	PTS	P	6-pyruvoyltetrahydropterin synthase	261640	A	Phenylketonuria due to PTS deficiency (3)	

Continued.

Table VII.1 *Continued.*

Location	Locus symbol	Status[a]	Title	MIM#[b]	Method[c]	Disorder(s)	Mouse locus
11q23	APOA1	C	Apolipoprotein A-I	107680	REa, RE, Fd, F, D	ApoA-I and apoC-III deficiency, combined (3); hypertriglyceridaemia, one form (3); hypoalphalipoproteinaemia (3); amyloidosis, Iowa type, 107680.0010 (3)	9(Apoa1)
11q23	APOC3	C	Apolipoprotein C-III	107720	REa, RE, F	Hypertriglyceridaemia (3)	
11q23	BRCA3	P	Breast cancer, 11; 22 translocation associated	600048	Ch	Breast cancer-3 (2)	
11q23	MLL, HRX, HTRX1	C	Myeloid/lymphoid, or mixed-lineage leukaemia; trithorax (Drosophila) homologue	159555	Ch, RE	Leukaemia, myeloid/lymphoid or mixed-lineage (2)	9(All1)
11q23	TCPT	L	Thrombocytopenia, Paris–Trousseau type (deletion 11q23 syndrome)	188025	Ch	?Thrombocytopenia, Paris–Trousseau type (2)	
11q23.1	PORC		Porphyria, acute, Chester type	176010	Fd	Porphyria, Chester type (2)	
11q24.1-q24.2	HMBS, P PBGD, UPS	C	Hydroxymethylbilane synthase	176000	S, D	Porphyria, acute intermittent (3)	9(Ups)
Chr.11	GIF		Gastric intrinsic factor	261000	Rea	Anaemia, pernicious, congenital, due to deficiency of intrinsic factor (1)	
12pter-p12	CD4	C	CD4 antigen (p55)	186940	REa, A	[CD4(+) lymphocyte deficiency] (2); {Lupus erythematosus, susceptibility to} (2)	6(Ly4)
12pter-p12	DRPLA	C	Dentatorubro-pallidoluysian atrophy	125370	Fd	Dentatorubro-pallidoluysian atrophy (3)	
12pter-q12	BCT1	C	Branched chain aminotransferase-1	113520	S	?Hyperleucinaemia-isoleucinaemia or hypervalinaemia (1)	
12p13.3	VWF, F8VWF	C	Coagulation factor VIII VWF (von Willebrand factor)	193400	A, REa, REb, Fd	von Willebrand disease (3)	6(Vwf)
12p13.3-p12.3	A2M	C	α-2-macroglobulin	103950	REa, A	Emphysema due to α-2 macroglobulin deficiency (1)	6(A2m)
12p13.3-p11.2	ACLS	L	Acrocallosal syndrome	200990	Ch	?Acrocallosal syndrome (2)	
12p13	C1R	C	Complement component-1, r subcomponent	216950	REa, Fd, RE, A	C1r/C1s deficiency, combined (1)	
12p13	C1S	C	Complement component-1, s subcomponent	120580	REa, Fd, RE, A	C1r/C1s deficiency, combined (1)	
12p13	KCNA1, AEMK	C	Potassium voltage-gated channel, shaker-related subfamily, member 1	176260	REa, Fd, A, H	Episodic ataxia/myokymia syndrome, 160120 (3)	6(Kcnal)
12p13	MPE	L	Malignant proliferation, eosinophil	131440	Ch	?Eosinophilic myeloproliferative disorder (2)	
12p13	TPI1	C	Triosephosphate isomerase-1	190450	S, D, R, REa	Haemolytic anaemia due to triosephosphate isomerase deficiency (3)	6(Tpi1)
12p12.1	KRAS2, RASK2	C	Kirsten rat sarcoma-2 viral (v-Ki-ras2) oncogene homologue	190070	REa, A, Fd	Colorectal adenoma (1); colorectal cancer (1)	6(Kras2)
12p12.1-p11.2	PTHLH	P	Parathyroid hormone-like hormone	168470	REa, A	?Humoral hypercalcaemia of malignancy (1)	6(Pthlh)
12q11-q13	KRT1	C	Keratin-1	139350	H, REa, A	Epidermolytic hyperkeratosis, 113800 (3); keratoderma, palmoplantar, nonepidermolytic (3)	15(Krt2)
12q11-q13	KRT2E	P	Keratin-2e	600194	Fd	Ichthyosis bullosa of Siemens, 146800 (3)	
12p11-q13	KRT5	P	Keratin-5	148040	A, Fd	Epidermolysis bullosa simplex, Dowling–Meara type, 131760 (3); Epidermolysis bullosa simplex, Koebner type, 131900 (3); Epidermolysis bullosa, Weber–Cockayne type, 131800 (3)	
12q11-q13	PPKB	P	Palmoplantar keratoderma, Bothnia type	600231	Fd	Palmoplantar keratoderma, Bothnia type (2)	
12q12-q13	CD63, MLA1	P	CD63 antigen (melanoma 1 antigen)	155740	REa, A	?Hermansky–Pudlak syndrome, 203300 (1)	
12q12-q14	VDR	P	Vitamin D (1,25-dihydroxyvitamin D3) receptor	277440	REa, A	Rickets, vitamin D-resistant (3); ?osteoporosis, involutional (1)	

Continued on p. 910.

Table VII.1 *Continued.*

Location	Locus symbol	Status[a]	Title	MIM#[b]	Method[c]	Disorder(s)	Mouse locus
12q13	AQP2	C	Aquaporin 2 (collecting duct)	107777	A	Diabetes insipidus nephrogenic, autosomal recessive (3)	
12q13.1-q13.2	DDIT3, GADD153, CHOP10	C	DNA-damage-inducible transcript-3	126337	REa, A, Ch	Myxoid liposarcoma (3)	
12q13.11-q13.2	COL2A1	C	Collagen, type II, α-1 polypeptide	120140	REa, A	Stickler syndrome, type I (3); SED congenita (3); Kniest dysplasia (3); achondrogenesis-hypochondrogenesis, type II (3); osteoarthrosis, precocious (3); Wagner syndrome, type II (3); SMED Strudwick type (3)	
12q13.2-q24.1	FEOM, CFEOM	P	Fibrosis of the extraocular muscles, congenital	135700	Fd	Fibrosis of the extraocular muscles, congenital (2)	
12q14	GNS, G6S	P	*N*-acetylglucosamine-6-sulphatase	252940	A, REa	Sanfilippo syndrome D (1)	
12q14	PDDR, VDD1	C	Pseudo-vitamin D dependency rickets 1	264700	Fd	Pseudo-vitamin D dependency rickets 1 (2)	
12q14-qter	PPD	P	4-hydroxyphenylpyruvate dioxygenase	276710	REa	Tyrosinaemia, type III (1)	
12q15	BABL, LIPO	C	Lipoma (breakpoint in benign lipoma)	151900	Ch	Lipoma, benign (2); ?multiple lipomatosis (2)	
12q21	CNA2	P	Cornea plana 2 (autosomal recessive)	217300	Fd	Cornea plana congenita, recessive (2)	
12q21.3-q22	HOS	P	Holt–Oram syndrome	142900	Fd	Holt–Oram syndrome (2)	
12q22	MGCT	P	Male germ cell tumour	273300	D, Ch	Male germ cell tumour (2)	
12q22-q23	HAL, HSTD	C	Histidine ammonia-lyase (histidase)	235800	REa, A	[Histidinaemia] (1)	10(Hstd)
12q22-qter	ACADS	P	Acyl-Coenzyme A dehydrogenase, C-2 to C-3 short chain	201470	REa	Acyl-CoA dehydrogenase, short-chain, deficiency of (3)	5(Bcd1)
12q22-qter	MODY3	P	Maturity-onset diabetes of the young, type III	600496	Fd	Maturity-onset diabetes of the young, type III (2)	
12q22-qter	NS1	P	Noonan syndrome 1	163950	Fd	Noonan syndrome-1 (2)	
12q23-q24.1	DAR	C	Darier disease (keratosis follicularis)	124200	Fd	Darier disease (keratosis follicularis) (2)	
12q24	SCA2	C	Spinocerebellar ataxia 2 (olivopontocerebellar ataxia 2, autosomal dominant)	183090	Fd	Spinocerebellar atrophy II (2)	
12q24.1	IFNG	C	Interferon, gamma	147570	REa, A	Interferon, immune, deficiency (1)	10(Ifg)
12q24.1	PAH, PKU1	C	Phenylalanine hydroxylase	261600	REa, A, Fd	Phenylketonuria (3); [hyperphenyl-alaninaemia, mild] (3)	10(Pah)
12q24.2	ALDH2	C	Aldehyde dehydrogenase-2, mitochondrial	100650	REa, A, H	Alcohol intolerance, acute (3); {?fetal alcohol syndrome} (1)	4(Aldh2)
Chr.12	LYZ	P	Lysozyme	153450	Rea	Amyloidosis, renal, 105200 (3)	
Chr.12	MVK, MVLK	P	Mevalonate kinase	251170	REa	Mevalonicaciduria (3)	
13q12	DFNB1	P	Deafness, neurosensory, autosomal recessive, 1	220290	Fd	Deafness, neurosensory, AR, 1 (2)	
13q12-q13	BRCA2	C	Breast cancer 2, early onset	600185	Fd	Breast cancer 2, early onset (2)	
13q12-q13	DMDA1	P	Duchenne-like muscular dystrophy, autosomal recessive	253700	Fd	Muscular dystrophy, Duchenne-like, autosomal (2)	
13q12.2-q13	MBS	L	Moebius syndrome	157900	Ch	?Moebius syndrome (2)	
13q14	D13S25, DBM	P	Disrupted in B-cell neoplasia	109543	D	Leukaemia, chronic lymphocytic, B-cell (2)	
13q14-q31	LESD	L	Letterer–Siwe disease	246400	Ch	?Letterer–Siwe disease (2)	
13q14.1	FKHR	P	Fork head (Drosophila) homologue 1 (rhabdomyosarcoma)	136533	Ch	Rhabdomyosarcoma, alveolar, 268200 (3)	
13q14.1-q14.2	RB1	C	Retinoblastoma-1	180200	Ch, F, Fd	Retinoblastoma (3); osteosarcoma, 259500 (2); bladder cancer, 109800 (3)	14(Rb1)
13q13.3-q21.1	ATP7B, WND	C	ATPase, Cu^{2+} transporting, β polypeptide	277900	F, Fd	Wilson disease (3)	
13q21.1-q32	CLN5	P	Ceroid-lipofuscinosis, neuronal-5	256731	Fd	Ceroid-lipofuscinosis, neuronal, variant late infantile form (2)	
13q22	EDNRB, HSCR2	C	Endothelin receptor type B	131244	REa, Ch, LD	Hirschsprung disease-2, 600155 (3)	

Continued.

Table VII.1 *Continued.*

Location	Locus symbol	Status[a]	Title	MIM#[b]	Method[c]	Disorder(s)	Mouse locus
13q32	PCCA	C	Propionyl Coenzyme A carboxylase, α polypeptide	232000	REa, D, A, Fd	Propionicacidaemia, type I or pccA type (1)	14(Pcca)
13q33	ERCC5, XPG	C	Excision-repair, complementing defective, in Chinese hamster, number 5	133530	S, A	Xeroderma pigmentosum, group G (3)	
13q34	DJS	L	Dubin–Johnson syndrome	237500	LD	?Dubin–Johnson syndrome (2)	
13q34	F7	C	Coagulation factor VII	227700	D	Factor VII deficiency (3)	
13q34	F10	C	Coagulation factor X	227600	D, A, REa	Factor X deficiency (3)	8(Cf10)
13q34	HHH	L	Hyperornithinaemia-hyperammonaemia-homocitrullinaemia syndrome	238970	D	?HHH syndrome (2)	
13q34	STGD2	P	Macular dystrophy with flecks, type 2	153900	Fd	Stargardt disease 2 (2)	
Chr.13	BRCD1	P	Breast cancer, ductal, suppressor-1	211410	D	Breast cancer, ductal (2)	
Chr.13	CPB2	P	Carboxypeptidase B2 (plasma)	212070	Psh	Carboxypeptidase B deficiency (1)	
14q	MPD1	P	Myopathy, distal 1	160500	Fd	Myopathy, distal (2)	
14q	SPG3	P	Spastic paraplegia-3	182600	Fd	Spastic paraplegia-3 (2)	
14q11.1-q13	HPE4	L	Holoprosencephaly-4, semilobar	142946	Ch	?Holoprosencephaly-4 (2)	
14q11.2	ICR2, LI	P	Ichthyosis congenita II, non-erythromatous lamellar ichthyosis	242300	Fd	Lamellar ichthyosis, autosomal recessive (2)	
14q11.2	TCRA	C	T-cell antigen receptor, α-polypeptide	186880	H, REa, A, REn	Leukaemia/lymphoma, T-cell (3)	14(Tcra)
14q11.2-q13	OPMD1	P	Oculopharyngeal muscular dystrophy-1	164300	Fd	Oculopharyngeal muscular dystrophy-1 (2)	
14q12	MYH7, CMH1	C	Myosin, heavy polypeptide-7, cardiac muscle β	160760	REa, RE, D, A	Cardiomyopathy, familial hypertrophic, 1, 192600 (3); ?central core disease, one form (3)	
14q13.1	NP	C	Nucleotide phosphorylase	164050	S, D	Nucleoside phosphorylase deficiency, immunodeficiency due to (3)	14(Np1,2)
14q21-q22	PYGL	P	Phosphorylase, glycogen, liver	232700	Reb	Glycogen storage disease VI (1)	12(Pyg1)
14q22-q23.2	SPTB	C	Spectrin, β, erythrocytic	182870	REb, F, H, REa, A, RE	Elliptocytosis-3 (3); Spherocytosis-1 (3)	12(Sptb1)
14q22.1-q22.2	GCH1	P	GTP cyclohydrolase 1	600225	Psh, A	Phenylketonuria, atypical, due to GCH1 deficiency, 233910 (1); dystonia, DOPA-responsive, 128230 (3)	
14q23-q24	ARVD	P	Arrhythmogenic right ventricular dysplasia	107970	Fd	Arrhythmogenic right ventricular dysplasia (2)	
14q24-qter	CTAA1	L	Cataract, anterior polar, 1	115650	Ch	?Cataract, anterior polar, I (2)	
14q24.3	AD3	C	Alzheimer disease-3	104311	Fd	Alzheimer disease-3 (2)	
14q24.3-q31	MJD	C	Machado–Joseph disease	109150	Fd	Machado–Joseph disease (2)	
14q24.3-q32.1	GALC	C	Galactosylceraminidase	245200	REa, A, H, Fd	Krabbe disease (3)	12(tw)
14q24.3-qter	SCA3	C	Spinocerebellar ataxia 3 (olivopontocerebellar ataxia 3, autosomal dominant)	183085	Fd	Spinocerebellar ataxia-3 (2)	
14q31	TSHR	C	Thyroid-stimulating hormone receptor	275200	REa, Fd, A	Hypothyroidism, non-goitrous, due to TSH resistance (3); thyroid adenoma, hyperfunctioning (3); Graves disease, 275000 (1); hyperthroidism, congenital (3)	12(Tshr)
14q32	CKBE	P	Creatine kinase, ectopic expression 1	23270	F	[Creatine kinase, brain type, ectopic expression of] (2)	
14q32	SIV	L	Situs inversus viscerum	270100	H	?Situs inversus viscerum (2)	12(iv)
14q32	USH1A, USH1	C	Usher syndrome-1A	276900	Fd	Usher syndrome, type 1A (2)	
14q32	VP, PPOX	P	Variegate porphyria (protoporphyrinogen oxidase)	176200	F	Porphyria variegata (2)	
14q32.1	AACT	C	α-1-antichymotrypsin	107280	REa, A, Fd, Ren	α-1-antichymotrypsin deficiency (3); cerebrovascular disease, occlusive (3)	
14q32.1	CBG	C	Corticosteroid-binding globulin	122500	A, REn	[Transcortin deficiency] (1)	
14q32.1	PCI, PLANH3	C	Protein C inhibitor (plasminogen activator inhibitor-3)	227300	Psh, REn	Protein C inhibitor deficiency (2)	
14q32.1	PI, AAT	C	Protease inhibitor (α-antitrypsin)	107400	F, S, A, D, EM, Fd	Emphysema-cirrhosis (3) haemorrhagic diathesis due to 'antithrombin' Pittsburgh (3); emphysema (3)	12(Aat)
14q32.1	TCL1	C	T-cell lymphoma-1	186960	Ch, RE	Leukaemia/lymphoma, T-cell (2)	

Continued on p. 912.

Table VII.1 *Continued.*

Location	Locus symbol	Status[a]	Title	MIM#[b]	Method[c]	Disorder(s)	Mouse locus
14q32.33	IGH@	C	Immunoglobulin heavy chain gene cluster		REa, A	?Combined variable hypo-gammaglobulinaemia (1)	12(Igh)
14q32.33	IGHR	L	Immunoglobulin heavy chain regulator	144120	F	?Hyperimmunoglobulin G1 syndrome (2)	
Chr.14	MPS3C	L	Sanfilippo disease, type IIIC	252930	Ch	?Sanfilippo disease, type IIIC (2)	
Chr.14	RMCH	P	Rod monochromacy	216900	Ch	Rod monochromacy (2)	
15q11	PWCR, PWS	C	Prader–Willi syndrome chromosome region	176270	Ch, D	Prader–Willi syndrome (2)	
15q11-q13	AHO2	L	Albright hereditary osteodystrophy-2	103581	D	?Albright hereditary osteodystrophy-2 (2)	
15q11-q13	ANCR	C	Angelman syndrome chromosome region	105830	Ch, D	Angelman syndrome (2)	
15q11-q13	ITO	L	Hypomelanosis of Ito	146150	Ch	?Hypomelanosis of Ito (2)	
15q11.1	SPG6	P	Spastic paraplegia 6	600363	Fd	Spastic paraplegia-6 (2)	
15q11.2-q12	OCA2, P, PED, D15S12	C	Oculocutaneous albinism II (pink-eye dilution (murine) homologue)	203200	D, REa, Fd	Albinism, oculocutaneous, type II (3); albinism, ocular, autosomal recessive (3)	7(p)
15q12	SNRPN	P	Small nuclear ribonucleoprotein polypeptide N	182279	REa, D	?Prader–Willi syndrome (1)	7(Snrpn)
15q14-q15	IVD	P	Isovaleryl Coenzyme A dehydrogenase	243500	Rea	Isovalericacidaemia (3)	
15q15	EPB42	C	Erythrocyte surface protein band 4.2	177070	A	Spherocytosis, hereditary, Japanese type (3); ?Hermansky–Pudlak syndrome, 203300 (1)	2(Epb4.2)
15q15	SORD	C	Sorbitol dehydrogenase	182500	S, H, A, REa	?Cataract, congenital (2)	2(Sdh1)
15q15.1-q21.1	LGMD2A	C	Limb girdle muscular dystrophy 2A (autosomal recessive)	253600	Fd, A	Muscular dystrophy, limb-girdle, type 2A (2)	
15q21	CDAN3, CDA3	P	Congenital dyserythropoietic anaemia, type III	105600	Fd	Dyserythropoietic anaemia, congenital, type III (2)	
15q21-q22	B2M	C	β-2-microglobulin	109700	S, D, H	Haemodialysis-related amyloidosis (1)	2(B2m)
15q21-q23	LIPC	C	Lipase, hepatic	151670	REa, A	Hepatic lipase deficiency (3)	9(Hl)
15q21.1	CYP19, ARO	C	Cytochrome P450, subfamily XIX (aromatization of androgens)	107910	REa, A, H	?Gynaecomastia, familial, due to increased aromatase activity (1); virilization, maternal and fetal, from placental aromatase deficiency (3)	9(Cyp19)
15q21.1	FBN1, MFS1	C	Fibrillin-1	134797	A, Fd	Marfan syndrome, 154700 (3)	2(Fbn1)
15q22	PML, MYL	P	Promyelocytic leukaemia, inducer of	102578	Ch, RE	Leukaemia, acute promyelocytic (2)	
15q22	TPM1, CMH3	C	Tropomyosin 1 α	191010	Fd	Cardiomyopathy, familial hypertrophic, 3, 115196 (3)	9(Tpm1)
15q22.3-q23	BBS4	P	Bardet–Biedl syndrome 4	600374	Fd, LD	Bardet–Biedl syndrome-4 (2)	
15q23-q24	HEXA, TSD	C	Hexosaminidase A (α-polypeptide)	272800	S, D, A	Tay–Sachs disease (3); GM2-gangliosidosis, juvenile, adult (3); [Hex A pseudodeficiency] (1)	9(Hexa)
15q23-q25	ETFA, GA2	P	Electron transfer flavoprotein, α-polypeptide	231680	REa, A	Glutaricaciduria, type IIA (1)	
15q23-q25	FAH	C	Fumarylacetoacetase	276700	A, REa	Tyrosinaemia, type I (3)	
15q26	IDDM3	P	Insulin-dependent diabetes mellitus 3	600318	Fd	Diabetes mellitus, insulin-dependent, 3 (2)	
15q26.1	BLM, BS	C	Bloom syndrome	210900	M, LD	Bloom syndrome (2)	
Chr.15	TSK, SSS	L	Stiff skin syndrome	184900	H	?Stiff skin syndrome (2)	2(Tsk)
Chr.15	XPF	L	Xeroderma pigmentosum, complementation group F	278760	M	?Xeroderma pigmentosum, type F (2)	
16pter-p13.3	HBA1	C	Haemoglobin α-1	141800	HS	Thalassaemias, α- (3); methaemoglobinaemias, α- (3); erythremias, α- (3); Heinz body anaemias, α- (3)	11(Hba)
16pter-p13.3	HBHR, ATR1	C	α-thalassaemia/mental retardation syndrome, type 1	141750	Fd, RE	α-thalassaemia/mental retardation syndrome, type I (1)	
16p13.31-p13.12	PKD1	C	Polycystic kidney disease-1 (autosomal dominant)	173900	F, Fd, REn	Polycystic kidney disease-1 (3)	?17(Pkd1)
16p13.3	CATM	P	Cataract, congenital, with microphthalmia	156850	Ch	Cataract, congenital, microphthalmia (2)	
16p13.3	PKDTS	P	Polycystic kidney disease, infantile severe, with tuberous sclerosis	600273	RE	Polycystic kidney disease, infantile severe, with tuberous sclerosis (3)	

Continued.

Table VII.1 *Continued.*

Location	Locus symbol	Status[a]	Title	MIM#[b]	Method[c]	Disorder(s)	Mouse locus
16p13.3	RSTS	C	Rubinstein–Taybi syndrome	180849	Ch	Rubinstein–Taybi syndrome (2)	
16p13.3	TSC2	C	Tuberous sclerosis-2 (tuberin)	191092	Fd, Ch, D, REn	Tuberous sclerosis-2 (2)	17(Tsc2)
16p13.3-p13.2	CDG1	P	Carbohydrate-deficient glycoprotein	212065	Fd	Carbohydrate-deficient glycoprotein syndrome (2)	
16p13.11	SAH	P	SA (rat hypertension-associated) homologue	145505	REa, A	{?Hypertension, essential} (1)	
16p13	HAGH, GLO2	C	Hydroxyacyl glutathione hydrolase; glyoxalase II	138760	S	[Glyoxalase II deficiency] (1)	
16p13	MEF, FMF	C	Mediterranean fever, familial	249100	Fd, LD	Familial Mediterranean fever (2)	
16p12.3	SCNN1B	P	Sodium channel, non-voltage-gated 1 β	177200	REa, FD	Liddle syndrome (3)	
16p12	CLN3, BTS	C	Ceroid-lipofuscinosis, neuronal-3, juvenile (Batten disease)	204200	F, Fd	Batten disease (2)	
16p11.2	SGLT2	P	Sodium-glucose transporter-2	182381	REa	?Renal glucosuria, 253100 (1)	
16q	SCA4	P	Spinocerebellar ataxia 4	600223	Fd	Spinocerebellar ataxia, type 4 (2)	
16q12-q13.1	PHKB	C	Phosphorylase kinase, β-polypeptide	172490	REa, A	?Phosphorylase kinase deficiency of liver and muscle, 261750 (2)	
16q12.1	TBS	L	Townes–Brocks syndrome	107480	Ch	?Townes–Brocks syndrome (2)	16q13-
q22.1	CES1, SES1	P	Carboxylesterase 1 (monocyte/ macrophage serine esterase 1)	114835	REa	?Monocyte carboxyesterase deficiency (1)	8(Ces1)
16q21	BBS2	P	Bardet–Biedl syndrome 2	209900	Fd	Bardet–Biedl syndrome 2 (2)	
16q21	CETP	P	Cholesteryl ester transfer protein, plasma	118470	REa, A	[CETP deficiency] (3)	
16q22	CBFB	C	Core-binding factor, β-subunit	121360	Ch	Myeloid leukaemia, acute, M4Eo subtype (2)	
16q22-q24	ALDOA	C	Aldolase A, fructose-bisphosphatase	103850	REa, REb, A	Aldolase A deficiency (3)	
16q22.1	CDH1, UVO	C	Cadherin 1 (E-cadherin, uvomorulin)	192090	REa, D, Ch	Endometrial carcinoma (3); ovarian carcinoma (3)	8(Um)
16q22.1	CTM	C	Cataract, Marner type	116800	F	Cataract, Marner type (2)	16q22.1
	LCAT	C	Lecithin-cholesterol acyltransferase	245900	F, LD, A, REa	Norum disease (3); fish-eye disease (3)	8(Lcat)
16q22.1-q22.3	TAT	C	Tyrosine aminotransferase, cytosolic	276600	REa, A, H, D	Tyrosinaemia, type II (3)	8(Tat)
16q24	APRT	C	Adenine phosphoribosyltransferase	102600	S, D	Urolithiasis, 2,8-dihydroxy-adenine (3)	8(Aprt)
16q24	CYBA	C	Cytochrome b-245, α-polypeptide	233690	REa, A	Chronic granulomatous disease, autosomal, due to deficiency of CYBA (3)	
16q24.3	GALNS, MPS4A	C	Galactosamine (*N*-acetyl)-6-sulphate sulfatase	253000	A, Psh	Mucopolysaccharidosis IVA (3)	
Chr.16	ATP2A1	P	ATPase, Ca^{2+} transporting, fast-twitch, 1	108730	REa	Brody myopathy (1)	
Chr.16	CTH	P	Cystathionase	219500	S	[Cystathioninuria] (1)	
17pter-p13	ASPA	P	Aspartoacylase (aminoacylase-2)	271900	A	Canavan disease (3)	
17pter-p12	GP1BA	P	Glycoprotein Ib, platelet, α-polypeptide	231200	A	Bernard–Soulier syndrome (1)	
17pter-p12	PLI	P	α-2-plasmin inhibitor	262850	Psh	Plasmin inhibitor deficiency (3)	
17p13.3	BCPR	L	Breast cancer-related regulator of TP53	113721	D	?Breast cancer (1)	
17p13.3	MDCR, MDS	C	Miller–Dieker syndrome chromosome region	247200	Ch, D	Miller–Dieker lissencephaly syndrome (2)	11(Mds)
17p13.1	TP53	C	Tumour protein p53	191170	REa, A, D	Colorectal cancer, 114500 (3); Li–Fraumeni syndrome (3)	11(Trp53)
17p12-q12	DFNB3	P	Deafness, autosomal recessive 3	600316	Fd	Deafness-3, neurosen-sory non-syndromic recessive (2)	
17p11.2	PMP22, CMT1A	C	Peripheral myelin protein-22	118220	Fd, D, A	Charcot–Marie–Tooth neuropathy, slow nerve conduction type Ia (3); neuropathy, recurrent, with pressure palsies, 162500 (3); Dejerine–Sottas disease, PMP22 related, 145900 (3)	11(Tr)
17p11.2	SMCR	C	Smith–Magenis syndrome chromosome region	182290	Ch	Smith–Magenis syndrome (2)	

Continued on p. 914.

Table VII.1 *Continued.*

Location	Locus symbol	Status[a]	Title	MIM#[b]	Method[c]	Disorder(s)	Mouse locus
17p	RP13	P	Retinitis pigmentosa-13	600059	Fd	Retinitis pigmentosa-13 (2)	
17q	PSS1	P	Psoriasis susceptibility, familial, 1	177900	Fd	Psoriasis susceptibility (2)	
17q11.2	NF1, VRNF, WSS	C	Neurofibromatosis, type 1 (neurofibromatosis, von Recklinghausen disease, Watson syndrome)	162200	Fd, EM, Ch, F	Neurofibromatosis, type I (3); Watson syndrome, 193520 (3)	
17q11.2	SLS	P	Sjogren–Larsson syndrome	270200	Fd, LD	Sjogren–Larsson syndrome (2)	
17q11.2-q24	MHS2	P	Malignant hyperthermia susceptibility 2	154275	Fd	Malignant hyperthermia susceptibility 2 (2)	
17q12	RARA	C	Retinoic acid receptor, α-polypeptide	180240	A, Ch	Leukaemia, acute promyelocytic (1)	11(Rara)
17q12-q21	KRT9, EPPK	C	Keratin-9	144200	Fd, Rea	Epidermolytic palmoplantar keratoderma (3)	
17q12-q21	KRT14, EBS3, EBS4	P	Keratin-14	148066	REa	Epidermolysis bullosa simplex, Koebnertype, 131900 (3); epidermolysis bullosa simplex, Dowling–Meara type, 131670 (3); epidermolysis bullosa simplex Weber–Cockayne type, 131800 (3)	
17q12-q21	PCHC1	P	Pachyonychia congenita 1 (Jackson–Lawler type)	167210	Fd	Pachyonychia congenita, Jackson–Lawler type (2)	
17q12-q21.33	ADL, DAG2, LGMD2D	C	Adhalin	600119	Psh, A	Muscular dystrophy, Duchenne-like, type 2 (3)	
17q21	ACAC, ACC	P	Acetyl-Coenzyme A carboxylase	200350	A	Acetyl-CoA carboxylase deficiency (1)	
17q21	BRCA1	C	Breast cancer-1, early onset	113705	Fd	Breast cancer-1, early onset (3); ovarian cancer, sporadic (3)	
17q21	G6PT	C	Glucose-6-phosphatase	232200	REa, REn	Glycogen storage disease, type I (3)	
17q21-q22	DDPAC	P	Disinhibition-dementia-Parkinsonism-amyotrophy complex	600274	Fd	Disinhibition-dementia-Parkinsonism-amyotrophy complex (2)	
17q21-q22	GALK1	C	Galactokinase-1	230200	S, Ch, R, C	Galactokinase deficiency (1)	11(Glk)
17q21-q22	KRT10	C	Keratin-10	148080	REa, A, REn	Epidermolytic hyperkeratosis, 113800 (3)	
17q21-q22	PENT, PNMT	C	Phenylethanolamine *N*-methyltransferase	171190	REa, Fd	?Hypertension, essential, 145500 (1)	
17q21-q22	SLC4A1, AE1, EPB3	C	Solute carrier family 4, anion exchanger, member 1 (erythrocyte membrane protein band 3, Diego blood group)	109270	REa, RE, Fd, A	[Acanthocytosis, one form] (3); [elliptocytosis, Malaysian-Melanesian type] (3); Spherocytosis, hereditary (3)	
17q21.3-q22	MPO	C	Myeloperoxidase	254600	REa, A, F, Ch, C	Myeloperoxidase deficiency (3)	11(Mpo)
17q21.31-q22.05	COL1A1	C	Collagen I, α-1 polypeptide	120150	C, M, A, REa	Osteogenesis imperfecta, 4 clinical forms, 166200, 166210, 259420, 166220 (3); Ehlers–Danlos syndrome, type VIIA1, 130060 (3); osteoporosis, idiopathic, 166710 (3)	11(Cola1)
17q21.32	ITGA2B, GP2B, CD41B	C	Integrin, α 2b (platelet glycoprotein IIb of IIb/IIIa complex, antigen CD41B)	273800	A, REb, REa, RE, F, LD	Glanzmann thrombasthenia, type A (3); thrombocytopenia, neonatal alloimmune (1)	
17q21.32	ITGB3, GP3A	C	Integrin, β-3 (platelet glycoprotein IIIa; antigen CD61)	173470	REa, REb, A, RE, F, LD	Glanzmann thrombasthenia, type B (3)	
17q22-q24	CSH1, CSA, PL	C	Chorionic somatomammotropin hormone-1	150200	REa, A	[Placental lactogen deficiency] (1)	13(Pl1)
17q22-q24	GH1, GHN	C	Growth hormone-1	139250	REa, A, Fd	Isolated growth hormone deficiency, Illig type with absent GH and Kowarski +type with bioinactive GH (3)	11(Gh)
17q23	DCP1, ACE1	C	Dipeptidyl carboxypeptidase-1 (angiotensin I converting enzyme)	106180	A, H, Fd	{Myocardial infarction, susceptibility to} (3)	
17q23	GAA	C	Glucosidase, acid α-	232300	S, A, D, C	Glycogen storage disease, type II (3)	
17q23-qter	APOH	C	Apolipoprotein H (β-2-glycoprotein I)	138700	Fd, Rea	[Apolipoprotein H deficiency] (3)	11(Apoh)

Continued.

Table VII.1 *Continued.*

Location	Locus symbol	Status[a]	Title	MIM#[b]	Method[c]	Disorder(s)	Mouse locus
17q23-qter	TOC, TEC	P	Tylosis with oesophageal cancer	148500	Fd	Tylosis with oesophageal cancer (2)	
17q23.1-q25.3	SCN4A, HYPP, NAC1A	C	Sodium channel, voltage-gated, type 4, α polypeptide	170500	REa, Fd	Hyperkalaemic periodic paralysis (3); paramyotonia congenita, 168300 (3); myotonia congenita, atypical acetazolamide-responsive (3)	
	CCA1	P	Cataract, congenital, cerulean type	115660	Fd	Cataract, congenital, cerulean type (2)	
17q24.3-q25.1	CMD1, SOX9, SRA1	C	Campomelic dysplasia-1 (sex reversal, autosomal, 1)	211970	Ch	Campomelic dysplasia with autosomal sex reversal (3)	11(Ts, Sox9)
17q25	ACOX	P	Acyl-Coenzyme A oxidase	264470	A	Adrenoleucodystrophy, pseudoneonatal (2)	
17q25	RSS	P	Russell–Silver syndrome	180860	Ch	Russell–Silver syndrome (2)	
18pter-q11	HPE1	L	Holoprosencephaly-1, alobar	236100	Ch	?Holoprosencephaly-1 (2)	
18p11.32	MCL	L	Multiple hereditary cutaneous leiomyomata	150800	Ch	?Leiomyomata, multiple hereditary cutaneous (2)	
18p11.2	MC2R	C	Melanocortin-2 receptor (ACTH receptor)	202200	A, Psh	Glucocorticoid deficiency, due to ACTH unresponsiveness (1)	18(Mc2r)
18q11-q12	LCFS2	L	Lynch cancer family syndrome II	114400	F	?Lynch cancer family syndrome II (2)	
18q11-q12	NPC	C	Niemann–Pick disease, type C	257220	Ch, H, Fd, M	Niemann–Pick disease, type C (2)	18(spm)
18q11.2-q12.1	TTR, PALB	C	Transthyretin (prealbumin)	176300	REa, A	Amyloid neuropathy, familial, several allelic types (3); [Dystransthyretinaemic hyperthyroxinaemia] (3); amyloidosis, senile systemic (3); carpal tunnel syndrome, familial (3)	18 (Palb)
18q21.1-q22	FEO	P	Familial expansile osteolysis	174810	Fd	Familial expansile osteolysis (2)	
18q21.3	BCL2	C	B-cell CLL/lymphoma-2	151430	Ch, RE, REn	Leukaemia/lymphoma, B-cell, 2 (2)	1(Bcl2)
18q21.3	FECH, FCE	C	Ferrochelatase	177000	A, Reb	Protoporphyria, erythropoietic (3); protoporphyria, erythropoietic, recessive, with liver failure (3)	
18q21.3	FVT1	P	Follicular lymphoma, variant translocation 1	136440	RE	Lymphoma/leukaemia, B-cell, variant (1)	
18q22-qter	MS1	L	Multiple sclerosis	126200	Fd, LD	{?Multiple sclerosis, susceptibility to} (2)	
18q22.1	GTS	L	Gilles de la Tourette syndrome	137580	Ch	?Tourette syndrome (2)	
18q23	CYB5	C	Cytochrome b5	250790	Psh, REa, A	Methaemoglobinaemia due to cytochrome b5 deficiency (3)	
18q23.3	DCC	C	Deleted in colorectal carcinoma	120470	D, RE	Colorectal cancer (3)	18(Dcc)
19p13.3	FUT6	P	Fucosyltransferase 6 (α (1,3) fucosyltransferase)	136836	Psh, REn	Fucosyltransferase-6 deficiency (3)	
19p13.3	HHC2, FHH2	P	Hypocalciuric hypercalcaemia-2	145981	Fd	Hypocalciuric hypercalcaemia, type II (2)	
19p13.3	TBXA2R	C	Thromboxane A2 receptor	188070	Psh, Fd, A, H	Bleeding disorder due to defective thromboxane A2 receptor (3)	10(Tbxa2r)
19p13.3	TCF3, E2A	C	Transcription factor-3 (E2A immunoglobulin enhancer-binding factors 12/E47)	147141	REa, A	Leukaemia, acute lymphoblastic (1)	
19p13.3-p13.2	AMH, HIF	P	Anti-Mullerian hormone	261550	REa, A	Persistent Mullerian duct syndrome (3)	10(Amh)
19p13.3-p13.2	ATHS, ALP	P	Atherosclerosis susceptibility (lipoprotein associated)	108725	Fd	{Atherosclerosis, susceptibility to} (2)	
19p13.3-q13.2	C3	C	Complement component-3	120700	F, S, A, Rea	C3 deficiency (3)	17(C3)
19p13.3-q13.2	EPOR	C	Erythropoietin receptor	133171	A, REa, H, Fd	[Erythrocytosis, familial], 133100 (3)	9(Epor)
19p13.2	GCDH	P	Glutaryl-Coenzyme A dehydrogenase	231670	REa, A	Glutaricacidaemia, type I (3)	
19p13.2	INSR	C	Insulin receptor	147670	REa, A, REb	Leprechaunism (3); diabetes mellitus, insulin-resistant, with acanthosis nigricans (3); Rabson–Mendelhall syndrome (3)	8(Insr)

Continued on p. 916.

Table VII.1 *Continued.*

Location	Locus symbol	Status[a]	Title	MIM#[b]	Method[c]	Disorder(s)	Mouse locus
19p13.2-p13.1	LDLR, FHC	C	Familial hypercholesterolaemia (LDL receptor)	143890	F, REa, A	Hypercholesterolaemia, familial (3)	9(Ldlr)
19p13.2-p13.1	LYL1	C	Lymphoblastic leukaemia derived sequence-1	151440	Ch, A	Leukaemia, T-cell acute lymphoblastoid (2)	8(Lyl1)
19p13.2-q13.3	LPSA, D19S381E	P	Oncogene liposarcoma (DNA Segment, single copy, expressed, probes MC15, MC6)	164953	A	Liposarcoma (1)	
19p13.1	RFX1	C	Regulatory factor (trans-acting) 1 (influences HLA class II expression)	600006	A	Severe combined immunodeficiency, HLA class II-negative type, 209920 (2)	
19p13	MHP1	C	Migraine, hemiplegic 1	141500	Fd	Migraine, hemiplegic-1 (2)	
19p	APCA, CAPA	P	Cerebellar ataxia, paroxysmal acetazolamide-responsive	108500	Fd	Cerebellar ataxia, paroxysmal acetazolamide-responsive 2)	
19p	EXT3	P	Exostoses (multiple) 3	600209	Fd	Exostoses, multiple, type 3 (2)	
19cen-q12	MANB	C	Mannosidase, alpha B, lysosomal	248500	S, Psh	Mannosidosis (1)	
19cen-q13.11	PEPD	C	Peptidase D (prolidase)	170100	S, F, H, Fd	Prolidase deficiency (3)	7(Pep4)
19cen-q13.2	AD2	C	Alzheimer disease-2 (late-onset)	104310	Fd	Alzheimer disease-2, late onset (2)	
19q12	CADASIL, CASIL	P	Cerebral autosomal dominant arteriopathy with subcortical infarcts and leukoencephalopathy	125310	Fd	Cerebral arteriopathy with subcortical infarcts and leukoencephalopathy (2)	
19q12	EDM1, MED	C	Epiphyseal dysplasia, multiple 1	132400	Fd	Epiphyseal dysplasia, multiple 1 (2)	
19q12	PSACH	C	Pseudoachondroplastic dysplasia	177170	Fd	Pseudoachondroplastic dysplasia (2)	
19q12-q13.1	NPHS1, CNF, NFC	P	Nephrosis 1, congenital, Finnish type	256300	Fd	Nephrosis, congenital, Finnish (2)	
19q13	BCL3	C	B-cell CLL/lymphoma-3	109560	Ch, S, H	Leukaemia/lymphoma, B-cell, 3 (2)	7(Bcl3)
19q13.1	GPI	C	Glucose phosphate isomerase; neuroleukin	172400	S, D, A	Haemolytic anemia due to glucosephosphate isomerase deficiency (3); hydrops fetalis, one form (1)	7(Gpi1)
19q13.1	RYR1, MHS, CCO	C	Ryanodine receptor-1 (skeletal)	180901	A, Fd, H	Malignant hyperthermia susceptibility-1, 145600 (3); central core disease, 117000 (3)	7(Ryr)
19q13.1-q13.2	AKT2	P	Murine thymoma viral (v-akt) homologue-2	164731	A	Ovarian carcinoma, 167000 (2)	
19q13.1-q13.2	BCKDHA, MSUD1	C	Branched chain keto acid dehydrogenase E1, α polypeptide	248600	REa, REb, A	Maple syrup urine disease, type Ia (3)	
19q13.1-q13.2	CORD2, CRD	P	Cone rod dystrophy 2 (autosomal dominant)	120970	Fd	Cone-rod retinal dystrophy (2)	
19q13.2	APOE	C	Apolipoprotein E	107741	F, REa, LD, A, Fd	Hyperlipoproteinaemia, type III (3)	7(Apoe)
19q13.2	APOC2	C	Apolipoprotein C-II	207750	REa, F, LD, A, Fd	Hyperlipoproteinaemia, type Ib (3)	
19q13.2-q13.3	DM	C	Dystrophia myotonica	160900	F, Fd	Myotonic dystrophy (3)	7(Dm)
19q13.2-q13.3	ERCC2, EM9	C	Excision repair cross complementing rodent repair deficiency, complementation group-2	126340	S, RE, M	Xerodermal pigmentosum, group D, 278730 (3)	7(Ercc2)
19q13.2-q13.3	HB1, PFHB1	P	Heart block, progressive familial, type I	113900	Fd	Heart block progressive familial, type I (2)	
19q13.2-q13.3	LIG1	C	Ligase I, DNA, ATP-dependent	126391	REa, A	DNA ligase I deficiency (3)	
19q13.2-q13.3	PVS	C	Polio virus sensitivity	173850	S, A, REa	{Polio, susceptibility to} (2)	9(Pvs)
19q13.3	ETFB	C	Electron transfer flavoprotein, β-polypeptide	130410	REa, A	Glutaricaciduria, type IIB (3)	
19q13.3	GYS1, GYS	C	Glycogen synthase	138570	REa, A	{Non-insulin dependent diabetes mellitus, susceptibility to} (2)	
19q13.32	LHB	C	Luteinizing hormone, β-polypeptide	152780	RE	Hypogonadism, hypergonadotropic (3); ?male pseudohermaphroditism due to defective LH (1)	7(Lhb)
19q13.4	RP11	P	Retinitis pigmentosa-11 (autosomal dominant)	600138	Fd	Retinitis pigmentosa-11 (2)	
Chr.19	BCT2	P	Branched chain aminotransferase-2	113530	S	?Hypervalinaemia or hyperleucineisoleucinaemia (1)	

Continued.

Table VII.1 *Continued.*

Location	Locus symbol	Status[a]	Title	MIM#[b]	Method[c]	Disorder(s)	Mouse locus
20pter-p12	PRNP, PRIP	C	Prion protein (p27–30)	176640	REa, REb, A	Creutzfeldt–Jakob disease, 123400 (3); Gerstmann–Straussler disease, 137440 (3); insomnia, fatal familial (3)	2(Prnp)
20p13	AVP, AVRP, VP	C	Arginine vasopressin (neurophysin II, antidiuretic hormone)	192340	REa, RE, Fd	Diabetes insipidus, neurohypophyseal, 125700 (3)	2(Avp)
20p12	BMP2, BMP2A	C	Bone morphogenetic protein-2	112261	H, REa, A	?Fibrodysplasia ossificans progressiva (1)	2(Bmp2a)
20p12-cen	THBD, THRM	C	Thrombomodulin	188040	REb, A	Thrombophilia due to thrombomodulin defect (3)	
20p11.2	AGS, AHD	C	Alagille syndrome (arteriohepatic dysplasia)	118450	Ch, D	Alagille syndrome (2)	
20p11	CST3	C	Cystatin C	105150	REa, A	Cerebral amyloid angiopathy (3)	2(Cst3)
20p	ITPA	C	Inosine triphosphatase-A	147520	S	[Inosine triphosphatase deficiency] (1)	2(Itp)
20q11.2	GHRF	C	Growth hormone releasing factor; somatocrinin	139190	REa, REb, Ch, Fd, A	?Isolated growth hormone deficiency due to defect in GHRF (1); gigantism due to GHRF hypersecretion (1)	
20q13	MODY1	C	Maturity-onset diabetes of the young, type I	125850	Fd	MODY, type I (2)	
20q13.1	PPGB, GSL, NGBE, GLB2	C	Protective protein for β-galactosidase	256540	S, A, Fd	Galactosialidosis (3)	2(Ppgb)
20q13.11	ADA	C	Adenosine deaminase	102700	S, D, REa, F, A, Fd	Severe combined immunodeficiency due to ADA deficiency (3); hemolytic anemia due to ADA excess (1)	2(Ada)
20q13.2	GNAS1, GNAS, GPSA	C	Guanine nucleotide-binding protein (G protein), α-stimulating activity	139320	REa, H, A, Fd	Pseudohypoparathyroidism, type Ia, 103580 (3); McCune–Albright polyostotic fibrous dysplasia, 174800 (3); somatotrophinoma (3)	2(Gnas)
20q13.2-q13.3	CHRNA4, EBN1	C	Cholinergic receptor, nicotinic, α polypeptide-4	118504	REa, REn, A, Fd	Epilepsy, benign neonatal, type I, 121200 (3)	2(Acra4)
20q13.2-q13.3	FA1, FA, FACA	P	Fanconi anaemia-1	227650	Fd	Fanconi anaemia-1 (2)	
20q13.31	PCK1	C	Phosphoenolpyruvate carboxykinase-1 (soluble)	261680	REa, A, Fd	?Hypoglycaemia due to PCK1 deficiency (1)	2(Pck1)
21q11.2	MST	L	Myeloproliferative syndrome, transient	159595	Ch	?Leukaemia, transient (2)	
21q21.3-q22.05	APP, AAA, CVAP	C	Amyloid β (A4) precursor protein	104760	REa, A, Fd, RE	Amyloidosis, cerebroarterial, Dutch type (3); Alzheimer disease, APP-related (3); schizophrenia, chronic (3)	16(App)
21q22.1	HCS	P	Holocarbyoxylase synthetase	253270	Psh, A	Multiple carboxylase deficiency, biotin-responsive (3)	
21q22.1	SOD1, ALS1	C	Superoxide dismutase-1, soluble	147450	S, D, Fd	Amytrophic lateral sclerosis, due to SOD1 deficiency, 105400 (3)	16(Sod1)
21q22.3	APECED	P	Autoimmune polyglandular disease, type I	240300		Autoimmune polyglandular disease, type I (2)	
21q22.3	CBFA2, AML1	C	Core-binding factor, runt domain, α subunit (aml1 oncogene)	151385	Ch, Fd	Leukaemia, acute myeloid (3)	
21q22.3	CBS	C	Cystathionine β-synthase	236200	S, D, A, Fd	Homocystinuria, B6-responsive and non-responsive types (3)	17(Cbs)
21q22.3	DSCR	C	Down syndrome (critical region)	190685	Ch	Down syndrome (1)	
21q22.3	EPM1	C	Epilepsy, progressive myoclonic 1	254800	Fd, LD	Epilepsy, progressive myoclonus (2)	
21q22.3	ITGB2, CD18, LCAMB, LAD	C	Integrin, β-2 (antigen CD18 (p95), lymphocyte function-associated antigen-1; macrophage antigen, β polypeptide)	116920	S, A, Fd	Leukocyte adhesion deficiency (1)	7(Ly15)
21q22.3	PFKL	C	Phosphofructokinase, liver type	171860	S, D, Fd	Haemolytic anaemia due to phosphofructokinase deficiency (1)	17(Pfk1)
22q11	CECR, CES	C	Cat eye syndrome	115470	Ch, A, D	Cat eye syndrome (2)	
22q11	CTHM	L	Conotruncal cardiac anomalies	217095	D	?Conotruncal cardiac anomalies (2)	

Continued on p. 918.

Table VII.1 *Continued.*

Location	Locus symbol	Status[a]	Title	MIM#[b]	Method[c]	Disorder(s)	Mouse locus
22q11	DGCR, DGS, VCF	C	DiGeorge syndrome chromosome region (velocardiofacial syndrome)	188400	Ch, D	DiGeorge syndrome (2); Velocardiofacial syndrome, 192430 (2)	
22q11	HCF2, HC2	C	Heparin cofactor II	142360	REb, REa	Thrombophilia due to heparin cofactor II deficiency (3)	
22q11	NAGA	C	Acetylgalactosaminidase, α-*N*- (α-galactosidase B)	104170	S, Ch	Schindler disease (3); Kanzaki disease (3)	
22q11.1-q11.2	GGT1, GTG	C	γ-glutamyltransferase-1	231950	A, S, F, RE	Glutathioninuria (1)	
22q11.12	GGT2	P	γ-glutamyltransferase-2	137181	REn	[Gamma-glutamyltransferase, familial high serum] (2)	
22q11.2-qter	SGLT1	P	Sodium-glucose transporter-1	182380	Rea	Glucose/galactose malabsorption (3)	
22q11.2-qter	TCN2, TC2	C	Transcobalamin II	275350	F, S, D	Transcobalamin II deficiency (3)	11(Tcn2)
22q11.21	BCR, CML, PHL	C	Breakpoint cluster region	151410	Ch, RE	Leukaemia, chronic myeloid (3)	10(Bcr)
22q12	EWSCR, EWS	C	Ewing sarcoma breakpoint region-1	133450	Ch	Ewing sarcoma (3); Neuroepithelioma (2)	
22q12.1-q13.2	TIMP3, SFD	C	Tissue inhibitor of metalloproteinase-3	188826	REa, A, Fd	Sorsby fundus dystrophy, 136900 (3)	
22q12.2	NF2	C	Neurofibromatosis-2 (bilateral acoustic neuroma)	101000	RE, F, Ch, D, Fd	Neurofibromatosis, type 2 (3); meningioma, NF2-related (3); Schwannoma, sporadic (3);	11(Nf2)
22q12.3-q13.1	PDGFB, SIS	C	Platelet-derived growth factor, β polypeptide (oncogene SIS)	190040	REa, Fd	Meningioma, SIS-related (3)	15(Pdgfb)
22q13-qter	ACR	P	Acrosin	102480	REa	?Male infertility due to acrosin deficiency (2)	15(Acr)
22q13.1	ADSL	C	Adenylosuccinate lyase	103050	S, REa, A	Adenylosuccinase deficiency (1); autism, succinylpurinaemic (3)	
22q13.1	CYP2P@, CYP2D, P450C2D	C	Cytochrome P450, subfamily IID	124030	F, Fd, Psh, A	{?Parkinsonism, susceptibility to} (1); Debrisoquine sensitivity (3)	15(Cyp2d)
22q13.1-qter	SFD	P	Sorsby fundus dystrophy	136900	Fd	Sorsby fundus dystrophy (2)	
22q13.31-qter	ARSA	C	Arylsulphatase A	250100	S, D	Metachromatic leukodystrophy (3)	15(As2)
22q13.31-qter	DIA1	C	Diaphorase (NADH); cytochrome b-5 reductase	250800	S, REa	Methaemoglobinaemia, enzymopathic (3)	15(Dia1)
Xpter-p22.32	GCFX, SS	P	Growth control factor, X-linked	312865	Fd	Short stature (2)	
Xpter-p22.2	CFND	L	Craniofrontonasal dysplasia	304110	Ch	?Craniofrontonasal dysplasia (2)	
Xp22.32	CSF2RA	C	Colony-stimulating factor-2 receptor, α, low-affinity (granulocyte-macrophage)	306250	A	Leukaemia, acute myeloid, 19(Csf2ra) M2 type (1)	
Xp22.32	STS, ARSC1, SSDD	C	Steroid sulphatase, microsomal	308100	F, S, D	Ichthyosis, X-linked (3); placental steroid sulphatase deficiency (3)	X,Y(Sts)
Xp22.31	DHOF, FODH	P	Dermal hypoplasia, focal	305600	Ch	Focal dermal hypoplasia (2)	
Xp22.3	ARSE, CDPX1, CDPXR	C	Arylsulphatase E	302950	D, Fd	Chondrodysplasia punctata, X-linked recessive (3)	
Xp22.3	KAL1, KMS, ADMLX	C	Kallmann syndrome-1 sequence	308700	F, Fd, D, REa, Ren	Kallman syndrome (3)	
Xp22.3	OA1	C	Ocular albinism-1, Nettleship–Falls type	300500	F, Fd	Ocular albinism, Nettleship-Falls type (3)	
Xp22.3	OASD	P	Ocular albinism and sensorineural deafness	300650	Fd	Ocular albinism with sensorineural deafness (2)	
Xp22.3-p22.1	AMELX, AMG, AIH1, AMGX	C	Amelogenin	301200	REa, A, Fd	Amelogenesis imperfecta (3)	X-(Amel)
Xp22.3-p21.1	NHS	C	Nance–Horan cataract-dental syndrome	302350	Fd	Nance–Horan syndrome (2)	?X(Xcat)
Xp22.3-p21.1	POLA	C	Polymerase (DNA directed), α	312040	S	?N syndrome, 310465 (1)	X(Pola)
Xp22.3-p22.1	RS	C	Retinoschisis	312700	F, Fd	Retinoschisis (2)	
Xp22.2	CMTX2	P	Charcot–Marie–Tooth disease,	302801	Fd	Charcot–Marie–Tooth neuropathy, X-linked-2, recessive (2)	
Xp22.2	FCPX, FCP	P	F-cell production	305435	F, Fd	Heterocellular hereditary persistence of fetal haemoglobin, Swiss type (2)	

Continued.

Table VII.1 *Continued.*

Location	Locus symbol	Status[a]	Title	MIM#[b]	Method[c]	Disorder(s)	Mouse locus
Xp22.2	HOMG, HSH, HMGX	P	Hypomagnesaemia, secondary hypocalcaemia	307600	X/A	Hypomagnesaemia, X-linked primary (2)	
Xp22.2	MLS, MITF	P	Microphthalmia with linear skin defects (microphthalmia-associated transcription factor)	309801	Ch	?Microphthalmia with linear skin defects (2)	
Xp22.2-p21.2	KFSD	P	Keratosis follicularis spinulosa decalvans	308800	Fd	Keratosis follicularis spinulosa decalvans (2)	
Xp22.2-p22.1	CLS	P	Coffin–Lowry syndrome	303600	Fd	Coffin–Lowry syndrome (2)	
Xp22.2-p22.1	HYP, HPDR1	C	Hypophosphataemia, vitamin D resistant rickets	307800	Fd	Hypophosphataemia, hereditary (2)	X(Hyp)
Xp22.2-p22.1	PDHA1, PHE1A	C	Pyruvate dehydrogenase, El-α polypeptide-1	312170	REa, A	Pyruvate dehydrogenase deficiency (3)	X(Pdha1)
Xp22.2-p22.1	PHK, PHKA2	C	Phosphorylase kinase deficiency, liver (glycogen storage disease type VIII)	306000	Fd, REa, A	Glycogen storage disease, X-linked hepatic (2)	
Xp22.2-p22.1	PRTS, MRXS1	P	Partington syndrome (mental retardation, X-linked, syndromic-1, with dystonic movements, ataxia, and seizures)	309510	Fd	Mental retardation, X-linked, syndromic-1, with dystonic movements, ataxia, and seizures (2)	
Xp22.2-p22.1	SEDL, SEDT	C	Spondyloepiphyseal dysplasia, late	313400	Fd	Spondyloepiphyseal dysplasia tarda (2)	
Xp22.11-p21.2	GDXY, TDFX, SRVX	P	Gonadal dysgenesis, XY female type	306100	F, Ch	Gonadal dysgenesis, XY female type (2)	
Xp22	AGMX2, XLA2, IMD6	P	Agammaglobulinaemia, X-linked 2 (with growth hormone deficiency)	300310	Fd	Agammaglobulinaemia, type 2, X-linked (2)	X(Xid)
Xp22	AIC	C	Aicardi syndrome	304050	X/A, Ch	Aicardi syndrome (2)	
Xp22	GY, HYP2	L	Hereditary hypophosphataemia II (gyro equivalent)	307810	H	?Hypophosphataemia with deafness (2)	X(Gy)
Xp22	MRX1	C	Mental retardation, X-linked-1, non-dysmorphic	309530	F, Fd, D	Mental retardation, X-linked-1, non-dysmorphic (2)	
Xp22-p21	PDR	P	Pigment disorder, reticulate	301220	Fd	Partington syndrome II (2)	
Xp21.3-p21.2	DAX1, AHC, AHX	C	DSS, AHC, X gene 1	300200	D, Fd	Adrenal hypoplasia, congenital, with hypogonadotrophic hypogonadism (3)	
Xp21.3-p21.2	GK	C	Glycerol kinase deficiency	307030	D, Fd	Glycerol kinase deficiency (2)	
Xp21.3-p21.2	RP6	L	Retinitis pigmentosa-6 (X-linked recessive)	312612	Fd	?Retinitis pigmentosa-6 (2)	
Xp21.2	DFN4	P	Deafness 4, congenital sensorineural	600203	Fd	Deafness 4, congenital sensorineural (2)	
Xp21.2	DMD, BMD	C	Dystrophin (muscular dystrophy, Duchenne and Becker types)	310200	X/A, Fd, D	Duchenne muscular dystrophy (3); Becker muscular dystrophy (3); Cardiomyopathy, dilated, X-linked (3)	X(Dmd)
Xp21.2-p21.1	XK	C	Kell blood group precursor	314850	F, D	McLeod phenotype (3)	
Xp21.1	CYBB, CGD	C	Cytochrome b-245, β-polypeptide	306400	F, D	Chronic granulomatous disease, X-linked (3)	X(Cybb)
Xp21.1	OTC	C	Ornithine transcarbamylase	311250	L, REa, A, D	Ornithine transcarbamylase deficiency (3)	X(spf; Otc)
Xp21.1	RP3	C	Retinitis pigmentosa-3 (X-linked recessive)	312610	Fd, D	Retinitis pigmentosa-3 (2)	
Xp21.1-q22	WTS, MRXS6	P	Wilson–Turner syndrome (mental retardation, X-linked, syndromic-6, with gynaecomastia and obesity)	309585	Fd	Mental retardation, X-linked, syndromic-6, with gynaecomastia and obesity (2)	
Xp21	GTD	L	Gonadotropin deficiency	306190	D	?Gonadotropin deficiency (2); ?cryptorchidism (2)	
Xp21	SRS, MRSR	P	Snyder–Robinson X-linked mental retardation syndrome	309583	Fd	Mental retardation, Snyder–Robinson type (2)	
Xp11.4	NDP, ND	C	Norrie disease (pseudoglioma)	310600	Fd, D	Norrie disease (3), Exudative vitreoretinopathy, X-linked, 305390 (3)	
Xp11.4-p11.23	AIED, OA2	C	Aland island eye disease (ocular albinism, Forsius–Eriksson type)	300600	F, D, Fd	Ocular albinism, Forsius–Eriksson type (2)	
Xp11.4-p11.23	PFC, PFD	C	Properdin P factor, complement	312060	Fd, REa, A	Properdin deficiency, X-linked (3)	X(Pfc)
Xp11.3	COD1, PCDX	C	Cone dystrophy-1 (X-linked)	304020	Fd	Progressive cone dystrophy (2)	
Xp11.3	CSNB1	I	Congenital stationary night blindness-1	310500	Fd	Night blindness, congenital stationary, type I (2)	
Xp11.3	RP2	C	Retinitis pigmentosa-2 (X-linked recessive)	312600	Fd	Retinitis pigmentosa-2 (2)	

Continued on p. 920.

Table VII.1 *Continued.*

Location	Locus symbol	Status[a]	Title	MIM#[b]	Method[c]	Disorder(s)	Mouse locus
Xp11.23	MAOA	C	Monoamine oxidase A	309850	Fd, REa, D, A, REn	Brunner syndrome (3)	X(Maoa)
Xp11.23-p11.22	NPHL2, DENTS	P	Nephrolithiasis 2, X-linked (Dent syndrome)	600248	Fd	Nephrolithiasis 2, X-linked (2)	
Xp11.23-p11.22	WAS, IMD2, THC	C	Wiskott–Aldrich syndrome	301000	Fd, X/A	Wiskott–Aldrich syndrome (3); Thrombocytopenia, X-linked, 313900 (3)	
Xp11.22	CLCK2	P	Chloride channel, voltage-gated, K2	600260	RE	?Dent disease, 310468 (2)	
Xp11.22	NPHL1, XRN	P	Nephrolithiasis 1 (X-linked)	310468	Fd	Nephrolithiasis, X-linked, with renal failure (2)	
Xp11.21	ALAS2, ASB, ANH1	C	Aminolaevulinate, δ-, synthase-2	301300	Ch, REa, A, Fd	Anaemia, sideroblastic/hypochromic (3)	
Xp11.21	FGDY, AAS	C	Faciogenital dysplasia (Aarskog–Scott syndrome)	305400	X/A, Fd	Aarskog–Scott syndrome (3)	
Xp11.21	IP1, IP	C	Incontinentia pigmenti-1, sporadic type	308300	X/A	Incontinentia pigmenti, sporadic type (2)	X(Td)
Xp11.2	RCCP2	P	Renal cell carcinoma, papillary	312390	Ch, S	Renal cell carcinoma, papillary, 2 (2)	
Xp11.2	SSRC, SSX	C	Sarcoma, synovial	312820	Ch, RE, A	Sarcoma, synovial (3)	Xp11
	MRXA	L	Mental retardation, X-linked non-specific, with aphasia	309545	Fd	?Mental retardation, X-linked non-specific, with aphasia (2)	
Xp11-q21	PRS, MRXS2	P	Prieto syndrome (mental retardation, X-linked, syndromic-2, with dysmorphism and cerebral atrophy)	309610	Fd	Mental retardation, X-linked, syndromic-2, with dysmorphism and cerebral atrophy (2)	
Xp11-q21.3	SHS, MRXS3	P	Sutherland–Haan syndrome (mental retardation, X-linked, syndromic-3, with spastic diplegia)	309470	Fd	Mental retardation, X-linked, syndromic-3, with spastic diplegia (2)	
Xp	CCT	L	Cataracts, congenital total	302200	Fd	?Cataract, congenital total (2)	
Xp	RTT, RTS	L	Rett syndrome	312750	Ch	?Rett syndrome (2)	
Xp	SMAX2	P	Spinal muscular atrophy, X-linked lethal infantile	600199	Fd	Spinal muscular atrophy X-linked lethal infantile (2)	
Xq11-q12	AR, DHTR, TFM, SBMA, KD	C	Androgen receptor (dihydrotestosterone receptor)	313700	S, Fd, REa, A	Androgen insensitivity, several forms (3); spinal and bulbar muscular atrophy of Kennedy, 313200 (3); prostate cancer (3); perineal hypospadias (3); breast cancer, male, with Reifenstein syndrome (3)	X(Tfm)
Xq11-q12	MRX2	L	Mental retardation, X-linked-2, non-dysmorphic	309540	Fd	?Mental retardation, X-linked-2, non-dysmorphic (2)	
Xq12-q13	ATP7A, MNK, MK	C	ATPase, Cu^{2+} transporting, α-polypeptide	309400	Fc, X/A, H	Menkes' disease (2)	X(Mnk)
Xq12-q13.1	DYT3	C	Torsion dystonia-Parkinsonism, Filipino type	314250	Fd	Torsion dystonia-Parkinsonism, Filipino type (2)	
Xq12-q21	JMS	P	Juberg–Marsidi syndrome	309590	Fd	Juberg–Marsidi syndrome (2)	
Xq12.2-q13.1	EDA, HED	C	Anhidrotic ectodermal dysplasia	305100	X/A, H, Fd	Anhidrotic ectodermal dysplasia (2)	X(Ta)
Xq13	ASAT	L	Anaemia, sideroblastic, with spinocerebellar ataxia	301310	Fd	?Anaemia, sideroblastic, with spinocerebellar ataxia (2)	
Xq13	IL2RG, SCIDX1, SCIDX, IMD4	C	Interleukin-2 receptor, γ	308380	Fd	Severe combined immunodeficiency X-linked, 300400 (3); combined immunodeficiency, X-linked, moderate, 312863 (3)	X(Il2rg)
Xq13	PGK1, PGKA	C	Phosphoglycerate kinase-1	311800	S, R, REb, Fd	Haemolytic anaemia due to PGK deficiency (3); myoglobinuria/haemolysis due to PGK deficiency (3)	X(Pgk1)
Xq13	PHKA1	C	Phosphorylase kinase, muscle, α-polypeptide	311870	REa, A, REn	Muscle glycogenosis (3)	X(Phka)
Xq13	RAD54, XH2, ATRX, ATR2	C	RAD54 (*Saccharomyces cerevisiae*)	600254	RE, Fd	α-thalassaemia/mental retardation syndrome, type 2, 301040 (3)	X(Xh2)
Xq13-q21	WWS	P	Wieacker–Wolff syndrome	314580	Fd	Wieacker–Wolff syndrome (2)	
Xq13-q22	MCS, MRXS4	P	Miles–Carpenter syndrome (mental retardation, X-linked, syndromic-4, with congenital contractures and low fingertip arches)	309605	Fd	Mental retardation, X-linked, syndromic-4, with congenital contractures and low fingertip arches (2)	

Continued.

Table VII.1 *Continued.*

Location	Locus symbol	Status[a]	Title	MIM#[b]	Method[c]	Disorder(s)	Mouse locus
Xq13.1	GJB1, CX32, CMTX1	C	Gap junction protein, β-1, 32 kD (connexin 32)	304040	REa, Fd	Charcot–Marie–Tooth neuropathy, X-linked-1, dominant, 302800 (3)	X(Gjb1)
Xq13.1	RPS4X, CCG2, SCAR	C	Ribosomal protein S4, X-linked	312760	A, REa, REn	Turner syndrome (1)	X(Rps4x)
Xq21	AHDS	P	Allan–Herndon–Dudley mental retardation syndrome	309600	Fd	Allan–Herndon syndrome (2)	
Xq21.1	POU3F4, DFN3	P	POU domain, class 3, transcription factor 4	600420	Fd, D, H, REn	Deafness, conductive, with stapes fixation, 304400 (3)	
Xq21.1-q21.31	CPX	C	Cleft palate and/or ankyloglossia	303400	Fd, D	Cleft palate, X-linked (2)	
Xq21.2	CHM, TCD	C	Choroideraemia	303100	Fd, LD, D, A, Ch, X/A	Choroideraemia (3)	
Xq21.3-q22	BTK, AGMX1, IMD1, XLA, AT	C	Bruton agammaglobulinaemia tyrosine kinase	300300	H, Fd, A	Agammaglobulinaemia, type 1, X-linked (3); ?XLA and isolated growth hormone deficiency, 307200 (3)	X(xid, Btp)
Xq21.3-q22	MGC1, MGCN	P	Megalocornea-1, X-linked	309300	Fd	Megalocornea, X-linked (2)	
Xq21.3-q22	PHP, GHDX	L	Panhypopituitarism, X-linked	312000	Fd	?Panhypopituitarism, X-linked (2)	
Xq22	COL4A5, ATS, ASLN	C	Collagen, type IV, alpha-5 polypeptide	303630	REa, A, Fd	Alport syndrome, 301050 (3); Leiomyomatosis-nephropathy syndrome, 308940 (1)	
Xq22	COL4A6	C	Collagen, type IV, α-6 polypeptide	303631	REn, A	Leiomyomatosis, diffuse (1); ?Alport syndrome, X-linked, type 2 (1)	
Xq22	GLA	C	Galactosidase, α	301500	S, R, A, Fd	Fabry disease (3)	X(Ags)
Xq22	PLP, PMD	C	Proteolipid protein; Pelizaeus–Merzbacher disease	312080	REa, A, Ch, R, Fd	Pelizaeus–Merzbacher disease (3); Spastic paraplegia 2, 312920 (3)	X-P1p(jp))
Xq22	TBG	C	Thyroxine-binding globulin	314200	REa, A	[Euthyroidal hyper- and hypothyroxinaemia] (1)	
Xq22-q24	PRPS1	C	Phosphoribosyl pyrophosphate synthetase-1	311850	S, R, REa, A	Phosphoribosyl pyrophosphate synthetase-related gout (3)	
Xq22-q28	AIH3	L	Amelogenesis imperfecta-3, hypomaturation or hypoplastic type	301201	Fd	?Amelogenesis imperfecta-3, hypoplastic type (2)	
Xq22.1	PIGA	P	Phosphatidylinositol glycan class A	311770	A	Paroxysmal nocturnal haemoglobinuria (3)	X(Piga)
Xq25	LYP, IMD5, XLP, XLPD	C	Lymphoproliferative syndrome	308240	Fd, D	Lymphoproliferative syndrome, X-linked (2)	
Xq25-q26	HTX1	P	Heterotaxy-1	306955	Fd	Heterotaxy, X-linked visceral (2)	
Xq25-q26.1	TAS	P	Thoracoabdominal syndrome	313850	Fd	Thoracoabdominal syndrome (2)	
Xq25-q27	PGS, MRXS5	P	Pettigrew syndrome (mental retardation, X-linked, with Dandy–Walker malformation, basal ganglia disease, and seizures)	304340	Fd	Mental retardation, X-linked, syndromic-5, with Dandy–Walker malformation, basal ganglia disease, and seizures (2)	
Xq26	CD40LG, HIGM1 IGM	C	CD40 antigen ligand (hyper-IgM syndrome)	308230	Fd, A, Psh	Immunodeficiency, X-linked, with hyper-IgM (3)	X(CD40l)
Xq26	GUST	P	Gustavson mental retardation syndrome (with microcephaly, optic atrophy, deafness)	309555	Fd	Gustavson syndrome (2)	
Xq26	SDYS, SGB	C	Simpson dysmorphia syndrome	312870	Fd	Simpson–Golabi–Behmel syndrome (2)	
Xq26	SHFM2, SHFD2	P	Split hand/foot malformation, type (ectrodactyly) 2	313350	Fd	Split hand/foot malformation, type 2 (2)	
Xq26-q27	BFLS	P	Borjeson–Forssman–Lehmann syndrome	301900	Fd	Borjeson–Forssman–Lehmann syndrome (2)	
Xq26-q27	HPT, HPTX, HYPX	P	Hypoparathyroidism	307700	Fd	Hypoparathyroidism, X-linked (2)	
Xq26-q27	POF1, POF	L	Premature ovarian failure-1	311360	Ch	Ovarian failure, premature (2)	
Xq26-q27.2	HPRT	C	Hypoxanthine phosphoribosyltransferase	308000	S, M, C, R, REa, Fd	Lesch–Nyhan syndrome (3); HPRT-related gout (3)	X(Hprt)
Xq26-qter	INDX	P	Immunoneurologic syndrome, X-linked, of Wood, Black, and Norbury	600486	Fd	Wood's neuroimmunologic syndrome (2)	
Xq26.1	OCRL, LOCR, OCRL1	C	Oculocerebrorenal syndrome of Lowe	309000	X/A, Fd	Lowe syndrome (3)	

Continued on p. 922.

Table VII.1 *Continued.*

Location	Locus symbol	Status[a]	Title	MIM#[b]	Method[c]	Disorder(s)	Mouse locus
Xq26.3-q27.1	ADFN, ALDS	P	Albinism-deafness syndrome	300700	Fd	Albinism-deafness syndrome (2)	
Xq27-q28	ANOP1	L	Anophthalmos-1 (with mental retardation but without anomalies)	301590	F	?Anophthalmos-1 (2)	
Xq27.1-q27.2	F9, HEMB	C	Coagulation factor IX (plasma thromboplastic component)	306900	REa, A, Fd, D, X/A, RE	Haemophilia B (3)	X(Cf9)
Xq27.3	FMR1, FRAXA	C	Fragile X mental retardation-1	309550	Ch, F, Fd, Re	Fragile X syndrome (3)	X(Fmr1)
Xq28	ALD	C	Adrenoleucodystrophy	300100	F, Fd, D	Adrenoleucodystrophy (3); adrenomyeloneuropathy (2)	X(Ald)
Xq28	AVPR2, DIR, DI1, ADHR	C	Arginine vasopressin receptor-2 (nephrogenic diabetes insipidus)	304800	Fd, S, REa, Psh	Diabetes insipidus, nephrogenic (3)	
Xq28	CBBM, BCM	C	Blue-monochromatic colour-blindness (blue cone monochromacy)	303700	F, Fd, RE	Colour blindness, blue monochromatic (3)	
Xq28	CDPX2, CPXD, CPX	L	Chondrodysplasia punctata-2, X-linked dominant (Happle syndrome)	302960	H	Chondrodysplasia punctata, X-linked dominant (2)	X(Bpa)
Xq28	DKC	C	Dyskeratosis congenita	305000	Fd	Dyskeratosis congenita (2)	
Xq28	EFE2, BTHS	C	Endocardial fibroelastosis-2 (Barth syndrome; cardioskeletal myopathy with neutropenia and abnormal mitochondria)	302060	Fd	Endocardial fibroelastosis-2 (2); Barth syndrome (2)	
Xq28	EMD, EDMD	C	Emery–Dreifuss muscular dystrophy	310300	F, Fd, H, REn	Emery–Dreifuss muscular dystrophy (3)	
Xq28	F8C, HEMA	C	Coagulation factor VIIIc, procoagulant component	306700	F, Fd, REa, A, RE	Haemophilia A (3)	X(Cf8)
Xq28	FRAXE, FMR2	P	Fragile site, X-linked, E	309548	Ch, REn	Mental retardation, X-linked, FRAXE type (3)	
Xq28	FRAXF	P	Fragile site, folic acid type, rare, fra(X)(q28)	600226	Ch, RE	Mental retardation, X-linked, FRAXF type (3)	
Xq28	G6PD, G6PD1	C	Glucose-6-phosphate dehydrogenase	305900	F, S, REb, RE	G6PD deficiency (3); Favism (3); haemolytic anaemia due to G6PD deficiency (3)	X(G6pd)
Xq28	HMS1, GAY1	L	Homosexuality, male	306995	Fd	[?Homosexuality, male] (2)	
Xq28	GCP, CBD	C	Green cone pigment	303800	F, RE, A, Fd	Colour-blindness, deutan (3)	X(Rsvp)
Xq28	IDS, MPS2, SIDS	C	Iduronate 2-sulphatase (Hunter syndrome)	309900	X/A, Fd, F, RE	Mucopolysaccharidosis II (3)	X(Ids)
Xq28	IP2	C	Incontinentia pigmenti-2 (familial, male-lethal type)	308310	Fd	Incontinentia pigmenti, familial (2)	X(?Str)
Xq28	L1CAM, CAML1, HSAS1	C	L1 cell adhesion molecule	308840	A, RE, H, Fd	Hydrocephalus due to aqueductal stenosis, 307000 (3); MASA syndrome, 303350 (3); spastic paraplegia, 312900 (3)	X(L1cam)
Xq28	MAFD2, MDX	L	Major affective disorder-2	309200	F	?Manic-depressive illness, X-linked (2)	
Xq28	MRSD, CHRS	P	Mental retardation-skeletal dysplasia	309620	Fd	Mental retardation-skeletal dysplasia (2)	
Xq28	MRX3	P	Mental retardation, X-linked-3	309541	Fd	Mental retardation, X-linked-3 (2)	
Xq28	MTM1, MTMX	C	Myotubular myopathy-1	310400	Fd	Myotubular myopathy, X-linked (2)	
Xq28	MYP1, BED	P	Myopia-1 (Bornholm eye disease)	310460	Fd	Myopia-1 (2); Bornholm eye disease (2)	
Xq28	OPD1	P	Otopalatodigital syndrome, type I	311300	Fd	Otopalatodigital syndrome, type I (2)	
Xq28	RCP, CBP	C	Red cone pigment	303900	F, RE, A, Fd	Colour-blindness, protan (3)	X(Rsvp)
Xq28	TKC, TKCR	C	Torticollis, keloids, cryptorchidism and renal dysplasia	314300	X/A	Goeminne TKCR syndrome (2)	
Xq28	WSN, BGMR	P	Waisman syndrome (basal ganglion disorder with mental retardation)	311510	Fd	Waisman parkinsonism-mental retardation syndrome (2)	
Yp11.3	TDF, SRY	C	Testis determining factor (sex-determining relation Y)	480000	Ch, Fd	Gonadal dysgenesis, XY type (3)	Yp(Tdy, Sry)

Table VII.2 The morbid anatomy of the human genome (alphabetically by disorder).

Disorder	Location
Aarskog–Scott syndrome (3)	Xp11.21
Abetalipoproteinaemia (3)	2p24
[Acanthocytosis, one form] (3)	17q21-q22
Acatalasaemia (3)	11p13
Acetyl-CoA carboxylase deficiency (1)	17q21
Achondrogenesis-hypochondrogenesis, type II (3)	12q13.11-q13.2
Achondroplasia, 100800 (3)	4p16.3
?Acrocallosal syndrome (2)	12p13.3-p11.2
?Acrofacial dysostosis, Nager type (2)	9q32
ACTH deficiency (1)	2p25
Acyl-CoA dehydrogenase, long chain, deficiency of (3)	2q34-q35
Acyl-CoA dehydrogenase, medium chain, deficiency of (3)	1p31
Acyl-CoA dehydrogenase, short chain, deficiency of (3)	12q22-qter
Adenylosuccinase deficiency (1)	22q13.1
Adrenal hyperplasia, congenital, due to 11-β-hydroxylase deficiency (3)	8q21
Adrenal hyperplasia, congenital, due to 17-α-hydroxylase deficiency (3)	10q24.3
Adrenal hyperplasia, congenital, due to 21-hydroxylase deficiency (3)	6p21.3
Adrenal hypoplasia, congenital, with hypogonadotropic hypogonadism (3)	Xp21.3-p21.2
Adrenocortical carcinoma, hereditary, 202300 (2)	11p15.5
Adrenoleucodystrophy (3)	Xq28
Adrenoleucodystrophy, pseudoneonatal (2)	17q25
Adrenomyeloneuropathy (2)	Xq28
[AFP deficiency, congenital] (1)	4q11-q13
Agammaglobulinaemia, type 1, X-linked (3)	Xq21.3-q22
Agammaglobulinaemia, type 2, X-linked (3)	Xp22
Aicardi syndrome (2)	Xp22
Alagille syndrome (2)	20p11.2
Albinism, brown, 203290 (1)	9p23
Albinism, ocular, autosomal recessive (3)	15q11.2-q12
Albinism, oculocutaneous, type IA (3)	11q14-q21
Albinism, oculocutaneous, type II (3)	15q11.2-q12
Albinism-deafness syndrome (2)	Xq26.3-q27.1
?Albright hereditary osteodystrophy-2 (2)	15q11-q13
Alcohol intolerance, acute (3)	12q24.2
Aldolase A deficiency (3)	16q22-q24
Aldosteronism, glucocorticoid-remediable (3)	8q21
Alkaptonuria (2)	3q2
Allan–Herndon syndrome (2)	Xq21
Alpha-1-antichymotrypsin deficiency (3)	14q32.1
Alpha-ketoglutarate dehydrogenase deficiency (1)	7p13-p11.2
Alpha-thalassaemia/mental retardation syndrome, type 1 (1)	16pter-p13.3
Alpha-thalassaemia/mental retardation syndrome, type 2, 301040 (3)	Xq13
Alport syndrome, 301050 (3)	Xq22
Alport syndrome, autosomal recessive, 203780 (3)	2q36
Alport syndrome, autosomal recessive, 203780 (3)	2q36-q37
?Alport syndrome, X-linked, type 2 (1)	Xq22
Alzheimer disease, APP-related (3)	21q21.3-q22.05
Alzheimer disease-2, late onset (2)	19cen-q13.2
Alzheimer disease-3 (2)	14q24.3
Amelogenesis imperfecta (3)	Xp22.3-p22.1
?Amelogenesis imperfecta-3, hypoplastic type (2)	Xq22-q28
[AMP deaminase deficiency, erythrocytic] (3)	11pter-p13
Amyloid neuropathy, familial, several allelic types (3)	18q11.2-q12.1
Amyloidosis, cerebroarterial, Dutch type (3)	21q21.3-q22.05
Amyloidosis, Finnish type, 105120 (3)	9q34
Amyloidosis, hereditary renal, 105200 (3)	4q28
Amyloidosis, Iowa type, 107680.0010 (3)	11q23

Continued on p. 924.

Table VII.2 *Continued.*

Disorder	Location
Amyloidosis, renal, 105200 (3)	Chr.12
{?Amyloidosis, secondary, susceptibility to} (1)	1q21-q23
Amyloidosis, senile systemic (3)	18q11.2-q12.1
Amyotrophic lateral sclerosis, juvenile (2)	2q33-q35
Amytrophic lateral sclerosis, due to SOD1 deficiency, 105400 (3)	21q22.1
?Anal canal carcinoma (2)	11q22-qter
Analbuminaemia (3)	4q11-q13
Androgen insensitivity, several forms (3)	Xq11-q12
?Anaemia, megaloblastic, due to DHFR deficiency (1)	5q11.2-q13.2
Anaemia, pernicious, congenital, due to deficiency of intrinsic factor (1)	Chr.11
?Anaemia, sideroblastic, with spinocerebellar ataxia (2)	Xq13
Anaemia, sideroblastic/hypochromic (3)	Xp11.21
Aneurysm, familial, 100070 (3)	2q31
Angelman syndrome (2)	15q11-q13
Angio-oedema, hereditary (3)	11q11-q13.1
Anhidrotic ectodermal dysplasia (2)	Xq12.2-q13.1
Aniridia (3)	11p13
Ankylosing spondylitis (2)	6p21.3
?Anophthalmos-1 (2)	Xq27-q28
Anterior segment mesenchymal dysgenesis (2)	4q28-q31
Antithrombin III deficiency (3)	1q23-q25
Apert syndrome, 101200 (3)	10q26
Apnoea, postanaesthetic (3)	3q26.1-q26.2
ApoA-I and apoC-III deficiency, combined (3)	11q23
Apolipoprotein B-100, ligand-defective (3)	2p24
[Apolipoprotein H deficiency] (3)	17q23-qter
Argininaemia (3)	6q23
Argininosuccinicaciduria (3)	7cen-q11.2
Arrhythmogenic right ventricular dysplasia (2)	14q23-q24
Aspartylglucosaminuria (3)	4q32-q33
Ataxia with isolated vitamin E deficiency, 277460 (3)	8q
Ataxia-telangiectasia (2)	11q22.3
{Atherosclerosis, susceptibility to} (2)	19p13.3-p13.2
{Atherosclerosis, susceptibility to} (2)	1q23-q25
?{Atherosclerosis, susceptibility to} (3)	8p21-p12
Atopy (2)	11q12-q13
Atransferrinaemia (1)	3q21
Atrial septal defect, secundum type (2)	6p21.3
Autism, succinylpurinaemic (3)	22q13.1
Autoimmune polyglandular disease, type I (2)	21q22.3
Bardet–Biedl syndrome 1 (2)	11q13
Bardet–Biedl syndrome 2 (2)	16q21
Bardet–Biedl syndrome 3 (2)	3p13-p12
Bardet–Biedl syndrome-4 (2)	15q22.3-q23
Bare lymphocyte syndrome, type I, due to TAP2 deficiency (1)	6p21.3
Barth syndrome (2)	Xq28
?Basal cell carcinoma (2)	9q31
Basal cell carcinoma (3)	5q13.3
Basal cell naevus syndrome (2)	9q31
Batten disease (2)	16q12
Becker muscular dystrophy (3)	Xp21.2
Beckwith–Wiedemann syndrome (2)	11pter-p15.4
Bernard–Soulier syndrome (1)	17pter-p12
{Beryllium disease, chronic, susceptibility to} (3)	6p21.3
3-β-hydroxysteroid dehydrogenase, type II, deficiency (3)	1p13.1
Biotinidase deficiency (1)	3p25
Bladder cancer, 109800 (3)	13q14.1-q14.2

Continued.

Table VII.2 *Continued.*

Disorder	Location
Bleeding disorder due to defective thromboxane A2 receptor (3)	19p13.3
Blepharophimosis, epicanthus inversus and ptosis (2)	3q22-q23
Bloom syndrome (2)	15q26.1
Borjeson–Forssman–Lehmann syndrome (2)	Xq26-q27
Bornholm eye disease (2)	Xq28
?Brachydactyly type E (2)	2q37
Brachydactyly-mental retardation syndrome (2)	2q37
Branchio-otorenal dysplasia (2)	8q13.3
?Breast cancer (1)	17p13.3
Breast cancer (1)	6q25.1
Breast cancer 2, early onset (2)	13q12-q13
Breast cancer, ductal (2)	1p36
Breast cancer, ductal (2)	Chr.13
Breast cancer, male, with Reifenstein syndrome (3)	Xq11-q12
Breast cancer-1, early onset (3)	17q21
Breast cancer-3 (2)	11q23
Brody myopathy (1)	Chr.16
Brunner syndrome (3)	Xp11.23
Burkitt lymphoma (3)	8q24.12-q24.13
Butterfly dystrophy, retinal (3)	6p21.1-cen
?C1q deficiency (1)	1p36.3-p34.1
?C1q deficiency (1)	1p36.3-p34.1
C1r/C1s deficiency, combined (1)	12p13
C1r/C1s deficiency, combined (1)	12p13
C2 deficiency (3)	6p21.3
C3 deficiency (3)	19p13.3-p13.2
C3b inactivator deficiency (1)	4q25
C4 deficiency (3)	6p21.3
C4 deficiency (3)	6p21.3
C5 deficiency (1)	9q34.1
C6 deficiency (1)	5p13
C7 deficiency (1)	5p13
C8 deficiency, type I (2)	1p32
C8 deficiency, type II (3)	1p32
C9 deficiency (1)	5p13
Campomelic dysplasia with autosomal sex reversal (3)	17q24.3-q25.1
Canavan disease (3)	17pter-p13
Carbamoylphosphate synthetase I deficiency (3)	2q33-q36
Carbohydrate-deficient glycoprotein syndrome (2)	16p13.3-p13.2
Carboxypeptidase B deficiency (1)	Chr.13
?Cardiomyopathy (1)	2q35
Cardiomyopathy, dilated, X-linked (3)	Xp21.2
Cardiomyopathy, familial dilated, with conduction defect (2)	1p11-q11
Cardiomyopathy, familial hypertrophic, 1, 192600 (3)	14q12
Cardiomyopathy, familial hypertrophic, 2, 115195 (3)	1q3
Cardiomyopathy, familial hypertrophic, 3, 115196 (3)	15q22
Cardiomyopathy, familial hypertrophic, 4 (2)	11p13-q13
?Carnitine acetyltransferase deficiency (1)	9q34.1
?Carnitine palmitoyltransferase I deficiency (2)	11q
Carnitine palmitoyltransferase II deficiency (3)	1p32
Carpal tunnel syndrome, familial (3)	18q11.2-q12.1
Cartilage-hair hypoplasia (2)	9p13-q11
Cat eye syndrome (2)	22q11
?Cataract, anterior polar, I (2)	14q24-qter
?Cataract, congenital (2)	15q15
Cataract, congenital, cerulean type (2)	CCA1
?Cataract, congenital total (2)	Xp

Continued on p. 926.

Table VII.2 *Continued.*

Disorder	Location
Cataract, congenital, with late-onset corneal dystrophy (3)	11p13
Cataract, congenital, with microphthalmia (2)	16p13.3
Cataract, Coppock-like (3)	2q33-q35
Cataract, Marner type (2)	16q22.1
Cataract, zonular pulverulent-1 (2)	1q2
Cavernous angiomatous malformations (2)	7q11-q22
CD3, ζ-chain, deficiency (1)	1q23-q25
[CD4(+) lymphocyte deficiency] (2)	12pter-p12
CD59 deficiency (3)	11p13
Central core disease, 117000 (3)	19q13.1
?Central core disease, one form (3)	14q12
Centrocytic lymphoma (2)	11q13
Cerebellar ataxia, paroxysmal acetazolamide-responsive (2)	19p
Cerebellar ataxia with retinal degeneration (2)	3p21.1-p12
Cerebral amyloid angiopathy (3)	20p11
Cerebral arteriopathy with subcortical infarcts and leucoencephalopathy (2)	19q12
Cerebrotendinous xanthomatosis (3)	2q33-qter
Cerebrovascular disease, occlusive (3)	14q32.1
Ceroid lipofuscinosis, neuronal-1, infantile (2)	1p32
Ceroid-lipofuscinosis, neuronal, variant late infantile form (2)	13q21.1-q32
Cervical carcinoma (2)	11q13
[CETP deficiency] (3)	16q21
Charcot–Marie–Tooth disease, type II (2)	1p36-p35
Charcot–Marie–Tooth disease, type IVA (2)	8q13-q21.1
Charcot–Marie–Tooth neuropathy, slow nerve conduction type Ia (3)	17p11.2
Charcot–Marie–Tooth neuropathy, slow nerve conduction type Ib, 118200 (3)	1q22
Charcot–Marie–Tooth neuropathy, X-linked-1, dominant, 302800 (3)	Xq13.1
Charcot–Marie–Tooth neuropathy, X-linked-2, recessive (2)	Xp22.2
Chloride diarrhoea, congenital (2)	7q31
Cholesteryl ester storage disease (3)	10q24-q25
?Chondrodysplasia punctata, rhizomelic (2)	4p16-p14
Chondrodysplasia punctata, X-linked dominant (2)	Xq28
Chondrodysplasia punctata, X-linked recessive (3)	Xp22.3
Choroideraemia (3)	Xq21.2
Chronic granulomatous disease, autosomal, due to deficiency of CYBA (3)	16q24
Chronic granulomatous disease due to deficiency of NCF-1 (3)	7q11.23
Chronic granulomatous disease due to deficiency of NCF-2 (1)	1q25
Chronic granulomatous disease, X-linked (3)	Xp21.1
{Chronic infections, due to opsonin defect} (3)	10q11.2-q21
Citrullinaemia (3)	9q34
Cleft palate, X-linked (2)	Xq21.1-q21.31
Cleidocranial dysplasia (2)	6q21
CMO II deficiency (3)	8q21
Cockayne syndrome-2, late onset, 216410 (2)	10q11
Coffin–Lowry syndrome (2)	Xp22.2-p22.1
Cohen syndrome (2)	8q22-q23
?Colon cancer (1)	7q22-q31.1
Colon cancer, familial non-polyposis, type 1 (3)	2p16-p15
Colour-blindness, blue monochromatic (3)	Xq28
Colour-blindness, deutan (3)	Xq28
Colour-blindness, protan (3)	Xq28
Colour-blindness, tritan (3)	7q31.3-q32
Colorectal adenoma (1)	12p12.1
Colorectal cancer (1)	12p12.1
Colorectal cancer, 114500 (3)	17p13.1
Colorectal cancer (3)	18q23.3
Colorectal cancer (3)	5q21

Continued.

Table VII.2 *Continued.*

Disorder	Location
Colorectal cancer (3)	5q21-q22
Colorectal cancer, familial non-polyposis type 2 (3)	3p21.3
Colton blood group (3)	7p14
Combined C6/C7 deficiency (1)	5p13
Combined immunodeficiency, X-linked, moderate, 312863 (3)	Xq13
?Combined variable hypogammaglobulinaemia (1)	14q32.33
Cone-rod retinal dystrophy (2)	19q13.1-q13.2
Congenital bilateral absence of vas deferens (3)	7q31.2
?Conotruncal cardiac anomalies (2)	22q11
Contractural arachnodactyly, congenital (3)	5q23-q31
Coproporphyria (3)	3q12
Cornea plana congenita, recessive (2)	12q21
Corneal dystrophy, combined granular/lattice type (2)	5q22-q33.3
Corneal dystrophy, Groenouw type I (2)	5q22-q33.3
Corneal dystrophy, lattice type I, 122200 (2)	5q22-q33.3
?Cornelia de Lange syndrome (2)	3q26.3
{Coronary artery disease, susceptibility to} (1)	6q27
Cortisol resistance (3)	5q31
CR1 deficiency (1)	1q32
?Craniofrontonasal dysplasia (2)	Xpter-p22.2
Craniosynostosis, type 1 (2)	7p21.3-p21.2
Craniosynostosis, type 2 (3)	5q34-q35
[Creatine kinase, brain type, ectopic expression of] (2)	14q32
Creutzfeldt–Jakob disease, 123400 (3)	20pter-p12
Crigler–Najjar syndrome, type I, 218800 (3)	Chr.2
Crouzon craniofacial dysostosis, 123500 (3)	10q26
?Cryptorchidism (2)	Xp21
?Cutis laxa, marfanoid neonatal type (1)	7q31.1-q31.3
[Cystathioninuria] (1)	Chr.16
Cystic fibrosis (3)	7q31.2
Cystinuria, 220100 (3)	2p21
Darier disease (keratosis follicularis) (2)	12q23-q24.1
Deafness 4, congenital sensorineural (2)	Xp21.2
Deafness, autosomal non-syndromic sensorineural, 2 (2)	1p32
Deafness, conductive, with stapes fixation, 304400 (3)	Xq21.1
Deafness, low-tone (2)	5q31-q33
Deafness, neurosensory, AR, 1 (2)	13q12
Deafness, non-syndromic, recessive, 2 (2)	11q13.5
Deafness-3, neurosensory non-syndromic recessive (2)	17p12-q12
Debrisoquine sensitivity (3)	22q13.1
Dejerine–Sottas disease, myelin P(0)-related, 145900 (3)	1q22
Dejerine–Sottas disease, PMP22 related 145900 (3)	17p11.2
?Dent disease, 310468 (2)	Xp11.22
Dentatorubro-pallidoluysian atrophy (3)	12pter-p12
Dentinogenesis imperfecta-1 (2)	4q13-q21
Denys–Drash syndrome (3)	11p13
Diabetes insipidus, nephrogenic (3)	Xq28
Diabetes insipidus nephrogenic, autosomal recessive (3)	12q13
Diabetes insipidus, neurohypophyseal, 125700 (3)	20p13
Diabetes mellitus, insulin-dependent, 3 (2)	15q26
Diabetes mellitus, insulin-dependent, 4 (2)	11q13
?Diabetes mellitus, insulin-dependent, 7 (2)	2q31
?Diabetes mellitus, insulin-dependent, neonatal (2)	Chr.6
?Diabetes mellitus, insulin-dependent-1 (2)	6p21.3
Diabetes mellitus, insulin-resistant, with acanthosis nigricans (3)	19p13.2
Diabetes mellitus, rare form (1)	11p15.5
Diastrophic dysplasia (3)	5q31-q34

Continued on p. 928.

Table VII.2 *Continued.*

Disorder	Location
?Dicarboxylicaminoaciduria, 222730 (1)	9p24
DiGeorge syndrome (2)	22q11
Diphenylhydantoin toxicity (1)	1p11-qter
{Diphtheria, susceptibility to} (1)	5q23
Disinhibition-dementia-Parkinsonism-amyotrophy complex (2)	17q21-q22
Distal arthrogryposis-1 (2)	9p21-q21
DNA ligase I deficiency (3)	19q13.2-q13.3
Dopamine-β-hydroxylase deficiency (1)	9q34
Down syndrome (1)	21q22.3
?Dubin–Johnson syndrome (2)	13q34
Duchenne muscular dystrophy (3)	Xp21.2
[Dysalbuminaemic hyperthyroxinaemia] (3)	4q11-q13
[Dysalbuminaemic hyperzincaemia] (3)	4q11-q13
Dysautonomia, familial (2)	9q31-q33
Dyserythropoietic anaemia, congenital, type III (2)	15q21
Dysfibrinogenaemia, α types (3)	4q28
Dysfibrinogenaemia, β types (3)	4q28
Dysfibrinogenaemia, γ types (3)	4q28
Dyskeratosis congenita (2)	Xq28
Dyslexia, specific, 2 (2)	6p21.3
Dysplasminogenaemic thrombophilia (1)	6q26
Dysprothrombinaemia (3)	11p11-q12
Dystonia, DOPA-responsive, 128230 (3)	14q22.1-q22.2
[Dystransthyretinaemic hyperthyroxinaemia] (3)	18q11.2-q12.1
?EEC syndrome (2)	7q11.2-q21.3
Ehlers–Danlos syndrome, type III (3)	2q31
Ehlers–Danlos syndrome, type IV, 130050 (3)	2q31
Ehlers–Danlos syndrome, type unspecified (3)	9q34.2-q34.3
Ehlers–Danlos syndrome, type VI, 225400 (3)	1p36.3-p36.2
Ehlers–Danlos syndrome, type VIIA1, 130060 (3)	17q21.31-q22.05
Ehlers–Danlos syndrome, type VIIA2, 130060 (3)	7q22.1
?Ehlers–Danlos syndrome, type X (1)	2q34
[Elliptocytosis, Malaysian-Melanesian type] (3)	17q21-q22
Elliptocytosis-1 (3)	1p36.2-p34
Elliptocytosis-2 (3)	1q21
Elliptocytosis-3 (3)	14q22-q23.2
Emery–Dreifuss muscular dystrophy (3)	Xq28
Emphysema (3)	14q32.1
Emphysema due to α-2-macroglobulin deficiency (1)	12p13.3-p12.3
Emphysema-cirrhosis (3)	14q32.1
Endocardial fibroelastosis-2 (2)	Xq28
Endometrial carcinoma (3)	16q22.1
Enolase deficiency (1)	1pter-p36.13
?Eosinophilic myeloproliferative disorder (2)	12p13
Epidermolysis bullosa dystrophica, dominant, 131750 (3)	3p21.3
Epidermolysis bullosa dystrophica, recessive, 226600 (3)	3p21.3
Epidermolysis bullosa, Herlitz junctional type, 226700 (3)	1q25-q31
Epidermolysis bullosa, Ogna type (2)	8q24
Epidermolysis bullosa simplex, Dowling–Meara type, 131670 (3)	17q12-q21
Epidermolysis bullosa simplex, Dowling–Meara type, 131760 (3)	12q11-q13
Epidermolysis bullosa simplex, Koebner type, 131900 (3)	12q11-q13
Epidermolysis bullosa simplex, Koebner type, 131900 (3)	17q12-q21
Epidermolysis bullosa simplex, Weber–Cockayne type, 131800 (3)	17q12-q21
Epidermolysis bullosa, Weber–Cockayne type, 131800 (3)	12q11-q13
Epidermolytic hyperkeratosis, 113800 (3)	12q11-q13
Epidermolytic hyperkeratosis, 113800 (3)	17q21-q22
Epidermolytic palmoplantar keratoderma (3)	17q12-q21

Continued.

Table VII.2 *Continued.*

Disorder	Location
Epilepsy, benign neonatal, type 2 (2)	8q
Epilepsy, benign neonatal, type I, 121200 (3)	20q13.2-q13.3
Epilepsy, juvenile myoclonic (2)	6p21.3
Epilepsy, partial (2)	10q
Epilepsy, progressive myoclonus (2)	21q22.3
Epilepsy, progressive, with mental retardation (2)	8pter-p22
Epiphyseal dysplasia, multiple 1 (2)	19q12
Epiphyseal dysplasia, multiple 2 (2)	1p32
Episodic ataxia/myokymia syndrome, 160120 (3)	12p13
Epithelioma, self-healing, squamous 1, Ferguson–Smith type (2)	9q31
?Erythraemia (1)	7q21
Erythraemias, α- (3)	16pter-p13.3
Erythraemias, β- (3)	11p15.5
Erythroblastosis fetalis (1)	1p36.2-p34
[Erythrocytosis, familial], 133100 (3)	19p13.3-p13.2
Erythrokeratodermia variabilis (2)	1p36.2-p34
[Euthyroidal hyper- and hypothyroxinaemia] (1)	Xq22
Ewing sarcoma (3)	22q12
Exertional myoglobinuria due to deficiency of LDH-A (3)	11p15.4
Exostoses, multiple, type 1 (2)	8q24.11-q24.13
?Exostoses, multiple, type 2 (2)	11p11-q11
Exostoses, multiple, type 3 (2)	19p
Exudative vitreoretinopathy, X-linked, 305390 (3)	Xp11.4
Fabry disease (3)	Xq22
Facioscapulohumeral muscular dystrophy 1A (2)	4q35
Factor H deficiency (1)	1q32
Factor V deficiency (1)	1q23
Factor VII deficiency (3)	13q34
Factor X deficiency (3)	13q34
Factor XI deficiency (3)	4q35
Factor XII deficiency (3)	5q33-qter
Factor XIIIA deficiency (3)	6p25-p24
Factor XIIIB deficiency (3)	1q31-q32.1
Familial expansile osteolysis (2)	18q21.1-q22
Familial Mediterranean fever (2)	16p13
?Fanconi anaemia (1)	1q42
Fanconi anaemia-1 (2)	20q13.2-q13.3
Favism (3)	Xq28
{?Fetal alcohol syndrome} (1)	12q24.2
?Fetal hydantoin syndrome (1)	1p11-qter
?Fibrodysplasia ossificans progressiva (1)	20p12
Fibromuscular dysplasia of arteries, 135580 (3)	2q31
Fibrosis of the extraocular muscles, congenital (2)	12q13.2-q24.1
Fish-eye disease (3)	16q22.1
[Fish-odour syndrome] (1)	1q
Fletcher factor deficiency (1)	4q35
{Fluorouracil toxicity, sensitivity to} (1)	1p22-q21
focal dermal hypoplasia (2)	Xp22.31
Fragile X syndrome (3)	Xq27.3
Friedreich ataxia (2)	9q13-q21.1
Fructose intolerance (3)	9q22
Fucosidosis (3)	1p34
Fucosyltransferase-6 deficiency (3)	19p13.3
Fukuyama type congenital muscular dystrophy (2)	9q31-q33
Fumarase deficiency (3)	1q42.1
Fundus flavimaculatus with macular dystrophy (2)	1p21-p13

Continued on p. 930.

Table VII.2 *Continued.*

Disorder	Location
G6PD deficiency (3)	Xq28
Galactokinase deficiency (1)	17q21-q22
Galactose epimerase deficiency (1)	1p36-p35
Galactosaemia (3)	9p13
Galactosialidosis (3)	20q13.1
[γ-glutamyltransferase, familial high serum] (2)	22q11.12
Gardner syndrome (3)	5q21-q22
Gaucher disease (3)	1q21
Gaucher disease, variant form (3)	10q21-q22
Generalized atrophic benign epidermolysis bullosa, 226650 (1)	10q24.3
Gerstmann–Straussler disease, 137440 (3)	20pter-p12
Gigantism due to GHRF hypersecretion (1)	20q11.2
?Gilbert syndrome, 143500 (1)	Chr.2
Glanzmann thrombasthenia, type A (3)	17q21.32
Glanzmann thrombasthemia, type B (3)	17q21.32
Glaucoma, primary open angle, juvenile-onset (2)	1q21-q31
Glioblastoma multiforme (2)	10p12-q23.2
Glucocorticoid deficiency, due to ACTH unresponsiveness (1)	18p11.2
Glucose/galactose malabsorption (3)	22q11.2-qter
Glutaricacidaemia, type I (3)	19p13.2
Glutaricacidaemia, type IIC (3)	4q32-qter
Glutaricaciduria, type IIA (1)	15q23-q25
Glutaricaciduria, type IIB (3)	19q13.3
Glutathioninuria (1)	22q11.1-q11.2
Glycerol kinase deficiency (2)	Xp21.3-q21.2
Glycogen storage disease III (1)	1p21
Glycogen storage disease IV (1)	3p12
Glycogen storage disease, type I (3)	17q21
Glycogen storage disease, type II (3)	17q23
Glycogen storage disease VI (1)	14q21-q22
Glycogen storage disease VII (3)	1cen-q32
Glycogen storage disease, X-linked hepatic (2)	Xp22.2-p22.1
?Glycoprotein Ia deficiency (2)	5q23-q31
[Glyoxalase II deficiency] (1)	16p13
GM1-gangliosidosis (3)	3p21.33
GM2-gangliosidosis, AB variant (3)	5q31.3-q33.1
GM2-gangliosidosis, juvenile, adult (3)	15q23-q24
Goeminne TKCR syndrome (2)	Xq28
Goitre adolescent multinodular (1)	8q24.2-q24.3
Goitre, congenital (3)	2p13
Goitre, non-endemic, simple (3)	8q24.2-q24.3
?Goldenhar syndrome (2)	7p
Gonadal dysgenesis, XY female type (2)	Xp22.11-p21.2
Gonadal dysgenesis, XY type (3)	Yp11.3
?Gonadotropin deficiency (2)	Xp21
Graves disease, 275000 (1)	14q31
Greig cephalopolysyndactyly syndrome, 175700 (3)	7p13
?Growth hormone deficient dwarfism (1)	7p15-p14
Gustavson syndrome (2)	Xq26
?Gynaecomastia, familial, due to increased aromatase activity (1)	15q21.1
Gyrate atrophy of choroid and retina with ornithinemia, B6 responsive or unresponsive (3)	10q26
Haemochromatosis (2)	6p21.3
Haemodialysis-related amyloidosis (1)	15q21-q22
Haemolytic anaemia due to ADA excess (1)	20q13.11
Haemolytic anaemia due to adenylate kinase deficiency (1)	9q34.1
Haemolytic anaemia due to bisphosphoglycerate mutase deficiency (1)	7q31-q34
Haemolytic anaemia due to G6PD deficiency (3)	Xq28

Continued.

Table VII.2 *Continued.*

Disorder	Location
Haemolytic anaemia due to glucosephosphate isomerase deficiency (3)	19q13.1
Haemolytic anaemia due to glutathione peroxidase deficiency (1)	3q11-q12
Haemolytic anaemia due to glutathione reductase deficiency (1)	8p21.1
Haemolytic anaemia due to hexokinase deficiency (1)	10q22
Haemolytic anaemia due to PGK deficiency (3)	Xq13
Haemolytic anaemia due to phosphofructokinase deficiency (1)	21q22.3
Haemolytic anaemia due to triosephosphate isomerase deficiency (3)	12p13
Haemophilia A (3)	Xq28
Haemophilia B (3)	Xq27.1-q27.2
Haemorrhagic diathesis due to stroke 'antithrombin' Pittsburgh (3)	14q32.1
Haemorrhagic diathesis due to PAI1 deficiency (1)	7q21.3-q22
Harderoporphyrinuria (3)	3q12
Heart block, progressive familial, type I (2)	19q13.2-q13.3
Heinz body anaemias, α- (3)	16pter-p13.3
Heinz body anaemias, β- (3)	11p15.5
Hepatic lipase deficiency (3)	15q21-q23
Hepatocellular carcinoma (1)	11p14-p13
?Hepatocellular carcinoma (1)	2q14-q21
Hepatocellular carcinoma (3)	4q32.1
Hereditary haemorrhagic telangiectasia, 187300 (3)	9q34.1
[Hereditary persistence of α-fetoprotein] (3)	4q11-q13
?Hereditary persistence of fetal hemoglobin (3)	11p15.5
?Hereditary persistence of fetal hemoglobin, heterocellular, Indian type (2)	7q36
?Hermansky–Pudlak syndrome, 203300 (1)	12q12-q13
?Hermansky–Pudlak syndrome, 203300 (1)	15q15
Heterocellular hereditary persistence of fetal hemoglobin, Swiss type (2)	Xp22.2
Heterotaxy, X-linked visceral (2)	Xq25-q26
[Hex A pseudodeficiency] (1)	15q23-q24
?HHH syndrome (2)	13q34
Hirschsprung disease, 142623 (3)	10q11.2
Hirschsprung disease-2, 600155 (3)	13q22
[Histidinaemia] (1)	12q22-q23
HMG-CoA lyase deficiency (3)	1pter-p33
Holoprosencephaly, type 3 (2)	7q36
?Holoprosencephaly-1 (2)	18pter-q11
?Holoprosencephaly-2 (2)	2p21
?Holoprosencephaly-4 (2)	14q11.1-q13
Holt–Oram syndrome (2)	12q21.3-q22
Homocystinuria, B6-responsive and non-responsive types (3)	21q22.3
Homocystinuria due to MTHFR deficiency (3)	1p36.3
[?Homosexuality, male] (2)	Xq28
HPFH, deletion type (3)	11p15.5
HPFH, non-deletion type A (3)	11p15.5
HPFH, non-deletion type G (3)	11p15.5
HPRT-related gout (3)	Xq26-q27.2
?Humoral hypercalcaemia of malignancy (1)	12p12.1-p11.2
Huntington disease (3)	4p16.3
Hydrocephalus due to aqueductal stenosis, 307000 (3)	Xq28
Hydrops fetalis, one form (1)	19q13.1
3-hydroxyacyl-CoA dehydrogenase deficiency (1)	Chr.7
Hyperbetalipoproteinaemia (3)	2p24
Hypercalcaemia, hypocalciuric, familial (3)	3q21-q24
Hypercholesterolaemia, familial (3)	19p13.2-p13.1
Hyperchylomicronaemia syndrome, familial (3)	8p22
Hyperglycinaemia, isolated non-ketotic, type I (3)	9p22
Hyperglycinaemia, non-ketotic, type II (1)	3p21.2-p21.1
?Hyperimmunoglobulin G1 syndrome (2)	14q32.33

Continued on p. 932.

Table VII.2 *Continued.*

Disorder	Location
Hyperkalaemic periodic paralysis (3)	17q23.1-q25.3
?Hyperleucinaemia-isoleucinaemia or hypervalinaemia (1)	12pter-q12
Hyperlipoproteinemia I (1)	8p22
Hyperlipoproteinaemia, type Ib (3)	19q13.2
Hyperlipoproteinaemia, type III (3)	19q13.2
Hyperoxaluria, primary, type 1 (3)	2q36-q37
Hyperphenylalaninaemia due to pterin-4a-carbinolamine dehydratase deficiency, 264070 (3)	10q22
[Hyperphenylalaninaemia, mild] (3)	12q24.1
[?Hyperproglucagonaemia] (1)	2q36-q37
Hyperproinsulinaemia, familial (3)	11p15.5
[Hyperproreninaemia] (3)	1q32
{?Hypertension, essential} (1)	16p13.11
?Hypertension, essential, 145500 (1)	17q21-q22
{Hypertension, essential, susceptibility to} (3)	1q42-q43
Hyperthyroidism congenital (3)	14q31
Hypertriglyceridaemia (3)	11q23
Hypertriglyceridaemia, one form (3)	11q23
?Hypervalinaemia or hyperleucine-isoleucinaemia (1)	Chr.19
Hypoalphalipoproteinaemia (3)	11q23
Hypobetalipoproteinaemia (3)	2p24
Hypocalcaemia, autosomal dominant (3)	3q21-q24
Hypocalciuric hypercalcaemia, type II (2)	19p13.3
[Hypoceruloplasminaemia, hereditary] (1)	3q21-q24
Hypochondroplasia, 146000 (3)	4p16.3
Hypofibrinogenaemia, γ types (3)	4q28
?Hypoglycaemia due to PCK1 deficiency (1)	20q13.31
Hypogonadism, hypergonadotropic (3)	19q13.32
?Hypogonadotropic hypogonadism due to GNRH deficiency, 227200 (1)	8p21-p11.2
Hypokalaemic periodic paralysis, 170400 (3)	1q32
Hypomagnesaemia, X-linked primary (2)	Xp22.2
?Hypomelanosis of Ito (2)	15q11-q13
?Hypomelanosis of Ito (2)	9q33-qter
Hypoparathyroidism, autosomal dominant (3)	11p15.3-p15.1
Hypoparathyroidism, autosomal recessive (3)	11p15.3-p15.1
Hypoparathyroidism, familial (2)	3q13
Hypoparathyroidism, X-linked (2)	Xq26-q27
?Hypophosphatasia, adult, 146300 (1)	1p36.1-p34
Hypophosphatasia, infantile, 241500 (3)	1p36.1-p34
Hypophosphataemia, hereditary (2)	Xp22.2-p22.1
?Hypophosphataemia with deafness (2)	Xp22
Hypoprothrombinaemia (3)	11p11-q12
?Hypospadias-dysphagia syndrome (2)	5p13-p12
Hypothyroidism, congenital (3)	2p13
Hypothyroidism, hereditary congenital (3)	8q24.2-q24.3
Hypothyroidism, non-goitrous (3)	1p13
Hypothyroidism, non-goitrous, due to TSH resistance (3)	14q31
Ichthyosis bullosa of Siemens, 146800 (3)	12q11-q13
?Ichthyosis vulgaris, 146700 (1)	1q21
Ichthyosis, X-linked (3)	Xp22.32
[Ii blood group, 110800] (1)	9q21
?Immotile cilia syndrome (2)	6p
Immunodeficiency, X-linked, with hyper-IgM (3)	Xq26
Incontinentia pigmenti, familial (2)	Xq28
Incontinentia pigmenti, sporadic type (2)	Xp11.21
[Inosine triphosphatase deficiency] (1)	20p
Insomnia, fatal familial (3)	20pter-p12
Insulin-dependent diabetes mellitus-2 (2)	11p15.5

Continued.

Table VII.2 *Continued.*

Disorder	Location
Interferon, α, deficiency (1)	9p21
Interferon, immune, deficiency (1)	12q24.1
?Isolated growth hormone deficiency due to defect in GHRF (1)	20q11.2
Isolated growth hormone deficiency, Illig type with absent GH and Kowarski type with bioinactive GH (3)	17q22-q24
Isovalericacidaemia (3)	15q14-q15
Jackson–Weiss syndrome, 123150 (3)	10q26
?Jacobsen syndrome (2)	11q
Juberg–Marsidi syndrome (2)	Xq12-q21
Junctional epidermolysis bullosa inversa (2)	1q31
Kallmann syndrome (3)	Xp22.3
Kanzaki disease (3)	22q11
[κ light chain deficiency] (3)	2p12
Keratoderma, palmoplantar, non-epidermolytic (3)	12q11-q13
Keratosis follicularis spinulosa decalvans (2)	Xp22.2-p21.2
3-ketothiolase deficiency (3)	11q22.3-q23.1
[Kininogen deficiency] (3)	3q26-qter
?Klippel–Feil syndrome (2)	5q11.2
Kniest dysplasia (3)	12q13.11-q13.2
Kostmann neutropenia, 202700 (3)	1p35-p34.3
Krabbe disease (3)	14q24.3-q32.1
?Lactase deficiency, adult, 223100 (1)	2q21
?Lactase deficiency, congenital (1)	2q21
Lactic acidosis due to defect in iron-sulphur cluster of complex I (1)	2q33-q34
?Lactoferrin-deficient neutrophils, 245480 (1)	3q21-q23
Lamellar ichthyosis, autosomal recessive (2)	14q11.2
Langer–Giedion syndrome (2)	8q24.11-q24.13
Laron dwarfism (3)	5p13-p12
?Laryngeal adductor paralysis (2)	6p21.3-p21.2
{Lead poisoning, susceptibility to} (3)	9q34
Leucocyte adhesion deficiency (1)	21q22.3
?Leiomyomata, multiple hereditary cutaneous (2)	18p11.32
Leiomyomatosis, diffuse (1)	Xq22
Leiomyomatosis-nephropathy syndrome, 308940 (1)	Xq22
Leprechaunism (3)	19p13.2
Lesch–Nyhan syndrome (3)	Xq26-q27.2
?Letterer–Siwe disease (2)	13q14-q31
Leukaemia, acute lymphoblastic (1)	19p13.3
Leukaemia, acute lymphoblastic (2)	9p22-p21
?Leukaemia, acute lymphocytic, with 4/11 translocation (3)	4q21
Leukaemia, acute myeloid (2)	9q34.1
Leukaemia, acute myeloid (3)	21q22.3
Leukaemia, acute myeloid, M2 type (1)	Xp22.32
Leukaemia, acute non-lymphocytic (2)	6p23
Leukaemia, acute pre-B-cell (2)	1q23
Leukaemia, acute promyelocytic (1)	17q12
Leukaemia, acute promyelocytic (2)	15q22
Leukaemia, acute T-cell (2)	11p13
Leukaemia, acute, T-cell (2)	11p13
Leukaemia, chronic lymphocytic, B-cell (2)	13q14
Leukaemia, chronic myeloid (3)	22q11.21
Leukaemia, chronic myeloid (3)	9q34.1
Leukaemia, myeloid/lymphoid or mixed-lineage (2)	11q23
Leukaemia, T-cell acute lymphoblastic (2)	11p15
Leukaemia, T-cell acute lymphoblastic (2)	9q34.3
Leukaemia, T-cell acute lymphoblastoid (2)	19p13.2-p13.1
Leukaemia, T-cell acute lymphocytic (2)	10q24

Continued on p. 934.

Table VII.2 *Continued.*

Disorder	Location
?Leukaemia, transient (2)	21q11.2
Leukaemia-1, T-cell acute lymphoblastic (3)	1p32
Leukaemia-2, T-cell acute lymphoblastic (3)	9q31
Leukaemia/lymphoma, B-cell, 1 (2)	11q13.3
Leukaemia/lymphoma, B-cell, 2 (2)	18q21.3
Leukaemia/lymphoma, B-cell, 3 (2)	19q13
Leukaemia/lymphoma, T-cell (2)	14q32.1
Leukaemia/lymphoma, T-cell (2)	2q34
Leukaemia/lymphoma, T-cell (3)	14q11.2
Liddle syndrome (3)	16p12.3
Li–Fraumeni syndrome (3)	17p13.1
Lipoamide dehydrogenase deficiency (3)	7q31-q32
Lipoma, benign (2)	12q15
Lipoprotein lipase deficiency (3)	8p22
Liposarcoma (1)	19p13.2-q13.3
Long QT syndrome-1 (2)	11p15.5
Long QT syndrome-2 (3)	7q35-q36
Long QT syndrome-3 (3)	3p24-p21
Lowe syndrome (3)	Xq26.1
{Lupus erythematosus, susceptibility to} (2)	12pter-p12
Lupus erythematosus, systemic, 152700 (1)	1q23
Lymphoma, B-cell (2)	3q27
Lymphoma, diffuse large cell (3)	3q27
Lymphoma/leukaemia, B-cell, variant (1)	18q21.3
Lymphoproliferative syndrome, X-linked (2)	Xq25
?Lynch cancer family syndrome II (2)	18q11-q12
?Lysosomal acid phosphatase deficiency (1)	11p12-p11
Machado–Joseph disease (2)	14q24.3-q31
Macrocytic anaemia of 5q- syndrome, refractory (2)	5q12-q32
Macrocytic anaemia refractory, of 5q- syndrome, 153550 (3)	5q31.1
[Macrothrombocytopenia] (1)	7q11.2
Macular dystrophy (3)	6p21.1-cen
Macular dystrophy, atypical vitelliform (2)	8q24
Macular dystrophy, dominant cystoid (2)	7p21-p15
Macular dystrophy, North Carolina type (2)	6q14-q16.2
Macular dystrophy, vitelliform type (2)	11q13
Male germ cell tumour (2)	12q22
?Male infertility due to acrosin deficiency (2)	22q13-qter
?Male infertility, familial (1)	11p13
?Male infertility, familial (1)	11p13
?Male pseudohermaphroditism due to defective LH (1)	19q13.32
Malignant hyperthermia susceptibility 2 (2)	17q11.2-q24
Malignant hyperthermia susceptibility-1, 145600 (3)	19q13.1
Malignant hyperthermia susceptibility-3, 154276 (3)	7q21-q22
Malignant melanoma, cutaneous (2)	1p36
?Manic-depressive illness, X-linked (2)	Xq28
Mannosidosis (1)	19cen-q12
Maple syrup urine disease, type 3 (3)	6p22-p21
Maple syrup urine disease, type Ia (3)	19p13.1-q13.2
Maple syrup urine disease, type II (3)	1p31
Marfan syndrome, 154700 (3)	15q21.1
Maroteaux–Lamy syndrome, several forms (3)	5q11-q13
MASA syndrome, 303350 (3)	Xq28
Mast cell leukaemia (3)	4q12
Maturity-onset diabetes of the young, type III (2)	12q22-qter
McArdle disease (3)	11q13
McCune–Albright polyostotic fibrous dysplasia, 174800 (3)	20q13.2

Continued.

Table VII.2 *Continued.*

Disorder	Location
McLeod phenotype (3)	Xp21.2-p21.1
Medullary thyroid carcinoma, 155240 (3)	10q11.2
Megalocornea, X-linked (2)	Xq21.3-q22
?Melanoma (1)	2p25.3
Melanoma (1)	9p21
Melanoma, cutaneous malignant (2)	9p21
?Melkersson–Rosenthal syndrome (2)	9p11
Membroproliferative glomerulonephritis (1)	1q32
Meningioma, NF2-related (3)	22q12.2
Meningioma, SIS-related (3)	22q12.3-q13.1
Menkes' disease (2)	Xq12-q13
Mental retardation, Snyder–Robinson type (2)	Xp21
Mental retardation, X-linked, FRAXE type (3)	Xq28
Mental retardation, X-linked, FRAXF type (3)	Xq28
?Mental retardation, X-linked non-specific, with aphasia (2)	Xp11
Mental retardation, X-linked, syndromic-1, with dystonic movements, ataxia, and seizures (2)	Xp22.2-p22.1
Mental retardation, X-linked, syndromic-2, with dysmorphism and cerebral atrophy (2)	Xp11-q21
Mental retardation, X-linked, syndromic-3, with spastic diplegia (2)	Xp11-q21.3
Mental retardation, X-linked, syndromic-4, with congenital contractures and low fingertip arches (2)	Xq13-q22
Mental retardation, X-linked, syndromic-5, with Dandy–Walker malformation, basal ganglia disease, and seizures (2)	Xq25-q27
Mental retardation, X-linked, syndromic-6, with gynaecomastia and obesity (2)	Xp21.1-q22
Mental retardation, X-linked-1, non-dysmorphic (2)	Xp22
?Mental retardation, X-linked-2, non-dysmorphic (2)	Xq11-q12
Mental retardation, X-linked-3 (2)	Xq28
Mental retardation-skeletal dysplasia (2)	Xq28
Mephenytoin poor metabolizer (3)	10q24.1-q24.3
Metachromatic leucodystrophy (3)	22q13.31-qter
Metachromatic leucodystrophy due to deficiency of SAP-1 (3)	10q21-q22
Metaphyseal chondrodysplasia, Murk Jansen type, 156400 (3)	3p22-p21.1
Metaphyseal chondrodysplasia, Schmid type (3)	6q21-q22.3
Methaemoglobinaemia due to cytochrome b5 deficiency (3)	18q23
Methaemoglobinaemia, enzymopathic (3)	22q13.31-qter
Methaemoglobinaemias, α- (3)	16pter-p13.3
Methaemoglobinaemias, β- (3)	11p15.5
Methylmalonicaciduria, mutase deficiency type (3)	6p21
Mevalonicaciduria (3)	Chr.12
?Microphthalmia with linear skin defects (2)	Xp22.2
Migraine, hemiplegic-1 (2)	19p13
Miller–Dieker lissencephaly syndrome (2)	17p13.3
?Mitochondrial complex I deficiency, 252010 (1)	11q13
MODY, one form (3)	11p15.5
MODY, type I (2)	20q13
MODY, type II, 125851 (3)	7p15-p13
?Moebius syndrome (2)	13q12.2-q13
?Monocyte carboxyesterase deficiency (1)	16q13-q22.1
Mucolipidosis II (1)	4q21-q23
Mucolipidosis III (1)	4q21-q23
Mucopolysaccharidosis Ih (3)	4p16.3
Mucopolysaccharidosis Ih/s (3)	4p16.3
Mucopolysaccharidosis II (3)	Xq28
Mucopolysaccharidosis Is (3)	4p16.3
Mucopolysaccharidosis IVA (3)	16q24.3
Mucopolysaccharidosis IVB (3)	3p21.33
Mucopolysaccharidosis VII (3)	7q21.11
Multiple carboxylase deficiency, biotin-responsive (3)	21q22.1

Continued on p. 936.

Table VII.2 *Continued.*

Disorder	Location
Multiple endocrine neoplasia I (1)	11q13
Multiple endocrine neoplasia IIA, 171400 (3)	10q11.2
Multiple endocrine neoplasia IIB, 162300 (3)	10q11.2
?Multiple lipomatosis (2)	12q15
{?Multiple sclerosis, susceptibility to} (2)	18q22-qter
Muscle glycogenosis (3)	Xq13
Muscular dystrophy, congenital, merosin-negative (2)	6q22-q23
Muscular dystrophy, Duchenne-like, autosomal (2)	13q12-q13
Muscular dystrophy, Duchenne-like, type 2 (3)	17q12-q21.33
Muscular dystrophy, limb-girdle, autosomal dominant (2)	5q22.3-q31.3
Muscular dystrophy, limb-girdle, type 2A (2)	15q15.1-q21.1
Muscular dystrophy, limb-girdle, type 2B (2)	2p16-p13
Myelodysplasia syndrome-1 (3)	3q26
Myelodysplastic syndrome, preleukaemic (3)	5q31.1
Myelogenous leukaemia, acute (3)	5q31.1
Myeloid leukaemia, acute, M4Eo subtype (2)	16q22
Myeloperoxidase deficiency (3)	17q21.3-q22
Myoadenylate deaminase deficiency (3)	1p21-p13
{Myocardial infarction, susceptibility to} (3)	17q23
Myoglobinuria/haemolysis due to PGK deficiency (3)	Xq13
?Myopathy, desminopathic (1)	2q35
Myopathy, distal (2)	14q
Myopathy due to phosphoglycerate mutase deficiency (3)	7p13-p12.3
?Myopathy due to succinate dehydrogenase deficiency (1)	1p22.1-qter
Myopia-1 (2)	Xq28
Myotonia congenita, atypical acetazolamide-responsive (3)	17q23.1-q25.3
Myotonia congenita, dominant, 160800 (3)	7q35
Myotonia congenita, recessive, 255700 (3)	7q35
Myotonic dystrophy (3)	19q13.2-q13.3
Myotubular myopathy, X-linked (2)	Xq28
Myxoid liposarcoma (3)	12q13.1-q13.2
?N syndrome, 310465 (1)	Xp22.3-p21.1
Nail-patella syndrome (2)	9q34.1
Nance–Horan syndrome (2)	Xp22.3-p21.1
Nemaline myopathy-1, 161800 (3)	1q22-q23
Neonatal alloimmune thrombocytopenia (2)	5q23-q31
Neonatal hyperparathyroidism, 239200 (3)	3q21-q24
Nephrolithiasis 2, X-linked (2)	Xp11.23-p11.22
Nephrolithiasis, X-linked, with renal failure (2)	Xp11.22
Nephronophthisis, juvenile (2)	2q13
Nephrosis, congenital, Finnish (2)	19q12-q13.1
Neuroblastoma (2)	1p36.2-p36.1
Neuroepithelioma (2)	22q12
Neurofibromatosis, type 2 (3)	22q12.2
Neurofibromatosis, type I (3)	17q11.2
Neuropathy, recurrent, with pressure palsies, 162500 (3)	17p11.2
Neutropenia, immune (2)	1q23
Neutropenia, neonatal alloimmune (1)	Chr.4
Niemann–Pick disease, type A (3)	11p15.4-p15.1
Niemann–Pick disease, type B (3)	11p15.4-p15.1
Niemann–Pick disease, type C (2)	18q11-q12
Night blindness, congenital stationary, type 3, 163500 (3)	4p16.3
Night blindness, congenital stationary, type I (2)	Xp11.3
Night blindness, congenital stationery, rhodopsin-related (3)	3q21-q24
{Non-insulin dependent diabetes mellitus, susceptibility to} (2)	19q13.3
Noonan syndrome-1 (2)	12q22-qter
Norrie disease (3)	Xp11.4

Continued.

Table VII.2 *Continued.*

Disorder	Location
Norum disease (3)	16q22.1
Nucleoside phosphorylase deficiency, immunodeficiency due to (3)	14q13.1
?Obesity (2)	7q31
?Ocular albinism, autosomal recessive (2)	6q13-q15
Ocular albinism, Forsius–Eriksson type (2)	Xp11.4-p11.23
Ocular albinism, Nettleship–Falls type (3)	Xp22.3
Ocular albinism with sensorineural deafness (2)	Xp22.3
Oculopharyngeal muscular dystrophy-1 (2)	14q11.2-q13
Oestrogen resistance (3)	6q25.1
Optic atrophy 1 (2)	3q28-qter
Optic nerve coloboma with renal anomalies, 120330 (3)	10q25
Ornithine transcarbamylase deficiency (3)	Xp21.1
Orofacial cleft (2)	6p24.3
Oroticaciduria (1)	3q13
OSMED syndrome, 215150 (3)	6p21.3
Osteoarthrosis, precocious (3)	12q13.11-q13.2
Osteogenesis imperfecta, 4 clinical forms, 166200, 166210, 259420, 166220 (3)	17q21.31-q22.05
Osteogenesis imperfecta, 4 clinical forms, 166200, 166210, 259420, 166220 (3)	7q22.1
?Osteopetrosis, 259700 (1)	1p21-p13
Osteoporosis, idiopathic, 166710 (3)	17q21.31-q22.05
?Osteoporosis, involutional (1)	12q12-q14
Osteosarcoma, 259500 (2)	13q14.1-q14.2
Otopalatodigital syndrome, type I (2)	Xq28
Ovarian cancer, serous (2)	6q26-q27
Ovarian cancer, sporadic (3)	17q21
Ovarian carcinoma, 167000 (2)	19q13.1-q13.2
Ovarian carcinoma (2)	9p24
Ovarian carcinoma (3)	16q22.1
Ovarian failure, premature (2)	Xq26-q27
Pachyonychia congenita, Jackson–Lawler type (2)	17q12-q21
?Paget disease of bone (2)	6p21.3
?Pallister–Hall syndrome (2)	3p25.3
Palmoplantar keratoderma, Bothnia type (2)	12q11-q13
Pancreatic lipase deficiency (1)	10q26.1
?Panhypopituitarism, X-linked (2)	Xq21.3-q22
Paraganglioma (2)	11q22.3-q23.2
Paramyotonia congenita, 168300 (3)	17q23.1-q25.3
Paraneoplastic sensory neuropathy (1)	1p34
Parathyroid adenomatosis 1 (2)	11q13
?Parietal foramina (2)	11p12-p11.12
{?Parkinsonism, susceptibility to} (1)	22q13.1
Paroxysmal nocturnal haemoglobinuria (3)	Xq22.1
Partington syndrome II (2)	Xp22-p21
Pelizaeus–Merzbacher disease (3)	Xq22
Pelviureteric junction obstruction (2)	6p
?Pendred syndrome (2)	8q24
PEO with mitochondrial DNA deletions (2)	10q
Perineal hypospadias (3)	Xq11-q12
Periodontitis, juvenile (2)	4q11-q13
Peroxisomal bifunctional enzyme deficiency (1)	3q26.3-q28
Persistent hyperinsulinaemic hypoglycaemia of infancy (2)	11p15.1-p14
Persistent Mullerian duct syndrome (3)	19p13.3-p13.2
Peters anomaly (3)	11p13
Pfeiffer syndrome, 101600 (3)	10q26
Pfeiffer syndrome, 101600 (3)	8p12-p11.2
Phenylketonuria (3)	12q24.1
Phenylketonuria, atypical, due to GCH1 deficiency, 233910 (1)	14q22.1-q22.2
Phenylketonuria due to dihydropteridine reductase deficiency (3)	4p15.31

Continued on p. 938.

Table VII.2 *Continued.*

Disorder	Location
Phenylketonuria due to PTS deficiency (3)	11q22.3-q23.3
Phaeochromocytoma (2)	1p
Phosphoribosyl pyrophosphate synthetase-related gout (3)	Xq22-q24
?Phosphorylase kinase deficiency of liver and muscle, 261750 (2)	16q12-q13.1
Piebaldism (3)	4q12
Pituitary hormone deficiency, combined (3)	3p11
PK deficiency haemolytic anaemia (3)	1q21
[Placental lactogen deficiency] (1)	17q22-q24
Placental steroid sulphatase deficiency (3)	Xp22.32
Plasmin inhibitor deficiency (3)	17pter-p12
Plasminogen activator deficiency (1)	8p12
Plasminogen deficiency, types I and II (1)	6q26
Plasminogen Tochigi disease (3)	6q26
Platelet α/δ storage pool deficiency (1)	1q23-q25
Platelet glycoprotein IV deficiency (3)	7q11.2
{Polio, susceptibility to} (2)	19q13.2-q13.3
Polycystic kidney disease, adult, type II (2)	4q21-q23
Polycystic kidney disease, autosomal recessive (2)	6p21.1-p12
Polycystic kidney disease, infantile severe, with tuberous sclerosis (3)	16p13.3
Polycystic kidney disease-1 (3)	16p13.31-p13.12
Polyposis coli, familial (3)	5q21-q22
Porphyria, acute hepatic (3)	9q34
Porphyria, acute intermittent (3)	11q24.1-q24.2
Porphyria, Chester type (2)	11q23.1
Porphyria, congenital erythropoietic (3)	10q25.2-q26.3
Porphyria cutanea tarda (3)	1p34
Porphyria, hepatoerythropoietic (3)	1p34
Porphyria variegata (2)	14q32
?Prader–Willi syndrome (1)	15q12
Prader–Willi syndrome (2)	15q11
Precocious puberty, male, 176410 (3)	2p21
{Pre-eclampsia, susceptibility to} (3)	1q42-q43
Progressive cone dystrophy (2)	Xp11.3
Prolactinoma, hyperparathyroidism, carcinoid syndrome (2)	11q13
Prolidase deficiency (3)	19cen-q13.11
Properdin deficiency, X-linked (3)	Xp11.4-p11.23
Propionicacidaemia, type I or pccA type (1)	13q32
Propionicacidaemia, type II or pccB type (3)	3q21-q22
Prostate cancer (3)	Xq11-q12
Protein C cofactor deficiency (3)	1q23
Protein C inhibitor deficiency (2)	14q32.1
Protein S deficiency (3)	3p11.1-q11.2
Protoporphyria, erythropoietic (3)	18q21.3
Protoporphyria, erythropoietic, recessive, with liver failure (3)	18q21.3
Pseudoachondroplastic dysplasia (2)	19q12
Pseudohermaphroditism, male, with gynaecomastia (3)	9q22
Pseudohypoaldosteronism (1)	4q31.1
Pseudohypoparathyroidism, type Ia, 103580 (3)	20q13.2
Pseudovaginal perineoscrotal hypospadias (3)	Chr.2
Pseudo-vitamin D dependency rickets 1 (2)	12q14
Pseudo-Zellweger syndrome (1)	3p23-p22
Psoriasis susceptibility (2)	17q
Pulmonary alveolar proteinosis, congenital, 265120 (3)	2p12-p11.2
Purpura fulminans, neonatal (1)	2q13-q14
?Pyridoxine dependency with seizures (1)	2q31
Pyropoikilocytosis (3)	1q21
Pyruvate carboxylase deficiency (1)	11q

Continued.

Table VII.2 *Continued.*

Disorder	Location
Pyruvate dehydrogenase deficiency (3)	Xp22.2-p22.1
Rabson–Mendenhall syndrome (3)	19p13.2
?Ragweed sensitivity (2)	6p21.3
Renal cell carcinoma (2)	3p14.2
Renal cell carcinoma (3)	3p26-p25
?Renal cell carcinoma, papillary, 1 (2)	1q21
Renal cell carcinoma, papillary, 2 (2)	Xp11.2
?Renal glucosuria, 253100 (1)	16p11.2
[Renal glucosuria] (2)	6p21.3
Renal tubular acidosis-osteopetrosis syndrome (3)	8q22
?Resistance/susceptibility to TB, etc. (1)	2q35
?Retinal cone dystrophy-1 (2)	6q25-q26
Retinitis pigmentosa, autosomal recessive (3)	3q21-q24
Retinitis pigmentosa, digenic (3)	11q13
Retinitis pigmentosa, digenic (3)	6p21.1-cen
Retinitis pigmentosa, peripherin-related (3)	6p21.1-cen
Retinitis pigmentosa-1 (2)	8p11-q21
Retinitis pigmentosa-10 (2)	7q31-q35
Retinitis pigmentosa-11 (2)	19q13.4
Retinitis pigmentosa-12, autosomal recessive (2)	1q31-q32.1
Retinitis pigmentosa-13 (2)	17p
Retinitis pigmentosa-14 (2)	6p21.3
Retinitis pigmentosa-2 (2)	Xp11.3
Retinitis pigmentosa-3 (2)	Xp21.1
Retinitis pigmentosa-4, autosomal dominant (3)	3q21-q24
?Retinitis pigmentosa-6 (2)	Xp21.3-p21.2
Retinitis pigmentosa-9 (2)	7p15.1-p13
Retinitis punctata albescens (3)	6p21.1-cen
Retinoblastoma (3)	13q14.1-q14.2
?Retinol binding protein, deficiency of (1)	10q23-q24
Retinoschisis (2)	Xp22.3-p22.1
?Rett syndrome (2)	Xp
Rhabdomyosarcoma (2)	11p15.5
Rhabdomyosarcoma, alveolar, 268200 (3)	13q14.1
Rhabdomyosarcoma, alveolar, 268220 (3)	2q35
Rh-null disease (1)	3cen-q22
?Rh-null haemolytic anaemia (1)	1p36.2-p34
Rickets, vitamin D-resistant (3)	12q12-q14
Rieger syndrome (2)	4q25-q27
Rippling muscle disease-1 (2)	1q41
Rod monochromacy (2)	Chr.14
?Rothmund–Thomson syndrome (2)	Chr.8
Rubinstein–Taybi syndrome (2)	16p13.3
Russell–Silver syndrome (2)	17q25
Saethre–Chotzen syndrome (2)	7p21
Salivary gland pleomorphic adenoma (2)	8q12
Salla disease (2)	6q
Sandhoff disease (3)	5q13
?Sanfilippo disease, type IIIC (2)	Chr.14
Sanfilippo syndrome D (1)	12q14
Sarcoma, synovial (3)	Xp11.2
Schindler disease (3)	22q11
?Schizophrenia (2)	5q11.2-q13.3
Schizophrenia, chronic (3)	21q21.3-q22.05
{?Schizophrenia, susceptibility to} (2)	3q13.3
Schizophrenia-3 (2)	6pter-p22
Schwannoma, sporadic (3)	22q12.2

Continued on p. 940.

Table VII.2 *Continued.*

Disorder	Location
Sclerotylosis (2)	4q28-q31
SED congenita (3)	12q13.11-q13.2
Segawa syndrome, recessive (3)	11p15.5
Selective T-cell defect (3)	2q12
Severe combined immunodeficiency due to ADA deficiency (3)	20q13.11
Severe combined immunodeficiency due to IL2 deficiency (1)	4q26-q27
Severe combined immunodeficiency, HLA class II-negative type, 209920 (2)	19p13.1
?Severe combined immunodeficiency, type I (1)	8q11
Severe combined immunodeficiency, X-linked, 300400 (3)	Xq13
Short stature (2)	Xpter-p22.32
?Sialidosis (2)	6p21.3
Sickle cell anaemia (3)	11p15.5
Simpson–Golabi–Behmel syndrome (2)	Xq26
?Situs inversus viscerum (2)	14q32
Sjogren–Larsson syndrome (2)	17q11.2
?SLE (1)	1q32
Small-cell cancer of lung (2)	3p23-p21
SMED Strudwick type (3)	12q13.11-q13.2
?Smith–Lemli–Opitz syndrome (2)	7p34-qter
Smith–Magenis syndrome (2)	17p11.2
Somatotrophinoma (2)	11q13
Somatotrophinoma (3)	20q13.2
Sorsby fundus dystrophy, 136900 (3)	22q12.1-q13.2
Sorsby fundus dystrophy (2)	22q13.1-qter
Spastic paraplegia 2, 312920 (3)	Xq22
Spastic paraplegia, 312900 (3)	Xq28
Spastic paraplegia 5A (2)	8p12-q13
Spastic paraplegia-3 (2)	14q
Spastic paraplegia-4 (2)	2p24-p21
Spastic paraplegia-6 (2)	15q11.1
Spherocytosis, hereditary (3)	17q21-q22
Spherocytosis, hereditary, Japanese type (3)	15q15
Spherocytosis, recessive (3)	1q21
Spherocytosis-1 (3)	14q22-q23.2
Spherocytosis-2 (3)	8p11.2
Spinal and bulbar muscular atrophy of Kennedy, 313200 (3)	Xq11-q12
Spinal muscular atrophy II (2)	5q12.2-q13.3
Spinal muscular atrophy III (2)	5q12.2-q13.3
Spinal muscular atrophy X-linked lethal infantile (2)	Xp
Spinocerebellar ataxia, type 4 (2)	16q
Spinocerebellar ataxia, type 5 (2)	11p11-q11
Spinocerebellar ataxia-1 (3)	6p23
Spinocerebellar ataxia-3 (2)	14q24.3-qter
Spinocerebellar atrophy II (2)	12q24
Split hand/foot malformation, type 2 (2)	Xq26
Split-hand/split-foot malformation, type 1 (2)	7q21.2-q21.3
Spondyloepiphyseal dysplasia tarda (2)	Xp22.2-p22.1
Stargardt disease 2 (2)	13q34
Stargardt disease 3 (2)	6cen-q14
Stargardt macular dystrophy (2)	1p21-p13
Startle disease, autosomal recessive (3)	5q32
Startle disease/hyperefplexia, autosomal dominant, 149400 (3)	5q32
Stickler syndrome, type 2 (2)	6p22-p21.3
Stickler syndrome, type I (3)	12q13.11-q13.2
Stickler syndrome, type II, 184840 (3)	6p21.3
?Stiff skin syndrome (2)	Chr.15
?Stomatocytosis I (1)	9q34.1

Continued.

Table VII.2 *Continued.*

Disorder	Location
Sucrose intolerance (1)	3q25-q26
Supravalvar aortic stenosis, 185500 (3)	7q11.2
?{Susceptibility to IDDM} (1)	11q13
{Susceptibility to measles} (1)	1q32
Tay–Sachs disease (3)	15q23-q24
Thalassaemias, α- (3)	16pter-p13.3
Thalassaemias, β- (3)	11p15.5
Thoracoabdominal syndrome (2)	Xq25-q26.1
Thrombocytopenia, neonatal alloimmune (1)	17q21.32
?Thrombocytopenia, Paris–Trousseau type (2)	11q23
Thrombocytopenia, X-linked, 313900 (3)	Xp11.23-p11.22
?Thrombophilia due to elevated HRG (1)	3q28-q29
Thrombophilia due to excessive plasminogen activator inhibitor (1)	7q21.3-q22
Thrombophilia due to heparin cofactor II deficiency (3)	22q11
Thrombophilia due to protein C deficiency (3)	2q13-q14
Thrombophilia due to thrombomodulin defect (3)	20p12-cen
Thromboxane synthase deficiency (2)	7q34
Thymine-uraciluria (1)	1p22-q21
Thyroid adenoma, hyperfunctioning (3)	14q31
Thyroid hormone resistance, 274300, 188570 (3)	3p24.3
Thyroid iodine peroxidase deficiency (1)	2p13
Thyroid papillary carcinoma (1)	10q11-q12
Thyrotropin-releasing hormone deficiency (1)	Chr.3
Torsion dystonia (2)	9q32-q34
Torsion dystonia-Parkinsonism, Filipino type (2)	Xq12-q13.1
Total anomalous pulmonary venous return (2)	4p13-q12
?Tourette syndrome (2)	18q22.1
?Townes–Brocks syndrome (2)	16q12.1
Transcobalamin II deficiency (3)	22q11.2-qter
[Transcortin deficiency] (1)	14q32.1
Treacher Collns mandibulofacial dysostosis (2)	5q32-q33.1
Trichorhinophalangeal syndrome, type I (2)	8q24.12
Triphalangeal thumb-polysyndactyly syndrome (2)	7q36
Trypsinogen deficiency (1)	7q32-qter
{?Tuberculosis, susceptibility to] (2)	2q
Tuberous sclerosis-1 (2)	9q34
Tuberous sclerosis-2 (2)	16p13.3
Turcot syndrome with glioblastoma, 276300 (3)	3p21.3
Turner syndrome (1)	Xq13.1
Tylosis with oesophageal cancer (2)	17q23-qter
Tyrosinaemia, type I (3)	15q23-q25
Tyrosinaemia, type II (3)	16q22.1-q22.3
Tyrosinaemia, type III (1)	12q14-qter
Urate oxidase deficiency (1)	1p22
Urolithiasis, 2,8-dihydroxyadenine (3)	16q24
Usher syndrome, type 1A (2)	14q32
Usher syndrome, type 1B (2)	11q13.5
Usher syndrome, type 1C (2)	11p15.1
Usher syndrome, type 2 (2)	1q32
Usher syndrome, type 3 (2)	3q21-q25
van der Woude syndrome (2)	1q32
Velocardiofacial syndrome, 192430 (2)	22q11
Venous malformations, multiple cutaneous and mucosal (2)	9p
Virilization, maternal and fetal, from placental aromatase deficiency (3)	15q21.1
Vitreoretinopathy, exudative, familial (2)	11q13-q23
Vitreoretinopathy, neovascular inflammatory (2)	11q13
{Vivax malaria, susceptibility to} (1)	1q21-q22

Continued on p. 942.

Table VII.2 *Continued.*

Disorder	Location
von Hippel–Lindau syndrome (3)	3p26-p25
von Willebrand disease (3)	12p13.3
Waardenburg syndrome, type 2A, 193510 (3)	3p14.1-p12.3
Waardenburg syndrome, type I (3)	2q35
Waardenburg syndrome, type III, 148820 (3)	2q35
Wagner syndrome, type II (3)	12q13.11-q13.2
Waisman Parkinsonism-mental retardation syndrome (2)	Xq28
?Walker–Warburg syndrome, 236670 (2)	9q31-q33
Watson syndrome, 193520 (3)	17q11.2
Werdnig–Hoffmann disease (2)	5q12.2-q13.3
Werner syndrome (2)	8p12-p11
{Wernicke–Korsakoff syndrome, susceptibility to} (1)	3p14.3
Wieacker–Wolff syndrome (2)	Xq13-q21
Williams–Beuren syndrome, 194050 (3)	7q11.2
Wilms' tumour (3)	11p13
Wilms' tumour, type 2 (2)	11p15.5
Wilson disease (3)	13q14.3-q21.1
Wiskott–Aldrich syndrome (3)	Xp11.23-p11.22
?Wolf–Hirschhorn syndrome, 194190 (3)	4p16.1
Wolf–Hirschhorn syndrome (2)	4p16.3
Wolfram syndrome (2)	4p
Wolman disease (3)	10q24-q25
Woods neuroimmunologic syndrome (2)	Xq26-qter
Wrinkly skin syndrome (2)	2q32
Xanthinuria (1)	2p23-p22
?Xeroderma pigmentosum (1)	1q42
Xeroderma pigmentosum, complementation group C (3)	3p25
Xeroderma pigmentosum, group B (3)	2q21
Xeroderma pigmentosum, group D, 278730 (3)	19q13.2-q13.3
Xeroderma pigmentosum, group G (3)	13q33
Xeroderma pigmentosum, type A (3)	9q34.1
?Xeroderma pigmentosum, type F (2)	Chr.15
?XLA and isolated growth hormone deficiency, 307200 (3)	Xq21.3-q22
Zellweger syndrome-1 (2)	7q11.23
Zellweger syndrome-2 (3)	1p22-p21
Zellweger syndrome-3 (3)	8q21.1

[ ], nondisease genes; { }, susceptibility genes.

Appendix VIII Mouse gene knock-out tables

The ability to engineer specific mutations in mice by mutating a specific gene, by homologous recombination in embryonic stem cells and then transferring the mutation into a developing mouse, represented a major breakthrough in mouse genetics (see e.g. refs 1–3). The most common point of this exercise has been to inactivate the targeted gene and to observe the phenotypic effects of the 'knock-out' on the developing mouse (e.g. refs 4–6).

Table VIII.1 includes the majority of targeted mouse mutants for which the exact gene mutation has been characterized. In several cases a targeted mutation has produced a phenotype similar to a previously studied mouse mutant, and has thus facilitated the molecular characterization of the previously established mutant [7–13].

The first column of Table VIII.1 gives the name of the protein product of the targeted gene (with abbreviated and/or alternative names in parentheses), or a description of the targeted genomic locus if it is not a coding sequence. If the entry is not a null mutant, this is also indicated by, for example, '*modification*' or '*partial*' (a 'leaky' mutation). The table is alphabetized by the first column.

In cases where more than one group has performed essentially the same mutation, the number of independent groups is indicated in the second column (e.g. ×2, ×3). Also noted in the second column is whether the mutant has been crossed with another mutant derived by gene targeting, and studied as a double mutant; this is indicated by a 'D', preceded by the number of different double mutants (e.g. D, 2D). The double mutants are described in Table VIII.2.

The third column gives a synopsis of the key aspects of the phenotype, generally those reported in the primary reference(s) listed in the fourth column. When embryonic lethality is associated with the mutation this is indicated by an 'e' followed by the day(s) of gestation when the mutants die. Perinatal lethality is defined as death within the first 24–48 h after birth, and neonatal lethality is death before weaning. In many cases the perinatal and neonatal lethality is highly variable and may depend on the genetic background or environment, and the reader is referred to the primary references for the precise details.

In a few instances, when more than one group has created essentially the same mutant, disparate results have been reported. Regardless of the reason for the disagreement, these findings are marked with an asterisk (*) in the third column. The fourth column gives the reference(s) that first described the mutant, while the fifth column lists subsequent reports describing the mutant, or utilizing the mutant as a tool for other experiments.

In certain cases, unexpected roles for a protein have been discovered. In other cases, genetic redundancy or compensation allows partial function despite the absence of a protein thought to be crucial. Out of the 263 knock-outs listed in Table VIII.1, only about 25% lead to lethality before or just after birth, with another 10% resulting in death during the first 3–6 weeks. Most null mutants, however, survive into adulthood. Only a dozen or so of the mutants are apparently normal.

The data for Tables VIII.1 and VIII.2 were originally compiled by E.P. Brandon, R.L. Izerda and G.S. McKnight and are reprinted here from *Current Biology* (**5**, 625–634; 758–765; 873–881; [corrigenda] 1073) with permission.

Listings of targeted mutations in mouse, pig, rat and *Drosophila* can be found in TBASE. WWW: http://www.gdb.org/Dan/tbase/tbase/.html

Table VIII.1 Published targeted mutations.

Protein/locus	×; D (see text)	Phenotype	Initial report(s)	Follow-ups
Ab1	×2	Perinatal lethality; multiple developmental defects; lymphopenia	14–16	17
Acetylcholine receptor, nicotinic, β_2 subunit		Loss of high-affinity nicotine binding in brain; abnormal avoidance learning	18	
Acrosin		No defects in fertilization	19	
Activin/inhibin βA	D	Perinatal lethal; whiskers and incisors absent; cleft palate	20	
Activin/inhibin βB	D	Eyelid defects; female reproductive defects	21	
Activin receptor type II (ActRcII)		Reduced FSH; reproductive defects	22	
Adenomatous polyposis coli (APC) protein *modification*		Postimplantation embryonic lethal; heterozygotes develop intestinal tumours	23	
Adenylyl cyclase type I		Changes in long-term potentiation and spatial learning	24	
Adhesion molecule on glia (AMOG) (β_2 subunit of Na,K-ATPase)		Neonatal lethal; neural degeneration	25	
Amyloid (β-) precursor protein *modification*		Increased incidence of corpus callosum agenesis; behavioural deficits	26	
Angiotensinogen		Hypotension	27	
Apolipoprotein AI		Reduced HDL cholesterol	28	29
Apolipoprotein B *modification*		Hypobetalipoproteinaemia; exencephalus and hydrocephalus	30	
Apolipoprotein B		e10–20 lethal; heterozygotes protected from diet-induced hypercholesterolaemia	31	
Apolipoprotein C-III		Hypotriglyceridaemia	32	
Apolipoprotein E	×2; D	Hypercholesterolaemia and atherosclerosis	33–35	36–43
Argininosuccinate synthetase (ASS)		Neonatal lethal; citrullinaemia; hyperammonaemia	44	
Asialoglycoprotein receptor, minor subunit HL-2		Decreased HL-1 expression in liver	45	
Atrial natriuretic peptide (ANP)		Salt-sensitive hypertension	46	
B-cell lineage-specific activator protein (BSAP) (*Pax5* gene)		Neonatal lethality; posterior midbrain morphological defects; B-cell development disrupted	47	
B7 (CD28 ligand)		Decreased co-stimulated response to alloantigen	48	
Bcl-2	×2	Neonatal lethality; lymphocytopenia; multiple growth defects; tremor; melanin synthesis defect; polycystic kidneys	49, 50	
Bcl-x		e13 lethal; neuronal and haematopoietic apoptosis	51	
Bmi-1		Haematopoietic defects; ataxia; seizures; posterior transformation	52	
Brain-derived neurotrophic factor (BDNF)	×2	Neonatal lethality; coordination deficiency; of sensory ganglia degeneration	53, 54	

Continued.

Table VIII.1 *Continued.*

Protein/locus	×; D (see text)	Phenotype	Initial report(s)	Follow-ups
Cadherin (E-cadherin)	×2	e4–5 lethal; trophectoderm and blastocoel do not form	55, 56	
Calcium-calmodulin-dependent protein kinase II α (α-CaMKII)		Deficient hippocampal long-term potentiation and long-term depression; impaired spatial learning; seizure prone; abnormal fear and pain responses	57, 58	59, 60
Casein (β-casein)		Reduced casein micelles; reduced protein in milk, reduced growth of pups	61	
CD2		No defects observed	62	63
CD4	×2; D	Decreased helper T-cell activity	64, 65	66–77
CD8-α (Lyt-2)	D	Absence of cytotoxic T cells	78	74–77, 79–86
CD8-β		Reduced thymic maturation of $CD8^+$ T cells	87	
CD18 *partial*		Mild granulocytosis; impaired immune responses	88	
CD23	×3	Defects in IgE regulation and IgE-mediated signalling	89–91	
CD28		Decreased T-cell response to lectins; decreased IL-2Rα, IgG1, and IgG2b	92	93
CD40		Defects in thymus-dependent humoral immunity	94	
CD40 ligand (CD40L)	×2	Defects in thymus-dependent humoral immunity	95, 96	
CD45 exon 6		Impaired T-cell maturation	97	98
Cellular retinoic acid binding protein (CRABP-I)		No defects observed	99	
Ciliary neurotrophic factor (CNTF)		Motor neuron degeneration; muscle weakness	100	
Collagen α (IX)		Non-inflammatory degenerative joint disease	101	
Collagen α (V) *modificationn*		Neonatal lethality; abnormalities in spine, skin and eyes	102	
Collagen (X)		No defects observed	103	
Connexin 43		Perinatal lethal; cardiac malformation	104	
Corticotropin-releasing hormone (CRH)		Decreased adrenal corticosterone release in response to stress; offspring of homozygous mother perinatal lethal due to lung dysplasia	105	
Creatine kinase		No burst activity in skeletal muscle	106	107, 108
Csk	×2	e10 lethal; no notochord; increased Src and Fyn activity	109, 110	
Cyclic AMP-responsive element binding protein (CREB) α and δ isoforms		Lack late phase of CA1 long-term potentiation; decreased long term memory; increase in CREM	111	112
Cystathionine β-synthase (CBS)		Neonatal lethality; growth retardation; abnormal hepatic morphology	113	
Cystic fibrosis trans-membrane regulator (CFTR)	×3	Neonatal lethality; meconium ileus; defective epithelial chloride transport	114–117	118–124

Continued on p. 946.

Table VIII.1 *Continued.*

Protein/locus	×; D (see text)	Phenotype	Initial report(s)	Follow-ups
Cystic fibrosis transmembrane regulator (CFTR) *partial*		Neonatal lethality; meconium ileus; defective epithelial chloride transport	125	126–128
Cytochrome b, phagocyte-specific oxidase, 91 kD subunit		Increased susceptibility to pathogens; model for X-linked chronic granulomatous disease		129
DNA methyltransferase		e10–12 lethal; defective expression of imprinted genes	130	131
DNA polymerase β *modification*		Demonstrates feasibility of tissue-specific disruption using Cre-*loxP* system	132	
Dopamine D1 receptor (D1R)	×2	Premature lethality without wet food; growth retardation; reduced dynorphin and substance P expression; hyperactivity; decreased rearing behaviour	133, 134	135
E2A	×2	Neonatal lethality; growth retardation; lack B cells	136, 137	
En-1		Perinatal lethal; multiple developmental defects	138	
En-2		Abnormal cerebellar foliation	139	140
Endothelin-1		Perinatal lethal; craniofacial abnormalities; high blood pressure in heterozygotes	141	
Endothelin-3		Aganglionic megacolon; white spotting of skin and coat; allelic with lethal spotting (*ls*)	12	
Endothelin-B receptor		Aganglionic megacolon; white spotting of skin and coat; allelic with piebald (*s*)	13	
Evx1 (even-skipped homologue)		Early postimplantation lethal	142	
Excision repair cross-complementing protein (ERCC-1)		Neonatal lethal; liver failure and aneuploidy	143	
Fc receptor g subunit		Pleiotropic effector cell defects	144	
Fgr	D	No defects observed	145	
Fibroblast growth factor 3 (FGF3) (int-2)		Tail and inner ear developmental defects	146	
Fibroblast growth factor 4 (FGF4)		e4–6 lethal; inner cell mass does not develop	147	
Fibroblast growth factor 5 (FGF5)		Long hair; allelic with angora (*go*)	11	
Fibroblast growth factor receptor 1 (FGFR-1)	×2	e7–9 lethal; abnormal mesoderm patterning	148, 149	
Fibronectin		e9–10 lethal; defects in mesoderm development	150	
FMR1 protein		Macroorchidism; hyperactivity	151	
Follistatin		Perinatal lethal; multiple developmental defects	152	
Fos	×2	Perinatal lethality; osteopetrosis; defects in gametogenesis and haematopoiesis	153, 154	155–160

Continued.

Table VIII.1 *Continued.*

Protein/locus	×; D (see text)	Phenotype	Initial report(s)	Follow-ups
Fumarylacetoacetate (FAH)		Perinatal lethal; hepatic dysfunction; allelic with lethal albino (*alf/hsdr*-1)	10	
Fyn ($p59^{fyn}$)	×2; 2D	Signalling defect in thymocytes but not peripheral T cells; impaired long-term potentiation; abnormal olfactory glomeruli and hippocampal morphology; suckling defect	17, 161, 162	163–167
Fyn $(p59)^{fynT}$		Signalling defective in thymocytes but not peripheral T cells	168	
GATA-2		e10–11 lethal; severe anaemia	169	
Glial fibrillary acidic protein (GFAP)		No defects observed	170	
Globin (β-globin b1)		Premature lethality; anaemia	171	
Glucocerebrosidase		Perinatal lethal; lysosomal storage defect; model for Gaucher's disease	172	
Glutamate receptor, metabotropic type 1 (mGluR1)	×2	Ataxia; impaired cerebellar long-term depression and conditioned eyeblink response; impaired long-term potentiation, spatial learning and context-dependent fear conditioning	173–175	
Glutamate receptor, NMDA type 1 (NMDAR1)	×2	Perinatal lethal; respiratory failure	176, 177	
Glutamate receptor, NMDA type ε (NMDAR ε) (NR2A)		Reduced CA1 long-term potentiation; spatial learning defect	178	
Granulocyte colony-stimulating factor (G-CSF)		Granulopoietic defects	179	
Granulocyte–macrophage colony-stimulating factor (GM-CSF)	×2	Pulmonary pathology; apparently normal haematopoiesis	180, 181	182
Granzyme B		Cytotoxic T-lymphocyte defect	183	
Growth-associated protein-43 (GAP-43)		Perinatal and neonatal lethality; abnormal pathfinding at the optic chiasm	184	
Hck	D	Phagocytosis impaired; increased Lyn activity	145	
Hepatic lipase		Mild dyslipidaemia	185	
Hepatocyte factor/scatter factor (HGF/SF)	×2	e13–16 lethal; placental defect; small liver	186, 187	
Hepatocyte nuclear factor 3b (HNF-3b)	×2	e10–11 lethal; disorganized node and notochord	188, 189	
Hepatocyte nuclear factor 4 (HNF-4)		e6 lethal; ectodermal cell death; impaired gastrulation	190	
Hexosaminidase (β-hexosaminidase α subunit)		Accumulation of ganglioside in central nervous system; model for Tay–Sachs disease	191	
Hox 11		No spleen	192	
Hox-A1 (Hox 1.6)	×2	Perinatal lethal; hindbrain reorganization; cranial nerve and inner ear defects	193, 194	195–197

Continued on p. 948.

Table VIII.1 *Continued.*

Protein/locus	×; D (see text)	Phenotype	Initial report(s)	Follow-ups
Hox-A2 (Hox 1.11)	×2	Perinatal lethal; homeotic transformation of rostral head	198, 199	
Hox-A3 (Hox 1.5)	D	Perinatal lethal; athymic; aparathyroid; throat, heart, arterial, and craniofacial abnormalities	200	
Hox-A4 (Hox 1.4)		Rib and sternal defects	201	
Hox-A5 (Hox 1.3)		Perinatal lethality; cervical and thoracic homeotic transformations	202	
Hox-A11 (Hox 1.9)		Homeotic transformations; skeletal malformations	203	
Hox-B4 (Hox 2.6)		Neonatal lethality; cervical homeotic transformation; sternal defects	204	
Hox-B4 (Hox 2.6) *truncation*		Cervical homeotic transformation	204	
Hox-B5 (Hox 2.1)		Rostral shift in shoulder girdle; homeotic transformation of vertebrae C6 through T1	205	
Hox-B6 (Hox 2.2)		Missing first rib; bifid second rib; homeotic transformation of vertebrae C6 through T1	205	
Hox-C8 (Hox 3.1)		Neonatal lethality; skeletal transformations	206	
Hox-D3 (Hox 4.1)	D	Transformations of anterior vertebrae (atlas and axis)	207	
Hox-D11 (Hox 4.6)	×2	Vertebral homeotic transformations; other skeletal abnormalities	208, 209	
Hox-D13 (Hox 4.8)		Skeletal alterations along all body axes; males infertile	210	
Hypoxanthine-guanine phosphoribosyl-transferase (HPRT) *correction*		Demonstrates germline transmission of a genetic correction introduced by homologous recombination in embryonic stemcells	211	
Ik (Ikaros gene products)		Neonatal lethality; reduced size; lymphocytes and lymphoid progenitors absent	212	
Immunoglobulin D	×2	Reduced number of mature B cells*	213, 214	
Immunoglobulin E		No defects observed	215	
Immunoglobulin E receptor α chain		Resistant to cutaneous and systemic anaphylaxis	216	
Immunoglobulin κ intron enhance		No Igκ rearrangement; slight reduction in splenic B cells	217	218
Immunoglobulin κ light chain	×2	Reduced number of B cells	218, 219	220
Immunoglobulin κ replaced with human constant region		B cells produce human–mouse chimeric κ-bearing antibodies	221	
Immunoglobulin μ membrane exon		Absence of B cells	222	223–225
Inhibin a	D	Gonadal tumours; both males and females sterile	226	227, 228
Insulin receptor substrate-1 (IRS-1)	×2	Reduced size; impaired glucose tolerance; decrease in insulin-, IGF-1- and IGF-2-induced glucose uptake	229, 230	

Continued.

Table VIII.1 *Continued.*

Protein/locus	×; D (see text)	Phenotype	Initial report(s)	Follow-ups
Insulin-like growth factor I (IGF-I)	×2; 2D	Perinatal lethality (background strain dependent); 60% normal birthweight; infertile; under-developed muscle tissue; lung defects	230–233	
Insulin-like growth factor II (IGF-II)	2D	60% normal birthweight; (heterozygotes with paternally inherited null alllele similar to homozygotes; gene imprinting implicated)	234	231, 232, 235–237
Insulin-like growth factor receptor 1 (IGF1R)	2D	Perinatal lethal; organ hypoplasia; respiratory failure; 45% normal birthweight	231, 232	238–240
Insulin-like growth factor receptor 2 (IGF2R) (mannose 6-phosphate receptor 300; MPR300)	×2	Perinatal lethal; 30% larger at birth; organ and skeletal abnormalities; (heterozygotes with maternally inherited null allele similar to homozygotes; gene imprinting implicated)	241, 242	
Insulin-promoter-factor-1 (IPF-1)		Neonatal lethal; no pancreas	243	
Integrin (α5 integrin)		e10–11 lethal; mesodermal defects	244	
Intercellular adhesion molecule-1 (ICAM-1)	×2	Leucocytosis; impaired inflammatory and immune responses	245, 246	
Interferon αβ receptor		Antiviral defence impaired	247	
Interferon γ		Multiple immune response defects	248	249–252
Interferon γ receptor		Multiple immune response defects	253	254–256
Interferon regulatory factor 1 (IRF-1)	×2	Decreased CD4⁻8⁺ T cells; impaired interferon γ response	257, 258	259–261
Interferon regulatory factor 2 (IRF-2)		Premature lethality; defects in haematopoiesis; immunocompromised	257	
Interleukin-1β-converting enzyme (ICE)		Decreased IL-1 production; resistance to endotoxic shock	262	
Interleukin-2 (IL-2)	D	Premature lethality; normal T-cell subset composition, but dysregulated immune system; inflammatory bowel disease	263	264, 265
Interleukin-2 receptor γ chain (IL-2Rγ)		Lymphopenia; absence of NK cells	266	
Interleukin-4 (IL-4)	×2; D	CD4⁺ (Th2)-produced cytokines reduced; serum IgG1 and IgE reduced	267, 268	269, 270
Interleukin-6 (IL-6)	×2	Higher bone turnover rate; no bone loss when ovariectomized; immune defects; reduced IgA-producing cells	271, 272	273–275
Interleukin-7 receptor (IL-7R)		Early lymphocyte expansion severely impaired	275	
Interleukin-8 receptor (IL-8R)		Lymphadenopathy and splenomegaly; increased B cells and neutrophils	277	
Interleukin-10 (IL-10)		Reduced growth; anaemia; chronic enterocolitis	278	
Invariant chain (Ii)	×2	MHC class II transport and function defective; reduced CD4⁺ T cells	279, 280	281, 282

Continued on p. 950.

Table VIII.1 *Continued.*

Protein/locus	×; D (see text)	Phenotype	Initial report(s)	Follow-ups
J_H–E_m immunoglobulin heavy chain (joining and enhancer regions)		Suppression of switch recombination at μ gene; absence of B cells	283	223, 284
J_H immunoglobulin joining region	×2	Absence of B cells	285, 286	
J_H replaced with rearranged V region		Rearranged V transgene expressed in all B cells	287	
Jun	×2	e11–16 lethal; impaired hepatogenesis"; defective fetal liver erythropoiesis*; oedema; decreased growth of embryonic fibroblasts	288, 289	
Keratin 8		e12 lethal; internal bleeding; abnormal fetal liver; a few mice survive to adulthood	290	291
Krox-20	×2	Perinatal lethal; defects in hindbrain and associated cranial sensory ganglia	292, 293	294
L14 s-type lectin		No defects observed	295	
Lactalbumin (α-lactalbumin)		Females can't nurse offspring; milk viscous; no lactose	296	
Lactalbumin (α-lactalbumin replaced by human lactalbumin		Demonstrates germline transmission of ES cells that have undergone double-replacement targeting	297	
λ5		Defective B cell development	298	223
Laminin(s-laminin/ laminin β2)		Neonatal lethal; proteinuria; neuromuscular junction defects	299	
Lck ($p56^{Lck}$)		Thymic atrophy; reduced $CD4^+ 8^+$ T cells; very few mature T cells; immunocompromised	300	301, 302
Leukaemia inhibitory factor (LIF)	×2	Decreased haematopoietic stem cells; deficient neurotransmitter switch *in vitro* but normal sympathetic neurons *in vivo*; blastocysts do not implant in homozygous mother	303, 304	305
Lipoxygenase (5-lipoxygenase)	×2	Resistance to certain inflammatory agents	306, 307	
LMP-7		Defects in MHC class I expression and antigen presentation	308	
Low density lipoprotein receptor (LDLR)	D	Hypercholesterolaemia; increased apoB-100	309	310, 311
Low-density lipoprotein receptor-related protein (LRP)		Embryonic lethal; failed implantation of embryos	312	313
Lymphoid enhancer factor 1 (LEF-1)		Neonatal lethal; defects in development of multiple organs	314	
Major histocompatibility complex class II Aα (MHC II Aα)		Decreased $CD4^+ 8^-$ T cells; immune defects	315	
Major histocompatibility complex class II Aβ (MHC II Aβ)	×2; D	Decreased $CD4^+ 8^-$ T cells; deficient cell-mediated immunity; some B-cell dysfunctions; inflammatory bowel disease	316, 317	318–328

Continued.

Table VIII.1 *Continued.*

Protein/locus	×; D (see text)	Phenotype	Initial report(s)	Follow-ups
Mammalian achaete scute homologue 1 (Mash-1)		Neonatal lethal; olfactory and autonomic neuron deficiency	329	
Mammalian achaete scute homologue 2 (Mash-2)		e10 lethal; placental defects; (heterozygotes with paternally inherited null allele similar to homozygotes; gene imprinting implicated)	330	331
Mannose 6-phosphate receptor 46 (MPR46) (cation-dependent mannose 6-phosphate receptor; CD-MPR)	×2	Defects in targeting/retention of lysosomal enzymes	332, 333	
Metallothionein I (MT-I) and metallothionein II (MT-II)	×2	Sensitive to heavy metal	334, 335	336
Microglobulin (β_2-microglobulin)	×2; D	Decreased $CD4^-8^+$ T cells	337–338	318, 323–328, 340–368
Mos	×2	Reduced female fertility; parthenogenesis	369, 370	
Mp1		Thrombocytopenia; increased serum thrombopoietin	371	
Msx-1		Cleft palate; tooth and craniofacial abnormalities	372	
Mullerian-inhibiting substance (MIS)		Males infertile due to development of female reproductive organs; Leydig cell hyperplasia	373	
Multiple drug resistance protein 1a (mdr1a)		Blood–brain barrier defect; drug sensitivity	374	
Multiple drug resistance protein 2 (mdr2)		Liver disease; lack of phospholipid secretion into bile	375	376
Myb		e15 lethal; defect in haematopoiesis	377	
Myc (c-myc)		e10 lethal; heart and neural tube abnormal	378	
Myc (N-myc)	×3	e10–12 lethal; development of several organs affected	379–381	382, 383
Myc (N-myc) *partial*		Perinatal lethal; lung defect	384	383
Myelin-associated glycoprotein (MAG)	×2	Oligodendrocyte abnormalities; subtle tremor; increased NCAM expression;	385, 386	
Myf-5 D		Neonatal lethal; incomplete rib development	387	388
Myf-5β-galactosidase *insertion*		Used to study Myf-5 expression during early development	389	390, 391
MyoD D		No obvious defects but reduced survival; increased Myf-5	392	
Myogenin	×2	Perinatal lethal; decreased skeletal muscle	393, 394	
Myristoylated alanine-rich C-kinase substrate (MARCKS)		Perinatal lethal; defects in brain development	395	
N-acetylglucosaminyl-transferase I (GlcNAc-TI)	×2	e10 lethal; defects in neural tube formation, vascularization, and determination of left–right asymmetry	396, 397	

Continued on p. 952.

Table VIII.1 *Continued.*

Protein/locus	×; D (see text)	Phenotype	Initial report(s)	Follow-ups
NAD+:protein (ADP-ribosyl) transferase (ADPRT)		Susceptible to skin disease	398	
Nerve growth factor (NGF)		Neonatal lethal; sensory and sympathetic neurons drastically reduced; forebrain ACh neurons still present	399	
Nerve growth factor receptor (NGFR) (low affinity, p75)		Decreased sensory innervation; secondary infections and ulceration	400	401, 402
Neural cell adhesion molecule (NCAM)		Small olfactory bulb; deficient spatial learning	403	
Neural cell adhesion molecule-180 (NCAM-180)		Defective migration of olfactory neurons; small olfactory bulb	404	
Neurofibromatosis type 1 gene (NF1)		e11–13 lethal; malformation of heart; hyperplasia of sympathetic ganglia; heterozygotes predisposed to tumours	405	406
Neurotrophin-3 (NT-3)	×3	Neonatal lethal; peripheral sensory and sympathetic neurons reduced; limb proprioceptive afferents absent	407–409	
NF-IL6		Defects in macrophage bactericidal and tumoricidal activities	410	
NF-κB p50 subunit		Multifocal defects in immune responses	411	
Nitric oxide synthetase, neuronal (nNOS)		Stomach hypertrophy; normal CA1 long-term potentiation	412	413, 414
Notch 1		e10–11 lethal; widespread cell death	415	
Oct-2		Perinatal lethal; decreased IgM+ B cells	416	
Oestrogen receptor		Females infertile; males have reduced fertility	417	
P_0		Hypomyelination; myelin degeneration; tremors fertility	418	
p53	×3; 2D	Spontaneous tumours; thymocytes resistant to apoptosis by radiation or etoposide	419–421	422–438
Parathyroid hormone-related peptide (PTHrP)		Perinatal lethal; abnormal chondrocyte and bone development	439	
Perforin	×3	Impaired CTL and NK cell function; unable to clear LCMV infection	440–442	443
Pim-1		Impaired response of early B cells to interleukin-7 and steel factor; impaired response of bone marrow-derived mast cells to interleukin-3	444	445–447
Plasminogen activator inhibitor-1		Mildly hyperfibrinolytic; more resistant to thrombosis	448, 449	
Platelet-derived growth factor B (PDGF B)		Perinatal lethal; kidney defect; haemorrhagic; erythroblastosis; macrocytic anaemia; thrombocytopenia	450	
Platelet-derived growth factor receptor β (PDGFbR)		Perinatal lethal; kidney defect; haemorrhagic; anaemic; thrombocytopenia	451	

Continued.

Table VIII.1 *Continued.*

Protein/locus	×; D (see text)	Phenotype	Initial report(s)	Follow-ups
Prion protein (PrP)		Resistant to scrapie; weakened $GABA_A$R-mediated fast inhibition; impaired long-term potentiation	452	453–459
Protein kinase Cγ (PKCγ)		Deficient long-term potentiation; impaired context-dependent fear conditioning	460–461	
Proteolipid protein (PRP/DM20)		Disrupted myelination; decreased axonal conduction velocity; behavioural changes	462	
PU.1		e16–18 lethal; defect in development of lymphoid and myeloid cells	463	
Rab3A		Induced synaptic depression increased	464	
Ras (N-Ras)		No defects observed	465	
Rbtn2		e10 lethal; absence of erythrocytes	466	
Recombination activation gene 1 (RAG-1)	×2	Absence of mature B and T lymphocytes	467, 468	322, 469–473
Recombination activation gene 2 (RAG-2)		Absence of mature B and T lymphocytes	474	475–479
RelB	×2	Multiorgan inflammation; haematopoietic defects	480, 481	
Ret		Neonatal lethal: kidney and enteric nervous system defects	482	
Retinoblastoma-1 (Rb-1)	×2; D	e12–15 lethal; neural and tumours in heterozygotes haematopoietic defects; pituitary	483, 484	436
Retinoblastoma-1 (Rb-1) *truncation*		e12–15 lethal; neural and haematopoietic defects	485	486, 487
Retinoic acid receptor α (RARα)	3D	Neonatal lethal; testicular degeneration	488	
Retinoic acid receptor α1 (RARα1)	×2; 2D	No defects observed	488, 489	
Retinoic acid receptor β2 (RARβ2)	3D	No defects observed	490	
Retinoic acid receptor γ (RARγ)	4D	Neonatal lethality; growth deficiency; glandular, skeletal, and cartilage defects	491	
Retinoic acid receptor γ2 (RARγ2)		No defects observed	492	
RXRα	×2; 2D	e13–16 lethal; eye, heart, and liver defects	492, 493	
Ryanodine receptor		Perinatal lethal; skeletal muscle defects; excitation–contraction uncoupled	494	
Selectin (L-selectin)		Defects in lymphocyte homing and leucocyte rolling and migration	495	
Selectin (P-selectin)		Defects in leucocyte behaviour; increased neutrophils	496	
Serotonin receptor 1B (5-HT 1B receptor)		Aggressive behaviour	497	

Continued on p. 954.

Table VIII.1 *Continued.*

Protein/locus	×; D (see text)	Phenotype	Initial report(s)	Follow-ups
SF-1 (*Ftz-F1*)		Neonatal lethal; lack adrenals; both genders have female internal genitalia only	498	499
sγ1 class switch region		Shutdown IgM–IgG class switch at that allele	500	
Src	2D	Osteopetrosis	501	17, 165, 166, 502–507
Srm		No defects observed	508	
Synapsin I		Increased paired pulse facilitation	509	
Synaptotagmin I		Perinatal lethal; synaptic transmission severely impaired in cultured neurons	510	
T-cell factor-1 (TCF-1)		Defect in thymocyte development	511	
T-cell receptor α (TCRα)	×2; D	Loss of thymic medullae; devoid of single positive thymocytes; no ab T cells; inflammatory bowel disease	471, 512	323, 473, 513–516
T-cell receptor β (TCRβ)	2D	Reduced % CD4+8+, and total number of thymocytes; inflammatory bowel disease	471	322, 469, 513, 514
T-cell receptor δ (TCRδ)	D	Absence of γδ T cells	517	513, 514
T-cell receptor η (TCRη)		Neonatal lethal; (partial knockout of Oct-1 on opposite strand)	518	
T-cell receptor ηϕ (TCRηϕ)		Lower birth rate; T cells develop normally; (partial knockout of Oct-1 on opposite strand)	519	
T-cell receptor ζ (TCRζ)	×3	Decreased CD4+8+ thymocytes and single positive T cells; low TCR expression	520–522	523
T-cell receptor ζη (TCRζη)		Decreased CD4+8+ thymocytes and single positive T cells; low TCR expression	524	
Tal-1 (SCL)		e9–10 lethal; haematopoietic defect	525	
Tau		Altered microtubules in small calibre axons	526	
Tek receptor tyrosine kinase (Tek RTK)		e8–9 lethal; decreased endothelial cells; cardiac defects	527	
Tenascin-C		No defects observed	528	529
Terminal deoxynucleotidyltransferase		(TdT)Decreased TCR diversity	530	
Thrombomodulin		e8–9 lethal; growth retardation	531	
Tissue plasminogen activator (tPA)	D	Impaired clot lysis	532	
Transforming growth factor α (TGFα)	×2	Hair follicle and eye defects; allelic with waved-1 (*wa-1*)	8, 9	
Transforming growth factor β1 (TGFβ1)	×2	Neonatal lethal; multifocal inflammatory disease	533, 534	535–540
Transporter associated with antigen processing 1 (TAP1)		MHC class I transport and function defective; lack CD4-8+)	541	542–546
Transthyretin and thyroid hormone		Decreased serum retinol, retinol-binding protein	547	548, 549
TrkA		Neonatal lethal; severe sensory and sympathetic neuropathies; decreased forebrain ACh neurons	550	

Continued.

Table VIII.1 *Continued.*

Protein/locus	×; D (see text)	Phenotype	Initial report(s)	Follow-ups
TrkB		Neonatal lethal; deficiencies in central and peripheral nervous system	551	
TrkC		No Ia muscle afferents; proprioception disrupted	552	
Tumour necrosis factor receptor 1 (TNF-R-1) (p55)	×2	Resistant to endotoxic shock; susceptible to *Listeria* infection	553, 554	555
Tumour necrosis factor receptor 2 (TNF-R-2) (p75)		Resistance to TNF-induced necrosis and death	556	
Tumour necrosis factor-β (TNF-β) (lymphotoxin)		No Peyer's patches or lymph nodes; increased IgM^+ B cells	557	
Urate oxidase		Neonatal lethality; hyperuricaemia; urate nephropathy	558	
Urokinase plasminogen activator (uPA)	D	Occasional fibrin deposition	532	
Vascular cell adhesion molecule-1 (VCAM-1)		e8–10 lethality; chorioallantoic fusion disrupted; surviving adults have elevated mononuclear leucocytes	559	
Vav		e4–7 lethal; possible trophoblast defect	560	
Vimentin		No defects observed	561	
Wilms' tumour protein 1 (WT-1)		e11 lethal; kidney apoptosis; gonadal, lung, and heart defects	562	
Wnt-1 (int-1)		Neonatal lethality; cerebellum and midbrain absent; severe ataxia; allelic with swaying (*sw*)	563, 564	565, 566
Wnt-3a		e10 lethal; no hind portion	567	
Wnt-4		Perinatal lethal; kidney defects	568	
Wnt-7a		Limb abnormalities; sterile	569	
Yes	2D	No defects observed	570	17, 165, 166, 496

Table VIII.2 Double knockouts.

Mutants crossed	Phenotype	Initial report(s)	Follow-up(s)
Activin/inhibin βA & Activin/inhibin βB	Perinatal lethal; whiskers and incisors absent; cleft palate; eyelid defects	20	
Apolipoprotein E (apoE) and low-density lipoprotein receptor (LDLR)	ApoB-48 and apoB-100 both elevated	311	
CD4 & CD8	Some cytotoxic T cells are still present	571	572
Fgr & Hck	Increased susceptibility to *Listeria* infection	145	
Fyn & Src	Neonatal lethal; reduced size; osteopetrosis	567	
Fyn & Yes	Premature lethality; degenerative renal changes	567	
Hox-A3 (Hox 1.5) and Hox-D3 (Hox 4.1)	Synergistic defects; atlas deleted	572	

Continued on p. 956.

Table VIII.2 *Continued.*

Mutants crossed	Phenotype	Initial report(s)	Follow-up(s)
Inhibin (a) and p53	Gonadal tumours	574	
Insulin-like growth factor I (IGF-I) and insulin-like growth factor II (IGF-II)	30% normal birthweight	231, 232	
Insulin-like growth factor receptor 1 (IGF1R) and insulin-like growth factor I (IGF-I)	Appear identical to IGF1R knockout	231, 232	
Insulin-like growth factor receptor 1 (IGF1R) and insulin-like growth factor II (IGF-II)	Perinatal lethal; 30% normal birthweight	231, 232	
Interleukin-2 (IL-2) and interleukin-4 (IL-4)	Paradoxical increase in T-cell proliferation	575	
Microglobulin (β_2-microglobulin) and major histocompatibility complex II (MHC II)	Depleted of $CD4^{+}8^{-}$ and $CD4^{-}8^{+}$ T cells	328, 576	318, 577
MyoD1 & Myf-5	Neonatal lethal; no skeletal muscle	578	
p53 & retinoblastoma-1 (Rb-1)	Decrease in the ectopic apoptosis in lens fibre cells that is observed in Rb-1 single mutants	436	
RARα & RARβ2	Skeletal malformations; homeotic transformations; middle ear ossicle fusions; eye, oesophago-tracheal, thymus, thyroid, parathyroid, heart, and urogenital defects	579, 580	
RARα & RARγ	e13–16 lethality; reduced size; exencephaly; middle ear ossicle, fusions; eye, thymus, thyroid, parathyroid, heart, umbilical, gland, and urogenital defects; craniofacial and skeletal malformations; homeotic transformations	579, 580	
RARγ & RXRα	e13–16 lethal; eye and heart defects	493	
RARα1 & RARβ2	Middle ear ossicle fusions; eye, oesophagotracheal, thymus, thyroid, parathyroid, heart, and urogenital defects	579, 580	
RARα1 & RARγ	Skeletal malformations; homeotic transformation; gland defects	579, 580	
RARβ2 & RARγ	Skeletal and cartilage malformations; homeotic transformations; eye, thyroid, gland, and urogenital defects	579, 580	
RARγ & RXRα	e13–16 lethal; eye and heart defects	493	
Src & Yes	Mostly neonatal lethal; reduced size; osteopetrosis	570	
T-cell receptor α (TCRα) and T-cell receptor β (TCRβ)	Reduced % $CD4^{+}8^{+}$ cells, and total number of thymocytes	471	
T-cell receptor β (TCRβ) and T-cell receptor δ (TCRδ)	Devoid of $CD4^{+}8^{+}$ cells and single positive thymocytes; inflammatory bowel disease	471	322, 573
Tissue plasminogen activator (tPA) and urokinase plasminogen activator (uPA)	Extensive fibrin deposition	532	

References

1 Capecchi, M.R. (1989) Altering the genome by homologous recombination. *Science* **244**, 1288–1292.

2 Ramirez-Solis, R., Davis, A.C. & Bradley, A. (1993) Gene targeting in embryonic stem cells. *Methods Enzymol.* **225**, 855–878.

3 Bronson, S.K. & Smithies, O. (1994) Altering mice by homologous recombination using embryonic stem cells. *J. Biol. Chem.* **269**, 2715–2718.

4 Joyner, A.L. & Guillemot, F. (1994) Gene targeting and development of the nervous system. *Curr. Opin. Neurobiol.* **4**, 37–42.

5 Yeung, R.S., Penninger, J. & Mak, T.W. (1994) T-cell development and function in gene-knockout mice. *Curr. Opin. Immunol.* **6**, 298–307.

6 Soriano, P. (1995) Gene targeting in ES cells. *Annu. Rev. Neurosci.* **18**, 1–18.

7 Thomas, K.R., Musci, T.S., Neumann, P.E. & Capecchi, M.R. (1991) Swaying is a mutant allele of the proto-oncogene Wnt-1. *Cell* **67**, 969–976.

8 Mann, G.B., Fowler, K.J., Gabriel, A., Nice, E.C., Williams, R.L. & Dunn, A.R. (1993) Mice with a null mutation of the TGF alpha gene have abnormal skin architecture, wavy hair, and curly whiskers and often develop corneal inflammation. *Cell* **73**, 249–261.

9 Luetteke, N.C., Qiu, T.H., Peiffer, R.L., Oliver, P., Smithies, O. & Lee, D.C. (1993) TGF alpha deficiency results in hair follicle and eye abnormalities in targeted and waved-1 mice. *Cell* **73**, 263–278.

10 Grompe, M., al Dhalimy, M., Finegold, M. *et al.* (1993) Loss of fumarylacetoacetate hydrolase is responsible for the neonatal hepatic dysfunction phenotype of lethal albino mice. *Genes Dev.* **7**, 2298–2307.

11 Hébert, J.M., Rosenquist, T., Götz, J. & Martin, G.R. (1994) FGF5 as a regulator of the hair growth cycle: evidence from targeted and spontaneous mutations. *Cell* **78**, 1017–1025.

12 Greenstein Baynash, A., Hosoda, K., Giaid, A. *et al.* (1994) Interaction of endothelin-3 with endothelin-B receptor is essential for development of epidermal melanocytes and enteric neurons. *Cell* **79**, 1277–1285.

13 Hosoda, K., Hammer, R.E., Richardson, J.A. *et al.* (1994) Targeted and natural (piebald-lethal) mutations of endothelin-B receptor gene produce megacolon associated with spotted coat color in mice. *Cell* **79**, 1267–1276.

14 Schwartzberg, P.L, Goff, S.P. & Robertson, E.J. (1989) Germ-line transmission of a c-abl mutation produced by targeted gene disruption in ES cells. *Science* **246**, 799–803.

15 Schwartzberg, P.L., Stall, A.M., Hardin, J.D. *et al.* (1991) Mice homozygous for the ablm1 mutation show poor viability and depletion of selected B and T cell populations. *Cell* **65**, 1165–1175.

16 Tybulewicz, V.L., Crawford, C.E., Jackson, P.K., Bronson, R.T. & Mulligan, R.C. (1991) Neonatal lethality and lymphopenia in mice with a homozygous disruption of the c-abl proto-oncogene. *Cell* **65**, 1153–1163.

17 Grant, S.G., O'Dell, T.J., Karl, K.A., Stein, P.L., Soriano, P. & Kandel, E.R. (1992) Impaired long-term potentiation, spatial learning, and hippocampal development in fyn mutant mice. *Science* **258**, 1903–1910.

18 Picciotto, M.R., Zoli, M., Lena C. *et al.* (1995) Abnormal avoidance learning in mice lacking functional high-affinity nicotine receptor in the brain. *Nature* **374**, 65–67.

19 Baba, T., Azuma, S., Kashiwabara, S. & Toyoda, Y. (1994) Sperm from mice carrying a targeted mutation of the acrosin gene can penetrate the oocyte zona pellucida and effect fertilization. *J. Biol. Chem.* **269**, 31845–31849.

20 Matzuk, M.M., Kumar, T.R., Vassalli, A. *et al.* (1995) Functional analysis of activins during mammalian development. *Nature* **374**, 354–356.

21 Vassalli, A., Matzuk, M.M., Gardner, H.A., Lee, K.F. & Jaenisch, R. (1994) Activin/inhibin beta B subunit gene disruption leads to defects in eyelid development and female reproduction. *Genes Dev.* **8**, 414–427.

22 Matzuk, M.M., Kumar, T.R. & Bradley, A. (1995) Different phenotypes for mice deficient in either activins or activin receptor type II. *Nature* **374**, 356–359.

23 Fodde, R., Edelmann, W., Yang, K. *et al.* (1994) A targeted chain-termination mutation in the mouse Apc gene results in multiple intestinal tumors. *Proc. Natl. Acad. Sci. USA* **91**, 8969–8973.

24 Wu, Z.-L., Thomas, S.A., Villacres, E.C. *et al.* (1995) Altered behavior and long-term potentiation in type I adenylyl cyclase mutant mice. *Proc. Natl. Acad. Sci. USA* **92**, 220–224.

25 Magyar, J.P., Bartsch, U., Wang, Z.Q. *et al.* (1994) Degeneration of neural cells in the central nervous system of mice deficient in the gene for the adhesion molecule on Glia, the beta 2 subunit of murine Na, K-ATPase. *J. Cell Biol.* **127**, 835–845.

26 Müller, U., Cristina, N., Li, Z.W. *et al.* (1994) Behavioral and anatomical deficits in mice homozygous for a modified beta-amyloid precursor protein gene. *Cell* **79**, 755–765.

27 Tanimoto, K., Sugiyama, F., Goto, Y. *et al.* (1994) Angiotensinogen-deficient mice with hypotension. *J. Biol. Chem.* **269**, 31334–31337.

28 Williamson, R., Lee, D., Hagaman, J. & Maeda, N. (1992) Marked reduction of high density lipoprotein cholesterol in mice genetically modified to lack apolipoprotein A-I. *Proc. Natl. Acad. Sci. USA* **89**, 7134–7138.

29 Li, H., Reddick, R.L. & Maeda, N. (1993) Lack of apoA-I is not associated with increased susceptibility to atherosclerosis in mice. *Arterioscler. Thromb.* **13**, 1814–1821.

30 Homanics, G.E., Smith, T.J., Zhang, S.H., Lee, D., Young, S.G. & Maeda, N. (1993) Targeted modification of the apolipoprotein B gene results in hypobetalipoproteinemia and developmental abnormalities in mice. *Proc. Natl. Acad. Sci. USA* **90**, 2389–2393.

31 Farese, R.J.J., Ruland, S.L., Flynn, L.M., Stokowski, R.P. & Young, S.G. (1995) Knockout of the mouse apolipoprotein B gene results in embryonic lethality in homozygotes and protection against diet-induced hypercholesterolemia in heterozygotes. *Proc. Natl. Acad. Sci. USA* **92**, 1774–1778.

32 Maeda, N., Li, H., Lee, D., Oliver, P., Quarfordt, S.H. &

Osada, J. (1994) Targeted disruption of the apolipoprotein C-III gene in mice results in hypotriglyceridemia and protection from postprandial hypertriglyceridemia. *J. Biol. Chem.* **269**, 23610–23616.
33 Plump, A.S., Smith, J.D., Hayek, T. *et al.* (1992) Severe hypercholesterolemia and atherosclerosis in apolipoprotein E-deficient mice created by homologous recombination in ES cells. *Cell* **71**, 343–353.
34 Zhang, S.H., Reddick, R.L., Piedrahita, J.A. & Maeda, N. (1992) Spontaneous hypercholesterolemia and arterial lesions in mice lacking apolipoprotein E. *Science* **258**, 468–471.
35 Piedrahita, J.A., Zhang, S.H., Hagaman, J.R., Oliver, P.M. & Maeda, N. (1992) Generation of mice carrying a mutant apolipoprotein E gene inactivated by gene targeting in embryonic stem cells. *Proc. Natl. Acad. Sci. USA* **89**, 4471–4475.
36 Plump, A.S., Scott, C.J. & Breslow, J.L. (1994) Human apolipoprotein A-I gene expression increases high density lipoprotein and suppresses atherosclerosis in the apolipoprotein E-deficient mouse. *Proc. Natl. Acad. Sci. USA* **91**, 9607–9611.
37 Palinski, W., Ord., V.A., Plump, A.S., Breslow, J.L., Steinberg, D. & Witztum, J.L. (1994) ApoE-deficient mice are a model of lipoprotein oxidation in athero genesis: demonstration of oxidation-specific epitopes in lesions and high titers of autoantibodies to malondialdehyde-lysine in serum. *Arterioscler. Thromb.* **14**, 605–616.
38 Nakashima, Y., Plump, A.S., Raines, E.W., Breslow, J.L. & Ross, R. (1994) ApoE-deficient mice develop lesions of all phases of atherosclerosis throughout the arterial tree. *Arterioscler. Thromb.* **14**, 133–140.
39 Popko, B., Goodrum, J.F., Bouldin, T.W., Zhang, S.H. & Maeda, N. (1993) Nerve regeneration occurs in the absence of apolipoprotein E in mice. *Neurochem.* **60**, 1155–1158.
40 Pászty, C., Maeda, N., Verstuyft, J. & Rubin, E.M. (1994) Apolipoprotein AI transgene corrects apolipoprotein E deficiency-induced atherosclerosis in mice. *J. Clin. Invest.* **94**, 899–903.
41 Reddick, R.L., Zhang, S.H. & Maeda, N. (1994) Atherosclerosis in mice lacking apo E: evaluation of lesional development and progression. *Arterioscler. Thromb.* **14**, 141–147 (published erratum appears in *Arterioscler. Thromb.* **14**, 389).
42 Zhang, S.H., Reddick, R.L., Burkey, B. & Maeda, N. (1994) Diet-induced atherosclerosis in mice heterozygous and homozygous for apolipoprotein E gene disruption. *J. Clin. Invest.* **94**, 937–945.
43 Linton, M.F., Atkinson, J.B. & Fazio, S. (1995) Prevention of atherosclerosis in apolipoprotein E-deficient mice by bone marrow transplantation. *Science* **267**, 1034–1037.
44 Patejunas, G., Bradley, A., Beaudet, A.L. & O'Brien, W.E. (1994) Generation of a mouse model for citrullinemia by targeted disruption of the argininosuccinate synthetase gene. *Somat. Cell Mol. Genet.* **20**, 55–60.
45 Ishibashi, S., Hammer, R.E. & Herz, J. (1994) Asialoglycoprotein receptor deficiency in mice lacking the minor receptor subunit. *J. Biol. Chem.* **269**, 27803–27806.
46 John, S.W.M., Krege, J.H., Oliver, P.M. *et al.* (1995) Genetic decreases in atrial natriuretic peptide and salt-sensitive hypertension. *Science* **267**, 679–681.
47 Urbánek, P., Wang, Z.Q., Fetka, I., Wagner, E.F. & Busslinger, M. (1994) Complete block of early B cell differentiation and altered patterning of the posterior midbrain in mice lacking Pax5/BSAP. *Cell* **79**, 901–912.
48 Freeman, G.J., Borriello, F., Hodes, R.J. *et al.* (1993) Uncovering of functional alternative CTLA-4 counter-receptor in B7-deficient mice. *Science* **262**, 907–909.
49 Veis, D.J., Sorenson, C.M., Shutter, J.R. & Korsmeyer, S.J. (1993) Bcl-2-deficient mice demonstrate fulminant lymphoid apoptosis, polycystic kidneys, and hypopigmented hair. *Cell* **75**, 229–240.
50 Nakayama, K., Nakayama, K., Negishi, I., Kuida, K., Sawa, H. & Loh, D.Y. (1994) Targeted disruption of Bcl-2 alpha beta in mice: occurrence of gray hair, polycystic kidney disease, and lymphocytopenia. *Proc. Natl. Acad. Sci. USA* **91**, 3700–3704.
51 Motoyama, N., Wang, F., Roth, K.A. *et al.* (1995) Massive cell death of immature haematopoietic cells and neurons in Bcl-x-deficient mice. *Science* **267**, 1506–1510.
52 van der Lugt, N.M.T., Domen, J., Linders, K. *et al.* (1994) Posterior transformation, neurological abnormalities, and severe haematopoietic defects in mice with a targeted deletion of the bmi-1 proto-oncogene. *Genes Dev.* **8**, 757–769.
53 Ernfors, P., Lee, K.F. & Jaenisch, R. (1994) Mice lacking brain-derived neurotrophic factor develop with sensory deficits. *Nature* **368**, 147–150.
54 Jones, K.R., Farinas, I., Backus, C. & Reichardt, L.F. (1994) Targeted disruption of the BDNF gene perturbs brain and sensory neuron development but not motor neuron development. *Cell* **76**, 989–999.
55 Larue, L., Ohsugi, M., Hirchenhain, J. & Kemler, R. (1994) E-cadherin null mutant embryos fail to form a trophectoderm epithelium. *Proc. Natl. Acad. Sci. USA* **91**, 8263–8267.
56 Riethmacher, D., Brinkman, V. & Birchmeier, C. (1995) A targeted mutation in the mouse E-cadherin gene results in defective preimplantation development. *Proc. Natl Acad. Sci. USA* **92**, 855–859.
57 Silva, A.J., Paylor, R., Wehner, J.M. & Tonegawa, S. (1992) Impaired spatial learning in alpha-calcium-calmodulin kinase II mutant mice. *Science* **257**, 206–211.
58 Silva, A.J., Stevens, C.F., Tonegawa, S. & Wang, Y. (1992) Deficient hippocampal long-term potentiation in alpha-calcium-calmodulin kinase II mutant mice. *Science* **257**, 201–206.
59 Chen, C., Rainnie, D.G., Greene, R.W. & Tonegawa, S. (1994) Abnormal fear response and aggressive behavior in mutant mice deficient for alpha-calcium-calmodulin kinase II. *Science* **266**, 291–294.
60 Stevens, C.F., Tonegawa, S. & Wang, Y. (1994) The role of calcium-calmodulin kinase II in three forms of synaptic plasticity. *Curr. Biol.* **4**, 687–693.
61 Kumar, S., Clarke, A.R., Hooper, M.L. *et al.* (1994) Milk composition and lactation of beta-casein-deficient mice. *Proc. Natl Acad. Sci. USA* **91**, 6138–6142.

62 Killeen, N., Stuart, S.G. & Littman, D.R. (1992) Development and function of T cells in mice with a disrupted CD2 gene. *EMBO J.* **11**, 4329–4336.
63 Evans, C.F., Rall, G.F., Killeen, N., Littman, D. & Oldstone, M.B. (1993) CD2-deficient mice generate virus-specific cytotoxic T lymphocytes upon infection with lymphocytic choriomeningitis virus. *J. Immunol.* **151**, 6259–6264.
64 Killeen, N., Sawada, S. & Littman, D.R. (1993) Regulated expression of human CD4 rescues helper T cell development in mice lacking expression of endogenous CD4. *EMBO J.* **12**, 1547–1553.
65 Rahemtulla, A., Fung-Leung, W.-P., Schilham, M.W. *et al.* (1991) Normal development and function of CD8+ cells but markedly decreased helper cell activity in mice lacking CD4. *Nature* **353**, 180–184.
66 Locksley, R.M., Reiner, S.L., Hatam, F., Littman, D.R. & Killeen, N. (1993) Helper T cells without CD4: control of leishmaniasis in CD4-deficient mice. *Science* **261**, 1448–1451.
67 Killeen, N. & Littman, D.R. (1993) Helper T-cell development in the absence of CD4-p56lck association. *Nature* **364**, 729–732.
68 Rajan, T.V., Nelson, F.K., Killeen, N. *et al.* (1994) CD4+ T-lymphocytes are not required for murine resistance to the human filarial parasite, *Brugia malayi*. *Exp. Parasitol.* **78**, 352–360.
69 Wallace, V.A., Rahemtulla, A., Timms, E., Penninger, J. & Mak, T.W. (1992) CD4 expression is differentially required for deletion of MLS-1a-reactive T cells. *J. Exp. Med.* **176**, 1459–1463.
70 Koh, D.R., Ho, A., Rahemtulla, A., Penninger, J. & Mak, T.W. (1994) Experimental allergic encephalomyelitis (EAE) in mice lacking CD4+ T cells. *Eur. J. Immunol.* **24**, 2250–2253.
71 Rahemtulla, A., Kundig, T.M., Narendran, A. *et al.* (1994) Class II major histocompatibility complex-restricted T cell function in CD4-deficient mice. *Eur. J. Immunol.* **24**, 2213–2218.
72 Battegay, M., Moskophidis, D., Rahemtulla, A., Hengartner, H., Mak, T.W. & Zinkernagel, R.M. (1994) Enhanced establishment of a virus carrier state in adult CD4+ T-cell-deficient mice. *J. Virol.* **68**, 4700–4704.
73 Law, Y.M., Yeung, R.S., Mamalaki, C., Kioussis, D., Mak, T.W. & Flavell, R.A. (1994) Human CD4 restores normal T cell development and function in mice deficient in murine CD4. *J. Exp. Med.* **179**, 1233–1242.
74 Rottenberg, M.E., Bakhiet, M., Olsson, T. *et al.* (1993) Differential susceptibilities of mice genomically deleted of CD4 and CD8 to infections with *Trypanosoma cruzi* or *Trypanosoma brucei*. *Infect. Immun.* **61**, 5129–5133.
75 Penninger, J.M., Wallace, V.A., Timms, E. & Mak, T.W. (1994) Maternal transfer of infectious mouse mammary tumor retroviruses does not depend on clonal deletion of superantigen-reactive V beta 14+ T cells. *Eur. J. Immunol.* **24**, 1102–1108.
76 Zhang, L., Shannon, J., Sheldon, J., Teh, H.S., Mak, T.W. & Miller, R.G. (1994) Role of infused CD8+ cells in the induction of peripheral tolerance. *J. Immunol.* **152**, 2222–2228.
77 Penninger, J.M., Neu, N., Timms, E. *et al.* (1993) The induction of experimental autoimmune myocarditis in mice lacking CD4 or CD8 molecules *J. Exp. Med.* **178**, 1837–1842 (published erratum appears in *J. Exp. Med.* **179**, 371).
78 Fung-Leung, W.-P., Schilham, M.W., Rahemtulla, A. *et al.* (1991) CD8 is needed for development of cytotoxic T cells but not helper T cells. *Cell* **65**, 443–449.
79 Wallace, V.A., Fung, L.W.P., Timms, E. *et al.* (1992) CD45RA and CD45RB high expression induced by thymic selection events. *J. Exp. Med.* **176**, 1657–1663.
80 Koh, D.R., Fung, L.W.P., Ho, A., Gray, D., Acha, O.H. & Mak, T.W. (1992) Less mortality but more relapses in experimental allergic encephalomyelitis in $CD8^{-/-}$ mice. *Science* **256**, 1210–1213.
81 Battegay, M., Moskophidis, D., Waldner, H. *et al.* (1993) Impairment and delay of neutralizing antiviral antibody responses by virus-specific cytotoxic T cells. *J. Immunol.* **151**, 5408–5414 (published erratum appears in *J. Immunol.* **152**, 1635).
82 Fung-Leung, W.-P., Louie, M.C., Limmer, A. *et al.* (1993) The lack of CD8 alpha cytoplasmic domain resulted in a dramatic decrease in efficiency in thymic maturation but only a moderate reduction in cytotoxic function of CD8+ T lymphocytes. *Eur. J. Immunol.* **23**, 2834–2840.
83 Chan, I.T., Limmer, A., Louie, M.C. *et al.* (1993) Thymic selection of cytotoxic T cells independent of CD8 alpha-Lck association. *Science* **261**, 1581–1584.
84 Fung-Leung, W.-P., Wallace, V.A., Gray, D. *et al.* (1993) CD8 is needed for positive selection but differentially required for negative selection of T cells during thymic ontogeny. *Eur. J. Immunol.* **23**, 212–216.
85 Olsson, T., Bakhiet, M., Hojeberg, B. *et al.* (1993) CD8 is critically involved in lymphocyte activation by a T. brucei brucei-released molecule. *Cell* **72**, 715–727.
86 Lin, T., Matsuzaki, G., Kenai, H. *et al.* (1994) Characteristics of fetal thymus-derived T cell receptor γδ intestinal intraepithelial lymphocytes. *Eur. J. Immunol.* **24**, 1792–1798.
87 Fung, L.W.P., Kundig, T.M., Ngo, K. *et al.* (1994) Reduced thymic maturation but normal effector function of $CD8^+$ T cells in CD8 beta gene-targeted mice. *J. Exp. Med.* **180**, 959–967.
88 Wilson, R.W., Ballantyne, C.M., Smith, C.W. *et al.* (1993) Gene targeting yields a CD18-mutant mouse for study of inflammation. *J. Immunol.* **151**, 1571–1578.
89 Fujiwara, H., Kikutani, H., Suematsu, S. *et al.* (1994) The absence of IgE antibody-mediated augmentation of immune responses in CD23-deficient mice. *Proc. Natl Acad. Sci. USA* **91**, 6835–6839.
90 Stief, A., Texido, G., Sansig, G., Eibel, H., Legros, G. & Venderputten, H. (1994) Mice deficient in CD23 reveal its modulatory role in IgE production but no role in T and B cell development. *J. Immunol.* **152**, 3378–3390.
91 Yu, P., Koscovilbois, M., Richards, M., Kohler, G. & Lamers, M.C. (1994) Negative feedback regulation of IgE synthesis by murine CD23. *Nature* **369**, 753–756.
92 Shahinian, A., Pfeffer, K., Lee, K.P. *et al.* (1993) Differential T cell costimulatory requirements in CD28-deficient mice. *Science* **261**, 609–612.
93 Green, J.M., Noel, P.J., Sperling, A.I. *et al.* (1994) Absence of B7-dependent responses in CD28-deficient mice. *Immunity* **1**, 501–508.

94 Kawabe, T., Naka, T., Yoshida, K. *et al.* (1994) The immune responses in CD40-deficient mice: Impaired immunoglobulin class switching and germinal center formation. *Immunity* **1**, 167–178.
95 Renshaw, B.R., Fanslow, W.I., Armitage, R.J. *et al.* (1994) Humoral immune responses in CD40 ligand-deficient mice. *J. Exp. Med.* **180**, 1889–1900.
96 Xu, J.C., Foy, T.M., Laman, J.D. *et al.* (1994) Mice deficient for the CD40 ligand. *Immunity* **1**, 423–431.
97 Kishihara, K., Penninger, J., Wallace, V.A. *et al.* (1993) Normal B lymphocyte development but impaired T cell maturation in CD45-exon6 protein tyrosine phosphatase-deficient mice. *Cell* **74**, 143–156.
98 Berger, S.A., Mak, T.W. & Paige, C.J. (1994) Leukocyte common antigen (CD45) is required for immunoglobulin E-mediated degranulation of mast cells. *J. Exp. Med.* **180**, 471–476.
99 Gorry, P., Lufkin, T., Dierich, A. *et al.* (1994) The cellular retinoic acid binding protein I is dispensable. *Proc. Natl Acad. Sci. USA* **91**, 9032–9036.
100 Masu, Y., Wolf, E., Holtmann, B., Sendtner, M., Brem, G. & Thoenen, H. (1993) Disruption of the CNTF gene results in motor neuron degeneration. *Nature* **365**, 27–32.
101 Fassler, R., Schnegelsberg, P.N., Dausman, J. *et al.* (1994) Mice lacking alpha 1 (IX) collagen develop noninflammatory degenerative joint disease. *Proc. Natl Acad. Sci. USA* **91**, 5070–5074.
102 Andrikopoulos, K., Liu, X., Keene, D.R., Jaenisch, R. & Ramirez, F. (1995) Targeted mutation in the col5a2 gene reveals a regulatory role for type V collagen during matrix assembly. *Nature Genet.* **9**, 31–36.
103 Rosati, R., Horan, G., Pinero, G.J. *et al.* (1994) Normal long bone growth and development in type X collagen-null mice. *Nature Genet.* **8**, 129–135.
104 Reaume, A.G., de Sousa, P.A., Kulkarni, S. *et al.* (1995) Cardiac malformation in neonatal mice lacking connexin43. *Science* **267**, 1831–1834.
105 Muglia, L., Jacobson, L., Dikkes, P. & Majzoub, J.A. (1995) Corticotropin-releasing hormone deficiency reveals major fetal but not adult glucocorticoid need. *Nature* **373**, 427–432.
106 van Deursen, J., Heerschap, A., Oerlemans, F. *et al.* (1993) Skeletal muscles of mice deficient in muscle creatine kinase lack burst activity. *Cell* **74**, 621–631.
107 van Deursen, J., Ruitenbeek, W., Heerschap, A., Jap, P., ter Laak, H. & Wieringa, B. (1994) Creatine kinase (CK) in skeletal muscle energy metabolism: a study of mouse mutants with graded reduction in muscle CK expression. *Proc. Natl Acad. Sci. USA* **91**, 9091–9095.
108 van Deursen, J., Jap, P., Heerschap, A., ter Laak, H., Ruitenbeek, W. & Wieringa, B. (1994) Effects of the creatine analogue beta-guanidinopropionic acid on skeletal muscles of mice deficient in muscle creatine kinase. *Biochim. Biophys. Acta* **1185**, 327–335.
109 Imamoto, A. & Soriano, P. (1993) Disruption of the csk gene, encoding a negative regulator of Src family tyrosine kinases, leads to neural tube defects and embryonic lethality in mice. *Cell* **73**, 1117–1124.
110 Nada, S., Yagi, T., Takeda, H. *et al.* (1993) Constitutive activation of Src family kinases in mouse embryos that lack Csk. *Cell* **73**, 1125–1135.
111 Hummler, E., Cole, T.J., Blendy, J.A. *et al.* (1994) Targeted mutation of the CREB gene: compensation within the CREB/ATF family of transcription factors. *Proc. Natl Acad. Sci. USA* **91**, 5647–5651.
112 Bourtchuladze, R., Frenguelli, B., Blendy, J., Cioffi, D., Schutz, G. & Silva, A.J. (1994) Deficient long-term memory in mice with a targeted mutation of the cAMP-responsive element-binding protein. *Cell* **79**, 59–68.
113 Watanabe, M., Osada, J., Aratani, Y. *et al.* (1995) Mice deficient in cystathionine beta-synthase: animal models for mild and severe homocyst(e)inemia. *Proc. Natl Acad. Sci. USA* **92**, 1585–1589.
114 Ratcliff, R., Evans, M.J., Cuthbert, A.W. *et al.* (1993) Production of a severe cystic fibrosis mutation in mice by gene targeting. *Nature Genet.* **4**, 35–41.
115 Snouwaert, J.N., Brigman, K.K., Latour, A.M. *et al.* (1992) An animal model for cystic fibrosis made by gene targeting. *Science* **257**, 1083–1088.
116 Clarke, L.L., Grubb, B.R., Gabriel, S.E., Smithies, O., Koller, B.H. & Boucher, R.C. (1992) Defective epithelial chloride transport in a gene-targeted mouse model of cystic fibrosis. *Science* **257**, 1125–1128.
117 O'Neal, W.K., Hasty, P., McCray, P.B.J. *et al.* (1993) A severe phenotype in mice with a duplication of exon 3 in the cystic fibrosis locus. *Hum. Mol. Genet.* **2**, 1561–1569.
118 Valverde, M.A., O'Brien, J.A., Sep'ulveda, F.V., Ratcliff, R., Evans, M.J. & Colledge, W.H. (1993) Inactivation of the murine cftr gene abolishes cAMP-mediated but not Ca(2+)-mediated secretagogue-induced volume decrease in small-intestinal crypts. *Pflugers Arch.* **425**, 434–438.
119 Cuthbert, A.W., Hickman, M.E., MacVinish, L.J. *et al.* (1994) Chloride secretion in response to guanylin in colonic epithelial from normal and transgenic cystic fibrosis mice. *Br. J. Pharmacol.* **112**, 31–36.
120 Hyde, S.C., Gill, D.R., Higgins, C.F. *et al.* (1993) Correction of the ion transport defect in cystic fibrosis transgenic mice by gene therapy. *Nature* **362**, 250–255.
121 Gabriel, S.E., Clarke, L.L., Boucher, R.C. & Stutts, M.J. (1993) CFTR and outward rectifying chloride channels are distinct proteins with a regulatory relationship. *Nature* **363**, 263–268.
122 Zhou, L., Dey, C.R., Wert, S.E., DuVall, M.D., Frizzell, R.A. & Whitsett, J.A. (1994) Correction of lethal intestinal defect in a mouse model of cystic fibrosis by human CFTR. *Science* **266**, 1705–1708.
123 Grubb, B.R., Pickles, R.J., Ye, H. *et al.* (1994) Inefficient gene transfer by adenovirus vector to cystic fibrosis airway epithelia of mice and humans. *Nature* **371**, 802–806.
124 Gabriel, S.E., Brigman, K.N., Koller, B.H., Boucher, R.C. & Stutts, M.J. (1994) Cystic fibrosis heterozygote resistance to cholera toxin in the cystic fibrosis mouse model. *Science* **266**, 107–109.
125 Dorin, J.R., Dickinson, P., Alton, E.W. *et al.* (1992) Cystic fibrosis in the mouse by targeted insertional mutagenesis. *Nature* **359**, 211–215.
126 Gray, M.A., Winpenny, J.P., Porteous, D.J., Dorin, J.R. & Argent, B.E. (1994) CFTR and calcium-activated chloride currents in pancreatic duct cells of a transgenic CF mouse. *Am. J. Physiol.* **266**, 6213–6221.
127 Dorin, J.R., Stevenson, B.J., Fleming, S., Alton, E.W.,

Dickinson, P. & Porteous, D.J. (1994) Long-term survival of the exon 10 insertional cystic fibrosis mutant mouse is a consequence of low level residual wild-type Cftr gene expression. *Mamm. Genome* **5**, 465–472.

128 Davidson, D.J., Dorin, J.R., McLachlan, G. *et al.* (1995) Lung disease in the cystic fibrosis mouse exposed to bacterial pathogens. *Nature Genet.* **9**, 351–357.

129 Pollock, J.D., Williams, D.A., Gifford, M.A.C. *et al.* (1995) Mouse model of X-linked chronic granulomatous disease, an inherited defect in phagocyte superoxide production. *Nature Genet.* **9**, 202–209.

130 Li, E., Bestor, T.H. & Jaenisch, R. (1992) Targeted mutation of the DNA methyltransferase gene results in embryonic lethality. *Cell* **69**, 915–926.

131 Li, E., Beard, C. & Jaenisch, R. (1993) Role for DNA methylation in genomic imprinting. *Nature* **366**, 362–365.

132 Gu, H., Marth, J.D., Orban, P.C., Mossmann, H. & Rajewsky, K. (1994) Deletion of a DNA polymerase beta gene segment in T cells using cell type-specific gene targeting. *Science* **265**, 103–106.

133 Xu, M., Moratalla, R., Gold, L.H. *et al.* (1994) Dopamine D1 receptor mutant mice are deficient in striatal expression of dynorphin and in dopamine-mediated behavioral responses. *Cell* **79**, 729–742.

134 Drago, J., Gerfen, C.R., Lachowicz, J.E. *et al.* (1994) Altered striatal function in a mutant mouse lacking D1A dopamine receptors. *Proc. Natl Acad. Sci. USA* **91**, 12564–12568.

135 Xu, M., Hu, X.-T., Cooper, D.C. *et al.* (1994) Elimination of cocaine-induced hyperactivity and dopamine-mediated neurophysiological effects in dopamine D1 receptor mutant mice. *Cell* **79**, 945–955.

136 Zhuang, Y., Soriano, P. & Weintraub, H. (1994) The helix-loop-helix gene E2A is required for B cell formation. *Cell* **79**, 875–884.

137 Bain, G., Robanus Maandag, E.C., Izon, D.J. *et al.* (1994) E2A proteins are required for proper B cell development and initiation of immunoglobulin gene rearrangements. *Cell* **79**, 885–892.

138 Wurst, W., Auerbach, A.B. & Joyner, A.L. (1994) Multiple developmental defects in Engrailed-1 mutant mice: an early mid-hindbrain deletion and patterning defects in forelimbs and sternum. *Development* **120**, 2065–2075.

139 Joyner, A.L., Herrup, K., Auerbach, B.A., Davis, C.A. & Rossant, J. (1991) Subtle cerebellar phenotype in mice homozygous for a targeted deletion of the En-2 homeobox. *Science* **251**, 1239–1243.

140 Millen, K.J., Wurst, W., Herrup, K. & Joyner, A.L. (1994) Abnormal embryonic cerebellar development and patterning of postnatal foliation in two mouse Engrailed-2 mutants. *Development* **120**, 695–706.

141 Kurihara, Y., Kurihara, H., Suzuki, H. *et al.* (1994) Elevated blood pressure and craniofacial abnormalities in mice deficient in endothelin-1. *Nature* **368**, 703–710.

142 Lubahn, D.B., Moyer, J.S., Golding, T.S., Couse, J.F., Korach, K.S. & Smithies, O. (1993) Alteration of reproductive function but not prenatal sexual development after insertional disruption of the mouse estrogen receptor gene. *Proc. Natl Acad. Sci. USA* **90**, 11162–11166.

143 Spyropoulos, D.D. & Capecchi, M.R. (1994) Targeted disruption of the even-skipped gene, evx1, causes early postimplantation lethality of the mouse conceptus. *Genes Dev.* **8**, 1949–1961.

144 McWhir, J., Selfridge, J., Harrison, D.J., Squires, S. & Melton, D.W. (1993) Mice with DNA repair gene (ERCC-1) deficiency have elevated levels of p53, liver nuclear abnormalities and die before weaning. *Nature Genet.* **5**, 217–224.

145 Takai, T., Li, M., Sylvestre, D., Clynes, R. & Ravetch, J.V. (1994) FcR gamma chain deletion results in pleiotrophic effector cell defects. *Cell* **76**, 519–529.

146 Lowell, C.A., Soriano, P. & Varmus, H.E. (1994) Functional overlap in the src gene family: inactivation of hck and fgr impairs natural immunity. *Genes Dev.* **8**, 387–398.

147 Feldman, B., Poueymirou, W., Papaioannou, V.E., DeChiara, T.M. & Goldfarb, M. (1995) Requirement of FGF-4 for postimplantation mouse development. *Science* **267**, 246–249.

148 Deng, C.-X., Wynshaw-Boris, A., Shen, M.M., Daugherty, C., Ornitz D.M. & Leder, P. (1994) Murine FGFR-1 is required for early postimplantation growth and axial organization. *Genes Dev.* **8**, 3045–3057.

149 Yamaguchi, T.P., Harpal, J., Henkemeyer, M. & Rossant, J. (1994) fgfr-1 is required for embryonic growth and mesodermal patterning during mouse gastrulation. *Genes Dev.* **8**, 3032–3044.

150 George, E.L., Georges-Labouesse, E.N., Patel-King, R.S., Rayburn, H. & Hynes, R.O. (1993) Defects in mesoderm, neural tube and vascular development in mouse embryos lacking fibronectin. *Development* **119**, 1079–1091.

151 The Dutch-Belgium Fragile X Consortium (1994) Fmr1 knockout mice: a model to study fragile X-mental retardation. *Cell* **78**, 23–33.

152 Matzuk, M.M., Lu, N., Vogel, H., Sellheyer, K., Roop, D.R. & Bradley, A. (1995) Multiple defects and perinatal death in mice deficient in follistatin. *Nature* **374**, 360–363.

153 Johnson, R.S., Spiegelman, B.M. & Papaioannou, V. (1992) Pleiotropic effects of a null mutation in the c-fos proto-oncogene. *Cell* **71**, 577–586.

154 Wang, Z.Q., Ovitt, C., Grigoriadis, A.E., Mohle, S.U., Ruther, U. & Wagner, E.F. (1992) Bone and haematopoietic defects in mice lacking c-fos. *Nature* **360**, 741–745.

155 Baum, M.J., Brown, J.J., Kica, E., Rubin, B.S., Johnson, R.S. & Papaioannou, V.E. (1994) Effect of a null mutation of the c-fos proto-oncogene on sexual behavior of male mice. *Biol. Reprod.* **50**, 1040–1048.

156 Hu, E., Mueller, E., Oliviero, S., Papaioannou, V.E., Johnson, R. & Spiegelman, B.M. (1994) Targeted disruption of the c-fos gene demonstrates c-fos-dependent and -independent pathways for gene expression stimulated by growth factors or oncogenes. *EMBO J.* **13**, 3094–3103.

157 Jain, J., Nalefski, E.A., McCaffrey, P.G. *et al.* (1994) Normal peripheral T-cell function in c-Fos-deficient mice. *Mol. Cell Biol.* **14**, 1566–1574.

158 Paylor, R., Johnson, R.S., Papaioannou, V., Spiegelman, B.M. & Wehner, J.M. (1994) Behavioral assessment of c-fos mutant mice. *Brain Res.* **651**, 275–282.

159 Okada, S., Wang, Z.Q., Grigoriadis, A.E., Wagner, E.F. & von Ruden, T. (1994) Mice lacking c-fos have normal haematopoietic stem cells but exhibit altered B-cell differentiation due to an impaired bone marrow environment. *Mol. Cell Biol.* **14**, 382–390.

160 Grigoriadis, A.E., Wang, Z.-Q., Cecchini, M.G. *et al.* (1994) c-Fos: a key regulator of osteoclast-macrophage lineage determination and bone remodeling. *Science* **266**, 443–447.

161 Yagi, T., Aizawa, S., Tokunaga, T., Shigetani, Y., Takeda, N. & Ikawa, Y. (1993) A role for Fyn tyrosine kinase in the suckling behaviour of neonatal mice. *Nature* **366**, 742–745.

162 Stein, P.L., Lee, H.M., Rich, S. & Soriano, P. (1992) pp59fyn mutant mice display differential signaling in thymocytes and peripheral T cells. *Cell* **70**, 741–750.

163 Umemori, H., Sato, S., Yagi, T., Aizawa, S. & Yamamoto, T. (1994) Initial events of myelination involve Fyn tyrosine kinase signalling. *Nature* **367**, 572–576.

164 Yagi, T., Shigetani, Y., Okado, N., Tokunaga, T., Ikawa, Y. & Aizawa, S. (1993) Regional localization of Fyn in adult brain; studies with mice in which fyn gene was replaced by lacZ. *Oncogene* **8**, 3343–3351.

165 Ignelzi, M.A.J., Miller, D.R., Soriano, P. & Maness, P.F. (1994) Impaired neurite outgrowth of src-minus cerebellar neurons on the cell adhesion molecule L1. *Neuron* **12**, 873–884.

166 Kiefer, F., Anhauser, I., Soriano, P., Aguzzi, A., Courtneidge, S.A. & Wagner, E.F. (1994) Endothelial cell transformation by polyomavirus middle T antigen in mice lacking Src-related kinases. *Curr. Biol.* **4**, 100–109.

167 Beggs, H.E., Soriano, P. & Maness, P.F. (1994) NCAM-dependent neurite outgrowth is inhibited in neurons from Fyn-minus mice. *J. Cell Biol.* **127**, 825–833.

168 Appleby, M.W., Gross, J.A., Cooke, M.P., Levin, S.D., Qian, X. & Perlmutter, R.M. (1992) Defective T cell receptor signaling in mice lacking the thymic isoform of p59fyn. *Cell* **70**, 751–763.

169 Tsai, F.Y., Keller, G., Kuo, F.C. *et al.* (1994) An early haematopoietic defect in mice lacking the transcription factor GATA-2. *Nature* **371**, 221–226.

170 Gomi, H., Yokoyama, T., Fujimoto, K. *et al.* (1995) Mice devoid of the glial fibrillary acidic protein develop normally and are susceptible to scrapie prions. *Neuron* **14**, 29–41.

171 Shehee, W.R., Oliver, P. & Smithies, O. (1993) Lethal thalassemia after insertional disruption of the mouse major adult beta-globin gene. *Proc. Natl Acad. Sci. USA* **90**, 3177–3181.

172 Tybulewicz, V.L., Tremblay, M.L., LaMarca, M.E. *et al.* (1992) Animal model of Gaucher's disease from targeted disruption of the mouse glucocerebrosidae gene. *Nature* **357**, 407–410.

173 Aiba, A., Chen, C., Herrup, K., Rosenmund, C., Stevens, C.F. & Tonegawa, S. (1994) Reduced hippocampal long-term potentiation and context-specific deficit in associative learning in mGluR1 mutant mice. *Cell* **79**, 365–375.

174 Aiba, A., Kano, M., Chen, C. *et al.* (1994) Deficient cerebellar long-term depression and impaired motor learning in mGluR1 mutant mice. *Cell* **79**, 377–388.

175 Conquet, F., Bashir, Z.I., Davies, C.H. *et al.* (1994) Motor deficit and impairment of synaptic plasticity in mice lacking mGluR1. *Nature* **372**, 237–243.

176 Forrest, D., Yuzaki, M., Soares, H.D. *et al.* (1994) Targeted disruption of NMDA receptor 1 gene abolishes NMDA response and results in neonatal death. *Neuron* **13**, 325–338.

177 Li, Y., Erzurumlu, R.S., Chen, C., Jhaveri, S. & Tonegawa, S. (1994) Whisker-related neuronal patterns fail to develop in the trigeminal brainstem nuclei of NMDAR1 knockout mice. *Cell* **76**, 427–437.

178 Sakimura, K., Kutsuwada, T., Ito, I. *et al.* (1995) Reduced hippocampal LTP and spatial learning in mice lacking NMDA receptor epsilon1 subunit. *Nature* **373**, 151–155.

179 Lieschke, G.J., Grail, D., Hodgson, G. *et al.* (1994) Mice lacking granulocyte colony-stimulating factor have chronic neutropenia, granulocyte and macrophage progenitor cell deficiency, and impaired neutrophil mobilization. *Blood* **84**, 1737–1746.

180 Stanley, E., Lieschke, G.J., Grail, D. *et al.* (1994) Granulocyte/macrophage colony-stimulating factor-deficient mice show no major perturbation of haematopoiesis but develop a characteristic pulmonary pathology. *Proc. Natl Acad. Sci. USA* **91**, 5592–5596.

181 Dranoff, G., Crawford, A.D., Sadelain, M. *et al.* (1994) Involvement of granulocyte–macrophage colony-stimulating factor in pulmonary homeostasis. *Science* **264**, 713–716.

182 Lieschke, G.J., Stanley, E., Grail, D. *et al.* (1994) Mice lacking both macrophage- and granulocyte–macrophage colony-stimulating factor have macrophages and coexistent osteopetrosis and severe lung disease. *Blood* **84**, 27–35.

183 Heusel, J.W., Wesselschmidt, R.L., Shresta, S., Russell, J.H. & Ley, T.J. (1994) Cytotoxic lymphocytes require granzyme B for the rapid induction of DNA fragmentation and apoptosis in allogeneic target cells. *Cell* **76**, 977–987.

184 Strittmatter, S.M., Fankhauser, C., Huang, P.L., Mashimo, H. & Fishman, M.C. (1995) Neuronal pathfinding is abnormal in mice lacking the neuronal growth cone protein GAP-43. *Cell* **80**, 445–452.

185 Homanics, G.E., de Silva, H.V., Osada, J. *et al.* (1995) Mild dyslipidemia in mice following targeted inactivation of the hepatic lipase gene. *J. Biol. Chem.* **270**, 2974–2980.

186 Schmidt, C., Bladt, F., Goedecke, S. *et al.* (1995) Scatter factor/hepatocyte growth factor is essential for liver development. *Nature* **373**, 699–702.

187 Uehara, Y., Minowa, O., Mori, C. *et al.* (1995) Placental defect and embryonic lethality in mice lacking hepatocyte growth factor/scatter factor. *Nature* **373**, 702–705.

188 Weinstein, D.C., Ruiz i Altaba, A., Chen, W.S. *et al.* (1994) The winged-helix transcription factor HNF-3 beta is required for notochord development in the mouse embryo. *Cell* **78**, 575–588.

189 Ang, S.L. & Rossant, J. (1994) HNF-3 beta is essential for node and notochord formation in mouse development. *Cell* **78**, 561–574.

190 Chen, W.S., Manova, K., Weinstein, D.C. *et al.* (1994)

Disruption of the HNF-4 gene, expressed in visceral endoderm, leads to cell death in embryonic ectoderm and impaired gastrulation of mouse embryos. *Genes Dev.* **8**, 2466–2477.

191 Yamanaka, S., Johnson, M.D., Grinberg, A. *et al.* (1994) Targeted disruption of the Hexa gene results in mice with biochemical and pathologic features of Tay–Sachs disease. *Proc. Natl Acad. Sci. USA* **91**, 9975–9979.

192 Roberts, C.W., Shutter, J.R. & Korsmeyer, S.J. (1994) Hox11 controls the genesis of the spleen. *Nature* **368**, 747–749.

193 Chisaka, O., Musci, T.S. & Capecchi, M.R. (1992) Developmental defects of the ear, cranial nerves and hindbrain resulting from targeted disruption of the mouse homeobox gene Hox-1.6. *Nature* **355**, 516–520.

194 Lufkin, T., Dierich, A., LeMeur, M., Mark, M. & Chambon, P. (1991) Disruption of the Hox-1.6 homeobox gene results in defects in a region corresponding to its rostral domain of expression. *Cell* **66**, 1105–1119.

195 Carpenter, E.M., Goddard, J.M., Chisaka, O., Manley, N.R. & Capecchi, M.R. (1993) Loss of Hox-A1 (Hox-1.6) function results in the reorganization of the murine hindbrain. *Development* **118**, 1063–1075.

196 Doll'e, P., Lufkin, T., Krumlauf, R., Mark, M., Duboule, D. & Chambon, P. (1993) Local alterations of Krox-20 and Hox gene expression in the hindbrain suggest lack of rhombomeres 4 and 5 in homozygote null Hoxa-1 (Hox-1.6) mutant embryos. *Proc. Natl Acad. Sci. USA* **90**, 7666–7670.

197 Mark, M., Lufkin, T., Vonesch, J.L. *et al.* (1993) Two rhombomeres are altered in Hoxa-1 mutant mice. *Development* **119**, 319–338.

198 Rijli, F.M., Mark, M., Lakkaraju, S., Dierich, A., Doll'e, P. & Chambon, P. (1993) A homeotic transformation is generated in the rostral branchial region of the head by disruption of Hoxa-2, which acts as a selector gene. *Cell* **75**, 1333–1349.

199 Gendron, M.M., Mallo, M., Zhang, M. & Gridley, T. (1993) Hoxa-2 mutant mice exhibit homeotic transformation of skeletal elements derived from cranial neural crest. *Cell* **75**, 1317–1331.

200 Chisaka, O. & Capecchi, M.R. (1991) Regionally restricted developmental defects resulting from targeted disruption of the mouse homeobox gene hox-1.5. *Nature* **350**, 473–479.

201 Horan, G., Wu, K., Wolgemuth, D.J. & Behringer, R.R. (1994) Homeotic transformation of cervical vertebrae in Hoxa-4 mutant mice. *Proc. Natl Acad. Sci. USA* **91**, 12644–12648.

202 Jeannotte, L., Lemieux, M., Charron, J., Poirier, F. & Robertson, E.J. (1993) Specification of axial identity in the mouse: role of the Hoxa-5 (Hox1.3) gene. *Genes Dev.* **7**, 2085–2096.

203 Small, K.M. & Potter, S.S. (1993) Homeotic transformations and limb defects in Hox A11 mutant mice. *Genes Dev.* **7**, 2318–2328.

204 Ramirez-Solis, R., Zheng, H., Whiting, J., Krumlauf, R. & Bradley, A. (1993) Hoxb-4 (Hox-2.6) mutant mice show homeotic transformation of a cervical vertebra and defects in the closure of the sternal rudiments. *Cell* **73**, 279–294.

205 Rancourt, D.E., Tsuzuki, T. & Capecchi, M.R. (1995) Genetic interaction between *hoxb-5* and *hoxb-6* is revealed by nonallelic noncomplementation. *Genes Dev.* **9**, 108–122.

206 Le Mouellic, H., Lallemand, Y. & Brulet, P. (1992) Homeosis in the mouse induced by a null mutation in the Hox-3.1 gene. *Cell* **69**, 251–264.

207 Condie, B.G. & Capecchi, M.R. (1993) Mice homozygous for a targeted disruption of Hoxd-3 (Hox-4.1) exhibit anterior transformations of the first and second cervical vertebrae, the atlas and the axis. *Development* **119**, 579–595.

208 Davis, A.P. & Capecchi, M.R. (1994) Axial homeosis and appendicular skeleton defects in mice with a targeted disruption of hoxd-11. *Development* **120**, 2187–2198.

209 Favier, B., Le Meur, M., Chambon, P. & Doll'e, P. (1995) Axial skeleton homeosis and forelimb malformations in Hoxd-11 mutant mice. *Proc. Natl Acad. Sci. USA* **92**, 310–314.

210 Doll'e, P., Dierich, A., LeMeur, M. *et al.* (1993) Disruption of the Hoxd-13 gene induces localized heterochrony leading to mice with neotenic limbs. *Cell* **75**, 431–441.

211 Thompson, S., Clarke, A.R., Pow, A.M., Hooper, M.L. & Melton, D.W. (1989) Germ line transmission and expression of a corrected HPRT gene produced by gene targeting in embryonic stem cells. *Cell* **56**, 313–321.

212 Georgopouos, K., Bigby, M., Wang, J.-H. *et al.* (1994) The Ikaros gene is required for the development of all lymphoid lineages. *Cell* **79**, 143–156.

213 Nitschke, L., Kosco, M.H., Kohler, G. & Lamers, M.C. (1993) Immunoglobulin D-deficient mice can mount normal immune responses to thymus-independent and -dependent antigens. *Proc. Natl Acad. Sci. USA* **90**, 1887–1891.

214 Roes, J. & Rajewsky, K. (1993) Immunoglobulin D (IgD)-deficient mice reveal an auxiliary receptor function for IgD in antigen-mediated recruitment of B cells. *J. Exp. Med.* **177**, 45–55.

215 Oettgen, H.C., Martin, T.R., Wynshaw, B.A., Deng, C., Drazen, J.M. & Leder, P. (1994) Active anaphylaxis in IgE-deficient mice. *Nature* **370**, 367–370.

216 Dombrowicz, D., Flamand, V., Brigman, K.K., Koller, B.H. & Kinet, J.P. (1993) Abolition of anaphylaxis by targeted disruption of the high affinity immunoglobulin E receptor alpha chain gene. *Cell* **75**, 969–976.

217 Takeda, S., Zou, Y.R., Bluethmann, H., Kitamura, D., Muller, U. & Rajewsky, K. (1993) Deletion of the immunoglobulin kappa chain intron enhancer abolishes kappa chain gene rearrangement in cis but not lambda chain gene rearrangement in trans. *EMBO J.* **12**, 2329–2336.

218 Zou, Y.-R., Takeda, S. & Rajewsky, K. (1993) Gene targeting in the Ig kappa locus: efficient generation of lambda chain-expressing B cells, independent of gene rearrangement in Ig kappa. *EMBO J.* **12**, 811–820.

219 Chen, J., Trounstine, M., Kurahara, C. *et al.* (1993) B cell development in mice that lack one or both immunoglobulin kappa light chain genes. *EMBO J.* **12**, 821–830.

220 Kim, J.Y., Kurtz, B., Huszar, D. & Storb, U. (1994) Crossing the SJL lambda locus into kappa-knockout mice reveals a dysfunction of the lambda 1-contain-

ing immunoglobulin receptor in B cell differentiation. *EMBO J.* **13**, 827–834.

221 Zou, Y.R., Gu, H. & Rajewsky, K. (1993) Generation of a mouse strain that produces immunoglobulin kappa chains with human constant regions. *Science* **262**, 1271–1274.

222 Kitamura, D., Roes, J., Kuhn, R. & Rajewsky, K. (1991) A B cell-deficient mouse by targeted disruption of the membrane exon of the immunoglobulin mu chain gene. *Nature* **350**, 423–426.

223 Ehlich, A., Schaal, S., Gu, H., Kitamura, D., Muller, W. & Rajewsky, K. (1993) Immunoglobulin heavy and light chain genes rearrange independently at early stages of B cell development. *Cell* **72**, 695–704.

224 Kitamura, D. & Rajewsky, K. (1992) Targeted disruption of mu chain membrane exon causes loss of heavy-chain allelic exclusion. *Nature* **356**, 154–156.

225 Beutner, U., Kraus, E., Kitamura, D., Rajewsky, K. & Huber, B.T. (1994) B cells are essential for murine mammary tumor virus transmission, but not for presentation of endogenous superantigens. *J. Exp. Med.* **179**, 1457–1466.

226 Matzuk, M.M., Finegold, M.J., Su, J.G., Hsueh, A.J. & Bradley, A. (1992) Alpha-inhibin is a tumour-suppressor gene with gonadal specificity in mice. *Nature* **360**, 313–319.

227 Matzuk, M.M., Finegold, M.J., Mather, J.P., Krummen, L., Lu, H. & Bradley, A. (1994) Development of cancer cachexia-like syndrome and adrenal tumors in inhibin-deficient mice. *Proc. Natl Acad. Sci. USA* **91**, 8817–8821.

228 Trudeau, V.L., Matzuk, M.M., Hach'e, R.J. & Renaud, L.P. (1994) Overexpression of activin-beta A subunit mRNA is associated with decreased activin type II receptor mRNA levels in the testes of alpha-inhibin deficient mice. *Biochem. Biophys. Res. Commun.* **203**, 105–112.

229 Tamemoto, H., Kadowaki, T., Tobe, K. *et al.* (1994) Insulin resistance and growth retardation in mice lacking insulin receptor substrate-1. *Nature* **372**, 182–186.

230 Araki, E., Lipes, M.A., Patti, M.E. *et al.* (1994) Alternative pathway of insulin signalling in mice with targeted disruption of the IRS-1 gene. *Nature* **372**, 186–190.

231 Liu, J.P., Baker, J., Perkins, A.S., Robertson, E.J. & Efstratiadis, A. (1993) Mice carrying null mutations of the genes encoding insulin-like growth factor I (Igf-1) and type 1 IGF receptor (Igf1r). *Cell* **75**, 59–72.

232 Baker, J., Liu, J.P., Robertson, E.J. & Efstratiadis, A. (1993) Role of insulin-like growth factors in embryonic and postnatal growth. *Cell* **75**, 73–82.

233 Powell, B.L., Hollingshead, P., Warburton, C. *et al.* (1993) IGF-I is required for normal embryonic growth in mice. *Genes Dev.* **7**, 2609–2617.

234 DeChiara, T.M., Efstratiadis, A. & Robertson, E.J. (1990) A growth-deficiency phenotype in heterozygous mice carrying an insulin-like growth factor II gene disrupted by targeting. *Nature* **345**, 78–80.

235 DeChiara, T.M., Robertson, E.J. & Efstratiadis, A. (1991) Parental imprinting of the mouse insulin-like growth factor II gene. *Cell* **64**, 849–859.

236 Christofori, G., Naik, P. & Hanahan, D. (1994) A second signal supplied by insulin-like growth factor II in oncogene-induced tumorigenesis. *Nature* **369**, 414–418.

237 Filson, A.J., Louvi, A., Efstratiadis, A. & Robertson, E.J. (1993) Rescue of the T-associated maternal effect in mice carrying null mutations in Igf-2 and Igf2r, two reciprocally imprinted genes. *Development* **118**, 731–736.

238 Coppola, D., Ferber, A., Miura, M. *et al.* (1994) A functional insulin-like growth factor I receptor is required for the mitogenic and transforming activities of the epidermal growth factor receptor. *Mol. Cell Biol.* **14**, 4588–4595.

239 Sell, C., Rubini, M., Rubin, R., Liu, J.P., Efstratiadis, A. & Baserga, R. (1993) Simian virus 40 large tumor antigen is unable to transform mouse embryonic fibroblasts lacking type 1 insulin-like growth factor receptor. *Proc. Natl Acad. Sci. USA* **90**, 11217–11221.

240 Sell, C., Dumenil, G., Deveaud, C. *et al.* (1994) Effect of a null mutation of the insulin-like growth factor I receptor gene on growth and transformation of mouse embryo fibroblasts. *Mol. Cell Biol.* **14**, 3604–3612.

241 Wang, Z.-Q., Fung, M.R., Barlow, D.P. & Wagner, E.F. (1994) Regulation of embryonic growth and lysosomal targeting by the imprinted *Igf2/Mpr* gene. *Nature* **372**, 464–467.

242 Lau, M.M.H., Stewart, C.E.H., Liu, Z., Bhatt, H., Rotwein, P. & Stewart, C.L. (1994) Loss of the imprinted IGF2/cation-independent mannose 6-phosphate receptor results in fetal overgrowth and perinatal lethality. *Genes Dev.* **8**, 2953–2963.

243 Jonsson, J., Carlsson, L., Edlund, T. & Edlund, H. (1994) Insulin-promoter-factor 1 is required for pancreas development in mice. *Nature* **371**, 606–609.

244 Yang, J.T., Rayburn, H. & Hynes, R.O. (1993) Embryonic mesodermal defects in alpha 5 integrin-deficient mice. *Development* **119**, 1093–1105.

245 Xu, H., Gonzalo, J.A., St, P.Y. *et al.* (1994) Leukocytosis and resistance to septic shock in intercellular adhesion molecule 1-deficient mice. *J. Exp. Med.* **180**, 95–109.

246 Sligh, J.E.J., Ballantyne, C.M., Rich, S.S. *et al.* (1993) Inflammatory and immune responses are impaired in mice deficient in intercellular adhesion molecule 1. *Proc. Natl Acad. Sci. USA* **90**, 8529–8533.

247 Muller, U., Steinhoff, U., Reis, L.F. *et al.* (1994) Functional role of type I and type II interferons in antiviral defense. *Science* **264**, 1918–1921.

248 Dalton, D.K., Pitts-Meek, S., Keshav, S., Figari, I.S., Bradley, A. & Stewart, T.A. (1993) Multiple defects of immune cell function in mice with disrupted interferon-gamma genes. *Science* **259**, 1739–1742.

249 Cooper, A.M., Dalton, D.K., Stewart, T.A., Griffin, J.P., Russell, D.G. & Orme, I.M. (1993) Disseminated tuberculosis in interferon gamma gene-disrupted mice. *J. Exp. Med.* **178**, 2243–2247.

250 Flynn, J.L., Chan, J., Triebold, K.J., Dalton, D.K., Stewart, T.A. & Bloom, B.R. (1993) An essential role for interferon gamma in resistance to *Mycobacterium tuberculosis* infection. *J. Exp. Med.* **178**, 2249–2254.

251 Graham, M.B., Dalton, D.K., Giltinan, D., Braciale, V.L., Stewart, T.A. & Braciale, T.J. (1993) Response to

influenza infection in mice with a targeted disruption in the interferon gamma gene. *J. Exp. Med.* **178**, 1725–1732.
252 Wang, Z.E., Reiner, S.L., Zheng, S., Dalton, D.K. & Locksley, R.M. (1994) CD4+ effector cells default to the Th2 pathway in interferon gamma-deficient mice infected with *Leishmania major*. *J. Exp. Med.* **179**, 1367–1371.
253 Huang, S., Hendriks, W., Althage, A. *et al.* (1993) Immune response in mice that lack the interferon-gamma receptor. *Science* **259**, 1742–1745.
254 Kamijo, R., Le, J., Shapiro, D. *et al.* (1993) Mice that lack the interferon-gamma receptor have profoundly altered responses to infection with Bacillus Calmette-Guérin and subsequent challenge with lipopolysaccharide. *J. Exp. Med.* **178**, 1435–1440.
255 Car, B.D., Eng, V.M., Schnyder, B. *et al.* (1994) Interferon gamma receptor deficient mice are resistant to endotoxic shock. *J. Exp. Med.* **179**, 1437–1444.
256 Kamijo, R., Shapiro, D., Le, J., Huang, S., Aguet, M. & Vilcek, J. (1993) Generation of nitric oxide and induction of major histocompatibility complex class II antigen in macrophages from mice lacking the interferon gamma receptor. *Proc. Natl Acad. Sci. USA* **90**, 6626–6630.
257 Matsuyama, T., Kimura, T., Kitagawa, M. *et al.* (1993) Targeted disruption of IRF-1 or IRF-2 results in abnormal type I IFN gene induction and aberrant lymphocyte development. *Cell* **75**, 83–97.
258 Reis, L., Ruffner, H., Stark, G., Aguet, M. & Weissmann, C. (1994) Mice devoid of interferon regulatory factor 1 (IRF-1) show normal expression of type I interferon genes. *EMBO J.* **13**, 4798–4806.
259 Kamijo, R., Harada, H., Matsuyama, T. *et al.* (1994) Requirement for transcription factor IRF-1 in NO synthase induction in macrophages. *Science* **263**, 1612–1615.
260 Kimura, T., Nakayama, K., Penninger, J. *et al.* (1994) Involvement of the IRF-1 transcription factor in antiviral responses to interferons. *Science* **264**, 1921–1924.
261 Tanaka, N., Ishihara, M., Kitagawa, M. *et al.* (1994) Cellular commitment to oncogene-induced transformation or apoptosis is dependent on the transcription factor IRF-1. *Cell* **77**, 829–839.
262 Li, P., Allen, H., Banerjee, S. *et al.* (1995) Mice deficient in IL-1b-converting enzyme are defective in production of mature IL-1b and resistant to endotoxic shock. *Cell* **80**, 401–411.
263 Schorle, H., Holtschke, T., Hunig, T., Schimpl, A. & Horak, I. (1991) Development and function of T cells in mice rendered interleukin-2 deficient by gene targeting. *Nature* **352**, 621–624.
264 Sadlack, B., Merz, H., Schorle, H., Schimpl, A., Feller, A.C. & Horak, I. (1993) Ulcerative colitis-like disease in mice with a disrupted interleukin-2 gene. *Cell* **75**, 253–261.
265 Kundig, T.M., Schorle, H., Bachmann, M.F., Hengartner, H., Zinkernagel, R.M. & Horak, I. (1993) Immune responses in interleukin-2-deficient mice. *Science* **262**, 1059–1061.
266 DiSanto, J.P., Muller, W., Guy-Grand, D., Fischer, A. & Rajewsky, K. (1995) Lymphoid development in mice with a targeted deletion of the interleukin 2 receptor γ chain. *Proc. Natl Acad. Sci. USA* **92**, 377–381.
267 Kopf, M., Le, G.G., Bachmann, M., Lamers, M.C., Bluethmann, H. & Kohler, G. (1993) Disruption of the murine IL-4 gene blocks Th2 cytokine responses. *Nature* **362**, 245–248.
268 Kuhn, R., Rajewsky, K. & Muller, W. (1991) Generation and analysis of interleukin-4 deficient mice. *Science* **254**, 707–710.
269 Kanagawa, O., Vaupel, B.A., Gayama, S., Koehler, G. & Kopf, M. (1993) Resistance of mice deficient in IL-4 to retrovirus-induced immunodeficiency syndrome (MAIDS). *Science* **262**, 240–242.
270 Schmitz, J., Thiel, A., Kuhn, R. *et al.* (1994) Induction of interleukin 4 (IL-4) expression in T helper (Th) cells is not dependent on IL-4 from non-Th cells. *J. Exp. Med.* **179**, 1349–1353.
271 Poli, V., Balena, R., Fattori, E. *et al.* (1994) Interleukin-6 deficient mice are protected from bone loss caused by estrogen depletion. *EMBO J.* **13**, 1189–1196.
272 Kopf, M., Baumann, H., Freer, G. *et al.* (1994) Impaired immune and acute-phase responses in interleukin-6-deficient mice. *Nature* **368**, 339–342.
273 Fattori, E., Cappelletti, M., Costa, P. *et al.* (1994) Defective inflammatory response in interleukin 6-deficient mice. *J. Exp. Med.* **180**, 1243–1250.
274 Ramsay, A.J., Husband, A.J., Ramshaw, I.A. *et al.* (1994) The role of interleukin-6 in mucosal IgA antibody responses *in vivo*. *Science* **264**, 561–563.
275 Libert, C., Takahashi, N., Cauwels, A., Brouckaert, P., Bluethmann, H. & Fiers, W. (1994) Response of interleukin-6-deficient mice to tumor necrosis factor-induced metabolic changes and lethality. *Eur. J. Immunol.* **24**, 2237–2242.
276 Peschon, J.J., Morrissey, P.J., Grabstein, K.H. *et al.* (1994) Early lymphocyte expansion is severely impaired in interleukin 7 receptor-deficient mice. *J. Exp. Med.* **180**, 1955–1960.
277 Cacalano, G., Lee, J., Kikly, K. *et al.* (1994) Neutrophil and B cell expansion in mice that lack the murine IL-8 receptor homolog. *Science* **265**, 682–684.
278 Kuhn, R., Lohler, J., Rennick, D., Rajewsky, K. & Muller, W. (1993) Interleukin-10-deficient mice develop chronic enterocolitis. *Cell* **75**, 263–274.
279 Bikoff, E.K., Huang, L.Y., Episkopou, V., van Meerwijk, J., Germain, R.N. & Robertson, E.J. (1993) Defective major histocompatibility complex class II assembly, transport, peptide acquisition, and CD4+ T cell selection in mice lacking invariant chain expression. *J. Exp. Med.* **177**, 1699–1712.
280 Viville, S., Neefjes, J., Lotteau, V. *et al.* (1993) Mice lacking the MHC class II-associated invariant chain. *Cell* **72**, 635–648.
281 Bonnerot, C., Marks, M.S., Cosson, P. *et al.* (1994) Association with BiP and aggregation of class II MHC molecules synthesized in the absence of invariant chain. *EMBO J.* **13**, 934–944.
282 Elliot, E.A., Drake, J.R., Amigorena, S. *et al.* (1994) The invariant chain is required for intracellular transport and function of major histoincompatibility complex class II molecules. *J. Exp. Med.* **179**, 681–694.
283 Gu, H., Zou, Y.R. & Rajewsky, K. (1993) Independent control of immunoglobulin switch recombination at

individual switch regions evidenced through Cre-lowP-mediated gene targeting. *Cell* **73**, 1155–1164.

284 Wagner, S.D., Williams, G.T., Larson, T. *et al.* (1994) Antibodies generated from human immunoglobulin miniloci in transgenic mice. *Nucleic Acids Res.* **22**, 1389–1393.

285 Chen, J., Trounstine, M., Alt, F.W. *et al.* (1993) Immunoglobulin gene rearrangement in B cell deficient mice generated by targeted deletion of the JH locus. *Int. Immunol.* **5**, 647–656.

286 Jakobovits, A., Vergara, G.J., Kennedy, J.L. *et al.* (1993) Analysis of homozygous mutant chimeric mice: deletion of the immunoglobulin heavy-chain joining region blocks B-cell development and antibody production. *Proc. Natl Acad. Sci. USA* **90**, 2551–2555.

287 Taki, S., Meiering, M. & Rajewsky, K. (1993) Targeted insertion of a variable region gene into the immunoglobulin heavy chain locus. *Science* **262**, 1268–1271.

288 Hilberg, F., Aguzzi, A., Howells, N. & Wagner, E.F. (1993) c-jun is essential for normal mouse development and hepatogenesis. *Nature* **365**, 179–181 (published erratum appears in *Nature* **366**, 368).

289 Johnson, R.S., van L.B., Papaioannou, V.E. & Spiegelman, B.M. (1993) A null mutation at the c-jun locus causes embryonic lethality and retarded cell growth in culture. *Genes Dev.* **7**, 1309–1317.

290 Baribault, H., Price, J., Miyai, K. & Oshima, R.G. (1993) Mid-gestational lethality in mice lacking keratin 8. *Genes Dev.* **7**, 1191–1202.

291 Baribault, H., Penner, J., Iozzo, R.V. & Wilson, H.M. (1994) Colorectal hyperplasia and inflammation in keratin 8-deficient FVB/N mice. *Genes Dev.* **8**, 2964–2973.

292 Schneider, M.S., Topilko, P., Seitandou, T. *et al.* (1993) Disruption of Krox-20 results in alteration of rhombomeres 3 and 5 in the developing hindbrain. *Cell* **75**, 1199–1214.

293 Swiatek, P.J. & Gridley, T. (1993) Perinatal lethality and defects in hindbrain development in mice homozygous for a targeted mutation of the zinc finger gene Krox20. *Genes Dev.* **7**, 2071–2084.

294 Topilko, P., Schneider-Maunoury, S., Levi, G. *et al.* (1994) Krox-20 controls myelination in the peripheral nervous system. *Nature* **371**, 796–799.

295 Poirier, F. & Robertson, E.J. (1993) Normal development of mice carrying a null mutation in the gene encoding the L14 S-type lectin. *Development* **119**, 1229–1236.

296 Stinnakre, M.G., Vilotte, J.L., Soulier, S. & Mercier, J.C. (1994) Creation and phenotypic analysis of alpha-lactalbumin-deficient mice. *Proc. Natl Acad. Sci. USA* **91**, 6544–6548.

297 Stacey, A., Schnieke, A., McWhir, J., Cooper, J., Colman, A. & Melton, D.W. (1994) Use of double-replacement gene targeting to replace the murine alpha-lactalbumin gene with its human counterpart in embryonic stem cells and mice. *Mol. Cell Biol.* **14**, 1009–1016.

298 Kitamura, D., Kudo, A., Schaal, S., Muller, W., Melchers, F. & Rajewsky, K. (1992) A critical role of lambda 5 protein in B cell development. *Cell* **69**, 823–831.

299 Noakes, P.G., Gautam, M., Mudd, J., Sanes, J.R. & Merlie, J.P. (1995) Aberrant differentiation of neuromuscular junctions in mice lacking s-laminin/laminin beta2. *Nature* **374**, 258–262.

300 Molina, T.J., Kishihara, K., Siderovski, D.P. *et al.* (1992) Profound block in thymocyte development in mice lacking p56lck. *Nature* **357**, 161–164.

301 Molina, T.J., Bachmann, M.F., Kundig, T.M., Zinkernagel, R.M. & Mak, T.W. (1993) Peripheral T cells in mice lacking p56lck do not express significant antiviral effector functions. *J. Immunol.* **151**, 699–706.

302 Penninger, J., Kishihara, K., Molina, T. *et al.* (1993) Requirement for tyrosine kinase p56lck for thymic development of transgenic gamma delta T cells. *Science* **260**, 358–361.

303 Escary, J.L., Perreau, J. Dum'enil, D., Ezine, S. & Brulet P. (1993) Leukaemia inhibitory factor is necessary for maintenance of haematopoietic stem cells and thymocyte stimulation. *Nature* **363**, 361–364.

304 Stewart, C.L., Kaspar, P., Brunet, L.J. *et al.* (1992) Blastocyst implantation depends on maternal expression of leukaemia inhibitory factor. *Nature* **359**, 76–79.

305 Rao, M.S., Sun, Y., Escary, J.L. *et al.* (1993) Leukaemia inhibitory factor mediates an injury response but not a target-directed developmental transmitter switch in sympathetic neurons. *Neuron* **11**, 1175–1185.

306 Chen, X.S., Sheller, J.R., Johnson, E.N. & Funk, C.D. (1994) Role of leukotrienes revealed by targeted disruption of the 5-lipoxygenase gene. *Nature* **372**, 179–182.

307 Goulet, J.L., Snouwaert, J.N., Latour, A.M., Coffman, T.M. & Koller, B.H. (1994) Altered inflammatory responses in leukotriene-deficient mice. *Proc. Natl Acad. Sci. USA* **91**, 12852–12856.

308 Fehling, H.J., Swat, W., Laplace, C. *et al.* (1994) MHC class I expression in mice lacking the proteasome subunit LMP-7. *Science* **265**, 1234–1237.

309 Ishibashi, S., Brown, M.S., Goldstein, J.L., Gerard, R.D., Hammer, R.E. & Herz, J. (1993) Hypercholesterolemia in low density lipoprotein receptor knockout mice and its reversal by adenovirus-mediated gene delivery. *J. Clin. Invest.* **92**, 883–893.

310 Ishibashi, S., Goldstein, J.L., Brown, M.S., Herz, J. & Burns, D.K. (1994) Massive xanthomatosis and atherosclerosis in cholesterol-fed low density lipoprotein receptor-negative mice. *J. Clin. Invest.* **93**, 1885–1893.

311 Ishibashi, S., Herz, J., Maeda, N., Goldstein, J.L. & Brown, M.S. (1994) The two-receptor model of lipoprotein clearance: tests of the hypothesis in 'knockout' mice lacking the low density lipoprotein receptor, apolipoprotein E, or both proteins. *Proc. Natl Acad. Sci. USA* **91**, 4431–4435.

312 Herz, J., Clouthier, D.E. & Hammer, R.E. (1993) LDL receptor-related protein internalizes and degrades uPA-PAI-1 complexes and is essential for embryo implantation. *Cell* **71**, 411–421 (published erratum appears in *Cell* **73**, 428).

313 Herz, J., Couthier, D.E. & Hammer, R.E. (1993) Correction: LDL receptor-related protein internalizes and degrades uPA-PAI-1 complexes and is essential for embryo implantation. *Cell* **73**, 428.

314 Van Genderen, C., Okamura, R.M., Farias, I. *et al.* (1994) Development of several organs that require inductive epithelial-mesenchymal interactions is impaired in LEF-1-deficient mice. *Genes Dev.* **8**, 2691–2703.
315 Kontgen, F., Suss, G., Stewart, C., Steinmetz, M. & Bluethmann, H. (1993) Targeted disruption of the MHC class II Aa gene in C57BL/6 mice. *Int. Immunol.* **5**, 957–964.
316 Grusby, M.J., Johnson, R.S., Papaioannou, V.E. & Glimcher, L.H. (1991) Depletion of CD4+ T cells in major histocompatibility complex class II-deficient mice. *Science* **253**, 1417–1420.
317 Cosgrove, D., Gray, D., Dierich, A. *et al.* (1991) Mice lacking MHC class II molecules. *Cell* **66**, 1051–1066.
318 Crump, A.L., Grusby, M.J., Glimcher, L.H. & Cantor, H. (1993) Thymocyte development in major histocompatibility complex-deficient mice: evidence for stochastic commitment to the CD4 and CD8 lineages. *Proc. Natl Acad. Sci. USA* **90**, 10739–10743.
319 Laufer, T.M., von, H.M.G., Grusby, M.J., Oldstone, M.B. & Glimcher, L.H. (1993) Autoimmune diabetes can be induced in transgenic major histocompatibility complex class II-deficient mice. *J. Exp. Med.* **178**, 589–596.
320 Markowitz, J.S., Auchincloss, H.J., Grusby, M.J. & Glimcher, L.H. (1993) Class II-positive haematopoietic cells cannot mediate positive selection of CD4+ T lymphocytes in class II-deficient mice. *Proc. Natl Acad. Sci. USA* **90**, 2779–2783.
321 Jevnikar, A.M., Grusby, M.J. & Glimcher, L.H. (1994) Prevention of nephritis in major histocompatibility complex class II-deficient MRL-lpr mice. *J. Exp. Med.* **179**, 1137–1143.
322 Mombaerts, P., Mizoguchi, E., Grusby, M.J., Glimcher, L.H., Bhan, A.K. & Tonegawa, S. (1993) Spontaneous development of inflammatory bowel disease in T cell receptor mutant mice. *Cell* **75**, 274–282.
323 Coles, M.C. & Raulet, D.H. (1994) Class I dependence of the development of CD4+ CD8- NK1.1+ thymocytes. *J. Exp. Med.* **180**, 395–399.
324 Matsushima, G.K., Taniike, M., Glimcher, L.H. *et al.* (1994) Absence of MHC class II molecules reduces CNS demyelination, microglial/macrophage infiltration, and twitching in murine globoid cell leukodystrophy. *Cell* **78**, 645–656.
325 Takahama, Y., Suzuki, H., Katz, K.S., Grusby, M.J. & Singer, A. (1994) Positive selection of CD4+ T cells by TCR ligation without aggregation even in the absence of MHC. *Nature* **371**, 67–70.
326 Kariv, I., Hardy, R.R. & Hayakawa, K. (1994) Altered major histocompatibility complex restriction in the NK1.1+Ly-6C(hi) autoreactive CD4+ T cell subset from class II-deficient mice. *J. Exp. Med.* **180**, 2419–2424.
327 Markowitz, J.S., Rogers, P.R., Grusby, M.J., Parker, D.C. & Glimcher, L.H. (1993) B lymphocyte development and activation independent of MHC class II expression. *J. Immunol.* **150**, 1223–1233.
328 Chan, S.H., Cosgrove, D., Waltzinger, C., Benoist, C. & Mathis, D. (1993) Another view of the selective model of thymocyte selection. *Cell* **73**, 225–236.
329 Guillemot, F., Lo, L.C., Johnson, J.E., Auerbach, A., Anderson, D.J. & Joyner, A.L. (1993) Mammalian achaete-scute homolog 1 is required for the early development of olfactory and autonomic neurons. *Cell* **75**, 463–476.
330 Guillemot, F., Nagy, A., Auerbach, A., Rossant, J. & Joyner, A.L. (1994) Essential role of Mash-2 in extraembryonic development. *Nature* **371**, 333–336.
331 Guillemot, F., Caspary, T., Tilghman, S.M. *et al.* (1995) Genomic imprinting of Mash2, a mouse gene required for trophoblast development. *Nature Genet.* **9**, 235–241.
332 Ludwig, T., Ovitt, C.E., Bauer, U. *et al.* (1993) Targeted disruption of the mouse cation-dependent mannose 6-phosphate receptor results in partial missorting of multiple lysosomal enzymes. *EMBO J.* **12**, 5225–5235.
333 Koster, A., Saftig, P., Matzner, U., von, F.K., Peters, C. & Pohlmann, R. (1993) Targeted disruption of the M(r) 46,000 mannose 6-phosphate receptor gene in mice results in misrouting of lysosomal proteins. *EMBO J.* **12**, 5219–5223.
334 Michalska, A.E. & Choo, K.H. (1993) Targeting and germ-line transmission of a null mutation at the metallothionein I and II loci in mouse. *Proc. Natl Acad. Sci. USA* **90**, 8088–8092.
335 Masters, B.A., Kelly, E.J., Quaife, C.J., Brinster, R.L. & Palmiter, R.D. (1994) Targeted disruption of metallothionein I and II genes increases sensitivity to cadmium. *Proc. Natl Acad. Sci. USA* **91**, 584–588.
336 Lazo, J.S., Kondo, Y., Dellapiazza, D., Michalska, A.E., Choo, K.H.A. & Pitt, B.R. (1995) Enhanced sensitivity to oxidative stress in cultured embryonic cells from transgenic mice deficient in metallothionein I and II genes. *J. Biol. Chem.* **270**, 5506–5510.
337 Zijlstra, M., Li, E., Sajjadi, F., Subramani, S. & Jaenisch, R. (1989) Germ-line transmission of a disrupted beta 2-microglobulin gene produced by homologous recombination in embryonic stem cells. *Nature* **342**, 435–438.
338 Zijlstra, M., Bix, M., Simister, N.E., Loring, J.M., Raulet, D.H. & Jaenisch, R. (1990) Beta 2-microglobulin deficient mice lack CD4-8+ cytolytic T cells. *Nature* **344**, 742–746.
339 Koller, B.H., Marrack, P., Kappler, J.W. & Smithies, O. (1990) Normal development of mice deficient in beta 2M, MHC class I proteins, and CD8+ T cells. *Science* **248**, 1227–1230.
340 Eichelberger, M., Allan, W., Zijlstra, M., Jaenisch, R. & Doherty, P.C. (1991) Clearance of influenza virus respiratory infection in mice lacking class I major histocompatibility complex-restricted CD8+ T cells. *J. Exp. Med.* **174**, 875–880.
341 Liao, N.S., Bix, M., Zijlstra, M., Jaenisch, R. & Raulet, D. (1991) MHC class I deficiency: susceptibility to natural killer (NK) cells and impaired NK activity. *Science* **253**, 199–202.
342 Bix, M., Liao, N.S., Zijlstra, M., Loring, J., Jaenisch, R. & Raulet, D. (1991) Rejection of class I MHC-deficient haemopoietic cells by irradiated MHC-matched mice. *Nature* **349**, 329–331.
343 Hou, S., Doherty, P.C., Zijlstra, M., Jaenisch, R. & Katz, J.M. (1992) Delayed clearance of Sendai virus in mice lacking class I MHC-restricted CD8+ T cells. *J. Immunol.* **149**, 1319–1325.
344 Sanjuan, N., Zijlstra, M., Carroll, J., Jaenisch, R. &

Benjamin, T. (1992) Infection by polyomavirus of murine cells deficient in class I major histocompatibility complex expression. *J. Virol.* **66**, 4587–4590.

345 Zijlstra, M., Auchincloss, H.J., Loring, J.M., Chase, C.M., Russell, P.S. & Jaenisch, R. (1992) Skin graft rejection by beta 2-microglobulin-deficient mice. *J. Exp. Med.* **175**, 885–893.

346 Pereira, P., Zijlstra, M., McMaster, J., Loring, J.M., Jaenisch, R. & Tonegawa, S. (1992) Blockade of transgenic gamma delta T cell development in beta 2-microglobulin deficient mice. *EMBO J.* **11**, 25–31.

347 Correa, I., Bix, M., Liao, N.S., Zijlstra, M., Jaenisch, R. & Raulet, D. (1992) Most gamma delta T cells develop normally in beta 2-microglobulin-deficient mice. *Proc Natl Acad. Sci. USA* **89**, 653–657.

348 Osorio, R.W., Ascher, N.L., Jaenisch, R., Freise, C.E., Roberts, J.P. & Stock, P.G. (1993) Major histocompatibility complex class I deficiency prolongs islet allograft survival. *Diabetes* **42**, 1520–1527.

349 Rodriguez, M., Dunkel, A.J., Thiemann, R.L., Leibowitz, J., Zijlstra, M. & Jaenisch, R. (1993) Abrogation of resistance to Theiler's virus-induced demyelination in H-2b mice deficient in beta 2-microglobulin. *J. Immunol.* **151**, 266–276.

350 Wicker, L.S., Leiter, E.H., Todd, J.A. *et al.* (1994) Beta 2-microglobulin-deficient NOD mice do not develop insulitis or diabetes. *Diabetes* **43**, 500–504.

351 Udaka, K., Marusic, G.S. & Walden, P. (1994) $CD4^+$ and $CD8^+$ $\alpha\beta$, and $\gamma\delta$ T cells are cytotoxic effector cells of beta2-microglobulin-deficient mice against cells having normal MHC class I expression. *J. Immunol.* **153**, 2843–2850.

352 van Meerwijk, J.P. & Germain, R.N. (1993) Development of mature $CD8^+$ thymocytes: selection rather than instruction? *Science* **261**, 911–915.

353 Desai, N.M., Bassiri, H., Kim, J. *et al.* (1993) Islet allograft, islet xenograft, and skin allograft survival in CD8+ T lymphocyte-deficient mice. *Transplantation* **55**, 718–722.

354 Coffman, T., Geier, S., Ibrahim, S. *et al.* (1993) Improved renal function in mouse kidney allografts lacking MHC class I antigens. *J. Immunol.* **151**, 425–435.

355 Markmann, J.F., Bassiri, H., Desai, N.M. *et al.* (1992) Indefinite survival of MHC class I-deficient murine pancreatic islet allografts. *Transplantation* **54**, 1085–1089.

356 Spriggs, M.K., Koller, B.H., Sato, T. *et al.* (1992) Beta 2-microglobulin-, $CD8^+$ T-cell-deficient mice survive inoculation with high doses of vaccinia virus and exhibit altered IgG responses. *Proc. Natl Acad. Sci. USA* **89**, 6070–6074.

357 Tarleton, R.L., Koller, B.H., Latour, A. & Postan, M. (1992) Susceptibility of beta 2-microglobulin-deficient mice to *Trypanosoma cruzi* infection. *Nature* **356**, 338–340.

358 Muller, D., Koller, B.H., Whitton, J.L., LaPan, K.E., Brigman, K.K. & Frelinger, J.A. (1992) LCMV-specific, class II-restricted cytotoxic T cells in beta 2-microglobulin-deficient mice. *Science* **255**, 1576–1578.

359 Hoglund, P., Ohl'en, C., Carbone, E. *et al.* (1991) Recognition of beta 2-microglobulin-negative (beta $2m^-$) T-cell blasts by natural killer cells from normal but not from beta $2m^-$ mice: nonresponsiveness controlled by beta 2m-bone marrow in chimeric mice. *Proc. Natl Acad. Sci. USA* **88**, 10332–10336.

360 Rajan, T.V., Nelson, F.K., Shultz, L.D., Koller, B.H. & Greiner, D.L. (1992) $CD8^+$ T lymphocytes are not required for murine resistance to human filarial parasites. *J. Parasitol.* **78**, 744–746.

361 Flynn, J.L., Goldstein, M.M., Triebold, K.J., Koller, B. & Bloom, B.R. (1992) Major histocompatibility complex class I-restricted T cells are required for resistance to Mycobacterium tuberculosis infection. *Proc. Natl Acad. Sci. USA* **89**, 12013–12017.

362 Glas, R., Franksson, L., Ohl'en, C. *et al.* (1992) Major histocompatibility complex class I-specific and Ñ restricted killing of beta 2-microglobulin-deficient cells by $CD8^+$ cytotoxic T lymphocytes. *Proc. Natl Acad. Sci. USA* **89**, 11381–11385.

363 Glas, R., Ohl'en, C., Hoglund, P. & Karre, K. (1994) The CD8+ T cell repertoire in beta 2-microglobulin-deficient mice is biased towards reactivity against self-major histocompatibility class I. *J. Exp. Med.* **179**, 661–672.

364 Apasov, S. & Sitkovsky, M. (1993) Highly lytic $CD8^+$, α β T-cell receptor cytotoxic T cells with major histocompatibility complex (MHC) class I antigen-directed cytotoxicity in beta 2-microglobulin, MHC class I-deficient mice. *Proc. Natl Acad. Sci. USA* **90**, 2837–2841.

365 Wells, F.B., Gahm, S.J., Hedrick, S.M., Bluestone, J.A., Dent, A. & Matis, L.A. (1991) Requirement for positive selection of γ δ receptor-bearing T cells. *Science* **253**, 903–905.

366 Lehmann-Grube, F., Lohler, J., Utermohlen, O. & Gegin, C. (1993) Antiviral immune responses of lymphocytic choriomeningitis virus-infected mice lacking $CD8^+$ T lymphocytes because of disruption of the beta 2-microglobulin gene. *J. Virol.* **67**, 332–339.

367 Bender, B.S., Croghan, T., Zhang, L. & Small, P.A.J. (1992) Transgenic mice lacking class I major histocompatibility complex-restricted T cells have delayed viral clearance and increased mortality after influenza virus challenge. *J. Exp. Med.* **175**, 1143–1145.

368 Fiette, L, Aubert, C., Brahic, M. & Rossi, C.P. (1993) Theiler's virus infection of beta 2-microglobulin-deficient mice. *J. Virol.* **67**, 589–592.

369 Hashimoto, N., Watanabe, N., Furuta, Y. *et al.* (1994) Parthenogenetic activation of oocytes in c-mos-deficient mice. *Nature* **370**, 68–71 (published erratum appears in *Nature* **370**, 391).

370 Colledge, W.H., Carlton, M.B., Udy, G.B. & Evans, M.J. (1994) Disruption of c-mos causes parthenogenetic development of unfertilized mouse eggs. *Nature* **370**, 65–68.

371 Gurney, A.L., Carver, M.K., De Sauvage, F.J. & Moore, M.W. (1994) Thrombocytopenia in c-mpl-deficient mice. *Science* **265**, 1445–1447.

372 Satokata, I. & Maas, R. (1994) Msx1 deficient mice exhibit cleft palate and abnormalities of craniofacial and tooth development. *Nature Genet.* **6**, 348–356.

373 Behringer, R.R., Finegold, M.J. & Cate, R.L. (1994) Mullerian-inhibiting substance function during mammalian sexual development. *Cell* **79**, 415–425.

374 Schinkel, A.H., Smit, J.J., van T.O. *et al.* (1994)

Disruption of the mouse mdr1a P-glycoprotein gene leads to a deficiency in the blood-brain barrier and to increased sensitivity to drugs. *Cell* **77**, 491–502.

375 Smit, J.J., Schinkel, A.H., Oude Elferink, R.P. *et al.* (1993) Homozygous disruption of the murine mdr2 P-glycoprotein gene leads to a complete absence of phospholipid from bile and to liver disease. *Cell* **75**, 451–462.

376 Mauad, T.H., van Nieuwkerk, C.M., Dingemans, K.P. *et al.* (1994) Mice with homozygous disruption of the mdr2 P-glycoprotein gene: a novel animal model for studies of nonsuppurative inflammatory cholangitis and hepatocarcinogenesis. *Am. J. Pathol.* **145**, 1237–1245.

377 Mucenski, M.L., McLain, K., Kier, A.B. *et al.* (1991) A functional c-myb gene is required for normal murine fetal hepatic haematopoiesis. *Cell* **65**, 677–689.

378 Davis, A.C., Wims, M., Spotts, G.D., Hann, S.R. & Bradley, A. (1993) A null c-myc mutation causes lethality before 10.5 days of gestation in homozygotes and reduced fertility in heterozygous female mice. *Genes Dev.* **7**, 671–682.

379 Sawai, S., Shimono, A., Hanaoka, K. & Kondoh, H. (1991) Embryonic lethality resulting from disruption of both N-myc alleles in mouse zygotes. *New Biol.* **3**, 861–869.

380 Stanton, B.R., Perkins, A.S., Tessarollo, L., Sassoon, D.A. & Parada, L.F. (1992) Loss of N-myc function results in embryonic lethality and failure of the epithelial component of the embryo to develop. *Genes Dev.* **6**, 2235–2247.

381 Charron, J., Malynn, B.A., Fisher, P. *et al.* (1992) Embryonic lethality in mice homozygous for a targeted disruption of the N-myc gene. *Genes Dev.* **6**, 2248–2257.

382 Sawai, S., Shimono, A., Wakamatsu, Y., Palmes, C., Hanaoka, K. & Kondoh, H. (1993) Defects of embryonic organogenesis resulting from targeted disruption of the N-myc gene in the mouse. *Development* **117**, 1445–1455.

383 Moens, C.B., Stanton, B.R., Parada, L.F. & Rossant, J. (1993) Defects in heart and lung development in compound heterozygotes for two different targeted mutations at the N-myc locus. *Development* **119**, 485–499.

384 Moens, C.B., Auerbach, A.B., Conlon, R.A., Joyner, A.L. & Rossant, J. (1992) A targeted mutation reveals a role for N-myc in branching morphogenesis in the embryonic mouse lung. *Genes Dev.* **6**, 691–704.

385 Montag, D., Giese, K.P., Bartsch, U. *et al.* (1994) Mice deficient for the myelin-associated glycoprotein show subtle abnormalities in myelin. *Neuron* **13**, 229–246.

386 Li, C., Tropak, M.B., Gerlai, R. *et al.* (1994) Myelination in the absence of myelin-associated glycoprotein. *Nature* **369**, 747–750.

387 Braun, T., Rudnicki, M.A., Arnold, H.H. & Jaenisch, R. (1992) Targeted inactivation of the muscle regulatory gene Myf-5 results in abnormal rib development and perinatal death. *Cell* **71**, 369–382.

388 Braun, T., Bober, E., Rudnicki, M.A., Jaenisch, R. & Arnold, H.H. (1994) MyoD expression marks the onset of skeletal myogenesis in Myf-5 mutant mice. *Development* **120**, 3083–3092.

389 Tajbakhsh, S. & Buckingham, M.E. (1994) Mouse limb muscle is determined in the absence of the earliest myogenic factor myf-5. *Proc. Natl Acad. Sci. USA* **91**, 747–751.

390 Buckingham, M. & Tajbakhsh, S. (1993) Expression of myogenic factors in the mouse: myf-5, the first member of the MyoD gene family to be trascribed during skeletal myogenesis. *C.R. Acad. Sci. III* **316**, 1032–1046.

391 Tajbakhsh, S., Vivarelli, E., Cusella, D.A.G., Rocancourt, D., Buckingham, M. & Cossu, G. (1994) A population of myogenic cells derived from the mouse neural tube. *Neuron* **13**, 813–821.

392 Rudnicki, M.A., Braun, T., Hinuma, S. & Jaenisch, R. (1992) Inactivation of MyoD in mice leads to up-regulation of the myogenic HLH gene Myf-5 and results in apparently normal muscle development. *Cell* **71**, 383–390.

393 Hasty, P., Bradley, A., Morris, J.H. *et al.* (1993) Muscle deficiency and neonatal death in mice with a targeted mutation in the myogenin gene. *Nature* **364**, 501–506.

394 Nabeshima, Y., Hanaoka, K., Hayasaka, M. *et al.* (1993) Myogenin gene disruption results in perinatal lethality because of severe muscle defect. *Nature* **364**, 532–535.

395 Stumpo, D.J., Bock, C.B., Tuttle, J.S. & Blackshear, P.J. (1995) MARCKS deficiency in mice leads to abnormal brain development and perinatal death. *Proc. Natl Acad. Sci. USA* **92**, 944–948.

396 Ioffe, E. & Stanley, P. (1994) Mice lacking N-acetylglucosaminyltransferase I activity die at mid-gestation, revealing an essential role for complex or hybrid N-linked carbohydrates. *Proc. Natl Acad. Sci. USA* **91**, 728–732.

397 Metzler, M., Gertz, A., Sarkar, M., Schachter, H., Schrader, J.W. & Marth, J.D. (1994) Complex asparagine-linked oligosaccharides are required for morphogenic events during post-implantation development. *EMBO J.* **13**, 2056–2065.

398 Wang, Z.-Q., Auer, B., Stingl, L. *et al.* (1995) Mice lacking ADPRT and ply(ADP-ribosyl)ation develop normally but are susceptible to skin disease. *Genes Dev.* **9**, 509–520.

399 Crowley, C., Spencer, S.D., Nishimura, M.C. *et al.* (1994) Mice lacking nerve growth factor display perinatal loss of sensory and sympathetic neurons yet develop basal forebrain cholinergic neurons. *Cell* **76**, 1001–1011.

400 Lee, K.F., Li, E., Huber, L.J. *et al.* (1992) Targeted mutation of the gene encoding the low affinity NGF receptor p75 leads to deficits in the peripheral sensory nervous system. *Cell* **69**, 737–749

401 Davies, A.M., Lee, K.F. & Jaenisch, R. (1993) p75-deficient trigeminal sensory neurons have an altered response to NGF but not to other neutrophins. *Neuron* **11**, 565–574.

402 Lee, K.F., Bachman, K., Landis, S. & Jaenisch, R. (1994) Dependence on p75 for innervation of some sympathetic targets. *Science* **263**, 1447–1449.

403 Cremer, H., Lange, R., Christoph, A. *et al.* (1994) Inactivation of the N-CAM gene in mice results in size reduction of the olfactory bulb and deficits in spatial learning. *Nature* **367**, 455–459.

404 Tomasiewicz, H., Ono, K., Yee, D. *et al.* (1993) Genetic deletion of a neural cell adhesion molecule variant (N-CAM-180) produces distinct defects in the central nervous system. *Neuron* **11**, 1163–1174.
405 Brannan, C.I., Perkins, A.S., Vogel, K.S. *et al.* (1994) Targeted disruption of the neurofibromatosis type-1 gene leads to developmental abnormalities in heart and various neural crest-derived tissues. *Genes Dev.* **8**, 1019–1029.
406 Jacks, T., Shih, T.S., Schmitt, E.M., Bronson, R.T., Bernards, A. & Weinberg, R.A. (1994) Tumour predisposition in mice heterozygous for a targeted mutation in Nf1. *Nature Genet.* **7**, 353–361.
407 Ernfors, P., Lee, K.F., Kucera, J. & Jaenisch, R. (1994) Lack of neurotrophin-3 leads to deficiencies in the peripheral nervous system and loss of limb proprioceptive afferents. *Cell* **77**, 503–512.
408 Tessarollo, L, Vogel, K.S., Palko, M.E., Reid, S.W. & Parada, L.F. (1994) Targeted mutation in the neutrophin-3 gene results in loss of muscle sensory neurons. *Proc. Natl Acad. Sci. USA* **91**, 11844–11848.
409 Farinas, I., Jones, K.R., Backus, C., Wang, X.Y. & Reichardt, L.F. (1994) Severe sensory and sympathetic deficits in mice lacking neurotropin-3. *Nature* **369**, 658–661.
410 Tanaka, T., Akira, S., Yoshida, K. *et al.* (1995) Targeted disruption of the NF-IL6 gene discloses its essential role in bacteria killing and tumor cytotoxicity by macrophages. *Cell* **80**, 353–361.
411 Sha, W.C., Liou, H.-C., Tuomanen, E.I. & Baltimore, D. (1995) Targeted disruption of the p50 subunit of NF-kappaB leads to multifocal defects in immune responses. *Cell* **80**, 321–330.
412 Huang, P.L., Dawson, T.M., Bredt, D.S., Snyder, S.H. & Fishman, M.C. (1993) Targeted disruption of the neuronal nitric oxide synthase gene. *Cell* **75**, 1273–1286.
413 Huang, Z., Huang, P.L., Panahian, N., Dalkara, T., Fishman, M.C. & Moskowitz, M.A. (1994) Effects of cerebral ischemia in mice deficient in neuronal nitric oxide synthase. *Science* **265**, 1883–1885.
414 O'Dell, T.J., Huang, P.L., Dawson, T.M. *et al.* (1994) Endothelial NOS and the blockade of LTP by NOS inhibitors in mice lacking neuronal NOS. *Science* **265**, 542–546.
415 Swiatek, P.J., Lindsell, C.E., del A.F.F., Weinmaster, G. & Gridley, T. (1994) Notch 1 is essential for post-implantation development in mice. *Genes Dev.* **8**, 707–719.
416 Corcoran, L.M., Karvelas, M., Nossal, G.J., Ye, Z.S., Jacks, T. & Baltimore, D. (1993) Oct-2, although not required for early B-cell development, is critical for later B-cell maturation and for postnatal survival. *Genes Dev.* **7**, 570–582.
417 Mansour, S.L., Goddard, J.M. & Capecchi, M.R. (1993) Mice homozygous for a targeted disruption of the proto-oncogene int-2 have develop mental defects in the tail and inner ear. *Development* **117**, 13–28.
418 Giese, K.P., Martini, R., Lemke, G., Soriano, P. & Schachner, M. (1992) Mouse P0 gene disruption leads to hypomyelination, abnormal expression of recognition molecules, and degeneration of myelin and axons. *Cell* **71**, 565–576.
419 Donehower, L.A., Harvey, M., Slagle, B.L. *et al.* (1992) Mice deficient for p53 are developmentally normal but susceptible to spontaneous tumours. *Nature* **356**, 215–221.
420 Livingstone, L.R., White, A., Sprouse, J., Livanos, E., Jacks, T. & Tlsty, T.D. (1992) Altered cell cycle arrest and gene amplification potential accompany loss of wild-type p53. *Cell* **70**, 923–935.
421 Clarke, A.R., Purdie, C.A., Harrison, D.J. *et al.* (1993) Thymocyte apoptosis induced by p53-dependent and independent pathways. *Nature* **362**, 849–852.
422 Harvey, M., McArthur, M.J., Montgomery, C.A.J., Bradley, A. & Donehower, L.A. (1993) Genetic background alters the spectrum of tumors that develop in p53-deficient mice. *FASEB J.* **7**, 938–943.
423 Harvey, M., McArthur, M.J., Montgomery, C.A.J., Butel, J.S., Bradley, A. & Donehower, L.A. (1993) Spontaneous and carcinogen-induced tumorigenesis in p53-deficient mice. *Nature Genet.* **5**, 225–229.
424 Harvey, M., Sands, A.T., Weiss, R.S. *et al.* (1993) *In vitro* growth characteristics of embryo fibroblasts isolated from p53-deficient mice. *Oncogene* **8**, 2457–2467.
425 Hursting, S.D., Perkins, S.N. & Phang, J.M. (1994) Calorie restriction delays spontaneous tumorigenesis in p53-knockout transgenic mice. *Proc. Natl Acad. Sci. USA* **91**, 7036–7040.
426 Kemp, C.J., Donehower, L.A., Bradley, A. & Balmain, A. (1993) Reduction of p53 gene dosage does not increase initiation or promotion but enhances malignant progression of chemically induced skin tumors. *Cell* **74**, 813–822.
427 Kemp, C.J., Wheldon, T. & Balmain, A. (1994) p53-deficient mice are extremely susceptible to radiation-induced tumorigenesis. *Nature Genet.* **8**, 66–69.
428 Lee, J.M., Abrahamson, J., Kandel, R., Donehower, L.A. & Bernstein, A. (1994) Susceptibility to radiation-carcinogenesis and accumulation of chromosomal breakage in p53 deficient mice. *Oncogene* **9**, 3731–3736.
429 Lotem, J. & Sachs, L. (1993) Haematopoietic cells from mice deficient in wild-type p53 are more resistant to induction of apoptosis by some agents. *Blood* **82**, 1092–1096.
430 Berges, R.R., Furuya, Y., Remington, L., English, H.F., Jacks, T. & Isaacs, J.T. (1993) Cell proliferation, DNA repair, and p53 function are not required for programmed death of prostatic glandular cells induced by androgen ablation. *Proc. Natl Acad. Sci. USA* **90**, 8910–8914.
431 Kastan, M.B., Zhan, Q., El-Deiry, W.S. *et al.* (1992) A mammalian cell cycle checkpoint pathway utilizing p53 and GADD45 is defective in ataxia-telangiectasia. *Cell* **71**, 587–597.
432 Lowe, S.W., Ruley, H.E., Jacks, T. & Housman, D.E. (1993) p53-dependent apoptosis modulates the cytotoxicity of anticancer agents. *Cell* **74**, 957–967.
433 Lowe, S.W., Schmitt, E.M., Smith, S.W., Osborne, B.A. & Jacks, T. (1993) p53 is required for radiation-induced apoptosis in mouse thymocytes. *Nature* **362**, 847–849.
434 Lowe, S.W., Jacks, T., Housman, D.E. & Ruley, H.E. (1994) Abrogation of oncogene-associated apoptosis

allows transformation of p53-deficient cells. *Proc. Natl Acad. Sci. USA* **91**, 2026–2030.

435 Symonds, H., Krall, L., Remington, L. *et al.* (1994) p53-dependent apoptosis suppresses tumor growth and progression in vivo. *Cell* **78**, 703–711.

436 Morgenbesser, S.D., Williams, B.O., Jacks, T. & DePinho, R.A. (1994) p53-dependent apoptosis produced by Rb-deficiency in the developing mouse lens. *Nature* **371**, 72–74.

437 Clarke, A.R., Gledhill, S., Hooper, M.L., Bird, C.C. & Wyllie, A.H. (1994) p53 dependence of early apoptotic and proliferative responses within the mouse intestinal epithelium following gamma-irradiation. *Oncogene* **9**, 1767–1773.

438 Purdie, C.A., Harrison, D.J., Peter, A. *et al.* (1994) Tumour incidence, spectrum and ploidy in mice with a large deletion in the p53 gene. *Oncogene* **9**, 603–609.

439 Karaplis, A.C., Luz, A., Glowacki, J. *et al.* (1994) Lethal skeletal dysplasia from targeted disruption of the parathyroid hormone-related peptide gene. *Genes Dev.* **8**, 277–289.

440 Lowin, B., Beermann, F., Schmidt, A. & Tschopp, J. (1994) A null mutation in the perforin gene impairs cytolytic T lymphocyte- and natural killer cell-mediated cytotoxicity. *Proc. Natl Acad. Sci. USA* **91**, 11571–11575.

441 Walsh, C.M., Matloubian, M., Liu, C.C. *et al.* (1994) Immune function in mice lacking the perforin gene. *Proc. Natl Acad. Sci. USA* **91**, 10854–10858.

442 Kagi, D., Ledermann, B., Burki, K. *et al.* (1994) Cytotoxicity mediated by T cells and natural killer cells is greatly impaired in perforin-deficient mice. *Nature* **369**, 31–37.

443 Lowin, B., Hahne, M., Mattmann, C. & Tschopp, J. (1994) Cytolytic T-cell cytotoxicity is mediated through perforin and Fas lytic pathways. *Nature* **370**, 650–652.

444 Laird, P.W., van der Lugt, N.M.T., Clarke, A. *et al.* (1993) In vivo analysis of Pim-1 deficiency. *Nucleic Acids Res.* **21**, 4750–4755.

445 Domen, J., van der Lugt, N.M.T., Acton, D., Laird, P.W., Linders, K. & Berns, A. (1993) Pim-1 levels determine the size of early B lymphoid compartments in bone marrow. *J. Exp. Med.* **178**, 1665–1673.

446 Domen, J., van der Lugt, N.M.T., Laird, P.W., Saris, C.J. & Berns, A. (1993) Analysis of Pim-1 function in mutant mice. *Leukemia* **7** (Suppl. 2), 108–112.

447 Domen, J., van der Lugt, N.M.T., Laird, P.W. *et al.* (1993) Impaired interleukin-3 response in Pim-1-deficient bone marrow-derived mast cells. *Blood* **82**, 1445–1452.

448 Carmeliet, P., Kieckens, L., Schoonjans, L. *et al.* (1993) Plasminogen activator inhibitor-1 gene-deficient mice. I. Generation by homologous recombination and characterization. *J. Clin. Invest.* **92**, 2746–2755.

449 Carmeliet, P., Stassen, J.M., Schoonjans, L. *et al.* (1993) Plasminogen activator inhibitor-1 gene-deficient mice. II. Effects on hemostasis, thrombosis, and thrombolysis. *J. Clin. Invest.* **92**, 2756–2760.

450 Levéen, P., Pekny, M., Gebre, M.S., Swolin, B., Larsson, E. & Betsholtz, C. (1994) Mice deficient for PDGF B show renal, cardiovascular, and haematological abnormalities. *Genes Dev.* **8**, 1875–1887.

451 Soriano, P. (1994) Abnormal kidney development and haematological disorders in PDGF beta-receptor mutant mice. *Genes Dev.* **8**, 1888–1896.

452 Bueler, H., Fischer, M., Lang, Y. *et al.* (1992) Normal development and behaviour of mice lacking the neuronal cell-surface PrP protein. *Nature* **356**, 577–582.

453 Bueler, H., Aguzzi, A., Sailer, A. *et al.* (1993) Mice devoid of PrP are resistant to scrapie. *Cell* **73**, 1339–1347.

454 Prusiner, S.B., Groth, D., Serban, A. *et al.* (1993) Ablation of the prion protein (PrP) gene in mice prevents scrapie and facilitates production of anti-PrP antibodies. *Proc. Natl Acad. Sci. USA* **90**, 10608–10612.

455 Collinge, J., Whittington, M.A., Sidle, K.C. *et al.* (1994) Prion protein is necessary for normal synaptic function. *Nature* **370**, 295–297.

456 Weissmann, C., Bueler, H., Fischer, M. & Aguet, M. (1993) Role of the PrP gene in transmissible spongiform encephalopathies. *Intervirology* **35**, 164–175.

457 Weissmann, C., Bueler, H., Fischer, M., Sauer, A. & Aguet, M. (1994) Susceptibility to scrapie in mice is dependent on PrPC. *Philos. Trans. R. Soc. Lond. B* **343**, 431–433.

458 Sailer, A., Bueler, H., Fischer, M., Aguzzi, A. & Weissmann, C. (1994) No propagation of prions in mice devoid of PrP. *Cell* **77**, 967–968.

459 Whittington, M.A., Sidle, K.C.L., Gowland, I. *et al.* (1995) Rescue of neurophysiological phenotype seen in PrP null mice by transgene encoding human prion protein. *Nature Genet.* **9**, 197–201.

460 Abeliovich, A., Paylor, R., Chen, C., Kim, J.J., Wehner, J.M. & Tonegawa, S. (1993) PKC gamma mutant mice exhibit mild deficits in spatial and contextual learning. *Cell* **75**, 1263–1271.

461 Abeliovich, A., Chen, C., Goda, Y., Silva, A.J., Stevens, C.F. & Tonegawa, S. (1993) Modified hippocampal long-term potentiation in PKC gamma-mutant mice. *Cell* **75**, 1253–1262.

462 Boison, D. & Stoffel, W. (1994) Disruption of the compacted myelin sheath of axons of the central nervous system in proteolipid protein-deficient mice. *Proc. Natl Acad. Sci. USA* **91**, 11709–11713.

463 Scott, E.W., Simon, M.C., Anastasi, J. & Singh, H. (1994) Requirement of transcription factor PU.1 in the development of multiple haematopoietic lineages. *Science* **265**, 1573–1577.

464 Geppert, M., Bolshakov, V.Y., Siegelbaum, S.A. *et al.* (1994) The role of Rab3A in neurotransmitter release. *Nature* **369**, 493–497.

465 Umanoff, H., Edelmann, W., Pellicer, A. & Kucherlapati, R. (1995) The murine N-ras gene is not essential for growth and development. *Proc. Natl Acad. Sci. USA* **92**, 1709–1713.

466 Warren, A.J., Colledge, W.H., Carlton, M.B., Evans, M.J., Smith, A.J. & Rabbitts, T.H. (1994) The oncogenic cysteine-rich LIM domain protein rbtn2 is essential for erythroid development. *Cell* **78**, 45–57.

467 Mombaerts, P., Iacomini, J., Johnson, R.S., Herrup, K., Tonegawa, S. & Papaioannou, V.E. (1992) RAG-1-deficient mice have no mature B and T lymphocytes. *Cell* **68**, 869–877.

468 Spanopoulou, E., Roman, C.A., Corcoran, L.M. *et al.* (1994) Functional immunoglobulin transgenes guide ordered B-cell differentiation in Rag-1-deficient mice. *Genes Dev.* **8**, 1030–1042.
469 Levelt, C.N., Mombaerts, P., Iglesias, A., Tonegawa, S. & Eichmann, K. (1993) Restoration of early thymocyte differentiation in T-cell receptor beta-chain-deficient mutant mice by transmembrane signaling through CD3 epsilon. *Proc. Natl Acad. Sci. USA* **90**, 11401–11405.
470 Lafaille, J.J., Nagashima, K., Katsuki, M. & Tonegawa, S. (1994) High incidence of spontaneous autoimmune encephalomyelitis in immunodeficient anti-myelin basic protein T cell receptor transgenic mice. *Cell* **78**, 399–408.
471 Mombaerts, P., Clarke, A.R., Rudnicki, M.A. *et al.* (1992) Mutations in T-cell antigen receptor genes alpha and beta block thymocyte development at different stages. *Nature* **360**, 255–231 (published erratum appears in *Nature* **360**, 491).
472 Godfrey, D.I., Kennedy, J., Mombaerts, P., Tonegawa, S. & Zlotnik, A. (1994) Onset of TCR-beta gene rearrangement and role of TCR-beta expression during CD3⁻CD4⁻CD8⁻ thymocyte differentiation. *J. Immunol.* **152**, 4783–4792.
473 Jacobs, H., Vandeputte, D., Tolkamp, L., de Vries, E., Borst, J. & Berns, A. (1994) CD3 components at the surface of pro-T cells can mediate pre-T cell development in vivo. *Eur. J. Immunol.* **24**, 934–939.
474 Shinkai, Y., Rathbun, G., Lam, K.P. *et al.* (1992) RAG-2-deficient mice lack mature lymphocytes owing to inability to initiate V(D)J rearrangement. *Cell* **68**, 855–867.
475 Chen, J., Lansford, R., Stewart, V., Young, F. & Alt, F.W. (1993) RAG-2-deficient blastocyst complementation: an assay of gene function in lymphocyte development. *Proc. Natl Acad. Sci. USA* **90**, 4528–4532.
476 Rodewald, H.R., Awad, K., Moingeon, P. *et al.* (1993) Fc gamma RII/III and CD2 expression mark distinct subpopulations of immature CD4⁻CD8⁻ murine thymocytes: *in vivo* developmental kinetics and T cell receptor beta chain rearrangement status. *J. Exp. Med.* **177**, 1079–1092.
477 Bottaro, A., Lansford, R., Xu, L., Zhang, J., Rothman, P. & Alt, F.W. (1994) S region transcription per se promotes basal IgE class switch recombination but additional factors regulate the efficiency of the process. *EMBO J.* **13**, 665–674.
478 Shinkai, Y., Koyasu, S., Nakayama, K. *et al.* (1993) Restoration of T cell development in RAG-2-deficient mice by functional TCR transgenes. *Science* **259**, 822–825.
479 Castigli, E., Alt, F.W., Davidson, L. *et al.* (1994) CD40-deficient mice generated by recombination-activating gene-2-deficient blastocyst complementation. *Proc. Natl Acad. Sci. USA* **91**, 12135–12139.
480 Weih, F., Carrasco, D., Durham, S.K. *et al.* (1995) Multi-organ inflammation and haematopoietic abnormalities in mice with a targeted disruption of RelB, a member of the NF-kappaB/Rel family. *Cell* **80**, 331–340.
481 Burkly, L., Hession, C., Ogata, L. *et al.* (1995) Expression of relB is required for the development of thymic medulla and dendritic cells. *Nature* **373**, 531–536.
482 Schuchardt, A., D'Agati, V., Larsson, B.L., Costantini, F. & Pachnis, V. (1994) Defects in the kidney and enteric nervous system of mice lacking the tyrosine kinase receptor Ret. *Nature* **367**, 380–383.
483 Jacks, T., Fazeli, A., Schmitt, E.M., Bronson, R.T., Goodell, M.A. & Weinberg, R.A. (1992) Effects of an Rb mutation in the mouse. *Nature* **359**, 295–300.
484 Clarke, A.R., Maandag, E.R., van Roon, M. *et al.* (1992) Requirement for a functional Rb-1 gene in murine development. *Nature* **359**, 328–330.
485 Lee, E.Y., Chang, C.Y., Hu, N. *et al.* (1992) Mice deficient for Rb are nonviable and show defects in neurogenesis and haematopoiesis. *Nature* **359**, 288–294.
486 Hu, N., Gutsmann, A., Herbert, D.C., Bradley, A., Lee, W.H. & Lee, E.Y. (1994) Heterozygous Rb-1 delta 20/+ mice are predisposed to tumors of the pituitary gland with a nearly complete penetrance. *Oncogene* **9**, 1021–1027.
487 Lee, E.Y., Hu, N., Yuan, S.S. *et al.* (1994) Dual roles of the retinoblastoma protein in cell cycle regulation and neuron differentiation. *Genes Dev.* **8**, 2008–2021.
488 Lufkin, T., Lohnes, D., Mark, M. *et al.* (1993) High postnatal lethality and testis degeneration in retinoic acid receptor alpha mutant mice. *Proc. Natl Acad. Sci. USA* **90**, 7225–7229.
489 Li, E., Sucov, H.M., Lee, K.F., Evans, R.M. & Jaenisch, R. (1993) Normal development and growth of mice carrying a targeted disruption of the alpha 1 retinoic acid receptor gene. *Proc. Natl Acad. Sci. USA* **90**, 1590–1594.
490 Mendelsohn, C., Mark, M., Dollé, P. *et al.* (1994) Retinoic acid receptor beta2 (RARbeta2) null mutant mice appear normal. *Dev. Biol.* **166**, 246–258.
491 Lohnes, D., Kastner, P., Dierich, A., Mark, M., LeMeur, M. & Chambon, P. (1993) Function of retinoic acid receptor gamma in the mouse. *Cell* **73**, 643–658.
492 Sucov, H.M., Dyson, E., Gumeringer, C.L., Price, J., Chien, K.R. & Evans, R.M. (1994) RXR alpha mutant mice establish a genetic basis for vitamin A signaling in heart morphogenesis. *Genes Dev.* **8**, 1007–1018.
493 Kastner, P., Grondona, J.M., Mark, M. *et al.* (1994) Genetic analysis of RXR alpha developmental function: convergence of RXR and RAR signaling pathways in heart and eye morphogenesis. *Cell* **78**, 987–1003.
494 Takeshima, H., Iino, M., Takekura, H. *et al.* (1994) Excitation-contraction uncoupling and muscular degeneration in mice lacking functional skeletal muscle ryanodine-receptor gene. *Nature* **369**, 556–559.
495 Arbones, M.L., Ord, D.C., Ley, K. *et al.* (1994) Lymphocyte homing and leukocyte rolling and migration are impaired in L-selectin deficient mice. *Immunity* **1**, 247–260.
496 Mayadas, T.N., Johnson, R.C., Rayburn, H., Hynes, R.O. & Wagner, D.D. (1993) Leukocyte rolling and extravasation are severely compromised in P selectin-deficient mice. *Cell* **74**, 541–554.
497 Saudou, F., Amara, D.A., Dierich, A. *et al.* (1994) Enhanced aggressive behavior in mice lacking 5-HT1B receptor. *Science* **265**, 1875–1878.
498 Luo, X., Ikeda, Y. & Parker, K.L. (1994) A cell-specific

nuclear receptor is essential for adrenal and gonadal development and sexual differentiation. *Cell* **77**, 481–490.

499 Ingraham, H.A., Lala, D.S., Ikeda, Y. *et al.* (1994) The nuclear receptor steroidogenic factor 1 acts at multiple levels of the reproductive axis. *Genes Dev.* **8**, 2302–2312.

500 Jung, S., Rajewsky, K. & Radbruch, A. (1993) Shutdown of class switch recombination by deletion of a switch region control element. *Science* **259**, 984–987.

501 Soriano, P., Montgomery, C., Geske, R. & Bradley, A. (1991) Targeted disruption of the c-src proto-oncogene leads to osteopetrosis in mice. *Cell* **64**, 693–702.

502 Thomas, J.E., Soriano, P. & Brugge, J.S. (1991) Phosphorylation of c-Src on tyrosine 527 by another protein tyrosine kinase. *Science* **254**, 568–571.

503 Boyce, B.F., Chen, H., Soriano, P. & Mundy, G.R. (1993) Histomorphometric and immunocytochemical studies of src-related osteopetrosis. *Bone* **14**, 335–340.

504 Lowe, C., Yoneda, T., Boyce, B.F., Chen, H., Mundy, G.R. & Soriano, P. (1993) Osteopetrosis in Src-deficient mice is due to an autonomous defect of osteoclasts. *Proc. Natl Acad. Sci. USA* **90**, 4485–4489.

505 Guy, C.T., Muthuswamy, S.K., Cardiff, R.D., Soriano, P. & Muller, W.J. (1994) Activation of the c-Src tyrosine kinase is required for the induction of mammary tumors in transgenic mice. *Genes Dev.* **8**, 23–32.

506 Thomas, J.E., Aguzzi, A., Soriano, P., Wagner, E.F. & Brugge, J.S. (1993) Induction of tumor formation and cell transformation by polyoma middle T antigen in the absence of Src. *Oncogene* **8**, 2521–2529.

507 Boyce, B.F., Yoneda, T., Lowe, C., Soriano, P. & Mundy, G.R. (1992) Requirement of pp60c-src expression for osteoclasts to form ruffled borders and resorb bone in mice. *J. Clin. Invest.* **90**, 1622–1627.

508 Kohmura, N., Yagi, T., Tomooka, Y. *et al.* (1994) A novel nonreceptor tyrosine kinase, Srm: cloning and targeted disruption. *Mol. Cell Biol.* **14**, 6915–6925.

509 Rosahl, T.W., Geppert, M., Spillane, D. *et al.* (1993) Short-term synaptic plasticity is altered in mice lacking synapsin I. *Cell* **75**, 661–670.

510 Geppert, M., Goda, Y., Hammer, R.E. *et al.* (1994) Synaptotagmin I: a major Ca sensor for transmitter release at a central synapse. *Cell* **79**, 717–727.

511 Verbeek, S., Izon, D., Hofhuis, F. *et al.* (1995) An HMG-box-containing T-cell factor required for thymocyte differentiation. *Nature* **374**, 70–74.

512 Philpott, K.L., Viney, J.L., Kay, G. *et al.* (1992) Lymphoid development in mice congenitally lacking T cell receptor αβ-expressing cells. *Science* **256**, 1448–1452.

513 Mombaerts, P., Arnoldi, J., Russ, F., Tonegawa, S. & Kaufmann, S.H. (1993) Different roles of αβ and γδ T cells in immunity against an intracellular bacterial pathogen. *Nature* **365**, 53–56.

514 Tsuji, M., Mombaerts, P., Lefrancois, L., Nussenzweig, R.S., Zavala, F. & Tonegawa, S. (1994) Gamma delta T cells contribute to immunity against the liver stages of malaria in αβ T-cell-deficient mice. *Proc. Natl Acad. Sci. USA* **91**, 345–349.

515 Wen, L., Roberts, S.J., Viney, J.L. *et al.* (1994) Immunoglobulin synthesis and generalized autoimmunity in mice congenitally deficient in αβ(+) T cells. *Nature* **369**, 654–658.

516 Viney, J.L., Dianda, L., Roberts, S.J. *et al.* (1994) Lymphocyte proliferation in mice congenitally deficient in T-cell receptor $\alpha\beta^+$ cells. *Proc. Natl Acad. Sci. USA* **91**, 11948–11952.

517 Itohara, S., Mombaerts, P., Lafaille, J. *et al.* (1993) T cell receptor delta gene mutant mice: independent generation of alpha beta T cells and programmed rearrangements of gamma delta TCR genes. *Cell* **72**, 337–348.

518 Ohno, H., Goto, S., Taki, S. *et al.* (1994) Targeted disruption of the CD3 eta locus causes high lethality in mice: modulation of Oct-1 transcription on the opposite strand. *EMBO J.* **13**, 1157–1165.

519 Koyasu, S., Hussey, R.E., Clayton, L.K. *et al.* (1994) Targeted disruption within the CD3 zeta/eta/phi/Oct-1 locus in mouse. *EMBO J.* **13**, 784–797.

520 Love, P.E., Shores, E.W., Johnson, M.D. *et al.* (1993) T cell development in mice that lack the zeta chain of the T cell antigen receptor complex. *Science* **261**, 918–921.

521 Liu, C.-P., Ueda, R., She, J. *et al.* (1993) Abnormal T cell development in CD3-zeta$^{-/-}$ mutant mice and identification of a novel T cell population in the intestine. *EMBO J.* **12**, 4863–4875.

522 Ohno, H., Aoe, T., Taki, S. *et al.* (1993) Developmental and functional impairment of T cells in mice lacking CD3 zeta chains. *EMBO J.* **12**, 4357–4366.

523 Shores, E.W., Huang, K., Tran, T., Lee, E., Grinberg, A. & Love, P.E. (1994) Role of TCR zeta chain in T cell development and selection. *Science* **266**, 1047–1050.

524 Malissen, M., Gillet, A., Rocha, B. *et al.* (1993) T cell development in mice lacking the CD3-zeta/eta gene. *EMBO J.* **12**, 4347–4355.

525 Shivdasani, R.A., Mayer, E.L. & Orkin, S.H. (1995) Absence of blood formation in mice lacking the T-cell leukemia oncoprotein tal-1/SCL. *Nature* **373**, 432–434.

526 Harada, A., Oguchi, K., Okabe, S. *et al.* (1994) Altered microtubule organization in small-calibre axons of mice lacking tau protein. *Nature* **369**, 488–491.

527 Dumont, D.J., Gradwohl, G., Fong, G.H. *et al.* (1994) Dominant-negative and targeted null mutations in the endothelial receptor tyrosine kinase, tek, reveal a critical role in vasculogenesis of the embryo. *Genes Dev.* **8**, 1897–1909.

528 Saga, Y., Yagi, T., Ikawa, Y., Sakakura, T. & Aizawa, S. (1992) Mice develop normally without tenascin. *Genes Dev.* **6**, 1821–1831.

529 Steindler, D.A., Settles, D., Erickson, H.P. *et al.* (1995) Tenascin knockout mice: barrels, boundary molecules, and glial scars. *J. Neurosci.* **15**, 1971–1983.

530 Gilfillan, S., Dierich, A., Lemeur, M., Benoist, C. & Mathis, D. (1993) Mice lacking TdT: mature animals with an immature lymphocyte repertoire. *Science* **261**, 1175–1178 (published erratum appears in *Science* **262**, 1957).

531 Healy, A.M., Rayburn, H.B., Rosenberg, R.D. & Weiler, H. (1995) Absence of the blood-clotting regulator thrombomodulin causes embryonic lethality in mice before development of a functional cardiovascular system. *Proc. Natl Acad. Sci. USA* **92**, 850–854.

532 Carmeliet, P., Schoonjans, L., Kieckens, L. *et al.* (1994)

Physiological consequences of loss of plasminogen activator gene function in mice. *Nature* **368**, 419–424.
533 Kulkarni, A.B., Huh, C.G., Becker, D. *et al.* (1993) Transforming growth factor beta 1 null mutation in mice causes excessive inflammatory response and early death. *Proc. Natl Acad. Sci. USA* **90**, 770–774.
534 Shull, M.M., Ormsby, I., Kier, A.B. *et al.* (1992) Targeted disruption of the mouse transforming growth factor-beta 1 gene results in multifocal inflammatory disease. *Nature* **359**, 693–699.
535 Christ, M., McCartney-Francis, N.L., Kulkarni, A.B. *et al.* (1994) Immune dysregulation in TGF-beta 1-deficient mice. *J. Immunol.* **153**, 1936–1946.
536 Geiser, A.G., Letterio, J.J., Kulkarni, A.B., Karlsson, S., Roberts, A.B. & Sporn, M.B. (1993) Transforming growth factor beta 1 (TGF-beta 1) controls expression of major histocompatibility genes in the postnatal mouse: aberrant histocompatibility antigen expression in the pathogenesis of the TGF-beta 1 null mouse phenotype. *Proc. Natl Acad. Sci. USA* **90**, 9944–9948.
537 Glick, A.B., Kulkarni, A.B., Tennenbaum, T. *et al.* (1993) Loss of expression of transforming growth factor beta in skin and skin tumors is associated with hyperproliferation and a high risk for malignant conversion. *Proc. Natl Acad. Sci. USA* **90**, 6076–6080.
538 Letterio, J.J., Geiser, A.G., Kulkarni, A.B., Roche, N.S., Sporn, M.B. & Roberts, A.B. (1994) Maternal rescue of transforming growth factor-beta 1 null mice. *Science* **264**, 1936–1938.
539 Glick, A.B., Lee, M.M., Darwiche, N., Kulkarni, A.B., Karlsson, S. & Yuspa, S.H. (1994) Targeted deletion of the TGF-beta 1 gene causes rapid progression to squamous cell carcinoma. *Genes Dev.* **8**, 2429–2440.
540 Hines, K.L., Kulkarni, A.B., McCarthy, J.B. *et al.* (1994) Synthetic fibronectin peptides interrupt inflammatory cell infiltration in transforming growth factor beta 1 knockout mice. *Proc. Natl Acad. Sci. USA* **91**, 5187–5191.
541 Van Kaer, L., Ashton-Rickardt, P.G., Ploegh, H.L. & Tonegawa, S. (1992) TAP1 mutant mice are deficient in antigen presentation, surface class I molecules, and CD4⁻8⁺ T cells. *Cell* **71**, 1205–1214.
542 Ashton-Rickardt, P.G., Bandeira, A., Delaney, J.R. *et al.* (1994) Evidence for a differential avidity model of T cell selection in the thymus. *Cell* **76**, 651–663.
543 Moriwaki, S., Korn, B.S., Ichikawa, Y., Van Kaer, L. & Tonegawa, S. (1993) Amino acid substitutions in the floor of the putative antigen-binding site of H-2T22 affect recognition by a gamma delta T-cell receptor. *Proc. Natl Acad. Sci. USA* **90**, 11396–11400.
544 Ashton-Rickardt, P.G., Van Kaer, L., Schumacher, T.N., Ploegh, H.L. & Tonegawa, S. (1993) Peptide contributes to the specificity of positive selection of CD8⁺ T cells in the thymus. *Cell* **73**, 1041–1049.
545 Ljunggren, H.G., Van Kaer, L., Ploegh, H.L. & Tonegawa, S. (1994) Altered natural killer cell repertoire in Tap-1 mutant mice. *Proc. Natl Acad. Sci. USA* **91**, 6520–6524.
546 Aldrich, C.J., Ljunggren, H.G., Van Kaer, L., Ashton-Rickardt, P.G., Tonegawa, S. & Forman, J. (1994) Positive selection of self- and alloreactive CD8⁺ T cells in Tap-1 mutant mice. *Proc. Natl Acad. Sci. USA* **91**, 6525–6528.
547 Episkopou, V., Maeda, S., Nishiguchi, S. *et al.* (1993) Disruption of the transthyretin gene results in mice with depressed levels of plasma retinol and thyroid hormone. *Proc. Natl Acad. Sci. USA* **90**, 2375–2379.
548 Wei, S., Episkopou, V., Piantedosi, R. *et al.* (1995) Studies on the metabolism of retinol and retinol-binding protein in transthyretin-deficient mice produced by homologous recombination. *J. Biol. Chem.* **270**, 866–870.
549 Palha, J.A., Episkopou, V., Maeda, S., Shimada, K., Gottesman, M.E. & Saraiva, M.J.M. (1994) Thyroid hormone metabolism in a transthyretin-null mouse strain. *J. Biol. Chem.* **269**, 33135–33139.
550 Smeyne, R.J., Klein, R., Schnapp, A. *et al.* (1994) Severe sensory and sympathetic neuropathies in mice carrying a disrupted Trk/NGF receptor gene. *Nature* **368**, 246–249.
551 Klein, R., Smeyne, R.J., Wurst, W. *et al.* (1993) Targeted disruption of the *trkB* neurotrophin receptor gene results in nervous system lesions and neonatal death. *Cell* **75**, 113–122.
552 Klein, R., Silos, S.I., Smeyne, R.J. *et al.* (1994) Disruption of the neurotrophin-3 receptor gene *trkC* eliminates Ia muscle afferents and results in abnormal movements. *Nature* **368**, 249–251.
553 Rothe, J., Lesslauer, W., Lotscher, H. *et al.* (1993) Mice lacking the tumour necrosis factor receptor 1 are resistant to TNF-mediated toxicity but highly susceptible to infection by *Listeria monocytogenes*. *Nature* **364**, 798–802.
554 Pfeffer, K., Matsuyama, T., Kundig, T.M. *et al.* (1993) Mice deficient for the 55 kd tumor necrosis factor receptor are resistant to endotoxic shock, yet succumb to *L. monocytogenes* infection. *Cell* **73**, 457–467.
555 Mackay, F., Rothe, J., Bluethmann, H., Loetscher, H. & Lesslauer, W. (1994) Differential responses of fibroblasts from wild-type and TNF-R55-deficient mice to mouse and human TNF-alpha activation. *J. Immunol* **153**, 5274–5284.
556 Erickson, S.L., De Sauvage, F.J., Kikly, K. *et al.* (1994) Decreased sensitivity to tumour-necrosis factor but normal T-cell development in TNF receptor-2-deficient mice. *Nature* **372**, 560–563.
557 De Togni, P., Goellner, J., Ruddle, N.H. *et al.* (1994) Abnormal development of peripheral lymphoid organs in mice deficient in lymphotoxin. *Science* **264**, 703–707.
558 Wu, X., Wakamiya, M., Vaishnav, S. *et al.* (1994) Hyperuricemia and urate nephropathy in urate oxidasedeficient mice. *Proc. Natl Acad. Sci. USA* **91**, 742–746.
559 Gurtner, G.C., Davis, V., Li, H., McCoy, M.J., Sharpe, A. & Cybulsky, M.I. (1995) Targeted disruption of the murine VCAM1 gene: essential role of VCAM-1 in chorioallantoic fusion and placentation. *Genes Dev.* **9**, 1–14.
560 Zmuidzinas, A., Fischer, K.D., Lira, S.A. *et al.* (1995) The vav proto-oncogene is required early in embryogenesis but not for haematopoietic development in vitro. *EMBO J.* **14**, 1–11.
561 Colucci, G.E., Portier, M.M., Dunia, I., Paulin, D., Pournin, S. & Babinet, C. (1994) Mice lacking vimentin develop and reproduce without an obvious phenotype. *Cell* **79**, 679–694.

562 Kreidberg, J.A., Sariola, H., Loring, J.M. *et al.* (1993) WT-1 is required for early kidney development. *Cell* **74**, 679–691.
563 McMahon, A.P. & Bradley, A. (1990) The Wnt-1 (int-1) proto-oncogene is required for development of a large region of the mouse brain. *Cell* **62**, 1073–1085.
564 Thomas, K.R. & Capecchi, M.R. (1990) Targeted disruption of the murine int-1 proto-oncogene resulting in severe abnormalities in midbrain and cerebellar development. *Nature* **346**, 847–850.
565 McMahon, A.P., Joyner, A.L., Bradley, A. & McMahon, J.A. (1992) The midbrain-hindbrain phenotype of Wnt-1$^-$/Wnt-1$^-$ mice results from stepwise deletion of *engrailed*-expressing cells by 9.5 days postcoitum. *Cell* **69**, 581–595.
566 Shimamura, K., Hirano, S., McMahon, A.P. & Takeichi, M. (1994) *Wnt-1*-dependent regulation of local E-cadherin and alphaN-catenin expression in the embryonic mouse brain. *Development* **120**, 2225–2234.
567 Takada, S., Stark, K.L., Shea, M.J., Vassileva, G., McMahon, J.A. & McMahon, A.P. (1994) Wnt-3a regulates somite and tailbud formation in the mouse embryo. *Genes Dev.* **8**, 174–189.
568 Stark, K., Vainio, S., Vassileva, G. & McMahon, A.P. (1994) Epithelial transformation of metanephric mesenchyme in the developing kidney regulated by Wnt-4. *Nature* **372**, 679–683.
569 Parr, B.A. & McMahon, A.P. (1995) Dorsalizing signal Wnt-7a required for normal polarity of D–V and A–P axes of mouse limb. *Nature* **374**, 350–353.
570 Stein, P.L., Vogel, H. & Soriano, P. (1994) Combined deficiencies of Src, Fyn, and Yes tyrosine kinases in mutant mice. *Genes Dev.* **8**, 1999–2007.
571 Schilham, M.W., Fung, L.W.P., Rahemtulla, A. *et al.* (1993) Alloreactive cytotoxic T cells can develop and function in mice lacking both CD4 and CD8. *Eur. J. Immunol.* **23**, 1299–1304.
572 Yeung, R., Penninger, J.M., Kündig, T.M. *et al.* (1994) Human CD4-major histocompatibility complex class II (DQw6) transgenic mice in an endogenous CD4/CD8-deficient background: reconstitution of phenotype and human-restricted function. *J. Exp. Med.* **180**, 1911–1920.
573 Condie, B.G. & Capecchi, M.R. (1994) Mice with targeted disruptions in the paralogous genes *hoxa-3* and *hoxd-3* reveal synergistic interactions. *Nature* **370**, 304–307.
574 Shikone, T., Matzuk, M.M., Perlas, E. *et al.* (1994) Characterization of gonadal sex cord-stromal tumor cell lines from inhibin-alpha and p53-deficient mice: the role of activin as an autocrine growth factor. *Mol. Endocrinol.* **8**, 983–995.
575 Sadlack, B., Kuhn, R., Schorle, H., Rajewsky, K., Muller, W. & Horak, I. (1994) Development and proliferation of lymphocytes in mice deficient for both interleukins-2 and -4. *Eur. J. Immunol.* **24**, 281–284.
576 Grusby, M.J., Auchincloss, H.J., Lee, R. *et al.* (1993) Mice lacking major histocompatibility complex class I and class II molecules. *Proc. Natl Acad. Sci. USA* **90**, 3913–3917.
577 Dierich, A., Chan, S.H., Benoist, C. & Mathis, D. (1993) Graft rejection by T cells not restricted by conventional major histocompatibility complex molecules. *Eur. J. Immunol.* **23**, 2725–2728.
578 Rudnicki, M.A., Schnegelsberg, P.N., Stead, R.H., Braun, T., Arnold, H.H. & Jaenisch, R. (1993) MyoD or Myf-5 is required for the formation of skeletal muscle. *Cell* **75**, 1351–1359.
579 Lohnes, D., Mark, M., Mendelsohn, C. *et al.* (1994) Function of the retinoic acid receptors (RARs) during development (I). *Development* **120**, 2723–2748.
580 Mendelsohn, C., Lohnes, D., Décimo, D. *et al.* (1994) Function of the retinoic acid receptors (RARs) during development (II). *Development* **120**, 2749–2771.

Appendix IX Chromosome aberrations associated with cancer

Tables IX.1–IX.10 refer to haematological malignancies (see Chapter 7); Tables IX.11 and IX.12 refer to solid tumours (see Chapter 8).

Table IX.1 Chromosome abnormalities in acute myeloid leukaemia (AML).

Chromosomal abnormality	Association with disease
der (1)t(1;7)(p11;p11)	Secondary AML mostly M4[a]
ins(3;3)(q26;q21q26)	
inv(3)(q21q26)	
t(3;3)(q21;q26)	Abnormal megakaryocytes and thrombocytosis
trisomy 4	M2 and M4 or M5
monosomy 5	Secondary AML
del(5q)	Secondary AML
t(6;9)(p23;q34)	M2 and M4 (see Fig. 7.6 in Chapter 7)
monosomy 7	Secondary AML
del(7q)	Secondary AML
trisomy 8	Myeloid disease
t(8;21)(q22;q22)	M2 with Auer rods, eosinophilia

Continued on p. 978.

Table IX.1 *Continued.*

Chromosomal abnormality	Association with disease
t(9;11)(p21;q23)	M5 mostly M5a
t(9;22)(q34;q11)	M1 and M2[b]
t(10;11)(p14;q13)	M4 and M5
del/t(11q)	M4, M5, mostly M5a
del/t(12p)	Secondary AML
t(15;17)(q22;q21)	M3 and M3v (see Fig. 7.8 in Chapter 7)
t(16;16)(p13;q22)	M4Eo
inv(16)(p13q22)	M6
del(16)(q22q24)	
del(20q)	
missing Y	Age-related phenomenon

[a]M1, M2, etc. are types of acute myeloid leukaemia; see Table IX.2.
[b]This translocation is mostly associated with CML/CGL but is also found in AML and ALL.

Table IX.2 French–American–British classification of acute myeloid leukaemia.

Classification	Description
M1	Myeloblastic without maturation
M2	Myeloblastic with maturation
M3	Promyelocytic (hypergranular)
M3 variant	Promyelocytic (hypo- or microgranular)
M4	Myelomonocytic
M4Eo	M4 with eosinophilia
M5a	Monoblastic
M5b	Promonocytic-monocytic
M6	Erythroblastic ($<30\%$ blasts if $>50\%$ erythrocytes)
M7	Megakaryoblastic
M0	Myeloblastic with minimal differentiation

Table IX.3 Chromosome changes in acute lymphoid leukaemia (ALL).

Chromosomal abnormality	Association (if any)
t(8;14)(q24;q32)	L3[a] poor prognosis
t(8;22)(q24;q11)	L3 poor prognosis
t(2;8)(p12;q24)	L3 poor prognosis
duplications of 1q	L3
t(1;19)(q23;p13)	Pre B-cell ALL L1
t(1;11)(p32;q23)	Pre B-cell ALL L1
t(1;14)(p32;q11)	T-lineage ALL
t(8;14)(q24;q11)	T-lineage ALL
t(10;14)(q24;q11)	T-lineage ALL
t(11;14)(p13;q11)	T-lineage ALL
t(7;14)(q35;q11)	T-lineage ALL
inv(14)(q11q32) and t(14;14)(q11;q32)	Adult T-cell leukaemia
i(6p)	
del(6q)	Common ALL, T- or B-lineage, intermediate prognosis

Continued.

Table IX.3 *Continued.*

Chromosomal abnormality	Association (if any)
dic(7;9)(p11;p11)	
i(7q)	
del(9)(p21)	ALL L1 and L2, T- or B-lineage
t(9;22)(q34;q11)	Poor prognosis, immature B-cell ALL
t(4;11)(q21;q23)	Poor prognosis, immature B-cell ALL
t(11;19)(q23;p13)	Biphenotypic acute leukaemia
t(9;11)(p22;q23)	Biphenotypic acute leukaemia
t(11;17)(q23;p13)	Biphenotypic acute leukaemia
del(12)(p11p13)	Common ALL L1 or L2
trisomy 6	
trisomy 8	
trisomy 18	
trisomy 21	
hyperdiploidy (>50 chromosomes)	Early B precursor ALL, favourable prognosis
near haploidy (<30 chromosomes)	Common ALL, very poor prognosis
hypodiploidy (30–39 chromosomes)	Mainly observed in adult ALL

[a]L3 is a classification type of ALL (see Table IX.4). Note the involvement of immunoglobulin gene loci (e.g. 14q32, *IGH*) and T-cell receptor gene loci (e.g. 14q11, *TCRA* and *TCRD*) in the breakpoints of chromosomes.

Table IX.4 French–American–British classification of acute lymphoid leukaemia.

Classification	Description
L1	Blasts are mainly small and relatively uniform in appearance. The nucleocytoplasmic ratio is high. Nuclei are predominantly round, nucleoli inconspicuous. Chromatin pattern diffuse, smaller blasts show some chromatin condensation. Cytoplasm is scanty and slightly to moderately basophilic. Some cytoplasmic vacuolation may be present.
L2	Blasts are larger than in L1, and more heterogeneous. The nucleocytoplasmic ratio is lower. Nuclei are more pleomorphic, with some nuclei being indented, cleft or irregular. Cytoplasm varies in amount and is often abundant. Cytoplasmic basophilia is variable. Cytoplasmic vacuolation may be present.
L3	Blasts are large and homogeneous. The nucleocyto-plasmic ratio is high, though not as high as in L1. Nuclei are predominantly round with a finely stippled chromatin pattern, and prominent nucleoli. Cytoplasm is strongly basophilic, and in at least some cells there is heavy vacuolation.

Table IX.5 Common chromosome changes in myelodysplastic syndromes (MDS)[a] excluding chronic myelomonocytic leukaemia (MML).

del(5)
monosomy 7
del(7)
trisomy 8
del(11)(q13q25) (see Fig. 7.9)
del(13)
del(20)(q11q13.3) or (q11q13.1)
–Y
del(12)(p11p13)

[a]See Table IX.6 for classification.

Table IX.6 French–American–British classification of myelodysplastic syndromes excluding chronic myelomonocytic leukaemia.

Disease	Blasts in marrow (%)	Other features
Refractory anaemia	<5	<15% ringed sideroblasts in nucleated red cells
Refractory anaemia with sideroblasts	<5	>15% ringed sideroblasts in nucleated red cells
Refractory anaemia with excess blasts	5–20	
Refractory anaemia with excess blasts in transformation (RAEBt)	21–29[a]	Also RAEBt if Auer rods present, irrespective of blast count or if ⩾5% blasts in blood

[a]A blast count of 30% is a diagnostic criteria for acute myeloid leukaemia.

Table IX.7 Common chromosome changes in myeloproliferative disorders (MPD)[a].

t(9;22)(q34;q11) (CML and CGL) (see Fig. 7.7)
+1q
–7, +der(1)t(1;7)(p11;p11)
monosomy 7
del(7q)
trisomy 8
trisomy 9
del(13q)
del(20)(q11q13.1) or (q11q13.3)

[a]Chronic myeloid leukaemia, chronic granulocytic leukaemia, chronic myelomonocytic leukaemia, polycythaemia rubra vera, essential thrombocythaemia, myelofibrosis. See Table IX.8 for classification.

Table IX.8 Classification of myeloproliferative disorders.

Disease	Major proliferative component
Chronic myeloid leukaemia	
Chronic granulocytic leukaemia	
Chronic myelomonocytic leukaemia	Myeloid activity predominates
Polycythaemia rubra vera	Red cell activity predominates
Essential thrombocythaemia	Platelet activity predominates
Myelofibrosis	Reactive marrow fibrosis predominates

Table IX.9 Chromosome abnormalities in lymphomas.

Chromosomal abnormality	Association in lymphoma
Chromosome 1 (structural changes, i.e. translocations, deletions, duplications, etc.)	25% of non-Hodgkin's lymphomas (NHL), often as secondary change
1p (structural changes)	T-cell lymphoma
1q (structural changes)	Diffuse large cell lymphoma
inv(2;2)(p13;p11.2p11.14)	
t(2;5)(p23;q35)	Malignant histiocytosis
trisomy 3	T-cell/Diffuse mixed large and small cell lymphoma
Chromosome 3 (structural changes)	25% of NHL/diffuse large cell lymphoma
t(3;14)(q27;q32)	Diffuse large cell lymphoma
t(3;4)(q27;p11)	
del(3p)	Immunoblastic lymphoma
t(4;16)(q26;p13)	T-cell lymphoma
6p (structural changes)	T-cell lymphoma
del(6q)	15% NHL, often as secondary change
i(6p)	Follicular small cleaved cell type lymphoma
trisomy 7	5–15% NHL/follicular large cell lymphoma
7p15p21 (structural changes)	
7q35q36	T-cell lymphoma
trisomy 8	Follicular, mixed small cleaved cell and large cell lymphomas
t(8;14)(q24;q32), t(2;8)(p12;q24), t(8;22)(q24;q11)	Most common translocations seen in Burkitt's lymphoma
t(10;14)(q24;q32)	B-cell lymphoma
del(11)(q23q25)	B-cell immunophenotype. Diffuse mixed small and large cell lymphoma
t(11;14)(q13;q32)	Small cell lymphocytic lymphoma and B-cell chronic lymphocytic leukaemia (CLL)
trisomy 12	Small cell lymphocytic lymphoma and B-cell CLL
t(14;18)(q32;q21)	Most common change in NHL. B-cell lymphomas of follicular morphology, frequent in small cleaved cell type
14q+	Most common change in NHL (includes the t(14;18) and t(11;14) as described above). Occurs in 50% cases
inv(14)(q11;q32)	T/B-cell lymphoma
17q21q25 (structural changes)	Follicular large cell lymphoma
trisomy 18	10–15% NHL, usually as a secondary change
del(22)(q11)	Occurring as a Philadelphia translocation, variable histological type

Note the involvement of immunoglobulin gene loci (e.g. 14q32, *IGH*) and T-cell receptor gene loci (e.g. 14q11, *TCRA* and *TCRD*) in the breakpoints.

Table IX.10 Chromosome changes in chronic lymphoproliferative disorders.

Chromosomal abnormality	Association
B lineage	
Rearrangements of chromosome 1, 14q+	Multiple myeloma
Rearrangements of chromosome 1, 14q+	Plasma cell leukaemia
+12, 14q+, del(13q)	Chronic lymphocytic leukaemia
14q+, t/del(12)(p12p13)	Prolymphocytic leukaemia
14q+, del(14q)	Hairy cell leukaemia
?	Waldenström's macroglobulinaemia
T lineage	
Rearrangements of chromosome 1, t/del(6p)	Cutaneous T-cell lymphoma, Sézary's syndrome, mycosis fungoides
inv(14)(q11q32), t/del(14)(q11)	Large granular lymphocytic leukaemia
14q+, 14q11, del(6q)	Adult T-cell leukaemia/lymphoma
14q11 (structural changes)	Prolymphocytic leukaemia

Table IX.11 Consistent chromosome rearrangements in solid tumours.

Tumour type	Chromosomal rearrangement	Genes
Alveolar rhabdomyosarcoma	t(2;13)(q35;q14)	*PAX3, FKHR*
	t(1;13)(p36;q14)	*PAX7, FKHR*
Breast adenocarcinoma	del(3)(p14p23)	
	del(1)(p13p36)	
	del(16)(q21q24)	
	del(6)(q21q27)	
	dmin, hsr	
Colorectal adenoma	+8, +13, +14	
	del(1)(p32–36)	
Colorectal adenocarcinoma	+13, −14, −18, +X	
	del(17)(p11p13)	
	del(8)(p11p23)	
	del(5)(q22q35)	
	del(10)(q22q26)	
Clear cell sarcoma	t(12;22)(q13;q12)	*ATF1, EWS*
Ewing's sarcoma and peripheral primitive neuroectodermal tumours (pPNET), Askin tumour	t(11;22)(q24;q12)	*FLI1, EWS*
	t(21;22)(q22;q12)	*ERG, EWS*
	t(7;22)(p22;q12)	*ETV1, EWS*
Ewing's sarcoma, rhabdomyosarcoma, Wilm's tumour	der(1)t(1;16)(q10–25;q10–24)	
Extraskeletal myxoid chondrosarcoma	t(9;22)(q22;q12)	*CHN/TEC, EWS*
Follicular thyroid adenoma	+5, +12	
	t(2;3)(q12q13;p14p15)	
Glioma	dmin	
Haemangiopericytoma	t(12;19)(q13;q13.3)	
Intra-abdominal small cell sarcoma	t(11;22)(p13;q12)	*WT1, EWS*
Lipoma	t(3;12)(q27q28;q13q14)	
	t/ins(1;12)(p32p34;q13q15)	
	t/ins(12;21)(q13q15;q21q22)	
	t(2;12)(p21p23;q13q14)	
	del(13)(q12;q22)	
	Ring chromosomes	
Malignant fibrous histiocytoma	add(19)(p13) changes in chromosomes 1, 3p, 11p	

Continued.

Table IX.11 *Continued.*

Tumour type	Chromosomal rearrangement	Genes
Malignant melanoma	t/del(1)(p12p22)	
	t(1;19)(q12;p13)	
	t/del(6q)/i(6p)	
	+7	
Meningioma	–22, del(22q)	
Myxoid liposarcoma	t(12;16)(q13;p11)	*CHOP, TLS/FUS*
Neuroblastoma	del(1)(p31p36)	
	hsr, dmin	
Nonpapillary renal cell carcinoma	–14, –17	
	del(3)(p11p22)	
	del(5)(q22q35)	
	t(3;5)(p13;q22)	
Non-small cell undifferentiated lung carcinoma	del(3)(p14p23)	
	del(15)(p10p11)	
	del(9)(p21p23)	
	del(17)(p11p15)	
	del(11)(p11p15)	
	del(1)(p32p36)	
	del(7)(p11p13)	
	hsr, dmin	
Ovarian adenocarcinoma	–13, –17, –18, –X	
	del(6)(q15q25)	
	del(11)(q11q15)	
	del(1)(q21q44)	
	del(1)(p31p36)	
	del(3)(p13p23)	
	del(9)(p22p24)	
Papillary renal cell carcinoma	+17	
	t(X;1)(p11;q21)	
Papillary thyroid carcinoma	inv(10)(q11.2q21)	*RET*, unknown
	t(10;17)(q11.2;q23)	
Primitive neuroectodermal tumours of central nervous system	i(17p)	
Retinoblastoma	Structural changes of 1	
	i(6p)	
	del(13)(q14)/-13	*RB1*
Salivary gland adenoma	t(3;8)(p21p23;q12)	
Small cell undifferentiated lung adenocarcinoma	–13	
	del(3)(p14p24)	
	del(1)(q32q44)	
	del(17)(p11p13)	
	del(5)(q13q33)	
	hsr, dmin	
Synovial sarcoma	t(X:18)(p11.2;q11.2)	*SSX1/SSX2, SYT*
Transitional cell bladder carcinoma	–9/del(9)(q1q34)	
	del(11)(p11p15)	
	del(6)(q21q25)	
	del(3)(p14p21)	
	del(10)(q24q26)	
	i(5)(p10)	
Testicular teratoma/seminoma	i(12p)	
Uterine leiomyoma	del(7)(q11.2q22;q31q32)	
	t(12;14)(q14q15;q23q24)	
	–22	
	+12	
Wilm's tumour	+12, +18	
	del(11)(p13p15)	*WT1*

Table IX.12 Gene amplifications associated with solid tumours.

Gene	Location of normal allele	Malignancy
AR	Xq11q13	Prostate carcinoma
C-MYC	8q24	Breast, colorectal, lung carcinoma and many other solid tumours
CCND1	11q13	Breast, oesophageal carcinoma, squamous carcinoma, and many other solid tumours
HST-1		
GST		
SEA		
EGFR	7p11p13	Squamous cell carcinoma, astrocytoma
ERBB-2	17q12	Breast, ovarian, gastric carcinoma,and many other solid tumours
GLI	12q13	Soft tissue sarcomas, glioma
SAS		
CDK-4		
MDM2		
HRAS	11p15	Bladder carcinoma
IGFR-1	15q25q26	Breast carcinoma
MYCL	1p32	Small cell lung carcinoma
MYB	6q22q23	Colorectal carcinoma
MYCN	2p24	Neuroblastoma, retinoblastoma, small cell lung carcinoma, alveolar rhabdomyosarcoma
PDGFRA	4q12	Glioblastoma
PDGFRB	5q33q35	Glioblastoma

Abbreviations, acronyms and glossary

AtDB the *Arabidopsis thaliana* Database
ABHG American Board of Human Genetics
ABRC *Arabidopsis* Biological Research Centre
ACeDB the *Caenorhabditis elegans* Database
ACMG American College of Medical Genetics
ADA adenosine deaminase
ADP adenosine-5′-diphosphate
AFLP amplified fragment length polymorphism
AFM Association Française contre les Myopathies (French Muscular Dystrophy Association)
8-AG 8-azaguanine
AIMS *Arabidopsis* Information Management System
ALL acute lymphocytic leukaemia
Alu-PCR type of IRS-PCR using primers corresponding to the *Alu* repeat sequence in the human genome
AMCA 7-amino-4-methylcoumarin-3-acetic acid
AML acute myeloid leukaemia
AMP adenosine 5′ monophosphate
AMPFLP amplified fragment length polymorphism
AMV avian myeloblastosis virus
anchored island in genomic mapping, a group of one or more clones linked together by anchors (localized sequences) they share
APC familial adenomatous polyposis coli
APH aminoglycoside phosphotransferase
APML acute promyelocytic leukaemia
approximate map genetic map in which the position of markers is shown as the range of intervals which a particular marker could occupy at framework support.
APRT adenine phosphoribosyltransferase
APS ammonium persulphate
ARC Association pour la Recherche sur la Cancer
ARMS amplification refractory mutation system
ARS autonomously replicating sequence (yeast)
ASHG American Society of Human Genetics
ASO allele-specific oligonucleotide
ATCC American Type Culture Collection
ATP adenosine triphosphate
BA 6-benzyladenine
BAC bacterial artificial chromosome
back-cross the cross from a hybrid to one of its parental strains/species
bacterial artificial chromosome (BAC) vector based on a bacterial F factor
BCIP 5-bromo-4-chloro-3-indolyl phosphate (X-phos)
b.p. boiling point
bp base pair
Bq becquerel
BrdU 5-bromodeoxyuridine
BSA bovine serum albumin
C-banding chromosome-banding technique that stains the constitutive heterochromatin
cAMP cyclic adenosine-5′,5′-monophosphate
CAPS codominant cleaved amplified polymorphic sequences
CAT chloramphenicol acetyl transferase
cccDNA covalently closed-circular DNA
CCD charge-coupled device
CD cytosine deaminase
CDA deoxycytidine deaminase
CDC Centers for Disease Control
CDGE constant denaturing gel electrophoresis

cDNA complementary DNA
centiMorgan unit of recombination frequency. One centiMorgan (1 cM) is equivalent to 1% recombinant offspring or a recombination fraction of 0.01.
CEPH Centre d'Etude du Polymorphisme Humain (Centre for the Study of Human Polymorphism)
CF cystic fibrosis
CGH comparative genomic hybridization
CGL chronic granulocytic leukaemia
CHIAS chromosome image analysing system
CHLC Cooperative Human Linkage Center
chromosome microdissection technique in which a specific region of a chromosome is manually isolated from a cell using specially designed microneedles
chromosome painting visualization of a whole chromosome by hybridization with chromosome-specific fluorescent probes
Ci Curie
CIAP/CIP calf intestine alkaline phosphatase
CISSH chromosomal *in situ* suppression hybridization
CLL chronic lymphocytic leukaemia
cM centiMorgan
CML chronic myelomonocytic (myeloid) leukaemia
CNRS Centre National de la Recherche Scientifique (National Centre for Scientific Research)
coefficient of coincidence ratio of the observed number of recombinants to the expected number of recombinants.
comparative genomic hybridization (CGH) technique for identifying gains (including genomic amplification) and losses of chromosomal material by the cohybridization of differentially labelled tumour and normal DNA to normal chromosomes
complex traits/diseases traits or diseases that are not inherited as simple Mendelian traits
comprehensive map genetic map in which markers are included in their most likely positions irrespective of the statistical support
Con A concanavalin A
contig a contiguous set of clones spanning a region without gaps
CORN Council of Regional Networks for Genetic Services
cosmid plasmid vectors of approximately phage λ size, which are introduced into *Escherichia coli* by *in vitro* packaging and infection as defective λ phage and circularize *in vivo*
cpm counts per minute
cR centiRays
CSSH chromosomal *in situ* suppression hybridization
Da dalton
DABCO 1,4-diazobicyclo(2.2.2.)octane
DAPI diamidino-2-phenylindole-dihydrochloride
DBE direct blotting eletrophoresis
dbEST database of expressed sequence tags
DCK deoxycytidine kinase
DDBJ DNA Data Bank of Japan
ddNTP dideoxyribonucleoside triphosphates
DEAE diethyl aminoethyl
degenerate oligonucleotide primed PCR (DOP-PCR) method for random amplification by PCR of short DNA fragments at frequently occurring priming sites within the genome using a primer that contains partially degenerate sequence
denaturing gradient gel electrophoresis technique for detecting mutations
DGGE denaturing gradient gel electrophoresis
DHFR dihydrofolate reductase
DKFZ Deutsche Krebsforschungs Zenter (Heidelberg, Germany)
DMD Duchenne muscular dystrophy
DMF dimethylformamide
DMSO dimethylsulphoxide
DNA deoxyribonucleic acid
DNA fingerprinting strictly, a method of identifying an individual by DNA analysis by typing a large number of hypervariable loci to obtain an individual-specific pattern.
DNase deoxyribonuclease
DNP 2,4-dinitrophenyl
dNTP deoxyribonucleoside triphosphosphate
DOE Department of Energy (USA)
DOP-PCR degenerate oligonucleotide primed PCR
dpm disintegrations per minute
dsDNA double-stranded DNA
dsRNA double-stranded RNA
DSS disuccinimidyl suberate
DTE direct transfer electrophoresis
DTT dithiothreitol
EBI European Bioinformatics Institute
EBV Epstein–Barr virus
EC European Commission
EDTA ethylenediaminetetraacetic acid
EEC European Economic Community
ELISA enzyme-linked immunosorbent assay
EMBL European Molecular Biology Laboratory (Heidelberg, Germany)
EMBO European Molecular Biology Organization
EMS ethyl methanesulphonate
ESSA European Scientists Sequencing Arabdopsis (project)
EST expressed sequence tag
EtBr ethidium bromide
EU European Union
EUCIB European Collaborative Interspecific Backcross (program) (mouse)
EUROFAN European Functional Analysis Network
EUROGEM European Gene Mapping Project (EC sponsored project)
eV electron volt
exon trapping cloning technique for isolating coding sequences using specialized vectors in which only DNA containing exons can be maintained
F farad
F1, F2, etc. first filial generation; second filial generation, etc.
FACS fluorescence-activated flow sorting (of chromosomes)
FAP familial adenomatous polyposis coli
FCS fetal calf serum
FdU fluorodeoxyuridine

FIGE field inversion gel electrophoresis
FISH fluorescence *in situ* hybridization
FITC fluorescein isothiocyanate
fluorochrome dye that fluoresces in a particular colour under UV
forward chromosome painting technique using chromosome paints prepared from normal chromosomes and applied to metaphases containing abnormal chromosomes to reveal the identity of the aberrant chromosomes
fp flash point
framework map genetic map in which the placement of individual loci has a statistical support of at least 1000 : 1
FTP, ftp file transfer protocol
G-banding Giemsa banding of chromosomes
GDB Genome Data Base (Baltimore, USA)
gDGGE genomic DGGE
genetic map a map showing the order and distance between polymorphic chromosomal markers, constructed by linkage analysis
genomic mismatch scanning technique for identifying regions of identity between DNA samples by solution hybridization and detection of heteroduplexes
GISH genomic *in situ* hybridization, chromosome painting
GM-CSF granulocyte or macrophage colony stimulating factor
gridding arraying clones from multiple microtitre plates at high density onto membranes. These can be used for screening all clones in parallel by hybridization.
h hour
heterozygosity (of a locus) the frequency in the population of heterozygotes at that locus, which is usually expressed as a percentage or a frequency value between 0 and 1.0
HGMP Human Genome Mapping Project (UK)
HGMW Human Gene Mapping Workshops
HGPRT/HPRT hypoxanthine guanine phosphoribosyltransferase
HHMI Howard Hughes Medical Institute (USA)
HNPCC hereditary nonpolyposis colon cancer
homoeologues genetically and evolutionarily related chromosomes from different genomes within a heterogenomic polyploid or from related species; they are capable of pairing among themselves.
HPLC high-performance liquid chromatography
HUGO Human Genome Organization
hypervariable locus locus with an exceptionally high degree of polymorphism
i.b.d identical-by-descent
ICRF Imperial Cancer Research Fund (UK)
IFGT irradiation and fusion gene transfer
IGD Integrated Genome Database (Heidelberg, Germany)
***in situ* hybridization banding (ISHB)** R-banding by hybridizing with labelled Alu sequences
inclusive map *see* comprehensive map
incomplete penetrance case where some carriers of a dominant gene (or homozygotes for a recessive gene) do not express the phenotype; e.g. for a disease gene with incomplete penetrance, some carriers do not express any symptoms of the disease at all
integrated genome map
interphase FISH analysis FISH-based technique for identifying specific rearrangements in nondividing cells using region-specific markers.
IPTG isopropyl-D-thiogalactoside
irradiation and fusion gene transfer (IFGT) fusion of an irradiated donor cell with a nonirradiated recipient cell line; the resultant hybrids contain many fragments of donor chromosomes
IRS interspersed-repeat sequence
IRS-PCR interspersed repetitive sequence PCR
ISCN International System of Cytogenetic Nomenclature
ISHB *in situ* hybridization banding
ITC isothiocyanate
kb/kbp kilobases/kilobase pairs (10^3 bases/base pairs)
kd, kDa kilodalton
KOAc potassium acetate
LA-PCR linker-adaptor PCR
LB (1) loading buffer; (2) Luria broth
LINE long interspersed repeat element
linkage map *see* genetic map
linker-adaptor PCR technique for production of complex chromosome-specific libraries from DNA ligated at each end to an adaptor oligonucleotide and amplified by PCR using a primer for the adaptor sequence.
LMP low melting point
lod score (*z*) method of determining whether two loci are linked. The $\log_{10}$ of the ratio of the odds on linkage between two loci at a given recombination fraction.
LOH loss of heterozygosity
LSB low salt buffer
LTR long-terminal repeat
mA milliampere
MALDI matrix-assisted laser desorption
Mb/Mbp megabases/megabase pairs (10^6 bases/base pairs)
MC multiple copy
MDS myelodysplastic syndromes
mg milligram
MGD Mouse Genome Database
microFISH FISH using a probe produced by chromosome microdissection and amplification by PCR
microcell hybrids somatic cell hybrids derived from the fusion of micronuclei (subnuclear packets containing a subset of the donor genomic chromosomes) with intact recipient cells
microcell-mediated chromosome transfer (MMCT) production of a hybrid somatic cell by fusion of a cell from one species with a microcell derived from another
microsatellite an array of tandemly repeated very short sequences of nucleotides
min minute
minisatellite an array of tandem repeats typically in the range 1–30 kb and composed of 8- to 100-bp repeats
MMCT microcell-mediated chromosome transfer
mol mole
mol. wt molecular weight

m.p. melting point
MPD myeloproliferative diseases
MRC Medical Research Council (UK)
mRNA messenger RNA
MTX methotrexate
multiplex sequencing a variant of the shotgun sequencing strategy in which a number of samples are pooled during processing and separated by hybridization detection at the end of the process.
MVR-PCR minisatellite variant repeat PCR
NASC Nottingham *Arabidopsis* Information Centre
NBT *N*-hydroxylsuccinimidyl
NCHGR National Center for Human Genome Research (USA)
NCI National Cancer Institute (USA)
NIH National Institutes of Health (USA)
NOR nuclear organizer region
NPT II neomycin phosphotransferase II
NRC National Research Council (USA)
NSF National Science Foundation (USA)
NTA nitrilotriacetic acid
NTP nucleoside triphosphate
ODC ornithine decarboxylase
OLA oligonucleotide ligation assay
OMIM Online Mendelian Inheritance in Man
ORF open reading frame
P1 artificial chromosome (PAC) cloning system based on the P1 cloning vector but using a circular recombinant DNA
PAC P1 artificial chromosome
PAGE polyacrylamide gel electrophoresis
PALA *N*-(phosphonacetyl)-L-aspartate
PASA PCR amplification of specific alleles
PBS phosphate-buffered saline
PBSA phosphate-buffered saline solution A
PCR polymerase chain reaction
PEG polyethylene glycol
penetrance the probability that an individual is affected given their genotype at the disease-causing locus.
PFGE pulsed field gel electrophoresis
PGD Plant Genome Database
pH hydrogen ion exponential
PHA phytohaemagglutinin
phagemid plasmid containing the origin of replication from a filamentous phage, which can be packaged into the phage capsid.
physical map an ordered sequence of overlapping cloned DNAs that span a genomic region
pI isoelectric point
PI propidium iodide
PIC polymorphism information content
picking inoculating single colonies into wells of microtitre plates, and growing each as individual cultures. If these cultures are frozen in suitable freezing medium, they serve as a permanent source of these clones from which unlimited copies can be made.
PMA phorbol myristate acetate
PMSF phenylmethylsulphonylfluoride
polymorphism genetic variation at a locus at which at least one in 50 unselected individuals has a variant allele; that is, the variant allele has a frequency greater than 0.01.
polymorphism information content (PIC) a measure of informativeness of a marker in linkage studies which takes into account the fact that half the progeny of matings of the type $A1A2 \times A1A2$ will also be heterozygous and therefore uninformative for linkage.
probe labelled nucleic acid of known sequence used to detect and identify complementary sequences by hybridization
PTS probe-tagged site
PWM pokeweed mitogen
Q-banding quinacrine banding of chromosomes
QTL quantitative trait loci
R-banding reverse banding of chromosomes
RACE rapid amplification of cDNA ends
radiation hybrid (RH) somatic hybrid cell produced by fusion of a cell from one species with a radiation-treated cell from another species, in which the chromosomes have been fragmented.
radiation hybrid mapping *see* RH mapping
radiation mapping mapping technique involving cell hybrids that is based on the fact that the probability that two loci will be separated by a radiation-induced break and be carried on different chromosomal fragments should be proportional to their distance apart
RAPD random amplified polymorphic DNA
RAPD-PCR method of detecting polymorphisms by scanning the genome by PCR for a number of arbitrarily primed polymorphic loci
RDA representiational difference analysis
RE restriction endonuclease
recombination fraction (RF) the proportion of the total number offspring that do not have a parental combination of alleles; that is, those that have a recombined pattern. RF = (number of recombinant offspring)/(number of recombinant offspring + number of nonrecombinant offspring)
representational difference analysis (RDA) technique for identifying restriction fragments present in one sample but missing from another
RF recombination fraction
RFH radiation fusion hybrid
RFLP restriction fragment length polymorphism
RFLV restriction fragment length variant
RGP Japanese Rice Genome Research Program
RH radiation hybrid
RH mapping radiation hybrid mapping, a somatic cell genetic technique in which DNA markers can be mapped relative to one another using IFGT. It is based on the fact that the likelihood of two markers being separated by a radiation-induced break in the DNA is a function of physical distance, so that markers closer together have a higher probability of coretention in any given hybrid than markers further apart.
RIL recombinant inbred line
RNA ribonucleic acid
RNase ribonuclease
rpm revolutions per minute
rRNA ribosomal RNA

RT reverse transcription/transcriptase
RT-PCR reverse transcription/PCR amplification procedure
s second
SBH sequencing by hybridization
SC single copy
SCAR sequence characterized amplified regions
SCE buffer sorbitol/ sodium citrate/EDTA
SDS sodium dodecyl sulphate
shotgun sequencing strategy for DNA sequencing in which the DNA is fragmented at random, the fragments sequenced, and the sequence then assembled in the correct order.
SIGMA System for Integrated Genome Map Assembly
SINE short interspersed repeat elements
SSB single-stranded binding (protein)
SSC sodium chloride/sodium citrate
SSCP single-stranded conformation polymorphism
SSCP analysis technique for detecting mutations
SSCT sodium chloride/sodium citrate/Tween or Triton-X-100
SSCTM sodium chloride/sodium citrate/Triton-X-100/dried milk
ssDNA single-stranded DNA
SSLP simple sequence length polymorphism
ssRNA single-stranded RNA
STA Science and Technology Agency (Japan)
STC buffer sorbitol/Tris-HCl/$CaCl_2$
STM scanning tunnelling microscopy
STR simple tandem repeat, i.e. di- tri- and tetranucleotide repeat loci
STS sequence-tagged site
synteny the situation where a set of genes is in the same order on the chromosome in different organisms
TAE buffer Tris/acetic acid/EDTA
TBE buffer Tris/boric acid/EDTA
TdT terminal deoxynucleotidyltransferase
TE buffer Tris/EDTA
TEMED *N,N,N′,N′*-tetramethyl-1,2-diaminoethane
6-TG 6-thioguanine
TGGE temperature gradient gel electrophoresis
TK thymidine kinase
T_m melting temperature
TPA the mitogen 12-*O*-tetradecanoylphorbol-13-acetate
TRAP tumour necrosis factor-related activation protein
TRITC Texas red isothiocyanate
tRNA transfer RNA
UAS upstream activating sequence
URF unidentified reading frame
UV ultraviolet
VNTR variable number tandem repeat
v/v volume/volume
w/v weight/volume
w/w weight/weight
WWW World Wide Web
X-GAL 5-bromo-4-chloro-3-indolyl-D-galactopyranoside
X-Phos *see* BCIP
YAC yeast artificial chromosome
YGSC Yeast Genetic Stock Center
z lod score

Index

Page numbers in *italic* type refer to figures and tables; those in **bold** refer to protocols